MATERIAL AND PROCESS APPLICATIONS
LAND, SEA, AIR, SPACE

SOCIETY FOR THE ADVANCEMENT OF
MATERIAL AND PROCESS ENGINEERING

26TH NATIONAL SAMPE SYMPOSIUM AND EXHIBITION

VOLUME 26

MATERIAL AND PROCESS APPLICATIONS
LAND, SEA, AIR, SPACE

Hyatt @ Los Angeles Airport
Los Angeles, California
April 28-30, 1981

Additional copies of this publication may be obtained from
Society for the Advancement of Material and Process Engineering

SAMPE National Business Office
P. O. Box 613
Azusa, California 91702

Responsibility for the contents and security clearance
of papers published herein rests solely upon the
authors and not upon SAMPE or any of its members.

SESSIONS AND SESSION CHAIRMEN

COMPOSITES--FIBERS, MATRIX & PROCESSING

      R. J. Palmer, McDonnell Douglas

ADVANCED QUALITY ASSURANCE

      P. Hodgetts, Rockwell International

ADVANCED TECHNOLOGY

      E. Newell, Rohr Industries, Inc.

COMPOSITES--FIBERS, MATRIX & PROCESSING

      R. J. Palmer, McDonnell Douglas

SOLAR ENERGY SYSTEMS

      H. Maxwell, Jet Propulsion Laboratory

ADVANCED QUALITY ASSURANCE

      P. Hodgetts, Rockwell International

ADVANCED TECHNOLOGY

      E. Newell, Rohr Industries, Inc.

COMPOSITE COMPONENT MANUFACTURING

      R. Rapson, Air Force Material Laboratories

      L. "Roy" Meade, Lockheed-Georgia Company

SPACE SHUTTLE--MANUFACTURING FABRICATION--MATERIALS/PROCESSES

      H. Clancy, Rockwell International

MATERIALS AND PROCESSES IN AVIONICS

      W. Tolley, Hughes Aircraft Company

BALLISTIC MISSILE--MATERIALS AND PROCESSES

      D. Sayles, Ballistic Missile Defense Advanced Technology

ADVANCED SPACECRAFT ANTENNAS TECHNOLOGIES

      R. Dumke, Hughes Aircraft Company

COMPOSITE COMPONENT MANUFACTURING

      R. Rapson, Air Force Material Laboratories

      L. "Roy" Meade, Lockheed-Georgia Company

SPACE SHUTTLE--MANUFACTURING FABRICATION--MATERIALS/PROCESSES

      H. Clancy, Rockwell International

MATERIALS AND PROCESSES IN AVIONICS

      W. Tolley, Hughes Aircraft Company

BALLISTIC MISSILE--MATERIALS AND PROCESSES

      D. Sayles, Ballistic Missile Defense Advanced Technology

ADVANCES IN ALUMINUM ALLOYS AND PROCESSES

      T. Croucher, Tom Croucher & Associates

<u>PREVIOUS PUBLICATIONS</u>

THE SCIENCE OF ADVANCED MATERIALS
AND PROCESS ENGINEERING SERIES

Documents in this series contain papers which have been presented at prior
National SAMPE Symposia and cover a wide range of Materials and Process
Technology.

<u>Symposium Proceedings</u>	Non-Member Price	Member Price
Volume 1 Filament Winding	$ 3.00	$ 1.00
Volume 8 Insulation-Materials & Processes for Aerospace & Hydrospace Applications	8.00	5.00
Volume 9 Joining of Materials for Aerospace Systems	8.00	5.00
Volume 10 Advanced Fibrous Reinforced Composites	8.00	5.00
Volume 11 Effects of the Space Environment on Materials	10.00	5.00
Volume 12 Advances in Structural Composites	8.00	5.00
Volume 13 Energistic Materials	8.00	5.00
Volume 14 Advanced Techniques for Materials Investigation & Fabrication	10.00	8.00
Volume 16 Materials '71	10.00	8.00
Volume 17 Materials Review for '72	10.00	8.00
Volume 25 The 1980's - Payoff Decade for Advanced Materials	55.00	50.00
Volume 26 Material and Process Applications--Land, Sea, Air, Space	60.00	55.00

NATIONAL SAMPE TECHNICAL CONFERENCE SERIES

Volumes in this series contain papers which have been presented at various
technical conferences which deal with specific subject matter related to a
narrow segment of the Materials and Process technology.

<u>Technical Conference Series</u>		
NSTC 1 Aircraft Structures & Materials Application	10.00	8.00
NSTC 6 Materials on the Move	30.00	20.00
NSTC 7 Materials Review '75	40.00	30.00
NSTC 8 Bicentennial of Materials	40.00	30.00
NSTC 9 Materials & Processes--In Service Performance	40.00	30.00
NSTC 11 New Horizons--Materials and Processes for the Eighties	55.00	50.00
NSTC 12 Materials 1980	60.00	55.00

SAMPE, P.O. Box 613, Azusa, CA 91702

PREFACE

Los Angeles Chapter, the original SAMPE Chapter, is pleased and
proud to be the host of this 26th National SAMPE Symposium and
Exhibition. As the second quarter century of these meetings starts,
we pause to consider the theme of this Symposium - "Material and
Process Applications - Land, Sea, Air, Space" - and consider the ever
widening scope of material and process engineering. Progress in this
area of engineering offers every one involved in it a continuing
series of challenges which can only be surmounted by an on-going
individual learning effort. It is hoped that this SAMPE Symposium
will provide a significant learning opportunity for those attending
the sessions or referring to this volume of preprints. This learning
opportunity can be dual in nature in that it allows the specialists to
add to their knowledge of their particular field while also allowing
them to broaden their horizons by viewing the accomplishments taking
place in allied fields.

We would like to express our thanks to the various Committee Chairmen
who have given so generously of their time to bring this Symposium
and Exhibition to a successful culmination as well as personnel from
the National Business Office whose experience and expertise immeasurably lightened our burdens. Particular thanks must be given to the
Session Chairmen who worked long and hard to get meaningful papers
and thus make the technical aspects of this Symposium significant to
a broad segment of our profession.

Eugene R. Crilly   John A. Van Hamersveld
General Chairman   Program Chairman
26th SAMPE Symposium and Exhibition 26th SAMPE Symposium and Exhibition

CONTENTS

NAMES AND AUTHORS

SPECIAL SESSIONS

The papers shown below covering the subject areas of <u>Composite Component Manufacturing</u> and <u>Metal Matrix Composites Application Payoff and Composites Fabrication Assessments</u> were presented at the 26th National SAMPE Symposium and Exhibition but are not contained herein due to Federal Regulations. The Federal Export Data Dissemination (FEDD) regulations limit the distribution of data in these subject areas to U.S. citizens only. Copies of these papers are available to qualified individuals upon written request to the author. Copies are <u>not</u> available through the S.A.M.P. E. National Business Office.

Composite Component Manufacturing

Title	Author(s)
"Manufacturing Development of Kevlar/ PMR-15 Composite Fairing for High Temperature Environments"	R. Kawai and A. Zsolnay, Douglas Aircraft Company, 3855 Lakewood Boulevard, Long Beach, CA 90846

Metal Matrix Composites Application Payoff and Composites Fabrication Assessments

"Application Overview"	F. Boensch, AFWAL/FIBAA, WPAFB, OH 45433
"High Performance Aircraft Airframes"	Brian T. Gannon, Rockwell Int'l/NAAD, 826 Lapham St, MB-60, El Segundo, CA 90245 and Robert W. Gordon, AFWAL/ FIBAA, WPAFB, OH 45433
"Cargo/Bomber Aircraft"	A. V. Hawley, Douglas Aircraft Co., MS 35-42, 3855 Lakewood Boulevard, Long Beach, CA 90846, and Bill L. Shelton, AFWAL/FIBAA, WPAFB, OH 45433
"Cargo/Bomber Aircraft"	J. R. Carroll, Lockheed-Georgia Company, D/72-77, Zone 450, 86 South Cobb Drive, Marietta, GA 30063, and Bill L. Shelton, AFWAL/FIBAA, WPAFB, OH 45433
"Missile Airframes"	G. J. Inukai, McDonnell Douglas Astronautics, B/106, L/2, P/D-10, P.O. Box 516, St. Louis, MO 63166, and Vern Johnson, AFWAL/FIBAA, WPAFB, OH 45433
"Fabrication Overview"	C. Ronald, AFML-LL/WPAFB, OH 45433
"Extrusion of Silicon Carbide Whisker Reinforced Aluminum Alloy"	C. B. Criner, J. Cook and J. Pickens, Silag, Inc., P.O. Drawer H, Greer, SC 29651
"Low Cost Hot Molding of Fiber Reinforced Aluminum Composites"	J. Henshaw, J. Cornie and G. Burt, Avco Specialty Materials, 2 Industrial Avenue, Lowell, MA 01851

SPECIAL SESSIONS

<u>Title</u>	<u>Author(s)</u>
"Manufacturing Fiber and Particulate Reinforced Metallic Composites"	W. Harrigan, DWA Composite Specialties, 21119 Superior Street, Chatsworth, CA 91311
"Metal Matrix Composites Fabrication Technology and Costs"	R. E. Fisher, Amercom, Inc., 8948 Fulbright Avenue, Chatsworth, CA 91311
"Fiber FP Reinforced Magnesium Casting Development for Helicopter Transmission Housings"	E. Dhinghra and H. Nusbaum, E. I. duPont, and R. Pinckney, Boeing Vertol, PO Box 16358, Philadelphia, PA 19142

FAST CURING HIGH PERFORMANCE EPOXY
RESINS FOR FILAMENT WINDING APPLICATIONS

Robert Edelman and Thomas P. Carter
Celanese Plastics & Specialties Company
Summit, New Jersey

Abstract

Two new low cost high performance resin systems have been developed which are fast curing, have good mechanical properties at 121°C and are suitable for processing by filament winding procedures. This unusual combination of filament winding processability, elevated temperature capability and rapid cure is believed to be unique to these resin systems.

The two resin systems differ primarily in the cure chemistry used. One system provides easier processability while the other has higher elongation. Both systems will undergo complete cure after only five minute exposure to 150°C. No post cure is necessary. The dry Tg of both cured resin systems is ca. 160°C. This is reduced to 135-155°C after seven day exposure to boiling water.

Carbon fiber as well as glass reinforced composites prepared with these resins give adequate elevated temperature mechanical properties even after extended environmental conditioning. Both systems have been successfully used to make filament wound constructions such as automotive composite drive shafts and leafsprings. No excessive exotherm has been observed in the making of these parts.

These resin systems are useful for the rapid production of automotive, industrial, aerospace or recreational parts where filament winding is suitable.

1. INTRODUCTION

Epoxy resin technology has grown substantially since the initial resin systems became commercially available almost 35 years ago. Systems are now available that fulfill a large variety of processing and property needs. However deficiencies still exist in epoxy formulations that limit the use of these versatile materials in certain applications.

Recently in our work related to carbon fiber composite applications a need arose for a resin system that would be suitable for the filament winding of a hybrid composite drive shaft. This application required a resin system with a unique combination of processability and mechanical property performance.

The material selected had to have filament winding processability. Viscosity at the application temperature should be less than 2000 cps (2 Pa.sec.), preferably 500-1000 cps (0.5-1.0 Pa.sec.). Pot life of the formulation should be a minimum of two hours. Mechanical property performance and environmental resistance had to be adequate at 121°C. Environmental resistance was defined by the auto maker as the maintenance of acceptable performance after immersion in boiling water or saturated salt water for 24 hours. The other physical properties required included adequate tensile strength (10,000 psi) and impact strength.

A key requirement was that the system had to cure rapidly to allow for high volume production. This was defined as five minutes or less at elevated temperature. During the resin cure excessive exotherm could not be tolerated. Finally the resin had to be low in cost and commercially available. It was felt that an epoxy material would be most likely to fulfill the requirements desired.

Filament winding resins have been reported that have many of the features outlined.[1,2,3,4] However none are available that fulfill all of the desired needs. Frequently cycloaliphatic epoxies and bis-phenol F diglycidyl ether systems have been combined with various curing agents to give systems with excellent processability and adequate thermal performance. Bis-phenol F diglycidyl ethers have also been used to achieve low viscosity. These systems are cured for long periods of time and are frequently not low in cost. Other systems use various Lewis acid catalysts to achieve rapid cure but do not have adequate elevated temperature performance or environmental resistance. The latter systems frequently produce excessive exotherm in rapidly cured parts of any reasonable thickness.

2. RESULTS AND DISCUSSION

2.1 Cast Resin Properties

Two resin systems were formulated to meet the requirements outlined. They provide the user with fast curing high performance capability at low cost. The two resin systems discussed below differ primarily in the cure chemistry used. They are designated 30-119 and 30-120 and both are two package systems. The former provides higher elongation while the latter has better handleability and hot wet performance.

Both systems will undergo complete cure after only five minute exposure to a temperature of ca.150-160°C. The Tg of both systems is about 160°C. This is reduced to 132°C for System 30-119 and 154°C for System 30-120 after a seven day water boil. The filament winding application temperature is 55°C for System 30-119 and 65°C for System 30-120. Typical cast resin properties are shown in Table I. Flexural strength and modulus at 121°C are 65% of the room temperature value for the 30-119 product. The 30-120 product has a higher flexural strength at room temperature and maintains 57% of this value at 121°C while its modulus at elevated temperature is 78% of the room temperature value. Tensile elongation at room temperature of the 30-119 system is higher than that of the 30-120 product.

In Table II cast resin properties are shown as a function of cure time for System 30-119. The maximum heat deflection temperature of 152°C was obtained just four minutes after reaching cure temperature.

2.2 Composite Properties

In order to obtain property information on composite laminates made with these resin systems flat panels were prepared from 0° filament wound sheets. The sheets were made from either Celion® 6000 carbon fiber or PPG 1062 "E" glass and then laid up in a panel configuration. Curing was done in a hot mold and press at 177°C for five minutes using contact pressure to replicate the filament winding construction. Composite properties obtained are shown in Tables III and IV. Flexural and interlaminar shear properties of the carbon fiber composites are slightly higher for System 30-119 while tensile strength is higher for System 30-120. Mechanical properties of glass fiber composites are very similar for both systems.

In Table V are shown flexural and interlaminar shear properties of System 30-119 after 24 hour conditioning in boiling saturated salt water. Flexural strength and interlaminar shear strength are the key composite properties affected by the nature of the interface. Since environmental conditioning exerts it most severe effect on composites at the fiber resin interface, these properties would be the most appropriate to monitor. Both properties show only minor reductions after this treatment.

Additional testing of composite specimens was also done after much more extensive environmental conditioning than a 24 hour treatment. This initial procedure can only assess the response of the resin system to limited environmental attack. It is well known from work done with aerospace epoxy resin systems that composite coupons need more extensive conditioning to reach an equilibrium condition.[5] In Table VI carbon fiber composite flexural and interlaminar shear properties were determined on both systems after a seven day water boil. Shear strength at 121°C is maintained at ca. 50% of the original level for the 30-119 system and at ca. 70% for the 30-120 system. "E" glass composites show similar performance levels for both resin systems after this severe conditioning treatment. Shear strength at 121°C was maintained at ca. 55% of the dry value for both systems (see Table VII). These property values are considered to be acceptable for the current design of the drive shaft.

2.3 Part Fabrication and Performance

A frequent problem with rapid cure systems, particularly in thick parts, is excessive exotherm during cure. Fast cure of these new systems does not result in problems with exotherming based on observations made during actual prototype fabrication. Drive shafts with thicknesses as great as 0.4 inches have been prepared in our laboratories with no exotherm problems. Leaf springs as thick as one inch have been prepared by other users of the resin systems with no exotherm problems. This thicker part needed a longer cure of ten minutes. Drive shafts prepared with the resin systems discussed have been subjected to a variety of impact, torque and elevated temperature environmental conditioning tests. No premature shaft failures occurred.

Drive shafts prepared with system 30-119 have been used on Mazda race cars during competition and have performed without any failures. These shafts were subjected to speeds greater than 100 miles per hour for periods of up to 17 hours.

3. CONCLUSIONS

Low cost, low viscosity, rapid curing epoxy resin systems have been formulated for the filament winding of carbon and glass fiber composite systems. Parts can be fabricated quickly with these systems without being subjected to degradative exotherm behavior.

Composites prepared from these systems show adequate flexural and shear properties at a temperature of 121°C even after extended environmental conditioning. The resin systems are suitable for "real world" production of quality automotive, industrial and aerospace filament wound parts.

It is clear from the composite data presented as well as the record of part performance that the 30-119 and 30-120 systems are uniquely suited for filament wound structures.

4. REFERENCES

1. T. T. Chiao, E. S. Jessup and L. Penn, "Screening of Epoxy

Systems For High Performance
Filament Winding Applica-
tions," Seventh National
SAMPE Technical Conference,
Albuquerque, N. M., October
1975.

2. J. A. Rinde, H. A. Newey and
I. L. Chiu, "A Cycloaliphatic
Epoxy Resin/Anhydride System
Usable up to 150°C," Compos-
ites Technology Review 2 (2),
21 (1979).

3. J. A. Rinde, I. L. Chiu, E.
T. Mones and H. A. Newey,
"2,5-Dimethyl-2,5-Hexanedi-
amine: A Promising New Curing
Agent for Epoxy Resins," Composites Technology Review 1
(2), 4 (1979).

4. Arnox® 3110 Product Literature.

5. E. L. McKague Jr., J. E.
Halkias and J. D. Reynolds,
"Moisture Diffusion in Composites: The Effect of Supersonic Service on Diffusion,"
Journal of Composite Materials
9, 2 (1975).

5. ACKNOWLEDGMENTS

The authors would like to acknowledge the assistance of Richard
Dzejak and Ray Steele who prepared
all of the castings and panels used
in this work.

Jim Sabo and Pat Carbone were
responsible for the preparation of
prototype parts made with the new
resin systems.

6. BIOGRAPHIES

Bob Edelman is a Research Associate
in the Structural Composites Group of
the Celanese Plastics & Specialties
Co. He holds a Ph.D. in Organic
Chemistry from Rutgers University.
For the last four years he has been involved in developing new resin systems for carbon fiber applications.
Prior to this period he worked in
the area of synthesis and modification of new polyacetals and poly-
esters. Bob holds numerous patents
and has given many presentations in
the area of new resin systems for
structural composites.

Thomas Carter holds a Ph.D. in
Organic Chemistry from the University of Texas at Austin. For the
past year he has held the position
of Senior Research Chemist with
the Structural Composites Group of
the Celanese Plastics & Specialties
Co. Project responsibilities include
matrix development for carbon fiber
reinforced composites. Prior to
his current assignment, Dr. Carter
was Group Leader for new product
development with the Emulsions
Division of Celanese Polymer
Specialties Company. He has both
patents and publications in the
areas of acrylic solution and
emulsion polymers.

Table I. Resin and Casting Properties[1]

	30-119	30-120
Viscosity, cps (Pa.sec.)	2,040 (2.0) at 55°C	950 (0.95) at 65°C
Gel Time, Min[2]	145 at 55°C (181g)	337 at 65°C (181g)
Tg, °C	160	156
Tg, °C (One Day Water Boil)	132	155
Tg, °C (Seven Day Water Boil)	132	154
Heat Deflection Temperature, °C	161	156
Flexural — Strength at 23°C, ksi (MPa)	15.9 (109.6)	20.2 (139.2)
Flexural — Modulus, ksi (MPa)	446 (3075)	457 (3151)
Flexural — Strength at 121°C, ksi (MPa)	10.1 (69.6)	11.5 (79.3)
Flexural — Modulus, ksi (MPa)	290 (2000)	355 (2448)
Tensile — Strength at 23°C, ksi (MPa)	11.1 (76.5)	10.7 (73.8)
Tensile — Elongation (%)	4.3	3.0
Tensile — Modulus, ksi (MPa)	472 (3254)	486 (3351)

(1) Cure: Five minutes at 180°C; No post cure
(2) Tecam Gel Timer

Table II. Cast Resin Properties As A Function Of Cure Time[1]

Resin	Cure Time (Min.)	Rockwell Hardness "M" Scale	Heat Deflection Temperature (°C)	Flexural Strength ksi (MPa)	Flexural Modulus ksi (MPa)
	4	108	152	17.0 (117)	455 (3137)
System 30-119	9	108		16.6 (114)	449 (3096)
	14	106		16.6 (114)	446 (3075)
	19	107	152	15.8 (109)	441 (3041)

(1) Curing was done in a circulating air oven at 184°C.

Table III. Composite Properties of 0°_1 Filament Wound
Celion® 6000/Epoxy System[1]

		Test Temperature (°C)	System 30-119	System 30-120
Flexural[2]	Strength ksi (MPa)	23	249 (1717)	221 (1524)
		121	167 (1151)	158 (1089)
		149	142 (979.1)	129 (889.5)
	Modulus msi (GPa)	23	18.1 (124.8)	18.6 (128.2)
		121	17.5 (120.7)	18.5 (127.6)
		149	17.2 (118.6)	17.4 (120.0)
Interlaminar Shear Strength, ksi (MPa)		23	13.4 (92.4)	12.1 (83.4)
		121	8.7 (60.0)	7.6 (52.4)
		149	7.2 (49.6)	6.6 (45.5)
Tensile[2]	Strength ksi (MPa)	23	239 (1648)	274 (1889)
	Elongation (%)	23	1.2	1.3
	Modulus msi (GPa)	23	20.1 (138.6)	21.1 (145.5)

(1) Contact mold cure for five minutes at 177°C.

(2) Fiber volumes are normalized to 62%.

Table IV. Composite Properties of 0° Filament Wound
"E" Glass/Epoxy System[1]

		Test Temperature (°C)	System 30-119	System 30-120
Flexural[2]	Strength ksi (MPa)	23	218 (1503)	219 (1510)
		121	147 (1014)	138 (951.5)
		149	111 (765.3)	112 (772.2)
	Modulus msi (GPa)	23	6.6 (45.5)	6.5 (44.8)
		121	6.3 (43.4)	5.5 (37.9)
		149	5.7 (39.3)	5.7 (39.3)
Interlaminar Shear Strength ksi (MPa)		23	12.0 (82.7)	12.6 (86.9)
		121	7.4 (51.0)	8.0 (55.2)
		149	4.0 (27.6)	6.7 (46.2)
Tensile[2]	Strength ksi (MPa)	23	179 (1234)	155 (1069)
	Elongation (%)	23	2.8	2.4
	Modulus msi (GPa)	23	6.8 (46.9)	6.5 (44.8)

(1) Contact mold cure for five minutes at 177°C.

(2) Fiber volumes are normalized to 62%.

Table VA. Effect of Saturated Salt Water Conditioning
on Celion® 6000/30-119 System Composite Properties[1]

		Test Temperature (°C)	Control	One Day Saturated Salt Solution Boil
Flexural[2]	Strength	23	268 (1848)	238 (1641)
	ksi (MPa)	121	190 (1310)	171 (1179)
		149	154 (1062)	128 (883)
	Modulus	23	20.2 (139)	19.4 (134)
	msi (GPa)	121	19.3 (133)	19.4 (134)
		149	18.8 (130)	18.0 (124)
Interlaminar Shear		23	14.7 (101)	13.6 (93.8)
Strength ksi (MPa)		121	9.8 (67.6)	8.3 (57.2)
		149	7.8 (53.8)	6.9 (47.6)

(1) Contact mold cure for 0.5 minutes at 177°C followed by 100 psi
(0.69 MPa) for 4.5 minutes.

(2) Fiber volumes are normalized to 62%.

Table VB. Effect of Saturated Salt Water Conditioning
on "E" Glass/30-119 System Composite Properties[1]

		Test Temperature (°C)	Control	One Day Saturated Salt Solution Boil
Flexural[2]	Strength	23	222 (1531)	204 (1407)
	ksi (MPa)	121	142 (979)	132 (910)
		149	110 (758)	97 (669)
	Modulus	23	6.4 (44.1)	6.3 (43.4)
	msi (GPa)	121	6.0 (41.4)	6.1 (42.0)
		149	5.8 (40.0)	5.6 (38.6)
Interlaminar Shear		23	13.3 (91.7)	12.7 (87.6)
Strength ksi (MPa)		121	8.4 (57.9)	7.5 (51.7)
		149	4.8 (33.1)	4.2 (28.9)

(1) Contact mold cure for 0.5 minutes at 177°C followed by 100 psi
(0.69 MPa) for 4.5 minutes.

(2) Fiber volumes are normalized to 62%.

Table VIA. Effect of Environmental Conditioning on
Celion® 6000/30-119 System Composite Properties[2]

		Test Temperature (°C)	Control	One Day Water Boil	Seven Day Water Boil
Flexural[2]	Strength ksi (MPa)	23	249 (1717)	213 (1469)	228 (1572)
		121	167 (1151)	120 (827.4)	132 (910.1)
		149	142 (979.1)		
	Modulus msi (GPa)	23	18.1 (124.8)	17.8 (122.7)	18.9 (130.3)
		121	17.5 (120.7)	16.7 (115.1)	17.6 (121.4)
		149	17.2 (118.6)		
Interlaminar Shear Strength ksi (MPa)		23	13.4 (92.4)	11.6 (80.0)	11.3 (77.9)
		121	8.6 (59.3)	4.7 (32.4)	4.0 (27.6)
		149	7.2 (49.6)		

(1) Distilled water environment.

(2) Fiber volumes are normalized to 62%.

Table VIB. Effect of Environmental Conditioning on
Celion® 6000/30-120 System Composite Properties[1]

		Test Temperature (°C)	Control	One Day Water Boil	Seven Day Water Boil
Flexural[2]	Strength ksi (MPa)	23	221	219 (1510)	208 (1434)
		121	158	139 (958.4)	126 (868.8)
		149	129		
	Modulus msi (GPa)	23	18.6	18.7 (128.9)	18.8 (129.6)
		121	18.5	18.2 (125.5)	18.1 (124.8)
		149	17.4		
Interlaminar Shear Strength ksi (MPa)		23	12.1	11.1 (76.5)	10.7 (73.8)
		121.	7.6	6.7 (46.2)	5.2 (35.9)
		149	6.6		

(1) Distilled water environment

(2) Fiber volumes are normalized to 62%.

Table VIIA. Effect of Environmental Conditioning on 1062 "E" Glass/30–119 System Composite Properties[1]

	Test Temperature (°C)	Control	One Day Water Boil	Seven Day Water Boil
Flexural[2] Strength ksi (MPa)	23	218 (1503)	193 (1331)	117 (806.7)
	121	149 (1027)	116 (800)	80 (551.6)
	149	111 (765)		
Flexural[2] Modulus msi (GPa)	23	6.6 (45.5)	6.9 (47.6)	6.5 (44.8)
	121	6.3 (43.4)	6.0 (41.4)	5.5 (37.9)
	149	5.7 (39.3)		
Interlaminar Shear Strength ksi (MPa)	23	12 (82.7)	11.2 (77.2)	9.6 (66.2)
	121	7.4 (51.0)	6.6 (45.5)	4.0 (27.6)
	149	4.0 (27.6)		

(1) Distilled water environment

(2) Fiber volumes are normalized to 62%.

Table VIIB. Effect of Environmental Conditioning on 1062 "E" Glass/30–120 System Composite Properties[1]

	Test Temperature (°C)	Control	One Day Water Boil	Seven Day Water Boil
Flexural[2] Strength ksi (MPa)	23	219 (1510)	198 (1365)	192 (1324)
	121	138 (951.5)	118 (813.6)	112 (772)
	149	112 (772.2)		
Flexural[2] Modulus msi (GPa)	23	6.5 (44.8)	6.4 (44.1)	6.6 (45.5)
	121	5.5 (37.9)	5.9 (40.7)	5.8 (40.0)
	149	5.7 (39.3)		
Interlaminar Shear Strength ksi (MPa)	23	12.6 (86.9)	11.7 (80.7)	10.1 (69.6)
	121	8.0 (55.2)	6.1 (42.1)	4.5 (31.0)
	149	6.7 (46.2)		

(1) Distilled water environment

(2) Fiber volumes are normalized to 62%.

AN AUTOMATED ULTRASONIC TESTBED:
APPLICATION TO NDE IN GRAPHITE/EPOXY MATERIALS

Jim F. Martin
Rockwell International Science Center
Thousand Oaks, California 91360

Abstract

This paper describes a computer-controlled ultrasonic imaging and analysis NDE facility developed at Rockwell International's Science Center, and the results obtained with samples of graphite/epoxy. The basic system consists of a minicomputer, a multiaxis microprocessor controller, a waveform digitizer, a contour following system with six degrees of freedom, and a full color display with 512 × 512 × 8 bit resolution. A review of the conceptual design of the test bed hardware and software will be included.

The extended data gathering capability of the system has been demonstrated on a variety of materials, but only results obtained with graphite/epoxy will be discussed in detail. Images representing B-scans and C-scans will be shown of both porosity and laminar defects. Results on quantitative definition of defect identity and size will be given.

1. INTRODUCTION

The project described here was funded partially by the Department of Defense and partially by Rockwell International. It started as an outgrowth of the DARPA/AFML Interdisciplinary Program for Quantitative Flaw Definition, and was designed to create a system of hardware and software which would allow careful laboratory testing of some of the concepts which emerged from the program. It was therefore, termed the "Testbed"; actual construction began in 1978 and the first images were ready by June of 1979. It is important to realize that the automatic scanning and system described below and used for this report is just the first step in the detection and characterization of flaws. Additional work in inversion algorithms for both bulk and surface defects is continuing with the objective of achieving the ability to completely characterize defects in structural materials.

Figure 1 shows a photograph of the instrumentation for the Testbed. The large immersion tank, containing the turntable base and equipped with an Automation Industries (AI) bridge and turntable with six-axis movement, is the dominant feature. To its right is the California Data Corporation (CDC) microprocessor controller which performs the direct control and tracking functions for the six stepping motors which move the bridge, orient the transducer, and rotate the turntable. These functions are in turn initiated by English character data blocks sent from the host minicomputer. This minicomputer, a Data General S/200, is in the electronics racks to the left of the tank. Also visible are the other key elements of the system which perform data acquisition and display. These are discussed in more detail below.

Figure 2 is a photograph from a closer distance to show the gimbal mounting. The gimbal assembly is positioned under water in this photograph; the transducer is located in the center of the mounting. Two of the stepping motors on the bridge are dedicated to rotating the transducer in this mount. It has two angular degrees of freedom; one angular rotation is unlimited, and the other rotation is limited to ±35 degrees. The samples in the photograph are not those reported on below.

Fig. 2 Close-up of gimbal-mounting.

The stepping motor step sizes are 0.1 degrees for the gimbal angles, as well as for the turntable. For movement in the three linear position axes, the step sizes are 0.001 inches (0.0254 mm). The actual accuracy in movement is about 0.2 degrees in the angular rotations and 0.002 inches (0.0508 mm) for small linear movements. For linear movements greater than a few inches

Fig. 1 Photograph of Testbed hardware.

(several centimeters), the positional accuracy is abut 0.010 inches
(0.254 mm).

The maximum speed in the movement of
any single linear axis is 3 inches
per second (7.62 cm/sec). In practice, much slower speeds of one-
tenth or one-twentieth this value
are used. It is important to realize that the physical parameters of
this tool have been optimized to the
purposes of laboratory research. In
a more practical field version, much
higher speeds could be employed, although some loss of accuracy would
be sustained.

2. HARDWARE

Figure 3 shows a schematic diagram
of the system of hardware, which is
discussed in more detail elsewhere.[1] The AI bridge actually
includes the transducer manipulator
assembly, and the CDC controller is
referred to as the microprocessor
controller subsystem. The latter is
based on a Z80 chip coupled with two
8085 chips. The transducer and
pulser/amplifier are manufactured by
Panametrics; in the results reported
here, three focussed transducers
were employed, operating at 2.25, 5,
and 15 MHz. The Biomation waveform
digitizing unit is capable of digitizing, storing, and transferring to
the S/200 the received and amplified
ultrasonic signal returning from the
sample. This can occur at repetition rates of up to 200 Hz. A sample
of such a waveform will be shown be-

low. The minicomputer peripherals
include two user terminals, a 50
MByte disk for data storage, a
Genisco programmable display processor, and a 9-track magnetic tape
unit which is not shown in Fig. 3.
The programmable display processor
drives a 512 × 512 RGB color video
monitor which displays a raster image of the received data; photographic hardcopy is currently obtained with an Image Resource Videoprint 5200. Color versions of the
monochromatic scan images appearing
in this report are available from
the author. The display processor
includes a hardware cursor and a
hardware image zoom and scroll. It
and its software are discussed in
more detail elsewhere.[2]

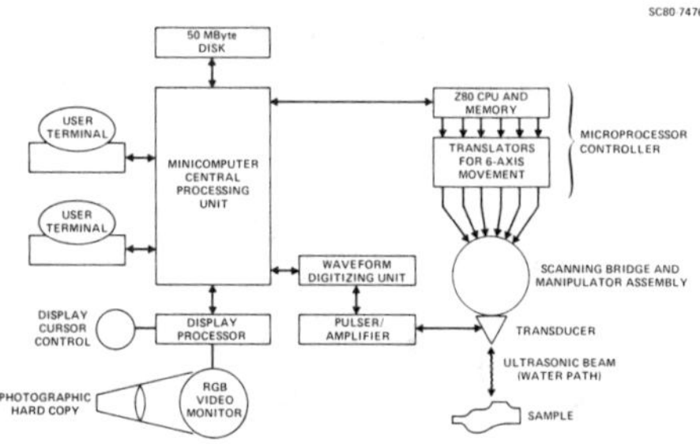

Fig. 3 Schematic diagram of the
hardware.

3. SOFTWARE

The software resides in two arenas:
the minicomputer and the microprocessor controller. The software resident in the minicomputer was designed and coded at the Science
Center. That resident in the microprocessor controller was designed

and coded by California Data
Corporation.

Figure 4 shows a schematic of the
multitasking data acquisition program. Four concurrent tasks are set
up at the start of the program and
their communication functions are
identified in the figure by numbers.
Task number one is the main program;
it allows user control by interact-
ing with the user and it sends com-
mands or data blocks to the microprocessor controller. Task number
two receives data from the microprocessor controller, reports it to
the user terminal, and when appropriate stores it on the disk. Task
number three controls the waveform
digitizer, and receives digitized
waveforms from it. It also processes the data and transfers the result for each waveform to task four,
which adds that data to a scan file
on the disk.

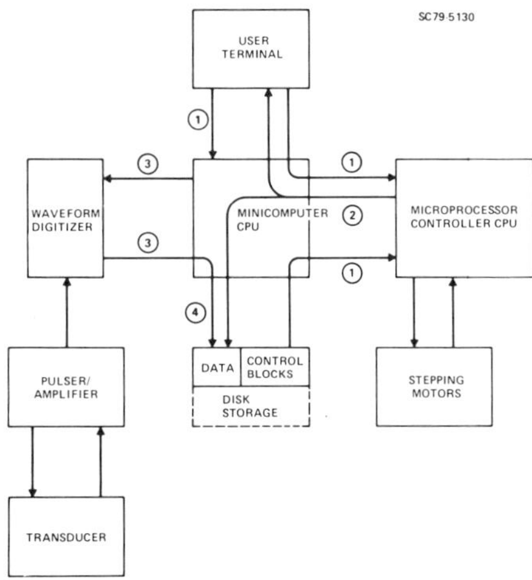

Fig. 4 Schematic diagram of the
scan control software.

The software in the microprocessor
controller is resident in EPROM.
Its function is to receive command
characters and data blocks from the
minicomputer, and translate these
into pulses to the stepping motors
and/or responses to the minicomputer. There are 13 command characters recognized by the microprocessor, and 13 response characters it
can generate, six of which are followed by signed integers indicating
position in thousandths of an inch
or tenths of a degree. Data is
passed to and from the Z80 in a
string of ASCII characters, followed
by a carriage return to denote the
end of each string. The command
characters include D, E, G, S, and
the letters X, Y, Z, A, B, and T. D
and E halt or continue whatever
movement is occurring. G and S obtain data on where the transducer is
in position and whether it is moving. The letters X, Y, Z, A, B, and
T refer to the six possible axes of
movement and allow querying for position or directing the controller to
reposition the transducer.

An example of a data block is "N1 F5
X1200 Z500 B910" which is a command
to "move from the present location
to X = 1.2 inches, Z = 0.5 inches,
and B = 91 degrees at a speed 5% of
the maximum". The data block is de-
coded and stored in a RAM memory until time to execute the move. Up to
256 separate data blocks may be
stored at one time. When a data
block is executed, the Z80

controller calculates the distance between the present location and the desired location, converts that difference into stepping motor pulses, and sends them out interleaved to provide maximum linearity in the move.

A unique feature of this system is its ability to follow curvilinear contours. If the data block includes two additional pieces of information, I7000 and K3200 for example, the Z80 will proceed to execute a move which folows a circular path, instead of a linear path. It calculates the circular path based on its current position, the desired position, and the I & K data which represent the X and Z position of the center of the circle. This allows the transducer to scan a part which has a convex or concave circular profile. If in addition, the sample has cylindrical symmetry, then the turntable can be rotated during the scan. In this way, a circular path is swept out and the entire part can be scanned with ultrasound despite its curvature. Since the controller can recall a sequence of steps, a part of rather complex profile can be inspected automatically.

4. SAMPLES

Prior to explaining the scan procedure it would be valuable to review the samples employed for this report. The geometry of both specimens in shown in Figs. 5 and 6. The

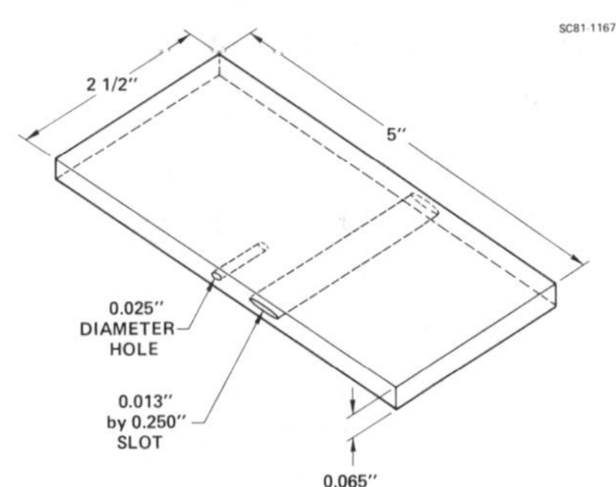

Fig. 5 Dimensions of the calibration specimen.

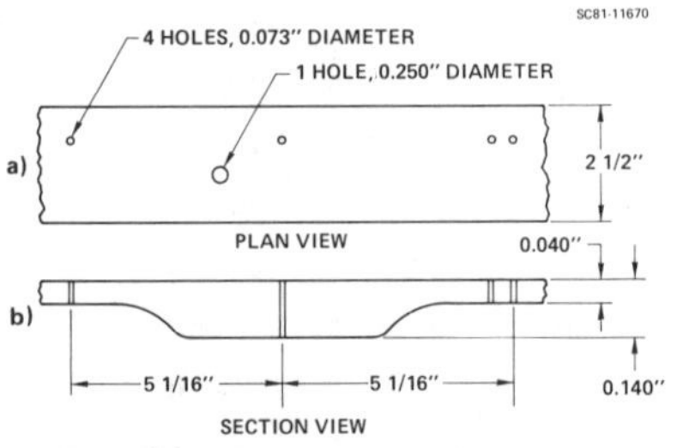

Fig. 6 Dimensions of the test specimen; a) plan view, b) section view.

material was a graphite/epoxy composite, composed of THORNEL T300 tape and fabric with compatible resins. The calibration piece, Fig. 5, was tested at the manufacturing plant and judged to be free of defects. Hence it could be used for comparison purposes. The test specimen was a structural element from the space shuttle called an "intercostal." Its dimensions are shown in Figs. 6a and 6b, although certain details of its geometry at its edges are not important to this report and are omitted from the figure. It is thicker in the center than at the two ends; the transition regions are

drawn more smoothly in Fig. 6b than
they occur in the specimen. It was
also tested at the plant, but was
judged to be defective. The tests
performed were ultrasound attenuation measurements, similar in nature
to the C-scan described below, although lacking the resolution shown
below. The objective of the scans
at the Science Center was to map out
the defective area in more detail
and attempt to identify the nature
of the defects ultrasonically.

5. SCAN PROCEDURE

The procedure for scanning the samples required only linear moves due
to their flat geometry. Prior to
the scan, a sequence of data blocks
was created and stored in a file on
the disk. Each data block was a
move in either X or Y, parallel to
the surface of the test specimen.
Figure 7 illustrates the experimental arrangement, showing the focussed
transducer, the ultrasonic beam
path, the sample, and the flat reflector plate. The dimension X is
shown by the arrows; the dimension Y
is perpendicular to X, and parallel
to the reflector plate. Two kinds
of scans were performed: in the jar-
gon of ultrasonic NDT, they were B-
scans and C-scans.

The C-scans covered a 14 inch by 6
inch (35.6 cm by 15.2 cm) area, including both the intercostal and the
calibration sample. Ultrasound
passed through the sample twice on
its way to the reflector plate and

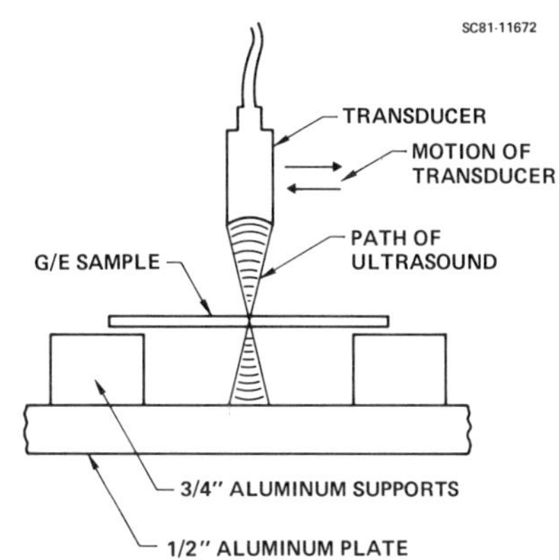

Fig. 7 Experimental arrangement of
the ultrasonic scan.

back. The transducer traversed a
raster pattern over the samples,
making large sweeps in X (14 inches
long, 35.6 cm) while sending and receiving ultrasound. At the end of
each sweep in X, a small move in Y
was made (0.048 inches, 1.22 mm) and
the transducer was swept in X again.
This was repeated until the entire
area was covered. Within a particular time gate designed to select
only the echo from the reflector
plate's front face, the peak amplitude was found and stored on disk.
Periodically, information was also
stored on the disk which identified
the current position of the
transducer.

After completion of the scan, a
display program was employed to
transfer to the display processor
the data stored on disk.

The peak amplitude at each (X, Y)
position was converted into an integer value in the range 1 to 16; this

integer value was then translated by
the display processor into a value
modulating the intensity of light
appearing on the CRT monitor, so
that higher intensity corresponded
to larger received ultrasonic
amplitude.

6. RESULTS

Figure 8 shows three C-scans; each
was performed with a different
transducer. The frequency used is
labelled in the picture. In each
picture, the calibration specimen
appears at the top and the intercostal at the bottom. A certain
amount of distortion in the pictures
is due to the curvature of the face
of the CRT: these photographs were
taken prior to the installation of
the Videoprint hardcopy system which
employs a flat faced CRT. Nevertheless, all the major features of each
specimen may be identified. The
calibration specimen is quite uniform, except for the artificial defects of the slot and edge-drilled
hole. Its support pieces are not
shown here. In the intercostal all
five drilled holes may be seen, as
well as the two regions where the
thickness of the sample changes from
0.140 inches (3.56 mm) to 0.040
inches (1.02 mm). The large patch
which is dark in the middle of the
intercostal to the right of the
large hole is the defective area; it
was suspected to contain porosity.
(For the purposes of this report,
porosity means a condition where the

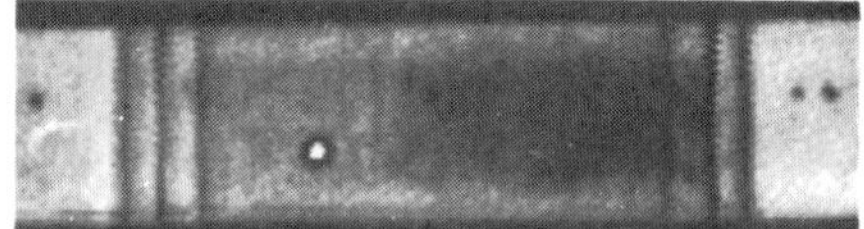

Fig. 8   C-scans; a) 2.25 MHz.

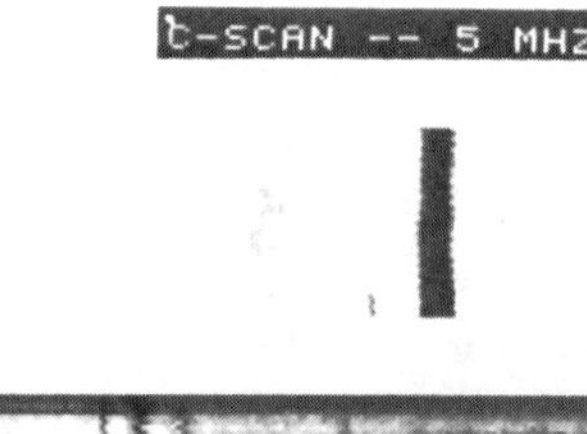

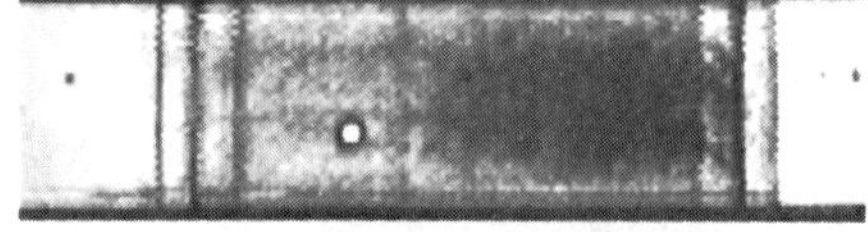

Fig. 8   C-scans; b) 5 MHz.

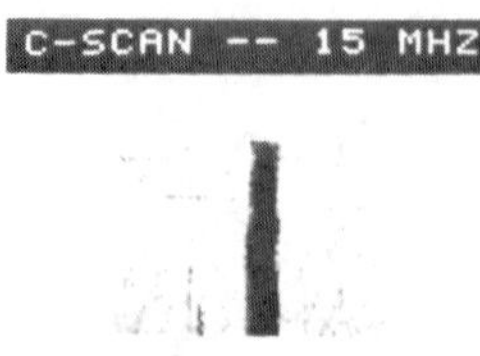

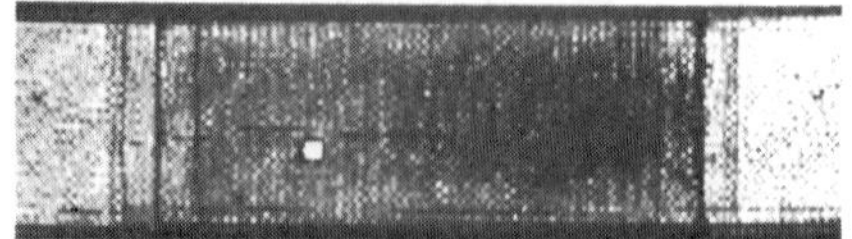

Fig. 8   C-scans; c) 15 MHz.

interstitial spaces in the fabric or
tape are not filled with resin.)
Since a delamination would be entirely black, as in the slot in the
calibration sample, it is reasonable
to conclude that the patch is not a
delamination.

As the frequency is increased, more
detail can be seen. At 15 MHz,
where even higher spatial resolution
can be expected, minor irregularities begin to appear in the calibration specimen, and the dark patch in
the intercostal is seen to contain
bright spots. These indicate that
in some tiny areas less than 0.050
inch by 0.025 inch in size (1.27 mm
by 0.64 mm), ultrasound is being
transmitted well. This is consistent with and supports the expecta-
tion that the defective area is a
patch containing porosity. Thus the
C-scans provided exact data on the
location of the patch and a map of
the various degrees of attenuation
occurring.

In order to analyze the sample further, B-scans were performed. To
help explain these, Fig. 9 shows a
typical ultrasonic waveform received
at a fixed X, Y position. The front
face and back face of the sample both
produce relatively large echoes. The
echoes occurring in the time interval between them are due to small
irregularities within the composite
and not electronic noise. In general, composites are acoustically noi-
sy and highly attenuating, so this
waveform is typical of that

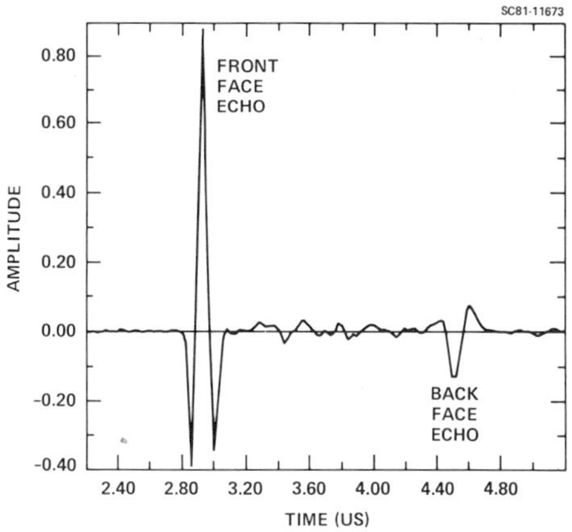

Fig. 9 Typical ultrasonic waveform
from the test specimen.

obtainable with even the highest
quality composite.

In a B-scan, the transducer is swept
in only one dimension. In this case
X was swept with Y fixed. The entire waveform collected at each X
position is stored on disk and the X
position recorded. Then, upon display, the absolute magnitude of the
signal appearing at each digital
time interval is coded into an intensity and displayed on the screen.
Figure 10 shows two B-scans of the
intercostal, one at 2.25 MHz and one
at 15 MHz. The X dimension extends
horizontally; the time dimension extends vertically and represents
depth in the material. Figures 10a
and 10b represent the left and right
half respectively of the B-scan at
2.25 MHz; Figs. 10c and 10d represent the left and right half respectively of the B-scan at 15 MHz. The
dominant features are the front and
back face echoes. These are well

19

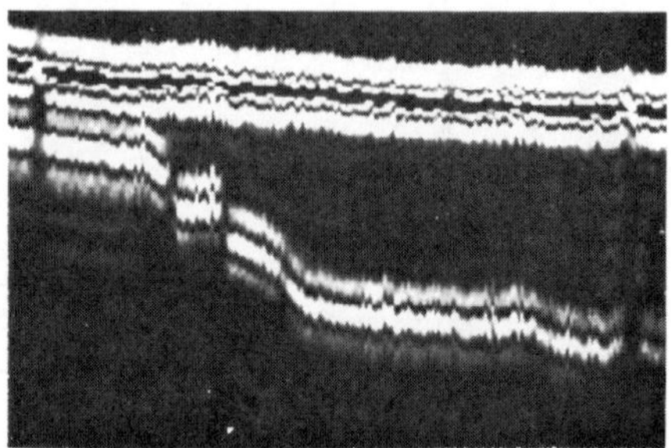

a) 2.25 MHz, left half.

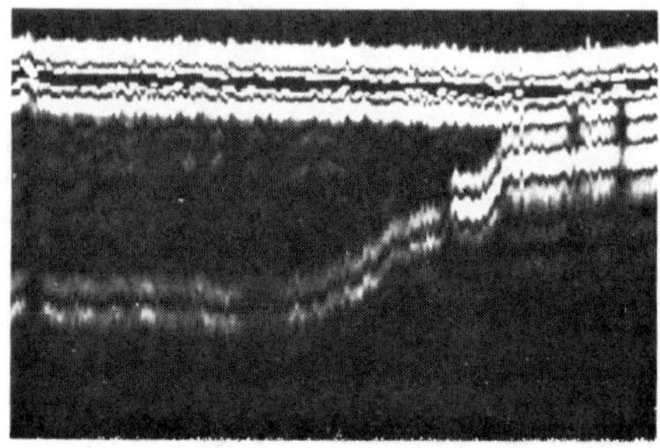

b) 2.25 MHz, right half.

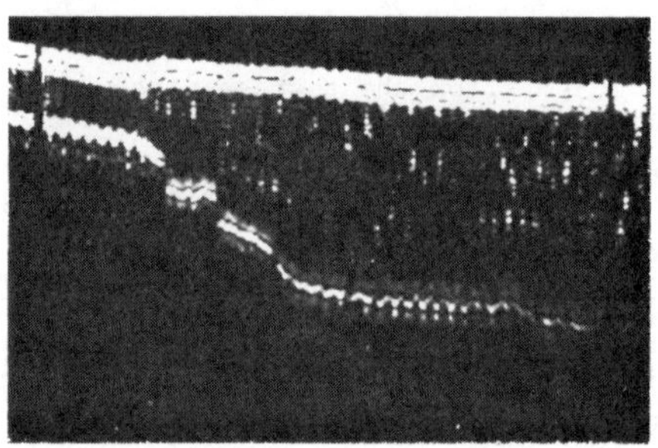

c) 15 MHz, left half.

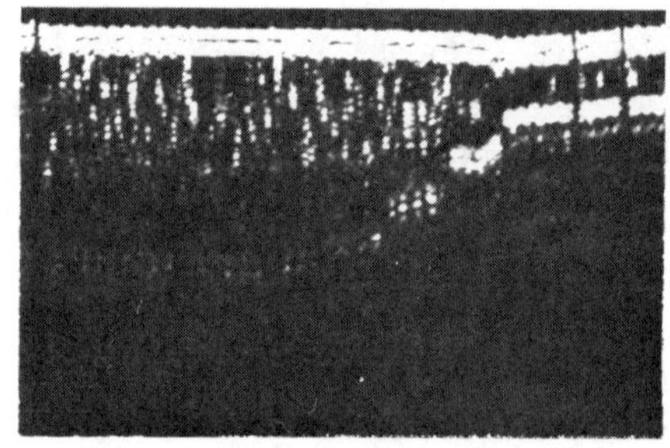

d) 15 MHz, right half.

Fig. 10   B-scans.

separated even in the thinner portion of the intercostal at 15 MHz.
At 2.25 MHz, these two echoes run
together and no detail can be seen
in the thinner portion of the sample. However, it is the thicker
portion of the intercostal which is
of interest.

The scans were oriented along a line
which sweeps out the four small
holes in the specimen; these are
visible in the scans as vertical
dark lines intercepting both the
front face and back face echoes.
This section was chosen because it
intersected both high quality and
low quality sections of the speci-

men. In all four figures, the center hole is visible, so there is a
small region of overlap between the
figures for reference purposes.

The defective area may be identified
in each of Figs. 10b and 10d by the
fact that the back face echo is
weaker than in Figs. 10a and 10c.
This is equivalent to the attenuation measurement available in the C-
scans. In addition, however, specific reflectors are visible within
the body of the composite. In Fig.
10a, these form a rather uniform
pattern which corresponds to the
pattern of layers within the composite. In Fig. 10b, however, there are

a few reflectors which are brighter.
In Figs. 10c and 10d the number of
reflectors visible is much larger;
again more appear in the defective
portion of the sample. Their size
in X can be estimated from a close-
up, or zoom, of the image in Fig.
10d which is not shown here for lack
of space. From this zoom, an estimate of the spatial extent in X of
the reflectors may be made. The entire scan includes waveforms taken
at each of 479 different points in
an X range of 12 inches (30.5 cm).
This corresponds to a spacing of 25
mils. In the closeup view, some of
the reflectors appear in only one
waveform; others extend across as
many as three waveforms. Therefore,
a rough estimate (ignoring the width
of the focal spot which is approximately 0.020 inches or 0.51 mm) of
the range of sizes of the reflectors
is 25 to 75 thousandths of an inch
(0.51 to 1.53 mm). Reflectors smaller than 0.025 inch in size in the X
dimension may exist; to detect these
a finer resolution scan would be required.

It will be noticed that there is not
a one to one correlation between reflectors appearing in the 2.25 MHz
B-scan and the 15 MHz B-scan because
the position of the two scans in Y
is slightly different.

Finally, Fig. 11 shows two portions
of a photograph of a cross section
of the intercostal taken after the
above results were obtained. The
specimen was cut in two along the

line marked by the 4 small holes and
one edge ground and polished. Figure 11a is a portion from the left
half with low attenuation and few
ultrasonic reflections; Fig. 11b is
from the right half with high attenuation and many reflectors. Note
that pores are extant in the fabric
of the composite and that they are
far more numerous and large in size
in Fig. 11b than in Fig. 11a.

The number and size range correspond
well to that indicated in the ultrasonic B-scans of Figs. 10c and 10d.
In Fig. 10c, there is an acoustically quiet region which corresponds
to the position of a bright ribbon
in Fig.11; that is the tape portion
of the lay-up. It is obscured in
Fig. 10d by the reflections from the
pores.

Due to the difficulty of obtaining a
saw cut which exactly matched the
line of the B-scan and due to the
small size of the defects it was not
expected that there would be a one
to one position correspondence between the reflectors appearing in
Fig. 10 and the pores appearing in
Fig. 11. This is especially true for
the 15 MHz data where the focal spot
is over 6 times smaller than that
for the 2.25 MHz scan. However, as
a final comment, it does appear that
the three or four brightest reflectors in the 2.25 MHz scan do match
up with specific pores visible in
Fig. 11.

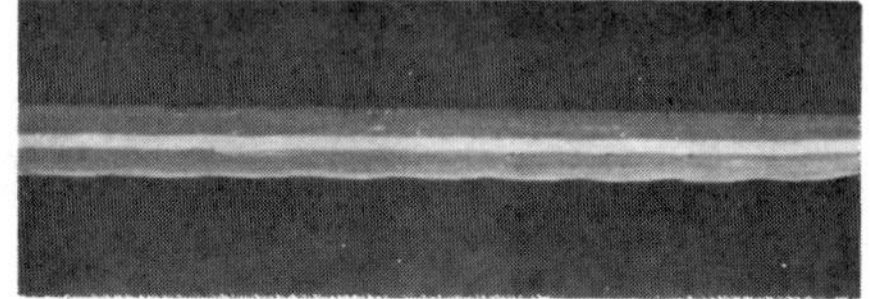

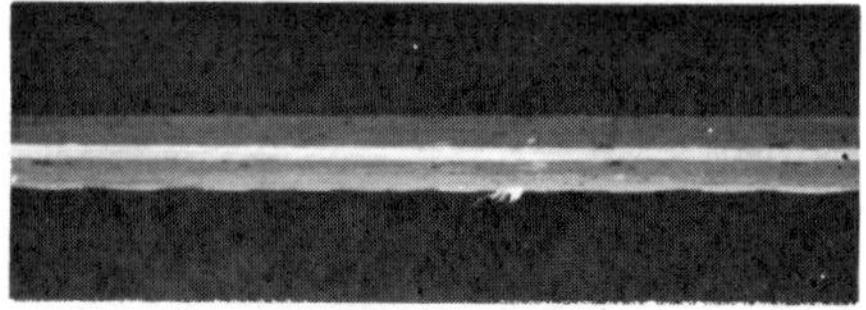

a) from the left half   b) from the right half

Fig. 11 Photograph of portions of the cross-section of the specimen after sectioning.

7. SUMMARY

A new automated ultrasonic scanning tank with an image display has been described which is designed for laboratory research and for demonstration and testing of new scientific techniques which may eventually be fielded. It has been successfully used to map out and partially identify defects in a graphite/ epoxy laminate of 0.140 inch (3.56 mm) thickness. The presence, quantity, and nature of the defects was confirmed by afterwards sectioning and photographing the specimen.

8. ACKNOWLEDGEMENTS

The aid of William J. Munn of Rockwell International was critical to the initiation and success of the study reported above. He provided samples, many hours of time, and the final effort of sectioning and photographing the sample. The permission of R.C. Addison, Science Center Project Manager for the Testbed, to use it for this study is gratefully acknowledged. More information on the Testbed program is available from the Science Center.[3]

9. REFERENCES

1. R.C. Addison et al, "Test Bed for Quantitative NDE," Proceedings of the DARPA/AF Review of Progress in Quantitative NDE, Rockwell International Science Center, in press.

2. J.F. Martin and C.C. Ruokangas, "Color Graphics: An Aid to Data Interpretation," Proceedings of the DARPA/AF Review of Progress in Quantitative NDE, Rockwell International Science Center, in press.

3. "Ultrasonic Test Bed for Quantitative NDE," Interim Technical Report No. 3, Rockwell International Science Center publication No. SC5164.261R, for Contract No. F33615-78-C-5164.

10. BIOGRAPHY

Jim Martin obtained a B.S. in Electrical Engineering from Iowa State University in 1965 and a Ph.D. in Physics from Massachusetts Institute of Technology in 1971. He worked as an experimental physicist in particle physics at M.I.T. and at Stanford University until 1978 when he joined Rockwell International. He has been deeply involved in the Testbed program since then, having

had overall responsibility for all
software created for it at the
Science Center. Dr. Martin is currently principal investigator for an
ultrasonic rail inspection program
and one task of the current DARPA/AF
program in quantitative NDE. He
also serves as program manager for a
large Science Center research program on advanced NDE techniques.
Author or co-author of over 30 technical papers, Dr. Martin is a member
of the Association for Nondestructive Testing, the American Physical
Society, the Institute of Electrical
and Electronic Engineers, Sigma Xi,
and Tau Beta Pi.

ADVANCED ULTRASONIC TESTING OF AEROSPACE STRUCTURES

Ronald J. Botsco
NDT Instruments, Inc.
Huntington Beach, CA 92649

Abstract

New ultrasonic techniques and associated instrumentation have been developed to aid with the nondestructive evaluation of complex aerospace structures from one accessible surface.

Ultrasonic impedance plane testing with the BondaScope (Patent No. 4,215,583) offers a simplified approach for evaluating bonded stuctures. The basic principles of this technique are described, as well as examples of its capability on a variety of bonded laminate, honeycomb and fibrous composite structures.

High resolution ultrasonics and its suitability for performing thickness gaging of complex/critical aerospace parts and materials are discussed. Typical results with the NovaScope are described for applications including turbine blades, radome rain erosion coatings, graphite composites and bondline thickness.

Keywords: Bondtesting, BondaScope, NovaScope, Thickness Gaging, Ultrasonics, Nondestructive Testing

1. INTRODUCTION

The complex compositions/shapes of and stringent integrity requirements for modern aerospace structures are obviously forcing considerable developments in nondestructive testing. Ultrasonic methods, in particular, have been undergoing major development in an attempt to meet these challenges.

Ironically, had it not been for the technical boom in micro-electronics, most of the advanced ultrasonic methods employed today would not have been possible. A good large-scaled example of this is the computer-based, automatic scan-recording/display systems which are rapidly appearing in virtually every major aerospace structures facility.

Micro-electronics is also responsible for the recent significant im-

provement in the capability/perfor-
mance levels of the smaller port-
able-type ultrasonic instrumentation. And there is little doubt
that the microprocessor will soon
become a dominant feature in many
types of portable equipment.

This specific paper concerns itself with two new and very diverse
ultrasonic test methods which are
especially well suited for inspecting aerospace structures. One
method utilizes the ultrasonic impedance plane (BondaScope) for inspecting bonded structures. The
other method involves high resolution pulsed ultrasound (NovaScope)
which finds particular use for the
thickness gaging of critical/com-
plex structures. Actual test results obtained on a wide variety
of typical structures are included
to demonstrate the capabilities and
versatilities of these ultrasonic
methods.

2. IMPEDANCE PLANE BONDTESTING

2.1 Basic Principles

Typical single-sided bondtesting
methods present an "incomplete"
(scaler-sensitive) readout or a
"complete" (vector-sensitive) readout which generally involves the
interpretation of changes in a
complicated waveform display. With
the present trend toward more complex bonded structures (for example, multi-layered laminates), the
above-mentioned methods are becoming increasingly difficult to
interpret.

In response to this situation, a
microprocessor-based instrument
named the BondaScope was developed.
See Figure 1.

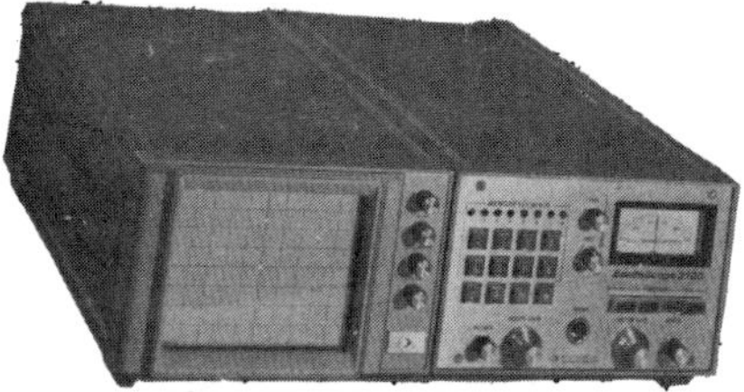

Figure 1. BondaScope 2100

The BondaScope operates on the ultrasonic impedance plane principle
and provides a more simplified vector-sensitive display (a dot on a
CRT monitor) which seems considerably easier for most operators to
interpret. It operates in a relatively low frequency range - normally less than 500 KHz.

Basically, this method consists of
monitoring the phase and amplitude
of a resonant single-element ultrasonic probe (transducer) which is
placed in contact with one surface
of the material under test. This
vector impedance information is displayed as a dot (the vector tip) on
the CRT, where phase changes cause
circumferential dot displacement
and amplitude changes cause radial
dot displacement. This type of
display is known as the ultrasonic
impedance plane.

When structural anomalies alter the

phase and amplitude of the continuous wave in the material, the impedance vector for the probe is
altered accordingly. The fortunate
result of this behavior is that detectable anomalies at different
depths in the bonded structure
possess characteristic dot locations (or addresses) on the CRT.
For comparison or reference purposes, the dot address for a well
bonded region of the material is
push-button nulled to the center of
the CRT display.

To further aid with interpretation,
the microprocessor also allows the
operator to program up to eight
anomaly dots from a material reference sample on the CRT. A meter is
included for obtaining quantative
phase and amplitude data - which
is particularly useful for measuring general material property
changes (in contrast to localized
anomalies).

2.2   Applications

2.2.1 _Metal Laminates_   The detection and depth location of unbonds
in multi-layered metal laminates is
the major area of application for
the impedance plane method. Figure
2 shows a typical response for a
five-ply (four bondline) adhesive-
ly-bonded aluminum laminate, where
each ply was 0.032" thick.

For successively deeper bondlines,
notice that the dot rotated count-
erclockwise to four distinct addresses on the CRT impedance plane.

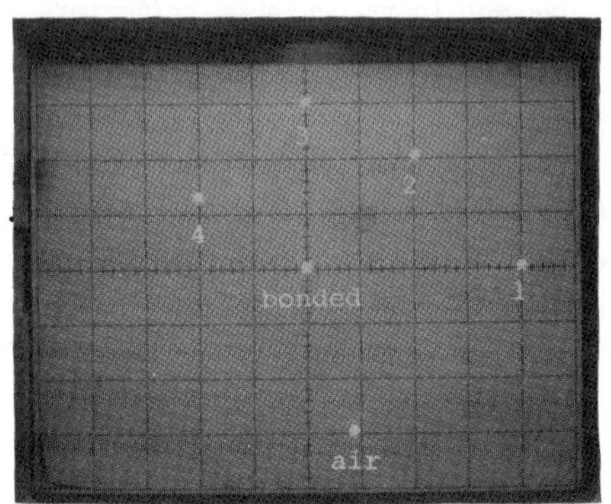

Figure 2. Unbonds At Four Different Bondline Depths In A Five-Ply
Aluminum Laminate.

The response to an unbond in a thick
2-ply aluminum laminate (each ply
0.250" thick) is shown in Figure 3.

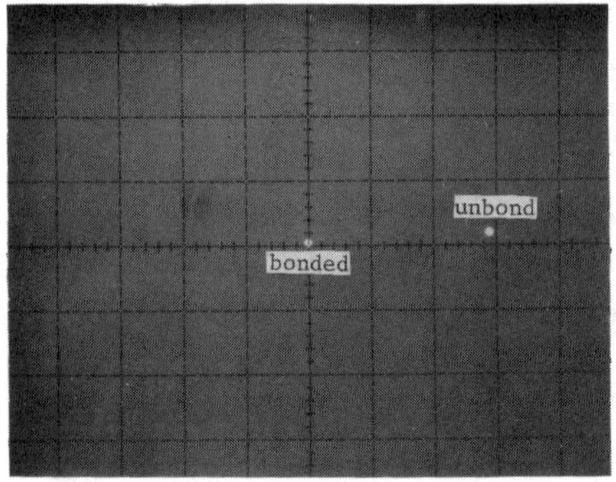

Figure 3. Response On A Thick Two-
Ply Aluminum Laminate (Each Ply
0.250"Thick).

These responses are typical for
many other types of multi-layered
laminates, including combinations
of metallic and non-metallic plies.

2.2.2 _Graphite Composites_   Figure 4
shows the impedance plane response
to unbonds at different depths in

a graphite-resin composite.

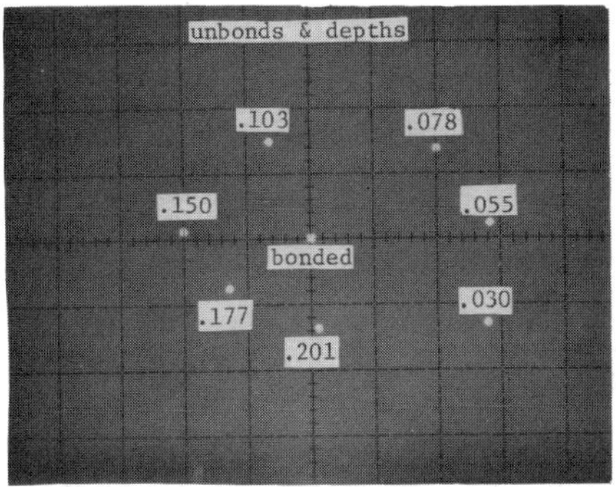

Figure 4. Response To Unbonds At Different Depths In A Graphite Composite.

This technique readily detects unbonds everywhere in graphite composites, except for in the last few layers on the opposite side from the test surface. The situation arises because the graphite layers are so thin (i.e., less than 9 mils) that the percentage change in the material thickness due to the bottom ply being unbonded, is immeasurably small. Of course, the situation is more apparent in thicker sections. The NovaScope, which is discussed later, is better suited to cope with this particular situation.

The impedance plane method is applicable for detecting unbonds between graphite layers bonded to other types of materials (such as aluminum, Kevlar, etc.), provided that the bondline depth isn't virtually located at the opposite surface.

In general, experience has indicated that ultrasonic thru transmission scanning is the best approach for inspecting simple sheets/plates of graphite composites. Single/side "bondtester methods" should be employed when one surface of the composite is accessible, or when the bondline(s) in question is not located too close to the opposite surface.

2.2.3 <u>Kevlar Structures</u> The discussion on graphite composites also applies for Kevlar composites because both are comprised of many thin layers bonded together. Figure 5 shows the response to three unbonds in a two-ply, Kevlar-steel laminate (each ply thickness was 0.093").

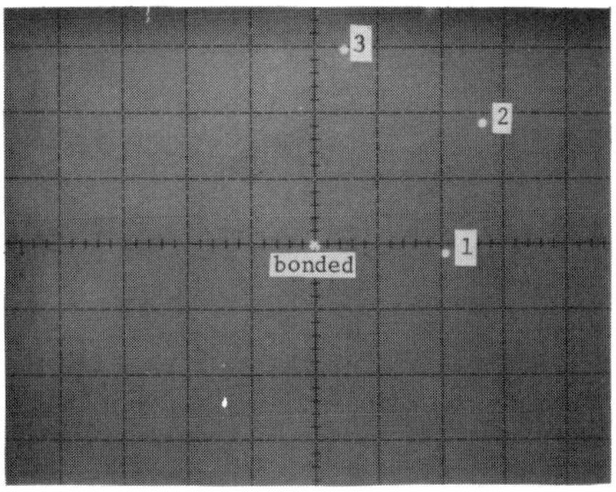

Figure 5. Unbonds Detected In A Kevlar-Steel Laminate.

The testing was performed from the Kevlar surface. The first two unbonds were located about 0.031" and 0.062" deep in the 0.093" thick Kevlar layer, while the third unbond was located in the bondline between the Kevlar composite and the steel. A variety of other laminates containing a Kevlar plate as a constituent have been successfully tested.

Preliminary work has also indicated that the impedance plane method is responsive to resin-rich and resin-starved areas in Kevlar. See Figure 6.

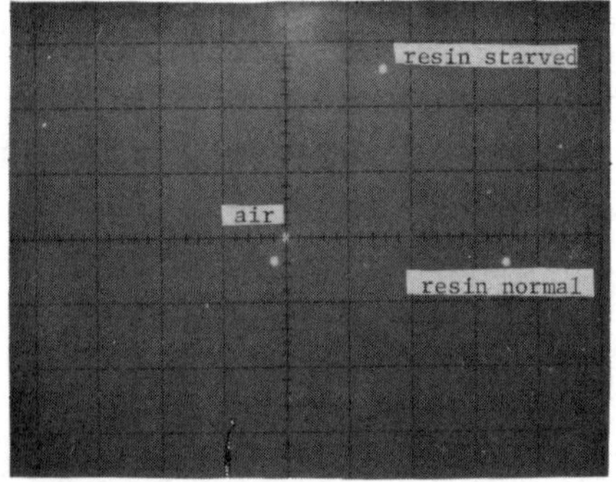

Figure 6. Detection Of Resin Ratio In Kevlar Laminate.

Such a response should be anticipated inasmuch as both the velocity and attenuation of the sound wave are theoretically affected.

2.2.4 Boron Fiber Composites Tests on boron fiber/resin matrix composites product results similar to those for graphite composites. The lower ultrasonic frequencies employed by the BondaScope minimize wave scattering from the fibers - which leads to a desirable low noise test background. Figure 7 shows the response to an unbond located between a 0.100"-thick boron/resin plate and an aluminum substrate. Testing was performed from the boron/resin surface.

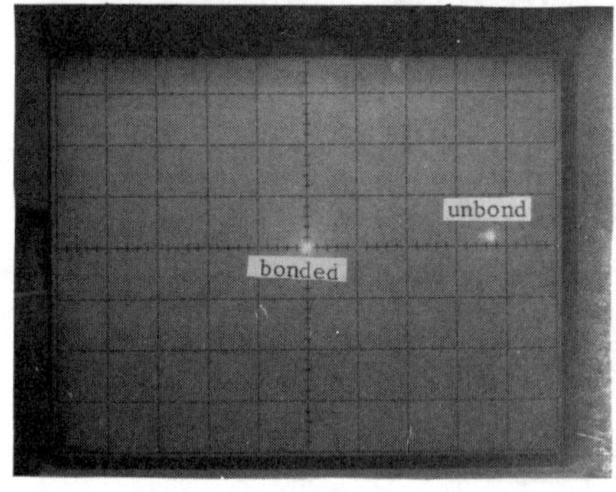

Figure 7. Unbond Response In A Boron/Resin - Aluminum Laminate.

2.2.5 Fiberglass The impedance plane method has been evaluated for detecting unbonds between two bonded fiberglass plates. Figure 8 shows the response to an unbond between two 1/4"-thick plates.

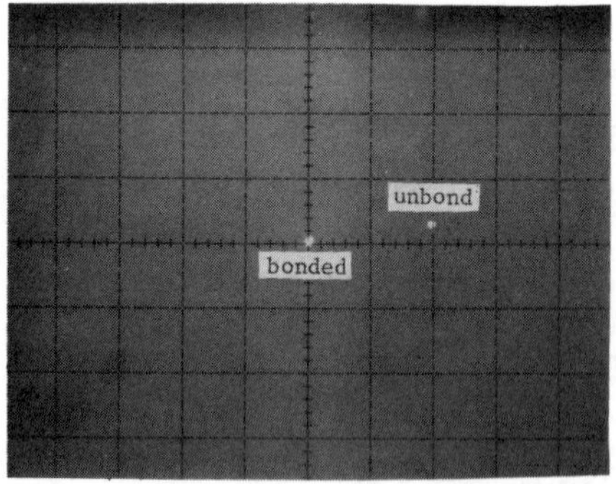

Figure 8. Unbond Detected Between Two 1/4"-Thick Fiberglass Plates.

Similar results have been obtained for thinner fiberglass plate laminates.

2.2.6 <u>Honeycomb</u> Both metallic and
non-metallic honeycomb structures
can be inspected for unbonds with
the impedance plane method. Figure 9 shows the response to near
and far side unbonds in an aluminum honeycomb panel (0.025" face
sheets X 3/4" core depth X 1/4"
cell size).

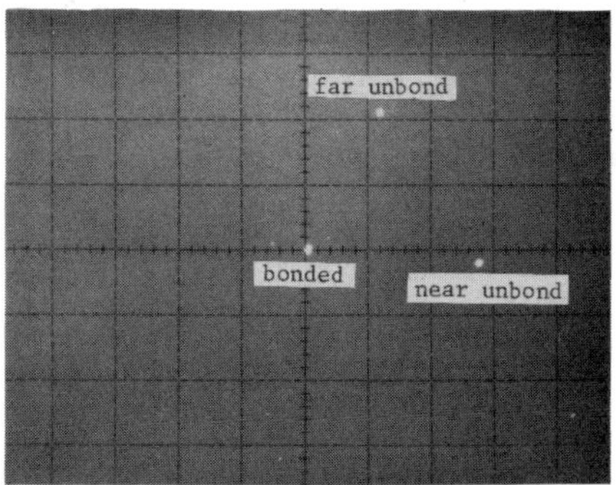

<u>Figure 9</u>. Detection Of Near And
Far Side Unbonds In An Aluminum
Honeycomb Panel.

Many times it is not possible to
detect far side unbonds in honeycomb panels. The success depends
upon the geometry and composition
of the panel.

Various combinations of Kevlar, aluminum core, Nomex core, graphite
and aluminum panels have been
tested.

3. HIGH RESOLUTION TECHNIQUES

3.1 <u>Basic Principles</u>

In contrast to the continuous-wave
low frequencies used for impedance
plane bondtesting, high resolution
ultrasonic methods employ short-
pulse, high frequency ultrasonic
energy. This method is essentially
an extension of the standard ultrasonic pulse-echo technique. Resolving power refers to the minimum echo period over which an instrument can clearly separate two
successive echoes.

High resolution instrumentation involves the use of a broadbanded
receiver, short-duration pulser,
high-speed CRT display, high-speed
logic circuits and, last but not
least, a well damped high frequency
transducer.

High resolution ultrasonics can be
used for both flaw detection and
thickness gaging applications. This
method is especially well suited for
the aerospace industry because of
the common use of thin structures.
High resolution equipment can clearly display the high-speed echo
patterns from such thin members in
a fashion similar to the way a
standard ultrasonic flaw detector
(tuned receiver, lower frequency)
displays A-Scan echo traces from
much thicker sections. Figure 10
shows a photo of the NovaScope 2000
which is a high resolution thickness gage. It provides a high-
speed scope display to assist in
qualifying the digital thickness
readout. A variety of controls permit the operator to handle many of
the complex ultrasonic signals obtained with both contact/delay

probes or immersion/squirter non-contact scanning test modes. Also a dual scope trace is used to show the time gate below the A-Scan.

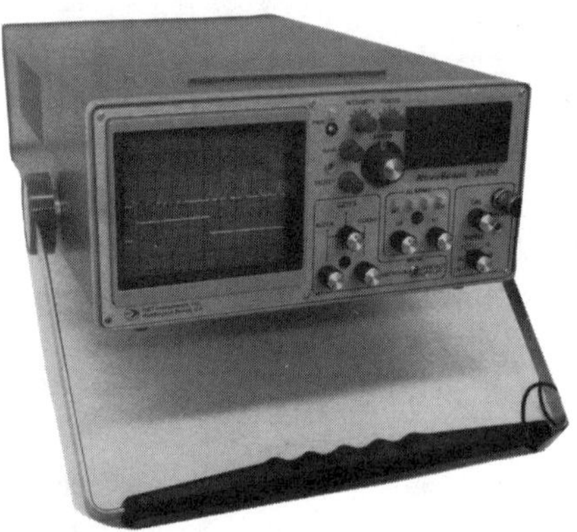

Figure 10. NovaScope 2000 High Resolution Thickness Gage.

The bandwidth of the NovaScope is well in excess of 35MHz, which permits the monitoring of echo patterns from some materials as thin as 0.005".

3.2   Applications

3.2.1 Metal Sheet And Tubing Highly contoured metal sheet products and small diameter tubing are prime applications for high resolution ultrasonic gaging, provided a CRT display is used to qualify the echo patterns. Figure 11 shows a good example for this test method where a wall thickness reading is made on the elbow of a small diameter thin-walled tube. The upper trace in the photo shows the high-speed A-Scan of the echo reverberation pattern from the thin tube wall. The lower trace shows the logic

circuit time gate, which measures the time between two echoes and ultimately sets the digital thickness readout value.

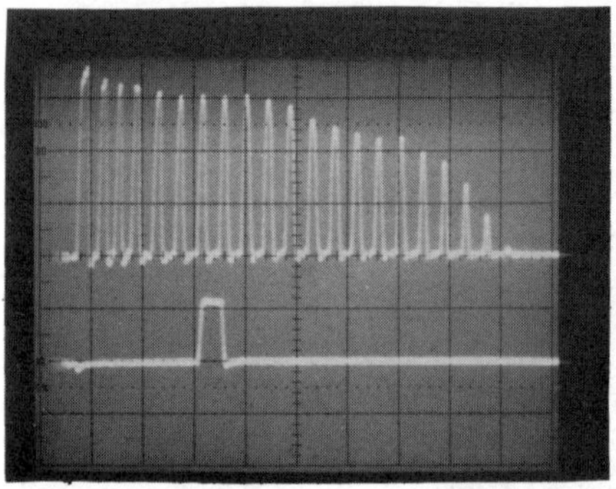

Figure 11. Thickness Gaging Of An 0.009" Wall On An Elbow Of A 5/8"OD Stainless Steel Tube.

Such tests can be performed by manual spot-checking or automatically on moving products (like sheet or tubing).

3.2.2 Turbine Blades Another major application for the NovaScope is to gage the wall thickness of jet engine turbine blades. This application involves a hollow,complex-contoured (airfoil) part manufactured from high-temperature alloys. The wall thickness of some of the more exotic blades is very thin - approaching 0.010".

This is a rather difficult ultrasonic application and the critical nature of reliable results can not be overemphasized. The approach requires the use of a highly focused ultrasonic beam, which is applied

via a "bubbler" water jet nozzle.
Normally the bubbler is in a stationary, upright position while
the operator manipulates the blade
on top of the nozzle.

Figure 12 shows a signal from a
turbine blade with a wall thickness of 0.015". Such measurements are virtually impossible
without the aid of the CRT display.

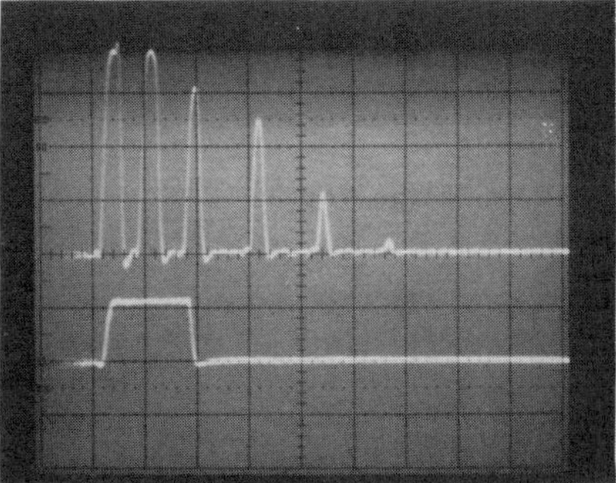

Figure 12. Gaging 0.015" Thick
Turbine Blade Wall.

3.2.3 Bondline Thickness It is obviously important to specify and
control adhesive thickness because
it affects the strength of a bonded
joint. It is presently common
practice to section and optically
evaluate test coupons to confirm
bondline thickness.

High resolution ultrasound with the
aid of a CRT, can many times offer
a quick nondestructive method for
gaging the bondline thickness of
the actual production structure.
The degree of applicability depends
upon the geometry and composition of

the laminate. A NovaScope trace
from a bondline thickness of 0.010"
between two aluminum plates is
shown in figure 13. Thousands of
bondline thickness measurements have
already been made on production aluminum laminates with the NovaScope.

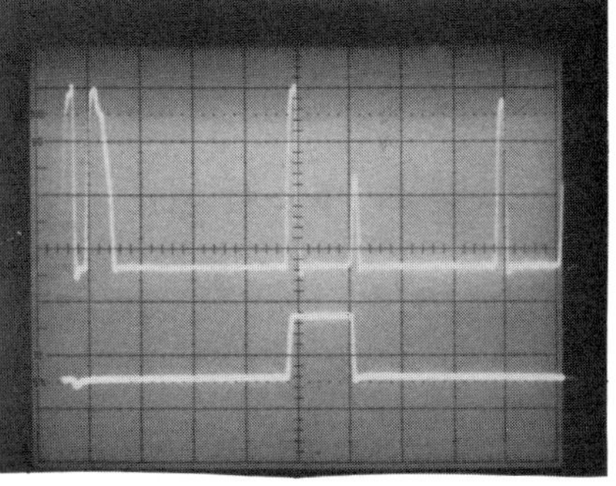

Figure 13. Gaging 0.010" Adhesive
Thickness In An Aluminum Laminate.

3.2.4 Graphite Composites The NovaScope is useful for detecting and
measuring the depth of unbonds
in graphite composites. It's also
being used for gaging the thickness of these composites.

Figure 14 shows the detection of an
unbond near the back surface of a
thick graphite composite. To detect
the echoes from such unbonds near
the back or front surfaces requires
high resolution techniques.

Thickness gaging precision is a
function of the uniformity of the
composites properties (which can
affect the ultrasonic velocity, and
thus, calibration). Experience has
shown that composite properties can

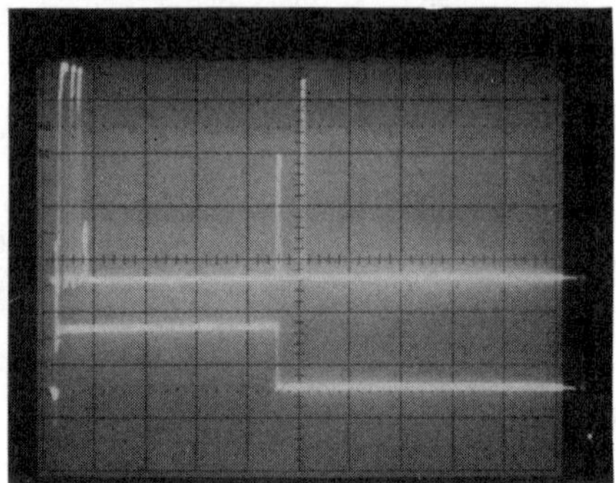

Figure 14. Detection Of An Unbond Near The Rear Surface Of A Graphite Composite With A Thickness Of 0.500".

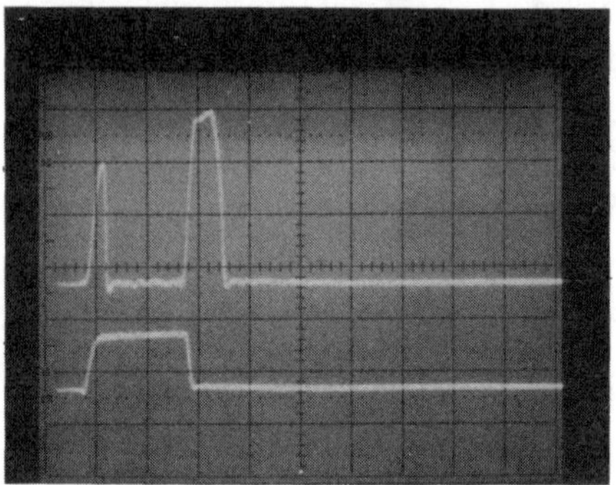

Figure 15. Gaging A Radome Rain Erosion Coating 0.005" Thick.

sometimes vary enough to cause appreciable errors in the thickness readout. Thus, considerable care should be exercised when attempting to make accurate ultrasonic measurements. The supplemental scope display occasionally reveals these undesirable property variations via changes in the echo amplitude (ultrasonic attenuation variations).

3.2.5 Radome Coatings It is common practice to protect aircraft radomes from rain erosion, etc. by coating them with certain elastomeric materials. The thickness of these thin coatings is important and frequently high resolution ultrasound can provide this information. Figure 15 shows the NovaScope display of a rain erosion coating having a thickness of 0.005". Coatings have been measured from approximately 0.003" to 0.017". The supplemental scope display adds reliability to these critical measurements; however, such coatings have been gaged without the benefit of a scope display.

4. CONCLUDING REMARKS

4.1 A rather wide variety of applications has been presented for two advanced ultrasonic test methods. To effectively use such methods at a production level requires accurate, standardized test procedures and the continual use of proper material reference standards. Close communication throughout industry is also important, especially because of the state-of-the-art situation for both the aerospace materials and the test methods.

<u>Biography</u>

Mr. Botsco is president and chairman of the board of NDT Instruments, Inc. His education includes an engineering degree from Case Institute of Technology and additional studies in business administration and science. He is a registered professional engineer in two fields,

a holder of patents and the author
of over fifty (50) papers on nondestructive testing. Mr. Botsco
is a member of ASNT and ASM. He
is an honorary fellow in ASNT and
was the recipient of their 1969
Achievement Award. He also holds
an ASNT Level III technical rating
in two categories and is listed in
a number of WHO's WHO publications.

26th National SAMPE Symposium
April 28-30, 1981

APPLICATIONS OF ELECTROMAGNETIC ACOUSTIC TRANSDUCERS (EMATs)

George A. Alers
University of New Mexico
Albuquerque, New Mexico 87131

Abstract

Ultrasonic inspections are often hampered by the necessity for liquid or grease couplants between the transducer and the part. Recently there have been advances in the design of transducers which excite and detect the ultrasonic waves by electromagnetic induction across an air gap and hence do not require any coupling medium. Examples of their use for high speed inspection of buried gas pipelines, railroad rails and artillery projectiles will be discussed. Other examples describe their use in inaccessible locations, in ultraclean environments and at temperatures of 800°F.

KEYWORDS

Electromagnetic transducers
Nondestructive testing
Ultrasonic
Inspection

1. INTRODUCTION

Ultrasonics is a rapidly growing method for inspecting materials because the high frequency, acoustic energy can penetrate the interior of very heavy sections, can detect very small flaws and can perform these functions in extremely short times. Furthermore, the electrical signal outputs of ultrasonic inspection systems readily lend themselves to process control and other automatic inspection devices. Unfortunately the full utilization of these systems is often restricted by the procedure used to inject and extract the ultrasonic energy from the material. Transducers for generating the ultrasonic waves as well as for detecting them must be connected to the part through a couplant fluid - usually water - which limits the range of speeds, temperatures, shapes and surface finishes that can be inspected. Other inspection methods such as X-rays or eddy currents are not so restricted because the sensing energy can be transferred across an air gap.

The purpose of this paper is to review the recent developments of a new class of ultrasonic transducer that operates by electromagnetic induction across an air gap and hence eliminates many of the restrictions. Because it is electromagnetic, it only operates on conducing materials (i.e. metals) and the air gap should be kept as small as possible. Furthermore, the forces that launch the ultrasonic waves and the displacements that must be measured in order to detect the waves are very small so the transducers must be engineered carefully in order to achieve the required sensitivity. Most of the applications to be discussed in the body of this paper are cases where conventional fluid coupled transducers could not be used at all. In the future, when more experience is gained in the electromagnetic technique, there will certainly be cases where both approaches could be used and the choice between them can be based upon questions of reliability of operation and information content in the electrical signals.

2. PRINCIPLES OF OPERATION

Figure 1 depicts the basic idea behind the electromagnetic acoustic transducer or EMAT.[1] In order to excite an ultrasonic wave in a metal, a wire carrying an alternating current

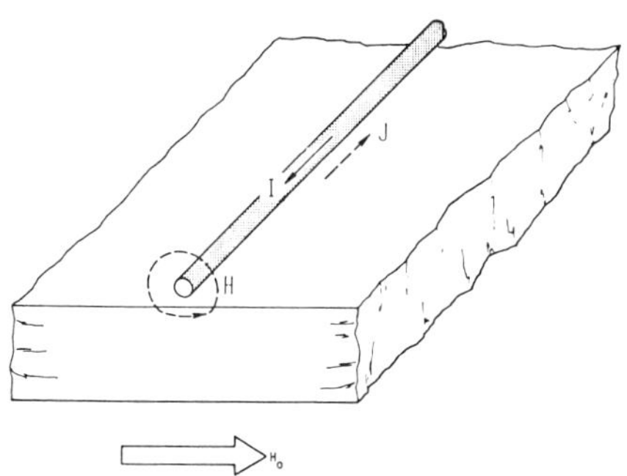

Fig. 1. Basic Form of an Electromagnetic Acoustic Transducer, EMAT.

I is held near enough to the metal to excite an eddy current J in the surface. By applying a large, static magnetic field H_o to this current, a force is generated just as in an electric motor. The direction of the force depends upon the orientation of the magnetic field relative to the eddy current. Thus, a field parallel to the surface generates a force normal to the surface and excites compressional or longitudinal waves in the metal. A field perpendicular to the surface exerts a tangential or shear force on the metal so shear waves can be launched. The use of an EMAT as a receiver involves the exact inverse of this generation process since an ultrasonic wave will move the surface of the metal in the presence of the static magnetic field and thereby produce an eddy current just as in an electric generator. This eddy current induces currents in the nearby wire which can be amplified and used to measure the amplitude of the ultrasonic wave in the metal.

If the metal is ferromagnetic there is another mechanism for launching and detecting ultrasonic waves which is based on the magnetostrictive response of the material.[2] In this case, an EMAT transmitter is formed when the alternating magnetic field H around the wire penetrates the metal surface and produces changes in the magnetization of the material under the wire. The strains associated with these magnetization changes launches an ultrasonic wave. Likewise, the strains associated with an ultrasonic wave impinging on the surface cause time dependent changes in the magnetization under the wire which induce currents in the wire to make an EMAT receiver.

The paper is organized by applications which utilize the various special features of EMATs. First, those cases where the product passes by the transducer at high speed are presented. Second, those cases where an array is used to achieve rapid inspection of a large area and third, those cases where special acoustic waves are used to interrogate inaccessible places are discussed. Then, examples are given of those cases where a couplant fluid cannot be used because of thermal or contamination constraints. The last application deals with the unique property of EMATs that they can measure stress levels in structures.

3. APPLICATIONS

3.1 High Speed Inspection

3.1.1 Buried Gas Pipelines. In seeking additional methods for inspecting buried gas pipelines, the American Gas Association sponsored a research program [3] to determine the feasibility of using ultrasonics for detecting corrosion and stress induced cracking in installed gas transmission pipelines. This required mounting the transducers on an unattended carriage which could be blown through the buried pipe by the gas at speeds of up to 30 mph

and over distances of up to 100 miles. By designing an EMAT consisting of a meander coil between the poles of a permanent magnet, Lamb type waves [4] were launched in a circumferential direction around the pipe as shown in Fig. 2. Three such EMAT units spaced at 120 degree intervals around the pipe proved

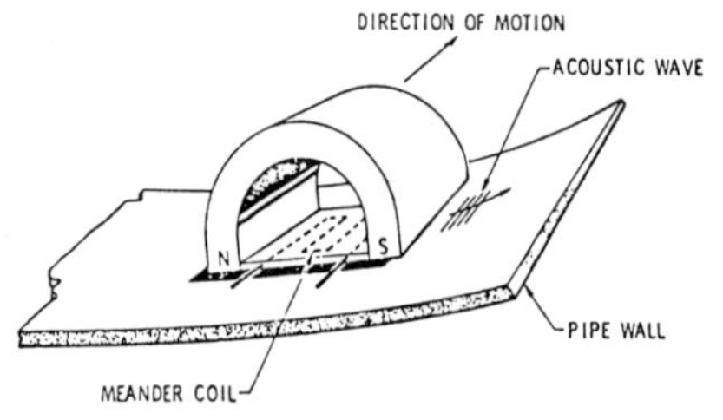

Fig. 2. Meander Coil EMAT for the Lamb Wave Inspection of Pipelines.

sufficient to detect defects by a pulse echo method. [5] The key feature in this inspection system was the fact that the EMATs could generate and detect Lamb waves when the pipe wall was moving past the transducer at high speed.

3.1.2 Railroad Track Inspection. One of the important causes of train disasters is the failure of the rail produced by the fatigue growth of microscopic defects within the rail. To find these flaws before they reach a critical size requires regular inspections of the rail while it is in place. In order to be economically acceptable, this inspection must be performed from a car driven down the track at speeds above 10 mph. Both magnetic and

ultrasonic methods are now used but
the inspection speed and reliability
of the ultrasonic method is limited
by the water filled wheel type
transducers currently employed. The
Department of Transportation is currently sponsoring a program to investigate the ability of EMATs to
replace these wheel transducers.[6]
Figure 3 shows a drawing of one concept being studied in which a large

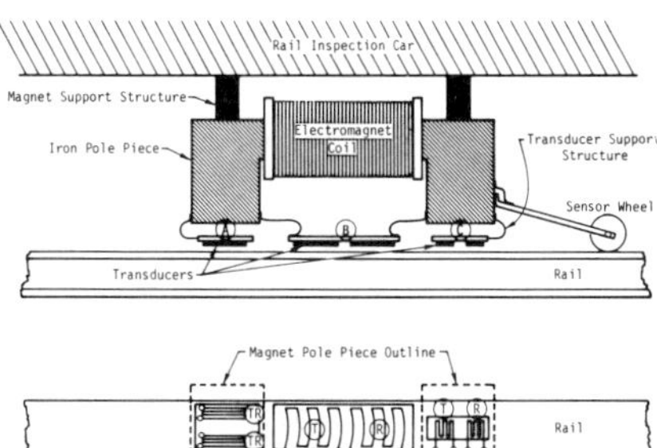

Fig. 3. Possible Arrangement of
EMATs for Rail Inspection.

electromagnet magnetizes a section
of rail so that an array of EMAT
coils can be distributed over the
rail head to give 100% coverage.
Coils A send high frequency acoustic
waves across the rail head to detect
longitudinal flaws in that region
while coils C send waves down into
the web to locate head-web separations and cracks at the bolt holes.
the large coils labeled B excite
wave modes that propagate along the
rail head to locate transverse
cracks.

3.2 Rapid Part Inspection

3.2.1 Artillery Projectile Inspec-
tion. Modern requirements on the
U. S. munitions manufacturers

demand that each article be free of
defects that could cause its premature detonation. As a result, each
high fragmentation artillery projectile must be 100% inspected for
cracks that are approximately 0.2
inches long and 0.02 inches deep.
At the anticipated production rates,
this ID and OD inspection of a
hollow object with compound curved
surfaces and variable wall thickness
must be completed in less than one
minute. Use of conventional piezoelectric transducers requires a
large number of transducer units
immersed in a water bath, each
critically aligned with respect to
the outer contours of the projectile.
A system using a single electromagnet and an array of nearly 50 EMAT
coils was constructed under the
sponsorship of the U. S. Army.[7]
By using a microprocessor to control
the entire electrical, mechanical
and acoustic process this experimental system was able to meet the
one minute inspection speed requirement and exhibited the necessary
sensitivity levels. This system did
not require any special cleaning or
immersion of the part and the inexpensive transducer coils did not
have to be maintained in precise
alignment over long periods of time.
Furthermore, the cost of replacing a
worn out EMAT coil transducer would
be much less than the cost of replacement of a single piezoelectric
transducer. Figure 4 shows a cross
section view of the projectile and
the array of EMATs used to achieve

100% coverage.

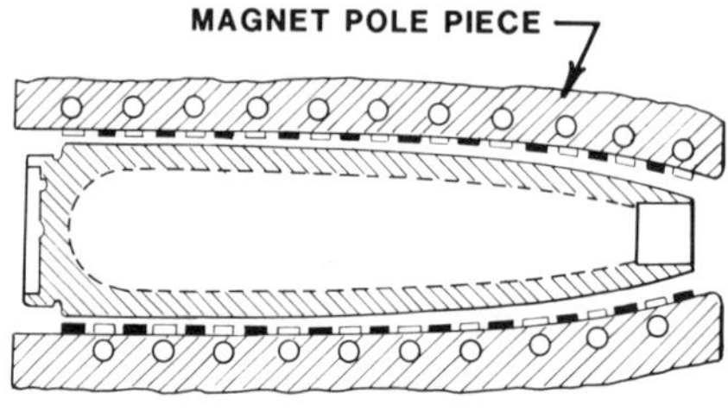

Fig. 4. EMAT Array for Inspecting Artillery Projectiles.

3.3 Inaccessible Locations

3.3.1 Heat Exchanger Tube Inspection. In nuclear reactor power plants, it is extremely important that there be no leaks in the heat exchangers which transfer the heat from the radioactive fluids in the reactor to the steam in the turbines. Thus, the long, curved, small diameter tubes in the heat exchangers must be inspected regularly to detect corrosion pits or fatigue cracks before they penetrate the tube walls. Eddy current and magnetic sensors attached to small probes that are pulled through the tubes have proven quite satisfactory at locating flaws in the regions between the tube supports, but in the immediate vicinity of the supports where stresses are concentrated and corrosion causing foreign matter collects, these methods become insensitive. The Electric Power Research Institute (EPRI) is sponsoring a program to determine the feasibility of using ultrasonic inspection for these locations.[8] EMATs turned out to be most

promising as the transducers for this application for two reasons: (1) their ability to couple acoustic energy in and out of the tubes was not critically dependent on a couplant fluid and precision alignment between the tube wall and the transducer; and (2) their ability to excite and detect special types of ultrasonic waves that were particularly well suited to detecting flaws near the supports. Figure 5 shows a drawing of the special, permanent

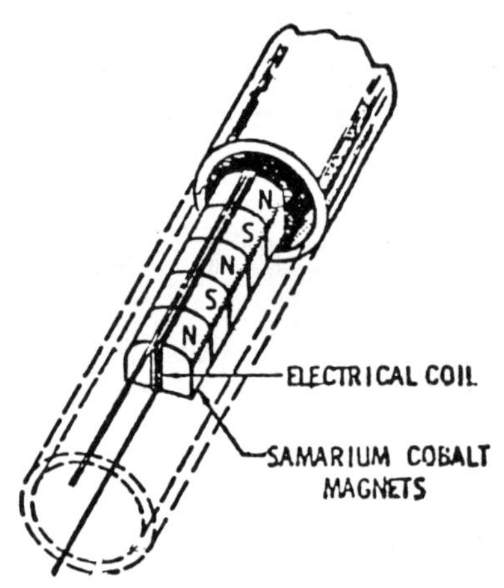

Fig. 5. EMAT for Inspecting Small Tubes.

magnet EMATs designed to slide easily through the 3/4" ID tubes. This arrangement excites a torsional wave in the tube which is insensitive to the tube support structure but which is very sensitive to those changes in the wall thickness which are produced by corrosion or cracking.

3.3.2 Cracks in Hidden Fastner Holes. Figure 6 shows the details of a common joint used in the construction of wings for an Air Force

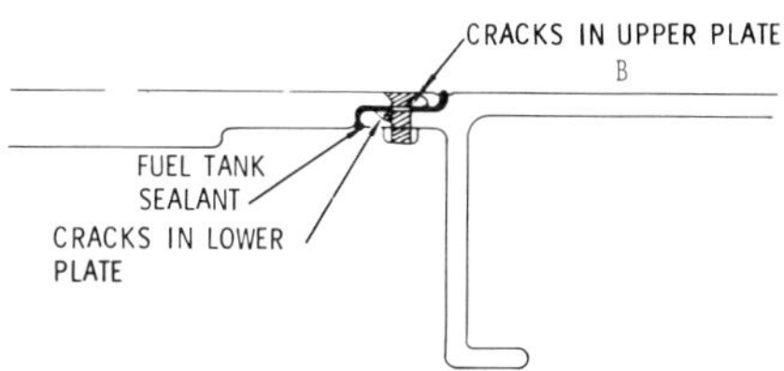

Fig. 6. Wing Structure Inspected by an EMAT at Point B.

cargo plane. In order to conserve weight, the wing skin serves both as a load bearing structural element as well as a seal for the fuel tanks. Hence, the individual parts of the wing have integral stiffeners and overlapping flanges for fasteners. Each fastener joint is filled with a sealant to prevent the leakage of fuel but this sealant also prevents ultrasonic energy from directly reaching the lower flange where fatigue cracks often develop. Thus, the inspection of the region around the fastner hole in the lower flange had to be achieved by guiding sound energy into the area from the exposed wing skin at the location B. In a special Air Force sponsored research study,[9] it was shown that EMATs similar to those used for inspecting the heat exchanger tubes could be used to excite a so-called "shear horizontal" wave (SH wave) in the plate at location B. Because the polarization of these particular shear waves is parallel to the corners at the stiffener intersection, reflections and mode conversion did not take place and the echo from the fastener hole could be clearly resolved. By observing the angular and frequency dependence of

this echo, it was possible to recognize the presence of a crack even when the fastener was in place.

3.4 No Couplant Environments

3.4.1 Aluminum Tube Welds. In order to assemble large vacuum systems for cryogenic and pure gas processing, it is necessary to make many welds in pipes made of very clean aluminum. Often the joints are inaccessible to X-ray testing so some other means of verifying the quality of the weld must be used. Conventional ultrasonic testing is possible, but the use of couplant fluids might contaminate the vacuum system and the inaccessibility of some joints make the achievement of good coupling unreliable. Small EMATs constructed with permanent magnets and coils specifically designed for exciting SH waves in the walls of the aluminum tubes proved to be capable of detecting small amounts of porosity, slag and foreign particles in the welds. An example of the results obtained in a research program sponsored by the Fluor Engineers and Constructors, Inc.[10] is shown in Figure 7.

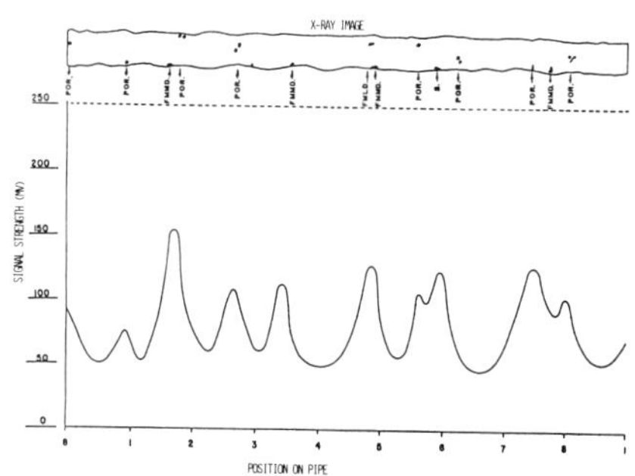

Fig. 7. Echo Amplitude Versus Location on a Weld.

Here the amplitude of the echo reflected by the total weld volume is plotted as a function of circumferential location of the transmit/receive EMAT around the pipe. A drawing of the X-ray photograph of this same weld is shown on the top of the figure with the film positioned so that it corresponds to the same circumferential locations as used by the EMAT. There exists an obvious correlation between the peaks in reflected ultrasonic energy and the existance of flaws in the weld.

3.4.2 Field Welds in Pipelines.
When a gas or liquid pipeline is constructed, the individual lengths of pipe must be welded together in the field where the environmental conditions can be so hot or cold that commonly used ultrasonic couplant fluids cannot be used. Furthermore, the rate at which the pipeline can be laid demands that automatic weld inspection methods be utilized to the fullest extent. Because the couplant used by conventional transducers does not lend itself to automation, a feasibility study was sponsored by a private company to show how EMATs might be applied to this problem. It was demonstrated that an EMAT made with a small electromagnet could be attached to a mechanical crawler that would automatically traverse the circumference of the pipe at the girth weld. The EMAT could be adjusted to excite a shear wave beam

that entered the pipe at an angle so it was possible to irradiate the region of the weld with acoustic energy as shown in Figure 8. By noting the time at which echos returned to the EMAT, it was possible to determine the location of flaws within the weld volume.

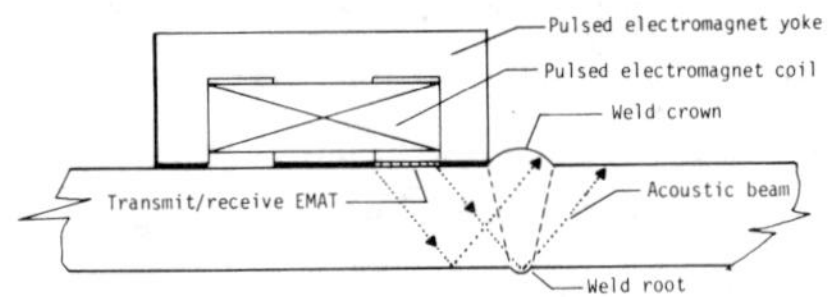

Fig. 8. Pulsed Magnet EMAT for Weld Inspection.

3.5.1 Hot Billet Inspection. A serious problem in the primary metals production industry is the detection and location of the casting "pipe" at the center of as-cast billets before they are rolled or forged into useful shapes. Visual inspection and mechanical cropping away of those parts of the billet that contain the "pipe" are now being used but when the surface conditions happen to obscure the pipe from visual detection, a considerable fraction of the production run can be ruined because it contains flaws rolled into the center of the product. Ultrasonic inspection is an obvious means for detecting the pipe but the requirement of a high temperature couplant fluid and the desire not to mar the surfaces of the billet make conventional approach either useless or very difficult. By making use of the fact that an EMAT can launch and detect ultrasonic waves across an air gap, it would appear that this class of transducer could solve the problem. To investigate this possibility, an EMAT was constructed out of heat resistant materials. By using a large pulsed electromagnet to generate the magnetic field and by winding an efficient coil to excite and detect bulk waves propagating normal to the metal surface, it was possible to detect a 1/4" diameter flat bottom hole at a distance of 8" in aluminum. Figure 9 shows photographs of the echo from the

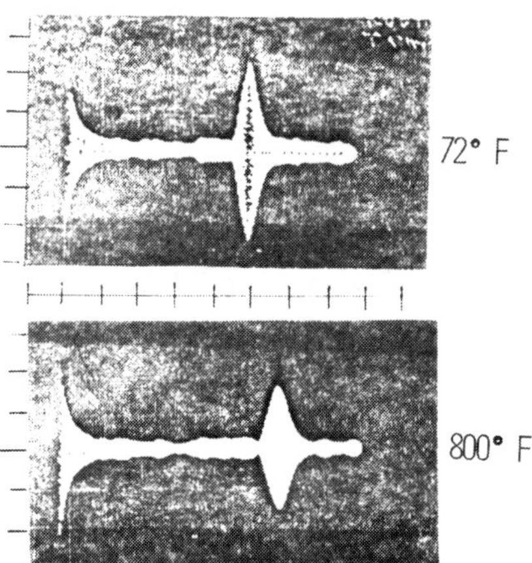

Fig. 9. Echos observed in a Cold and Hot Aluminum Block. (100mv/div, 10 μsec/div.)

bottom of an 8" thick aluminum block at both room temperature and at 800°F (430°C). Only a small decrease in signal strength occurred when the sample was heated to the elevated temperature.

3.6 Stress Measurement

3.6.1 Detection of Bending in Pipelines. When a pipeline must be laid across unstable soils such as on the ocean bottom, in fault zones, through permafrost soils and past mined out areas subject to subsidence, it is quite possible that a shift in the supporting ground will subject the pipeline to a serious bending stress. To detect this situation, a pipeline operator can try to maintain an accurate survey of the pipe's location but a far simpler method would be to monitor the stress level in the pipe wall from a carriage transported through the line by the fluid in the pipe. EMATs have shown a potential for performing this function because in steel type structures the efficiency of transduction is dependent on the state of stress. This relationship between the amplitude of sound waves generated by an EMAT and the external stress arises from the fact that in steel magnetostriction plays an important role in the ultrasonic wave generation mechanism and the magnetostrictive coupling coefficients depend upon the state of stress. If an electromagnet is used to supply the magnetic field to an EMAT coil held next to a steel member, the amplitude of the ultrasonic wave generated is a function of both the current in the electromagnet and the stress as shown in Figure 10. Data similar to that shown in Figure 10 were obtained with EMATs mounted on the inside of a gas transmission pipe when the pipe was subjected to a bending stress.[11] Thus, EMATs mounted on a carriage moving through a pipeline could record curves like those shown

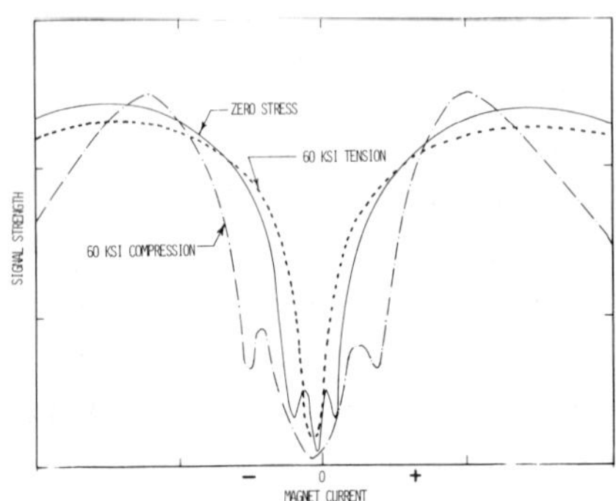

Fig. 10   EMAT Signal Strength Versus Electromagnet Current in Iron When Under Stress.

in Figure 10 at various locations. A subsequent analysis of the shape of these curves could be used to deduce the stress level present at each location. A more detailed investigation of this and other approaches is now being sponsored by the American Gas Association.[12]

3.7 Thickness Measurement

A unique feature of some EMAT coil designs is their ability to control the type of Lamb wave excited in sheet or plate materials.[4] Some of these Lamb wave modes have the property that their velocity of propagation depends upon the thickness of the plate so that a measurement of the time for this type of wave to traverse a fixed distance can be used to deduce the average value of the thickness over this distance. Since the noncontact feature of EMATs allows them to be used on hot or rapidly moving materials, this ability to measure thickness should be useful in many manufacturing operations involving sheet, plate or tubular materials.

4. CONCLUSIONS

The electromagnetic transducer, although weaker than its competitor, the piezoelectric transducer, has several features that give it a special role to play in ultrasonic nondestructive testing. Most obvious among these are those applications where a liquid couplant cannot be used because of environmental constraints (high speed motion, ultraclean conditions, inaccessible locations, extreme temperatures, etc.). Less obvious are those cases where special magnetic or acoustic phenomena permit unique measurements to be performed. The detection of stress in ferromagnetic metals and the use of Lamb wave dispersion for thickness measurement are two such examples on which only feasibility experiments have been performed. In order to apply EMATs effectively, the ultrasonic requirements of the problem should be clearly defined so that the most appropriate type of sound wave can be chosen and the optimum choice of magnetic field and EMAT coil configuration can be made. Then the type of magnet can be picked depending upon the space available or the electric power available to the system. In the future, as more applications are developed, all of these engineering choices will be better defined and specific uses can be quickly implemented.

5. REFERENCES

1. R. B. Thompson, "Electromagnetic Noncontact Transducers," 1973 Ultrasonics Symposium Proceedings, pg.385 (N. Y. IEEE, 1973).

2. R. B. Thompson, "A Model for the Electromagnetic Generation of Ultrasonic Guided Waves in Ferromagnetic Metal Polycrystals," IEEE Trans. Sonics and Ultrasonics Vol. SU25, No. 1, Jan. 1978.

3. Rockwell International Science Center, "Ultrasonic Inspection of In-place Gas Pipelines," Contract PR73-57, American Gas Association, Arlington, VA 1972-1976.

4. R. B. Thompson, "A Model for Electromagnetic Generation of Rayleigh and Lamb Waves," IEEE Trans. Sonics and Ultrasonics, SU-20, pg. 340, 1973.

5. R. B. Thompson, G. A. Alers and M. A. Tennison, "Application of Direct Electromagnetic Lamb Wave Generation to Gas Pipeline Inspection," Proc. IEEE Ultrasonics Symposium 1972.

6. Rockwell International Science Center, "Ultrasonic Inspection of Railroad Rails Using Electromagnetic Transducers," DOT-FR-9143 Dept. of Transportation, Federal Railway Administration, Washington, D. C., 1980.

7. Rockwell International Science Center, "Rapid Ultrasonic Inspection of Artillery Projectiles," Contract DAAK10-77-C-2020 ARRADCOM, Dover, NJ 1979.

8. Rockwell International Science Center, "Evaluation of Electromagnetic-Acoustic Concepts of Inspecting Steam Generator Tubes," Contract EPRI NP-519, Project 698-1, Electric Power Research Institute, Palo Alto, CA 1977.

9. Rockwell International Science Center, "Inspection of Wing Lap Joints for Second Layer Cracks with EMATs," Contract F33615-80-C-5004 DARPA and AFML.

10. Rockwell International Science Center, "Ultrasonic Inspection of Girth Welds in Aluminum Pipes by Non-contact Transducers," Contract 466666-9-K008, Fluor Engineers and Constructors, Inc., Irvine, CA 1979.

11. G. A. Alers and R. B. Thompson, "Detection of Stress in Buried Gas Pipelines," Review of Progress in Quantitative NDE La Jolla Meeting, July 1979, (In Press) DARPA-AFML Contract F33615-74-C-5180.

12. University of New Mexico, "Stress Measurement in Buried Gas Pipelines," Contract PR 155-132, American Gas Association, Arlington, VA 1981.

6. BIOGRAPHY

George A. Alers earned his B.S. in physics at Rice University and his M.S. and Ph.D. also in physics from the State University of Iowa in Iowa City, Iowa. In his early professional career, he was associated with the metallurgy department of the Westinghouse Electric Company's research laboratory working on

problems of plastic flow and fracture in metals. Later, he performed
basic research in the elastic properties of metals using ultrasonic
techniques at the Scientific Laboratory of the Ford Motor Company.
More recently, he has focused on
the development of advanced techniques for nondestructive testing
at the Science Center of Rockwell
International. Now associated with
the New Mexico Engineering Research
Institute of the University of New
Mexico, his work emphasizes the
application of non-contacting transducers to ultrasonic nondestructive
testing.

A NEW IMPACT MODIFIED AND HEAT RESISTANCE PHENOLIC

Dr. M. D. Bertolucci
Manager-Research and Development
Genal Products Section
General Electric Company

ABSTRACT

The excellent heat and solvent resistance of phenolic molding compositions is well known in the reinforced plastics industry. Recently glass fibers have also been incorporated to improve impact strength. Substantial increases in unit part cost ($¢/in^3$) result, however, due to the increased raw material cost and high specific gravity of the final composition. This paper describes a unique mineral and thermoplastic elastomer modified phenolic composition, which unlike previous thermoplastic modified thermosets, is significantly impact modified and has improved heat resistance over the unmodified matrix.

1. INTRODUCTION

Plastics engineers have long recognized thermoset composites as the most heat resistant plastics available on the market. The most adaptable thermoset to various high speed molding technologies has been the two stage phenol/formaldehyde or novolac type. Also known generically as phenolics, these compositions with their inherent solvent resistance, and good dielectric properties are stiff and therefore brittle. This paper describes a unique thermoplastic elastomer modified phenolic which retains all those properties generally attributed to heat resistant grade materials but with significantly improved impact strength.

It is shown that a glass fiber and mineral filled phenolic matrix modified with a reactive thermoplastic elastomer retains: the stiffness (> 1 million psi modulus) and strength (Flexure strength > 12000 psi) of medium heat resistant grade phenolics (specific gravity range 1.50-1.60) but with improved izod impact (0.6 ft. lbs/in notch) and heat resistance unparalleled in the industry (HDT $> 260^{o}C$ and Underwriters Laboratories Continuous Use Temperature Index of $180^{o}C$).

2. THERMAL MECHANICAL PROPERTIES

The need for improved impact grades of phenolic is clearly demonstrated by our search for low cost metal replacements in the automotive and electrical appliance markets. Key properties in this search are dimensional stability or stiffness, impact resistance and heat resistance. Secondary properties such as injection moldability, smooth surface appearance, paintability or platability and economy are also important. The unique combination of fillers and elastomer incorporated in the formulation described as MX582 satisfies to a large degree all these requirements.

Table 1 illustrates the property profile of a basic glass reinforced phenolic molding composition (PMC), MX 100, and the result of replacing mineral filler with a proprietary impact modifying system. Sample MX 582 contains an optimum level of this impact modifier. Flexural strength and modulus and tensile strength for MX 582 are seen to be typical for PMC in the 1.55 gravity range. Quite atypical, however, is the excellent drop ball and notched izod impact strength coupled with a heat deflection temperature in excess of 500°F.

Table 2 shows the stability of the enhanced flexural strength resulting from the proprietary matrix modification versus the MX 100 control formulation and a cellulosic fiber modified control, MX 200, following thermal aging at 350°F. Moreover, the data show that a modest post bake substantially improves the flexure strength of the MX 582 composition and stabilizes it to further more severe thermal aging conditions. The flexure strength retention of 170% of the original value following 100 hours exposure to 450°F is quite remarkable. These data, in addition to those for a typical cellulosic filled PMC, are illustrated graphically in Figure 1.

Further illustration of the unique heat resistance of the MX 582 composition is its long term flexure strength retention at temperatures well above its 180°C Continuous Use Thermal Index (U.L. 746B). Figure 2 shows that after heat aging over 1500 hours MX 582 retains 100% of its original flexural strength at 205°C and 118% at 190°C. As shown in Figure 3 comparable thermal aging of typical 160°C Thermal Index PMC results in flexural strength loss to below 60% in less than 1000 hours at 190°C.

A weight loss profile for the elastomer modified matrix and an industry leading heat resistant PMC is shown in Figure 4 as a function of time at 205°C exposure temperature. It is clear the onset of the thermal degradation which begins near 600 hours at these temperatures for the non asbestos heat resistance grade does not occur in the test interval for the MX 582 composition.

Table 3 summarizes the thermal-mechanical and electrical properties of the elastomer modified composite. Also shown are the resultant changes following a dimensionally stabilizing post bake of 5 hours at 350°F. Without significant exception all properties are maintained or enhanced.

Perhaps the most striking performance characteristic of the thermoplastic elastomer mineral reinforced composite is the heat deflection temperature measured under a 264 psi load. Figure 5 illustrates the deflection of a 1/2 x 1/2 x 5" bar as a function of temperature. It is noted that up to nearly 400°F or higher depending on the flow range of the product the matrix begins to recover against the applied force. The resistance to further distortion out to the limit of the silicone oil bath safe operating range prevents an accurate ultimate HDT to be calculated.

3. EXPERIMENTAL

The following procedures were used to produce and characterize the performance profile of modified phenol/formaldehyde matrixes reinforced with mineral fillers.

3.1 Heat resistance
Flexure strength (1) following thermal aging at elevated temperatures was used as a major criterion for resistance to thermal degradation of mechanical properties. Forced air circulating ovens with ± 2°F temperature control were used throughout.

Injection molded specimens were obtained from a 5 cavity mold which produced both 1/4" and 1/2" bars, tensile dog bone and both 1/8" thick 4" and 2" diameter discs. Specimens

molded with a 175 ton New Britain
press cured 5 seconds over minimum
time to completely cure the 1/2" x
1/2" x 5" bar (typically 35-45 seconds).

3.2. <u>Impact strength</u>
Both izod (2) and drop ball (3)
impact tests were used to evaluate
respectively 1" x 1/2" x 1/2" bar
and 1/8" x 4" dia. plate impact
strength. The drop ball test can
be viewed as a modified Gardner
test which typifies practical impact encounters. A 1/2 lb weight
is repeatedly dropped onto a disc
supported at its rim from succeed-
ingly greater heights beginning at
1" and increasing in 1" intervals.
Subsequent strength is recorded as
the height at which catastrophic
failure results.

Injection molded specimens were
used for all impact tests.

4. SUMMARY

An elastomer modified novolac resin
has been shown to form a unique
impact modified matrix for mineral
reinforced molding compositions.
The injection moldable formulation
demonstrates substantially improved
heat resistance over conventional
phenolics as measured by retention
of both impact and flexural strength
following thermal aging. When used
in conjunction with a crosslinking
agent the proprietary impact modification system increases resistance to thermal oxidation and reduces the rate of deflection under
low applied forces. These properties are reflected in the U.L.
Thermal Index of 180^{0}C and a heat
distortion temperature greater than
500^{0}F.

5. REFERENCES

(1) ASTM method D-790
(2) ASTM method D-256A
(3) General Electric Co. test
     method #1360

6. ACKNOWLEDGEMENTS

The technical contributions and direct support of Mr. J. R. Bartolo-
mucci and Mr. H. G. Oles, Jr. in
the development of this unique concept for impact reinforcement of
phenolic molding materials is greatly appreciated.

7. BIOGRAPHY

Dr. Michael D. Bertolucci is currently Manager of Research and Development for the GENAL Products
Section of General Electric Company.
Prior to his G.E. experiences, he
spent five years with Union Carbide's R&D organization in Bound
Brook, New Jersey. A Ph.D. physical
chemist from California Institute
of Technology in Pasadena, Dr.
Bertolucci has authored papers and
patents in the areas of surface
chemistry, organic chemistry, reinforced thermoplastics and thermosets. A text book entitled "Symmetry and Spectroscopy" co-authored
with Dr. D. C. Harris recently published by Oxford University Press
rounds out a breadth of interests
paralleled only by his enthusiasm
for applied research.

EFFECT OF IMPACT MODIFIER

Sample Designation	MX-100	MX-582
Thermoplastic Elastomer	No	Yes
Specific Gravity	1.74	1.55
Tensile Strength (psi)	9000	7500
Flexure Strength (psi)	15000	12200
Flexure Modulus (psi)	2.0×10^6	1.2×10^6
1/2# Drop Ball (Inches)	15	25
Izod Notched Ft-Lbs/In.	0.5	0.6
HDT @ 264 psi °F	390	>550

HEAT AGED FLEXURE STRENGTH RETENTION

Designation	MX-100	MX-582	MX-200
Impact Modifier	Glass Fiber	Proprietary System	Cellulosic Fiber
Specific Gravity (D792A)	1.75	1.55	1.40
Flexure Strength (psi)			
As Molded	15200	12200	9120
Post Baked 5 Hrs/350°F	17770	15300	—
5 Hrs/350°F + 4 Hrs/450°F	16530	17850	2890
5 Hrs/350°F + 100 Hrs/450°F	15200	20600	—

FIGURE 1

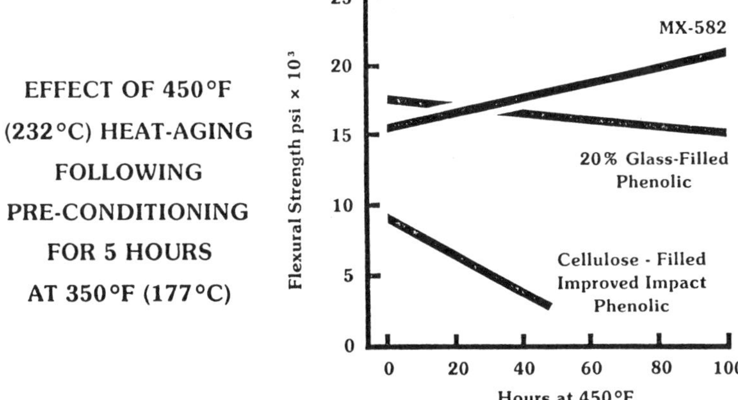

FIGURE 2

FLEXURE STRENGTH RETENTION
MX-582
(180°C THERMAL INDEX)

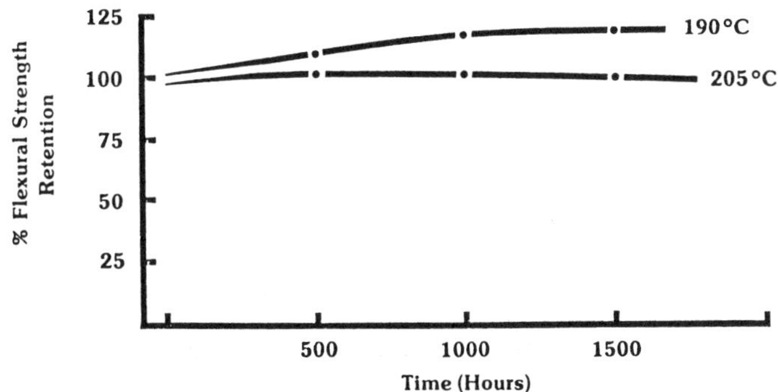

FLEXURE STRENGTH RETENTION
MINERAL FILLED HEAT RESISTANT GRADE
(160°C THERMAL INDEX)

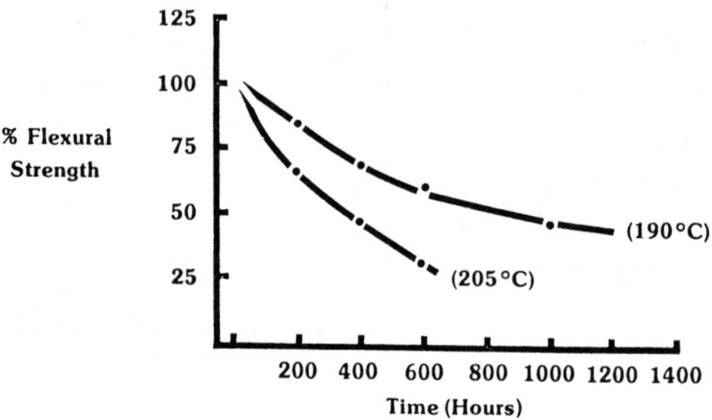

205°C WEIGHT LOSS PROFILE
MX-582 vs MINERAL FILLED HEAT RESISTANT GRADE

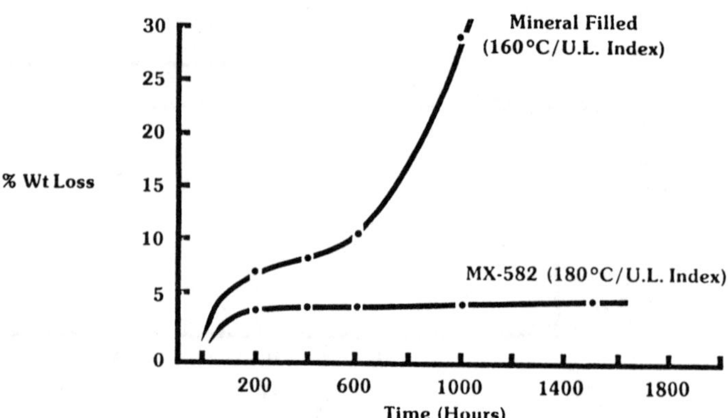

POST BAKED PROPERTY PROFILE MX-582

	As Molded	5 Hrs/350°F
Tensile Strength (psi)	7500	9500
Flexure Strength (psi)	12200	15300
Flexure Modulus (psi)	1.2×10^6	1.3×10^6
Compressive Strength (psi)	16000	21500
Hardness, Rockwell "M"	85	95
Drop Ball Impact (½# Wt)	25″	24″
Izod, Notched Ft Lbs/In.	0.6	0.6
HDT @ 264 psi	⪀ 550°F	Indeterminant
Arc Resistance, Sec.	130	177
CTI, Volts	175	190
Dielectric Strength S/T (vpm)		
40/23/50 (Dry)	395	400
96/35/90 (Wet)	420	430
Volume Resistivity (ohm-cm)		
40/23/50 (Dry)	1×10^{11}	1×10^{12}

FIGURE 5

COMPARISON OF HEAT
DEFLECTION
TEMPERATURE (HDT)
BEHAVIOR OF MX-582
TYPICAL
HEAT RESISTANT,
AND 20 PERCENT
GLASS-FILLED GRADES

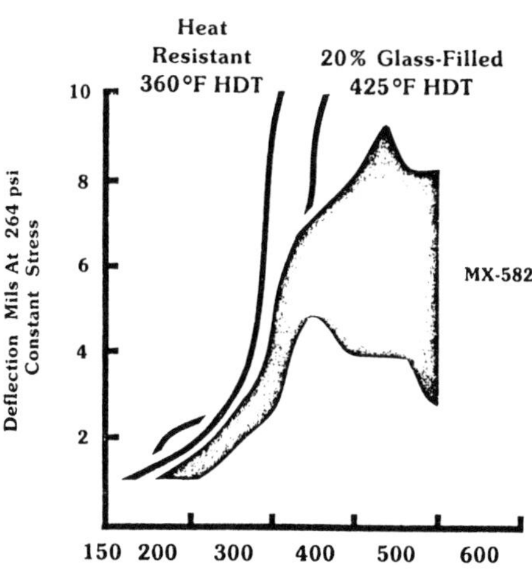

26th National SAMPE Symposium
April 28-30, 1981

HIGH-PERFORMANCE MULTIFUNCTIONAL CORROSION INHIBITOR FOR AIRCRAFT

M. Khobaib
Systems Research Laboratories, Inc.
2800 Indian Ripple Road
Dayton, OH 45440

C. T. Lynch
Air Force Wright Aeronautical Laboratories
Materials Laboratory, AFWAL/MLLN
Wright-Patterson Air Force Base, OH 45433

Abstract

Extensive research has been conducted on the development of a water-soluble non-toxic inhibitor to be used in the lower bay, galley, and related areas of large aircraft. A large number of inhibitor formulations have been tested for their effectiveness in reducing aqueous corrosion in these areas of aircraft. Synthetic urine consisting of more than twenty aggressive ingredients has been prepared in the laboratory to simulate the bilge solution. The corrosion behavior of this synthetic urine has been compared with that of several specimens of natural urine. Anodic and cathodic potentiodynamic-polarization and immersion test results have shown quantitative comparability with the corrosion behavior of natural urine. Based upon extensive prior experience with borax-nitrite inhibitor formulations, corrosion tests have been conducted on numerous formulations consisting of water-soluble non-toxic ingredients such as borax, nitrite, phosphate, and silicate. Some commercial inhibitors also have been screened using fast electro-chemical techniques. Corrosion-fatigue and crack-growth studies on aluminum alloys with inhibited and uninhibited urine solutions have also been conducted. Several of the formulations show promise as effective inhibitors.

Results of anodic, cathodic, and linear polarization and immersion tests on high-strength aluminum alloys, steel, and copper will be discussed.

1. INTRODUCTION

During the past several years, a considerable number of studies have been conducted at several levels within the Air Force and at the National Bureau of Standards[1] regarding the total cost of corrosion prevention and control for aircraft. The inescapable conclusion has been that total corrosion costs in terms of life-cycle management and maintenance of aircraft represent an intolerable burden to the Air Force in maintaining force effectiveness at a reasonable cost to the taxpayer. The direct costs of corrosion maintenance in the field and at the depot level for major aircraft systems have been estimated to exceed 750 million dollars a year, with the total corrosion costs including facilities estimated at more than one billion dollars a year. One of the major cost items for depot repair has been due to corrosion in the lower surfaces of particularly larger aircraft, such as the C-130 and C-141A. Generally inaccessible and hard to inspect, these areas (referred to as "bilge" areas) act as traps for dirt and moisture which accelerate corrosion and result in major corrosion damage that requires large-scale structural repair. In addition to lower cargo bay areas, hot spots for corrosion are found in the galley and latrine areas. Warner Robins Air Logistics Center Corrosion Management Office has often cited the need to investigate the development of corrosion inhibitors for this type of application that would be simple to apply or use. Bilge inhibitors in package form have already been reported in use by the Royal Air Force in Great Britain[2], but they are reported to contain chromates which would not be satisfactory from a toxicity standpoint for the Air Force. Some commercial use has been made of water-displacement compounds, but these have limited life, are difficult to apply, and are toxic in application. The magnitude of the cost of bilge-area rework is indicated from recent figures for depot work on the C-141A, on which 1182 hours were required per aircraft for a cost in excess of two million dollars per year for the entire C-141A force.

Although chromate-based[3,4] corrosion inhibitors have been widely used to combat corrosion of ferrous and non-ferrous metals and alloys, the use of chromates has recently been the subject of ecological concern. The present investigation was carried out to find alternatives to chromates, such as a borax-nitrite-based inhibitor. The value of borax-nitrite as a corrosion inhibitor has long been recognized.[5,6] Earlier work[7] has shown a borax-nitrite combination to be very effective in controlling general corrosion as well as crevice corrosion of high-strength steels. However, this combination was not found to be effective against the corrosion of other ferrous and nonferrous metals and alloys. The main objective of the present study was the development of a nontoxic multifunctional corrosion inhibitor (which would be effective against corrosion of other typical aerospace ferrous and nonferrous metals and alloys, for example, aluminum and copper alloys) to be incorporated into the hot-spot corrosion-prone areas of aircraft.

The initial basis for the current work was provided in a prior investigation[8] to develop nontoxic, multifunctional, water-soluble corrosion inhibitors for use in the Air Force automatic rinse facility for fighter aircraft at MacDill Air Force Base, Florida. This rinse facility was placed in operation in 1979 to remove corrosion-causing contaminants such as sea salt routinely through a rinse procedure to reduce corrosion maintenance costs. Several hundred inhibitor compounds and formulations were studied with regard to their effect upon electrochemical behavior (general corrosion, galvanic corrosion, crevice corrosion) and corrosion fatigue. As a result, a new formulation has been developed, which in lab tests has proven its effectiveness for use in the lower bay and galley areas of aircraft. This mixture contains no chromate, is biodegradable, and offers important advantages over chromate-based combinations.

2. GENERAL CONSIDERATIONS FOR INHIBITORS

Several commercial inhibitors are available for various service applications such as cooling-tower circuits, central heating systems, automotive radiators, and water-desalination plants. These formulations are normally combinations of several classes of inhibitor compounds, some functioning as anodic inhibitors and others as cathodic inhibitors. Commercial experience has shown that such combinations are often more effective due to some synergistic[9] effect. Unfortunately, most of them are optimized for a specific application.

A substantial number of corrosion-preventive compounds (called CPC's) are also available commercially. These are based upon the use of water displacing compounds, protective organic films, and surfactant agents in a solvent vehicle. Their chief disadvantages appear to be limited effective life, requirement of clean and uncorroded surfaces, cost, and toxicity of the carrier solvents.

The results of work conducted by the Air Force have demonstrated an encouraging inhibition effect of borax-nitrite systems upon high-strength steels;[10] chromates have not been found to be so effective in the presence of chloride ions.[11] The promising results of the borax-nitrite combination were observed in crack-growth experiments--both in static tests and cyclic corrosion fatigue tests. However, this combination itself was not effective in inhibiting the corrosion of high-strength aluminum alloys, copper, and other alloys used in aircraft structures. But the encouraging results obtained on high-strength steels served as the basis for further exploration.

In order to systematize the development of new inhibitor formulations and current commercial products, it is desirable to develop some guidelines for inhibitor selection for further experimental screening. In Table I some of the more important considerations and possible compound types are listed. Most of these are obvious considerations, but a few are particularly important for aerospace or other applications where high-strength alloys are utilized. These considerations require ability to retard or eliminate hydrogen embrittlement, stress-corrosion cracking, and corrosion fatigue which can lead to catastrophic failure in such alloys.[10]

The first and foremost requirement in the screening of the inhibitors was the question of toxicity. All the inhibitor formulations which were obviously toxic (based upon literature information)[12] were eliminated. Chromates, aniline, and arsenic additions were obvious examples.

The use of the guidelines in Table I and previous work on the development of rinse inhibitors for aircraft[8] led to a list of inhibitor compounds and formulations for subsequent experimental screening. Special solutions were developed, which are discussed in the experimental section, to synthetically reproduce the most aggressive media expected to be present in the lower bay areas of large aircraft.

3. EXPERIMENTAL

Since the inhibitor was being developed for aircraft galley and low bay areas, the materials chosen to be tested were those commonly used in aircraft structural alloys. Three high-strength aluminum alloys, 2024-T3, 7075-T6 and 7050, along with high-strength steel, 4130, 4340, cast iron, and 70-30 brass were obtained from a local supplier (Jorgenson Steel, Dayton, OH).

Standard 60 × 30 × 3.125 mm test coupons were used for immersion tests on aluminum alloys. Smaller rectangular sheets with dimensions 75 × 25 × 3.125 mm were used for high-strength steel, brass, and cast iron. These were mechanically polished with emery paper up to 400, cleaned thoroughly in acetone and alcohol, and in several instances finally degreased with petroleum ether. A hole, nearly 5 mm in diam, was made close to one end and the specimens were suspended by means of a fish line (nylon thread). The maximum duration of these tests has been up to 30 months, but most of the test specimens were immersed for 90 days (3 mo.).

The working electrodes for the electrochemical tests were 25-mm-square pieces which were carefully mounted in resin and were tapped with 3-48 thread for attaching to the electrode holder. All electrochemical tests were carried out in accordance with ASTM standard G5-72, "Standard Recommended Practice for Standard Reference Method for Making Potentiostatic and Potentiodynamic Standardization Measurements." The measurements were conducted by means of an automated PAR unit consisting of a corrosion cell, potentiostat/galvanostat, log converter, programmer, and X-Y recorder.

A series of systematic tests were conducted which involved the screening of 1) anodic inhibitors (single component, multicomponent); 2) cathodic inhibitors (single component, multicomponent); 3) combination of anodic and cathodic inhibitors; and 4) multifunctional systems containing the anodic-cathodic inhibitors and other components such as film formers, chelating agents, and wetting agents. The screening of a large number of combinations was conducted by the potentiodynamic polarization technique and weight-loss methods.

In the immersion tests, the weight loss per unit of surface area of the specimens in different electrolytes was converted to mpy (mils per year). The percentage inhibitive efficiencies were not calculated because the final selection of the inhibitor was decided, based on the visual observation (where there was no change in surface appearance) and polarization results. In some cases, pieces of aluminum, high-strength steels, brass, and cast iron were suspended together in one electrolyte to check the effectiveness of inhibitors against interfering ions. Finally, the effectiveness of the inhibitor for metallic parts prone to galvanically coupled conditions was also examined.

Low-cycle corrosion-fatigue tests[13] were conducted to determine the effectiveness of the inhibitor formulations in retarding crack growth. The modified rinse inhibitor, with the addition of 125 ppm Richonate 1850, was chosen for this purpose. Compact-tension plane-strain fracture-toughness specimens (Al 7075-T6) were used to determine the crack-growth rate in the presence of uninhibited and inhibited

natural and synthetic urine. A detailed description of the corrosion-fatigue tests on aluminum alloys is given in Ref. 14. Sinusoidal tension-tension cycling was used at a frequency of 0.1 Hz.

The crack-length-vs.-number-of-cycles data were converted to fatigue-crack-growth rates (da/dN) using a computer program.[15] Seven to eleven data points were fitted to a second-order polynomial, and the derivative (da/dN) was then obtained for the middle data point. This process was then repeated over the entire range of data. The da/dN-vs.-ΔK curves were then constructed from the data for tests in uninhibited and inhibited solutions.

Commercial inhibitor solutions and aerosols were obtained from the manufacturers or commercial vendors. Reagent-grade chemicals and distilled water were used to make solutions, with the exception of the use of tap water for typical hard-water simulation. The most aggressive solution used was a synthetic urine developed to simulate the aggressive corrosion behavior of natural urine. The composition of this synthetic solution, consisting of the most aggressive components of natural urine, is given in Table II. The corrosive behavior of the synthetic urine was found to closely approximate that of natural urine and was used in most subsequent testing. The polarization behavior of synthetic urine is found to be almost identical to that of natural urine on Al 7075-T6, as shown in Fig. 1.

4. RESULTS AND DISCUSSION

From the hundreds of polarization and immersion tests that have been conducted in this investigation, a number of representative results have been selected for analysis in this paper. It is important to understand that optimizing inhibitor formulations for aggressive environments such as a urine (or synthetic substitute) requires many more experiments than will be discussed here.

Initially the new formulations were tested with the alloy Al 7075-T6. Those formulations that gave encouraging results in terms of low corrosion currents and reasonably wide potential ranges for passive (or pseudo-passive) regions were tested through experiments with other alloys such as high-strength steel and brass. It was found that aggressive environmental effects that were inhibitive in corrosion of the high-strength aluminum alloys were generally protective toward other aerospace alloys such as high-strength steels. The converse of this, however, was generally not true. Figure 2 is the anodic polarization plot of Al 7075-T6 in the synthetic-urine solution with the rinse inhibitor[8] added. The values of Tafel slopes and corrosion current calculated from linear polarization tests are given in Table III. Concurrent with the polarization tests, immersion experiments were conducted. These results are given in Table IV. These results show that only limited protection is provided by the rinse inhibitor in this aggressive environment. The anodic-polarization curve (Fig. 2) indicates a very small passive region. However, the current density corresponding to the break-down potential (pitting potential) has been significantly lowered as compared to that in synthetic urine.

Since the synthetic (and natural) urine contains nearly 1% sodium chloride and since sodium nitrate is known to provide inhibition against chloride ions, several formulations with higher concentrations of nitrites and nitrates were used in the development process. The immersion results in Table IV suggest that better protection is provided by formulation Nos. 2 and 3. This is mainly the effect of higher nitrite and nitrate concentrations, as indicated earlier. Long-term immersion resulted, however, in the pitting of

aluminum and steel. This is thought to be due to the local breakdown of the passive layer by ions present in the complex chemistry of the synthetic urine. At the same time the corrosion rates as shown in Table III decreased only by a small amount and there was only slight improvement in the breakdown (of passivity) potential.

Several modifications of rinse-inhibitor formulations[8] by additions of small concentrations of film formers, chloride absorbers, and a number of chelating agents were tested. Figure 3 shows the effect of the addition of low concentrations of isoproylamine, Triton X-114, and Estersulf. The anodic polarization curves exhibit a large passive region, and the linear polarization results shown in Table III represent marked lowering of the corrosion current. The immersion test results on Al 7075-T6 were found to be excellent for some of these formulations. There was no visible corrosion, upon close inspection after six to ten months of immersion. No weight loss could be detected and was reported as negligible. This is consistent with the observation that amines, along with silicates and nitrites,[16,17] provide passivation for aluminum. Vermilyea, _et al._,[18] has also reported that satisfactory passivation is provided to aluminum by salts of organic acids (sulphonic acid, stearic acid, etc.).

Problems were encountered when pieces of aluminum, steel, and brass were immersed together in the synthetic urine. Some of the formulations which exhibited excellent inhibition to aluminum alloys failed to protect the metals under these conditions. Table V gives immersion results for selected formulations. There was evidence that copper ions were dissolving out from the brass and interfering with the inhibiting effect of these formulations.

When different metallic materials are present simultaneously in one aggressive electrolyte, it becomes necessary to use a combination of different inhibitors in order to provide protection from the effect of interfering ions on all the materials present. For the combination of steel and copper or copper and aluminum, in which the corrosion of steel--and particularly of aluminum--is frequently increased by copper ions, corrosion of the copper itself may not be significant. In view of this, the concentration of mercaptabenzothiazole or benzothiazole was increased in some formulations. ZnSO4 was also included in several formulations to provide additional protection to steel. The immersion results of some of these combinations are shown in Table V. Figure 4 contains the anodic polarization curves for aluminum, brass, and steel. The weight-loss measurements combined with the polarization results show that very effective inhibition is provided by 0.35 w/o borate + 0.2 w/o nitrite + 0.2 w/o nitrate + 0.01 w/o silicate + 50 ppm phosphate + 50 ppm MBT + 125 ppm Richonate 1850. Weight-loss results clearly demonstrate the very improved protection provided by this formulation, compared to some of the commercial formulations called out in Table V.

Certain formulations which are effective for general corrosion inhibition actually accelerate corrosion in crevice situations and are crack accelerators. In order to evaluate the effectiveness of the inhibitors in all practical applications, low-cycle corrosion-fatigue tests were conducted. Figure 5 illustrates the corrosion-fatigue behavior of Al 7075-T6 in synthetic urine and synthetic urine inhibited by the formulation developed as a result of the present study. The air result is shown for comparison. The reduction in the fatigue-crack-growth rate due to

the addition of the inhibitor can be clearly observed. It is interesting to note that the crack-growth rate is two to three times lower in Region II, as compared to the air value, and five to six times lower than that obtained in uninhibited synthetic urine. The difference in the crack-growth rates in air and inhibited synthetic urine may be due to the fact that the air value is not a vacuum value and, in general, is higher. This suggests that the introduction of the inhibitor nearly eliminates the environmental influence upon the crack growth of this alloy. For applications involving high-strength aerospace alloys, it is important to establish this inhibition or elimination of the environmental acceleration of crack-growth rates in the corrosion-inhibiting medium.

5. CONCLUSIONS

A synthetic-urine solution has been formulated and found to experimentally reproduce the corrosion behavior of natural urine which is considered to be the most aggressive corrosive environment in the lower-cargo bay, galley, and comfort-station areas of large aircraft.

Extensive polarization and immersion experiments have been conducted to determine and optimize the effectiveness of various inhibitor systems in aggressive media.

High-strength aluminum alloys are inhibited successfully in a rinse-inhibitor formulation containing small amounts of isopropylamine. In the presence of other metals such as iron and copper, this inhibitor system loses effectiveness. Additions of chelating agents and surfactants restore inhibitor effectiveness.

A borax-nitrite-based inhibitor with 0.35 w/o borate, 0.2 w/o nitrite, 0.2 w/o nitrate, 0.01 w/o silicate, 50 ppm phosphate, 50 ppm MBT, and 125 ppm

Richonate 1850 offered excellent
protection to a wide variety of
aerospace alloys.

Environmental effects upon crack-
growth rates of aluminum alloys
were eliminated, reducing the rates
in corrosion fatigue to those
obtained in air.

6. REFERENCES

1. L. H. Bennett, J. Kruger, R. L.
Parker, E. Passaglia, C. Reimann,
A. W. Ruff, and H. Yakowitz,
Economic Effects of Metallic
Corrosion in the United States,
Parts I and II, NBS Special
Publication 511-1,2 (U.S. Dept
of Commerce, National Bureau
of Standards, Washington, D.C.).

2. Data presented at AFLC Corrosion
Managers Conference, WRALC,
Robins AFB, GA, October 1975.

3. Corrosion Inhibitors (C. C.
Nathan, ed.)(National Associa-
tion of Corrosion Engineers,
Houston, TX, December 1974).

4. The Corrosion and Oxidation of
Metals (Second Supplementary
Volume) (U.R. Evans, ed.)(Edward
Arnold Publishers, Ltd., U.K.,
1976).

5. J. Green and D. B. Boies, Cor-
rosion Inhibiting Compositions.
U.S. Patent No. 2,815,328,
December 3, 1957.

6. J. T. Bregman and D. B. Boies,
Corrosion 14, 275 (1958).

7. K. Bhansali, C. T. Lynch, F.
Vahldiek, and R. Summit, "Effect
of Multifunctional Inhibitors on
Crack Propagation Rates of High
Strength Steels in Corrosive
Environments," Prepared discuss-
ion presented at the Conference
on the Effect of Hydrogen on
Mechanical Behavior of Materials,
Moran, WY, 7-11 September 1975.

8. M. Khobaib, "Materials Evalu-
ation; Part II - Development of
Corrosion Inhibitors," AFML-TR-
79-4127 (Air Force Materials
Laboratory, Wright-Patterson
AFB, OH, September 1979).

9. F. N. Speller, Discussion, Proc.
ASTM 36 (Part 2), 695 (1936).

10. C. T. Lynch, F. Vahldiek, K. J.
Bhansali, and R. Summit,
"Inhibition of Environmentally
Enhanced Crack Growth Rates in
High Strength Steels," Symposi-
um on Environment Sensitive
Fracture of Engineering Materi-
als, AIME Meeting, Chicago, IL,
26 October 1977. Proceedings
Volume, AIME, 1979, pp. 639-658.

11. C. T. Lynch, K. J. Bhansali, and
P. A. Parrish, "Inhibition of
Crack Propagation of High-
Strength Steels through Single-
and Multi-Functional Inhibitors,"
AFML-TR-76-120 (Air Force
Materials Laboratory, Wright-
Patterson AFB, OH, 1976).

12. N. I. Sax, Dangerous Properties
of Industrial Materials (Second
Edition) (Reinhold Publishing
Corp, NY, 1963).

13. "Tentative Method of Test for
Plane-Strain Fracture Toughness
of Metallic Materials, E 399-
70T," Annual Book of ASTM Standards, Part 31, (American
Society for Testing and Materials, Philadelphia, PA, July
1971).

14. M. Khobaib, C. T. Lynch,
Accepted for publication in
Corrosion.

15. N. E. Ashbaugh, "Mechanical
Property Testing and Materials
Evaluation and Modeling," AFML-
TR-79-4127, Part I (Air Force
Materials Laboratory, Wright-
Patterson AFB, OH, September
1979).

16. G. Schick, Mater. Perf. 23
    (February 1975).

17. A. H. Roebuck and T. R.
    Pritchett, Mater. Perf. 16
    (June 1966).

18. D. A. Vermilyea, J. F. Brown,
    and D. R. Ochae, J. Electro-
    chem. Sci. 783 (June 1970).

7. BIOGRAPHIES

Dr. M. Khobaib is a Senior
Metallurgist with Systems Research
Laboratories, Inc. (SRL), Dayton,
OH. He received his Ph.D. degree
in Materials Science from the
University of Connecticut in 1974
and served there for two years as a
Postdoctoral Fellow. He received
his M.S. degree from the Indian
Institute of Technology, Kanpur,
India, in 1969 and his B.Sc.Engg.
degree from Regional Institute of
Technology, Jamshedpur, India, in
1965. Currently he is involved in
research on corrosion fatigue,
stress corrosion, and corrosion
inhibition (patent pending). Prior
to joining SRL, Dr. Khobaib was a
Research Associate at the Construction Engineering Research Laboratory, U.S. Army Corps of Engineers,
Champaign, IL, where he developed
design guidelines and criteria for
cathodic protection of coastal
structures. He is a member of ASM,
AIME, and NACE, and has authored
more than twenty publications and
presentations.

Dr. C. T. Lynch received his B.S.
degree in 1955 in Chemistry from
the George Washington University,
his M.S. degree in 1957, and Ph.D.
degree in 1960 in Analytical
Chemistry from the University of
Illinois. From 1960 to 1962 he was
an Air Force officer at Wright-
Patterson AFB. Since 1962 he has
served in research and management
positions with the AFWAL Materials
Laboratory and is currently Senior
Scientist for Environmental Effects,
responsible for planning and initi-
ating R&D in the fields of corrosion prediction, prevention, and
control. He has conducted and
managed a wide variety of materials
efforts including the development
of new refractory inorganic
materials, metal-matrix composites,
and multifunctional non-toxic
corrosion inhibitors. He was
responsible for developing new
methods of synthesis of refractory
materials from organometallic compounds and was co-developer of the
new refractory ceramic "Zyttrite."
He has taught chemistry and
lectured on corrosion at Wright
State University in Dayton, OH.
Dr. Lynch is the author of
approximately 200 publications and
presentations, holds 13 patents,
and has co-authored one book. He
is the Editor of the CRC Handbook
Series on Materials Science, now
in its forth volume.

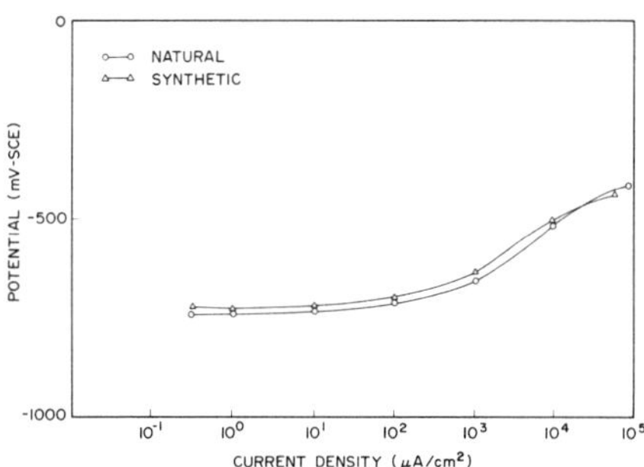

Fig. 1. Comparison on Anodic Polarization Behavior of Al 7075-T6
in Synthetic and Natural
Urine.

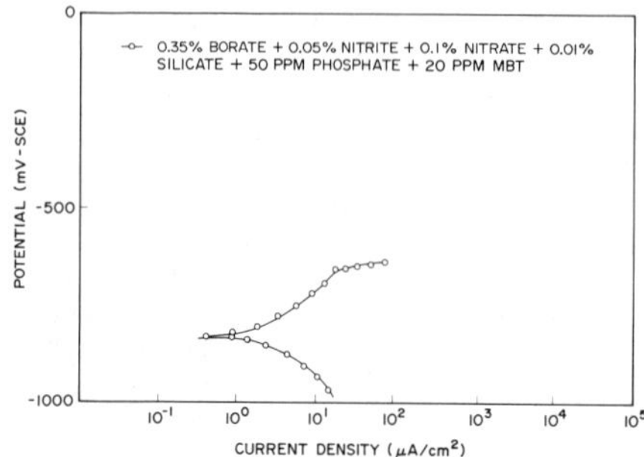

Fig. 2. Anodic and Cathodic
Polarization Behavior of
Al 7075-T6 in Synthetic
Urine Inhibited by Rinse
Formulation.

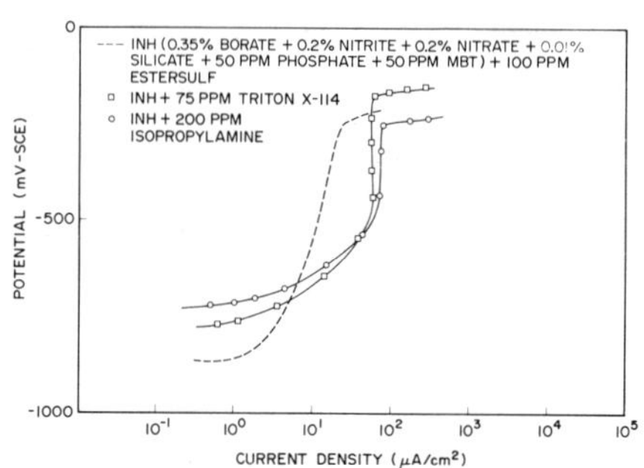

Fig. 3. Effect of Addition of
Estersulf, Triton X-114,
and Isopropylamine upon
the Anodic Polarization of
Al 7075-T6.

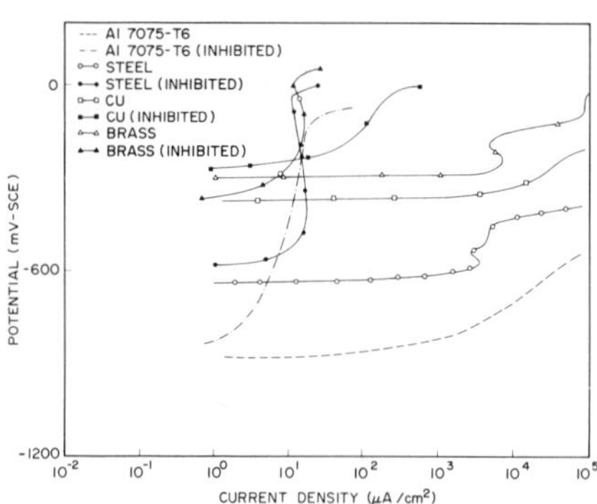

Fig. 4. Anodic Polarization Behavior
of Al 7075-T6, Steel,
Copper, and Brass in
Inhibited Synthetic Urine.

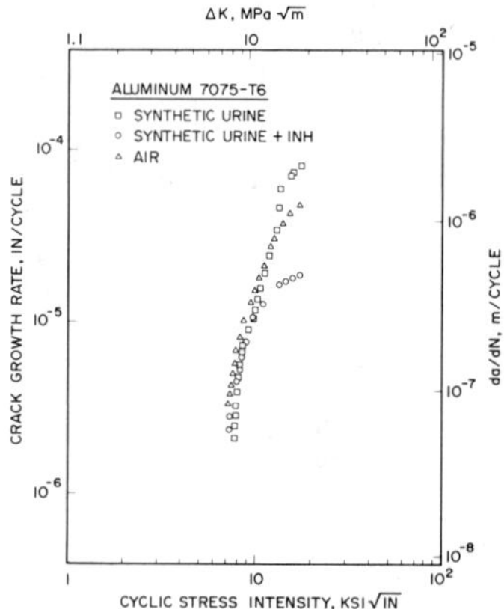

Fig. 5. Comparison of Crack-Growth
Rates of Al 7075-T6 in
Air and Inhibited and
Unhibited Synthetic Urine.

TABLE I
INHIBITORS

A. GENERAL CONSIDERATIONS

    1. Multifunctional
            Cathodic
            Anodic
            Chloride Absorbers
            Buffers
    2. Solubility Range
    3. Influence on Hydrogen Entry Rates
    4. Toxicity
    5. Cost

B. COMPOUNDS

    1. Cathodic: Polyphosphate, Zinc, Silicate
    2. Anodic: Orthophosphate, Chromate, Ferrocyanide, Nitrite
    3. Combinations: Polyphosphate-Chromate
                       Polyphosphate-Ferrocyanide
                       Borax-Nitrite
                       Benzoate-Nitrite
                       Silicate-Chromate
    4. Film Formers: Emulsified or Soluble Oils
                       Octadecylamine
                       Long-Chain Amines
                       Alcohols and Carboxylic Acids

C. GENERAL CONSIDERATIONS

    1. Stress Corrosion and Corrosion Fatigue
    2. Special Bilge Environments
    3. Long-Term Effectiveness
    4. Method of Application

TABLE II
INGREDIENTS OF SYNTHETIC URINE

	(wt in gm/liter)
Urea	20.60
5-Hydroxyindoleacetic Acid	0.0045
Uric Acid	0.052
Glucuronic Acid	0.431
Oxalic Acid	0.031
Citric Acid	0.462
Glycolic Acid	0.042
Creatine	0.0721
Guanidinoacetic Acid	0.027
Formic Acid	0.013
Glucose	0.072
Ammonium Sulfate	4.00
Potassium Phosphate	0.175
Potassium Chloride	0.0100
Potassium Bromide	0.008
Sodium Chloride	10.00
P-Cresol	0.087
Creatinine	1.500
Acetone	0.0001
Hydroxyquinoline-2 Carboxylic Acid	0.0028
Potassium Sulfate	0.134

TABLE III
TAFEL SLOPES AND CORROSION CURRENTS
IN DIFFERENT ELECTROLYTES
OF Al 7075-T6

ELECTROLYTE (wt%)	TAFEL SLOPES (mV/decade)		i_{corr} ($\mu A/cm^2$)
	ba	bc	
Synthetic Urine	100	135	6.24
Synthetic Urine + Rinse Inhibitor	120	160	3.42
Synthetic Urine + Rinse Inhibitor + 0.15 Nitrite + 0.1 Nitrate (INH)	160	130	2.44
Synthetic Urine + INH + 75 ppm Triton X-114	100	150	0.78
Synthetic Urine + INH + 100 ppm Estersulf	120	90	1.08
Synthetic Urine + INH + 100 ppm Isopropylamine	120	80	1.41
Synthetic Urine + INH + 125 ppm Richonate 1850	100	125	0.544
Synthetic Urine + INH + 125 ppm Richonate + 500 ppm $ZnSO_4$ + 50 ppm MBT	120	135	0.63

TABLE IV

IMMERSION TEST RESULTS ON Al 7075-T6

TYPE NO.	ELECTROLYTE (wt%)	mpy	pH INITIAL	pH FINAL	TIME OF EXPOSURE	SURFACE APPEARANCE
1	Synthetic Urine	0.23	5.6	5.8	3 Mo.	Nearly 20% surface area pitted.
2	Natural Urine	0.18	5.8	6.0	3 Mo.	Nearly 25% surface area pitted.
3	Synthetic Urine + Rinse Inhibitor	0.12	8.30	8.25	3 Mo.	Several pits on the surface.
4	Synthetic Urine + Rinse Inhibitor + 0.15 Nitrite + 0.1 Nitrate	0.078	8.35	8.35	3 Mo.	Surface clean, very light corrosion.
5	Synthetic Urine + Rinse Inhibitor + 0.25 Nitrite + 0.2 Nitrate	Negl.	8.40	8.35	3 Mo.	Clean as original. No sign of corrosion.
6	No. 4 + 25 ppm Triton X-114	Negl.	8.30	8.30	3 Mo.	Clean as original. No sign of corrosion.
7	No. 4 + 75 ppm Estersulf	Negl.	8.35	8.30	3 Mo.	Clean as original. No sign of corrosion.

TABLE V
TEST RESULTS ON Al, BRASS, AND STEEL
SPECIMENS IMMERSED TOGETHER

SPECIMEN	ELECTROLYTE (wt%)	pH INITIAL	pH FINAL	TIME OF EXPOSURE	SURFACE APPEARANCE
Al 7075-T6	0.35 Borate + 0.2 Nitrite + 0.2 Nitrate + 0.01 Silicate + 50 ppm Phosphate + 25 ppm MBT (INH) + 250 ppm Isopropylamine in Synthetic Urine.	8.45	8.30	3 Mo.	50% area pitted.
Brass				3 Mo.	Slowly dissolving into solution.
Steel				3 Mo.	Several pits.
Al 7075-T6	INH + 25 ppm Triton X-114 in Synthetic Urine.	7.75	7.60	3 Mo.	Several large pits.
Brass				3 Mo.	Clean but dull.
Steel				3 Mo.	Pitted – dull.
Al 7075-T6	INH + 25 ppm Triton X-114 + 100 ppm $ZnSO_4$ in Synthetic Urine.	7.65	7.60	3 Mo.	Clean and shiny.
Brass				3 Mo.	Clean.
Steel				3 Mo.	One corner pitted.
Al 7075-T6	INH + 75 ppm MBT + 125 ppm Richonate 1850 in Synthetic Urine.	8.25	8.20	3 Mo.	Clean and shiny as original.
Brass				3 Mo.	Clean, as original.
Steel				3 Mo.	Clean, as original.
Al 7075-T6	INH + 50 ppm Estersulf in Synthetic Urine.	8.30	8.20	3 Mo.	Clean, few pits.
Brass				3 Mo.	Clean.
Steel				3 Mo.	Clean, but several fine pits.
Al 7075-T6	AML Guard in Synthetic Urine.	6.90	6.95	3 Mo.	Nearly 20% area badly pitted.
Brass				3 Mo.	Nearly 20% area badly pitted.
Steel				3 Mo.	More than 50% area badly pitted.
Al 7075-T6	1% Boeshield T-9 in Synthetic Urine.	5.10	5.15	3 Mo.	Several large pits and number of small pits.
Brass				3 Mo.	Dull, one pit.
Steel				3 Mo.	Several large pits and number of small dark pits.

MICRODIELECTROMETRY: A NEW METHOD FOR IN SITU CURE MONITORING

Norman F. Sheppard, Steven L. Garverick, David R. Day,
and Stephen D. Senturia
Department of Electrical Engineering and Computer Science,
and Center for Materials Science and Engineering
Massachusetts Institute of Technology
Cambridge, Massachusetts

ABSTRACT

Integrated circuit technology has been used to develop a miniature dielectric cure monitor probe that combines small size with built in amplification to achieve good sensitivity down to 1 Hz, making the probe more sensitive to physical property changes than conventional dielectrometers operating at 1000 Hz. This paper describes the design and operation of the micro-dielectrometer chip, and the model used to determine the complex dielectric constant of the material under study. Two typical applications are illustrated by experiments. First, results of a post-cure study of a Versamide 140, Epon 828 mixture are presented and compared to the results of a similar study using a conventional parallel plate capacitor. In a second experiment, several probes were placed in a 100 ml mold and used to monitor the cure of a sample of the same resin system. The data show the nonuniformity of cure within the sample, and can be explained on the basis of gelation and the temperature dependence of the dielectric properties as determined in the post-cure experiment.

Keywords: Dielectrometry, cure monitoring, epoxy, integrated circuit

1. INTRODUCTION

As a result of the widespread use of a large variety of reactive thermosetting resins, a need exists for the ability to study the chemical and mechanical thermoset properties both as a function of cure and after cure. Conventional means of analysis for degree of conversion include reactive group analysis, FTIR, and DSC, while bulk properties can be monitored through torsional braid analysis, viscosity, and dielectrometry. All of these methods can be cumbersome, and this has inspired the development of a simple cure monitoring integrated circuit.

The new microdielectrometer 'chip' measures the low frequency dielectric properties of a resin both during and after cure, as in conventional dielectric techniques. The integrated sensors present an advantage in that they are small and can be placed in various locations of a large test sample, or can serve as implants for long-term property monitoring. The small mass enables quick thermal equilibration for isothermal or ramped measurements on small resin samples. The microdielectrometer chip also has the capability of monitoring much lower frequencies than conventional dielectrometry (down to 1 Hz or below, as opposed to 100 Hz). At these lower frequencies the dielectric response is more representative of the mechanical properties and therefore presents a viable alternative to torsional braid analysis. In addition, 'on chip' signal amplification eliminates the need for electrical shielding and increases sensitivity.

In this paper we will discuss the device structure and explain its operation. Initial results will be presented for a thermal ramp of a cured thermoset and compared to the equivalent results from a conventional parallel plate arrangement. Also to be presented are the data from a large sample during cure containing multiple chip sensors at various locations within the bulk. These results will be explained on the basis of gelation and post cure properties as a function of temperature.

2. DEVICE DESIGN

The present microdielectrometer chip evolved from the charge-flow transistor (CFT).[1] A charge-flow transistor consists of a field effect transistor (FET) with a gap in the gate over the critical channel region of the transistor which is filled by the material to be studied. The turn on characteristics of the transistor are determined by the flow of charges from the gate electrode through the material of interest to the channel region. The step waveform response of the CFT has been shown to change as a function of thermoset resin cure and can be explained through suitable modeling of the resin.[2]

The original CFT design was difficult to calibrate, and yielded an output that required nonlinear large signal analysis for interpretation. For this reason, a new sensor has been developed which can be analyzed with linear device models, and which permits accurately calibrated measurements to be made using differential techniques.[3] In this design, the device response is balanced against a reference FET so that all device characteristics are cancelled, leaving only the desired response of the material under study.

The new chip is 75 mils square, and employs a large area interdigitated capacitor as the sensing element. A sinusoidal voltage applied to one electrode, the "driven electrode" (see Fig. 1), transfers charge through the material under study to the opposite electrode, "the sensing electrode". As with standard parallel plate capacitors, the current due to this charge transfer could be measured with a capacitance bridge or dielectrometer using shielded cables and other appropriate measures to prevent noise. Noise problems are largely eliminated in the new chip by attaching the low side of the interdigitated capacitor (the sensing electrode) directly to the gate of a depletion-mode FET operating in its linear region. The presence of charge on this gate produces a voltage that controls the current flowing between the drain and source terminals of the transistor. The drain to source current resulting from a sinusoidal driving voltage contains the same information about charge transfer through the material as the conventional capacitor current, but is about 10^6 times larger. This on-chip amplification enables measurement of dielectric properties down to 1 Hz, a frequency at which capacitor currents are too small to be measured in a conventional manner.

3. MEASUREMENT CIRCUIT

The function of the measurement circuit (which is described in detail in Reference 3) is to measure the drain to source current of the sensing FET, and from it, to infer the sensing gate voltage which is directly related to the charge transfer from the driven gate to the sensing gate. To do this, a second depletion mode FET, identical in all dimensions to the sensing FET, is included on the chip as a reference. The gate of the reference FET is driven by an off chip current comparator so that the currents in both FETs are equal. Since the device geometries are identical, the voltage applied by the current comparator to the reference FET is necessarily equal to the voltage on the floating gate of the sensing FET. This measurement technique provides an output voltage equal to that on the floating gate and eliminates any current offset due to temperature variations, enabling the chip to function at temperatures up to 250°C.

During an experiment, the driven gate voltage and the reference voltage are simultaneously sent to a HP3575A gain-phase meter which determines the ratio of their amplitudes and relative phase shift. The digitized information is logged in real time by a HP85 desktop computer. The HP85 also controls a HP3325A function generator, which sets the frequency and amplitude of the driven gate voltage. This system allows for real time measurement as a function of frequency, temperature, and cure. The system is also capable of taking measurements on multiple devices through automatic switching, enabling cure data to be obtained from various regions within a sample simultaneously.

4. TRANSFER FUNCTION MODEL

The transfer function for the planar interdigitated capacitor is somewhat more complicated than that for a conventional parallel plate geometry. The dependence of the amplitude and phase of the sensing gate voltage on the complex dielectric constant of the resin (permittivity ε', and loss factor ε'') can be modeled by a one-dimensional distributed RC circuit, as illustrated in Fig. 1. The bracketed elements represent a differential element of the distributed system. C_A is proportional to ε', and R_A is proportional to $1/\omega\varepsilon''$, where ω is the angular frequency. Thus, $\tan\delta = 1/\omega R_A C_A$. C_T represents the capacitance between the plane of the electrode and the ground plane beneath the silicon dioxide, and C_G represents the total gate capacitance of the sensing FET and sensing electrode.

The exact transfer function for this model is given in the Appendix. The calibration of the device depends on C_T, C_G, and the separation betweeen electrodes, all of which can be controlled by device design and processing conditions. With this information, plus a single semi-empirical thickness parameter, d, there is a unique relation of ε' and ε'' to the magnitude and phase of the sensing gate voltage relative to the driven gate voltage.

5. EXPERIMENTAL

Initial experimentation has been carried out on an equal volume mix of Versamide 140, a linear aliphatic amide, and EPON 828 (DGEBA). This system was chosen because of its reasonably short cure time in 100ml batches at room temperature and the realitively low glass-rubber transition in the post cured material.

Two types of experiments were performed. For the study of fully cured material, samples were prepared by placing small drops of the resin on the chip sensing area and then curing at 180°C for 10 minutes. A parallel plate capacitor was also filled with the resin and cured under the same conditions. After cure, dielectric measurements were monitored at three frequencies (10, 100, and 1000 Hz) during a temperature ramp from 60°C at 4 deg/min.

In a second experiment to demonstrate the cure monitoring ability, several chips were placed at various levels within an empty mold (5.5 x 5.5 cm). An integrated temperature sensor (Analog Devices AD590) was included alongside each sensing chip for accurate temperature tracking during the cure. After the epoxy was mixed and poured into the mold, automatic measurements were initiated. The dielectric response and temperature were monitored during cure as a function of three frequencies (10, 100, and 1000 Hz).

6.0 RESULTS

6.1 Fully Cured Material

The dielectric behavior of the cured material as measured in the parallel plate capacitor is indicated in Fig. 2. The ε' data exhibit a frequency dependent transition which we attribute to dipole relaxation associated with the glass-rubber transition.[4] The ε'' behavior exhibits a steeply rising term on which are superimposed small peaks that one expects from the dipole relaxation. This suggests the existence of an activated ohmic conductivity of the following form:[4]

$$\sigma(T) = \sigma_o \exp(-E_C/kT) \qquad (1)$$

Where $\sigma(T)$ is the conductivity and E_C is the activation energy. The conductivity observed in the parallel plate data, which is most likely due to ionic conduction, is found to have an activation energy of 6.5 kcal/mole.

The relative amplitude (gain in decibels) and phase response of the microdielectrometer for fully cured material as a function of temperature is shown in Fig. 3. Using the relations in the Appendix, ε' and ε'' values were obtained from these data, and are shown in Fig. 4. The ε' values show excellent agreement with the parallel plate data, but there is a significant difference in the ε'' values. The conductance peaks correspond closely to those expected from normal dipole relaxation, but the strong ionic conductance observed in the parallel plate experiment appears to be missing. As a test, a conduction term of the form of Eq. 1 was added to the microdielectrometer data, resulting in the graphs of Fig. 5, which correspond quite closely with the parallel plate results of Fig. 3. Thus it appears that the planar electrode technique is sensitive only to dielectric loss contributions from dipole relaxation and is immune to ohmic conductivity, at least for this material and under these conditions. This behavior is presently not understood, but may be of great benefit in observing relaxation phenomena that are normally obscured by high ohmic conductivities.

6.2 <u>Multiple</u> <u>Chip</u> <u>Cure</u> <u>Monitoring</u>

During the cure of the large epoxy sample containing multiple sensors, the exothermic reaction caused the temperature to rise from 20°C to 130°C just after gelation and then finally settle back down to 20°C. As a result of the positioning of the sensors in the mold, different temperature profiles and dielectric responses were observed.

Raw data from the devices at the bottom and center of the mold are shown in Fig. 6. The temperature profiles indicate a slightly faster rise in temperature in the center of the mold than at the bottom, as expected.

The sharp discontinuity in the gain and phase data near 100 minutes is attributed to the gel point, for two reasons. First, the slope of the temperature profile starts decreasing at this point, and within a few minutes, the actual temperature starts to decrease, indicating a sharp reduction in reaction rate. Crosslinking reactions are known to be inhibited after gelation as a result of the formation of an infinite network which greatly impedes the diffusion of reactants to reaction sites. Second, the behavior of the gain and phase after the occurance of the sharp discontinuity are nearly identical to that of the

fully cured material shown in Fig. 2. Using the phase versus temperature data of the fully cured material, the phase has been replotted as a function of the temperature profile from Fig. 6. In the resulting curves (Fig. 7), the phase matches very closely to that observed in the mold cure at times beyond the gelation point. This similarity in behavior indicates that the peaks that occur after gelation are a result of the temperature passing above and then back down through the relaxation temperature of the cured material.

As a result of the sharp discontinuity in gain and phase at gelation, differences in gel point times for the various sensor locations are easily determined. In this particular sample the middle of the resin gelled five minutes before the bottom, which was only two centimeters away. This result stresses the difference in reaction rates which can occur from variation in mold geometry and temperature.

7. DISCUSSION

The microdielectrometer chip has been shown to yield results in general agreement with the results of conventional parallel plate dielectrometry. Advantages of the chip lie in its small size, making implantation and measurements on small samples possible. On-chip amplification also eliminates the need for electrical shielding. In addition, the chip has frequency monitoring ability down to 1 Hz. Low frequency monitoring is useful for observing long relaxation times which occur far into cure and which are generally difficult to measure.

Further improvements in the chip are planned, including the addition of the current comparator and temperature sensor to the integrated device structure.

8. ACKNOWLEDGEMENT

This research was sponsored in part by the Office of Naval Research. Device fabrication was carried out in the M.I.T. Microelectronics Laboratory, a Central Facility of the Center for Materials Science and Engineering, which is supported in part by the National Science Foundation under Contract DMR-78-24185. Polymer samples were obtained from the Army Materials and Mechanics Research Center. Some of the measurement equipment was purchased under NSF Contract ENG-7717129.

9. REFERENCES

1. S.D. Senturia, C.M. Sechen, and J.A. Wishneusky, Appl. Phys. Lett., $\underline{30}$, 106 (1977).

2. S.D. Senturia, N.F. Sheppard, S.Y. Poh, and H.R. Appelman, Polymer Eng. and Sci., in press.

3. S.L. Garverick and S.D. Senturia, 1980 IEDM Technical Digest, paper 26.6, p. 685 (1980)

4. P. Hedvig, _Dielectric Spectroscopy of Polymers_, Wiley, New York, (1977).

10. BIOGRAPHIES

Norman F. Sheppard is presently a graduate student in the M.I.T. Department of Electrical Engineering and Computer Science, having received an S.B. (1978) and S.M. (1979) in Chemical Engineering from M.I.T.

Steven L. Garverick received an S.B. (1979) and S.M. (1980) from the M.I.T. Department of Electrical Engineering and Computer Science. He is presently a Staff Member at the M.I.T. Lincoln Laboratory, Lexington, Massachusetts.

David. R. Day is a Postdoctoral Associate in the M.I.T. Department of Electrical Engineering and Computer Science, having received an S.B. (1975), S.M. (1976), and Ph.D. (1980) from Case Western Reserve University in Macromolecular Science.

Stephen D. Senturia is a Professor of Electrical Engineering at M.I.T., having received an S.B. in Physics from Harvard (1961), and a Ph.D. in Physics from M.I.T. (1966).

APPENDIX: TRANSFER FUNCTION MODEL

For a voltage $V_o e^{j\omega t}$ applied to the driven gate of the model of Figure 2, the complex voltage at the sensing electrode is of the form

$$V_{FG} = \frac{V_o}{A + jB}$$

The magnitude of the sensing electrode voltage will be

$$|V_{FG}| = \frac{V_o}{\sqrt{A^2 + B^2}}$$

The phase shift relative to the driving voltage will be

$$V_{FG} = -\tan^{-1}\left(\frac{B}{A}\right)$$

where $A = \cosh(m'x_o)\cos(m''x_o)$
$+ a\sinh(m'x_o)\cos(m''x_o)$
$- b\cosh(m'x_o)\sin(m''x_o)$

$B = \sinh(m'x_o)\cos(m''x_o)$
$+ b\sinh(m'x_o)\cos(m''x_o)$
$+ a\cosh(m'x_o)\sin(m''x_o)$

and

$$a = \frac{\omega R_A C_G (m'' + \omega m' R_A C_A)}{(1+\omega^2 R_A^2 C_A^2)(m'^2 + m''^2)}$$

$$b = \frac{\omega R_A C_G (m' - \omega m'' R_A C_A)}{(1+\omega^2 R_A^2 C_A^2)(m'^2 + m''^2)}$$

$$m' = \sqrt{\frac{\omega R_A C_T}{2}\frac{(\sqrt{1+\omega^2 R_A^2 C_A^2} + \omega R_A C_A)}{(1+\omega^2 R_A^2 C_A^2)}}$$

$$m'' = \sqrt{\frac{\omega R_A C_T}{2}\frac{(\sqrt{1+\omega^2 R_A^2 C_A^2} - \omega R_A C_A)}{(1+\omega^2 R_A^2 C_A^2)}}$$

Expressed in terms of permittivity, ε' and loss factor, ε'', and a thickness parameter, d,

$$R_A = 1/\omega\varepsilon''d \qquad C_A = \varepsilon'd$$

The constants used in evaluating these expressions were:

C_G = capacitance parameters of sensing electrode = 7.3×10^{-15} F

C_{ox} = capacitance of insulating oxide per unit area = 6.4×10^{-9} F/cm^2

x_o = distance between driven and sensing electrodes = 12.5 microns

d = thickness parameter = 12.5 microns

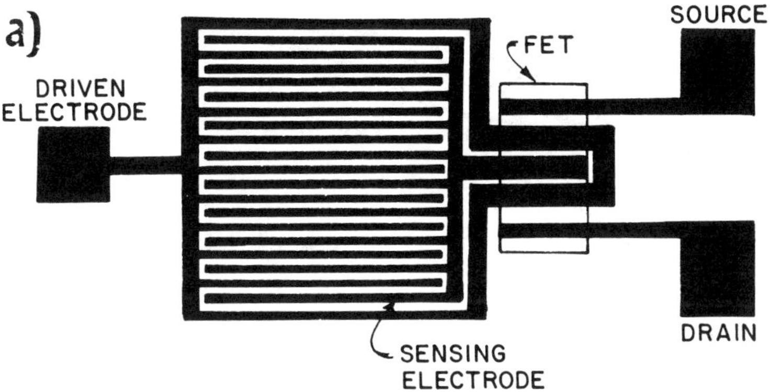

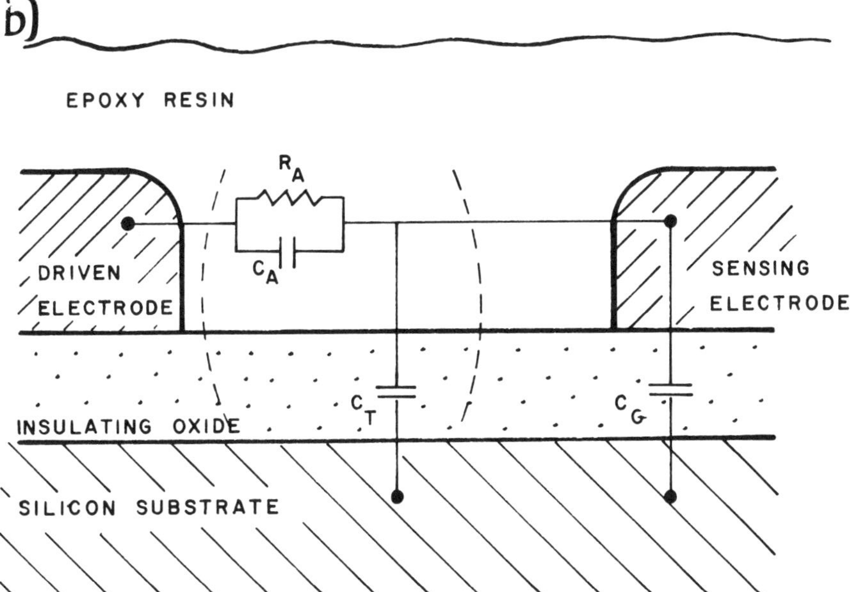

Fig.1 a) Schematic top view of the active portion of the microdielectrometer chip. b) Schematic cross-section of the transfer function
model.

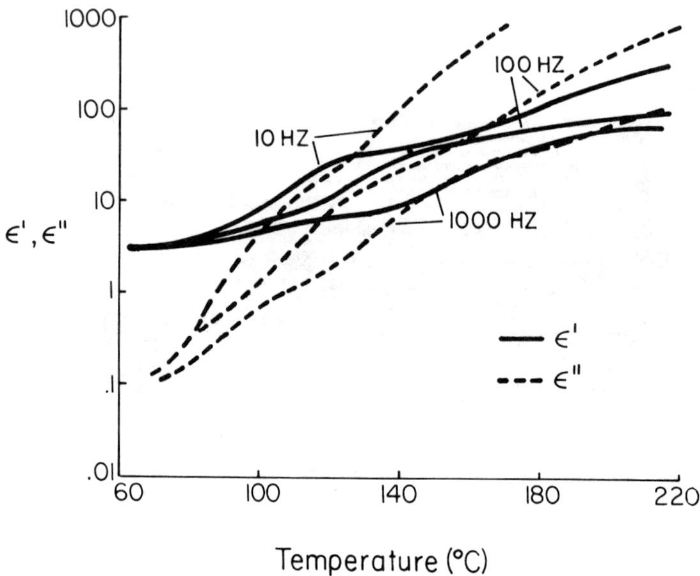

Fig. 2 Dielectric spectra of fully cured epoxy from parallel plate
capacitor measurement.

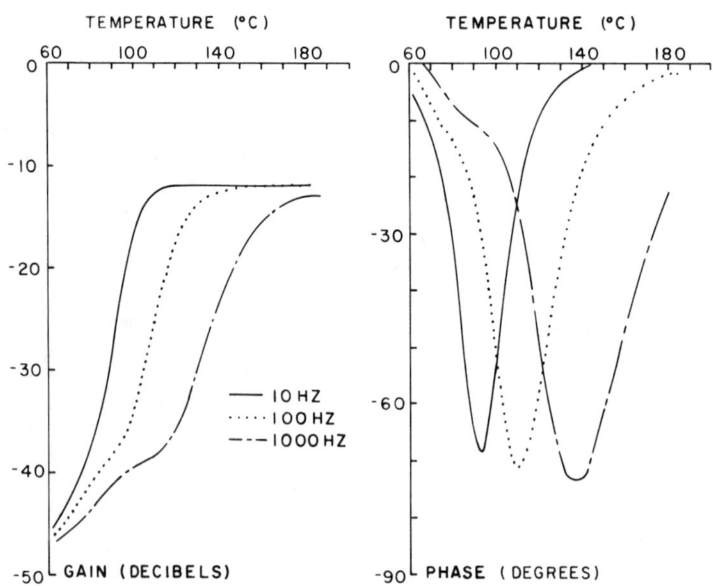

Fig. 3 Raw data (gain and phase) for fully cured epoxy as a function of
temperature from the microdielectrometer chip.

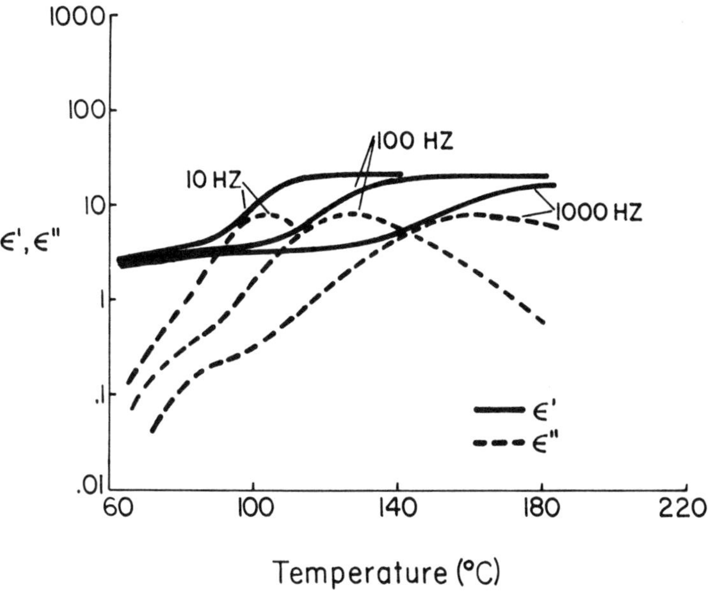

Fig. 4 Dielectric spectra of fully cured epoxy calculated from the data of
Fig. 3 using the transfer function in the Appendix.

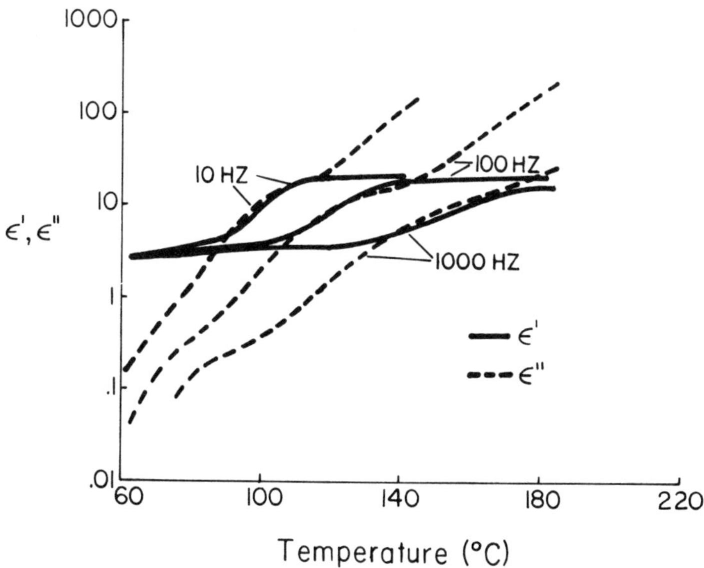

Fig. 5 Dielectric spectra of fully cured epoxy calculated from Fig. 4 with
the addition of an ohmic conductivity (Eq. 1).

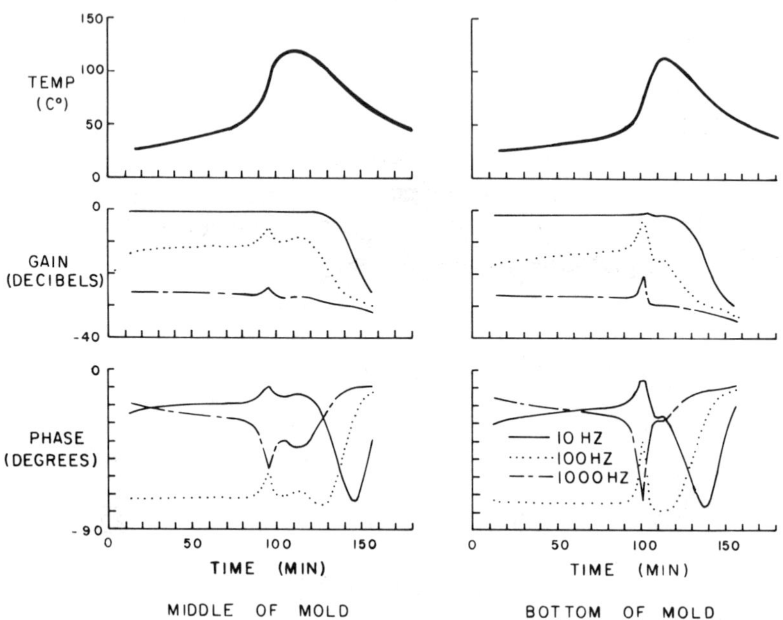

Fig. 6 Gain and phase response of microdielectrometer chips located in the bottom and center of a 5.5x5.5 cm cylindrical mold during epoxy cure.

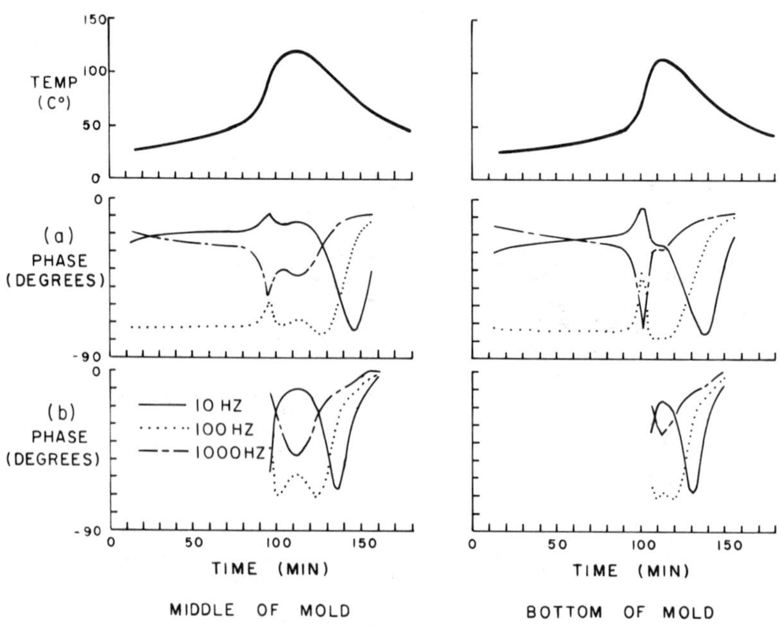

Fig. 7 Comparison of a) measured phase response of microdielectrometer during epoxy cure to b) the corresponding phase response measured for the fully cured material at equivalent temperatures.

26th National SAMPE Symposium
April 28-30, 1981

MONITORING CURE OF LARGE AUTOCLAVE MOLDED PARTS BY DIELECTRIC ANALYSIS

J. Chottiner, Z. N. Sanjana
Westinghouse Electric Corporation
R&D Center
Pittsburgh, PA 15235
M. R. Kodani, K. W. Lengel, G. B. Rosenblatt
Westinghouse Electric Corporation
Marine Division
Sunnyvale, CA 94088

Abstract

A study was undertaken to determine the feasibility of using dielectric analysis as a means of monitoring and controlling cure of large closures during autoclave molding. In dielectric analysis the dissipation factor and capacitance of the sample is continuously monitored as a function of time and frequency. Dissipation factor profiles were established for the suppliers' recommended cure cycle and for modified cure cycles. Good reproducibility was obtained in dissipation factor profiles on subsequent scaling up to production size (7 ft x 20 ft) autoclaves. Good correlation was also observed during production runs of full-scale closures. The effects of cure variables on the dissipation factor profiles and on the mechanical properties of the prepared laminates were analyzed for extent of correlation. Results of this study show:

(1) dielectric analysis can be used to monitor autoclave cure of composites, and (2) within limits, process control may be feasible.

"Keywords": dielectric analysis; autoclave molding; epoxy; glass; dissipation factor; capacitance

1. INTRODUCTION

The launch tube closure assembly for the Fleet Ballistic Missile (FBM) Launch System is currently autoclave molded from a phenolic/asbestos prepreg material. Because of the recognized carcenogenicity of asbestos, there has been an active program aimed at finding a suitable replacement material. The most promising of these materials are epoxy/ high silica glass cloth prepregs.[1]

These launch tube closures are domed-shaped structures approximately 6 ft in diameter at the base and slightly over 2 ft high at the apex. The lay-up procedure, autoclave and post bake cycle, finishing operations and qualification testing combine to form a costly and lengthly process. Improved quality control during processing is an ongoing objective. The work described in this paper is an effort to incorporate dielectric analysis as a quality control tool in production.

A useful cure monitoring technique should be capable of: (1) interrogating the material within the autoclave or press, (2) giving a response which is sensitive to changes in the incoming material and to changes that accompany cure, and (3) giving the information continuously and immediately so that, if necessary, process changes can be made based on the response. Most of the published work in cure monitoring indicates that some form of electrical measurement is the most feasible way to meet the requirements. May[2], Hudson[3], and Martin[4] have discussed the use of an automatic dielectrometer to continuously monitor the dissipation factor and the capacitance of the sample during cure. Yokota[5] has discussed the relative advantages of using ac voltage and phase angle measurements. Crabtree[6] has discussed dc resistance and voltage measurements during autoclave cure of a composite.

The technique discussed here is an ac technique in which an automatic dielectrometer is used to continuously monitor the dissipation fac-

tor (DF or tan δ) and capacitance (C) of a sample as a function of time, temperature and frequency of the sample. Both observed parameters are sensitive to dipolar motion within the matrix of the composite which is sensitive to the degree of advancement or cure of the resin. The use of dielectric analysis to follow the degree of polymerization and cure kinetics has been discussed by one of the authors[7] and in the papers referenced earlier[2-4].

2. EXPERIMENTAL

The closure application requires an unusual combination of properties including the ability to withstand substantial stresses from either side. It must, however, be readily sheared by a linear shaped charge just prior to launch of the missile. A family of epoxy/high silica glass cloth prepreg materials appear to possess the desired combination of properties.[1] Four materials from this family were selected for prototype closure evaluation and cure monitoring. These materials are Fiberite 934/Siltemp 82, Fiberite 934/Refrasil C100-48, Fiberite 30854/Refrasil C100-48 and Ferro CE321/Refrasil C100-48.

While dissipation factor (DF) and capacitance (C) can be measured by any ac bridge, continuous measurements are most conveniently done using an automatic dielectrometer. The dielectrometer is a self-balancing bridge with a frequency range of from 0.1 to 1.0 kHz and is often referred to by its acronym -

Audrey. It is manufactured by
Tetrahedron Associates.

The measured sample consisted of
one (for Refrasil C100-48) or two
(for Siltemp 82) plies of prepreg
in the middle of the stack. Each
electrode consisted of copper foil
to which leads had been soldered
and the effective area was one
square inch. One of the electrodes
was "blocked", i.e., separated from
the sample by one layer of 0.001"
thick polyimide film. The use of
the "blocking" film permits the
D.F. response to exhibit frequency
dependent relaxation peaks associated with the melt and flow of the
resin, and with the gelation and
cure process.[7] Figure 1 shows
the layup of the test panel. Total
number of plies in the stack was
selected to provide a shell thickness of 0.220" $\pm$ .02". Figure 2
shows the panel placed in an autoclave. To reduce noise it was
found beneficial, in some experiments, to ground the dielectrometer to the autoclave wall. For
these preliminary experiments we
selected a frequency of 1 kHz to
reduce noise to a minimum.

When experimenting with a full-
scale shell, it was decided to
incorporate the electrodes on the
flange of the shell where some
machining was necessary that would
subsequently remove the encapsulated electrode assembly. Figure 3
shows the molded shell with the
electrodes on the flange.

The standard autoclave cure cycles
for Ferro CE321 and Fiberite 934
and 30854 resin systems are shown in
Tables I and II. Dielectric analyses were run with the standard cure
cycles and a number of modified
cure cycles, including applying the
pressure at points based on the
dissipation factor or capacitance
profiles. The specific modifications employed are shown in Tables
III through V. The effect of cycle
modification was determined by
testing autoclave molded panels
(which were monitored using dielectrometry) for properties pertinent
to the closure application. These
properties are shown in Tables III
through V. Standard ASTM test
specimens and procedures were
employed.

A minimum of three full-scale closures were molded from each of the
four candidate materials. The molding cycle employed was selected on
the basis of the results obtained
in molding flat panels. A flat
panel was also molded in the autoclave while the closure itself was
being molded. Both the closure and
the flat panel were equipped with
electrodes and connected to the
dielectric analysis equipment.

3. RESULTS AND DISCUSSION
Figure 4 shows the profiles for DF
and C for the cure of Ferro CE321
resin on Refrasil C100/48 high
silica glass using the manufacturer's recommended cure cycle
(Table I). The figure also indi-

cates the ten experiments performed and the points at which pressure was applied (by means of circled numbers). The numbers also refer to the panel numbers in Table III which gives mechanical test data for each panel.

As the panel was heated up at some temperature, DF went through a maximum and then declined to a minimum. The temperature of this maximum (for convenience called the "melt" peak) is a function of the frequency of observation and the degree of advancement of the prepreg.[8] The maximum in C indicates the point in time beyond which the reaction begins to reduce the mobility of the polar groups. After this point the reaction (cure) proceeds rapidly resulting in a sharp and continuous decrease in C. Simultaneously the DF rises rapidly and becomes invarient with time as the rate of reaction ceases.

Several experiments repeating the standard cycle were performed in a laboratory autoclave and showed the results to be quite repeatable. Thus we were able to perform the ten modifications shown in Table III on panels cured in a production autoclave. One modification consisted of using a higher cure temperature which is shown in Fig. 5. An interesting departure from Fig. 4 occurs when the final cure temperature is reached. The DF and C profiles continued to show a significant rate of change until cool down.

The sensitivity of dielectric measurements to frequency is also shown in Fig. 5. During the cool down, a series of peaks were obtained. The frequency dependencies of the dielectric response of polymers during cure has been previously discussed.[7] It should also be noted that in this paper whenever we mention "melt" or "gel" peak it is only for convenience. The peaks result from relaxation phenomena associated with softening and flow of the resin and gelation of the resin. The occurrence of the peaks does not define the point in time when softening or gelation occurs because the occurrence of a peak is also frequency dependent.

The data in Table III show a rather large scatter in mechanical properties but it does appear that pressure application at the "melt" peak (Panels 2 and 4) resulted in higher mechanical properties. Pressure application past the capacitance peak (Panels 3 and 9) appears to result in poorer properties. Unfortunately, the autoclaves were manually controlled and the temperature rate of rise could not be controlled as well as desired, thus contributing to the inconsistencies in the data. Improved properties were noted for Panel 6 where a higher pressure was used. Panels 7 and 8 were cured simultaneously to evaluate the effect of different bleeder systems.

Fiberite 934 is a 350°F curing resin containing a tetrafunctional epoxy resin cured with an aromatic amine. Figure 6 shows the "standard" cure profile for the resin on Siltemp 82 high silica glass. The standard cure is defined in Table II. Pressure application points for the 12 experiments with this material are shown in Fig. 6 and the modifications are shown in Table IV.

As the material is heated, a "melt" peak in DF is noted as with the Ferro CE321 material. Temperature at which this peak occurs can be used to define the advancement of the prepreg. As the heat-up continues, DF drops to a minimum which coincides with a maximum in the C response. This should be in the region of minimum viscosity and Panels 3 and 8 had pressure applied at this point. Towards the end of the hold, DF went through a maximum and then as the temperature was increased to 350°F it increased again and went through a final maximum. For convenience we define the latter as "gel" peak. After this maximum, the reaction proceeds rapidly as noted by the large drop in DF. At 350°F, the reaction continues although at a much slower rate. In Fig. 7 we see the changes in the profile resulting from the slowest heat-up rate used (1.5°F/ min compared to 2.0°F/min in Fig. 6). The principal difference is the increase in breadth of the peaks. Due to increased staging,

the first peak associated with gelation was much larger and more completely defined.

The data of Table IV shows considerable scatter but four observations can be made: (a) Applying pressure past the "gel" peak as in Panel 4 lowers the mechanical properties. (b) As expected, lower pressure resulted in poorer mechanicals. (c) Applying the pressure at the DF minimum which is earlier than recommended appears to improve mechanical properties, and (d) The slow heat-up rate of Panel 11 did not adversely affect mechanicals.

Prepregs of both Fiberite 30854 and 934 resin systems on Refrasil C100-48 high silica glass were obtained and cured in pairs in an autoclave at the same time. Dielectrometer readings were obtained from both materials concurrently by switching between the two, and a typical result for a standard cure cycle is shown in Fig. 8. The standard cure is identical for both materials and is shown in Table II. The numbers in Fig. 8 correspond to various modifications as noted in Table V. The 30854 resin is basically similar to 934 except it is modified to increase flexibility. The essential similarity in chemistry of the two is exhibited by the close resemblance of the two profiles in Fig. 8. Only one important difference is noted and that is at the end of cure the flexibilized 30854 displays a higher DF than does the more rigid

934. The enhanced flexibility
allows greater dipolar mobility at
any temperature and therefore a
higher DF.

The results of mechanical tests are
given in Table V for the 30854
system. It is seen that late pressure application at the "gel" peak
is disadvantageous as shown in
Panel 4. Note particularly the
high void content. Applying pressure very early at the
"melt" peak as for example Panel 6
appears to be disadvantageous.

Full-scale shells and test panels
were simultaneously molded, in the
same autoclave. Both shells and
panels were monitored by switching
the dielectrometer between the two.
Figure 9 shows this for the Fiberite 934 material. It is interesting to note the similarity between
the profiles for the panel and the
shell and the similarity to Figs.
6 and 7. This program is continuing and all production shells are
being monitored. The shells are
being evaluated but testing is not
yet complete. This evaluation
includes the ability of the shells
to withstand internal and external
pressures and to be cleanly cut-
through by the explosive charge.

4. CONCLUSIONS

1. Dielectric analysis can be used
    to monitor autoclave cure of
    composites.
2. The dissipation factor and
    capacitance responses of the
    polymer composites are quite
    reproducible when materials and
    processes are not changed, thus
    making process control feasible.
3. The responses that we have cho-
    sen to study are very sensitive
    to the temperature and the rate
    of change of temperature at the
    moment of observation, particu-
    larly in the pre-gelation period
    of the cure. This adds to the
    difficulty of instituting pro-
    cess control unless tempera-
    tures can be very accurately
    programmed.
4. The dissipation factor profile
    obtained during cure can be used
    to define a range of time in
    which pressure must be applied,
    and also may be used to define
    an optimum pressure application
    point.

5. ACKNOWLEDGMENT

The authors wish to express their
appreciation to HITCO and especially
to Mr. Jack Novak of that Company
for their assistance in providing
facilities and consultation during
the course of this investigation.
The assistance of J. H. Testa and
R. L. Selby of Westinghouse R&D is
also deeply appreciated.

6. REFERENCES

1. J. Chottiner, G. B. Rosenblatt
   and K. W. Lengel, "Substitute
   Composites for FBM Launch System
   Closures", Technical Proceedings,
   35th Annual Conference, The
   Society of Plastics Industries,
   Feb. 1980.

2. C. A. May, Pro. SAMPE Symp.,
   V. 20, p. 108 (1975).

3. D. Hudson, Composites, p. 247
   (Nov. 1974).

3. D. Hudson, Composites, p. 247
   (Nov. 1974).

4. B. G. Martin, Mater. Eval.,
   V. 34, p. 49 (March 1976).

5. M. J. Yokota, Proc. SAMPE Symp.,
   V. 22, p. 416 (1977).

6. D. J. Crabtree, Proc. SAMPE
   Symp., V. 22, p. 636 (1977).

7. Z. N. Sanjana, "Science and
   Technology of Polymer Processing", ed. N. P. Suh and
   N. Sung, p. 827, MIT Press
   (1979).

8. Z. N. Sanjana, SAMPE Jnl.,
   V. 16, p. 5, (Jan. 1980).

7. BIOGRAPHIES

J. Chottiner and Z. N. Sanjana are
senior engineers at the Westinghouse R&D Center, Pittsburgh, PA.
Both are active in the evaluation
and application of high performance composites.

M. R. Kodani, K. W. Lengel and
G. B. Rosenblatt are with the
Westinghouse Marine Division in
Sunnyvale, CA and have all been
active in the design, manufacturing
and testing of FBM closures.

TABLE I – STANDARD AUTOCLAVE CURE
FOR FERRO CE321 EPOXY RESIN SYSTEM

1. Apply full vacuum and raise
   temperature to 200°F at
   2°-5°F per minute.

2. Apply 100 psi pressure at
   200°F, vent bag and continue
   raising temperature to 275°F
   ($\pm$ 5°F) at 2°-5°F per minute.

3. Hold at 275°F ($\pm$ 5°F) for
   two hours.

4. Cool under pressure to below
   175°F.

TABLE II – STANDARD AUTOCLAVE
CURE CYCLE FOR FIBERITE 934
AND 30854 EPOXY RESIN SYSTEMS

1. Apply full vacuum and hold
   at room temperature for
   30 minutes.

2. Maintain full vacuum
   throughout entire cure cycle.

3. Raise temperature to 250°F
   (+5, -10°F) at 2°-5°F per
   minute.

4. Hold at 250°F for 15 $\pm$ 5
   minutes.

5. Apply 100 psi pressure.

6. Hold at 250°F and 100 psi
   for 45 $\pm$ 5 minutes.

7. Raise temperature to 350°F
   (+10°, -0°F) at 2°-5°F
   per minute.

8. Hold at 350°F for two hours
   $\pm$ 15 minutes.

9. Cool under pressure to below
   175°F.

TABLE III - PROPERTIES OF FERRO CE321/REFRASIL C100-48 PANELS
MOLDED DURING DIELECTRIC ANALYSIS EXPERIMENTS

Panel No.	Cure Cycle Modification	Flexural		Tensile		Compressive		Thick. (mils)	Sp. Gr.	Resin Content (%)	Void Content (%)
		Str. (ksi)	Mod. (Msi)	Str. (ksi)	Mod. (Msi)	Str. (ksi)	Mod. (Msi)				
1	Standard	31.1	2.6	20.3	2.9	48.8	3.4	232	1.66	35.5	3.0
2	Apply Pressure at "Melt" Peak	40.2	2.9	24.8	3.2	57.1	3.6	221	1.68	33.2	3.0
3	Apply Pressure at C Peak	36.0	2.8	23.4	3.1	51.0	3.2	228	1.67	34.5	2.9
4	Apply Pressure at "Melt" Peak	36.3	3.0	23.1	3.0	55.6	3.5	223	1.68	32.9	3.1
5	Cure at 325°F	33.8	2.9	23.3	3.0	53.2	3.7	212	1.69	31.5	3.3
6	Cure at 200 psi	40.8	3.1	24.5	3.4	56.8	3.6	216	1.69	31.0	3.5
7	Standard	34.4	3.0	19.9	3.3	55.0	3.5	220	1.71	31.7	2.0
8	Standard	33.4	3.1	21.8	2.9	55.6	3.4	220	1.68	32.3	--
9	Apply Pressure Past C Peak	29.9	2.7	21.2	2.6	--	--	233	1.68	33.6	2.8
10	Standard	39.2	3.3	25.0	3.3	--	--	218	1.70	31.0	3.0

TABLE IV - PROPERTIES OF FIBERITE 934/SILTEMP 82 PANELS
MOLDED DURING DIELECTRIC ANALYSIS EXPERIMENTS

Panel No.	Cure Cycle Modification	Flexural		Tensile		Compressive		Sp. Gr.	Resin Content (%)
		Str. (ksi)	Mod. (Msi)	Str. (ksi)	Mod. (Msi)	Str. (ksi)	Mod. (Msi)		
1	Standard	34.0	3.5	22.1	4.6	51.1	3.8	1.68	28.1
2	Apply Pressure at "Gel" Peak	35.4	3.4	22.1	3.9	57.1	3.7	1.66	29.5
3	Apply Pressure at D.F. Minimum	38.1	3.4	21.6	4.1	59.5	3.6	1.67	29.5
4	Apply Pressure 10 Min. Past "Gel" Peak	27.5	2.7	17.4	3.4	38.9	3.0	1.48	29.7
5	30 psi Pressure	27.1	2.9	20.6	3.7	53.0	3.6	1.58	29.7
6	Standard	33.9	3.3	23.8	3.7	62.2	3.5	1.67	--
7	Apply Pressure at End of 250°F Hold	32.3	3.3	22.1	3.9	63.3	3.6	1.67	29.8
8	Apply Pressure at D.F. Minimum	35.0	3.2	23.2	3.5	59.8	3.6	1.67	29.5
9	Apply Pressure at End of 250°F Hold	33.8	3.3	24.0	4.1	62.2	3.8	1.67	30.1
10	35 psi Pressure	33.1	3.2	22.1	4.0	55.2	3.6	1.61	29.0
11	Slow Heat-up (1.5°F/Min.)	31.6	3.2	23.8	4.0	60.5	3.8	1.66	29.2
12	Standard	36.0	3.3	24.7	4.2	61.0	3.5	1.68	29.3

TABLE V - PROPERTIES OF FIBERITE 30854/REFRASIL C100-48 PANELS
MOLDED DURING DIELECTRIC ANALYSIS EXPERIMENTS

Panel No.	Cure Cycle Modification	Flexural		Tensile		Compressive		Thick. (mils)	Sp. Gr.	Resin Content (%)	Void Content (%)
		Str. (ksi)	Mod. (Msi)	Str. (ksi)	Mod. (Msi)	Str. (ksi)	Mod. (Msi)				
1	Standard	27.7	2.6	15.5	2.8	41.1	2.7	227	1.65	33.4	4.6
2	Apply Pressure at "Melt" Peak	28.9	2.8	19.0	3.3	41.6	2.8	224	1.67	31.0	4.7
3	Apply Pressure at Capacitance Peak	31.3	2.8	18.9	3.3	42.1	2.8	223	1.67	31.5	4.4
4	Apply Pressure at "Gel" Peak	31.4	2.5	16.9	3.2	31.6	2.3	246	1.51	31.3	13.7
5	Standard	32.5	2.9	19.3	3.6	37.6	2.6	231	1.63	31.3	6.8
6	Apply Pressure at "Melt" Peak	26.7	2.7	17.4	3.2	40.8	2.6	230	1.64	33.0	5.4
7	Apply Pressure at Capacitance Peak	27.8	2.7	18.3	3.4	43.3	3.1	227	1.66	32.7	4.4
8	Apply 150 psi at Capacitance Peak	31.7	2.9	20.6	4.1	45.9	3.3	220	1.68	30.8	5.0
9	Apply 150 psi at "Melt" Peak	27.0	2.8	17.5	3.6	39.1	3.2	225	1.67	32.2	4.0
10	Apply 150 psi at Capacitance Peak	32.8	2.9	20.8	3.6	44.6	3.4	214	1.68	29.3	4.2

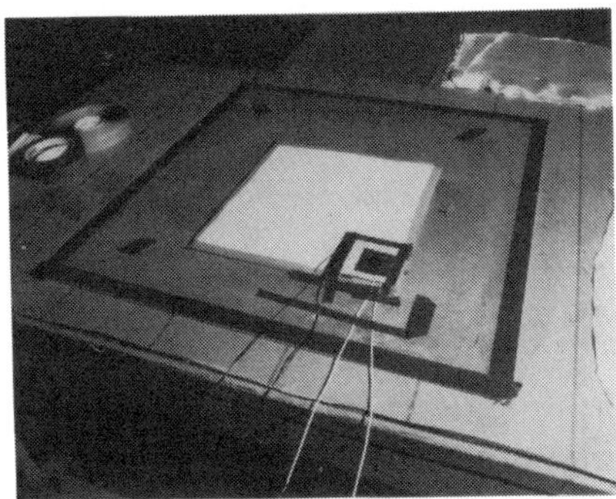

Fig. 1 - Lay-up of panel showing top electrode and thermocouple.

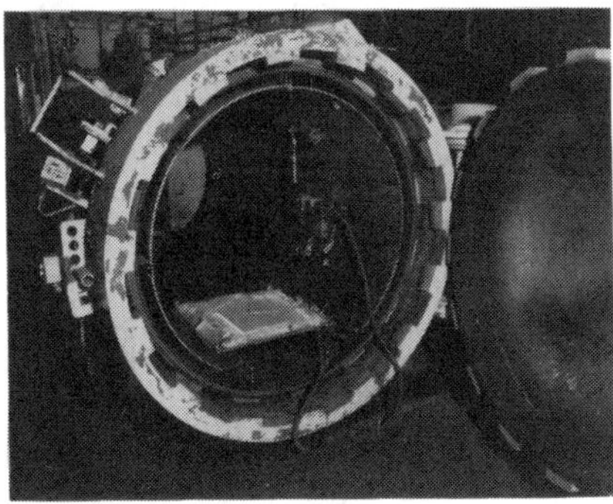

Fig. 2 - Vacuum bagged panel assembly in autoclave.

Fig. 3 - Full-scale shell showing electrodes placed on flange.

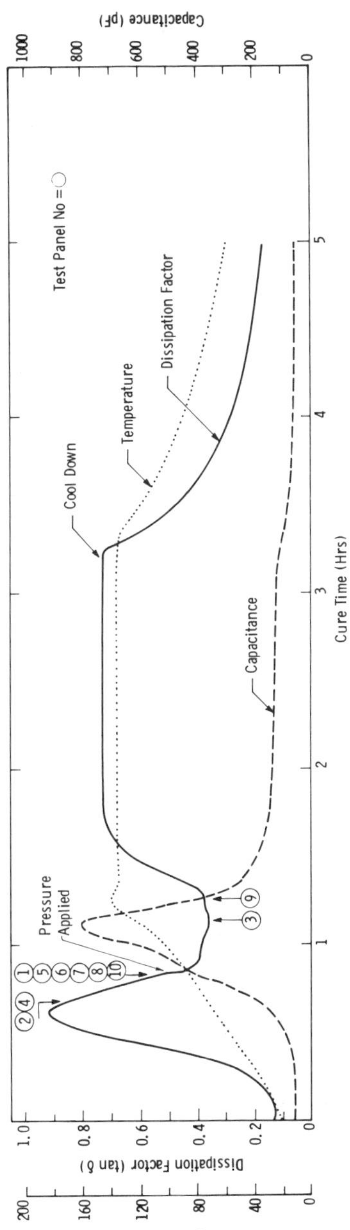

Fig. 4 – Dielectrometer profiles for Ferro CE 321 using manufacturer's recommended cure cycle. Fequency = 1 kHz

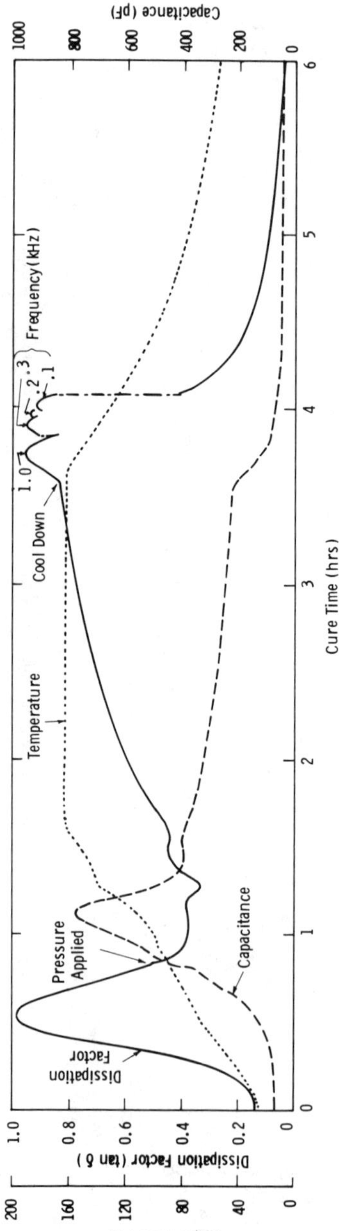

Fig. 5 — Dielectrometer profiles for Ferro CE 321 cured at 325°F (163°C). Frequency = 1 kHz except where noted

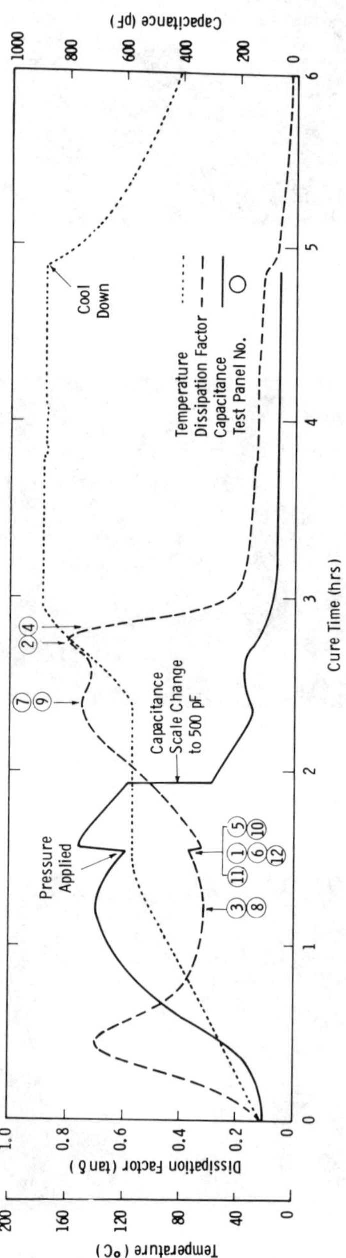

Fig. 6 — Dielectrometer profiles for Fiberite 934 using manufacturer's recommended cure cycle. Frequency = 1 kHz

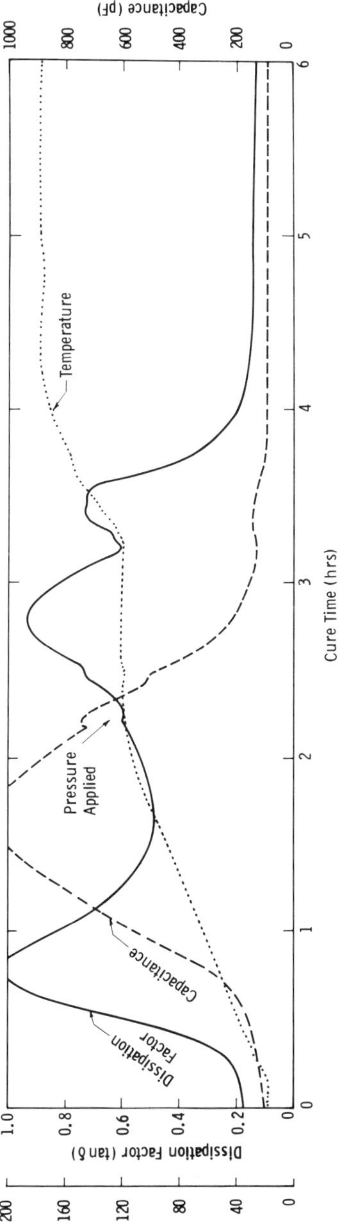

Fig. 7 — Dielectrometer profiles for Fiberite 934 using very slow (1.5°F/min) heat-up rate in cure cycle. Frequency = 1 kHz

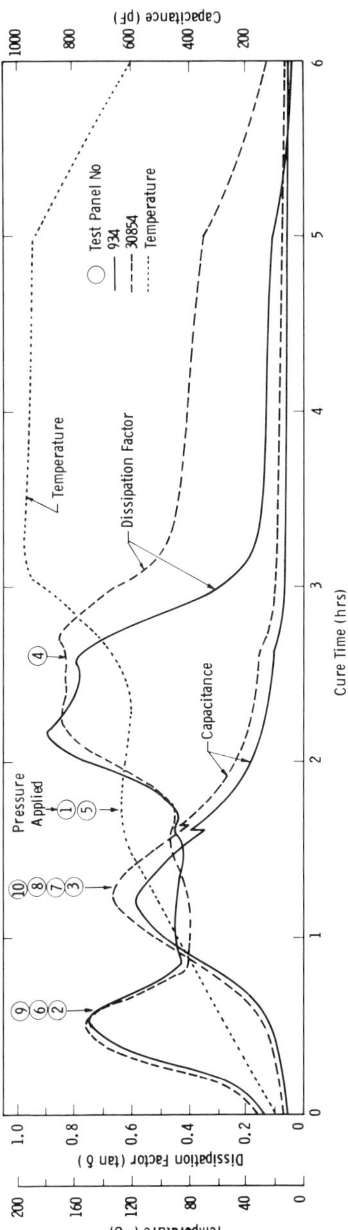

Fig. 8 — Dielectrometer profiles for Fiberite 934 and 30854 using manufacturer's recommended cure cycle. Frquency = 1 kHz

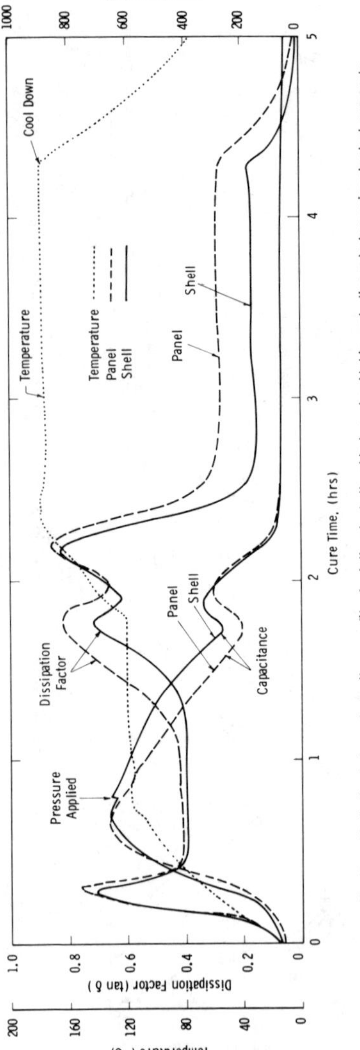

Fig. 9 — Dissipation factor and capacitance profiles for full-scale shell and test panel molded in production autoclave using standard cure cycle (Fiberite 934). Frequency = 1 kHz

26th National SAMPE Symposium
April 28-30, 1981

EFFECT OF STORING B-STAGED GRAPHITE/POLYIMIDE ON FABRICATION OF STRUCTURAL ELEMENTS

Robert M. Baucom
NASA-Langley Research Center
Hampton, VA 23665
and
Paul W. Kidder
Kentron International, Inc.
Hampton, VA 23666

Abstract

Graphite/polyimide structural elements have been successfully fabricated by forming and curing B-staged Celion 6000®/PMR-15 and Celion 6000/LaRC-160 laminates in heated matched metal tooling. The B-staged laminates were subjected to storage for various periods of time up to six months at ambient shop conditions and high humidity prior to fabrication into structural elements to determine the processability of the graphite/polyimide after exposure to these conditions. Changes in processability were monitored by measuring flexure and interlaminar shear strengths, glass transition temperature, moisture content, and formability of the B-staged laminates. No siginificant change in the processability of either Celion 6000/PMR-15 or Celion 6000/LaRC-160 was recorded indicating that after B-staging the graphite/polyimide laminates can be stored for extended periods of time in normal shop and high humidity environments and then be fabricated into defect-free structural elements in heated platen presses.

®Celion is a registered trademark of Celanese Corporation

1. INTRODUCTION

The storage of addition-type polyimide polymers at low and ambient temperatures has resulted in variations in processing response of the resin and variations in the final properties of the material. This problem has restricted the development of a single processing cycle suitable for curing freshly mixed material and material that has been stored in refrigerated and shop environments. Several studies have been directed toward the identification of the mechanisms that contribute to instability of these polyimide systems. The PMR-15 polyimide resin system developed at NASA-Lewis has been investigated and it was found that impurities of the diester of benzophenonetetracarboxylic acid (BTDE) were formed during storage in ethanol solvent.[1] This study concluded that higher ester impurities of BTDE increased with time during ambient and 5°C (41°F) storage of BTDE solutions. In addition, graphite reinforced PMR-15 composite laminates fabricated with resin made from aged BTDE exhibited lower mechanical properties than laminates fabricated with resin formulated with freshly mixed BTDE.

Since the storage of BTDE in the presence of ethanol was shown to cause the formation of undesirable higher ester compounds and the mechanical properties of graphite/PMR-15 made from aged PMR-15 resin were lower than mechanical properties of laminates made from unaged PMR-15 a study was undertaken to establish the effect of removing the ethanol by B-staging this system prior to processing. Celion 6000/PMR-15 was B-staged and then stored for up to six months in both a shop environment and 100 percent relative humidity chamber prior to processing into cured laminates and structural elements. The effect of staging under these conditions prior to processing was determined by measuring flexure and interlaminar shear strength of test specimens machined from the laminates and structural elements. A second addition-type polyimide composite system, Celion 6000/LaRC-160 was included for comparison.

2. MATERIAL AND SPECIMEN DESCRIPTION

The Celion 6000 graphite reinforced PMR-15 and LaRC-160 polyimide pregs were fabricated in-house on a conventional rotating drum filament winding machine. The solvent content for each of the as-wound prepreg systems ranged from 33 to 36 weight percent. Following the winding operation, radiant heat lamps were positioned near the surface of the prepreg sheets to remove the excess solvent. The prepreg sheets were heated to approximately 66°C (150°F) for 30 minutes to reduce the solvent content to approximately 15 weight percent. The prepreg was then removed from the drum and representative 7.6cm x 7.6cm (3 in. x 3 in.) squares of the prepreg were cut from each edge and the center for resin, fiber, and volatile content determinations (table 1). The prepreg sheets were then cut and laid up into appropriate orientation required for the fabrication of laminates.

In order to monitor the effect of B-staging and exposure on the graphite/polyimide systems a series of 16 ply unidirectional interlaminar shear panels were fabricated and tested. The interlaminar shear panels were assembled by cutting and stacking 16 plies of prepreg in a unidirectional orientation. The structural element chosen to demonstrate the fabricability of the B-staged and exposed laminates was a hat shaped stringer element. The laminates for these elements were assembled by cutting and stacking 9 plies of prepreg in a $\pm 45_2$, 0, $\overline{+}45_2$ orientation. This fiber orientation was chosen as being representative of bias ply filament reinforcement found in many stringer element designs where loads are transferred from the stringer to the skin by shear.

3. HOT FORMING PROCESS DEVELOPMENT

3.1 B-Stage Development

Contractual[2] and in-house[3] cure cycle development studies have shown that the solvent content for PMR-15 and LaRC-160 polyimide pre-preg systems must be below 3 weight percent to prevent the generation of excessive voids and delaminations during final cure. Solvent reduction or B-staging is achieved by heating the laminate under vacuum for a period of time.

During the development of B-stage cycles for Celion 6000/PMR-15 and Celion 6000/LaRC-160 handleability problems were encountered with laminates B-staged to the 3 weight percent solvent level. These laminates were found to be extremely difficult to handle during routine fabrication due to individual ply separations, loose pieces in a single ply and dry, powdery resin areas. The handleability was significantly improved by reducing the residual solvent content to below 1 weight percent. The laminates staged to this solvent content level were found to be very durable and presented no unusual handling problems. The B-stage cycles that were developed for this study are shown in figure 1. The low vacuum level of .3kPa (0.1 in. Hg) during the B-staging cycle for the Celion 6000/LaRC-160 system was found to be necessary in order to prevent excessive flow in the heat-up portion of the cycle.

3.2 Laminate Conditioning

B-staged Celion 6000/PMR-15 and Celion 6000/LaRC-160 laminates were conditioned in "normal" shop and high humidity environments for extended periods of time prior to hot forming into structural elements to determine if these systems would remain stable and retain their processability. The B-staged laminate conditioning schedule is shown in table 2. The "normal" shop environment is defined as a typical composite lay-up and assembly area in the Langley Research Center's Composite Model and Development Shop. The humidity and temperature in this area is maintained at 50 ± 8 percent and 18-24°C (65-75°F), respectively. The laminates scheduled for exposure to the high humidity environment were placed in a chamber which had been modified to maintain a temperature of 35-41°C (95-105°F) and 100 percent relative humidity. At specified intervals, the B-staged laminates were removed and processed into structural elements and interlaminar shear panels.

3.3 Hot Forming Cycle Development

After conditioning in either the shop or high humidity environments for various times, the laminates were removed, weighed, and prepared for hot forming and curing. A single ply of Teflon coated style

104 fiberglass release cloth was placed on each side of the laminate, followed by a sheet of 0.12mm (.005 in.) polyimide film on each side (figure 2). The structural element laminates were then formed in the sequence shown schematically in figure 3. The laminate was placed in the open die (figure 4) which was heated to 260°C (500°F). At this temperature the polyimide matrix resin melts and becomes viscous. The upper and lower halves of the die were then gradually closed (figures 5 and 6), and pressure was applied to the laminate as the temperature was increased. After the die halves were closed, 2.07MPa (300 psi) of pressure was applied over a period of about 2 minutes. At the end of this pressurization, the dies were heated to 330°C (625°F) and the pressure and temperature were maintained for one half hour to fully form and cure the hat section element (figure 8). The hot form cure cycle shown in figure 7 was used to cure the Celion 6000/ PMR-15 and Celion 6000/LaRC-160 structural elements and interlaminar shear panels.

The completeness of cure of each formed Celion 6000/PMR-15 and Celion 6000/LaRC-160 laminate was determined by measuring the glass transition temperature with conventional TMA equipment. In all cases the glass transition temperature, T_g, of the laminates cured by the cycle shown in figure 7 exceeded 316°C (600°F). A very modest increase in T_g of approximately 3°C (5°F) was obtained by subjecting the laminates to the post cure cycle shown in figure 9. Glass transition temperatures for both the hat section elements and interlaminar shear laminates before and after post cure are shown in figure 10.

4. HOT FORMED PANEL TESTING AND EVALUATION

A total of 28 hot formed hat section elements were fabricated and evaluated in this study. An equal number of interlaminar shear panels were fabricated and evaluated. Each of the elements and panels were subjected to nondestructive testing and evaluation.

4.1 Ultrasonic Evaluation

Each of the hot formed panels was ultrasonically examined for delaminations and/or voids. The equipment utilized for these examinations consisted of a 5 MHz pulse-echo transducer, water immersion tank, scanning bridge, and signal monitoring devices. Prior to scanning each laminate the equipment was calibrated to standards capable of detecting defects as small as 0.25 mm (0.010 in.) in diameter. Typical ultrasonic c-scans of hat section elements are shown in figure 11. The c-scans were developed from hat section elements that had been cut at the bends of the element. The residual flat sections were placed down in the same order as the cutting sequence for scanning. No voids or delaminations were found in any of the c-scans for the hat section elements or interlaminar shear laminates fabricated in this study. The dark anomaly shown in the lower center of the c-scan for the Celion 6000/PMR-15 hat section represents a depression in the element where the thermocouple was located.

4.2 Microscopic Examination

Representative samples of the hot formed and cured hat section elements and laminates were machined and prepared for microscopic examination. The samples were viewed at 126x in an optical microscope to look for fiber distribtuion, resin rich areas, and microscopic voids. Typical photomicrographs of hat section elements fabricated from B-staged laminates after exposure are shown in figures 12 and 13. The fiber distribution represented in these photomicrographs are typical of good quality graphite/polyimide laminates. In addition, no significant voids were noted, which verifies the findings in the ultrasonic c-scans of these samples. No anomalies were detected in the microscopic examination of the remaining elements and laminates in this study.

4.3 Interlaminar Shear Tests

The laminates prepared for interlaminar shear property determination of Celion 6000/PMR-15 and Celion 6000/LaRC-160 were machined into shear samples and tested according to the procedures outlined in ASTM-D2344-76. The interlaminar shear strength of Celion 6000/PMR-15 and Celion 6000/LaRC-160 fabricated after B-stage and exposure to 50 ± 8 and 100 percent relative humidity for times up to six months is shown in figures 14 and 15. The interlaminar shear strength data for each graphite/polyimide system exhibited similar decreases in shear strength (approximately 15 percent) as a function of exposure time. There appears to be no significant difference in interlaminar shear strength between Celion 6000/PMR-15 and Celion 6000/LaRC-160 for exposures up to six months.

4.4 Flexure Tests

In the initial screening effort for baseline properties, ASTM-D-790 flexure specimens were machined from the flange, ramp, and cap segments of the hat section elements. The flexure strength of specimens taken from all of these areas were within a scatter band of ±5 percent of the average. Therefore, to monitor the flexure strength of elements subsequently fabricated in this study, flexure test specimens were taken only from the flanges of the hat section elements.

As outlined in section 2. the graphite fiber orientation in the hat section elements was selected to represent a typical lay-up pattern for skin stringer elements. The flexure strength obtained from the hat section flange samples fabricated with this lay-up pattern ($\pm45_2$, 0, $\mp45_2$) was used to generate data to indicate property changes with exposure. The data presented in

figures 16 and 17 indicate that
Celion 6000/PMR-15 can be B-staged,
stored for six months at 50 or 100
percent relative humidity and then
be fabricated into hat section elements with a modest (approximately 9
percent) drop in flexure strength.
A more significant drop in flexure
strength (approximately 24 percent)
was recorded for Celion 6000/LaRC-
160 after being tested following the
same exposure. The only known difference in the Celion 6000/PMR-15
and Celion 6000/LaRC-160 test specimens was a slightly lower resin content (approximately 3 percent) for
the latter system.

5. CONCLUDING REMARKS

A process has been developed which
eliminates a variety of problems
associated with graphite/polyimide
prepreg storage and conversion to
structural shapes. After B-staging,
the essentially solvent-free (less
than 1 percent) graphite/polyimide
laminates can be stored in "normal"
shop and high humidity environments
at ambient temperature up to six
months without changing the processability of the material. Ultrasonic
c-scans of the as-fabricated hat
section elements and interlaminar
shear panels indicated no detectable
voids or delaminations. These findings were verified by examining
cross-sections of the as-fabricated
graphite/polyimide elements and
panels. A drop in the mechanical
properties of the test samples was
recorded as the B-stage storage time
prior to processing was increased.

6. REFERENCES

1. Lauver, R. W.; and Vannucci, R.
D.: Characterization of PMR Poly-
imides- Correlation of Ester Impurities with Composite Materials.
24th National SAMPE Conference,
Volume 24, Book 1, 1979, pp. 522-
532.

2. Leahy, J. D.; and Frost, R. K.:
Development and Demonstration of
Manufacturing Processes for Fabri-
cating Graphite/LaRC-160 Polyimide
Structural Elements, (7th Quarterly
Progress Report for NAS1-15371),
March 1980. (Available as NASA CR-
159303.)

3. Baucom, Robert M.: LaRC Fabrication Development. Graphite/
Polyimide Composites, NASA CP 2079,
pp. 19-37, February 1979.

7. BIOGRAPHIES

Mr. Robert M. Baucom is a Materials
Engineer in the Materials Processing
and Applications Branch at NASA-
LaRC. Mr. Baucom is a Mechanical
Engineering graduate from North
Carolina State University, Raleigh,
North Carolina. He has over 17
years experience in metals and
resin matrix composites research.
His current responsibilities
include the development of advanced
fabrication techniques for graphite
fiber reinforced polyimides in support of the CASTS and SCR programs
at Langley. He is the author of 25
technical papers in the fields of
metals and advanced composites
research.

Mr. Paul W. Kidder is an Engineering Specialist with Kentron Inter-

national, Inc., at the Hampton
Technical Center. He is a graduate
of Wentworth Institute, Boston,
Massachusetts and has over 27
years experience in the design,
development, and fabrication of composite structures. His experience
has included development of methods
and processes in support of advanced
materials research programs, and
tooling and manufacture of composite
material products for aero-hydro-
space applications.

TABLE 1. MATERIAL DESCRIPTION

GRAPHITE FIBER	POLYIMIDE RESIN	FIBER AREAL WEIGHT, gms/m^2 (oz/yd^2)	RESIN SOLIDS, weight percent	VOLATILE CONTENT, weight percent	FIBER VOLUME FRACTION percent	PLY THICKNESS, mm. (in.)
Celion 6000	PMR-15	185 (5.47)	33	10.3	61	0.17 (.0067)
Celion 6000	LaRC-160	176 (5.20)	36	12.5	59	0.16 (.0064)

TABLE 2. GRAPHITE/POLYIMIDE CONDITIONING SCHEDULE

GRAPHITE/POLYIMIDE SYSTEM	B-STAGE EXPOSURE		
	RELATIVE HUMIDITY, percent	TEMPERATURE, °C (°F)	TIME, days
Celion 6000/PMR-15	50 ±8	18-24 (65-75)	0,1,7,14,30,120,180
	100	35-41 (95-105)	"
Celion 6000/LaRC-160	50 ±8	18-24 (65-75)	"
	100	35-41 (95-105)	"

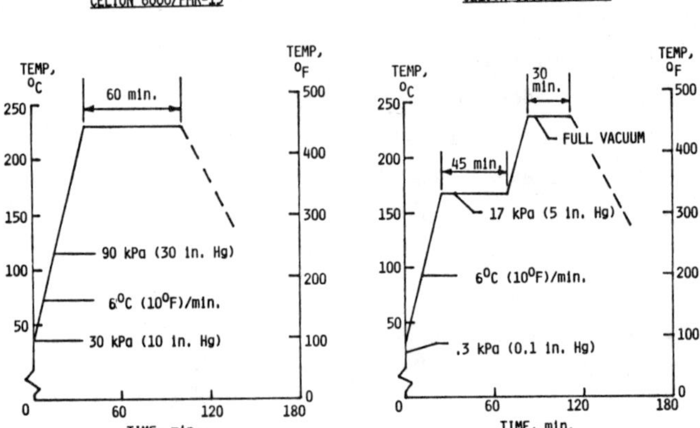

Figure 1. B-stage cycles for Celion 6000/PMR-15 and Celion 6000 LaRC-160.

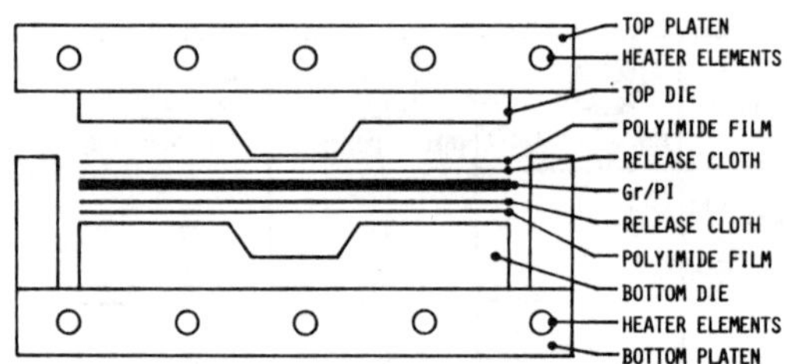

Figure 2. Fabrication assembly for hot formed hot section elements.

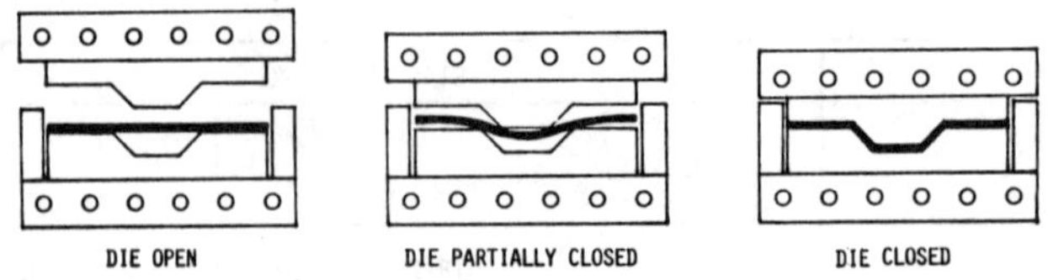

Figure 3. Hot section element fabrication sequence

Figure 4. Graphite/polyimide hot forming equipment.

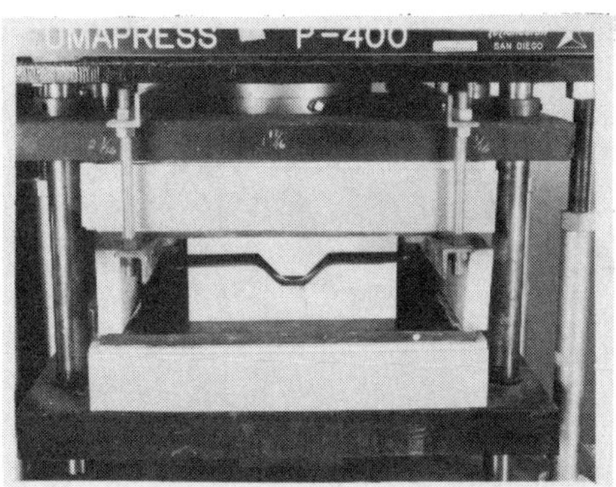

Figure 5. Partially formed graphite/polyimide hot section element.

Figure 6. Fully formed graphite/polyimide hot section element.

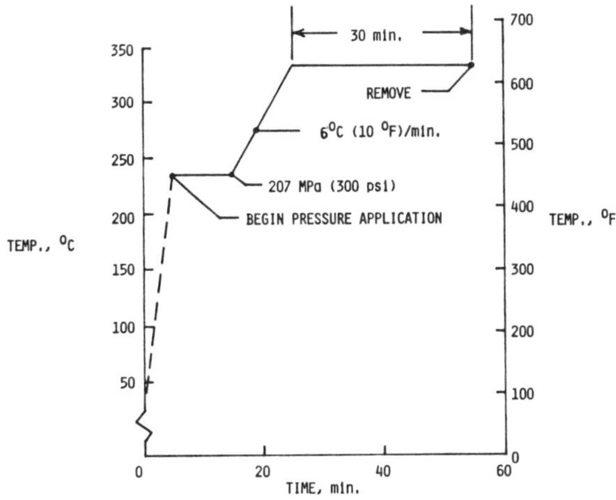

Figure 7. Forming cycle for Celion 6000/PMR-15 and Celion 6000/LaRC-160.

Figure 8. Graphite/polyimide hot section elements.

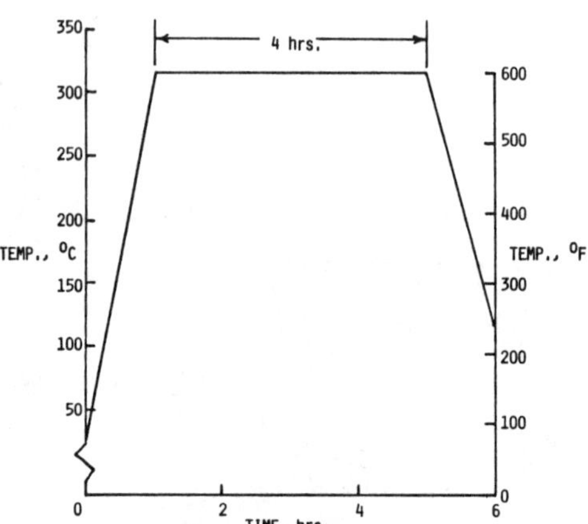

Figure 9. Post cure cycle for Celion 6000/PMR-15 and Celion 6000/LaRC-160.

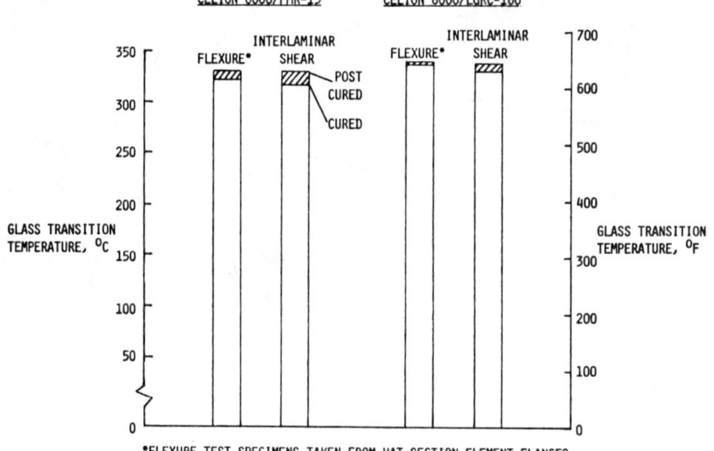

Figure 10. Glass transition temperatures for Celion 6000/PMR-15 and Celion 6000/LaRC-160.

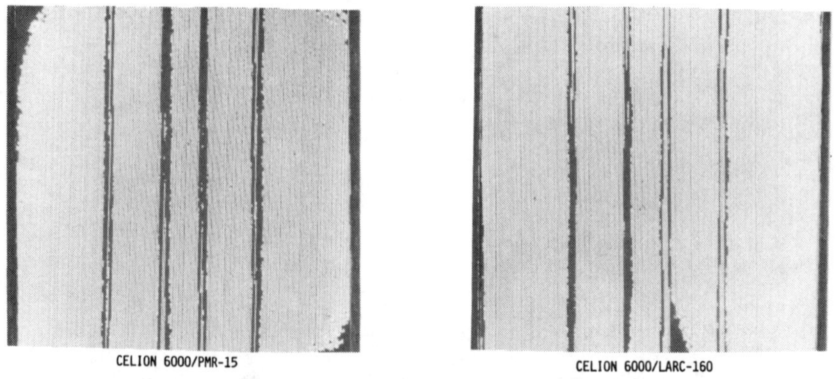

Figure 11. Ultrasonic c-scan photographs of hat section elements.

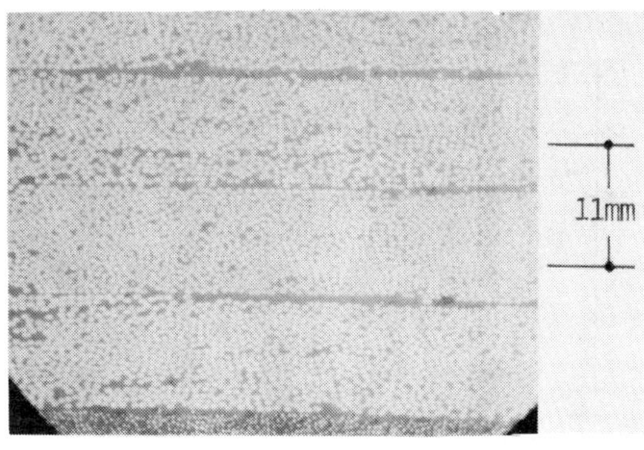

B-STAGE/6 MONTHS EXPOSURE AT 50 PERCENT R.H.

Figure 12. Photomicrograph of cured Celion 6000/PMR-15 hot section elements.

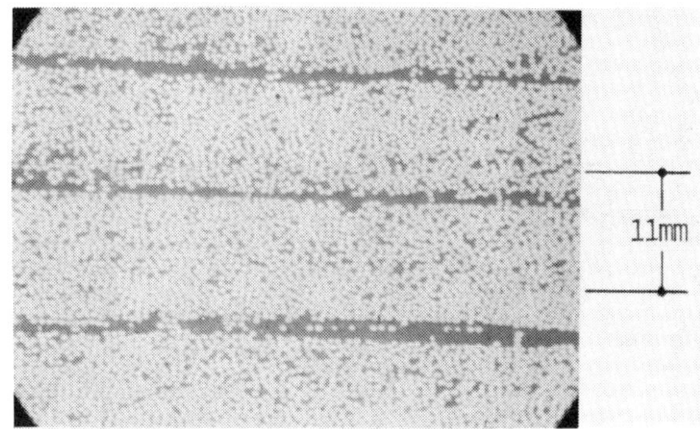

B-STAGE/6 MONTHS EXPOSURE AT 50 PERCENT R.H.

Figure 13. Photomicrograph of cured Celion 6000/LaRC-160 hot section elements.

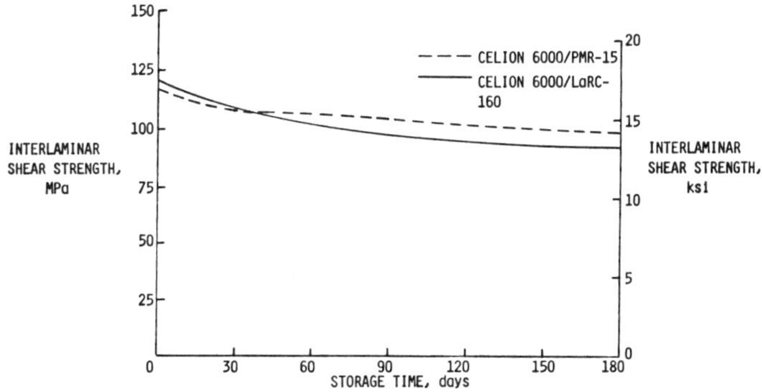

Figure 14. Interlaminar shear strength of unidirectional graphite/polyimide fabricated
from B-staged laminates stored at 100 percent relative humidity.

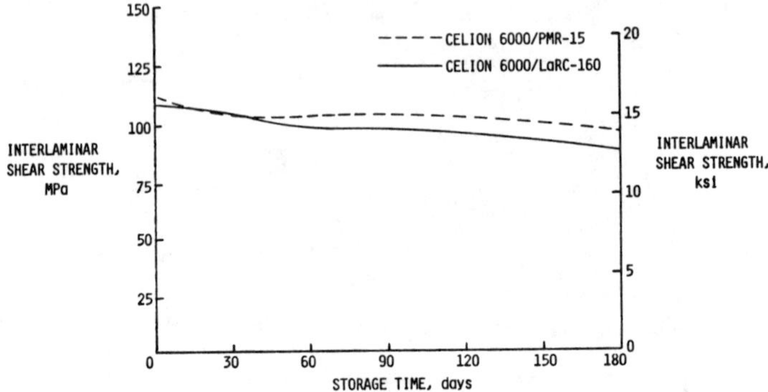

Figure 15. Interlaminar shear strength of unidirectional graphite/polyimide fabricated from B-staged laminates stored at 50 percent relative humidity.

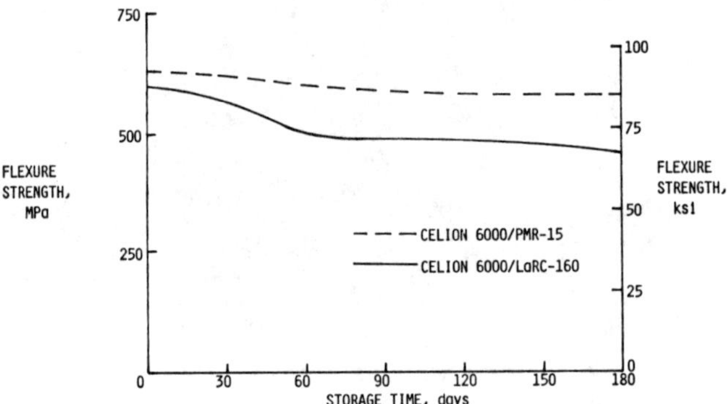

Figure 16. Flexure strength of $\pm45_2$, 0, $\mp45_2$ graphite/polyimide fabricated from B-staged laminates stored at 50 percent relative humidity.

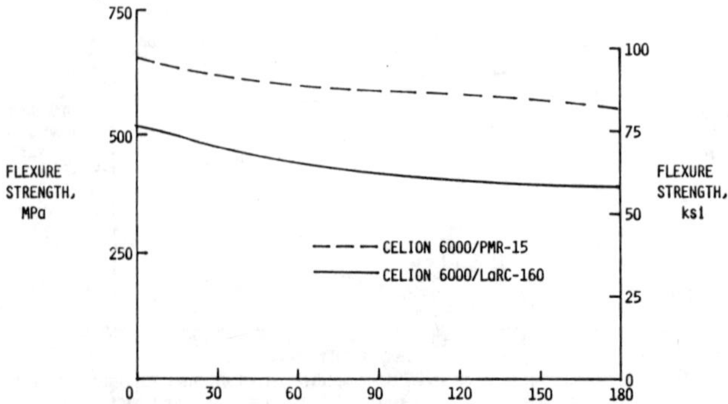

Figure 17. Flexure strength of $\pm45_2$, 0, $\mp45_2$ graphite/polyimide fabricated from B-staged laminates stored at 100 percent relative humidity.

A LOW-COST, EFFICIENT AND
DURABLE LOW-TEMPERATURE SOLAR COLLECTOR
Timothy P. ODonnell
Solar Materials, Inc.
P.O. Box 284
Altadena, California

Abstract

A relatively new synthetic elastomer material for low-temperature solar thermal applications is described. The cost nature, efficiency and durability of this material is highlighted. This material, specially formulated ethylene-propylene-diene-monomer (EPDM), used in a patented solar collector design (SolaRoll) has been considered a significant advancement of the state-of-the-art in the solar energy field. Cost/application experience of Solar Materials, Inc. with this product is discussed. This material/product, SolaRoll, is presented here as a viable alternate energy approach in lieu of the ever increasing costs of standard energy.

1. INTRODUCTION

Consumer energy costs are rising at an alarming rate. There have been significant yearly increases in prices paid by industrial and commercial establishments, homeowners, etc. for space and water heating needs in the last several years. Experts expect this trend to continue.

With all fuel costs increasing solar energy has reemerged as a reasonable energy alternative to consider. Solar hot water heating was fairly popular in the early 1900's in Florida and Southern California. Low-temperature solar thermal systems are proving once again to be a good investment. To date, well over ten million square feet of low-temperature solar collectors have been manufactured.

One unique solar collector system/material is a product called Sola-Roll. SolaRoll is the trademark of solar components manufactured by Bio Energy Systems, Inc. in New York. SolaRoll is a low-cost, efficient, durable and flexible collector material made of ethylene-propylene-diene-monomer (EPDM). This material has been used in solar systems to heat domestic water, pools, spas, homes by radiant energy, air space and various other applications.

    2. MATERIAL DESCRIPTION

EPDM or ethylene propylene rubber compounds are synthetic elastomers. EPDM elastomers combine superior ozone, good heat and oxygen resistance and very good low temperature properties to produce a compound with excellent overall age resistance. These properties have made EPDM compounds ideally suited for outdoor applications, especially solar system use.

Ethylene-propylene-diene-monomer is a terpolymer produced in several variations, the principal variant being diene used in the system. Dicyclopentadiene, ethylidene morbornene and 1, 4-hexadiene are the types usually selected as the non-conjugated diene component.

For example:

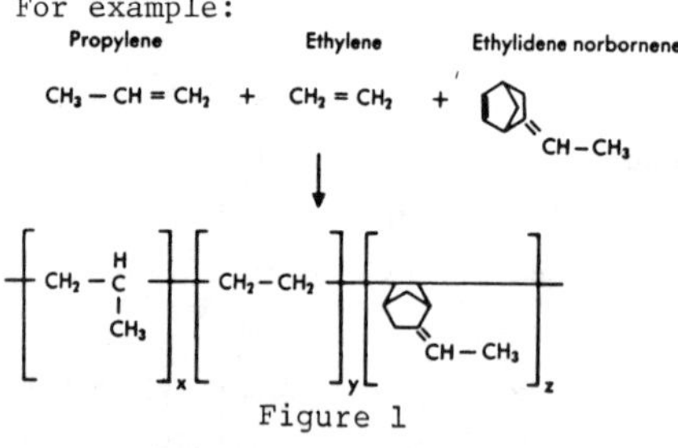

Figure 1

The advent of the Ziegler-Natta catalysts (mixtures of aluminum alkyls and titanium halides), about 1955, enabled the synthesis of sterospecific polymers, including the long sought for duplication of natural rubber, cis-1,4-polyisoprene. These polymerizations are carried out in hydrocarbon solvents, in the absence of air or moisture.

The Ziegler-Natta catalytic systems can also be used to prepare copolymers of ethylene and propylene (3:1), which are noncrystallizing elastomers exhibiting excellent mechanical properties. The absence of unsaturation makes these elastomers exceptionally suitable for age-resistant uses (e.g. solar collectors). However, in order to enable sulfur vulcanization, these copolymers are generally prepared with a few per cent of unsaturated units (nonconjugated dienes) and these are known as ethylene-propylene terpolymers (EPDM, for "ethylene-propylene diene monomer"). The unsaturated terpolymers of ethylene and propylene also have good heat resistance because their double bonds are not in the main chain and their reaction does not involve scission of the main chain. Although the nominal concentration of atmosphere ozone is only about 0.007 parts per one hundred million parts of air (or 7×10^{-5} PPM) laboratory tests have been conducted at concentrations as high as 10,000 parts ozone per hundred million parts of air (100 PPM). Specially formulated EPDM remains immune to ozone attack at its surface double bonds. Bond damage can cause cracking, hardening or tack.

The following table summarizes EPDM properties. The data has been provided by DuPont. Other manufacturers include Exxon, Goodrich and Uniroyal.

102

Table 1

EPDM Properties

Tensile Strength	Good to excellent	Maximum values exceed 3000 psi [211 kg/sq cm] (ASTM D-412-64T)
Hardness	Broad range	Durometer A values of 40 to 90 (ASTM D-2240-64T)
Tear Strength	Good	Nominally 200 lb./in. [35.6 kg/cm] (ASTM Die B)
Compression Set	Good to excellent	Nominally 10-25% (22 hrs. at 158°F. [70°C.]). With special compounding, values of 5% are possible. (ASTM D-395-61, Method B)
Hot Stress-Strain Properties	Good to excellent	Nominally maintains 60% of its room temperature tensile strength and elongation when tested at 212°F. (100°C.)
Resilience	Broad range with low degree of temperature dependence	Yerzley values of 30 to 80% (ASTM D-945-59)
Electrical Properties	Excellent	Dielectric constant of 3, power factor of 1%. Suitable for high voltage applications in wet or dry environments.
• **Resistance to:**		
Abrasion	Good to excellent	Up to 300 index by NBS test (ASTM D-1630-61)
Weathering	Excellent	No crazing or discoloration after 3 years' exposure in Florida.
Ozone	Excellent	No cracking even after 500 hours at 10,000 pphm, 20% extension (ASTM D-1149-64)
Heat	Excellent	Continuous service at temperatures up to 300°F. (149°C.). With special compounding, intermittent service to 400°F. (204°C.).
Cold	Excellent	Continuous service as low as −60°F. (−51°C.). Special compounds flexible down to −90°F. (−68°C.) (ASTM D-1043-61T)
Steam	Excellent	
Flame	Degree of flame resistance depends on compounding.	
Chemicals (Immersed to 28 days at 75°F. [24°C.])		
Acids	Good	
Bases	Excellent	
Petroleum Oils	Poor	
Polar Materials (Water, phosphate esters, ketones, alcohols and glycols.)	Good to excellent	

Bar chart — RETENTION, % vs. AT 302F (150C); bars for TENSILE STRENGTH and ELONGATION at 70 HOURS, 7 DAYS, 14 DAYS, and 28 DAYS.

The SolaRoll material (EPDM) is extruded into 4.4 inch wide mats. Each mat has six small tubes alternating with thin webbing. A cross section view of the SolaRoll tube mat is shown in Figure 1.

.731   .068   .203 DIA   .338 DIA   4.390

(Dimensions in inches)

The mat is available in rolls 600 feet in length which equals about 220 square feet of collector area. The continuous-length nature of the mat allows an installer to cut, assemble and construct systems with precisely the length and width desired.

The SolaRoll absorber mat will adhere to any clean building surface

with the use of thermosetting con-
struction-grade mastic adhesive.
Shingles, tile, brick, concrete,
wood, metal, insulation board-just
about anything can be used as a
mounting surface. In most cases
nails, screws, or other fastening
hardware are not required.
The webbing between the circulation
tubes can easily be stripped away
at any point along the mat segment.
In order to attach these tubes to
manifolds, webbing is stripped
from both ends of each mat segment.
Using a lubricant, Teflon jamb
sleeves are inserted into the end
of each tube. The tube ends are
then thumb-pressed into specially
designed holes in manifolds made of
copper or ABS plastic. The result
is a completely watertight connection that will withstand well over
50 psi of pressure. See figure
below.

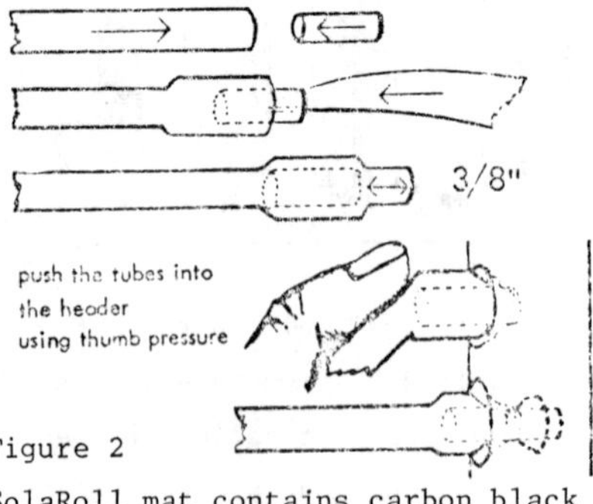

Figure 2

SolaRoll mat contains carbon black
throughout. Therefore there is no
separate coating to chip or delaminate. Carbon black acts to increase the materials solar absorp-
tivity, mechanical/abrasion

strength, thermal conductivity and
crosslinking efficiency.

3. COST

The low cost of the monomers used in
EPDM makes these elastomers especially attractive commercially. Small
quantity purchases of SolaRoll re-
tail at about $1.30 per linear ft.
or about $3.50 per square foot. All
cost figures in this paper express
1980 dollars. This compares with
the least expensive all copper solar
absorber panels costing about $4.50
a square foot. Generally, copper
collector panels are much more expensive. Purchase of SolaRoll in
bulk results in significant reduction in the costs listed above.
The simplicity and ease of installation plus the low-cost of EPDM
extensions keep the price of Sola-
Roll systems relatively low. Solar
Materials, Inc. has recently installed unglazed solar systems (including plumbing, pumps, controls etc)
for pools at about $8 - $9 a square
foot. The cost of a site built collector (single glass glazing), and
all other necessary components for
a single family (4-5) domestic hot
water system may average about $3200.
Consumer cost is drastically reduced
from this level by a combination of
all of the following: Federal tax
credit - 40%, California State tax
credit to be bring credit to the 55%
level and utility credits of about
$240 a year for four years. This
utility credit is in addition to the

federal and state income tax credits. The final cost of a $3200 solar system is then <u>only $480</u>.

4. EFFICIENCY

SolaRoll collectors have been tested using standard test procedures set by the American Society of Heating, Refrigeration and Air-Conditioning Engineers (ASHRAE), ASHRAE 93-77, and Housing and Urban Development (HUD), HUD 4930.2. In test conducted by the University of Conneticuts' Solar Energy Evaluation Center a SolaRoll collector operated at a higher average efficiency than did most of sixteen other commercially available collectors. Figure 3 shows the efficiency curves for just the Sola-Roll while Figure 4 shows the relationship of all collectors tested.

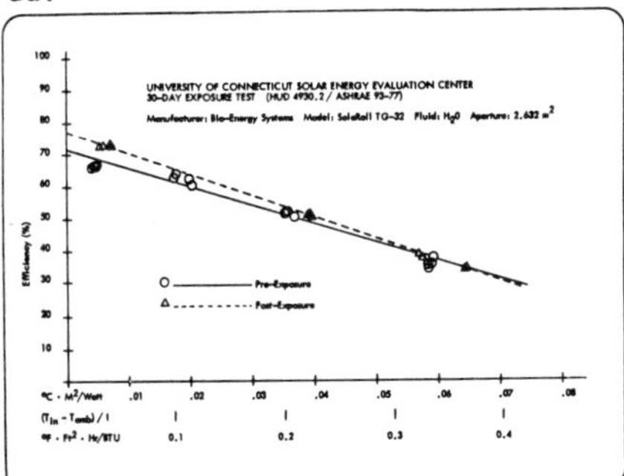

Figure 3

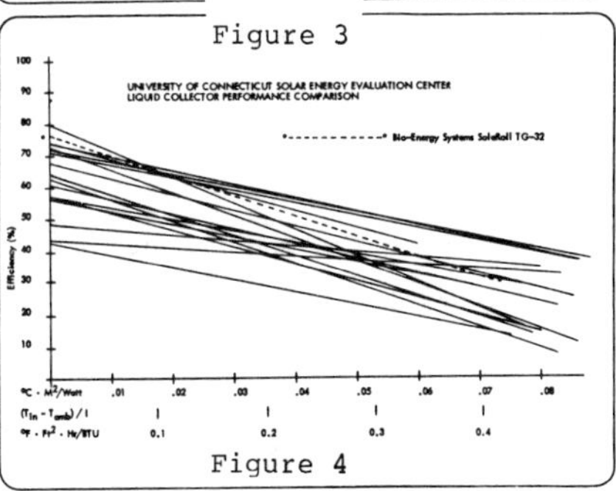

Figure 4

Given the intrinsically low thermal conductivity of EPDM the above results may seem surprising. S. Krulick of Bio Energy Systems, Inc. has suggested several reasons for high efficiency. They were: 1. Absolute thermal conductance of an absorber material in a collector isn't crucial because of the diffuse nature of the incoming solar radiation. A relatively small amount of heat is transferred a small distance, 2. The mat tubes are very close together – only 3/4 inch apart, compared with four or five inches on a typical copper absorber plate. SolaRoll absorber mats therefore carry a greater amount of water through the collector. With this design there is a greater collector surface to water mass ratio meaning more heat is collected, and 3. The material has less mass than metals. It heats up faster but admittedly cools down faster when the sun goes behind a cloud. The net result is likely a pick up of more heat.

Many feel that the stand alone criterium of cost and efficiency are inadequate in comparing collectors. A meaningful value is the Btu's delivered over the dollars spent for a collector. SolaRoll has one of the highest Btu/$ ratios.

5. DURABILITY

SolaRoll has been specially formulated to resist UV degradation and last in solar service well over twenty years. In weatherability, this product appears to be comparable to that of the best specialty elastomers. Comparable EPDM samples were exposed (no shading) for more than ten years in Florida. The samples were free from surface crazing and showed high retention of original physical properties as shown in the following DuPont data (Table 2)

Table 2
EPDM Florida Exposure

	Original	After 1 year	After 5 years	After 10 years
Tensile strength, psi	2900	3000	3000	3000
Elongation, %	560	490	360	350
Hardness, Durometer A	63	69	72	80
Surface Condition	—	No crazing	No crazing	No crazing

Water has little effect on the properties of this material. SolaRoll or vulcanized EPDM hydrocarbon rubber show minor loss in tensile strength and elongation upon being exposed to boiling water. Tests also show a very low degree of water absorption.

A main forte of this material is its ability to withstand environmental extremes. It is resistant to heat up to 375°F and intermittenly to 400°F. Cold temperatures down to about -60°F are no problem. The collector material has been designed to handle water freeze/thaw cycles with no detrimental effects.

Chemical resistance of EPDM elastomeric materials have been well characterized. Over one hundred chemicals have been evaluated in contact with EPDM by DuPont. In solar service, SolaRoll is virtually impervious to corrosion Solar installations with this product have been in existance well over seven years. Material degradation has not been evident.

As an indication of durability, this material can be buried in concrete for radiant floor heating. It is virtually impervious to concrete and will easily withstand the shifting, expansion and contraction common to this flooring. In this manner, radiant heating can prevent ice and snow build-up on outside surfaces.

6. SUMMARY

SolaRoll, a specially formulated EPDM solar collector material, is a low-cost, efficient and durable alternative energy approach. The ratio of Btu's delivered per dollars expended for a SolaRoll collector is extremely high. Tests/analyses conducted for and by the National Bureau of Standards (NBS) confirmed this fact. Equivalent collector areas (ft.^{2}) of a brand name copper collector and SolaRoll were installed and monitored during 1978-79. SolaRoll exhibited a Btu/$ ratio over three and one-half times that of the copper collector. In fact, in a report to the Department of Energy (DOE) the NBS considered this product a "significant advance of the state-of-the-art" which could "provide a much needed powerful impetus toward conversion to solar energy." It continued, "its economics put it well within the realm of good consumer acceptance." Sola-Roll was the first flat plate collector to be recommended (by NBS) to DOE for support.

Solar Materials, Inc. has searched for a low-cost reliable collector material. To date, the product discussed in this paper has performed to our satisfaction in a variety of applications. For certain applications the material is backed by a full five year warranty with ten additional years of a limited warranty.

BIOGRAPHY

Timothy P. O'Donnell is a co-founder and executive officer of Solar Materials, Inc., P.O. Box 284, Altadena, California, 91001, USA.

Mr. O'Donnell has a Bachelor of Science degree in Metallurgical Engineering from California Polytechnic State University, San Luis Obispo, CA. He was employed as a Scientist Associate at Lockheed-California's Rye Canyon Research Plant working on sensitive aircraft failure analysis cases. He has also had over five years as a Materials Specialist/Senior Engineer at the Jet Propulsion Laboratory, Pasadena, California. At JPL he has been involved in and provided materials engineering support to numerous spacecraft and energy projects including: Voyager, Low-cost Solar Array project (photovoltaics) and Solar Thermal Power Systems (high temperature solar). His technical society memberships include: SAMPE - Los Angeles Chapter member, American Ceramic Society and Alternate Consumer Energy Society.

A QUANTITATIVE METHOD FOR PHOTOVOLTAIC ENCAPSULATION SYSTEM OPTIMIZATION

Alexander Garcia III
Spectrolab, Inc.
12500 Gladstone Avenue
Sylmar, CA 91342

Charles P. Minning
Hughes Aircraft Company
Culver City, CA 90230

Edward F. Cuddihy
Jet Propulsion Laboratory
4800 Oak Grove Drive
Pasadena, CA 91109

ABSTRACT

This paper describes a set of analytical methods which have been developed to enable quantitative analysis of encapsulation system designs for terrestrial photovoltaic modules. Design factors determined most important include: encapsulant thickness and modulus, emissivity of module surface, ribs on substrate modulus, and AR coatings.

1. INTRODUCTION

The design of encapsulation systems for flat plate photovoltaic modules requires the fulfillment of conflicting design requirements. For example, structural requirements favor a thick layer of encapsulant between the cell and cover glass (i.e. the load-bearing member) of a glass superstrate module; on the other hand, optical and thermal requirements favor a thin encapsulant layer for this type of module. In the past, design tradeoffs to meet these requirements have been carried out in a cut-and-try fashion with resultant heavy investments in time and money.

The goals of these investigations were basically three-fold: (1) systematize the mathematical models to be used in each of the technical areas with special attention devoted to determination of module performance sensitivity to thickness and physical properties of candidate encapsulants, (2) establish trends that will be useful to the module designer, and (3) establish design priorities; that

*The research described in this paper was supported by the Jet Propulsion Laboratory of the California Institute of Technology and was sponsored by The Department of Energy through an inter-agency agreement with NASA.

is, identify items that exert the strongest influences on module electrical power output, cost, and safety.

2. OVERALL MODULE DESIGN

A baseline geometry for module construction was developed for the analysis. The baseline module was 48" x 48" square and consisted of 121, 4" x 4" square solar cells. Two families of module designs were considered. The superstrate design family is based on a transparent front cover, which also acts as the load bearing member of the module. The substrate is based on an opaque backing material which is the load bearing member.

3. STRUCTURAL ANALYSIS

3.1 Deflection Stress Sensitivity Studies

Mechanical loading such as wind and snow pressure, acceleration due to earthquakes, and imposed twist due to support settlement or misalignment will generate stress in the module. As the module deflects under load, the resulting strain in the load bearing layer will transfer through the encapsulant to the solar cells.

In order to determine the sensitivity of solar cell stress to encapsulant modulus and thickness, the analysis must account for the strain relief through the thickness of the module provided by the encapsulant layer. The modes of strain relief are transverse shear flexibility and thickness stretch. Classical lamination theory, which is based upon simple bending theory, does not account for these modes of strain relief. Also, deflection of the module due to wind pressure loading can be large relative to the thickness of the load-bearing layer. This necessitates the use of nonlinear, large deflection theory in order to accurately predict stress and deflection and to avoid overly conservative design.

In order to reduce time and cost, a relatively simplistic approach was used. The analysis utilized a two-dimensional, finite-element model wherein one-half of a cell and the additional layers through the thickness of the module were modeled. Small defelction theory was assumed to be valid in the localized region of a single cell. Finite-element analyses utilizing the MSC/NASTRAN structural analysis computer program were performed.

The model consisted of rectangular plate elements with symmetric boundary conditions along the center of the cell and free edge conditions along the imaginary cut plane between adjacent cells, as shown in Figure 1. The model permitted easy modification of the encapsulant thickness and modulus. Enforced displacements were applied to the load bearing layer and the resulting ratio of strain in the

cell to strain in the load bearing
layer was determined. Utilizing
the strain ratio for a given configuration, solar cell stress was
calculated for a desired working
stress in the load bearing layer.

The results of the sensitivity
studies for the glass superstrate
and wood product substrate are
summarized in Figures 2 and 3, in
that order. Solar cell stress is
plotted as a function of encapsulant thickness and modulus of
elasticity. The encapsulant thickness refers to the thickness
between the solar cell and the
load bearing layer. The analysis
assumed an equal thickness above
and below the cell. In each of
the figures the 8000 PSI solar
cell fatigue-allowable is indicated by a dotted line.

In Figure 2, the glass superstrate,
it is seen that for low values of
encapsulant modulus, the cell
stress decreases as the modulus
decreases or as the encapsulant
thickness increases. The encapsulant provides strain relief
through the thickness of the module. For ethylene vinyl acetate
(EVA) (E = 1000 PSI), approximately
5 mils are required to maintain
cell stress below the 8000 PSI
fatigue limit. For encapsulant
moduli above 50 KSI, no strain
relief is provided. For a modulus
of 250 KSI, the cell stress
increases as the encapsulant
thickness increases. For a stiff

encapsulant the cell stress increases as the distance between the
cell and the load bearing layer
(glass) increases.

In Figure 3 cell stress is plotted
for a ribbed plate and for an
unstiffened plate using EVA encapsulant. About 15 mils of encapsulant are required when no ribs
are present. Stiffening ribs
reduce the strain in the plate
which reduces the strain transferred to the solar cell. With
ribs, cell stress is not sensitive
to the thickness of EVA encapsulant.

3.2 Thermal/Stress Sensitivity Studies

Thermal/stress analyses were performed in order to determine the
sensitivity of solar cell stress
to the encapsulant modulus and
thickness. The analyses utilized
the same two-dimensional MSC/NASTRAN
model previously described. In
addition, a $5\frac{1}{2}$-cell, two dimensional model was developed in order
to determine whether cell stress
varied significantly as a function
of position in the module. In
all cases studied the thermal loading consisted of a 100°C uniform
temperature excursion. Room temperature material properties,
invariant with temperature, were
utilized in the analysis. This
assumption is violated in the real
world since the modulus of polymeric encapsulant materials increases with decreasing temperature
and decreases with increasing

temperature. However, the intent
of the analysis was to predict
trends and therefore the assumption
of linear, elastic, temperature
invariant properties was deemed
acceptable.

Analysis of the five and one half
cell model showed that cell stress
at the edge of the module was
approximately ten percent higher
than at the center of the module.
Therefore, in the subsequent sensitivity analysis, free edge conditions were imposed at the boundary of the one-half cell model.

Figure 4 presents the results for
the glass superstrate design. For
encapsulant modulus less than 10
KSI, strain relief is provided
and cell stress decreases as the
encapsulant thickness increases.
For EVA about 1 mil is sufficient
to maintain cell stress below the
5000 PSI thermal fatigue limit.
For a modulus of 250 KSI the encapsulant activity stresses the cell
due to its own expansion and the
cell stress increases as the encapsulant thickness increases.

The thermal/stress sensitivity for
a wood product substrate design is
given in Figure 5. In this case
cell stress is not critical to
thermal loading except for relatively stiff encapsulants (E > 50
KSI). For a modulus of 250 KSI
the encapsulant stresses the cell
due to its own expansion.

4. ELECTRICAL ANALYSIS

Electrical safety is an important
factor that must be incorporated
into the design of photovoltaic
modules. Of primary concern here
is the determination of material
thicknesses in the encapsulation
system to withstand at least 3000
volts DC (in 1000 volt system) before the occurrence of electrical
breakdown.

Electrical breakdown in dielectric
material is a complex phenomenon
which is not well understood. For
instance, values of electrical
properties such as dielectric
constant, dielectric strength, and
volume resistivity are known to be
strongly affected by humidity and
temperature. In addition, manufacturing defects such as cracks
and bubbles will also have a deleterious effect on the ability of a
module to withstand high voltage
stresses.

A detailed analysis that takes all
of the above factors into account
was not done in the present study.
Instead, a simple series capacitance model was developed for determining the electric field strength
in each material layer of the encapsulation system.

For a two-layer system, such as the
front cover and pottant above cell
combination found in substrate
designs, the following expressions
apply for the electric field apply,

$(E_1$ front cover and E_2 pottant).

$$E_1 = \frac{V_o}{t_1 + \left(\frac{\gamma_1 t_2}{\gamma_2}\right)} \qquad E_2 = \frac{V_o}{\left(\frac{\gamma_2 t_1}{\gamma_1}\right) + \gamma_2}$$

t_1 = front cover thickness

t_2 = pottant thickness

γ_1 = front cover dielectric constant

γ_2 = pottant dielectric constant

V_o = applied voltage

The electrical safety design used in this study is that the electric field strength in each layer <u>does not</u> exceed the dielectric strength, S_k, of that layer. In other words,

$$E_k < S_k$$

for all layers when V_o = 3000 volts DC.

A comparison between the results of these calculations with those of the previous section indicates that <u>for defect-free materials</u> (i.e., encapsulants and front covers) the encapsulant thickness required to satisfy both structural and electrical insolation requirements can be met by satisfying the structural requirement alone. However, these results are tentative because no material is defect free and humidity is known to adversely affect both the structural properties and the dielectric strength of organic materials.

5. OPTICAL ANALYSES

The primary purpose of the optical analysis is to predict the effective transmittance of an encapsulation system and indicate design directions for maximizing it. Effective transmittance is the ratio of the power output of an encapsulated cell to the power output of a bare cell. Hence, the spectral response of the cell is accounted for. To determine the effective transmittance requires calculating a weighted average transmittance of the encapsulation system where the weighting corresponds to the spectral response of the solar cell.

Also required of the optical analysis is the fraction of the total solar energy incident on an array absorbed in each layer of the encapsulation system. This information is used in the thermal analysis of the encapsulation system.

The following is a brief summary of the analytical procedure.

1. Required Information

a) Solar cell spectral power conversion efficiency or relative spectral response.

b) Thicknesses of encapsulation layers.

c) Absorption coefficients

d) Reflectivity of intercell areas.

e) Reflectivities of outer encapsulant layer and solar cell if they have been specially treated to reduce reflections. Otherwise, they can be calculated from indices of refraction.

2. Based upon the wavelength dependence of the above quantities, decide on the number of spectral bands that the solar spectrum should be divided into. Determine average values of the above quantities for each of these spectral bands.

3. Solve the radiosity-irradiation network for each of the spectral bands.

a) Calculate encapsulant layer transmittances.

b) Calculate reflectivities if not specified.

c) Solve for radiosities.

d) Solve for fraction of incident energy absorbed into each layer and surface.

4. Calculate the effective transmittance of the encapsulation system from equation.

5. Calculate the fraction solar irradiation absorbed in each layer and surface for input to the thermal analysis.

6. THERMAL ANALYSIS

To determine a realistic power prediction for a given module configuration, the steady state temperature at which the cell operates must be known. The results of the optical analyses are fed into an analysis program utilizing the CINDA 3G* thermal analyzer program.

A thermal network is defined describing the various heat flow paths as functions of material properties and dimensions. These transfers are analogous to an electrical resistance network. A typical network is shown below.

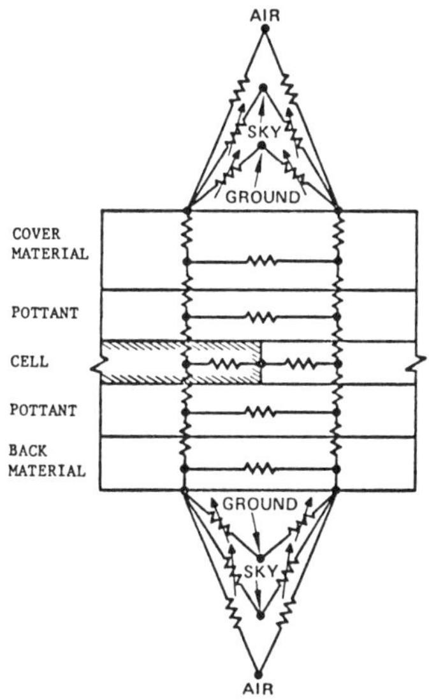

The results of the optical program are then incorporated into the thermal program to determine the cell power output. Table 1 is a summary of thermal/optical runs for the superstrate design. EVA/CG refers to a pottant material consisting of ethylene vinyl acetate and Crane Glas, a non-woven fiberglass matt. Open circuit conditions are those when there is no load on the solar cell. Power generation condition represents a loaded solar cell. Cell output power is for a single 4" x 4" square solar cell.

---

*Chrysler Improved Numerical Differential Analyzer, 3rd Generation

Figure 6 shows the energy balance
of a typical module under load.

7. CONCLUSIONS

The results of the thermal, optical,
structural, and electrical isolation analyses described in this
paper indicate that the following
items are major factors in the
design of terrestrial photovoltaic
modules:

1. For defect-free materials,
minimum encapsulation thicknesses
are determined primarily by structural considerations. Module deflection due to wind loads is the
primary determinant of encapsulant
thickness for superstrate modules.
On the other hand, both wind loads
and temperature excursions are
equally important in the determination of encapsulant thickness for
substrate modules.

2. Cell temperature is not strongly affected by encapsulant thickness or thermal conductivity.

3. Enissivity of module surfaces
exerts a significant influence on
cell temperature.

4. Encapsulants should be elastomeric.

5. Ribs are required on substrate
modules.

6. Aluminum is unsuitable as a
substrate material.

7. Antireflection coating is
required on cell surfaces.

8. Presence of Crane Glas in

encapsulant does not significantly
affect optical transmittance of
encapsulation system.

It was found that material defects
would strongly influence the
minimum encapsulant thickness required for electrical isolation.
In addition, it was found that
structural, optical, and electrical properties are heavily influenced by the presence of moisture.

For a more detailed treatment of
this research, the final report
for Phase I of the Design Analysis
and Test Verification of Advanced
Encapsulation Systems Contract
955567 with the Jet Propulsion
Laboratory should be obtained
from the Low-Cost Solar Array Data
Center.

| Cell Data | | Glass Data | | | Encapsulant | | | Calculated Results[2] | | |
| | | | | | | | | Cell Temp, °C | | |
Type[1]	AR - Coating Data	Iron Content	Thickness, mil	AR - Coating Data	Material	Thickness, mil	Backside Emissivity	Open Circuit	Power Generation	Cell Output Power, w
SC - non text	None	no	125	None	EVA/CG	5	0.92[3]	53.6	49.0	1.33
SC - non text	None	no	125	None	EVA/CG	10	0.92	54.1	49.5	1.32
SC - non text	None	no	125	None	EVA/CG	10	0.58[4]	57.2	52.4	1.30
SC - non text	None	no	125	None	EVA/CG	15	0.92	54.4	49.9	1.31
SC - non text	None	no	187.5	None	EVA/CG	10	0.92	54.5	50.0	1.31
SC - non text	None	no	250	None	EVA/CG	10	0.92	54.9	50.4	1.30
SC - non text	None	medium	125	None	EVA/CG	10	0.92	54.3	50.0	1.27
SC - non text	None	high	125	None	EVA/CG	10	0.92	54.9	50.8	1.19
SC - texturized	None	no	125	None	EVA/CG	10	0.92	58.2	53.2	1.47
PC - non text	None	no	125	None	EVA/CG	10	0.92	54.1	51.4	0.78
A - non text	None	no	125	None	EVA/CG	10	0.92	54.1	52.4	0.50
SC - non text	None	no	125	Opt. for air/glass; air side only	EVA/CG	10	0.92	54.7	50.1	1.35
SC - non text	None	no	125	Opt. for air/glass; both sides	EVA/CG	10	0.92	54.2	49.7	1.33
SC - non tex'	Opt. for air/silicon	no	125	None	EVA/CG	10	0.92	58.5	53.4	1.53
SC - non text	Opt. for encapsulant/silicon	no	125	None	EVA/CG	10	0.92	59.2	53.9	1.56
SC - non text	TiO$_2$	no·	125	None	EVA/CG	10	0.92	58.3	53.2	1.52
SC - non text	Opt. for encapsulant/silicon	no	125	Opt. for air/glass; both sides	EVA/CG	10	0.92	59.4	54.2	1.57
SC - texturized	TiO$_2$	no	125	None	EVA/CG	10	0.92	59.3	54.3	1.52

Notes: (1) SC - single crystal silicon (efficiency = 16.3%)
PC - polycrystalline silicon (efficinecy = 9.9%)
A - amorphous silicon (efficiency = 5.8%)

(2) Operational Environment

T_{air} = 20°C
T_{sky} = -5.2°C
T_{ground} = 20°C
Ground Emissivity = 0.9
Tilt Angle = 37°
Insulation = Air Mass 1.5 (normal incidence at 37° tilt)

(3) Aluminum Foil with white Mylar on outside surface

(4) Aluminized Mylar; Mylar is 0.5 mil thick

Table 1. SUMMARY OF THERMAL/OPTICAL COMPUTER RUNS FOR SUPERSTRATE DESIGN

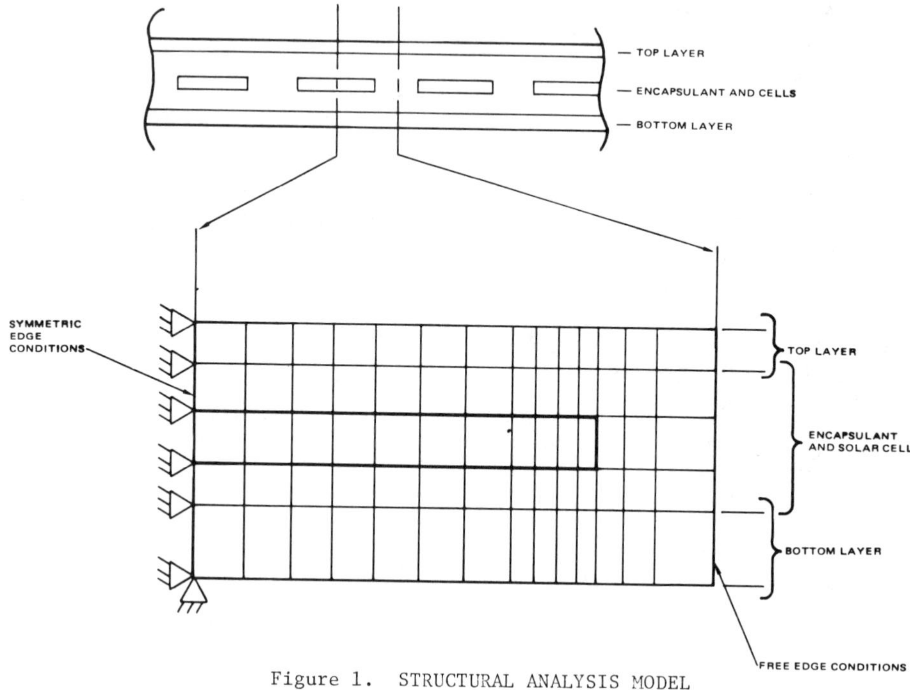

Figure 1.   STRUCTURAL ANALYSIS MODEL

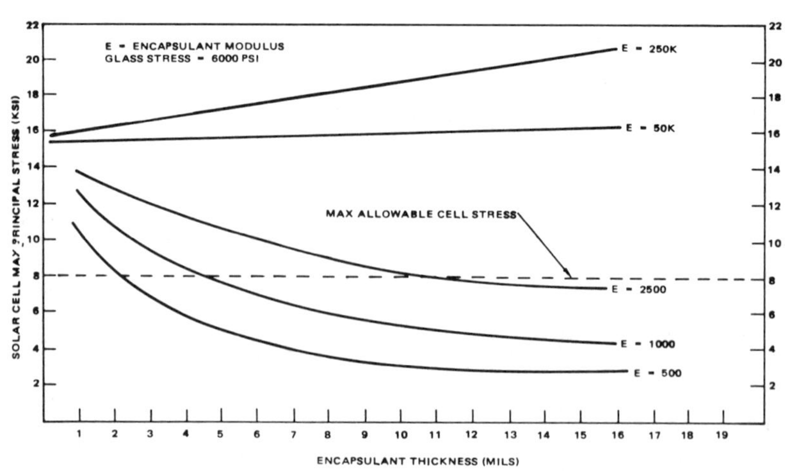

Figure 2.   DEFLECTION/STRESS RESULTS GLASS SUPERSTRATE

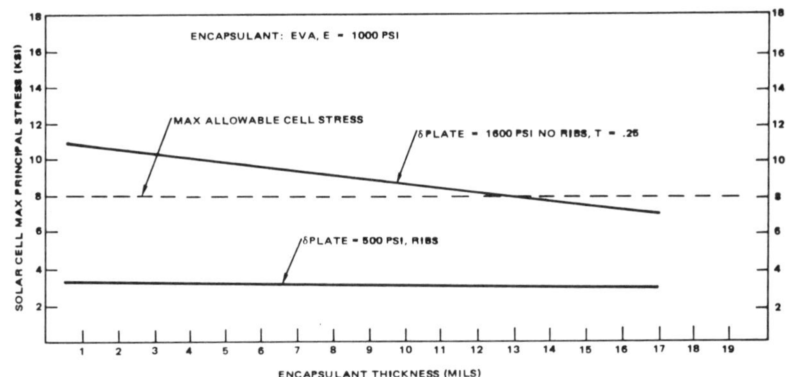

Figure 3.   DEFLECTION/STRESS RESULTS WOOD PRODUCT SUBSTRATE

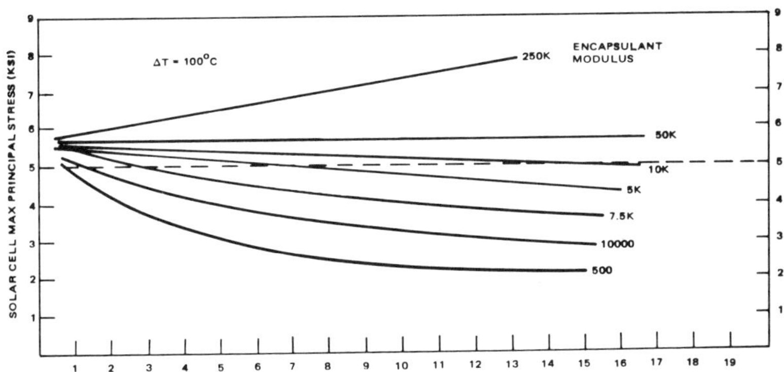

Figure 4.   THERMAL/STRESS RESULTS GLASS SUPERSTRATE

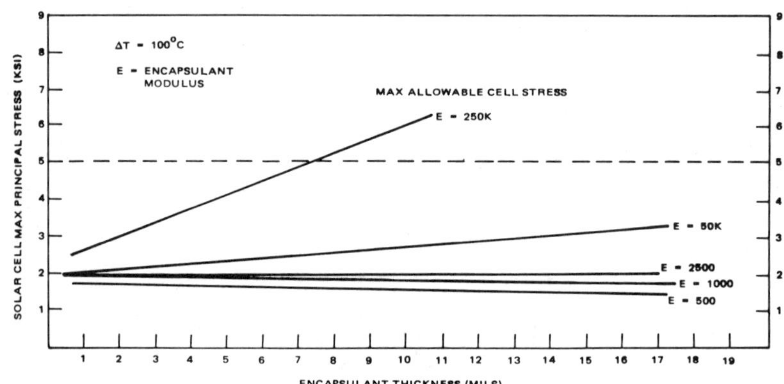

Figure 5.   THERMAL/STRESS RESULTS WOOD PRODUCT SUBSTRATE

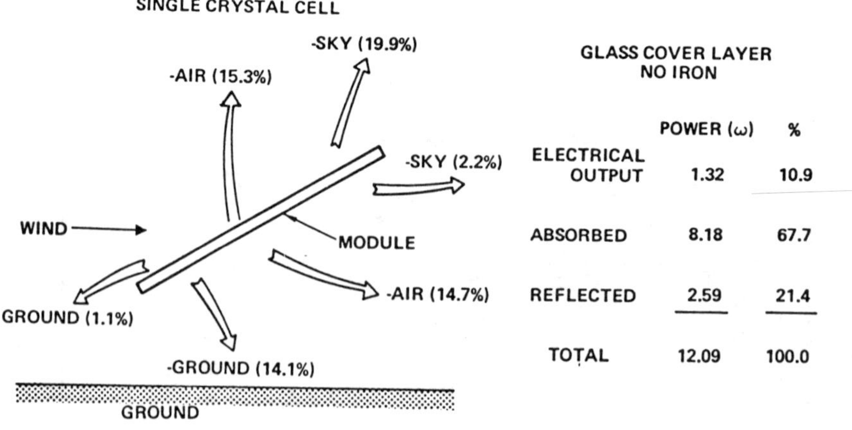

Figure 6.   TYPICAL MODULE ENERGY BALANCE

26th National SAMPE Symposium
April 28-30, 1981

MOLYBDENUM-TIN AS A SOLAR CELL METALLIZATION SYSTEM

D. W. Boyd and C. Radics
Jet Propulsion Laboratory
Pasadena, California 91103

Abstract

The objective of the new method of metallization for silicon solar cells described in this paper is to produce a low-cost, mechanically strong and stress-free ohmic contact. The contact, silk screened by using an ink consisting predominately of molybdenum tri-oxide (MoO_3) and tin, is subjected to a firing cycle characterized by a preheat period in a neutral atmosphere up to the melting point of MoO_3, and a final heat period in a reducing atmosphere of forming gas. MoO_3 has a relatively low melting point (795°C) and is easily reduced to metallic molybdenum which forms an ohmic contact with the silicon solar cell surface while also becoming readily wetted by the tin.

The advantages of this method are that the contact is purely metallic (free of frit), employs common metals, has a relatively closely matched coefficient of thermal expansion with silicon (Si), is easy to solder with low-melting alloys and can be applied by an automated silk-screen process and fired at temperatures compatible with standard inks. Additionally, the proposed method can be readily incorporated in a manufacturing line since it does not require specialized equipment or facilities. The savings resulting from the use of this metallization system are attributable to reductions in material costs.

1. INTRODUCTION

The Department of Energy (DOE), responding to this country's need for development of alternate sources of energy, has set up a significant Research and Development Program (R&D). DOE's goal is to achieve economically competitive electric power production via direct solar energy conversion utilizing solar cells. The Low-Cost Solar Array (LSA) Project at the Jet Propulsion Laboratory, part of DOE's R&D program, is developing low cost solar energy conversion using flat-plate solar cell modules.

A solar cell module is the basic unit for converting sunlight directly into electric power (Figure 1). It is made of several individual solar

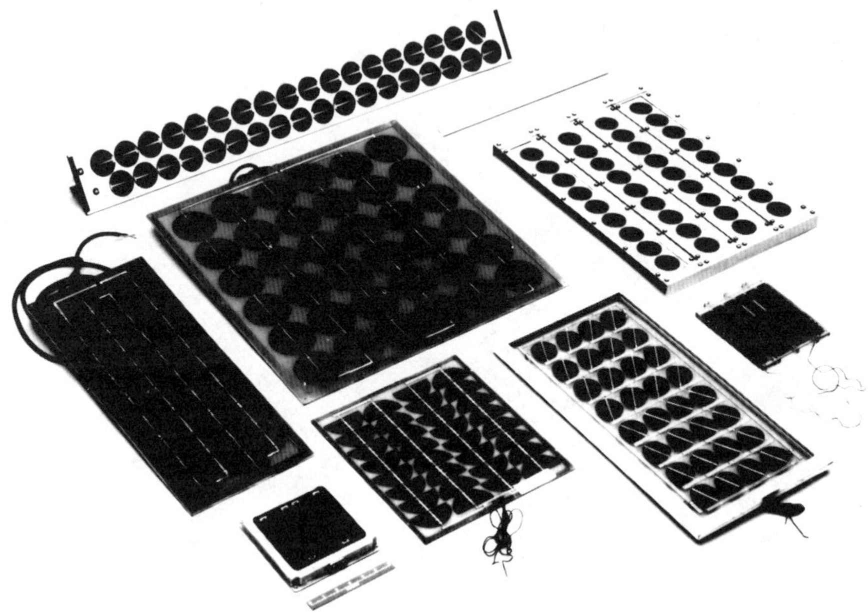

Figure 1.   Flat Plate Solar Cell Modules

cell subunits interconnected electrically and mounted into a rigid frame. Typically, the cells in a module are covered by a transparent layer that gives environmental protection.

The cost goal for the solar cell modules, in 1980 dollars, is set at $.70 per peak watt capacity for 1986 (Figure 2).

Several materials are candidates for use as solar cells. Silicon (Si) is the most widely used for practical reasons, e.g., relative ease of production, highest operating efficiency obtained (Figure 3), relatively low cost and abundance in nature. Virtually all solar cells manufactured in the world today are made of Si.[1]

2.   CELL MANUFACTURE

Highly purified semiconductor or nearly semiconductor grade Si mate-

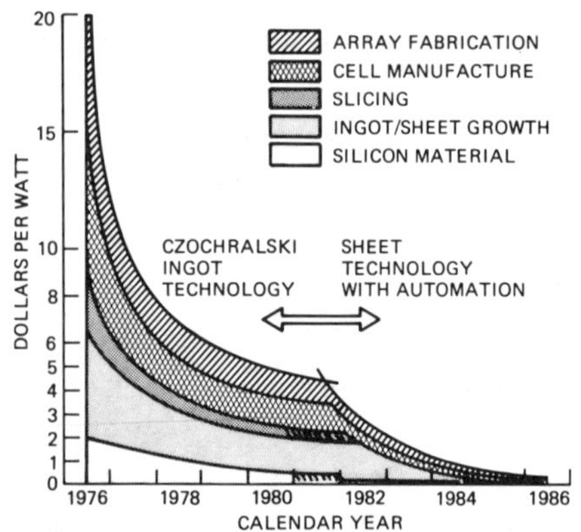

Figure 2.   Low-Cost Silicon Solar Array Project, Terrestrial Solar Array Price Goals

rial is needed as starting material for solar cells. This is converted into sheet form mostly by the relatively complex process of seeded crystal growth from the melt followed by diamond blade sawing of the resulting ingots into sheet material. Sheet material is cut to suitable

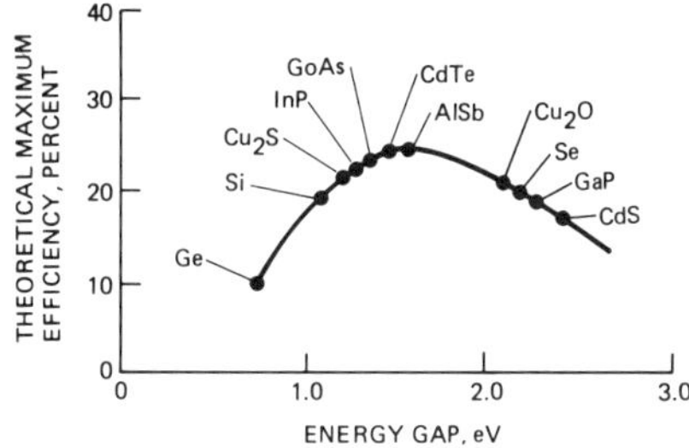

Figure 3. Theoretical maximum conversion efficiency of semiconductors as a function of bandgap energy. (IEEE Spectrum, Feb. 1980)

size subunits that are etched and cleaned, then diffused to form the carrier collecting active layer or p-n junction. Finally, ohmic electrical contacts are provided to both the front and back of the cell blanks, resulting in solar cells. These cells are then mounted into the modules (Figure 4).

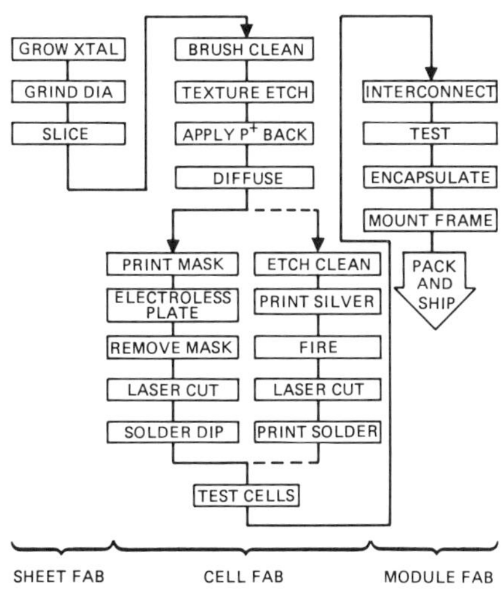

Figure 4. Low-Cost Silicon Solar Array Project, Manufacturing Sequence

3. CELL METALLIZATION

The formation of reliable, ohmic, low loss and low cost metal contacts on solar cells is a critical process step in cell manufacturing. One of the most reliable contacts in current use is made by vacuum evaporation of layers of titanium, palladium and silver (Ti-Pd-Ag). This system is well known, highly reproducible, compatible with soldering techniques, but very costly. One lower cost metallization system is produced by electroless or electrolytic plating. A more commonly used low cost metallization is produced by screen printing a metal powder-glass frit ink on the surface of the Si. At the elevated temperatures reached during a subsequent firing process, the glass frit melts and provides the bond between the Si surface and the conductive metal powder.[2] Several modified systems are under investigation to both lower the cost of materials and reduce process complexity. A technique, developed by Dr. M. Macha, et al, at Sollos, Inc, under a Department of Energy and Jet Propulsion Laboratory (JPL) contract[3] appears to successfully accomplish both these goals by utilizing a molybdenum-tin (MoSn) alloy for the metal contacts.

4. THE Mo-Sn SYSTEM

The ink used in this system is formulated from MoO_3 with Sn powder and a trace amount of titanium resinate. Resistive losses of the resulting contacts are low because the ink contains no frit.

During the Mo-Sn metallization process, the ink (78% Sn:22% MoO_3) is

screen printed on the solar cell
blank, then dried and initially
fired in an inert atmosphere (N_2) to
melt the MoO_3 (MP = 795°C) and then
fired in forming gas ($N_2 + H_2$) to
reduce the MoO_3 to Mo metal (Figure 5). The resulting Mo is highly
reactive which facilitates the Mo-Si
bonding. Sn readily wets Mo thereby
forming a convenient soldering base
for subsequent interconnections.

Preliminary environmental test data
are currently available and are presented in Section 7 of this paper.

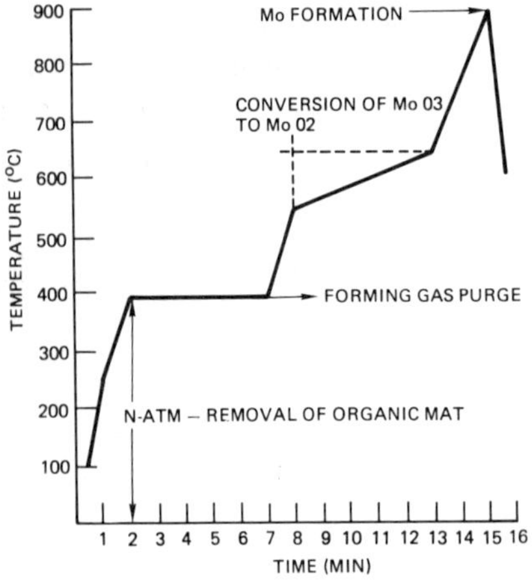

Figure 5. Thermal Cycle for Conversion of MoO_3 to Mo

5. SYSTEM ADVANTAGES AND DISADVANTAGES

5.1 System Advantages

Cost and performance features of the
system are currently being evaluated.
Apparent advantages identified to
date are:

(1) Both Mo and Sn are low in
cost compared to silver and
other metallization systems.

(2) The coefficient of expansion
of Mo is close to that of Si
(3.1 vs 4.2 cm/cm/°C x 10^{-6})
thereby minimizing stress
problems, a common cause of
metal contact failure in Si
devices.

(3) The system is
(a) Frit free which mini-
mizes resistive losses,
(b) Compatible with
soldering;
(c) Implementable with
low cost state-of-the-
art equipment;
(d) Easy to automate;
(e) Readily adaptable to
mass production.

5.1 System Disadvantages

The major disadvantage is that the
Mo-Sn ink contains titanium resinate
(in trace amounts) which is not
readily available for procurement.

6. COST ANALYSIS

An analysis of costs associated with
the Mo-Sn metallization system indicates that, by comparison to similar
metallization systems, appreciable
cost savings are possible. Assuming
a continuous manufacturing process
that uses automatic screen printers
and a conveyor belt furnace, the
only significant difference in the
resource costs of screen printing
Mo-Sn metallization versus any other
metallization is attributable to
material.

Solar Array Manufacturing Industry
Cost Standards (SAMICS) analysis at
JPL, based on mass production of

enough cells to produce a 15 MW output, indicates that material costs for a Mo-Sn system are 0.05 to 0.1 times as costly as the material for a silver system when the same metallization grid pattern is used for both. Other non-silver systems have been compared to the Mo-Sn system but none appear to be less costly.

7. MECHANICAL AND ELECTRICAL PERFORMANCE DATA[3]

7.1 Mechanical Performance

The contact was evaluated for the mechanical characteristics of adherence and solderability. The contact had good adherence as determined by scratch tests and lead pull tests (Table 1). The pull test was also a successful measure of solderability. Evaluation of the Mo-Sn system under environmental stresses included humidity exposure and subjection to temperature extremes. No mechanical breakage was observed after these tests.

7.2 Electrical Performance

Minimal electrical change was observed in Mo-Sn metallized cells after environmental testing. In general, evaluation of the I-V characteristics of these cells indicate that such cells are electrically equivalent to silver titanium (Ag-Ti) metallized cells (Table 1). Electrical performance of Mo-Sn cells is dependent on the metallization firing cycle. Peak temperature of the cycle, necessary to reduce MoO_3 to Mo to facilitate ohmic bonding, degrades cell electrical performance. However, subsequent heat treatment at a low temperature improves cell I-V characteristics to the quality of cells metallized at low temperatures only.

8. CONCLUSIONS

It has been experimentally demonstrated that ohmic contact to silicon solar cells can be achieved by a Mo-Sn metallization system formed by thermal reduction of an MoO_3-Sn solution mixed with trace amount of titanium resinate.

Low temperature annealing of cells following the Mo-Sn metallization firing cycle is essential for optimization of cell electrical and mechanical properties. Electrical (I-V curve) properties are enhanced by low temperature annealing while mechanical properties (metallization adhesion and solderability) are enhanced by the peak temperature of

Table 1. Comparison of Mo-Sn Screened, Ag Screened and Ni Plated Contacts

SOURCE	TYPE	METALLIZATION	Voc (V)	$\frac{I_{sc}\ (mA)}{cm^2}$	PULL STRENGTH (g)
MANUFACTURER "A"	P/N	Ni-PLATED	0.56	28	200 +
MANUFACTURER "B"	N/P	Ag-SCREENED	0.54	25	0 (CONTRACT SEPARATION)
SOLLOS, INC.	P/N	Mo/sn SCREENED	0.56	26	200 +

the firing cycle. The characteristics of cells metallized with the Mo-Sn system appear to be comparable with the characteristics of cells metallized with standard methods (plated Ni and screened Ag).

The Mo-Sn process uses standard silk screening equipment. The advantage of this new method is primarily attributable to cost savings in material based on the ink formulation.

9. BIOGRAPHIES

Donald W. Boyd is a process engineer for Photovoltaic Production Processes and Equipment activity on the DOE/JPL Low-Cost Solar Array Project. For four years on this project, he has directed the technical efforts of subcontractors during their investigations of solar module performance improvements based on improved manufacturing process sequences and improved solar cell geometry and metallization systems.

During this last year, he has also chaired a committee for the evaluation of cost savings attributable to individual process step improvements in commonly utilized solar module manufacturing process sequences.

Mr. Boyd holds a B.S. in Civil Engineering from Howard University (1958), an M.S. in Systems Engineering from West Coast University (1981), and an M.A. in Business Administration from University of Southern California (1970).

Charles Radics received his degree in Chemistry (MS) from the University Babes-Bolyai, Cluj, Roumania in 1961. Between 1961 and 1969, he was process engineer in various chemical plants in Roumania. Between 1969 and 1971, he was chemist at Teledyne Semiconductors, Hawthorne, California. 1971 to 1975, he was process engineer for Wacker Chemical Corp., Los Angeles, California. His specialization is silicon materials preparation and wafers processing. From 1976 he has been with Jet Propulsion Laboratory, Pasadena, California. He is involved in the materials development and characterization for the Low Cost Terrestrial Photovoltaic activity of the Low-Cost Solar Array Project.

Credit Statement: The research described in this paper was carried out at the Jet Propulsion Laboratory, California Institute of Technology, and was sponsored by U.S. Department of Energy through an agreement with NASA.

REFERENCES

1. H.J. Hovel: Solar Cells. Willardson-Beer, Ed: Semiconductors and Semimetals, Vol. 11 – Acad. Press N.Y. 1975.

2. Photovoltaics Review. Special issue of IEEE-Spectrum, Feb. 1980.

3. M. Macha, SOLLOS, Inc.: Low Cost Solar Cell Metallization. Final Report DOE/JPL Contract No. 954882, 1980.

26th National SAMPE Symposium
April 28-30, 1981

APPLICATIONS OF POLYMER EXTRUSION
TECHNOLOGY TO COAL PROCESSING

D. W. Lewis
Jet Propulsion Laboratory
California Institute of Technology
Pasadena, California

Abstract

"Keywords": Coal, Plastics, Extrusion, Control, Twin-Screw, Coal Pump

Large deposits of high-volatile bituminous coals exhibit an intriguing plastic property when heated to a temperature in the range of 780 to 900°F (416 to 482°C). Renewed research interest in the plastic state of coal has resulted in several innovative energy applications but development has been blocked by an inability to control the continuous processing of large quantities of plastic coal. Experiments at the Jet Propulsion Laboratory and its contractors have shown that instabilities occur with large single-screw extruders but are avoided when an intermeshing twin-screw extruder is used. Also discussed is how the release of volatiles from the hot coal have contributed to the unstable operations in the single-screw extruder and how this is a significant complication when compared to the extrusion of polymers. The improved mixing achieved by the intermeshing screws of the twin extruder is sufficient to result in a steady controlled process.

1. INTRODUCTION

There is an increasing recognition of the great value of the United States' coal resources to this nation and the entire industrial world. Surprisingly, there is much about this common material that remains unknown. Its properties vary considerably with location, age, and the existing conditions during its formation from prehistoric peat. Younger coals (or older peat) have significantly different physical and chemical compositions from the oldest anthracite coals which have been almost completely converted to carbon. There is an intriguing plastic property of many of the middle-aged bituminous coals. Upon heating, these coals exhibit a plasticity very similar to polyethylene for a few minutes but then

begin to polymerize to form a coke
with properties more like a thermosetting plastic. Plastic coal can
be extruded, pelletized or molded
using common plastics technology
and equipment. It can even be
sprayed as a mist like water. During the past four years my col-
leagues and I at the Jet Propulsion
Laboratory (JPL) have been experimenting with the plastic state of
coals.* We have characterized
some of its intrinsic properties
and are developing the techniques
which will make useful commercial
applications possible. Kushida
summarized the investigations at
JPL and its contractors in the
final report of the Coal Pump
Development Task.[1] This paper
will discuss one of the more critical questions addressed in this
work and how it was successfully
answered. Almost a decade ago
Oldaker observed clogging, belching
and control instabilities in his
attempts to continuously extrude
coal.[2] Since then others have
observed these same control problems.[3,4] Until this problem
could be resolved the long list of
attractive applications of plastic
coal would remain a set of laboratory curiosities.

2. PLASTIC-STATE COAL

A preliminary experiment at JPL
using a small piston and die device
may best demonstrate the plastic

characteristic of coal. In this
experiment, a machined one-inch
diameter cylinder of coal (a Utah
bituminous) was inserted into a
cylindrical die, heated to 734°F
(390°C) and pressurized to approximately 5000 psi by a hydraulically
actuated piston. A stream of fine
coal particles was extruded through
a 0.020 inch diameter orifice in
the die into the atmosphere. Un-
heated coal would not extrude even
at much higher pressures.
Studies of several high volatile
bituminous coals were conducted to
acquire a better understanding of
the flow properties of coals at
high temperature and pressure.[5]
It was generally found that upon
heating coal to a transition temperature it becomes plastic enough to
demonstrate flow properties. Within a few seconds a large decrease in
viscosity occurs (can be several
orders of magnitude) followed by a
more gradual increase due to molecular cross-linking or coking. The
time required for coking varies
from 2 minutes to 30 minutes, or
greater, depending on the coal and
the temperature.

3. CONTINUOUS EXTRUSION
Probably the most significant difference between the continuous extrusion of coal and the extrusion
of a thermoplastic polymer is that
volatiles are continuously being
released from the coal. Most ex-

*This work was initially sponsored by the National Aeronautics and Space
Administration (NASA), Energy Systems Division and subsequently funded by
the U.S. Department of Energy (DOE), Fossil Energy Programs.

planations of the plastication of coal indeed require the release of volatiles to achieve the plastic state. It is also this presence of volatiles that causes the erratic performance observed in coal extrusion. An interesting and important observation is that reports of control problems were always associated with the use of a single-screw plasticating extruder. There are at least four phenomena observed when volatiles are released while plasticising coal that almost preclude a steady-state operation in a simple single-screw extruder. By using an intermeshing twin-screw extruder these problems can be avoided. Stable operations have been demonstrated on a corotational twin but there would appear to be no fundamental problem with counter rotating screws.

3.1 <u>Dragflow</u> - Before discussing the phenomena resulting from the generation of volatiles, please permit a few tutorial words about how a single-screw extruder works. The flow of material through the serpentine channel of the screw and the buildup of pressure behind the die is caused by the material having a greater drag on the inner surface of the barrel than on the screw surfaces. This condition does not need to be met for all stations along the screw for material can be pushed along by upstream pumping, but it must occur somewhere. In the early stations along the screw (the feed zone) this drag is due to frictional forces which under normal conditions generates the pressure and heat to melt the material. After the material is melted the drag is developed by viscous shear. If for some reason these frictional or viscous shear forces are not developed the screw will turn within the barrel like a solid shaft. In the same way that a threaded fastener will not advance in a hole with stripped threads the single-screw extruder will not work without a favorable "dragflow" condition.

3.2 <u>Solids feeding</u> - The portion of screw between the inlet feed port and the zone where melting occurs is known as the feed section. When the screw channel is full, the material is compacted rapidly by increasing frictional forces provided that the coefficient of friction is sufficient. As the coal is heated by compaction and frictional drag, the hot volatiles which are released backstream toward the lower pressure of the feed port and condense on the incoming coal. This contributes to the rapid heating of the coal but unfortunately this also lubricates the coal. The resulting loss of frictional drag causes erratic feeding and eventually the material flow will stop. A circulation is also set up where the condensate is carried forward to be revaporized and returned to condense again on the incoming coal. The amount of condensate increases as more coal is melted and eventually the entire feed section of the screw will be saturated.

In tests using a 1½ inch (screw diameter) extruder it was possible to vent these volatiles and operate as long as desired but these techniques would not scale up to a larger 2½ inch extruder. The feed section of the screw would fill with condensate over periods ranging from 30 minutes to 4 hours depending on the feed rate, temperature distribution and venting configuration. Another phenomena (which may be another form of condensate lubrication) was demonstrated by Chung in an innovative screw simulator.[6] The relative motion between the coal and the barrel surface was maintained but the barrel was represented by a heated rotating steel cylinder. A coal sample was pressed against the drum with an instrumented holder. The frictional force was measured for various temperatures, pressures and conditions of the coal. As indicated in Fig. 1 at 400°F (204°C) the coefficient of friction abruptly dropped to a very small value. In this example the frictional drag force would be lost at a temperature 385°F (213°C) below the melting temperature of the coal.

3.3 <u>Melt pumping</u> - Downstream of the melting zone in a single-screw extruder the dragflow is generated by viscous shear. A very low effective viscosity will result in low die pressures and erratic flow. As shown in Fig. 2 the viscosity of coal quickly drops by a significant amount as it is plasticized. The result of this phenomenon is a loss of pumping in the melting zone and immediately downstream of it. A pool of low viscosity melt will build up and stagnate for several diameters downstream of the melt zone. In time (several minutes) the viscosity will increase due to coking but flow may not resume if there is inadequate mixing. Subsequently melted lower viscosity coal may flow over the thickening stagnate region rather than mix with it in the channel. Eventually the overflow path becomes so long and thin that the coal cokes before making the transit over it. When this occurs the screw stops turning or jams. A fourth phenomenon, foaming, occurs in plastic coal which can lower the effective viscosity and prevent a build up of a steady die pressure. Volatiles continue to evolve in the melt and form small gas bubbles at first but soon coalese to form large gas pockets. This progression can be observed in the solidified coal remaining in the screw after an immediate shutdown. Once formed these gas pockets are difficult to disperse and cause erratic spitting and belching as the material exits the die. If the melt is maintained above the vapor pressure of the majority of the volatiles (above 800 psi) this effect can be minimized. Careful temperature control and good mixing also help.

4. THE ANSWER

Persistent attempts were made to understand and compensate for these effects using two sizes of single-screw extruders. The work with the smaller 1½ inch extruder was successful but at the 2½ inch scale satisfactory control was not realized. The problem of large-scale continuous processing of plastic coal remained. Ultimately, the answer was to minimize the "dragflow" problems by changing to an intermeshing twin-screw extruder. Vogt reports that stable processing conditions were realized using a 57 mm twin-corotating screw extruder.[7] This type of extruder has a feature that each surface is wiped by the passage of the flight of the other screw. Hence, surface accumulations are wiped off and mixing is more thorough since there is a transfer of the material from one screw to the other. The solid transport still occurs by a frictional drag mechanism; but because of the wiping action, there is considerably more latitude for frictional variations than in a single-screw extruder and virtually no stagnation of the flow. In 6 hours of continuous operation using a high volatile Pittsburgh No. 8 coal with a steady throughput of 350 lb/hr there was no evidence of build-up of condensate in the feed section. It took about two hours of extruder time to "home in" on the stable extruder configuration. More was accomplished in these two hours with the twin-screw than was accomplished in about 1½ years of constant struggle with the single-screw extruder.

5. WHAT NEXT?

With the demonstration that plastic coal, even the high volatile Pittsburgh No. 8, can be processed with standard equipment and at a scale typical of much larger machines the door is now open for a large variety of applications. Caution is in order since the demonstration was limited in scope and duration but research into gasification, direct liquefaction, coal combustion, and coal feeding applications is now on a much firmer footing.

In applications for coal gasification the desirable characteristics of plastic coal extrusion are; 1) the ability to pump directly into gasifiers operating above 1000 psi, 2) the formation of a fine reactive mist, and 3) a positive metering of the coal feed. This latter characteristic is particularly important for the control and safe operation of pressurized oxygen-fed gasifiers. The undesirable characteristics are; 1) it is a new unproven technique, 2) only coals that will plasticise can be used, 3) for most coals the feed system should be purged upon shutdown to prevent coking within the plumbing and 4) in the spray application the nozzle is probably located within the reactor where it might be inaccessible. Wear data

for plastic coal is also sparce but
initial indications are that it is
a lubricant. Economic studies comparing plastic coal extrusion with
other feeding techniques show it to
be less expensive particularly when
the coal preparation and handling
equipment are included in the comparisons.

Combustion applications will probably take advantage of plastic coal
spray. Generally it will be less
expensive to pulverize coal and
blow it through standard injectors
for most applications. Advantages
may be realized where the handling
of pulverized coal is particularly
difficult such as with boilers and
kilns retrofitted to use coal.
Pressurized fluid bed combustion or
gasification processes may take advantage of the ability to blend very
fine material, such as calcium carbonate, with coal in the extruder
while it is a liquid yet control the
size of the extrudate to a coarser
mesh for the fluid bed.

For the direct liquefaction or hy-
drogenation of coal the application
is more direct. The liquid coal
can be extruded directly into a
stream of hydrogen donor solvent
as configured in current developments. The extruder would replace
the coal preparation, slurry formulation and pumping to the 2,000 to
3,000 psi system pressure. Experiments are currently in progress at
JPL to determine if the coal can be
hydrogenated directly in the liquid
state thus performing the entire
liquefaction within the extruder and
die.

Non-energy related applications may
also exist for plastic coal. For
example, there has been some discussion (no experiments) about
using coal to replace epoxy as a
polymer to pump under concrete sections of highways to re-align them
when they have settled unevenly.
There may be other applications
once it is realized that coal is the
most plentiful and least expensive
polymeric material known.

6. REFERENCES

1. Kushida, R., et al., _Coal Pump
Development, Phase III Final Report_
document 5030-479, Jet Propulsion
Laboratory, Pasadena, California,
July 1980.

2. Oldaker, E.C. and Gillmore,D.W.,
_The Hot Extrusion of Coal - A Very
Sticky Problem_, proceedings of the
Institute for Briquetting and Agglo-
meration, Vol. 13, Morgantown, West
Virginia, approx. 1970.

3. Furman, A.H., _Pressurized Feeding Of The GEGAS System_, presentation to 81st American Institute of
Chemical Engineers Conference,
Kansas City, Mo., April 1976.

4. Schatz, W.J., et al., _Coal Pump
Development, Phase I Feasibility
Report_, document 5030-235, p. 5-27,
Jet Propulsion Laboratory, Pasadena,
Calif., Sept. 1978.

5. Kushida, R., et al., _Coal Pump
Development, Phase II Interim Re-
port_, document 5030-460, p. 9-25,

Jet Propulsion Laboratory, Pasadena, Calif., June 1978.

6.   ibid, p. 11-1

7.   ibid, p. 6-1

BIOGRAPHY

Don W. Lewis is currently the manager of the Coal Technology Programs at the Jet Propulsion Laboratory, California Institute of Technology, Pasadena, California. A graduate of Pepperdine University (1950) in mathematics and the University of Southern California (1956) in mechanical engineering, he has spent most of his professional life in rocket and spacecraft development at North American Aviation (4 yrs.) and the Jet Propulsion Laboratory (26 yrs.). During the past 6 years his attention has focused on coal technology particularly coal conversion techniques. He directed the experimental work with the 2½ inch single-screw and 57 mm twin-screw extruders referenced in this paper.

The research described in this paper was carried out at the Jet Propulsion Laboratory, California Institute of Technology, and was sponsored by NASA and the DOE through an agreement with NASA.

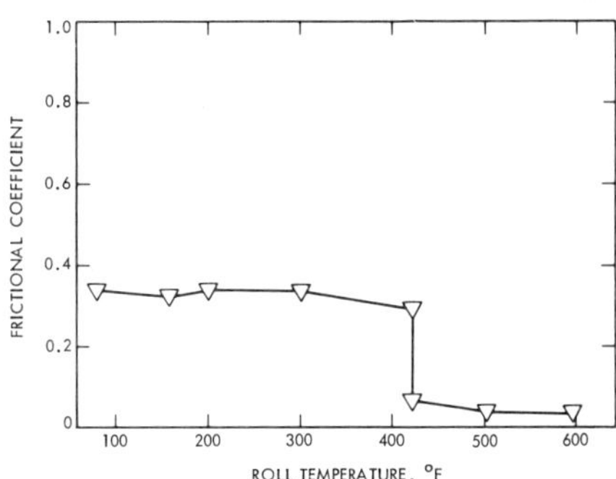

Fig. 1 Frictional Coefficient of Coal. Pittsburgh No. 8 at Room Temperature, 8140 Mesh, Roll Speed 2 in/sec.

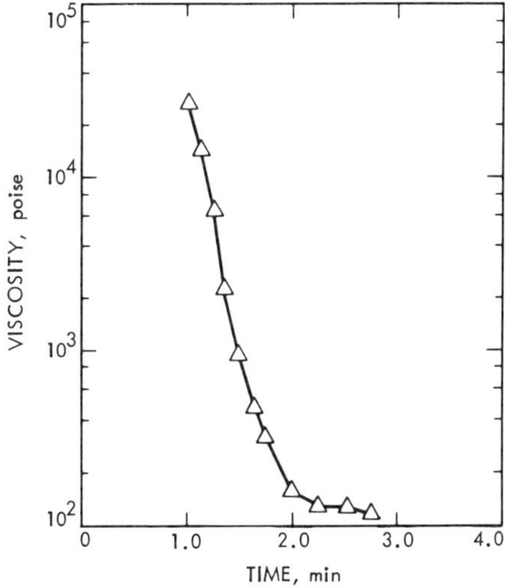

Fig. 2 Viscosity vs. Time for Pittsburgh No. 8 at 785°F (418°C).

131

TECHNICAL APPLICATIONS FOR THE
PERSONAL COMPUTER

Robert E. Sanders
Tulsa Division, Laboratories,
Rockwell International

Abstract

A small personal computer has evolved into a low-cost, yet sophisticated device which can be effectively and economically used in the laboratory for data calculations. Selection and use of such a computer are reviewed with programming examples in Basic.

1.   INTRODUCTION

There seems to be no end of the calculations in testing and development laboratories. The recent emphasis on product liability and the added requirement for more quality control have increased the testing and, consequently, the number of required calculations. In the past, laboratories have relied on a variety of instruments to perform the repetitive calculations. Initially, of course, the slide rule was used to perform these calculations, but was replaced to some degree by adding machines and slow-cumbersome mechanical calculators. These have in turn been replaced by pocket calculators in recent years. Some of these pocket calculators are very sophisticated and even include the ability to store programs on a variety of magnetic cards. A number of laboratories also added computer facilities to help in data acquisition and reduction. These required computer specialists and were often too expensive or not available for use on small programs.

In the past three to four years, gigantic improvements in electronic technology have allowed manufacture of "suitcase" size computers that have the power of most of the room-size computers of the past. Some computers have even been produced on a single electronic chip. The "suitcase" computers were initially assembled from a variety of kits and sometimes required continuous attention to maintain working order. As in any evolution, the kit reliability improved, and some of the "kits" even became available as completely assembled, operational units. Thus the "Personal Computer" evolved. Much of the initial use of the "Personal Computer" was game playing, which may still be a major application. This usage may have created a "toy" image of the personal computer; this image seems to persist although the personal computer is nonetheless a very sophisticated device. The game playing and programming are a very important learning method as the games often require computer simulations of real life situations and complicated programs. Many recent publications reflect applications of the small personal computer in system descriptions [1-3] and with programs [4-5]. Technical societies are also devoting a significant amount of time to laboratory computers.[6] In its present state of development, the personal computer or microcomputer provides a low cost method of reducing laboratory data. They have the advantage over the programmable calculator in that programs can be changed easily and can be designed to allow a wide range of user interaction.

2. COMPUTER SELECTION

There is a wide variety of personal computers available at this time. Each of the units has a unique combination of features, as well as a unique price. Some of the more widely used small computer systems are:

Texas Instruments TI 99/4
Radio Shack TRS-80 Models I, II, and III
Heath Kit H-19
Apple II and III
Atari 400/800
Southwest Technical Products
Commodore Pet
Ohio Scientific C4P, C8P, and C3
Rockwell Aim 65

It seems that buying a personal computer nowadays is somewhat like buying a car; choices depend greatly on personal preference. While in the past kit packages were very popular, the current trend seems to be toward assembled units at about the same price as kits.

In selection of a general usage laboratory computer, the following factors were considered:

Cost
Completely assembled
Portable
Upper and lower case letters
Standard screen display
Numeric key pad
Available programs
Compatability

Cost was perhaps a primary consideration in selecting the laboratory computer as it was envisioned that several would be required in different laboratory areas. The unit should also be completely assembled and relatively portable. Next it should have upper and lower case type available as future applications in word processing were considered. A standard display at least (32 lines x 64 characters) and compatability with other equipment, printers, plotters, modems, etc. was also desired. Initially, a small, lesser known computer was obtained and used for about 6 months for conducting timing studies and making the majority of the calculations. This unit has since been replaced because of the problems with data retrieval from the disk drives, which could not be corrected. The replacement was a Radio Shack TRS-80 Model III (Figure 1), which is being operated with a cassette recorder until reliability of the disk drives for this model is assured. Operation of the cassette recorder as a program and data storage device seems to work well because of the high rate of data transfer (1500 baud).

3. APPLICATION

Introduction of the computer to the laboratory was initially greeted with some skepticism and, possibly, distrust. However, the need to perform a large number of calculations on a new material evaluation program provided an opportunity to evaluate the computer as a laboratory tool. An

FIGURE 1.   TRS-80 COMPUTER SYSTEM

estimated 3500 calculations were in the program. A large percentage of these calculations were previously performed on a large Nova mainframe computer that was not available for use at the time of the new material evaluation. Programs were written in Basic language for the calculations using a menu type display to provide for the specific desired program, Figure 2. The operator selects the desired program by entering the program number; the program then guides the operator through the input process to provide the computer with the required data to make the calculatons and print the results.

Figure 3 is the Basic program, which provides menu display and Figure 4 is the Basic program for calculation of the flexural strength and modulus for a laminate, which is one of the more involved calculations. Prior to

134

using the computer, the calculations were made with a desk calculator with a single memory storage. Table I is a listing of the time saved per calculation for typical calculatons performed in one section of the laboratory. Although the individual saving for some of the simpler calculations is small, the saving is significant when the total number of calculations per year are factored into the calculations. Also because the computer does the data manipulation, the chance of errors is reduced. The savings in time on a group of typical applications is expected to pay for the computer in a little over a year.

wide variety of complex calculations. The calculations performed by the computer are expected to provide printed copies of the data in report format. Automatic data storage as calculations are performed, use of the stored data to prepare historical trend or quality control charts, and advanced statistical analysis as well as word processing, are also planned. For the most part, engineers have an aversion to performing an analysis of variance or even a least square fit of data if the data must be hand manipulated with a calculator. The small computer offers a solution to these aversions.

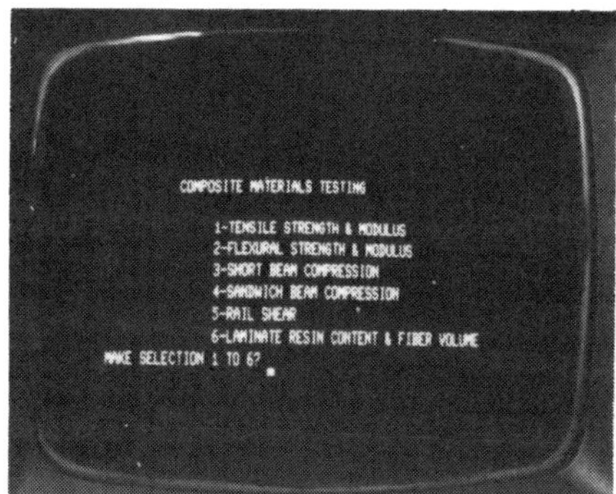

FIGURE 2. MENU DISPLAY FOR COMPOSITE MATERIALS DATA CALCULATIONS

4. CONCLUSIONS

The personal computer or microcomputer has emerged as a powerful low cost tool for the laboratory. Programs are becoming available or can be easily written to perform a

REFRENCES

1. Miller, D. L., "Computers, Glass Fiber Composites Are State-of-the-Art in Sports Equipment Design, "Platics Design & Processing November 1980, pp. 23-27.

2. Teschler, L., "Home Computers Find a New Home," Machine Design January 1980, pp. 20-24.

3. __________, "Portec's Comprehensive Quality Program," Quality, September 1980, pp. 61-65.

4. John, D. G. and K. Hladky, "BASIC Computer Program for Calculation of Corrosion Rate," Materials Performance, November 1980, pp. 42-45.

5. Kelter, P. B. and J. D. Carr, "Microcomputer Compatable Method of Resolving Rate Constants in Mixed First-and Second-Order Kinetic Rate Laws," Analytical Chemistry, Vol. 51, No. 11, September 1979, pp. 1828-34

6. American Chemical Society Symposium, 1979 Fall Annual Meeting, "Choosing a Laboratory Computer System."

BIOGRAPHY

Robert E. Sanders received a B.S. in Chemistry from Ohio State University in 1949. He has been employed by Rockwell International for the past 30 years in a variety of positions at the Columbus Division and the Tulsa Division. He is currently Supervisor, Laboratory Services for the Tulsa Division Laboratories. He is a member of the American Association for the Advancement of Science, and the National Management Association as well as a Certified Manger.

TABLE I ESTIMATED TIME SAVINGS WITH COMPUTERIZED CALCULATIONS

CALCULATION TYPE	MINUTES PER CALCULATIONS	CALCULATIONS PER YEAR	ANNUAL SAVINGS (MIN)
Tensile Strength & Modulus	0.15	300	45
Flexural Strength & Modulus	3.28	208	682
Sandwich Beam Compression	5.00	600	3000
Standard Deviation	0.58	960	554
Rivet Shear Strength	0.22	960	214

ESTIMATED TOTAL 4495
75 Manhours

```
10 CLS
20 PRINT:PRINT:PRINT
30 PRINT TAB(10) "COMPOSITE MATERIALS TESTING":PRINT
40 PRINT TAB(15) "1-TENSILE STRENGTH & MODULUS"
50 PRINT TAB(15) "2-FLEXURAL STRENGTH & MODULUS"
60 PRINT TAB(15) "3-SHORT BEAM COMPRESSION"
70 PRINT TAB(15) "4-SANDWICH BEAM COMPRESSION"
80 PRINT TAB(15) "5-RAIL SHEAR"
90 PRINT TAB(15) "6-LAMINATE RESIN CONTENT & FIBER VOLUME"
100 INPUT "MAKE SELECTION 1 TO 6"; A
105 CLS:PRINT:PRINT
110 ON A GOTO 200,300,400,500,600,700
```

Figure 3. Listing of Basic Menu Program


```
400  CLS:PRINT:PRINT
405 REM PROGRAM TO CALCULATE LAMINATE FLEXURAL
410 REM STRENGTH & MODULUS - REV A 4-11-80
415 REM ROCKWELL-TULSA-FOR TRS-80III
420 CLS:PRINT:PRINT:PRINT
425 INPUT"ENTER TEST MACHINE FACTOR";K
430 DIM P(20),W(20),T(20),F(20),M(20)
435 INPUT "ENTER NUMBER OF SAMPLES";N: DIM P1(N)
440 LET T1=0: T2=0
445 FOR X=1 TO N
450 PRINT: PRINT "SAMPLE # ";X
455 INPUT "            WIDTH       ";W(X)
460 INPUT "            THICKNESS   ";T(X)
465 INPUT "            FLEX LOAD   ";P(X)
470 INPUT "            MOD  LOAD   ";P1(X)
475 LET F(X)=(3.75*P(X))/(W(X)*T(X)*T(X))
480 LET M(X)= (K*P1(X))/(W(X)*T(X)*T(X)*T(X))
485 LET T1=T1+F(X)
490 LET T2=T2+M(X)
495 NEXT X: PRINT: PRINT
500 PRINT TAB(15)"FLEX LOAD"; TAB(30) "MOD LOAD"
505 PRINT
510 FOR X=1 TO N
515 PRINT TAB(15) INT(F(X));
520 PRINT TAB(30) (INT(M(X)/10[4)/100);"E6"
525 NEXT X
530 LET A1=T1/N: A2=T2/N: PRINT
535 PRINT TAB(5) "AVERAGE="; INT(A1);
540 PRINT TAB(30) (INT((A2)/10[4)/100)
545 PRINT:PRINT
550 INPUT "RUN MORE FLEX SAMPLES?, Y/N"; Q$
555 IF Q$="Y" GOTO 430
560 END
```

Figure 4. Listing of Laminate Flexural Calculation Program

ULTRASONIC EXTENSOMETRY FOR DETERMINING
BOLT PRELOAD IN HEAVY INDUSTRY - FROM
PETROCHEMICAL TO REACTORS

Donald C. Erdman
Consultant
Raymond Engineering, Inc.
Middletown, Ct.

Abstract

Use of ultrasonic extensometers has found wide application for bolt preload determination in airborne and aerospace applications where elongation measurement accuracy is often required to .0001". Experience has now been gained in heavy industry with fasteners up to 12 feet long, often on studs and bolts with relatively rough head surfaces. Here, accuracy may be reduced to .001 inch, a figure far better than available from torque wrenches. This paper describes some of these heavy industry applications.

1. INTRODUCTION

When bolts, studs, or other fasteners must be stressed to up to 80% of yield, it is important that a measuring technique be used that can be accurate to around 5%. The ultrasonic method uses the familiar pulse echo technique often found in thickness gages, but corrected for the fact that the velocity of ultrasound is a function of stress. Elongation of the fastener is a convenient measure of the preload acquired when fasteners are tightened. The accuracy of measurement is not affected by the length of the fastener. Thus, one inch long bolts used in aircraft turbine engines can be tested as well as 16 foot long columns in an extrusion press. Often the actual elongation is not as important as the equalization of elongation among four columns of a press, or in a bolt circle. Fig. 1 shows the path taken by the ultrasound in a bolt. The short pulse of ultrasound reflects from the far end of the bolt and returns to the transducer, permitting an elapsed time measurement that is a function of elongation.

2. APPLICATIONS

Figs. 2 and 3 show bolts being test on a bridge in Belgium, where bolts were failing in fatigue. Interest was both on the preload

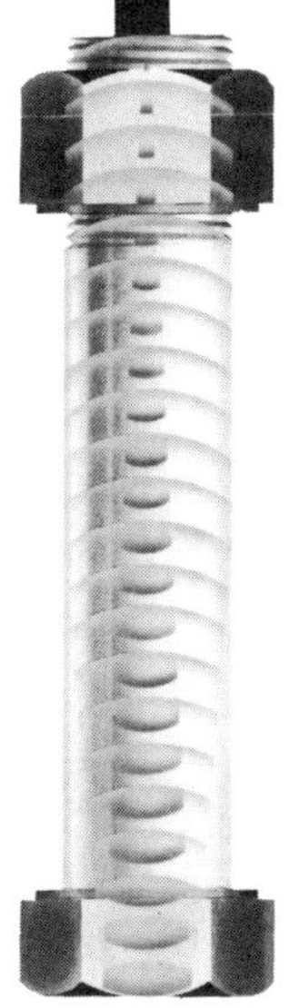

Fig. 1

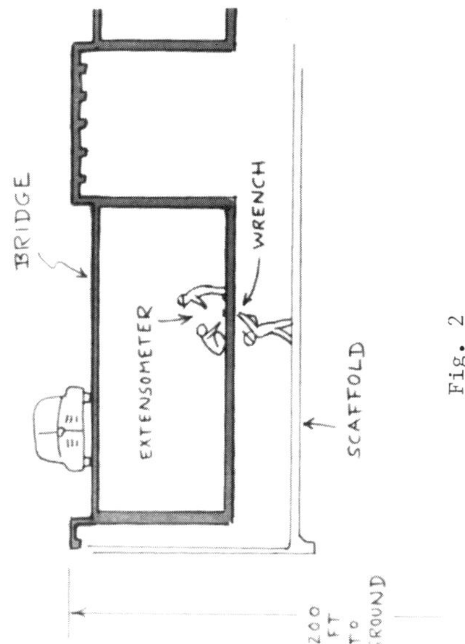

Fig. 2

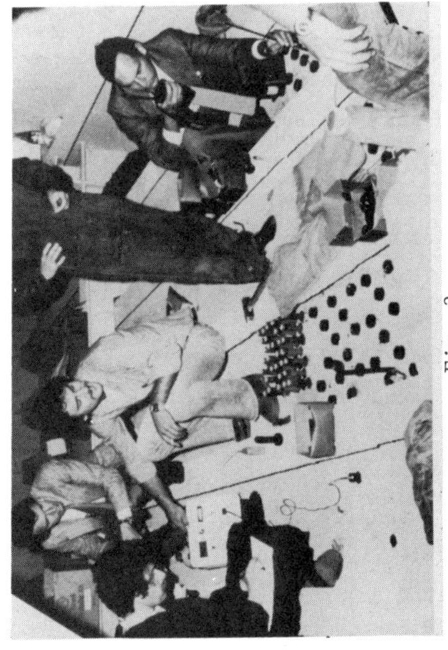

Fig. 3

existing in old bolts and upon replacing these with new bolts with
the proper preload. The old bolts
required noting the stretch length
versus the unloaded length.

Petrochemical plants have both end
closures and flange joints that require even preloading of the bolts
or studs. This is particularly
true in hydrogen crackers. Figs.
4, 5, and 6 show some of the chemical and refinery applications.
Fig. 7 is a typical joint that may
have 30 or more fasteners, each of
which must produce the same preload. Heat exchangers present
another application for careful
control of bolt preload. Fig. 8
gives some idea of the quantity of
fasteners involved. As an example
of a critical use of the ultrasonic
extensometer, there is a requirement that all nuclear subs in for
repair have their manway and
access hatches on the heat exchangers secured by bolts that have
been checked for preload with the
ultrasonic extensometers. These
bolts must keep the hatches from
leaking for many years. If records
are kept of the preload on individual bolts, occurance of creep
is easily noted.

Another area where accurate preload
is essential is on off-shore oil
rigs. Figs. 9 and 10 show the
method of application of the

extensometer. The instrument is
kept in the diver's hut. A 300 ft.
coax cable extends under water to
the transducer. The tie bolts holding the leg collar to the brace are
often 6 ft. long. The fact that
the bolts are continuously threaded
does not affect the accuracy of the
reading.

The method works well in insuring
even tightening on the studs that
secure the cover on nuclear reactor
pressure vessels. Some remarkably
long and crooked tie rods have been
inspected. An example is the roof
support rods in coal mines. This
can be done if the accessible end
is squared off neatly, and if there
are not serious grip marks on the
rods. These grip marks cause
spurious reflections that may be
interpreted by the instrument as
echoes from the far end of the tie
rod.

Accuracy of measurement depends
upon several controllable factors.
In short fasteners, parallelness
of the end surfaces is important,
as is the surface finish. Newer
techniques permit inspection of
bolts having external grade mark-
ings and a shape as severe as a
carriage head. Temperature of the
fastener will affect the velocity
of ultrasound as well as the actual
length, due to thermal expansion.
If all fasteners in a bolt circle

Fig. 4

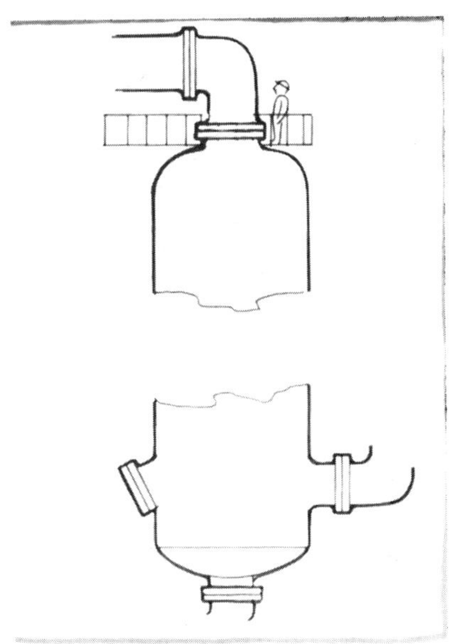

Fig. 6

Fig. 5

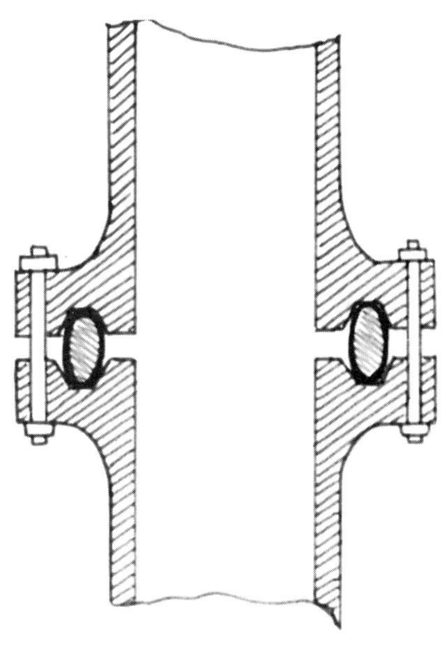

Fig. 7

Fig. 8

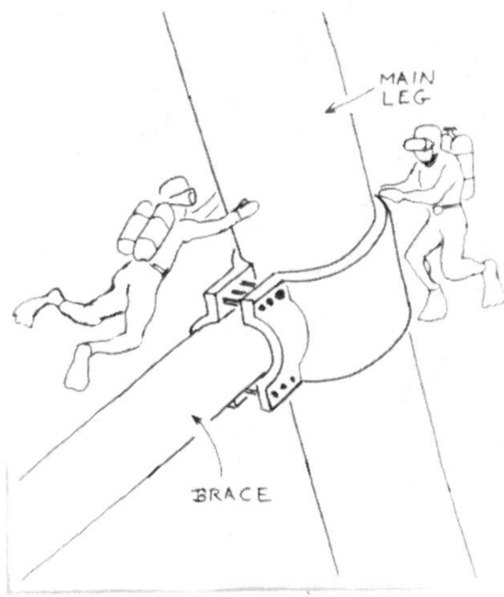

Fig. 10

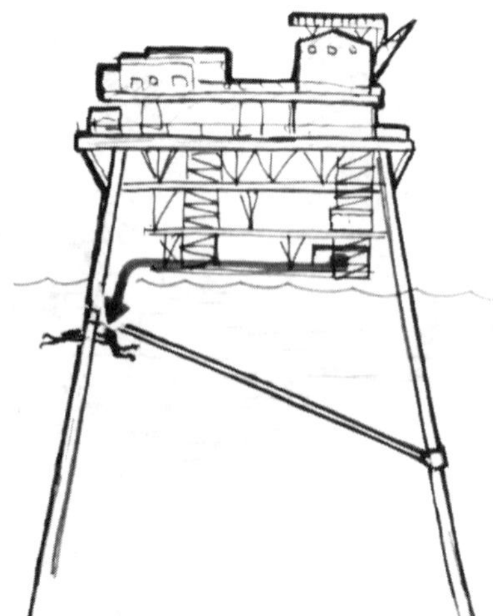

Fig. 9

are at the same temperature, equal preloads are readily obtained even if the actual total preload may be in slight error. Formulas are available for calculating the effect of temperature on the measurement.

If exact preloads provided by a new fastener must be observed, test specimens should be subjected to tensile tests using precision dial gages or linear displacement transducers. In the case of bolts used for the flange joints on the Space Shuttle main engines, standards of each size bolts are provided for the operator for calibrating the instrument. Once the effect of the second and third order Lame' and Murnaghan constants is known, a simple multiplying factor can be applied to elongation to compensate for the stress-velocity changes. It is often quite practical to use two reference specimens of slightly different lengths as back-up standards. The above precision calibration is not often required; only where preloads are approaching yield. Internal calibration adjustments are provided in the extensometer to compensate for velocity changes in various alloys.

3. SUMMARY

Illustrations are given where precision preload determination can be obtained in bolts, studs, and tie rods. Both relative preloads among fasteners in a bolt circle and accurate preload on individual fasteners are practical. Accuracy obtained will vary from as high as 1% to as low as 5%, depending upon the preparation of the specimen, and upon the care taken in calibration of the instrumentation.

4. REFERENCES

1. "Investigation of Threaded Fastener Structural Integrity", R. K. Swanson and John Barton, Southwest Research Institute, San Antonio, Texas. NASA report on Contract No. NAS9-15140. Available from NTIS N78-10473. Work performed Oct. 1977. This is a complete study of several preload or elongation measuring techniques.

2. "Using Ultrasonics to Measure Bolt Tension - A State of the Art Report", John H. Bickford, Vice-President, Raymond Engineering, Inc., Middletown, Conn. Published in Society of Manufacturing Engineers, Dearborn, Mich. 1976. Available from Raymond Engineering, Inc. 217 Smith Street, Middletown, Ct., phone 203/632-1000.

3. "Measuring Fastener Strain", Donald C. Erdman, 1179 Romney Drive, Pasadena, Calif. 91105. Presented 1973 to Air Transport Association of America. Copies available from author. Phone 213/792-4508.

5. BIOGRAPHY

Donald C. Erdman. Graduated in
Physics, Pomona College, 1938.
Built first immersed inspection UT
instrumentation while working at
MIT during World War II. Designed
the widely used Immerscope, B-Scan,
many of the large UT scanning
systems, and the Nanoscope.
Invented the ultrasonic extensometer
in conjunction with H. McFaul at
Douglas Aircraft in 1970. Consultant to Aerojet on Polaris A1, A2,
and A3. Consultant on B70, C5A,
Northrop new design, and presently
half time with Raymond Engineering,
Inc. President of Erdman
Instruments, Inc., working on ground
support equipment for testing Space
Shuttle main engines.

THICKNESS MEASUREMENT METHODS FOR
THIN MULTILAYER METAL FILMS

G. M. Light, Ph.D.
G. P. Singh, Ph.D.
Research and Development
Quality Assurance Systems and Engineering Division
Southwest Research Institute
San Antonio, Texas
and
F. D. McDaniel, Ph.D.
Department of Physics
North Texas State University
Denton, Texas

Abstract

Two experimental techniques--
characteristic X-ray fluorescence
and high frequency, pulse-echo
ultrasonics--have been used to
determine layer thicknesses ranging
from about 0.0005 inch (0.0127 mm)
to 0.005 inch (0.127 mm). Thin
layers of stainless steel, tin,
gold, and lead were examined.
Tests were conducted on free standing foils as well as foils in contact with thick copper backings.
The characteristic X-ray fluorescent studies were conducted using
Co_{27}^{57} and Cd_{48}^{109} sources with
strengths in the range of 0.1
millicurie. The ultrasonic test
frequency used in the study was 300
MHz. Equipment description and
results of this investigation are
presented in the paper.

Keywords: NDE thickness
measurement, thin films, X-ray
fluorescence, ultrasonic.

1. INTRODUCTION

Recent discussions with representa-
tives from industry indicate a need
for a nondestructive testing (NDT)
technique for measuring the thickness of thin multilayer metal films.
The thickness of the individual
layers range from 0.0005 inch
(0.0127 mm) to approximately 0.005
inch (0.127 mm); and they are
usually deposited on a thick substrate in a symmetric manner, as
shown in Figure 1.

The usual NDT practice for determining the thickness of thin films
is gamma ray absorption or eddy current techniques. The gamma ray
absorption technique only provides
information about total through-wall

thickness and consequently would
not be helpful in evaluating
individual layer thickness for
samples similar to the one shown in
Figure 1. Additionally, the layers
often have similar electrical
conductivity, thus preventing the
use of eddy current.

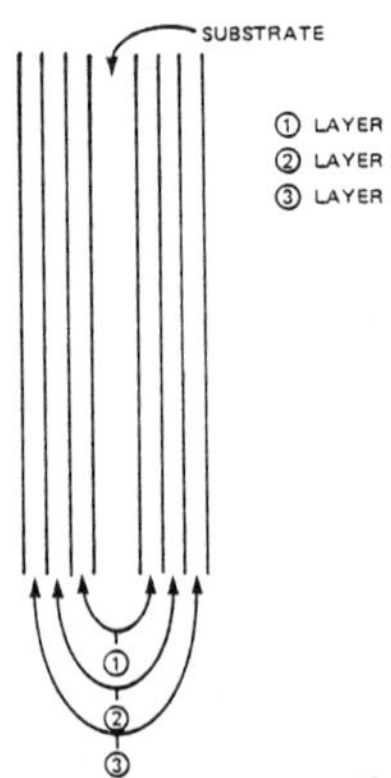

FIGURE 1. DIAGRAM OF MULTILAYER
METAL SAMPLE

With these obstacles in mind, the
impetus of this work was to develop
NDT techniques that would work in
the thickness range of interest and
not be effected by the previously
mentioned problems for layers
similar to those shown in Figure 1.
Two techniques were studied: a
very high frequency ultrasonic
pulse/echo examination (UT) method
and a gamma ray radioactive source
induced X-ray fluorescent technique. The purpose was to study
the feasibility and useful thickness range of both methods and,
finally, to compare the usefulness
of both techniques for industrial
applications.

Thickness measurements were made
for stainless steel, tin, gold, and
lead foils. Foils were used in
lieu of multilayered plated specimens due to classification of the
plating processes involved in their
fabrication. However, it was felt
that the foils would serve to demonstrate the capabilities of the
techniques.

2. ULTRASONIC TECHNIQUE

2.1   Background

Conventional, ultrasonic frequencies as high as 30 MHz have
been used to measure thickness as
of material layers. The traditional technique makes use of the
time-of- flight difference between
the front surface and the back surface echoes; and, with knowledge of
the ultrasonic velocity in the
sample, the time-of-flight can be
related to thickness. The measured
thickness can be described by the
expression:

$$\text{Thickness} = \frac{\text{velocity}}{2} \times (\text{measured}$$

time between front and
back echoes)

or

$$d = \frac{V}{2} \Delta t \qquad (1)$$

The lower limit of the thickness
range measurable by this technique
is primarily a function of transducer frequency and damping. To
estimate the transit time of a
reflected acoustic signal, it is

necessary to have good temporal separation between the signals produced by the front and back surfaces of the layer. The transient response of the transducer is shortened by incorporating acoustic damping, but the practical minimum response achievable is approximately one and a half periods of the center frequency that satisfies the inequality given by

$$\text{Freq.} > \frac{3}{4} \frac{(\text{Longitudinal Vel.})}{(\text{Layer Thickness})} \qquad (2)$$

Using this criterion, i.e., assuming an impulse duration of three halves the period, a transducer having a frequency on the order of 250 MHz would be needed to measure the thickness of a 0.0005 inch (0.0127 mm) film such as gold (velocity = 3.24 x 105 cm/sec).

Recent work by B. T. Khuri-Yakub[1,2] and G. S. Kino[1] demonstrated the usefulness of 200 MHz transducers for evaluation of ceramics. During these studies, the transducer was coupled to the ceramic with a thin gold foil; and the authors observed that the thickness of the gold foil could be measured. With these results in mind, it was decided to develop such a system and use it to measure thin film thicknesses of various materials.

2.2 Experimental Description

A 300 MHz ZnO transducer was fabricated by Dr. B. T. Khuri-Yakub for use in this experiment. The transducer consisted of a ZnO film approximately 8 microns thick, which was sputtered onto a sapphire buffer rod approximately 0.25 inch (6.4 mm) long. A thin layer of gold (approximately 1000 angstroms thick) was placed between the sapphire and ZnO to provide one of the electrodes. The transducer is schematically shown in Figure 2.

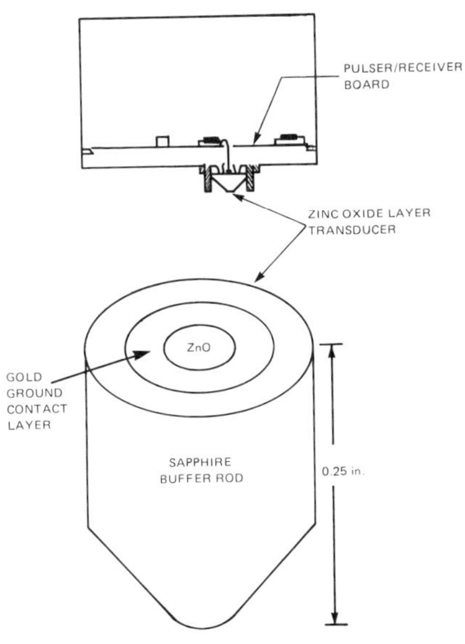

FIGURE 2. SCHEMATIC OF 300 MHz TRANSDUCER

The pulser for the transducer was mounted in the transducer housing in close proximity to the transducer (see Figure 2), and the connection between the ZnO transducer element and the pulsing circuit was made with a 0.002 inch (0.05 mm) gold wire. The pulser/receiver circuit with associated protection diodes is shown schematically in Figure 3.

The equipment set up for the experiment is shown in Figure 4. The pulser/receiver circuit in the

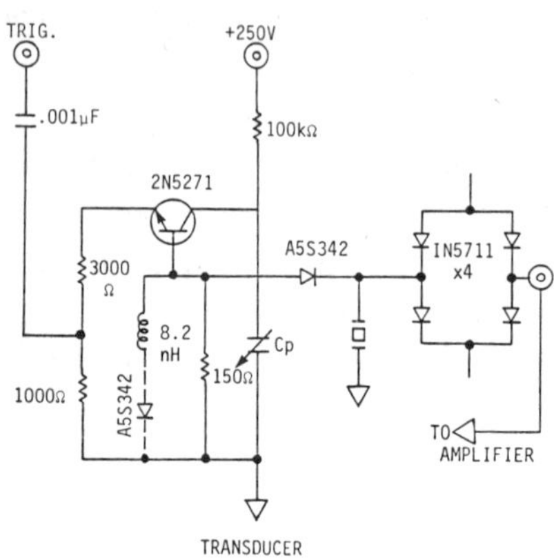

FIGURE 3.   SCHEMATIC OF THE 300
MHz PULSER/RECEIVER CIRCUIT

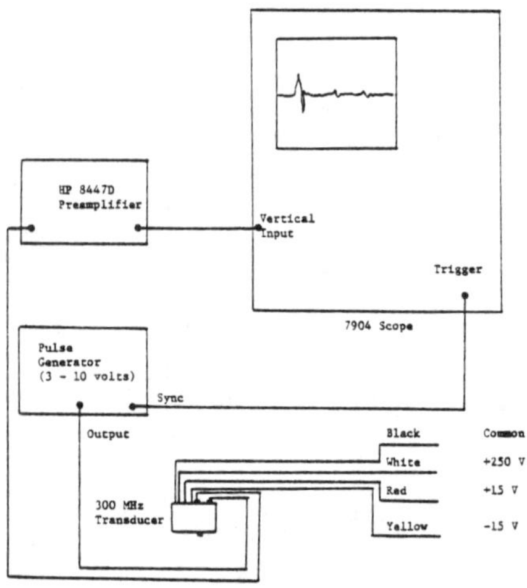

FIGURE 4.   EQUIPMENT SCHEMATIC
USED FOR 300 MHz SYSTEM

transducer housing required a
±15 and a +250 volt power supply.
The trigger source used in the
system was an Interstate model F44
voltage sweep generator. An
HP8447D amplifier (applicable over
the frequency range of 0.1 to 1300
MHz) was used, and the output
served as the y-axis input to the
Tektronix 7904 scope, which had
7A16, 7B71, and 7B70 plug ins. The

trigger repetition rate was 55 KHz,
and the measured center frequency
of the transducer was found to be
283 MHz ±20 MHz.

The thicknesses of lead, tin, and
302 stainless steel foils were
measured both ultrasonically and
mechanically (vernier caliper).
The difference in time between the
front and back echoes was measured
to within 4 nanoseconds, and the
thickness calculated using Equation 1. A typical oscilloscope
trace is shown in Figure 5. The
thicknesses for various lead, tin,

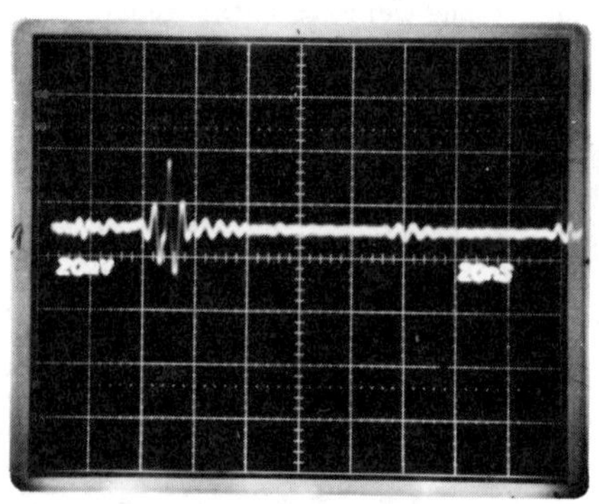

FIGURE 5.   SCOPE TRACE FOR
0.0045-INCH (0.11-mm)
LEAD FOIL

and 302 stainless steel foils
measured by the ultrasonic technique and the vernier caliper are
compared in Table 1, and graphically represented in Figures 6
and 7.

The results of the ultrasonic and
vernier caliper measurements are in
good agreement for all of the foil
materials studied up to a thickness
of about 0.0045 inch (0.11 mm).
For thicknesses greater than

TABLE 1.   THICKNESS MEASUREMENT DATA

	Specimen No.	Vernier Caliper Error = 0.0005 inch	Ultrasonic Error = 0.002 inch
		In units of 10^{-3} inch	In units of 10^{-3} inch
For Lead Foils	(1)	0.8	0.9 (0.7)*
Assumed Velocity = 0.85×10^5 in./sec	(2)	1.5	1.6 (1.4)
	(3)	2.3	(2.2)
	(4)	2.8	2.6 (2.6)
	(5)	3.0	2.6
	(6)	4.5	4.2
	(7)	5.0	4.3 (4.3)
For 302 Stainless Steel Foils	(1)	1.2	1.3
	(2)	2.0	2.0
Assumed Velocity = 2.23×10^5 in./sec	(3)	3.0	2.7
	(4)	5.0	4.5
For Tin Foil	(1)	1.5	1.4
Assumed Velocity = 1.31×10^5 in./sec			

*The data in parenthesis are data taken with the same equipment by a different operator.

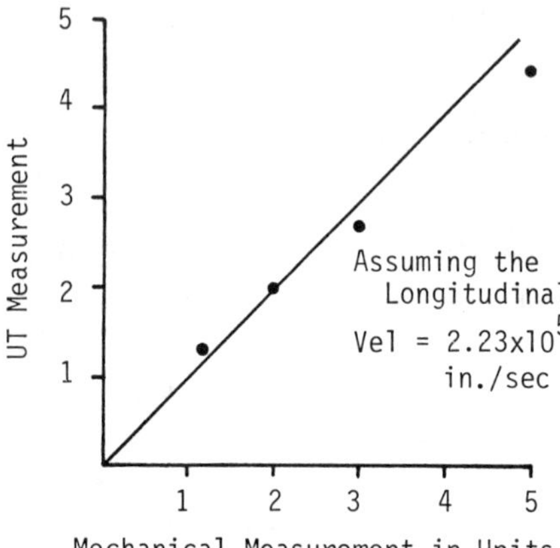

FIGURE 6. ULTRASONIC THICKNESS
MEASUREMENT FOR LEAD COMPARED
TO THICKNESS OBTAINED WITH
VERNIER CALIPER

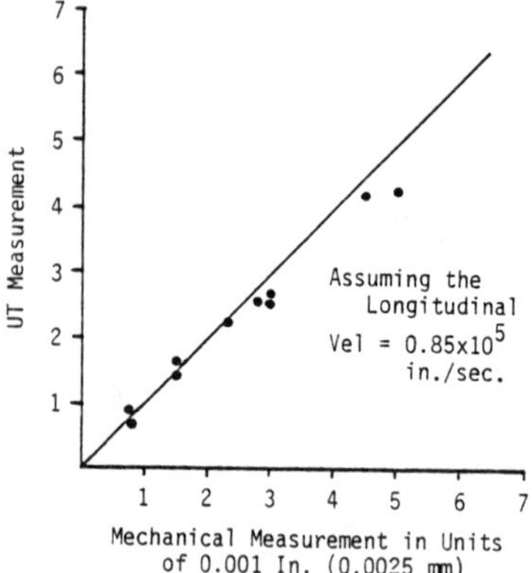

FIGURE 7. ULTRASONIC THICKNESS
MEASUREMENT FOR STAINLESS
STEEL COMPARED TO THICK-
NESS OBTAINED WITH
VERNIER CALIPER

than 0.005 inch (0.127 mm), the results show larger errors. The cause of this behavior is not well understood at this time. Possible sources of this error could be diffraction effects or a variation in longitudinal velocity as a function of thickness.

If the thickness of a foil greater than 0.004 inch (0.102 mm) thick is assumed to be known (i.e., by vernier caliper measurement) and the relationship shown in Equation 1 is rearranged to solve for the velocity, the velocity calculated is about 10 percent greater than the published value.[3] This could be attributed to the "rolling method" by which the test foils were fabricated. If these velocity values are used to calculate the thickness from the ultrasonic time-of-flight data, the thinner films are over-estimated while the thicker films are estimated well within the experimental error.

3. RADIOACTIVE SOURCE-INDUCED
X-RAY FLUORESCENCE TECHNIQUE

3.1 Background

For many years X-ray fluorescence has been used for trace element detection and analysis. This technique can be used to identify not only the elemental composition of thin layers, but also their thicknessses. In this work, measurements were made of gold and lead thicknesses ranging from 0.001 inch (0.025 mm) to 0.014 inch (0.36 mm).

The use of X-ray fluorescence for thickness measurement was suggested by Hamos.[4] According to his theory, if the initial intensity of the incident radiation (I_o), the incident and exit angles (θ_1 and θ_2, respectively), the absorption coefficient of the material (μ), the density of the material (ρ), and a characteristic of the material called the excitation constant (Q) are known, then the intensity of the emitted radiation from a film of thickness x can be calculated from the expression

$$I = \frac{QI_0}{\mu^1} \cdot [1-\exp(-\mu^1 \rho x)] \qquad (3)$$

where $\mu^1 = \mu (\sec \Theta_1 + \sec \Theta_2)$

3.2 Experimental Description

The equipment needed for this technique included a multichannel analyzer (MCA), an amplifier, and an intrinsic germanium (Ge) solid-state X-ray detector and radio-active source. The experimental setup is schematically shown in Figure 8. A Tracer Northern MCA with a Tektronix 5110 scope was used to acquire the data and perform preliminary analysis of the source-induced X-ray fluorescent spectra. The detector system included a liquid-nitrogen-cooled intrinsic Ge detector and an ORTEC model 440A amplifier.

A cadmium 109 (Cd_{48}^{109}) source excited the target materials and resulted in the emission of characteristic X-rays. The Cd_{48}^{109} source was a ring source with a strength of 0.71 millicuries. [The half life of Cd_{48}^{109} is 453 days, and it produces an 88 keV (kiloelectron volt) gamma ray.] Once the data were acquired, they were transferred to computer files should further analyses be required.

The detector geometry used in this experiment is shown in Figure 9. The materials making up the target holder were low atomic number (Z) elements to prevent the fluoresced X-rays from these materials from interfering with the target X-rays, i.e, the characteristic X-rays of the target holder were of much lower energy than the target characteristic X-rays.

Also, the mass of the material surrounding the target was kept to a minimum to reduce the effects of Compton scattering. The theoretical value for Compton scattering at 180 degrees for an 88 keV

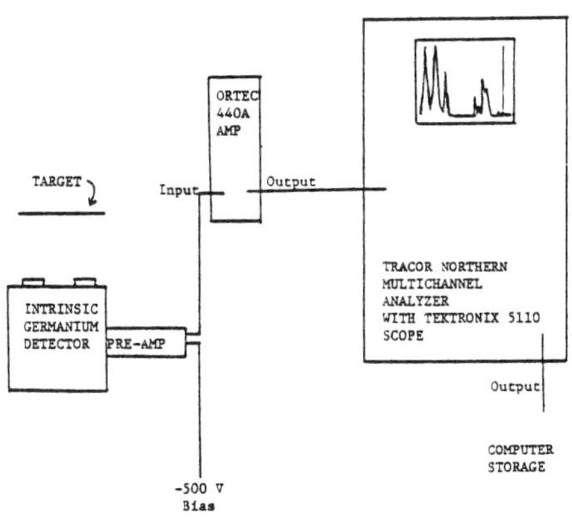

FIGURE 8. EXPERIMENTAL EQUIPMENT SETUP FOR THE SOURCE-INDUCED X-RAY FLUORESCENT TECHNIQUE FOR MEASURING THICKNESS

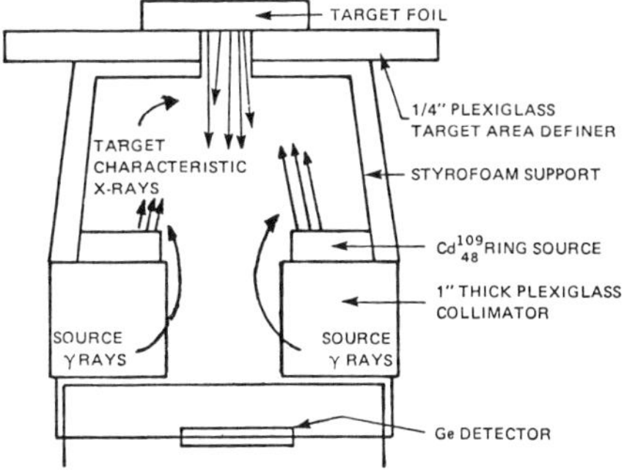

FIGURE 9. DETECTOR GEOMETRY

gamma ray beam is approximately
66 keV. The energies of the gold
$K\alpha_1$, $K\alpha_2$, $K\beta_1$, and $K\beta_2$ X-rays are
68.8, 67.0, 78.0, and 80.2 keV,
respectively; and the lead K_1,
$K\alpha_2$, $K\beta_1$, and $K\beta_2$ X-ray energies
are 74.9, 72.8, 84.9, and 87.3 keV,
respectively. Thus, one can see
the importance of not having a
large intensity peak at 66 keV to
interfere with the analysis of the
characteristic X-ray peaks.

The Ge detector system was set up
and calibrated using americium
(Am_{95}^{241}) and cadmium (Cd_{48}^{109})
sources. The known energy lines
emitted by these sources are 13.5,
17.8, 20.8, 22.1, 25.0, 26.4, 59.5,
and 88.0 keV. The detector system
is linear; and, therefore, these
data points provided a very good
experimental calibration line.

By rearranging Equation 2 to solve
for the excitation constant Q and
utilizing the experimentally
observed $K\alpha$ yields in the equation,
the value of Q can be determined;
and a curve predicting $K\alpha$ X-ray
yields versus film thickness can be
developed. The value of Q was
calculated from the $K\alpha$ yield from
each foil and compared. These Q
values are listed in Table 2. As
can be seen, some variations in the
Q values exist, which may be due to
nonuniform thickness of target
foils.

Figure 10 shows the $K\alpha$ X-ray
yields obtained for foils of known
thicknesses. These curves represent

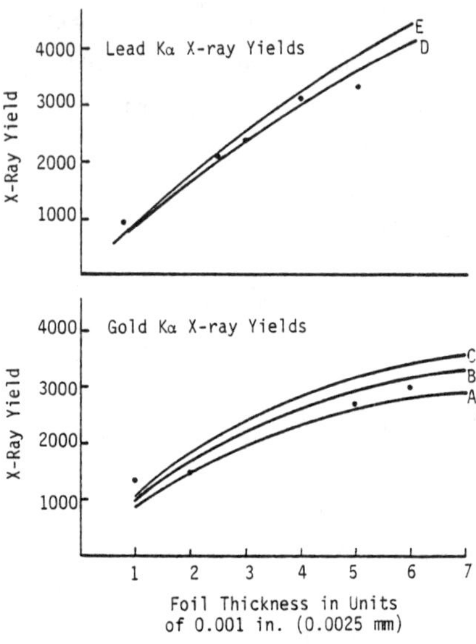

FIGURE 10. EXPERIMENT X-RAY YIELDS
COMPARED TO THEORETICAL PREDICTIONS
AS A FUNCTION OF TARGET THICKNESS

predictions of $K\alpha$ X-ray intensity
versus foil thickness for various
values of Q. Curve A represents
the prediction made using Q = 3.9 x
10^{-6} cm^2/g, which is the average of
the Q values obtained from the
0.002-, 0.005-, and 0.006-inch
(0.051-, 0.127-, and 0.152-mm)
thick foil X-ray yields. Curve B
represents similar prediction made
using Q = 4.45 x 10^{-6} cm^2/g, which
was obtained by adding the Q from
the 0.001-inch (0.025-mm) foil to
the previous Q values and taking
that average. Curve C was obtained
by adding to the previous Q values
the Q from the 0.0085-inch
(0.216-mm) foil, which yielded a
Q = 4.76 x 10^{-6} cm^2/g. All three
curves are in good agreement with
the experimental data.

Similar curves were generated for
the lead foils. Curve D represents
the average of excitation constants
obtained from the 0.0025-, 0.003-,
0.004-, and 0.005-inch (0.064-,
0.076-, 0.102-, 0.127-mm) thick
foils. This Q was 5.85 x 10^{-6}
cm^2/g. Curve E is similar to
Curve D except the Q value was
obtained with the addition to the Q
from the 0.0008-inch (0.02-mm)
foil, and thus the Q was
6.21 x 10^{-6} cm/^{2}/g.

The values used to generate these
curves were as follows. The
densities were 19.3 and 11.35 g/cm^3
for gold and lead, respectively;
and the attenuations for the Kα
X-ray lines were 3.06 cm^2/g and
1.4 cm^2/g for gold and lead,
respectively.

If the measured Kα and Kβ yields
are plotted as a function of foil
thickness, a very linear response
is observed for both gold and lead
for foil thicknesses from 0.001
(0.025 mm) to approximately 0.009
inch (0.229 mm) as shown in Figure 11. (These data have been
corrected for background.) Each
data point took 200 seconds
(3.3 minutes) to acquire and
represents an average thickness
value over an area of approximately
0.5-inch (12.7-mm) square.

Data were taken for both individual
foils and the same foils placed on
a 0.03-inch (0.76-mm) copper substrate. The results of the K-shell

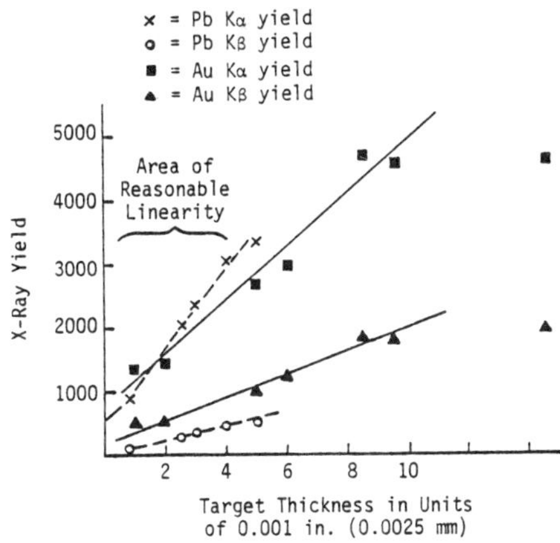

FIGURE 11.   X-RAY YIELDS AS A
FUNCTION OF TARGET THICKNESS

X-ray yields for individual foils
versus the foils on copper substrate were within experimental
error. In addition, data were
taken using a Co$_{27}^{57}$ source, which
has an energy of 121 and 136 keV
and a half life of 270 days. However, the strength of the source
was only microcuries; and, consequently the data acquisition time
was too long to be practical for
this experiment. But since the
energy is well above K-shell excitation energy, this source would be
applicable for such work if the
strength of the source were
greater.

4.   COMPARISON OF METHODS

The two techniques discussed have
been shown to be useful for non-
destructively measuring the thickness of foils in the range of 0.001
to 0.005 inch (0.025 to 0.127 mm).
Both techniques have unique characteristics, making them suitable for
particular examinations, and also

some disadvantages; these must be considered when determining which technique is best suited for a given task.

The X-ray fluorescent technique uses standard equipment, and the sample can be tested in air. The ultrasonic technique, on the other hand, consists of state-of-the-art pulser/receiver and amplifier electronics; and the sample must be subjected to some type of ultrasonic coupling mechanism (such as water, which was used in this experiment). The X-ray technique takes about 200 seconds to acquire each data point, which represents an area average thickness value. On the other hand, the ultrasonic technique is almost instantaneous and represents a specific point thickness value.

The X-ray technique requires the use of radioactive sources and calibration foils. The ultrasonic system requires that the ultrasonic velocity in the sample material be known.

The major disadvantages are that the X-ray technique requires a radioactive source (which means special handling procedures), and the ultrasonic technique requires the use of very high frequency electronics (which are not off-the-shelf items).

5. ACKNOWLEDGEMENT

The authors would like to thank K. Sandstrom and J. Ball for their help in developing the high frequency pulser/receiver electronics. Also, thanks are extended to Dr. B. T. Khuri-Yakub and Mr. J. L. Jackson for their many useful comments.

TABLE 2. EXCITATION CONSTANTS (Q) ESTIMATED FROM X-RAY YIELDS FROM VARIOUS FOILS

Mat'l	Foil Thickness Units of 0.001 inch	Q cm^2/g
Gold	1.0	5.0×10^{-6}
	2.0	3.7×10^{-6}
	5.0	4.0×10^{-6}
	6.0	4.1×10^{-6}
	8.5	6.0×10^{-6}
	9.5	5.6×10^{-6}
Lead	0.8	7.9×10^{-6}
	2.5	6.1×10^{-6}
	3.0	5.9×10^{-6}
	4.0	6.0×10^{-6}
	5.0	5.4×10^{-6}

REFERENCES

1. B. T. Khuri-Yakub and G. S. Kino, "High Frequency Pulse Echo Measurements of Ceramics and Thin Layers," 1976 Ultrasonic Symposium Proceedings, IEEE Cat. No. 76 CH1120-5SU (1976), pp. 964-66.

2. Private communication, June 1980.

3. Nondestructive Handbook, Edited by R. C. McMasters (New York: The Ronald Press Co., 1963).

4. Liven Hamos, Arkiv f. Matematik, Astronomi, och Fysik 31A, Article No. 25 (1941).

BIOGRAPHIES

Glenn M. Light, Ph.D.
Senior Research Physicist
Research and Development

B.S. in Physics, McMurry College, 1972; M.S. in Atomic and Nuclear Physics, North Texas State University, 1974; and Ph.D. in Atomic and Nuclear Physics, North Texas State University, 1978.

Dr. Light has over six years of experience in the fields of atomic and nuclear physics, including extensive experimentation in L-shell and K-shell X-ray production cross sections using ion bombardment. At SwRI, Dr. Light is currently investigating new methods of NDT testing, and working with adaptive learning networks and the Institute's Schlieren imaging system.

Gurvinder P. Singh, Ph.D.
Senior Research Engineer
Research and Development

B.S. in Aeronautical Engineering, Panjab Engineering College, India, 1974; M.S. in Mechanical Engineering, Drexel University, 1976; and Ph.D. in Mechanical Engineering and Mechanics, Drexel University, 1979.

Dr. Singh has several years of experience in the acoustic wave propagation, transducer design, digital signal processing, and pattern recognition fields. He has been engaged in the design of specialized transducers for ultrasonic testing and research applications in the nuclear and medical areas.

F. D. McDaniel, Ph.D.
Department of Physics
North Texas State University

B.S. at Memphis State University, 1977; M.S. at University of Georgia, 1968; and Ph.D. at University of Georgia, 1971.

Dr. McDaniel is presently an assistant professor at North Texas State University. He has worked in the areas of atomic and nuclear physics for fifteen years and has published 55 papers in this field. In addition to his teaching, Dr. McDaniel is a consultant to the Physics Division of Oak Ridge National Laboratory operated by Union Carbide Corporation for the Department of Energy.

FIBER FELTS AS LOW DENSITY
STRUCTURAL MATERIALS
John V. Milewski
Stephen E. Newfield
Los Alamos Scientific Laboratory
Los Alamos, New Mexico

Abstract

Short fiber felts structures can be made which provide improvements in properties over foams. In applications where resistance to compression set or stress relaxation are important, bonded fiber felts excel due to the flexing of individual fibers within their elastic limit. Felts of stainless steel and polyester fibers were prepared by deposition from liquid slurries. Compressive properties were determined as a function of felt parent material, extent of bonding, felt density, and length-to-diameter (L/D) ratio of starting fibers.

1. INTRODUCTION

By felting short fibers of various materials and subsequently bonding the fibers together at their intersections, low density materials can be made that have interesting physical properties. The ultimate properties obtained are a function of several variables, including: fiber material; bonding medium; and length-to-diameter (L/D) ratio of fiber used.

A laboratory-scale technique was developed for felting and coating fibers. Some physical properties have been determined for sample felts as functions of L/D, fiber material and extent of felt coating. In layers up to 12.7 mm thick, such felts could be used as cushions with attractive properties. By selecting fibers such as carbon or stainless steel, the stress relaxation and compression set failure of the cushion could be greatly reduced, since individual brittle fibers could flex within their elastic limit without deformation.

2. FELT FABRICATION

Fibers in various metal, ceramic and polymeric materials are becoming increasingly available commercially. Figure 1 shows four-μm-diam stainless steel fibers chopped into one-mm in length. The fibers are frequently supplied in ribbon-like bundles, held lightly together with an organic sizing material.

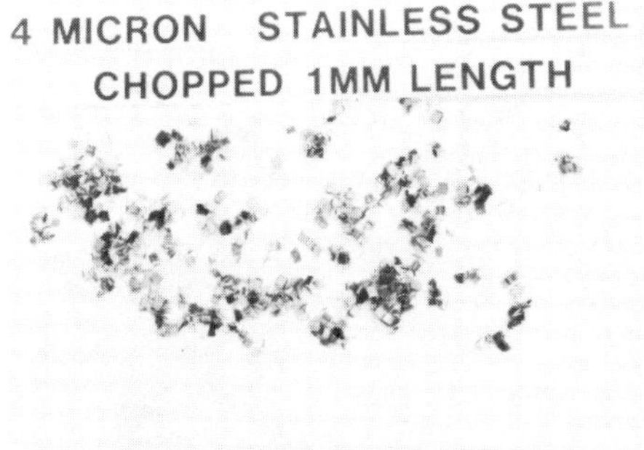

Figure 1. Fibers Bundles.

The fiber bundles were separated into individual fibers by vigorous agitation in hot water. The suspending liquid was then changed to isopropanol. Figure 2 shows the individual fibers as they were being agitated in the alcohol.

Figure 2. Fibers in alcohol.

In Figure 3, a Buchner funnel with a fritted glass filter is shown inverted into the agitated suspension. A vacuum drew the fiber-liquid suspension into the funnel, and the fibers deposited on the filter as a felt.

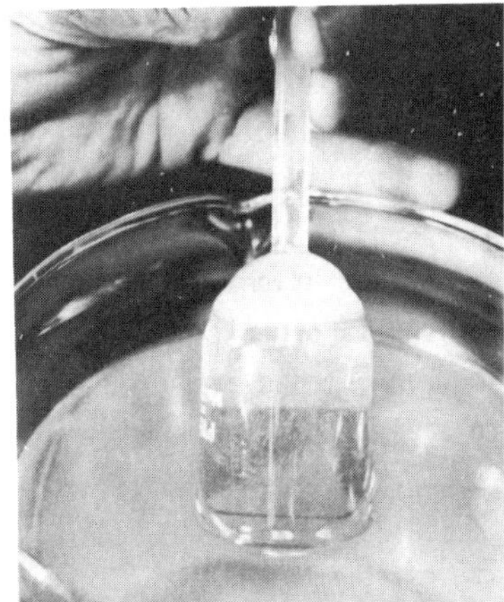

Figure 3. Inverted Buchner funnel with vacuum.

Figure 4 shows a deposited felt on the filter.

Figure 4. Felt on fritted filter. In order to bond the fibers together, a water-soluble resin solution of hair spray was drawn thourgh the felt and it was allowed to dry. A felt prepared in this manner will be referred to as "bonded fibers," since the "coated" felts will be subjected to further coating procedures.

A coating was added by drawing an
epoxy resin solution with curing
agent through the felt. This
leaves the felt wet with solution. The solvent evaporates
and the epoxy cures. Coating extent can be controlled by solution concentration.
The types of organic coatings
which can be used are quite
extensive. For tensile strength,
strong resins such as polycarbonate could be applied to inorganic felts by solution and
subsequently heat treated to
form very strong hot-melt
bonds.[1] Elastomeric coatings
could be used where rubbery
linkage of fibers is needed.
A conductive felt could be
electroplated with a metal
coating. The potential for
such materials composites will
be the focus of future work.
Figure 5 shows various materials
which have been felted by the
technique described.

Figure 5. Various felted
materials.

Figure 6 shows hemispherical shell
geometries.

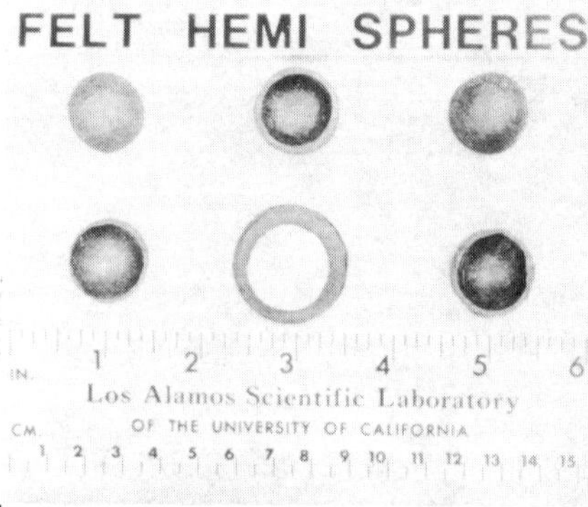

Figure 6. Hemisphrical felts.

3. PROPERTIES TESTING

The assembly shown in Figure 7 was
used to measure compressive properties of several felts.

Figure 7. Compression test
assembly.

The white felt in the center is
compressed through a recording
load cell against a fixed post.
Figure 8 shows a typical chart
recording of 5.0 gram increments

of load. The compressive data
will be presented later to help
show the effects of the felt
construction variables.
The reproducibility of felts
was determined by measuring the
volume of solids content for
several samples of stainless
steel and polyester fiber felts.

Figure 10 shows similar data for
polyester felts from fibers of
different L/D's. Percent solids
volume (and therefore density)
becomes lower as the L/D becomes
greater.

COMPRESSION TESTING

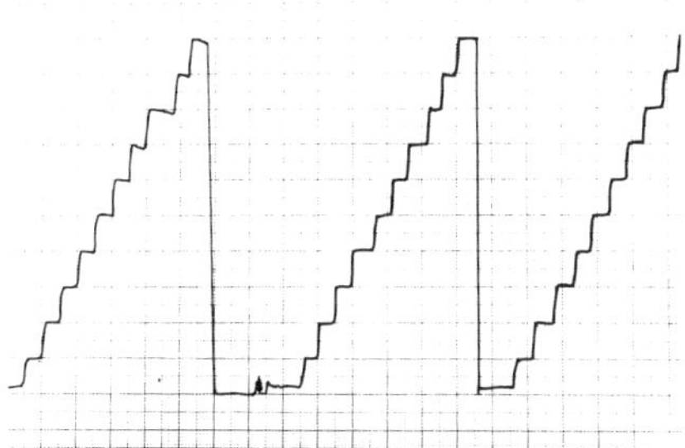

Figure 8.

Figure 9 shows the spread of data
for stainless steel felts made
from fibers 1000 μm long x 4 μm
diam, an L/D of 250.

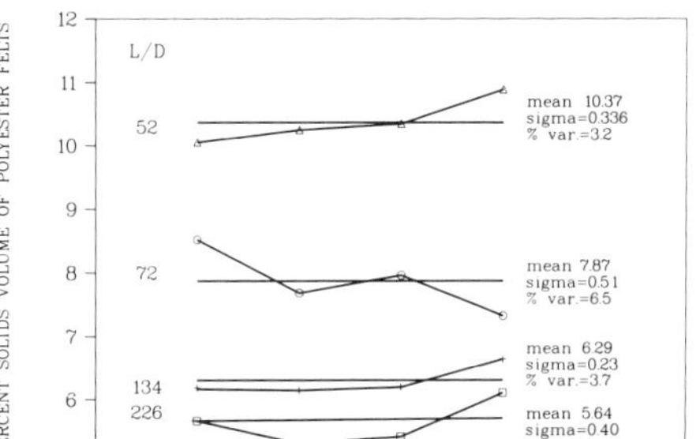

Figure 10.

Figure 11 shows the relationship
of L/D to density, expressed as
percent solids volume. This re-

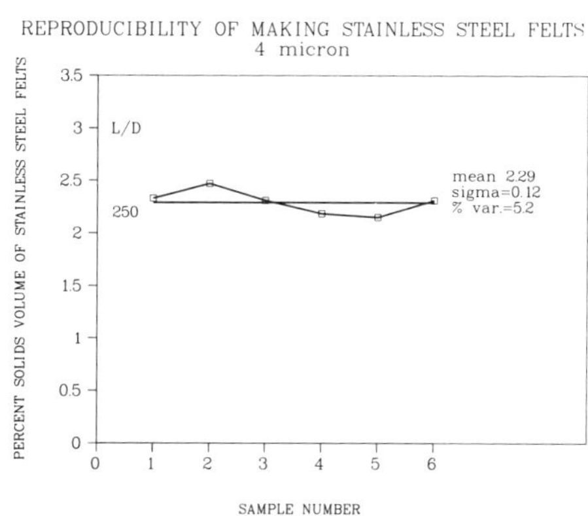

Figure 9.

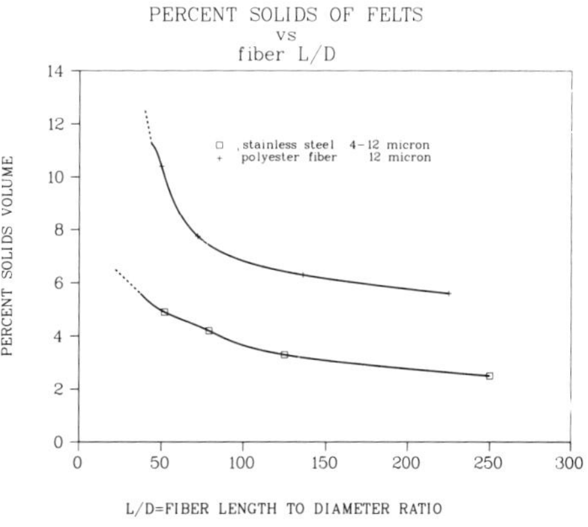

Figure 11.

sult is not unexpected, with
shorter fiber showing greater
density, insofar as shorter
fibers can be expected to
orient and pack more, with
fewer bridging interactions.
Figure 12 shows the expected
linear relationship of coating
added to coating solution concentration. The coating medium
for this work was Epon $^{(*)}$828,
an epicholorohydrin - bis-phenol-
A epoxy, cured with diethylene-
triamine.

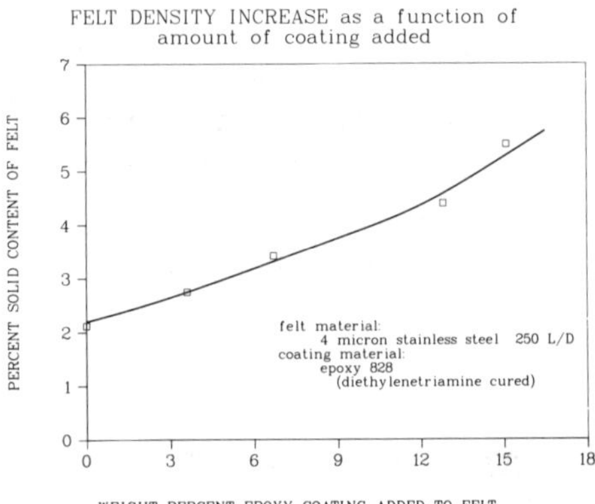

Figure 13.

The amount of epoxy pickup as a
function of L/D shown in Figure 14
is not what one would intuitively
expect. Smaller L/D ratios, or
shorter fibers with more surface
area per unit weight, might be
expected to accept more coating.

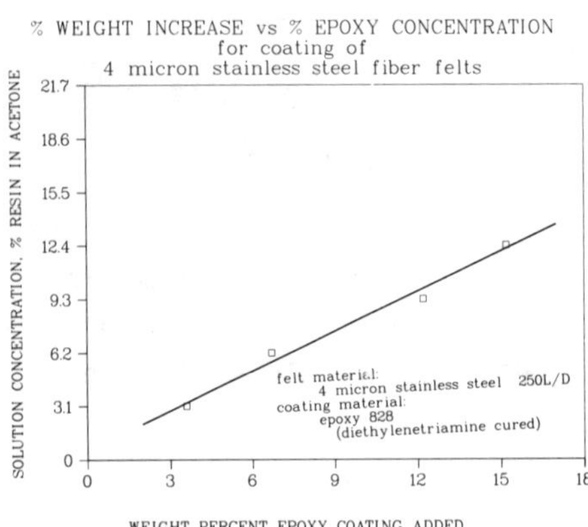

Figure 12.

The trend is toward higher percent solids content as the
weight percent of added epoxy
increases as shown in Figure 13.
The additional epoxy coating
simply fills available space
within the fiber matrix.

*"Epon" resins trademark of
Shell Chemical Co.

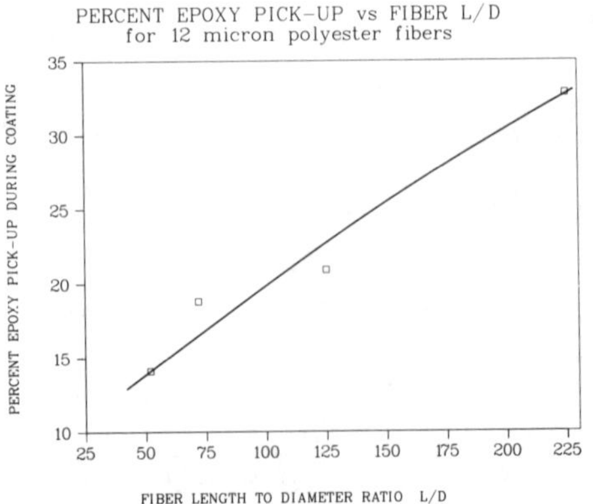

Figure 14.

However, the coating webs at fiber intersections, can be seen in Figs. 15 and 16. Longer fibers will have more intersections, which helps to explain the greater coating pickup data.

Uncoated polyester felts of various L/D's were compared in compression. Figure 17 shows that compressive strength increases for felts made from fibers of decreasing L/D.

Figure 15. Webbing at intersections.

Figure 16. Concentration at intersections.

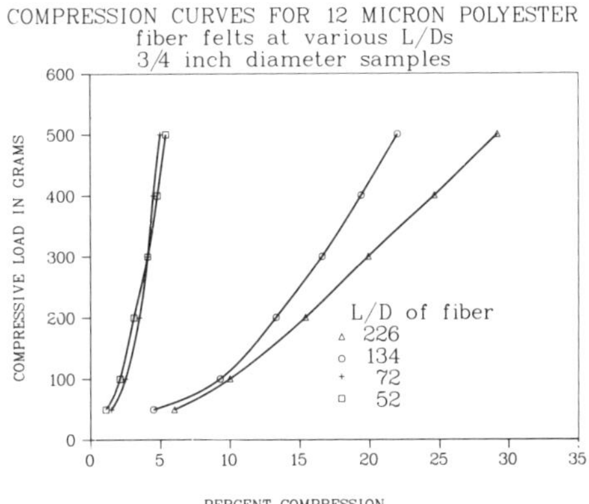

Figure 17.

In Figure 18, the effects of coating on two separate L/D fiber ratios are graphed. The coating ties ends and intersections together, causing fibers to flex rather than orient to the stress.

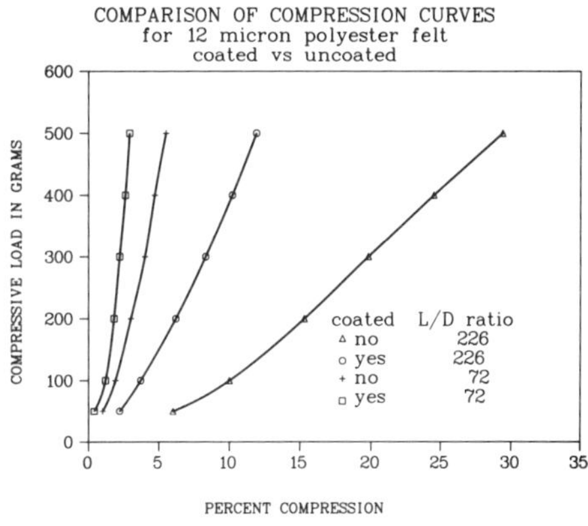

Figure 18.

Figure 19 shows plots for stainless steel felts with three
different epoxy coating contents.
By increasing the epoxy content
15%, compression for a given
load is reduced at least 50%,
yet total solids percent only
increase by 3.

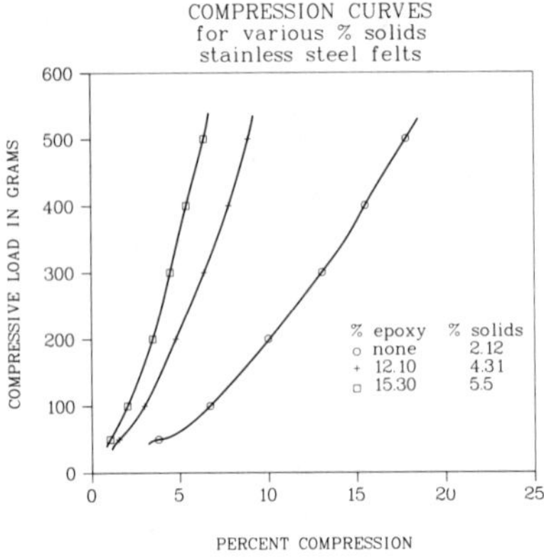

COMPRESSION CURVES
for various % solids
stainless steel felts

Figure 19.

By keeping the compressive load
constant and varying fiber L/D,
the curves in Figure 20 were
generated. The solid line was
drawn smoothly through the data.
However, the dashed line might
be a real condition. As fiber
length increases, there may be
a point at which a beam effect
is observed. Once a rod buckles
in the center its flexual properties abruptly become nonlinear, which may be what is
observed in the dashed line.

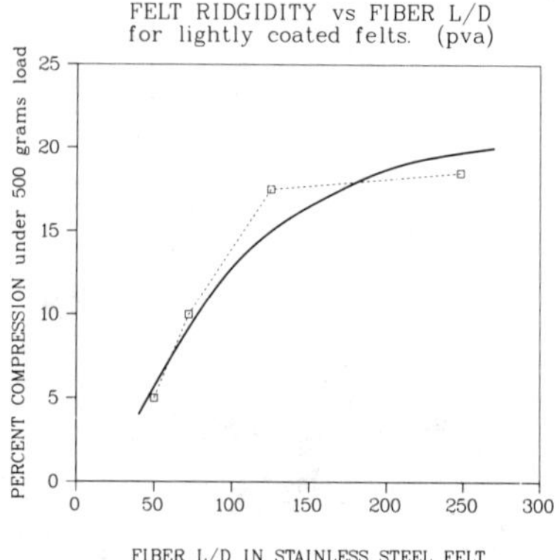

FELT RIDGIDITY vs FIBER L/D
for lightly coated felts. (pva)

Figure 20.

Figure 21 shows percent compression as a function of percent
solids. Increasing the solids
content of a felt lowers the
compression distortion under a
given load.

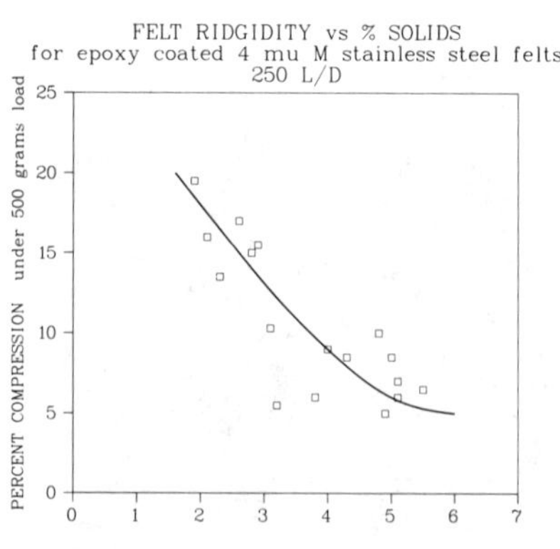

FELT RIDGIDITY vs % SOLIDS
for epoxy coated 4 mu M stainless steel felts
250 L/D

Figure 21.

4. TAILORING OF PROPERTIES

Fiber felting offers a method for tailoring the end properties needed in a low density structure. Density, for example, can be tailored in one or more of three ways. Selection of the fiber materials is one way to adjust density. The second way is to select an L/D to achieve a desired density. The smaller the L/D the higher the composite density (see Figure 10). The third way to adjust density is to vary the amount of coating added to the felt, as shown in Figure 13. Adjustments to other properties of a structural felt can also be made by changes in fiber material, L/D of fibers, and quantity and nature of the coating. There are trade-offs in designing properties. For example, for a given fiber material, in order to increase compressive strength it will be necessary to increase density. However, with the broad range of fibers commercially available, it might be possible to achieve the desired compressive strength at low density by selecting a higher-modulus fiber.

5. CONCLUSIONS

Low density structural composites with custom-tailored end properties can be made by felting and bonding fibers. Such structures can take advantage of the materials properties of the fiber, as well as the three-dimensional and low density properties of the matrix. Tailoring of properties is effected by varying the fiber material, L/D of fibers, and coating of the fibers.

The felting technique is applicable for making felts up to 12.7-mm thick reproducibly. Such structures are useful in compressive cushion applications, where the flexing of individual zero-creep fibers limits stress relaxation and compression set. The low density feature is attractive where weight or mass must be limited.

6. REFERENCES

1. "Polycarbonate Resin as a Hot Melt Adhesive," Stephen E. Newfield and E. P. Ehart, SPE 24, Dec. 1968.

BIOGRAPHIES

Stephen E. Newfield is the Section Leader of the Plastics Section of the Materials Technology Group of the Los Alamos Scientific Laboratory. He holds a B.S. in Chemistry from New Mexico State University and an M.S. in Electrical Engineering Computer Science from the University of New Mexico. He is involved in elastomers and polymers development, as well as composites and injection moldable ceramics and metals. He directs a group which provides service and development in most areas of elastomers, composites, adhesives, coatings, and conventional plastics.

John V. Milewski has over 31 years of industrial experience in the field of high strength and high

temperature materials and composites engineering. He has specialized in plastics and fiber reinforced composites, whiskers, and
other higher strength-fiber development and manufacturing, and
recently has become involved in
composite formulations and packing
concepts for filler-fiber combinations. Dr. Milewski has given
numerous technical presentations
throughout the world on the efficient use of high strength fibers.
He is the coeditor of the "Handbook
of Fillers and Reinforcements" for
Plastics and has over 20 technical
publications in this field, and
has been granted 18 patents.
Dr. Milewski is a licensed Professional Engineer in the state of
New Jersey, and is a member of
several professional societies.

27th National SAMPE Symposium
April 28-30, 1981

A MULTI-PURPOSE THERMOPLASTIC POLYIMIDE

Anne K. St. Clair and Terry L. St. Clair
NASA-Langley Research Center
Hampton, VA 23665

ABSTRACT

A linear thermoplastic polyimide, LARC-TPI, has been characterized and developed for a variety of high-temperature applications.
In its fully imidized form, this new material can be used as an
adhesive for bonding metals such as titanium, aluminum, copper,
brass, and stainless steel. LARC-TPI is being evaluated as a
thermoplastic for bonding large pieces of polyimide film to produce flexible, 100% void-free laminates for flexible circuit
applications. The further development of LARC-TPI as a potential molding powder, composite matrix resin, high-temperature
film and fiber will also be discussed.

1. INTRODUCTION

Linear aromatic polyimides are presently under consideration for applications on future aircraft and spacecraft. As a class, linear polyimides are attractive to the aerospace
industry because of their toughness
and flexibility, remarkable thermal
and thermo-oxidative stability, radiation and solvent resistance, light-
ness of weight, and excellent mechanical and electrical characteristics over a wide temperature range.
However, the processing of these materials is tedious compared to other
engineering plastics. A soluble
polyamic acid precursor must first
be dissolved in a high-boiling solvent, and later converted to the

polyimide by heating to an extremely
high temperature which causes evolution of water and residual solvent.
The preparation of void-free moldings, large-area laminates or adhesive joints with such a material becomes exceedingly difficult.

The use of meta-substituted aromatic
diamines in the preparation of linear polyimides has been a major advancement toward improving the processability of these polymers and
lowering the glass transition temperature.[1] A further advancement
has been the synthesis of these polyimides in a nontoxic, low-boiling
ether solvent.[2] The adhesive lap-
shear strength of a meta-oriented

polyimide was improved 2 to 5-fold by synthesis in bis(2-methoxyethyl)-ether(diglyme).[3,4] More recently, this material was used as an adhesive for bonding ultrathin polyimide film in the proposed NASA-Solar Sail Program.[5] The small-area sail joints (0.64 cm overlap) allowed an easy escape for volatiles during processing. However, the need still exists for a high-temperature adhesive that can bond larger areas without the evolution of volatiles and the entrapment of voids.

LARC-TPI is a linear thermoplastic polyimide which is presently being developed at NASA-Langley Research Center as an adhesive for the large-area bonding of metals and films. This material is a linear polymer which contains a meta-substituted diamine and is prepared in an ether solvent. Unlike conventional polyimides, LARC-TPI can be imidized and freed of volatiles at a reasonably low temperature (230°C) and then be processed as a thermoplastic. In an effort to exemplify the multiple uses of this material, this paper reports the preliminary development of LARC-TPI as an adhesive, molding powder, composite matrix resin, high-temperature film and fiber.

2. EXPERIMENTAL

2.1 <u>Preparation of LARC-TPI</u>

The monomers used in the preparation of LARC-TPI were 3,3',4,4'-benzo-phenone tetracarboxylic dianhydride (BTDA) and 3,3'-diaminobenzophenone (3,3'-DABP). The BTDA was used as received from Gulf Chemicals[*], m.p. 215°C, after drying overnight in a vacuum oven at 140°C. The 3,3'-DABP was obtained from Ash Stevens, m.p. 150°C, and used as received.

The polymerization was performed in a closed vessel at 20°-25°C at a concentration of 15% solids by weight in reagent grade diglyme. The 3,3'-DABP (21.2 grams) was slurried in 302.6 grams of diglyme by mechanical stirring. To this mixture, BTDA (32.2 grams) was added and stirring was continued for at least four hours to insure the complete reaction of monomers. If the polyamic acid tended to precipitate after an appreciable viscosity had been reached, small amounts of ethanol were added to redissolve the polymer.

The resin thus prepared was used directly for coating adhesive substrates, as a composite matrix resin in the prepregging of graphite fibers, and for spinning fibers. The resin could be cast as a thin film onto glass and heated to 300°C for one hour to prepare a flexible, self-supporting high-temperature film. The LARC-TPI resin was also precipitated as a solid by pouring into rapidly stirring water or methanol; and a molding powder was made by drying this solid and heating to complete imidization.

2.2 Characterization

The inherent viscosity of the polyamic acid solution was obtained at a concentration of 0.5% in dimethylacetamide at 35°C. Thermomechanical properties of LARC-TPI were obtained by torsional braid analysis (TBA). Glass braids were coated with a 7% polymer solution and preheated to 220°C in air before obtaining TBA spectra. Glass transition temperatures (T_gs) of various films, composites, and moldings were measured by thermomechanical analysis (TMA) on a DuPont 943 Analyzer in static air at a temperature program of 5°C/min. Thermograms of films were obtained by thermogravimetric analysis (TGA) by heating at a rate of 2.5°C/min. in static air. Melt-flow properties of LARC-TPI molding powder were observed by use of a parallel plate rheometer accessory with the DuPont 943 Thermomechanical Analyzer in static air at 5°C/min.

2.3 Adhesive Bonding

A LARC-TPI adhesive scrim cloth was prepared for bonding 2.54 cm wide 6Al-4V titanium adherends cleaned by the Pasa Jell 107 acid surface treatment. The resin was coated onto a 112 E-glass (A-1100 finish) cloth and heated for one hour at 100°C and 1/2 hr. at 200°C after each coat. Enough coats were applied to produce a final cloth thickness of 0.25-0.31 mm. Single lap-shear specimens were prepared by sandwiching the cloth between primed adherends using a 1.27 cm overlap. The specimens were bonded as follows:

(1) RT to 325°C at 7°C/min., apply 1.38 MPa (200 psi) at 280°C,
(2) Hold 5 min. at 325°C, and
(3) Cool under pressure.

Lap-shear tests were performed on a TT-6 Instron Testing Machine according to ASTM D-1002.

2.4 Film Laminating Process

LARC-TPI was used as an adhesive to bond thin polyimide Kapton® H-film measuring 0.025-0.076 mm in thickness. The Kapton film was cleaned with ethanol, primed with a thin coat of LARC-TPI adhesive resin, and gradually heated in air to 220°C for one hour to remove excess solvent and complete imidization. The primed film was then thermoplastically bonded to another primed or unprimed sheet of Kapton film as follows:

(1) Place sandwich between platens preheated to 232°C,
(2) Apply .345-2.07 MPa (50-300 psi),
(3) Heat to 343°C and hold 5 min., and
(4) Cool under pressure.

Thin film laminates were also prepared by placing a self-supported adhesive film of LARC-TPI staged to 220°C between two sheets of Kapton film and proceeding according to the above process. Metal-containing laminates were made by placing conductive metal sheets or foils between primed sheets of Kapton and bonding as described above.

2.5 Preparation of Molded Discs

Molded discs were prepared from LARC-TPI powder which had been imidized at 220°C. Approximately 20 grams of molding powder was placed in a 5.72 cm diameter steel mold and cured by the following cycle:

(1) Heat to 200°C without top on mold,

(2) Insert top and apply 6.89 MPA (1000 psi),

(3) Heat to 260°C, and

(4) Gradually cool under pressure.

Fracture toughness testing of the 0.16 cm thick discs was performed at the Naval Research Laboratory, Washington, D.C. Round compact tension test specimens were prepared from the LARC-TPI discs and tested for G_{I_c} (the opening-mode strain energy release rate).[6] Flexural strength and Izod impact strength were measured according to ASTM D-790 and D-256, respectively.[7]

2.6 Fabrication of Composites

Diluted LARC-TPI resin (5% solids) was applied as an initial coat to drum-wound Celion® 6000 graphite fibers in an effort to ensure good wetting of the fibers. The regular resin at 15% solids content was applied to the fibers in three subsequent coats. The prepreg was air-dried on the rotary drum, cut into 7.6 cm by 17.8 cm segments and stacked into a 12-ply unidirectional preform. The preformed billet was heated under full vacuum from room temperature to 204°C and held for four hours. Upon cooling to room temperature under vacuum, the consolidated billet was placed in a matched-metal mold and laminated in a hydraulic press according to the following cycle:

(1) RT to 335°C at 5°C/min.,

(2) Apply 4.14 MPa (600 psi) at 200°C,

(3) Hold 1/2 hour at 335°C, and

(4) Cool under pressure.

Mechanical tests on the composite were performed on an Instron Testing Machine TT-6. Flexural strength and modulus was determined according to ASTM D-790. Short beam shear strength was determined using a span-to-thickness ratio of 4.

2.7 Spinning of Fibers

Fibers were spun from a 15% solids LARC-TPI polyamic acid solution into a methanol coagulation bath at room temperature. Approximately 0.69 MPa

(100 psi) nitrogen was used to pressurize the solution through a teflon
spinneret with an orifice measuring
.15 mm in diameter. Yellow fibers
were spun onto a 7.62 cm diameter
teflon roller rotating at 25 rpm
which was located in the coagulation
bath. The fibers were imidized by
soaking for 24 hours in a 50/50 py-
ridine/acetic anhydride solution and
were then dried either in air or in
a forced air oven at 150°C. Single-
fiber tensile properties were measured at room temperature.

3. RESULTS AND DISCUSSION

3.1 Resin Chemistry and Properties

LARC-TPI was synthesized according
to the reaction scheme shown in Figure 1. Reaction of the monomers in
the ether solvent diglyme yielded a
highly viscous polyamic acid which
had an inherent viscosity of 0.70
dl/g. The thermal imidization of
this polyamic acid resin resulted in
a linear high molecular weight polyimide which could be processed as a
thermoplastic due to its unique molecular structure.

According to the TBA spectrum in Figure 2, LARC-TPI exhibited a relatively low T_g of 229°C after thermal imidization at 220°C in air. Upon further heating at 3°C/min. to 350°C,
the polymer had a T_g of 266°C as demonstrated by the TBA spectrum in
Figure 2 on cooling from 350°C. LARC-
TPI can be processed after imidizing

at a reasonably low temperature and
then heated further to increase its
final use temperature.

LARC-TPI is a very thermo-oxidative-
ly stable polymer as evidenced by
the thermograms shown in Figure 3.
The dynamic TGA curve obtained on
film heated to 300°C in air showed
essentially no weight loss prior to
400°C. Weight loss of the material
did not exceed 2-3% after isothermal
aging at 300°C for 550 hours.

Because it is processable as a thermoplastic after imdization and thermally stable at high temperatures,
LARC-TPI shows potential for a variety of applications (Table I).

3.2 Structural Adhesive Properties

LARC-TPI is currently a candidate
for the large-area bonding of an experimental graphite composite wing
panel in the NASA-Supersonic Cruise
Research (SCR) Program, Figure 4.
The materials in this program are being screened as structural adhesives
on both titanium and polyimide/
graphite composite adherends. Preliminary bonding results on LARC-TPI
were generated both in-house and in
studies at Boeing Aerospace Company.[8] LARC-TPI exhibited excellent
lap shear strengths on titanium at
room temperature; and the elevated
temperature strengths were above minimum strengths set by the SCR
program. The increase in bond
strength after aging at 232°C was

indicative of a post-cure effect
which is characteristic of LARC-TPI
adhesive. Although lap-shear specimens were not post-cured after bonding in the present study, a moderate
post-cure would advantageously increase the T_g and would therefore
increase the use temperature of the
adhesive.

The ability of LARC-TPI as a structural adhesive in bonding polyimide/
graphite and polyimide/glass composites has been demonstrated and reported elsewhere.[3] Composite and
composite-to-titanium lap shear
strengths averaged approximately
20.7 MPa (3000 psi), but failures
occurred in the composites rather
than in the adhesive.

3.3 Film Laminating

A need exists in the aerospace industry for reliable flexible electrical
circuitry that can withstand extreme
temperature variations and retain
flexibility. Problems to date have
been due partially to the presence
of voids in film laminates caused
by volatiles generated by the adhesive and/or the inherent rigidity of
some adhesives. Because it is both
flexible and imidized prior to bonding, LARC-TPI shows much potential
as a high-temperature adhesive for
laminating large ares of polyimide
film.

A film-laminating process has been
developed whereby films primed with
a thin coat of LARC-TPI adhesive are
bonded together using temperature
and pressure (Figure 5). As an alternate process, LARC-TPI polyamic
acid adhesive film may be imidized
by heating prior to being sandwiched
between polyimide film. When using
either process to produce flexible
circuits, a conductive metal may be
interposed between layers of the
polyimide film. Metal-containing
laminates have been made using aluminum, brass, copper and stainless
steel sheets or foils.

Large-area polyimide film laminates
were made which varied in size from
77.4 cm^2 to 645 cm^2. A laminate was
also prepared using 8 plies of polyimide film. Larger areas and thicker laminates can be made using the
process in Figure 5 as long as the
entire laminating area receives an
even distribution of temperature and
pressure.

In this study, over forty Kapton
film laminates were prepared, all of
which were clear yellow, flexible,
and 100% void-free. Conventional T-
peel (ASTM D-1876) and 180°-peel
(ASTM D-903) testing of the film
laminates was attempted, but no data
was generated on the adhesive. Failures occurred due to tearing of the
Kapton film indicating that the
strength of the LARC-TPI exceeds
that of the Kapton film.

In addition to peel strength, the
success of a flexible film laminate

for circuit applications is determined by its ability to withstand a dip in hot solder. Most laminates will blister (due to the adhesive) after just a few seconds. LARC-TPI laminates, however, have withstood a full 10 second dip in 288°C solder without blistering.[7]

3.4 Moldings

Molded discs were prepared from LARC-TPI molding powder by heating to 260°C under 6.89 MPa (1000 psi) pressure. Properties of the moldings are listed in Table II. Discs measuring 5.72 cm in diameter and 0.16 cm in thickness had a density of approximately 1.40 g/cm^3. Discs were flexible but low in T_g when heated only to 260°C during molding with no post cure. Higher T_gs in the range of 275°C have been achieved for moldings pressed at higher temperatures.[7]

LARC-TPI moldings used for measuring flexural strength were in the form of 12.7 cm x 1.27 cm x 0.64 cm bars, and a span of 10.2 cm was employed.[7] The deflection to break for these specimens was twice that displayed by other known commercial polyimides.

3.5 Composites

Once imidized, conventional linear polyimides ordinarily do not have enough flow during molding to be used as composite matrix resins. Because of its low T_g and unique meta-orien-ted molecular structure, however, LARC-TPI has the necessary flow properties and has demonstrated the ability to form composites with graphite fibers. The cure cycle used to mold B-staged LARC-TPI/Celion 6000 billets into composites is displayed in Figure 6. The 12-ply laminates press-molded by this cycle had a final thickness of 1.73 mm. Cross-sections of the laminates were examined by an optical microscope and found to have no lines of demarcation between the resin and fibers.

Properties of the LARC-TPI/Celion 6000 composites are listed in Table III. An average of two specimens each was used to obtain flexural strength and modulus with a ±5% variation in the data. The short beam shear strength represents an average strength of six specimens with a ±10% data variation. The low value for flexural modulus may be attributed to the high resin content (40.4% by weight) of the composite.

3.6 Films and Fibers

Linear polyimides are a prime choice for films and coatings in aerospace applications due not only to their thermal stability, but because of their toughness and flexibility.[12] Tough and flexible films were easily made from LARC-TPI amic acid solution cast onto glass plates. Properties shown in Table IV are characteristic of films air-cured 1 hour each at 100°,200°, and 300°C having a final

thickness of approximately 0.025 mm
(1 mil). Tensile properties (Table
IV) reported elsewhere[10] were reasonable in view of the fact that they
were recorded on unoriented films
having had no special thermal treatment following the cure cited above.

The direct wet-spinning of fibers
from a LARC-TPI polyamic acid solution has been demonstrated by the
procedure outlined in Figure 7. After
soaking in a pyridine/acetic anhydride bath for 24 hours, the fibers
were found by infrared spectroscopy
to be completely imidized. Cross-
sections of the fibers were examined
by optical microscopy and found to be
rounded in shape and to have solid
cores. Although the strengths obtained for these fibers are low compared to commercial high-strength fibers, preliminary results show
strengths (tenacity = 1.33-1.46 g/d)
to be comparable with those of many
common textile fibers.

4. CONCLUSIONS

LARC-TPI is a linear aromatic polyimide which shows potential as a
thermoplastic for a variety of high-
temperature applications. It is a
high molecular weight polymer which
is flexible, tough, and thermooxida-
tively stable at elevated temperatures (300°C). Because of its unique
meta-oriented molecular structure,
LARC-TPI may be processed as a thermoplastic after imidization and the
removal of solvent. After process-

ing, it can be further heated to increase the final use temperature.

LARC-TPI is being evaluated primarily
as an adhesive for bonding metals,
composites, and films. As a structural adhesive, this new material
has demonstrated adequate adhesive strengths for certain advanced
aircraft applications. LARC-TPI is
also being evaluated as an adhesive
in the large-area bonding of polyimide film to produce flexible, 100%
void-free laminates for flexible
circuit applications. LARC-TPI film
laminates have withstood a dip in
288°C solder and are totally resistant to peel forces.

In addition to being used as an adhesive, this resin can be precipitated to form a molding powder and
thermoplastically molded at 260°C.
Graphite composites have been prepared using LARC-TPI as a matrix
resin with promising results. LARC-
TPI also shows potential as a material for making high-temperature
films and for spinning high-temperature fibers.

5. ACKNOWLEDGEMENTS

The authors are indebted to Mr.
Philip Robinson, Mr. Robert Ely, and
Mr. Spencer Inge of NASA-Langley Re-
search Center, Hampton, VA for their
excellent technical assistance. We
also offer our appreciation to Mr.
James Tyeryar, NASA-Langley Research
Center, Hampton, VA for his

assistance in the spinning of fibers.

6. REFERENCES

1. V. L. Bell, U.S. Patent 4,094,
    862, June 13, 1978.
2. D. J. Progar, V. L. Bell, and
    T. L. St. Clair, U.S. Patent
    4,065,345, December 27, 1977.
3. T. L. St. Clair and D. J. Progar,
    ACS Polymer Preprints, 16(1),
    538 (1975).
4. D. J. Progar and T. L. St. Clair,
    Preprints, 7th National SAMPE
    Technical Conference,7,53 (1975).
5. A. K. St. Clair, W. S. Slemp,
    and T. L. St. Clair, Adhesives
    Age, Vol. 22(1), 35 (1979).
6. R. Y. Ting and R. L. Cottington,
    ACS Organic Coatings and Plastic
    Preprints, 41, 405 (1979).
7. R. T. Traskos, Special Projects
    Group, Lurie R&D Center, Rogers
    Corporation, Rogers, Connecti-
    cut 06263. Private communication.
8. C. L. Hendricks and S. G. Hill,
    Materials & Process Group,
    Boeing Aerospace Company,
    Seattle, Washington 98124.
    Private communication.
9. T. L. St. Clair, A. K. St. Clair,
    and E. N. Smith, Structure-
    Solubility Relationships in
    Polymers, Chpt. 15, pp. 199-215,
    Academic Press, Inc., New York
    (1977).
10. V. L. Bell, B. L. Stump, and
     H. Gager, J. Polym. Sci. Polym.
     Chem. Ed., 14, 2275 (1976).
11. L. T. Taylor, V. C. Carver, and
     A. K. St. Clair, ACS Organic
     Coatings and Plastic Preprints,
     40(1), 49 (1979).
12. A. K. St. Clair, Shell Polymers,
     Vol. 4(2), 50 (1980).

7. BIOGRAPHY

Anne St. Clair received a B.A. degree in Chemistry from Queens College
in 1969 and an M.S. in Chemistry
from Virginia Polytechnic Institute
and State University in 1972. From
1972 to September 1977, she was employed as a Research Associate at
NASA-Langley Research Center. She
recently joined the Polymer Group
at Langley as an Aerospace Technologist in the field of high temperature polymers.

Terry St. Clair received his B.S.
degree in Chemistry from Roanoke
College in 1965 and his Ph.D. in
Organic Chemistry from Virginia
Polytechnic Institute and State
University in 1972. From 1972
to 1975, he served as a Post-
doctoral research associate at
NASA-Langley. In 1975, he was employed by NASA-Langley as an Aerospace Technologist where he has
been responsible for research
and development of high temperature adhesives and composites.

TABLE I. APPLICATIONS FOR LARC-TPI

(1) Structural adhesive for metals & composites

(2) Adhesive for laminating polyimide film

(3) Molding Powder

(4) Composite Matrix Resin

(5) High-Temperature Film

(6) High-Temperature Fiber

TABLE II. PROPERTIES OF LARC-TPI MOLDINGS

Density	1.40 g/cm^3
	1.37 g/cm^{3}*
Adhesive fracture energy	662 J/m^2
Beginning flow temperature of molding powder	225°C
Maximum flow temperature of molding powder	255°C
T_g of discs molded at 260°C	225°C
(molded at >260°C)	275°C*
Flexural strength	159 MPa (23 ksi)*
Deflection to break	1.10 cm (0.435 in)*
Notched Izod Impact Strength (.64 cm x 1.27 cm bar)	21.4 N-m/m (0.4 ft.lb/in)*

*Data provided by R. T. Traskos, Rogers Corporation, Rogers, Connecticut

TABLE III. PROPERTIES OF LARC-TPI/CELION 6000 COMPOSITES

Resin Content	40.4%
Density	1.55 g/cm^3
T_g (no post cure)	244°C
Flexural Strength	1600 MPa (232 ksi)
Flexural Modulus	100 GPa (14.5 x 10^3 ksi)
Short Beam Shear Strength at room temperature	69 MPa (10.0 ksi)
Short Beam Shear Strength at 177°C	43 MPa (6.3 ksi)

TABLE IV. LARC-TPI HIGH-TEMPERATURE FILM PROPERTIES

T_g (film cured at 300°C)	259°C
Solubility in organic solvents[9]	Insoluble
Tensile Strength[10]	136 MPa (19.7 ksi)
Yield Strength (@2%)[10]	119 MPa (17.3 ksi)
Elongation[10]	4.8%
Tensile Modulus[10]	3720 MPa (540 ksi)
Polymer decomposition temperature	570°C
Volume Resistivity[11]	2.0 x 10^{16} ohms-cm

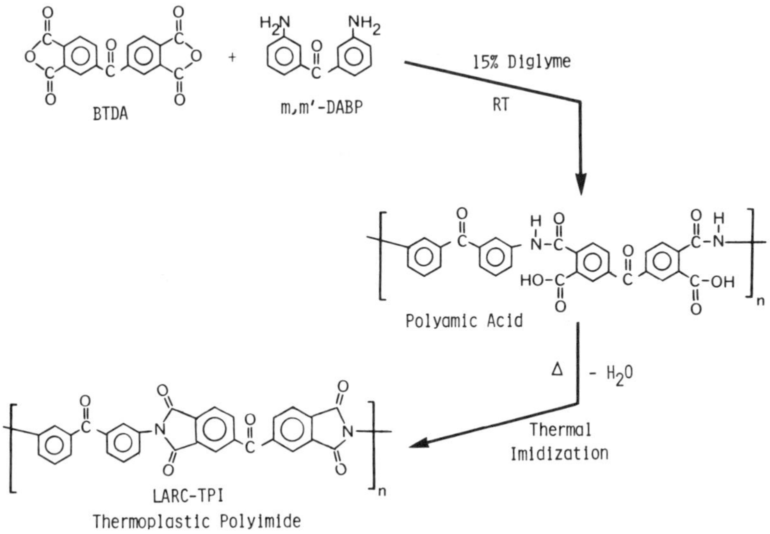

Figure 1. LARC-TPI Chemistry.

175

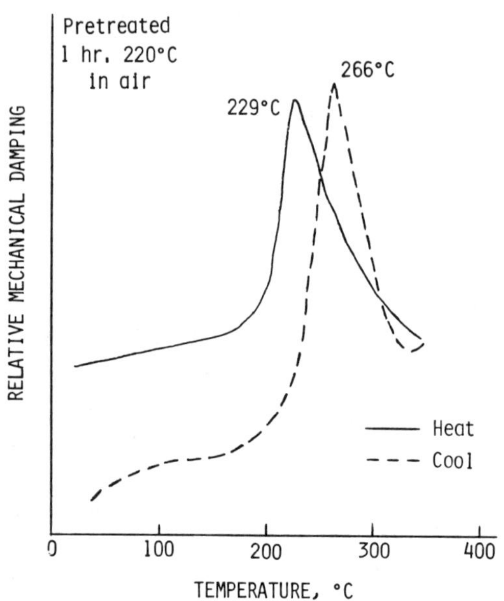

Figure 2.   Torsional Braid Analysis of LARC-TPI.

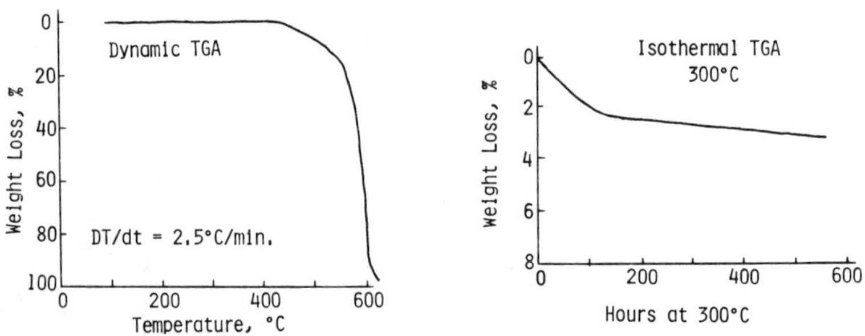

Figure 3.   Thermograms of LARC-TPI film.

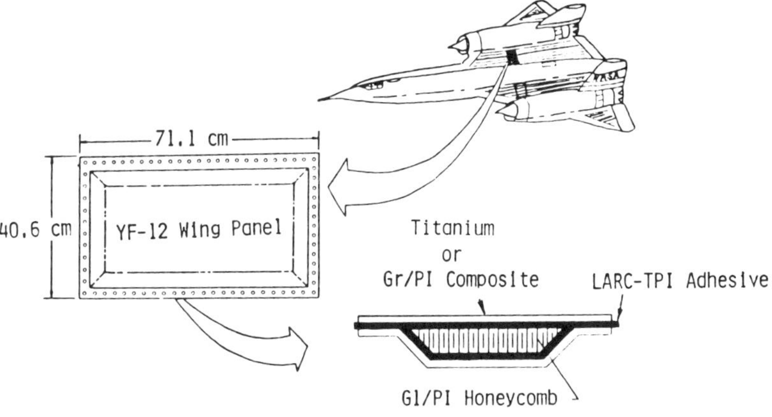

DATA FROM	Ti/Ti LAP SHEAR STRENGTH, MPa (psi)		Ti/Ti LAP SHEAR STRENGTH AFTER 3000 hrs at 232°C MPa (psi)
	RT	232°C	232°C
BOEING AEROSPACE	36.5 (5300)	13.1 (1900)	20.7 (3000)
NASA-LANGLEY	41.4 (6000)	17.9 (2600)	—

Figure 4. Structural adhesive application of LARC-TPI.

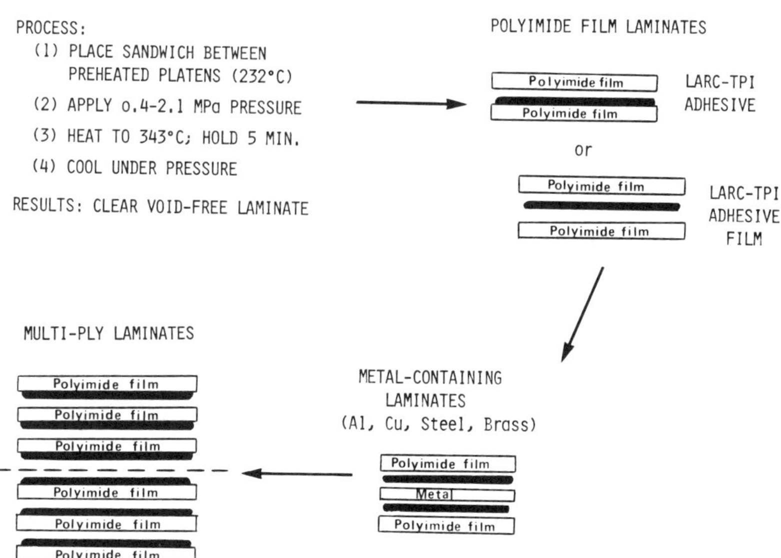

Figure 5. Process for preparing high-temperature film laminates.

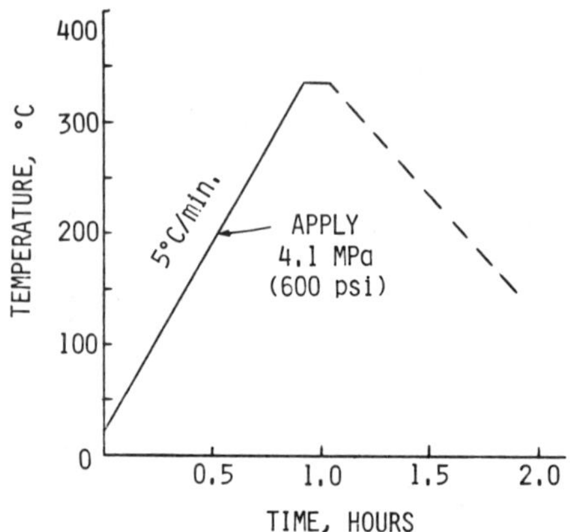

Figure 6. Cure cycle for LARC-TPI/Celion composites.

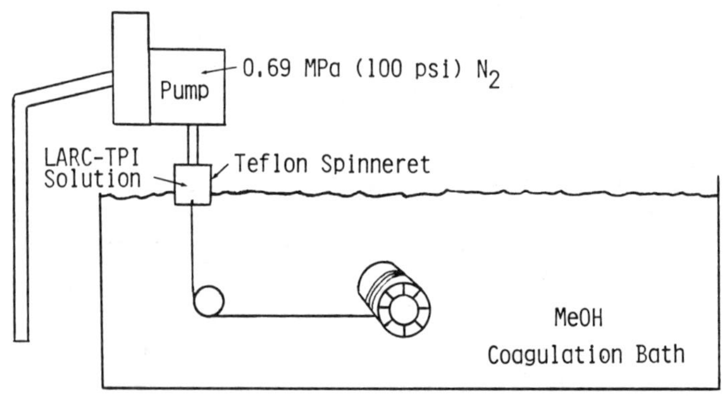

PROCEDURE: (1) Polymer solution pressurized through a 0.15 mm
(6 mil) spinneret onto a 7.62 cm roller.

(2) Fibers soaked for 24 hrs in 50/50 pyridine/
acetic anhydride.

(3) Fibers dried in air or oven @ 150°C.

PRELIMINARY
RESULTS: Fiber Denier 23-38

Fiber Tenacity 1.33-1.46 g/d

Figure 7. Spinning of LARC-TPI fibers.

ELECTRICAL CHARACTERISTICS OF
CARBON/GRAPHITE FIBER COMPOSITES
John Delmonte
Consultant, Delsen Testing Laboratory

Abstract

The electrical characteristics of carbon/graphite fiber composites are examined. These advance composites have been utilized for their structural qualities, particularly in aircraft applications. They possess interesting qualities as semi-conductors and volume resistivities are reported for some panels after temperature and humidity exposure. Applications are reviewed, particularly where electrical resistivity is important. Influence of ionic contaminants on the fibers oxidation resistance at high temperatures is noted. Application problems related to lightning strikes, carbon fiber release, electromagnetic shielding, etc., are reviewed. Attention is also directed to the stability of antenna configurations utilizing carbon/graphite fiber composites with very low coefficients of thermal expansion.

* * *

Carbon/graphite fiber composites have been the subject of numerous publications in recent years because of their good structural qualities. Aerospace applications have received the focus of this attention, accompanied by noteworthy uses in the sporting goods industry and automotive industry. Except for certain applications noted in this paper, carbon/graphite fiber composites in the electrical industry have not been significant. It is generally recognized that carbon and graphite fibers are semi-conductors of electricity and hence are not good candidates as insulative materials. In fact, where carbon-fiber fabrics were introduced commercially in the middle 1960's, there were attempts to utilize the materials as conductors in electrical heating panels, replacing in part metallic electrical resistance wires (i.e., "Nichrome" and others). These panels did not fare too well, and interest diminished.

Attention has been redirected

more recently towards the use of carbon/graphite fabrics and yarns for electrical heating (see Reference 1), though these instances are limited because of basic problems. Two of the more serious are: (1) carbon/graphite fibers do not possess good high temperature thermal stability when there are traces of certain ionic contaminants such as sodium ions on their surface, and (2) establishment of good electrical contact with connective wires or metallic components is not readily achieved with conventional devices. McMahon and Gibbs have reported on the influences of ionic contaminants on the temperature stability of carbon and graphite fibers in an oxidative atmosphere (2, 3). Some of their data appear in the accompanying Table 1, which reflects weight loss as a function of time and temperature. "T-300," which generally appears as a woven fabric, is now available in a low Na-ion grade.

This data is relevant in that electrical resistive heating elements must reach relatively high temperatures to transfer heat rapidly and efficiently through the composite to the exterior surfaces requiring elevated temperatures. Under sustained heating it is likely that the carbon/graphite fibers will degrade and lose weight and effectiveness, though less rapidly when encased in a resin matrix. It should be noted however, that there are recent grades of carbon/graphite fibers which are processed for low ionic contamination – and these should be selected for electrical resistive heating, together with a resin matrix that does not contribute traces of ionic contamination which accelerate degradation at very high temperatures.

In spite of these short-comings,

Table 1

TYPICAL OXIDATIVE STABILITY OF SELECTED CARBON FIBERS
(Data by Gibbs and by McMahon)

Carbon/Graphite Fiber	After 700 Hours In Air at 600° F.	After 3 Hours In Air at 932° F.
High Modulus Fibers		
GY-70 (Celanese)	---	0.1%
HMS (Hercules)	.08%	3.3%
VS-Pitch (U.C.C.)	---	0.6%
High Strength Fibers		
"Celion" 6000 (Celanese)	3.5 – 4.5%	---
HT-S (Hercules)	1.2%	5 – 30%
"Fortafil" 3 and 5 (Great Lakes)	< 1.0%	---
"Thornel" 300 (U.C.C.)	58%	---

there are noteworthy electrical
characteristics which offer potential when examining carbon/graphite
fiber composite applications. Some
of these qualities have been
examined by the writer in a recent
book on carbon/graphite fiber composites (Reference 4). Intrinsic
qualities of carbon/graphite composites which set them apart from
other advanced composites are as
follows:

1. Their specific stiffness (ratio
    of modulus of elasticity to
    density) is superior to other
    advanced composites

2. Fatigue resistance is superior

3. Coefficient of thermal expan-
    sion is very low with uni-
    directional fiber tapes - a
    zero C.T.E. may be achieved.

There are processing advantages,
which though not unique to carbon/
graphite composites, are nevertheless used to prepare composites.

a. Carbon/graphite fibers combined
    with glass fibers, polyaramids,
    boron fibers, and others form
    versatile hybrids.

b. Complicated one-piece assemblies
    are prepared through combina-
    tions of prepregs and tapes of
    carbon/graphite fibers placed
    in special tooling, and sub-
    jected to heat (for cure) and
    pressure from expandable rubber
    mandrels.

Because electrical data on
carbon/graphite composites does not
usually appear, we have run basic
dielectric properties on panels of
graphite fibers (60% vol.) and
epoxy resin binders, which are
reported in Table 2. As would be
expected, the semi-conductive
properties of the panels of
"T-300"/epoxy composites are
evident in their low values of
volume resistivity, performed per
ASTM-C-611. Data shows little
change in volume resistivity values
after high humidity and also high
temperature exposure. Meaningful
measurements of dielectric constant, dissipation factor at 60 to
1 megacycle, arc resistance, and
dielectric strength were not possible on this particular semi-
conductive composite - due to its
high volume concentration of
carbon/graphite fibers.

If high resistivity, high
dielectric strength, high arc
resistance, low loss factor and low
dielectric constants are required,
there are much better insulative
materials. The development
requirements for a new product may
place an important priority on high
dielectric constant (capacitative
devices for example); high loss
factors (molding compounds for
rapid dielectric heating); low
volume resistivity (reduce static
charges) and semi-conductive
properties (electro-static
shielding qualities); just to cite
a few examples where semi-
conductive features have advantages, and small amounts of fine
chopped carbon/graphite fibers
introduced with advantage.

Table 2

VOLUME RESISTIVITY (OHM-CM)
(T-300/Epoxy Panels)

<u>Conditioning</u>

As Received	.0068 (Aver.)
After 48 hours at 175° C.	.0063 (Aver.)
After 48 hours at 175° C. Plus 72 hours at 23° C. and 96% R.H.	.0065 (Aver.)
As Received Plus 72 hours at 23° C. and 96% R.H.	.0065 (Aver.)

<u>Resistivity</u>

Carbonization in an inert atmosphere, followed by graphitization at higher temperatures in an inert atmosphere produce a marked decrease of electrical resistivity on the carbon/graphite fibers. This fact is well known and typical data are illustrated in Figure 1. The changes are most

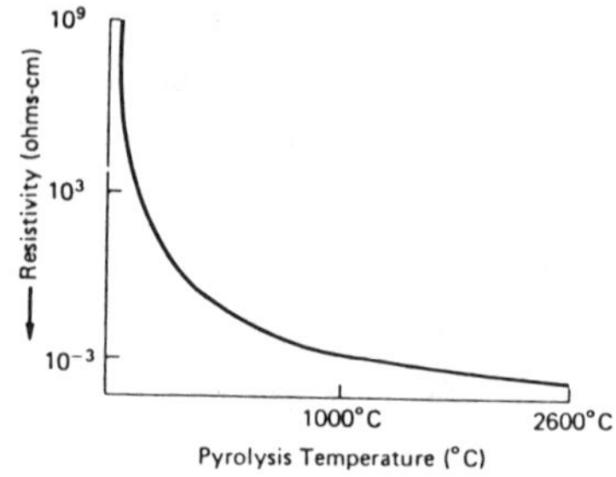

Pyrolysis temperature versus resistivity.

pronounced below 1000° C (carbonization in inert atmospheres). The reduction in resistivity is dramatic, as graphitization effects a more ordered crystalline structure among the carbon atoms. The lowest C.T.E. and highest moduli of elasticity are associated with the well graphitized fiber (the ultra-high modulus fibers).

On aircraft structures there is continuing concern for the effects of lightning strikes on graphite fiber composites used on wing structures, and exterior surfaces. These results have been reported (see Reference 5, for example). General conclusions suggest that while laminated graphite panels do not have as good electrical conductivity as aluminum alloy or titanium, they will dissipate the high voltage and amperage surges of lightning adequately without catastrophic failure. There are design details which should be considered and referral to original sources is urged. Electrical connections between metallic components and carbon/graphite fiber composites are best served by electrically conductive adhesives, in lieu of or at least augmenting mechanical devices.

Contact resistance between contiguous fibers or overlapping strands of woven fabrics of carbon and graphite fibers are subject to processing variables. The fibers have sizing agents or chemical treatments on their surface to enhance the bonding of the resin matrix. These measures enhance the interlaminar shear strength of

182

the composite. The fiber surface treatment should not be overlooked when seeking optimum ways of enhancing electrical conductivity of the composite. Pressures and cure temperatures during processing of carbon/graphite composites will have an influence insofar as they affect the contact resistances of the fibers to one another.

For structural effectiveness, carbon/graphite fiber composites are produced with many types of fiber orientation with the help of unidirectional fibers. Because of the anisotropic character of the physical properties, it is reasonable to expect fiber alignments to influence the electrical conductivity as well as the extent that electrical charges transfer from one bundle of fibers to another. When isotropic panels are prepared with woven carbon/graphite broad goods, uniformity in electrical conductivity may be expected.

X-Ray Transparency

Some manufacturers have recognized the value of utilizing the high specific modulus and strength of carbon/graphite graphite composites in the preparation of x-ray diagnostic tables and couches. Basic requirements include transparency to x-rays and the stiffness (resistance to bending) associated with a high modulus material. Glass fiber laminates do not possess adequate stiffness for this structural consideration, and while they have superior tensile strength, they defer to the higher moduli of carbon/graphite composites. Of commercially available fibers from all materials, the graphite fibers possess the lowest specific gravity concomitant with electrical conductivity. For x-ray tables, the light weight and high stiffness of carbon/graphite fiber composites provides a good structural basis.

X-ray diagnosis of reclined patients is served best when the patient can be kept still, the source of x-rays moved about the patient and the receptor film positioned under the structural supporting member. Present electrically conductive aluminum alloys do not permit this procedure.

Environmental Influences

The electrical properties of carbon/graphite composites may be affected by extremes of environmental exposure. Void content of the composite, coupling agents on the carbon/graphite fibers, and the chemical nature and cure history of the resin matrix binding the fibers are significant influences. Another mitigating factor arises from the differences in the coefficient of thermal expansion (C.T.E.) of the carbon/ graphite fiber and the encapsulating resin matrix. Under thermal cycling, fine micro-cracks in the resin matrix have been observed for some systems. The micro-cracks would be expected to contribute to increased moisture absorption and further derogation of electrical insulative characteristics. The low C.T.E. of the carbon/graphite fiber composites have achieved a minimum geometric surface distortion of large antennas designed for outer space (see Reference 6).

Release of Carbon Fibers

In recent years, attention has been called to the release of carbon fibers into the atmosphere, at the time of catastrophic fires which decompose the resin matrix binding the fibers. Obviously, this is of concern if the fine fibers drift about and short-circuit sensitive components of control apparatus or computers. Presumably, designs of such apparatus anticipate random exposure to fumes and particulate matter and should preclude this from happening.

Operation at high temperatures above the Tg of the resin matrix tends to reduce resin content at the surface of the composite and increase the concentration of carbon fibers at the outside surfaces. Improved resin matrices and selective hybrids (glass fibers and carbon/graphite fibers) have reduced this problem significantly. High char strength resin systems also appear to reduce carbon fiber release.

There are many other aspects of the utilization of carbon/graphite fiber composites and their adaptation to electrical applications. With the increasing maturity and understanding of the advantages and disadvantages of these products, further significant applications are evolving.

Bibliography

1. Bushman, E. F., and Bushman, G. R. "Carbon Fiber Tooling." 33rd Annual Technical Conference, SPI, RP/C Institute, Section 21F, Washington, D.C., February 1978.

2. Gibbs, H. H., Wendt, R. D., and Wilson, F. C. "Carbon Fiber Structural and Stability Studies." 33rd Annual Technical Conference, SPI, RP/C Institute, Section 24F, Washington, D.C., February 1978.

3. McMahon, P. E. "Thermal Oxidative Resistance of Carbon Fibers." SAMPE Symposium, Vol. 23, p. 150, Anaheim, 1978, and ASTM-STP-658, pages 254-266 (1978).

4. Delmonte, John. "Technology of Carbon and Graphite Fiber Composites." Van Nostrand Reinhold, 452 pages, New York, 1981.

5. English, F. C. "Advanced Composite Aileron for L1011 Transport Aircraft." L.R. 29056 - NAS1-15069, April 1979.

6. Haydostian, H., et al. "Lightweight Composite Spacecraft Antennas." SAMPE Journal, Vol. 13, page 14, September/ October 1977.

Biography

John Delmonte is an established consultant in the plastics field. He is currently chairman of the Board of Directors of Delsen Testing Labs and has also written and taught extensively on plastics. Some of Mr. Delmonte's many accomplishments include serving as

president of Furane Plastics, Inc.
for 25 years, developing the first
course in plastics ever taught in
the U.S., publishing five books
and 400 articles on plastics, and
acquiring 15 U.S. patents. He is
a member of the Plastics Hall of
Fame (Society of Plastics
Industry), and is active in the
Society of Plastics Engineers
(SPE), American Chemical Society
(ACS), American Society of Testing
Materials (ASTM), and Society for
Advancement of Materials and
Process Engineering (SAMPE).

ISOTHERMAL SHAPE ROLLING OF TITANIUM ALLOYS

A. Saith, R. Zandonella, J. R. Woodward and A. G. Metcalfe
Solar Turbines International
San Diego, California

Abstract

Isothermal rolling is an emerging process based on use of local electrical resistance heating with molybdenum alloy rolls or dies. Shaping of the metal can be accomplished by a roll or die configuration that may be used in conjunction with stationary tooling. The process is particularly well suited for rolling of titanium alloys. Several rolling processes are currently under development under DOD sponsorship. A process is being established for manufacture of a T-section component for the F-18 aircraft, starting from rectangular barstock. A SQUARE BEND process is being established where a radiused bend in sheet or plate is ironed by a roll to form a square external corner with an internal filleted corner. Sheet Rolling and Ring Rolling processes have been developed. Roll forge dies are used in another process to manufacture compressor blades for the T55 engine. The cost savings potential is derived primarily from the ability of the processes to make net sections and avoid excessive chip making. This factor is becoming increasingly important with the rapid rise in material costs and increases in lead times.

Keywords: Isothermal rolling, titanium, T-section, SQUARE BEND, sheet rolling, ring rolling, blade forging.

1. INTRODUCTION

The basic isothermal rolling concept is illustrated in Figure 1.

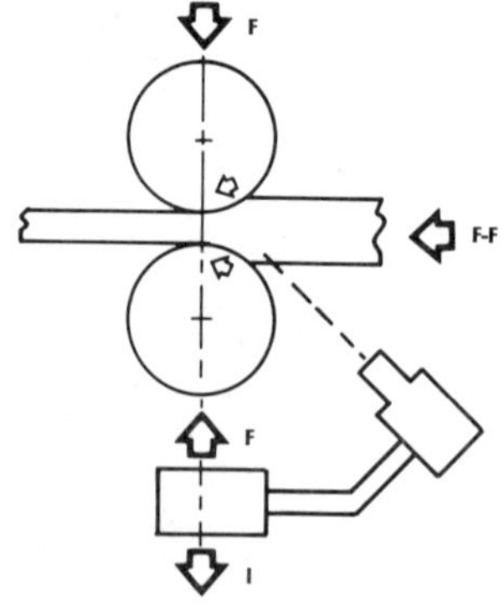

Figure 1. Basic Isothermal Rolling Concept

Molybdenum tooling, in this case rolls, serves also as electrodes. The current I heats the workpiece

and enough of the roll to form a
travelling hot zone. Shaping
occurs under the combined action
of the squeeze force F and a feed
force FF. The feed force FF is
essential for large reductions
that may exceed 95% reduction in
a single pass. The process may
be temperature controlled as
shown. The basic process can be
modified to use one roll operating
against fixed tooling, to use
shaped rolls, split rolls, roll-
forge dies, or combinations of
these. Isothermal rolling evolved
from the Continuous Seam Diffusion
Bonding process, one application
of which was the manufacture of
seals for the CF-6 engine. Figure
2 shows the production set-up at
Kelsey-Hayes for this application.

Figure 2. Continuous Seam Diffusion
Bonding Machine at Heintz

2. CUSTOM FABRICATION OF
F-18 SECTIONS

The F-18 aircraft uses large
amounts of titanium alloy stringers
and longerons that are machined
from heavy section extrusions.
The heavy section permits the
designer to thicken or widen the

flange or stem of a typical structural section. On the other hand,
the extrusion must have adequate
section to provide for the maximum
dimensions, even if this maximum
section is required in less than
10% of the length. Also the extrusion is typically made 0.25/0.30
inch oversize on all dimensions
with the result that the buy-to-fly
ratio may exceed 10:1.
Substantial cost reduction may be
achieved by isothermal rolling of
net thickness T-sections, thereby
eliminating the usual excess on all
dimensions. Additional metal may
be added by electron beam welding
for a second rolling pass to produce "custom" sections. This
additional rolling restores the
transformed beta structure of the
as-welded section shown in Figure
3 to the alpha-beta structure
shown in Figure 4.

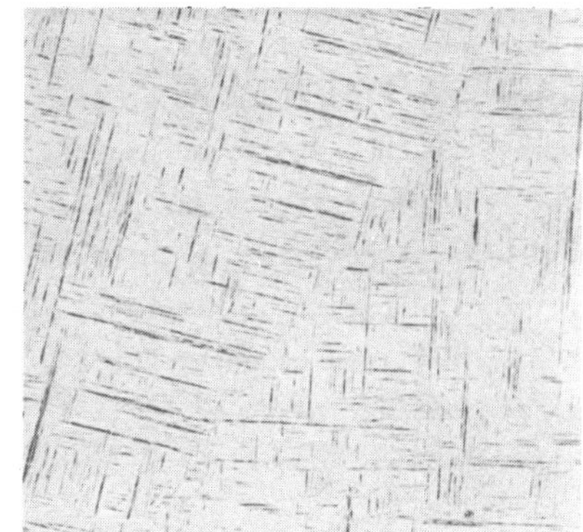

Figure 3. As Welded
(Magnification: 250X)

Consider a typical longeron for the
F-18 where the flange requirement
of 4.54 inch at one location demands a 5 inch wide extrusion

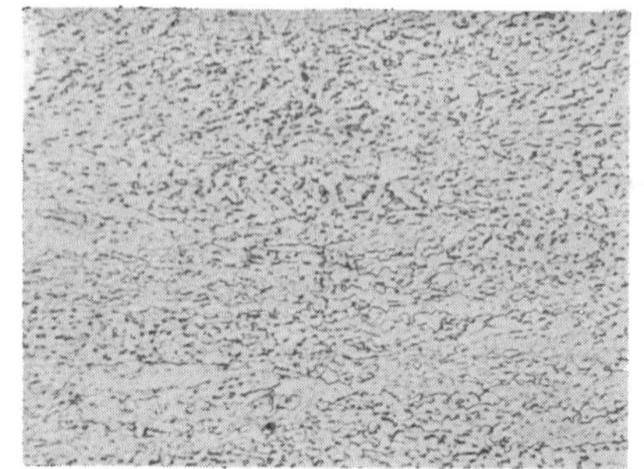

Figure 4. Rolled Weldments
(Magnification: 250X)

throughout. It is estimated that this part weighs 61.6 pounds per ship set, but requires 250 pounds of extrusion to manufacture. The blank for isothermal rolling will weigh 76 pounds for a savings of 174 pounds. The cost of this additional material plus that of chip making is over $4400. The exact cost of isothermal rolling cannot be established until the process has been demonstrated, but it is evident that very large cost savings will be realized once isothermal rolling is available. The Fuel Bay Floor Support shown in Figure 5 has been selected as the first F-18 component to serve as the demonstration article for this program. Custom thickening and widening will not be necessary for this component. Also, the part is neither fatigue nor maintenance critical, so that only static equivalence in properties will need to be demonstrated between extruded and isothermally rolled components. This should greatly facilitate initial acceptance of the process for F-18 components.

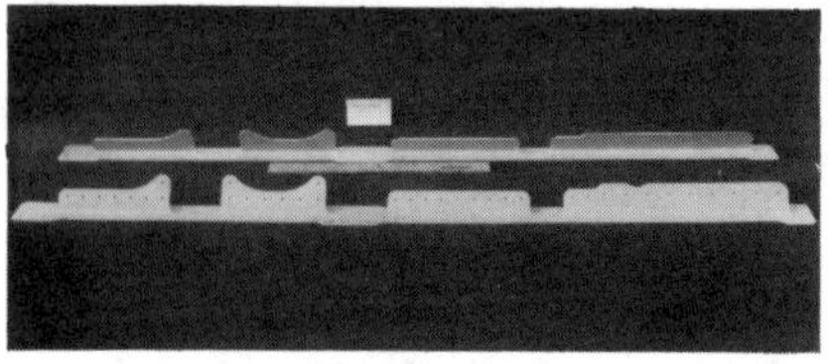

Figure 5. Fuel Bay Floor Supports

The basic two-step method for rolling T-sections starting with rectangular barstock is shown in Figure 6. Cross-sections from the three shapes of the process are shown in Figure 7. Figure 8 shows a 0.5 x 1.0 inch section of Ti6Al4V before and after the web rolling pass. Figure 9 shows T-sections produced by subsequent two-pass flange rolling. Typical rolling parameters for each operation are:

Stem Rolling

Squeeze force (lb)	40,000
Feed force (lb)	2,000
Current (amps)	16,000
Speed (ipm)	1.0

Flange Rolling

Squeeze force (lb)	48,000
Feed force (lb)	7,500
Current (amps)	28,000
Speed (ipm)	1.0

These sections are given stress relief annealing and straightening at 1350°F to produce T's as shown in Figure 10.

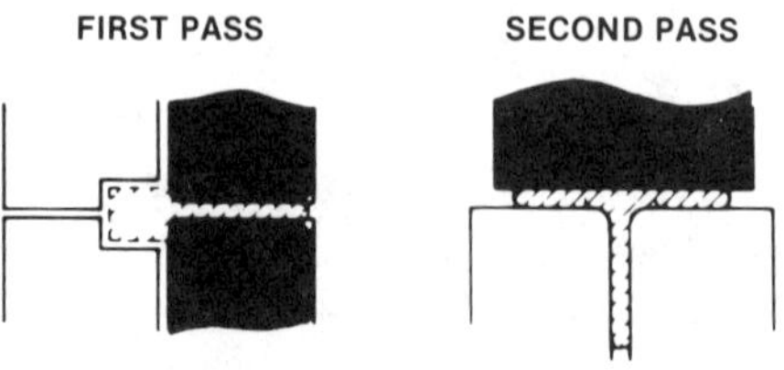

Figure 6. Two Pass Rolling of T-sections

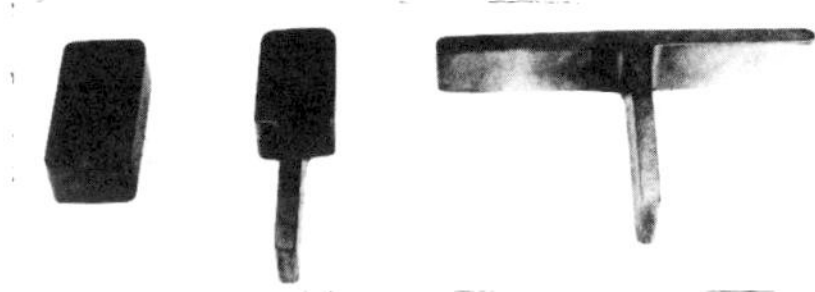

Figure 7. Starter Stock, Rolled Web, and Rolled Flange

Figure 8. Typical Rolled Web

Figure 9. Second Roll Pass to Form Flange

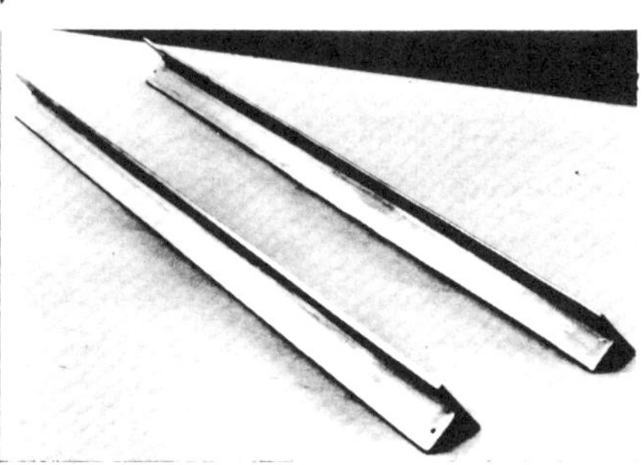

Figure 10. T-sections After Straightening

This program is being conducted under U.S. Navy sponsorship. Properties of processed material are being evaluated by Northrop Corporation.

3. SQUARE BEND

The SQUARE BEND process is also being developed under U.S. Navy sponsorship. Previous work had been limited to 0.062 inch thick Ti6Al4V sheet, and a scale-up to 0.20 inch plate in Ti6:6:2 alloy was required for the F-14 rear wing beam.

Figure 11 shows brake formed parts in 6:6:2 alloy with an approximately 5t bend. These parts are clamped into simple tooling as shown in Figure 12, and rolled isothermally. Rolled lengths of channel are shown in Figure 13. Control of bend quality is currently under study. An as-rolled bend is shown in Figure 14. At the present time, two passes are required to obtain the full bend. Typical SQUARE BEND rolling parameters are:

Squeeze force (lb)	6,000
Current (amps)	4,500
Speed (ipm)	3 in/min.

4. RING ROLLING

The purpose of this project was to develop manufacturing methods to produce isothermally rolled Ti6Al4V titanium alloy engine rings. Work was sponsored by the U.S. Air Force, and was performed in co-operation with the Heintz Division of Kelsey-Hayes, the Aircraft Engine Group of General Electric and Rohr Industries. Ring blanks were rectangular in cross-section, with a small machined shoulder, and were rolled two at a time, back-to-back. A key feature of

Figure 11.   Press Brake Formed Sections

Figure 12.   SQUARE BEND Rolling Set-up

Figure 13.   Channels Rolled by SQUARE BEND Process

the program was the development of a continuous force feeder to prevent slippage. Three L-section F-101 engine rings rolled on this program are shown in Figure 15. A variety of "hat" sections produced

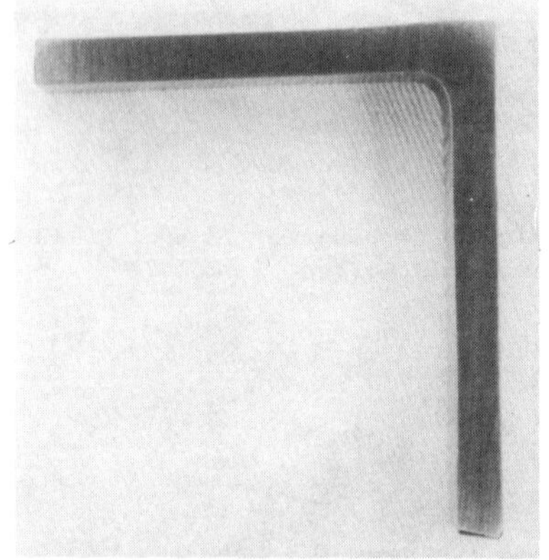

Figure 14.   Section of SQUARE BEND in Ti6:6:2 Alloy

Figure 15.   Isothermally Rolled F-101 Engine Rings

from Inco 718 and Hastelloy X on an Air Force program under sub-contract from General Electric are shown in Figure 16, and demonstrate the versatility of the process.

5.   BLADE FORGING

A program is being conducted under U.S. Army sponsorship to apply isothermal roll forging to the manufacture of compressor blades. The Avco T55 second stage compressor blade was selected for demonstration. Several blades made in

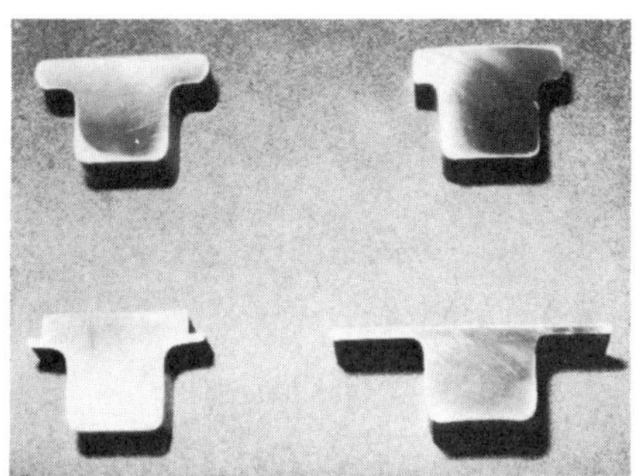

Figure 16. Engine Ring Preforms

this program are shown in Figure 17. The material is AM-350 alloy, but Ti6Al4V was used for proofing, and performed well. Two phases of this program have been completed.

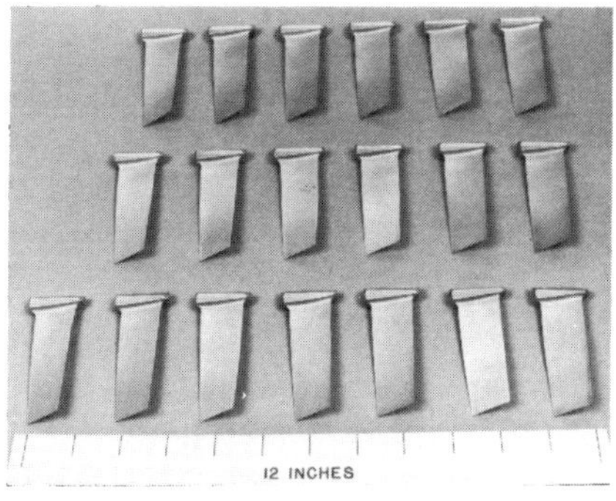

Figure 17. Finished T55 Blades

In the first phase, the feasibility of blade manufacture was demonstrated using the isothermal rolling mill; blades were manufactured to within a tolerance of 0.010 inch. In the second phase, dimensional control was tightened using an improved manufacturing process in a special machine of the knuckle rolling type. The machine permits root upsetting and airfoil rolling in the same operation. A microprocessor is utilized to accomplish the transition from root injection to airfoil rolling. Final coining and twisting are performed in a separate hot press forging operation. Phase III of the program has been initiated with the objective of producing engine quality blades. Substantial cost savings are anticipated due to a reduction in the number of operations, automation of the process by microprocessor, and the ability to roll forge thin trailing edges.

6. SHEET ROLLING

A high temperature sheet rolling process is being developed with U.S. Air Force sponsorship under subcontract. Although the objective of this program is directed to high temperature superalloy materials, Ti6Al4V titanium alloy has been used for tool proofing with excellent results. Figure 18 shows strip rolling in progress. One-eighth inch thick Ti6Al4V starter stock, two inches wide, was reduced in gage to various thicknesses between 0.025 inch and 0.048 inch in a single pass. Typical results are shown in Figure 19. A subsequent change in process control and technique allowed thicknesses of 0.012 inch and 0.009 inch to be produced in a single pass. Remarkably, these strips were free of internal defects and could readily be brought to a microstructural condition suitable for SPF/DB fabrication. The process appears to be a breakthrough for production of sheet and foil from titanium alloys. One potential area of application is metal matrix composites.

Figure 18. Strip Rolling in Progress

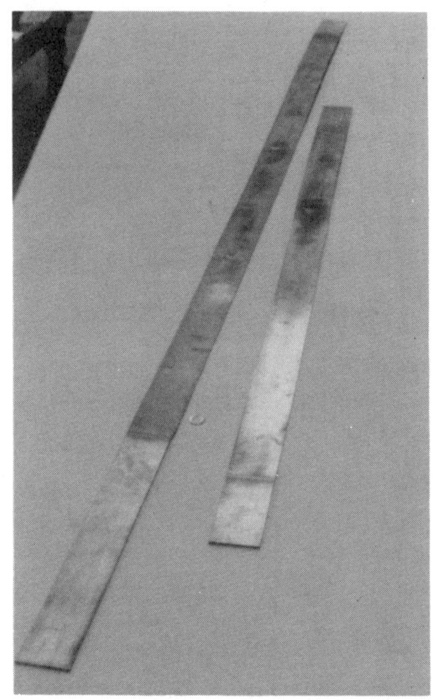

Figure 19. Typical Ti6Al4V Rolled Strip

7. VANESTOCK ROLLING

Isothermal rolling is particularly suited to single pass rolling of vanestock starting with simple barstock. The process has been demonstrated on a variety of materials, including Ti6Al4V titanium alloy.

8. POWDER COMPACTION

The ability to compact titanium alloy powder directly into shapes by isothermal rolling is currently being examined.

9. SUMMARY

In summary, isothermal rolling has been successfully used for manufacturing a variety of structural shapes and promises to be a major new production technology. Cost savings are realized by single pass rolling, thus eliminating intermediate processing between passes, and by net shape rolling, eliminating chip formation. The process is particularly suitable for titanium alloys.

Arun Saith is a Research Staff Engineer in the Advanced Manufacturing Technology Research (AMTEC) Group at Solar Turbines International. He has a Doctor of Engineering in Mechanical Engineering from the University of California at Berkeley. His specialty is manufacturing process development.

Rene Zandonella is a Research Engineer with the AMTEC Group. He has a B.S. in Engineering from Harvey Mudd College. He is currently involved with isothermal metal working process development.

James Woodward is the Manager of the AMTEC Group. His M.S. degree is from San Diego State University in Mechanical Engineering, with an A.E. degree from Northrop Aeronautical Institute. He has 14 patents in his field of material sciences.

Dr. Arthur Metcalfe is the Associate Director of Technology Planning. He has a PhD from Cambridge University, England. He has been associated with isothermal metal working technology from the start at Solar.

26th National SAMPE Symposium
April 28-30, 1981

COMPOSITE APPLICATIONS ON
BOEING COMMERCIAL AIRCRAFT

Vere S. Thompson
Boeing Commercial Airplane Company
Seattle, Washington

ABSTRACT

The Boeing Company has been using composite structures of fiberglass with plastic matrix on their commercial airplanes since the early 1960's. These structures have provided excellent engineering, manufacturing and service experience. The favorable background led to the application of advanced composites on the Model 767 and Model 757. Graphite, Kevlar and Graphite/Kevlar hybrids were applied to surfaces such as rudder, elevator, spoilers and ailerons, gear doors, fixed trailing edges, fairing, engine cowls. A weight savings of approximately 1250 pounds was achieved on the Model 767 by using advanced composites. This paper will describe the benefits of composite structures on commercial aircraft and our structural fiberglass experience. The Model 767/757 production readiness and manufacturing efforts for advanced composite structures will also be covered.

BENEFITS OF COMPOSITE STRUCTURES

Composite structures offer a number of advantages and benefits. Probably the most significant is a 25%-30% weight savings. For a Model 747 this translates into annual fuel savings of over $750,000 if graphite composites are utilized for secondary and primary structural applications. Another significant advantage of composite structures is a lower manufacturing cost. As an example, the Model 737 fiberglass rudder panel was approximately 25% the cost of a mechanically fastened aluminum structure. Another typical trade showed that a Model 727 structural fiberglass elevator control tab was about 65% the cost of the bonded aluminum metal honeycomb structure. Composite structures have also been a benefit from a commercial airplane standpoint because of their broad range of design possibilities, superior fatigue and sonic vibration resistance and improved durability.

STRUCTURAL FIBERGLASS EXPERIENCE
Structural fiberglass components have been used on our commercial aircraft since the Model 707. Over 25% of the wetted area on the Model 747 is structural fiberglass. The use of structural fiberglass on our airplanes has gone from only 200 sq. ft. on the Model 707 to over 10,000 sq. ft. on the Model 747. See Figure 1. The applications have expanded with each new airplane model. On the Model 707, structural fiberglass was used only for the radome and a few wing closure panels. Structural fiberglass components are currently used for fairings, rudder and elevator control surfaces, wing leading and trailing edge panels and inlet ducts, along with radomes and closure panels. The Model YC-14 expanded the list of applications further when hybrids of graphite and

structural fiberglass were utilized for doors, flaps and fairings.

A. Materials
Early parts were made by tailoring the glass fabric to the required configuration and pouring the liquid resin onto the fabric, spreading and sweeping the resin to impregnate the fabric, vacuum bagging the part and tool, and curing in an oven or autoclave. This wet layup method is very labor intensive.

The development of pre-impregnated fiberglass/epoxy materials ("prepregs") was a major step forward. Prepregs provide a relatively uniform amount of resin to the glass fabric and with proper processing yield consistent part properties. They also allow the co-cure of sandwich panels. The prepreg materials used through the early 1960's required a 350°F cure, some "bleeding" to remove excess resin and voids, and a separate adhesive layer for joining to honeycomb core. The processing requirements were essentially the same as those required for today's graphite composite systems. Development efforts over the years have evolved a prepreg system which Boeing uses for most of its composite parts. This system cures at 250°F and does not require a separate adhesive layer for joining to honeycomb core. These advantages result in lower material costs, less labor, reduced flow time, and reduced energy

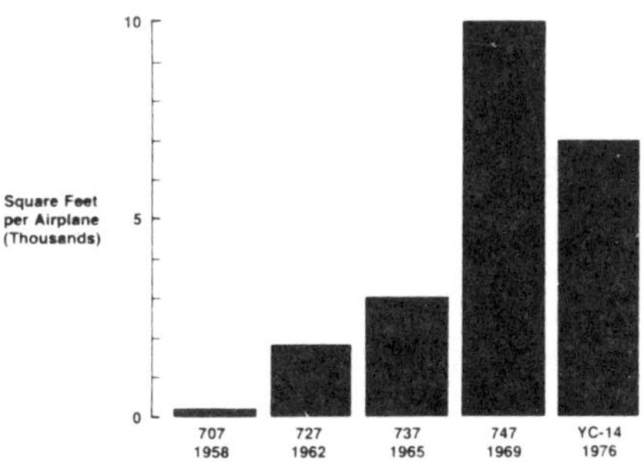

FIGURE 1
STRUCTURAL FIBERGLASS GROWTH

requirements.

Unidirectional material forms have been used very sparingly for airplane structural parts at Boeing. This is due in part to the much higher layup costs of unidirectional tapes as compared to woven fabrics.

Initial honeycomb core materials used by Boeing were polyester and nylon-phenolic for radomes. Later, heat resistant phenolic (HRP) core was used. The latest Boeing commercial applications use Nomex core due to a one pound per cubic foot weight savings compared to HRP core. The HRP core is still being used on earlier models and in localized areas requiring higher load carrying ability.

   B. Typical Construction
Most of our structural fiberglass applications involve sandwich panel type construction (Figure 2), in which honeycomb core is used for a weight and cost effective method of panel stabilization. A typical construction consists of lower and upper fiberglass skins and doublers and honeycomb core. Tedlar is normally applied to the upper skin as a moisture barrier. This sandwich construction has been in use since 1962.

It should be noted in most cases a separate adhesive layer is not required for joining the surface layers to the honeycomb core and the moisture barrier film is laid up and cured as an integral part of the component.

A wide variety of configurations are being produced with varying contour, size and thickness, depending on application. Areas with high localized loading are met using titanium inserts. Other requirements, such as antenna ground planes or lightning strike protection, are satisfied by bonding aluminum foil on inner surfaces, aluminum flame spray on outer surfaces and special finishing.

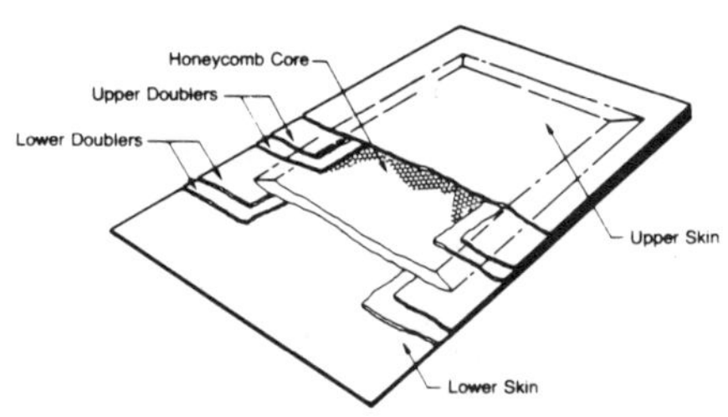

FIGURE 2
TYPICAL CONSTRUCTION

   C. Processing
The production flow (Figure 3) for fabricating structural fiberglass parts involves prepreg and core preparation, layup on molds, auto-clave cure, trim and exterior surface finishing. The prepreg is received in rolls and then "kits" prepared of the skin, doubler and filler plies. Kitting is the cutting and storing together of prepreg patterns for a specific part until the production schedule requires that part. Steps required for preparing honeycomb core details include slicing, forming,

196

splicing, machining and potting.

Skin, doubler and filler ply details
are oriented on a mold according to
the drawing, along with the honeycomb core detail. Autoclaves are
the principal curing method. The
flexibility of the autoclave is
most compatible with the production
part mix and curing schedule. After
cure, the parts are trimmed to size
and painted.

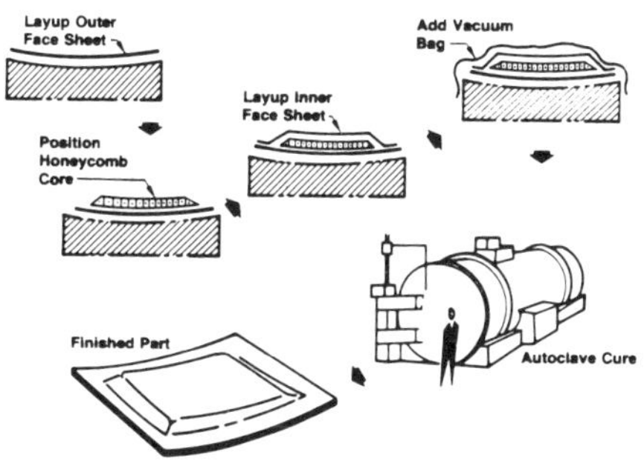

**FIGURE 3
PROCESSING STEPS**

ADVANCED COMPOSITE EFFORTS

Structural fiberglass composites
provided an excellent stepping-
stone to graphite composites. The
structural fiberglass approaches
for design to cost, manufacturing/
engineering interface, materials,
tooling, production and inspection
were a technology base for moving
into graphite secondary aircraft
structure. The cost and service
experience of producing and flying
approximately 9,000,000 sq. ft. of
structural fiberglass in commercial
operations further expanded this
technology base for graphite
composites.

A. Production Readiness Program
Before starting production of
advanced composites for the Model
767 and Model 757 aircraft we under-
took a readiness program to design,
fabricate and, in some cases, fly
advanced composite components for
our current Model 727, 737 and 747
aircraft. These components involved
the 727 Elevator, 727 Engine Cowl
Panel, 737 Spoilers, 737 Horizontal
Stabilizer and 747 Outboard Aileron.

1. 727 Graphite/Epoxy Elevator
Initial fabrication of Model 727
composite elevators was accomplished
in a NASA/Boeing contract. This
was followed by a Boeing funded
effort to build a number of elevators in order to evaluate different
materials and to obtain flight
service experience.

The Model 727 composite elevator,
Figure 4, consists of upper and
lower sandwich panels which are
mechanically fastened to a front
and rear spar of laminate construction. The ribs are sandwiched
honeycomb construction. Construction of the sandwich panels is
essentially the same as was used
for structural fiberglass.

By using composites on the Model
727 elevator, a weight savings of
150 pounds or approximately 29%
was achieved.

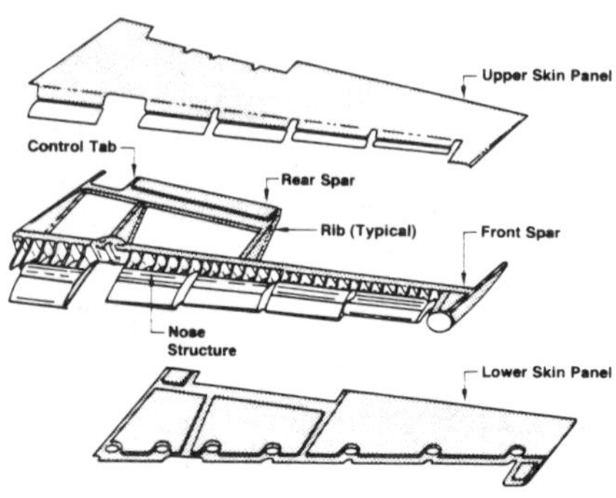

FIGURE 4
727 GRAPHITE/EPOXY ELEVATOR

2. 727 Graphite-Kevlar/Epoxy
    Hybrid Engine Cowl

Boeing and Rohr teamed to fabricate ten shipsets of hybrid cowl panels for the Model 727. The design is Kevlar with graphite utilized to reinforce core beams at the latch locations. The use of advanced composites for the engine cowl **panels** resulted in a 21% weight savings.

The cowl panels are currently in flight service on commercial aircraft without any problems being encountered. One airplane has accumulated over 3700 hours of flight time.

3. 737 Graphite Epoxy Spoiler

A NASA/Boeing contract to design, build and fly 140 spoilers for the Model 737 is one of the most significant portions of the production readiness program. This contract provides important data on the service experience of advanced composites. The spoilers have been in service on seven commercial air-

lines through the world for seven years.

The current bonded aluminum spoilers were redesigned using graphite/epoxy skins. Figure 5. The aluminum honeycomb and aluminum spar and actuator fittings were retained from the metal design. The use of graphite/epoxy skins resulted in a 16% weight savings.

After seven years, the spoilers have accumulated over 1,418,000 hours of total fleet service. All the airlines are very pleased with the performance of these spoilers.

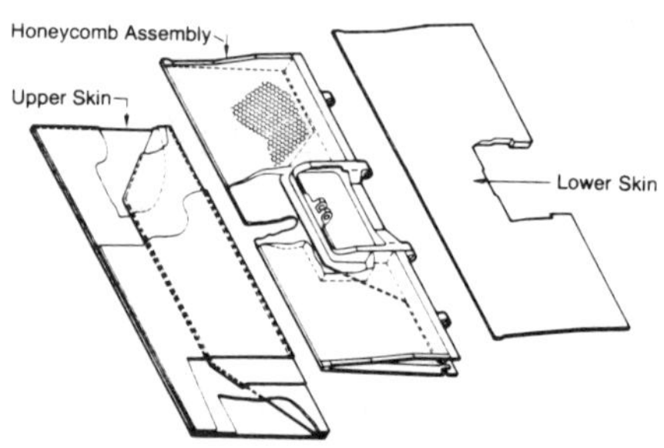

FIGURE 5
737 GRAPHITE/EPOXY SPOILER

4. 737 Graphite/Epoxy
    Horizontal Stabilizer

The graphite/epoxy horizontal stabilizer box was another NASA/Boeing contract. This contract involved the fabrication of 5 1/2 shipsets of stabilizer boxes. This was the first application of graphite advanced composites to primary structures on Boeing aircraft.

The stabilizer box, Figure 6,

consisted of "I" stiffened laminate
lower and upper skins, which are
mechanically fastened to front and
rear spars of laminate design. The
ribs use honeycomb sandwich construction and are very similar to
those fabricated for the Model 727
graphite/epoxy elevator. A 116
pound or 22% weight savings was
achieved by utilizing graphite/
epoxy for this application.

The first production box is
currently in ground fatigue testing.
The first flight of the stabilizers
on Boeing aircraft was in October
1980.

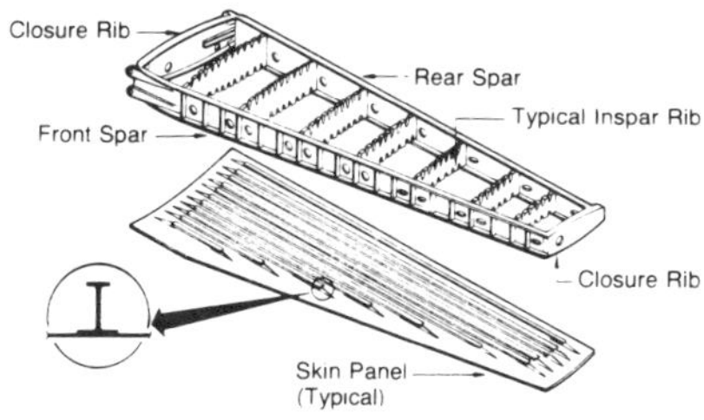

FIGURE 6
737 GRAPHITE/EPOXY STABILIZER BOX

5. 747 Graphite/Epoxy Aileron
Another part of the Production
Readiness Program involved the fabrication of two shipsets of graphite
/epoxy ailerons for the Model 747.
The design of these ailerons is
very similar to that of the Model
727 elevator in which graphite/
epoxy honeycomb panels are mechanically fastened to a laminate front
spar and graphite/epoxy sandwich
ribs.

A 90 pound or 26% weight savings
was achieved by using composites
for the Model 747 aileron.

B. 767/757 Advanced Composite
Production

As illustrated in Figure 7, 30% of
the wetted area on the Model 767
will be advanced composite
structures involving graphite or
graphite-Kevlar hybrids. Approximately 70% of advanced composite
structure is graphite-Kevlar hybrid
while 30% is all graphite. The use
of advanced composites on the Model
767 has enabled a weight savings of
over 1250 pounds.

Graphite applications involve the
elevator and rudder, which include
the panels, ribs and spars. The
outboard and inboard spoilers, and
outboard and inboard ailerons are
also graphite/epoxy applications.
Graphite fabric and Nomex honeycomb
are used for most of these applications.

The applications of advanced
composites on the Model 757 are
essentially the same as the Model
767. Preliminary design studies are

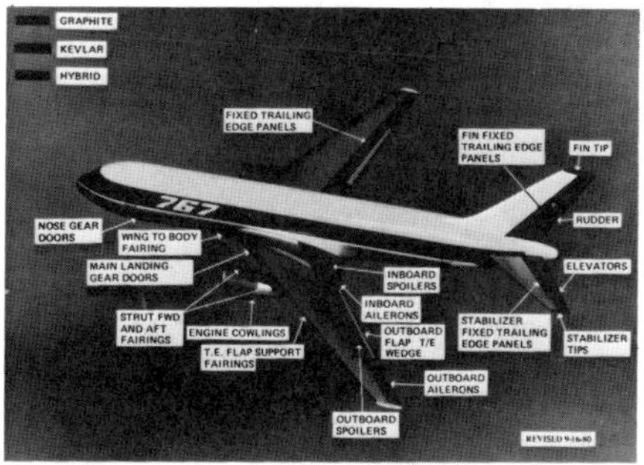

FIGURE 7

underway on the Model 757 to investigate other applications of advanced composites.

1. Elevators and Rudder
The Model 767 elevators and rudder consist of all graphite cover panels, ribs and spars. The cover panels are comprised of honeycomb core sandwich with graphite fabric skins. The honeycomb provides stiffening in non-attach areas and is removed in areas where the covers are fastened to ribs and spars. The ribs and spars are also honeycomb sandwich. Figures 8 and 9 show a Model 767 lower rudder cover panel and the rudder front spar, respectively. Assembly of the rudder and elevator components is accomplished with non-corroding fasteners (titanium or steel).

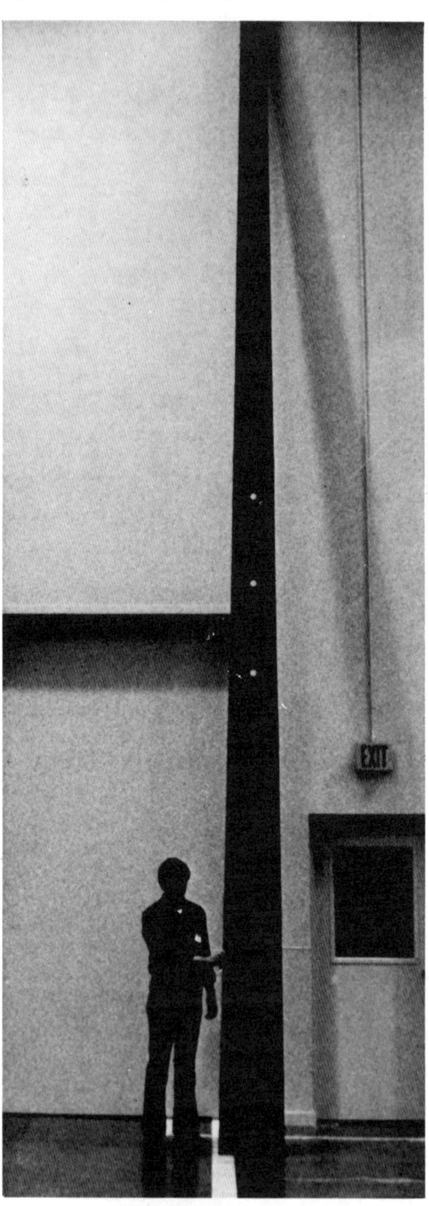

FIGURE 9
767 RUDDER FRONT SPAR

FIGURE 8
767 RUDDER PANEL

2. Spoilers and Ailerons
The spoilers are full depth honeycomb with thick graphite buildups and integral spars in attach areas.

200

The outboard spoiler has a pocket
area aft of the leading edge to
accommodate an actuator fitting.
The inboard spoiler has all attachments and fittings in the front
spar. The outboard aileron, Figure
10, is also full depth honeycomb
with graphite skins and a graphite
spar.

    3. Hybrid Applications
The graphite-Kevlar applications
include flaps, ducts, trailing edge
panels on the wing, stabilizer and
vertical fin, engine cowls, fairings
and main and nose gear doors. The
ratio of graphite to Kevlar varies
significantly for these components.
Graphite tape or fabric and Kevlar
fabric are all utilized. These
structures are primarily honeycomb
sandwich with laminated edge bands
where the panels are attached.
Figure 11 shows a wing trailing edge
closure panel during fabrication.
The typical construction of the
hybrid components is similar to that
illustrated in Figure 2 for
structural fiberglass.

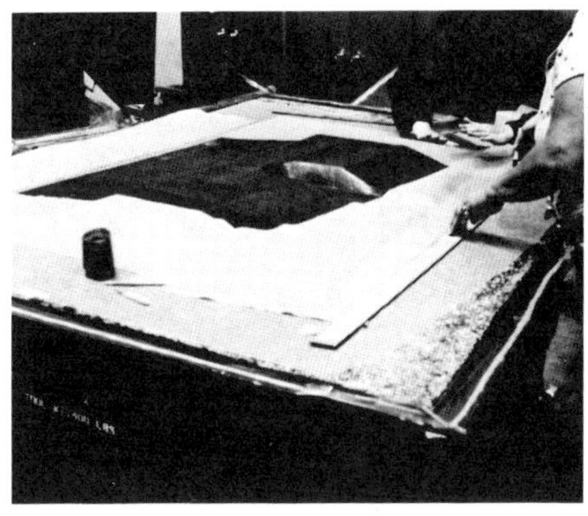

FIGURE 11
FABRICATION OF WING T.E. PANEL

COMPOSITE SUMMARY
Boeing has had excellent results
using composites of structural
fiberglass on our current aircraft.
In investigating advanced composites
for the Model 767 and 757, we found
no technical show stoppers that
would preclude their use. We are
finding that the future for advanced
composites will grow from our
current secondary structure applications to wing and fuselage primary
structures. Advanced composites are
the materials of the future.

FIGURE 10
767 OUTBOARD AILERON

26th National SAMPE Symposium
April 28-30, 1981

EF-111A GRAPHITE/EPOXY HORIZONTAL STABILIZER TRAILING EDGE

J.A. Suarez and S.J. Dastin
Grumman Aerospace Corporation
Bethpage, New York

Abstract

A production-suitable, vacuum-bag-cure manufacturing method for integrally stiffened graphite/epoxy sandwich structure has been verified for the EF-111A Horizontal Stabilizer Trailing Edge. Fabrication and static test of a structurally complete EF-111A Horizontal Stabilizer Trailing Edge has demonstrated the feasibility of constrained, vacuum pressure, oven-processing techniques for single integral cure exact molded (SICEM) assemblies. Cost and producibility analyses have been performed on the integrally-cured trapezoidal stiffener concept to determine production benefits and to project realistic total costs.

1. INTRODUCTION

Recent development efforts sponsored by the Air Force, coupled with industry-sponsored independent research, have established the feasibility of composite skin stabilization techniques with improved durability and maintainability as alternatives to full-depth aluminum honeycomb structures. It logically followed as a next step to demonstrate and validate low-cost and innovative skin stabilization manufacturing methods for composite secondary structure which are cost-competitive with full-depth honeycomb structure. A program to demonstrate constrained, vacuum pressure, oven processing techniques for single-cure assemblies was initiated. A production-suitable, vacuum-bag cure manufacturing method for integrally stiffened graphite/epoxy sandwich structure was verified for the EF-111A Horizontal Stabilizer Trailing Edge (Fig. 1).

The USAF/Grumman EF-111A is the latest tactical aircraft to be dedicated specifically to electronic warfare (EW) and the most modern, fully integrated, supersonic EW aircraft designed to date. This component provided an existing large-area (14.2 sq ft) aluminum honeycomb sandwich design for which considerable data was available. Also, due to its long service life on the F-111, significant maintenance experience existed. The selected component provided generic cost and weight comparison to full-depth honeycomb bonded aluminum structures for high-performance aircraft. Cost and producibility analyses were performed on the integrally-cured full-depth trapezoidal stiffener concept to determine production benefits and project realistic acquisition and life-cycle costs.

2. STRUCTURAL CONFIGURATION

To achieve a minimum weight (14% savings over the metal skin honeycomb-panel design) and low cost, the EF-111A trailing edge (Fig. 2) has been designed as a one-piece, integral, full-depth trapezoidal corrugation sandwich that reduces the assembly procedure to an absolute minimum.

The basic criteria for the design is as follows:

- Ply orientations restricted to the 0°, 90°, and ±45° family

R80-2149-001(T)

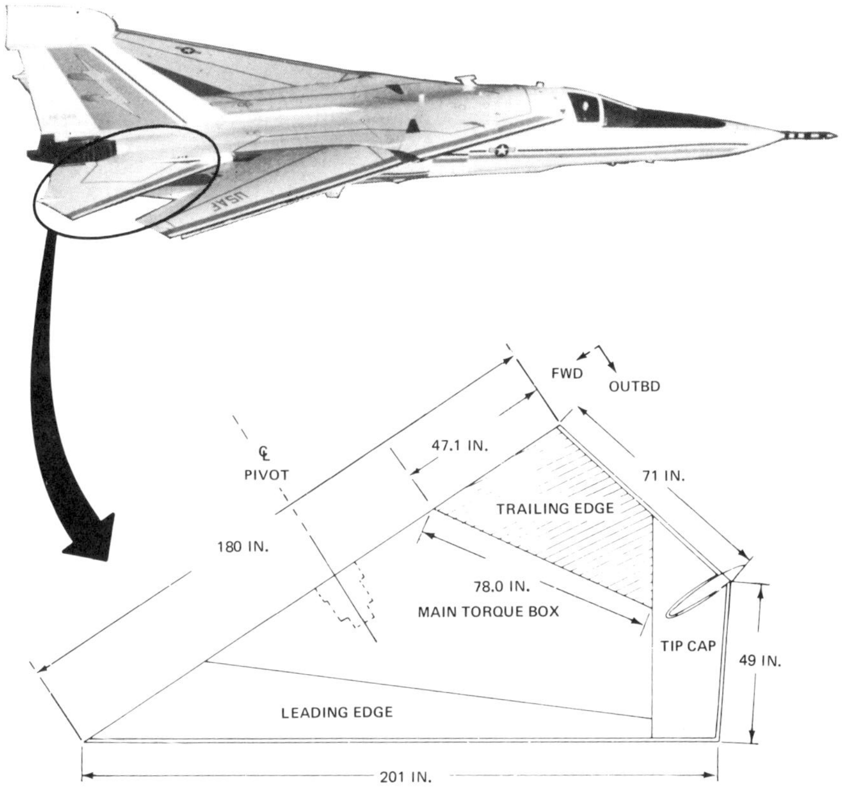

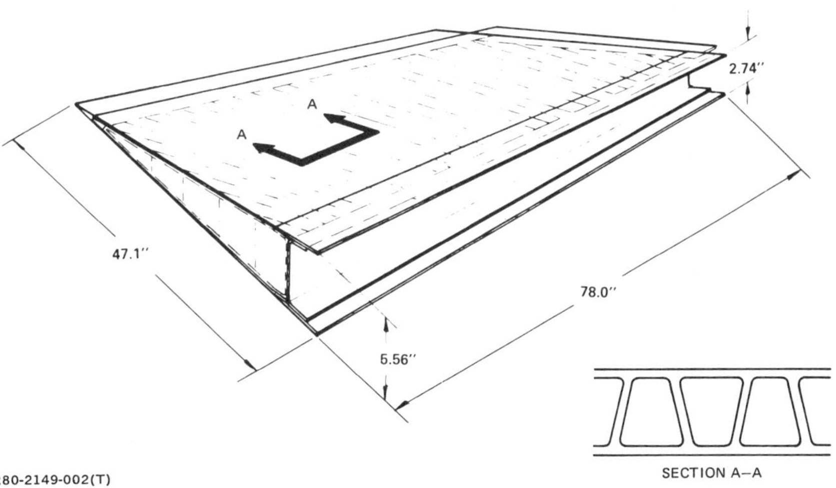

Fig. 1 EF-111A Horizontal Stabilizer Trailing Edge

R80-2149-002(T)

Fig. 2 EF-111A Horizontal Stabilizer Trailing Edge Demonstration Component

- No less than 10% and at least one 0° and one 90° ply in the cover skin to provide filament-controlled load capability

- Two outermost plies of every laminate must be a pair of ±45° for impact-damage consideration

- All laminates balanced and symmetrical about their midplane to minimize bending-stretching coupling effects

- All stiffener and frame webs composed primarily of ±45° plies to minimize their axial stiffness (hence their induced axial load) and provide maximum shear stiffness to minimize panel transverse shear deformation

- The two air-passage skins must meet or exceed the EI and GJ of the baseline aluminum skins

- The transverse strength of the skin panels must be sufficient to distribute normal air loads to the stiffener webs.

To minimize the number of tooling mandrels, the trapezoidal stiffeners are perpendicular to the trailing edge (arrowhead) reference plane. The inboard closure rib is fabricated along with the skins and stiffeners. To satisfy the fit, form, and function requirement of the composite replacement structure, it is attached to the main torque box and tip cap in the same manner as the aluminum structure.

The vertical shear load is transferred from the trailing edge to the main torque box by shear pins, the point of load application being on the web of the front beam of the trailing edge structure. This requires additional plies to be added to the front beam web at the pin location. The shear pins at the outboard closure rib-to-tip cap attachment are in the rib, which is included in the cured assembly, along with a fiberglass/epoxy corrosion barrier. The cover loads are transferred from the trailing edge skin to the main torque box, and from the tip cap to the trailing edge via mechanical attachments, the baseline bolt patterns being preserved. An aluminum splice plate is used due to close fastener spacings along the rear beam of the main torque box; again, a fiberglass/epoxy corrosion barrier is provided. The stiffener spacing was selected to minimize the skin thickness (i.e., to be as thin as possible and yet maintain the EI and GJ of the baseline component). An aluminum extrusion is used for the trailing edge arrowhead to reduce tooling costs and prevent graphite/epoxy brooming problems.

3. STRUCTURAL DESIGN

In the existing metal honeycomb trailing edge, the honeycomb core supports the local air load and provides a shear path to carry air load to the rear beam of the structural box. In the composite structure, the corrugations replace the honeycomb core and provide the core support functions.

Normal air load on the composite structure is resisted by skin panels spanning from web-to-web of the corrugations. The skin panel normal load is supported by the vertical webs; these transmit the load forward to the forward closure member from which it is transferred to the rear beam web of the structural box through shear pins. Moments are resisted by upper and lower cover loads transferred through bolt bearing and shear into the covers of the structural box.

The skin panels experience spanwise loading in carrying normal loading through bending plus tension, and chordwise loading in developing the cover loads balancing the moment due to the air load. Because of the material property reductions due to temperature and humidity, the covers are compression-critical (loads are fully reversible) and are analyzed for buckling at temperature under combined compression and shear loading. The panels are checked for maximum fiber strain under in-plane loading plus normal pressure. Maximum fiber strains are limited to 3700 μin./in. at ultimate load to provide a damage tolerant structure for low energy impact. Covers are nonbuckling to ultimate load. Cover panel laminates vary from $0_4/90_2/\pm45$ in the forward portion of the trailing edge to $0_2/90_2/\pm45_2$ at the aft end.

Vertical webs must resist normal air loading and transfer their load by shear. The webs are designed to be nonbuckling to ultimate load and are analyzed for in-plane shear and compression along the long edge of the web. Web laminates are $90_4/\pm45_2$ in the forward third of the trailing edge and $90_2/\pm45_2$ in the aft two-thirds. The lighter laminate is used as the height of the corrugation decreases, making the section less sensitive to shear buckling.

Cover laminates are locally reinforced where the composite trailing edge ties to the structural box cover. The composite is checked for bearing and shearout from fastener loads adjusted for peaking due to the presence of shear pins.

4. DESIGN ALLOWABLES

When vacuum pressure, oven-cure graphite/epoxy was first considered for the EF-111A Horizontal Stabilizer trailing edge, a review showed a lack of sufficient mechanical property data. Preliminary industry data looked promising but did not constitute a sufficient base for mechanical property characterization. As a result, a limited test program was initiated to establish the necessary design properties and, at the same time, demonstrate producibility of

the selected fabrication method. The basic program consisted of the following:

- Establishment of material and process specifications

- Selection of a conservative, yet representative, long-term moisture absorption criteria

- Coupon static tests to establish design values with environmental and vacuum-pressure-cure knockdown factors.

The strength and stiffness design values used in designing the trailing edge component were established from the mechanical property tests. Figure 3 shows these design values as ratios of dry strength at temperature of autoclave-cured laminates.

5. TOOLING

A tool feasibility and cost tradeoff study was performed to determine the tool design concept and tooling material to be used for the most critical tooling requirement, i.e., the skin stiffener mandrels. The candidates were Kirksite and eutectic salt.

The eutectic salt mandrel selected for the mandrel tooling approach for this program was based on the following:

Vacuum-Pressure-Cured Graphite/Epoxy Laminates[b]		
Property	Ratio of VPC to Autoclave-Cured, Dry Laminate at RT	Ratio of VPC Wet Laminate to Autoclave-Cured, Dry Laminate at 431°K (317°F)[a]
$F_t{}^*/F_t$	0.88	0.81
$F_c{}^*/F_c$	0.50	0.49
$F_s{}^*/F_s$	0.90	0.49[c]
$E_t{}^*/E_t$	1.00	0.91
$E_c{}^*/E_c$	0.87	0.77[c]
G^*/G	0.91	0.74[c]

(a) Moisture absorption is 1% — 1.4%

(b) Per-ply thickness of vacuum-pressure-cured laminate is 0.0133 cm (0.00525 in.)

(c) Estimated

* VPC laminate property

R80-2149-003(T)

Fig. 3 Design Values for Vacuum-Pressure-Cured Graphite/Epoxy

- Lower research, test, development, and evaluation (RTD&E)
  program costs

- Higher coefficient of thermal
  expansion, allowing for greater
  compaction of graphite laminates

- No possibility of part damage due
  to metal mandrel removal

- Greater flexibility in the engineering design, resulting in a more
  structurally efficient integrally-
  molded structure.

The eutectic salt material currently
in use at Grumman has a coefficient
of thermal expansion of 40×10^{-6}
in./in./°F, or slightly more than
twice that of Kirksite. The material
is reusable if properly reground to
maintain composition uniformity.
The nitrate salt mixture used to
fabricate the required mandrel tools
is Thermocast 320X; its properties
are listed in Figure 4. The salt is
melted in a heated pot provided with
a stirrer, cast to shape and trimmed
to length (see Figure 5). This material provides additional mandrel
expansion during the vacuum bag
pressure cure and, therefore,
greater compaction of the graphite
laminates. In addition, since the
material is melted out during post-
cure of the part, the front beam can
be molded into the assembly, making
for a completely integrally molded
structure with no need for secondary
attachment with fasteners. This
approach incurs a cost to recast mandrels for every unit built.

The eutectic salt mandrel fabrication
procedure is as follows:

- Solvent-clean the mandrel tools
  and apply two coats of Frecote 33
  release agent (or equivalent).
  Bake the release into the tool surfaces for 1 hour at 400°F. Hand-
  buff tool surface and apply one
  coat of Frecote 33 after each
  casting operation

- Preheat aluminum mandrel casting
  tools to 235° + 10°F at the start
  of the Thermocast 320X salt pouring
  operation

- Follow standard Grumman procedures for mixing and pouring
  the required 320X salt, taking
  care to prevent splash droplets
  on the sides of the mold (a metal
  funnel or trough may be used).
  The temperature of the eutectic
  mixture during pouring must be
  400°F ± 10°F

- To minimize warpage after salt has
  solidified, lay the assembly down
  horizontally with the top cover
  up. Allow the salt mandrels to
  cool down to ≤ 125°F in the fixture (using fans to facilitate assembly cool-down)

COLOR	GRAY
FORM	POWDER
MELTING & POURING TEMP	350 – 450°F
HEAT RESISTANCE	320°F
COEFF OF LINEAR EXPANSION	40×10^{-6} IN./IN./°F
COMPRESSIVE STRENGTH @ 76°F	18,000 PSI
COMPRESSIVE STRENGTH @ 320°F	2800 PSI
SOFTENING POINT	330 – 340°F
WATER SOLUBILITY RATE @ 140°F	5 – 10 MIN/LB 1.1 GM/MIN/CM^2
SPECIFIC VOLUME (SOLID CASTING)	12.9 – 13.3 CU IN./LB
DENSITY (SOLID CASTING)	128.5 – 132.5 LB/CU FT
LATENT HEAT OF FUSION	35 BTU/LB AVERAGE
SPECIFIC HEAT	0.33 – 0.7 BTU/LB°F
HEAT CONDUCTIVITY	SIMILAR TO WATER
DECOMPOSITION TEMP	APPROX 1200°F

R80-2149-004(T)

Fig. 4 Properties of Thermocast 320X

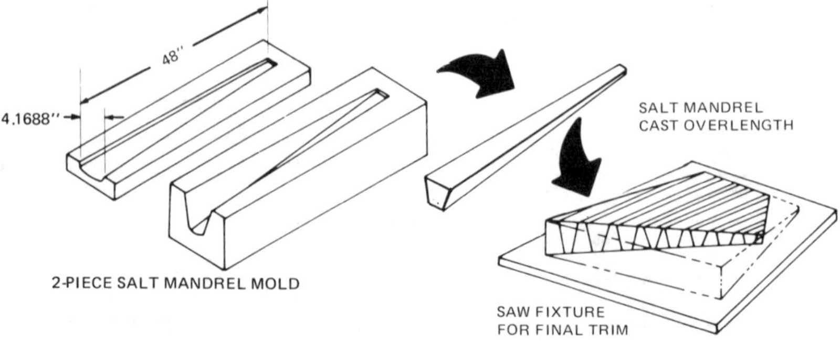

R80-2149-005(T)

Fig. 5 Eutectic Salt Mandrel Fabrication Sequence

- Inspect the completed salt mandrels and repair any pitting, seams, or surface irregularities (hand-sanding and a thermocast 320X/water filler may be used). Dimensionally check the mandrels for flatness (±0.020 out of plane).

  Finally, X-rayed the first six (6) mandrels to establish a process confidence level. Spray individual mandrels with acrylic lacquer to seal out moisture

- Minor cuts or breaks up to 2 sq in. may be repaired using EA 934 A/B or equivalent adhesive. Larger breaks or through cracks will be dispositioned per engineering direction.

Two mold forms were fabricated using cast aluminum tooling plate. Figure 6 shows a view of one of the 48-in.-long aluminum mold forms for casting the eutectic salt mandrels. Note the machined steps for ply drop-off.

Fig. 7 Steel Curing Fixture Consisting of Base Plate and Rails (Caul Plate Not Shown)

baseplate and rails. A 1-in.-thick steel caul plate (not shown) completes the curing fixture.

Each set of eutectic salt mandrels is cut and milled. The salt mandrels are first placed in the holding fixture, held with the aid of clamping blocks, and inscribed using a template. The individual mandrels were then cut with a band saw and routed using a form end mill cutter. Figure 8 shows the inboard salt mandrels in the curing fixture.

Fig. 6 View of Open Aluminum Mold Form Showing Machined Steps

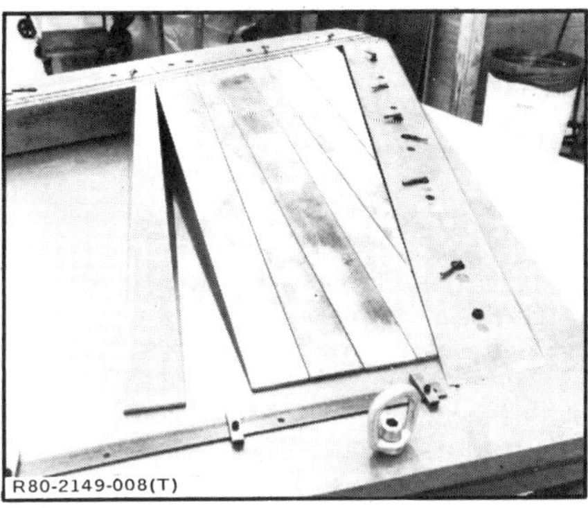

Fig. 8 Partial View of Curing Fixture with Eutectic Salt Mandrels in Place

The curing fixture design was also critical because of the thin gage of the skins and the size and mass of the mandrels. A horizontal laydown fixture was used in this program. This tooling method for assembly curing is a production method, and is an effective adaptation of the tooling used for the existing metal honeycomb trailing edge. Figure 7 shows the steel curing fixture consisting of

6. MATERIALS AND PROCESS REQUIREMENTS

The EF-111A horizontal stabilizer trailing edge was produced from 12-in.-wide uniaxial graphite/epoxy tape; this material form offers the best tradeoff of material and layup cost, and process compatibility. The selection of fiber is not significant since Hercules AS-1, Thornel T-300, and Celion 6000 all have adequate strength and stiffness properties. The critical item, however, is the resin applied to these fibers. Material with lack of tack or low flow must be avoided, since it is not compatible with non-autoclave-curing. Also to be avoided is high resin content prepreg, which results in resin-rich parts or use of excessively bulky bleeder systems. The bleeder systems are not only a high-cost center, but introduce a source of wrinkles and dimensional errors.

For this program, 3501-5A/AS-1 prepreg with a resin content of 37 + 2/-3% was specified. The standard Grumman bleeder system was modified to accommodate the reduced resin content graphite/epoxy prepreg and the vacuum-bag molding procedure. No bleeder was used for the web stiffeners, and a 50% reduced bleeder package (6:1; prepreg: 116 cloth) was utilized over the covers and front beam. The bleeder/breather system is described in Fig. 9.

7. MANUFACTURING SEQUENCE

Grumman's manufacturing approach to producing an integrally stiffened EF-111A horizontal stabilizer trailing edge was influenced by the interaction between the low resin content graphite/epoxy prepreg, the eutectic salt tooling concept, the integrally stiffened structural design, and the non-autoclave curing process. In the selected eutectic salt mandrel manufacturing concept, the structure is cured as a one-piece assembly-skins, stiffeners, inboard rib, front beam, trailing edge, and outboard closure rib using vacuum bag pressure. The salt mandrel tools are melted out during post-cure through holes in the front beam and the mixture is reclaimed for future mandrel tool fabrication. Figure 10 depicts the manufacturing sequence.

The manufacturing sequence consists of four phases: 1) detail part preparation; 2) assembly onto curing fixture; 3) cure/postcure/NDT

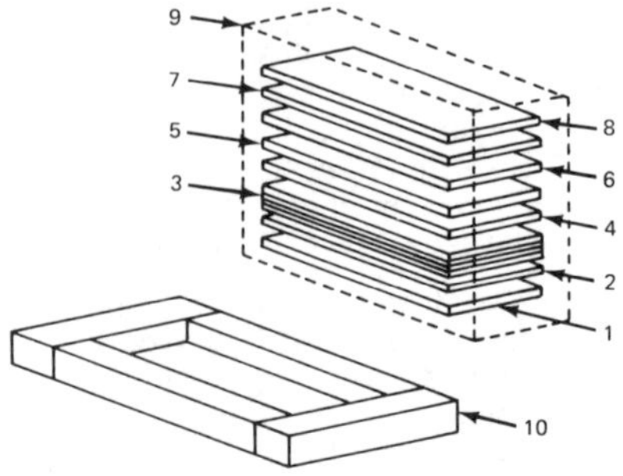

NOTES:

1. FLAT STEEL TOOLING PLATE COATED WITH MOLD RELEASE — FRECOTE 33 (OR EQUIVALENT) BAKED AT 350°F FOR 1 HOUR.

2. FIBERGLASS REINFORCED SILICONE RUBBER PAD (.020"). NOT REQUIRED FOR PROCESS VERIFICATION PANELS.

3. PREPREG LAYUP.

4. TEFLON IMPREGNATED CLOTH — TX1040 OR EQUIVALENT.

5. STYLE 116 OR 120 FIBERGLASS IN 6:1 (PREPREG:BLEEDER) RATIO ROUNDING OFF TO THE NEXT LARGER NUMBER (EXCEPT NO BLEEDER REQUIRED FOR WEB STIFFENERS).

6. TWO (2) PLIES OF 181 FIBERGLASS OVER ENTIRE COMPOSITE LAYUP.

7. .020" FIBERGLASS REINFORCED SILICONE RUBBER PAD AND AN ENGINEERED 1.0" STEEL CAUL PLATE. ALUMINUM CAUL PLATE (0.125" THICK) SHALL BE USED FOR PROCESS VERIFICATION PANELS ONLY.

8. THREE (3) PLIES OF STYLE 1000 FIBERGLASS BREATHER CLOTH.

9. NYLON BAGGING FILM.

10. PERIMETER DAM ("CORPRENE" OR EQUIVALENT METAL DAM). DAM TO BE ADJACENT TO LAYUP WITH A MAXIMUM GAP OF 1/16". PERIMETER DAM SHALL BE WITHIN NYLON FILM BAG; SHOWN SEPARATELY FOR CLARITY.

R80-2149-009(T)

Fig. 9 Bleeder/Breather Procedure For Vacuum Bag Molded Gr/Ep Laminates

inspection; and 4) final finish, as shown in Fig. 11. The front beam plies are laid up and stacked in the flat, and then vacuum-formed over a male tool which forms the front beam. The lightening holes are cut using a template. The integral stiffeners are designed to have only two layup orientations; thus, the required layup is cut from large uncured layups. The cut patterns are designed so that the layups are split and staggered at opposite corners of the mandrels to expedite layup. The mandrels are coated with room curing epoxy resin to minimize moisture pickup and then treated with flurocarbon mold release. A single ply of style 104

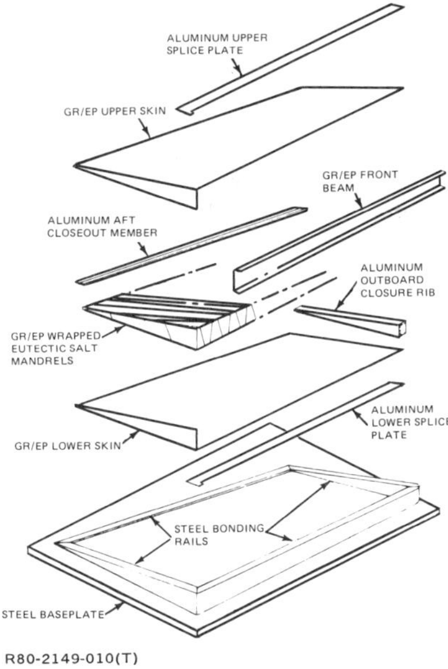

Fig. 10 Manufacturing Sequence

fiberglass/epoxy scrim is added on the graphite/epoxy in the radius

areas to hold the fiber orientation during transfer. The graphite/epoxy prepreg plies are then laid up on the mandrels and inspected (see Fig. 12). The upper and lower skin layups are produced using the ply-on-mylar technique and vacuum-compacted to assist in handling. The aluminum outboard closure rib is machined from plate, pretreated for

Fig. 12 Salt Mandrel Wrapped With the First Layer (45°) of 12-inch Graphite/Epoxy Tape

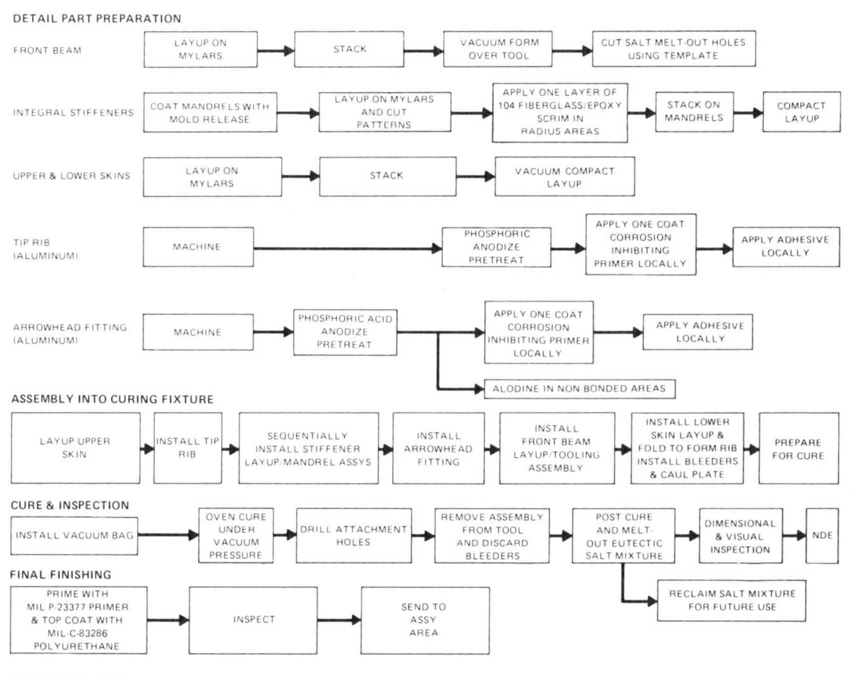

Fig. 11 Fabrication Process

209

bonding, phosphoric acid anodized, and then coated with corrosion-inhibiting primer. A similar procedure is followed with the two aluminum splice plates used to attach the trailing edge to the main torque box and the 6061 aluminum arrowhead.

The assembly of the trailing edge into the curing fixture begins with the placement of the aluminum upper splice plate. A layer of Metlbond 329 adhesive is applied to the plate, followed by a corrosion isolation layer of F161/7781 fiberglass/epoxy. The upper skin is then laid up with its bleeder system on the tool surface. Next, the pretreated outboard closure rib is also covered with a layer of Metlbond 329 adhesive and a layer of fiberglass/epoxy for corrosion isolation, and located in the fixture. The stiffener mandrel/layup assemblies are laid up, sequentially working inboard. Next, the front beam mandrel/layup assembly is installed. The 6061 arrowhead fitting is installed with a layer of Metlbond 329 adhesive, as well as a layer of fiberglass/epoxy for corrosion isolation. The lower skin is then added to the assembly and folded over to form part of the inboard closure rib web. Installation of the lower splice plate (in the same manner as the upper splice plate), lower skin bleeder/breather system, and caul plate finishes the layup process. After vacuum-bag-pressure cure, lightening holes, 1-in. in diameter, were drilled in the front beam between webs. Attachment holes for the 2024-T3 aluminum bearing plates on the front beam were also drilled. Figure 13 shows a plan-view of the trailing edge component with drilled lightening and attachment holes after post-cure. The eutectic salt melted out during post-cure. Some leftover salt in contact with the graphite/epoxy was water-washed out before installing the front beam bearing plates (7) using NAS1672-3L-3 fasteners. The component was located in the EF-111A horizontal stabilizer trailing edge assembly fixture (Fig. 14) and the center pilot holes for the shear pins opened to $0.4375 \begin{smallmatrix} +0.003 \\ -0.000 \end{smallmatrix}$ in. diameter (Fig. 15).

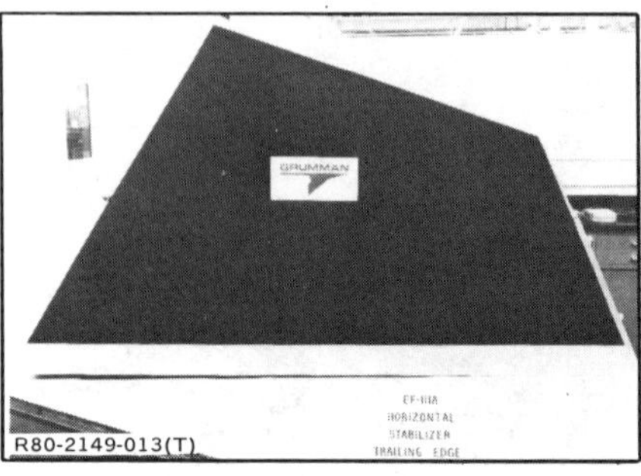

Fig. 13 Plan View of EF-111A Graphite/Epoxy Horizontal Stabilizer Trailing Edge

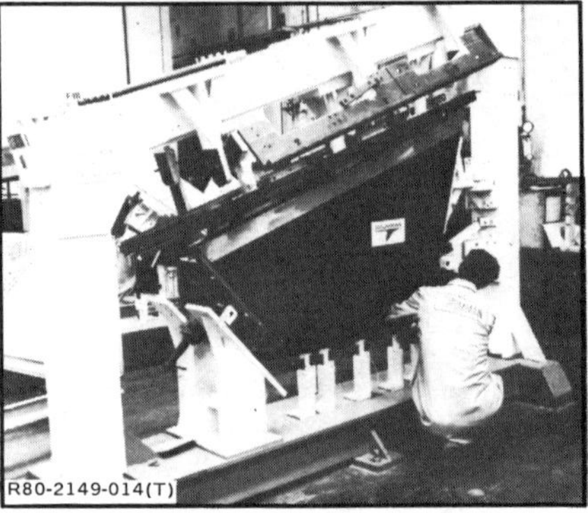

Fig. 14 EF-111A Horizontal Stabilizer Trailing Edge in Assembly Fixture

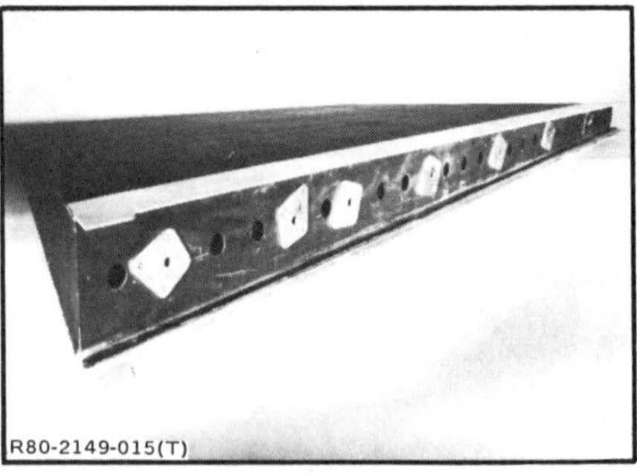

Fig. 15 View Showing Front Beam Of EF-111A Trailing Edge With Shear Pin Holes

In addition to the above, 24 1/4-in. diameter drain holes were drilled in the lower covers of the production demonstration components. These holes also allow for visual internal inspection of the trailing edges using fiberoptics. After fabrication, the components were attached to an existing stabilizer torque box (Fig. 16) for form and fit checks.

Fig. 16 Graphite/Epoxy EF-111A Trailing Edge Attached To Stabilizer Torque Box

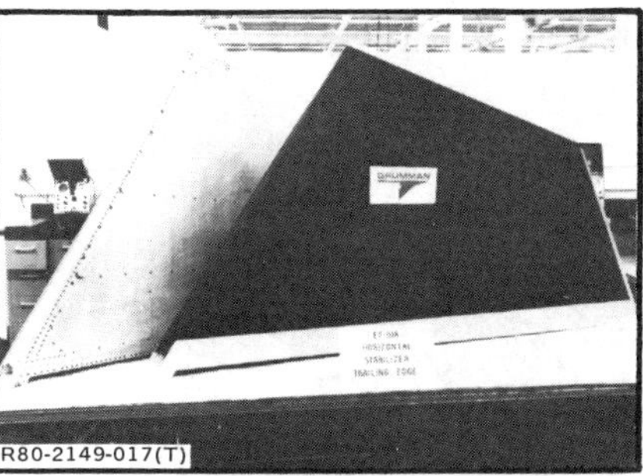

Fig. 17 Graphite/Epoxy And Metal EF-111A Horizontal Stabilizer Trailing Edges Side By Side

A quality control/nondestructive evaluation program plan was implemented. Ultrasonic evaluation was performed using the planned in-the-field contact resonance technique. Results were satisfactory. Selected sections of the substructure (radii of the mandrel plies) were also X-rayed and found to be well-compacted.

Skin compaction was checked by measuring thicknesses around the drain holes with a point micrometer. Twenty measurements were made; all were within tolerance. The front beams were also checked by measuring thicknesses of specimens cut from the web (1-in. diameter holes for salt extraction). The measurements were all within tolerance. In addition, the average horizontal shear strength value of 8.6 ksi for this $(0_2/90_2/\pm45_6)$ laminate compares favorably with the specification-required minimum of 7.0 ksi for a 0_{15t} laminate.

The average weight of the three graphite/epoxy trailing edges fabricated was compared to the weight of two existing metal trailing edges. The average weight of the metal components was 33.3 lb. The average composite trailing edge weighed 28.7 lb, a weight savings of 14% over the metal structure. Figure 17 shows graphite/epoxy and metal EF-111A horizontal stabilizer trailing edges side-by-side.

8. STATIC TEST

The critical HT-4 design condition was selected for static test of the EF-111A horizontal stabilizer trailing edge. The surface temperature associated with condition HT-4 is 317°F.

In addition, the predicted end-of-life laminate moisture content for the trailing edge is approximately 1.2% by weight. In order to conduct the test at ambient laboratory temperature and humidity conditions, the loads associated with condition HT-4 were increased by the following factors:

- Test at RT versus 317°F: increase loads by 1.25

- Test with specimen in essentially a dry condition versus containing 1.2% moisture: increase loads by an additional 1.25.

For the test, therefore, condition HT-4 loads were increased by a factor of 1.56.

The trailing edge was bolted to a dummy horizontal stabilizer rear beam and the completed assembly cantilevered from structural steel "L" frames (Fig. 18). Test loads were applied to the trailing edge surface by means of a whiffletree/tension pad system. Thirty-five 5 in. by 5 in. pads and three 4 in. by 4 in. pads were bonded to the compression surface. The pads were then connected together using aluminum angle and channel sections and steel eyebolt/barrel assemblies terminating at four load introduction points. Calibrated load links were installed in series with four hydraulic loading cylinders. Prior to test, the trailing edge test article was instrumented with 10 three-circuit rosette strain gages, and six deflection transducers.

Fig. 18 EF-111A Horizontal Stabilizer Trailing Edge Test Setup

The test specimen was subjected to the following test program:

- Test No. 1 - Functional test to 40% design ultimate

- Test No. 2 - Design limit load

- Test No. 3 - Failing load run.

For Test No. 1, the specimen was loaded in 5% increments of design ultimate to 40% design ultimate, and returned to zero load in 10% increments. During Test No. 2, the loads were applied in 5% increments to design limit and returned to zero in 10% increments. In Test No. 3, the specimen was loaded in 10% increments of design ultimate to design limit, returned to zero load, loaded in 10% increments of design ultimate to ultimate load, and then loaded at a constant rate to failure.

The EF-111A horizontal stabilizer trailing edge successfully withstood design ultimate load in the critical HT-4 design condition. Failure occurred at 135% of design ultimate load (DUL) in the compression cover. The flatwise tension skin disbond failure, which was not catastrophic, initiated in the outboard arrowhead tip bond area at Stabilizer Station 45.77. Crack initiation of the bonded splice at Stabilizer Station 27.00 was also observed. The predicted failure was in this region at 142% DUL. Therefore, if the load had not been applied using tension pads, the failure may not have initiated in the outboard arrowhead tip bond area and the failure load may have been closer to the predicted value of 142% DUL. Figures 19 and 20 show closeups of the disbond failure.

9. ACQUISITION AND LIFE CYCLE COSTS

9.1 Acquisition Cost Analysis

A cost analysis was performed to assess the benefits derived from a composite trailing edge versus a standard aluminum honeycomb assembly. The acquisition cost comparison was based on the cumulative average of the three fabricated composite units. The use of the eutectic salt mandrel tooling concept, 12-in.-wide graphite/epoxy tape with a semi-

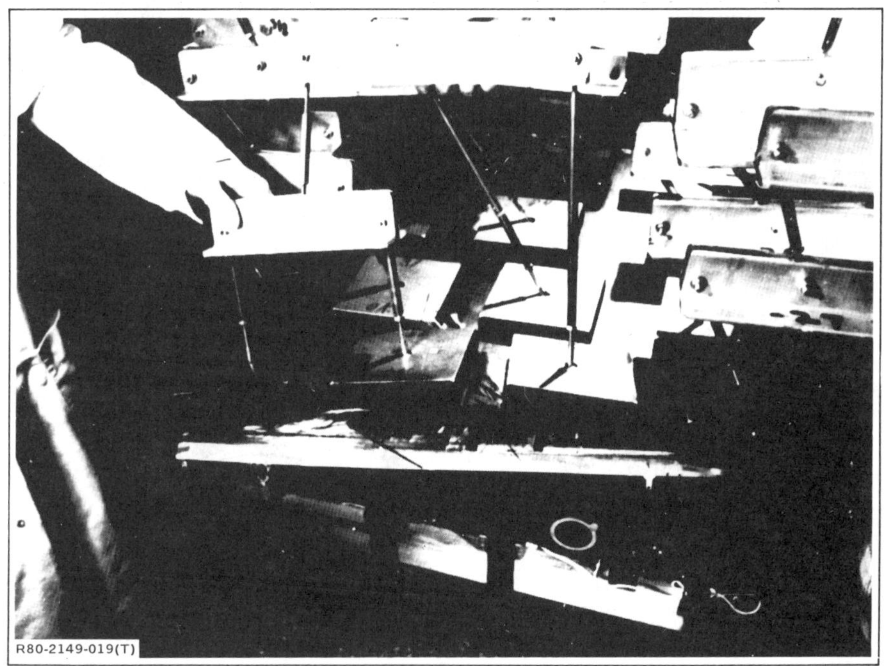

Fig. 19 Details of Aft-Inboard Edge Failure

Fig. 20 View of Aft-Outboard Edge Failure

213

automatic manufacturing process, shows a savings of 12% over the aluminum honeycomb design, Fig. 21.

9.2 Life-Cycle Cost Analysis

A life-cycle cost analysis was accomplished (Fig. 22) and shows a savings of 49% for vacuum-pressure-curing over a typical aluminum honeycomb design. The acquisition cost groundrules were the same but, in addition, assumed for Operations and Support (O&S) costs that no replacement of aluminum honeycomb trailing edges would occur over the life span of 20 years and that 20% of the aluminum O&S maintenance apply for the composite trailing edges. The 20% O&S maintenance on composite trailing edges was assumed by purging the metal data of most corrosion-induced problems. O&S cost was based on Grumman A-6A trailing edges and F-111 stabilizer torque boxes.

Using the Air Force equation for Return on Investment:

$$ROI = \frac{\text{Savings} - \text{Implementation Cost}}{\text{Implementation Cost}}$$

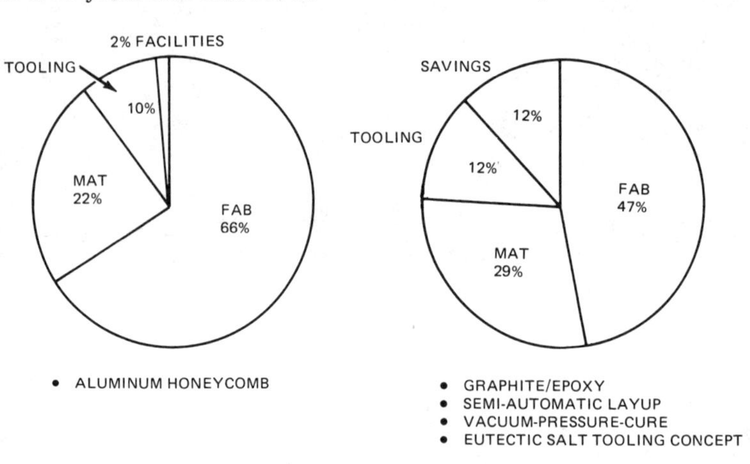

Fig. 21 Acquisition Cost Comparison (Based on Cumulative Average of 3 Units)

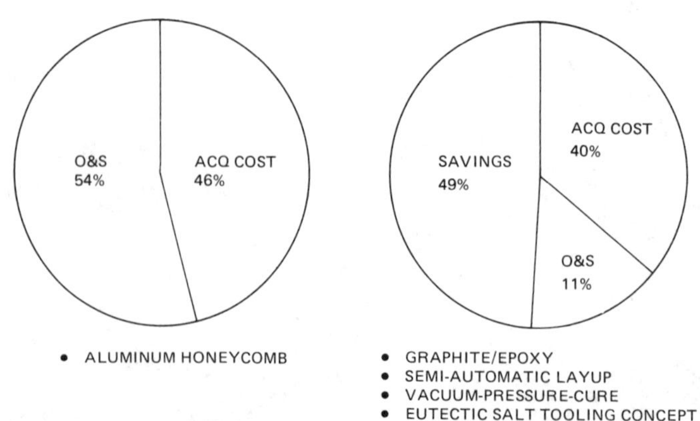

Fig. 22 Life-Cycle Cost Comparison (Based on Cumulative Average of 3 Units)

The ROI, based on the cumulative average of three units and assuming no aluminum honeycomb replacement spares, is 172%.

10. ACKNOWLEDGEMENT

The work described was performed for the Air Force Materials Laboratory under Contract F33615-78-C-5234. Mr. H.S. Reinert, Jr., AFWAL/MLTN, is the Air Force Project Engineer.

11. BIOGRAPHIES

Mr. Suarez holds an A.B. from Columbia College and a B.S., M.S., and Professional Degree in Engineering Mechanics from Columbia University. He is presently Group Leader in the Structural Mechanics Section at Grumman Aerospace Corporation, responsible for Advanced Composite development work. Mr. Suarez is a member of the American Society of Civil Engineers and the American Society of Mechanical Engineers.

Mr. Dastin holds a B.Ch.E. from City University of N.Y. He is presently Manager of Advanced Composites at Grumman Aerospace Corporation responsible for both in-house and contractural development of advanced composites structures. Mr. Dastin is a member of SAMPE and SPI.

ADVANCED COMPOSITES APPLICATIONS
IN RCA SATELLITES

Raj N. Gounder
RCA Astro-Electronics
Princeton, New Jersey   08540

Abstract

Application of advanced composites to the past and present RCA-built satellites is discussed. RCA satellites are built for a variety of missions including navigation, communication, military, meteorological, scientific and other missions. The drives for the application of advanced composites to satellite hardware come from light weight, high stiffness, high strength, thermal stability, RF transparency, and electrical behaviors of advanced fiber reinforced composites. Advanced fiber reinforced composite systems employed in satellite hardware include Kevlar/epoxy, graphite/epoxy, as well as glass/epoxy. Application of advanced composites to specific hardware such as antenna reflectors, antenna feed tower, waveguides and feed horns, multiplex microwave filters, precision mounting instrument platforms, solar panel substrates and structural subsystem hardware is discussed. The overall material selection criteria, design methods, and fabrication techniques are briefly described.

1.   INTRODUCTION

Since its inception in 1958, RCA Astro-Electronics has successfully been involved in the design and development of satellite systems for a large variety of missions. The missions include navigation, communication, military, meteorological, scientific and other (Figure 1). NAVSAT and NOVA are two examples of RCA satellites for navigational missions. RCA designed communications satellites include Satcom series of satellites for RCA Americom and Anik B for Telesat Canada. DMSP Series of satellites form the backbone of defense meteorological systems. The civilian weather satellites include TIROS and NOAA series of satellites. In the scientific area RCA has designed and built a long list of satellites including Ranger, SERT, Explorer and Dynamic Explorer.

The structures for the various satellite systems have difficult strength, weight, alignment and stiffness requirements. The limitations of the selected launch vehicle imposes strict size and weight constraints on the satellite structures. Advanced lightweight materials systems with superb specific strength and specific stiffness characteristics are required to meet this launch vehicle imposed weights requirements. The drive for the application of advanced materials with excellent specific strength and stiffness characteristics also come from the loads imposed on the satellite structure and other subsystem hardware during the satellite mission environment. Satellite structures are required to have proper stiffness to keep resonant frequencies separated from each other during launch and separated from any on-orbit excitations caused by rotating equipment. Advanced fiber reinforced composites with tailorable stiffness characteristics offer great potential to meet these satellite structural frequency requirements. Satellite structures also must keep the various instruments and control equipment properly aligned during the life of the satellite in order to provide the required pointing accuracies of the various instruments. This means the satellite structure and other subsystem hardware should be constructed of materials which are dimensionally stable under the mission environments of extreme temperatures, temperature distribution, radiation, vacuum and other imposed loads. The excellent coefficient of thermal expansion characteristics of advanced composite materials offer great potentials for designing dimensionally stable structures. In this paper, we shall briefly review selected applications of advanced composite materials to specific satellite hardware.

2. PROPERTIES OF ADVANCED COMPOSITES

Fig. 2 compares the stiffness and strength properties of the various types of advanced reinforcement fibers with the conventional materials. It can be readily seen that the stiffness and strength of advanced fibers are much superior to those of conventional materials. These superior properties of advanced fibers are being capitalized by reinforcing conventional materials such as organics and metals with these fibers. In a unidirectional prepreg or a lamina, these fibers are aligned parallel in a matrix so that the fiber reinforcement represents approximately 50 percent of the fiber properties, neglecting the properties of the matrix. Table 1 and Fig. 3 compare the unidirectional lamina properties of several epoxy matrix composites with the more conventional structural materials. Table 1 compares the structural mechanical properties (stiffness and strength) of advanced composites with those of conventional materials for equivalent volumes. The advantages of composites become even more evident when these properties are compared on an equivalent weight basis. The specific modulus (modulus/density) and the specific strength (strength/density) characteristics of advanced composites and conventional materials are compared in Fig. 3. The figure illustrates how composites can provide structural characteristics equivalent to metals at much less weight, or superior characteristics at equivalent size and weight. It is readily apparent from Fig. 3 that the unidirectional high modulus graphite/epoxy, for example is five times as stiff and three and a half times as strong as 7075-T6 aluminum, on an equivalent weight basis. As shown in Table 1, the advanced composites also exhibit excellent thermal characteristics. Both the Kevlar/epoxy and graphite/epoxy unidirectional composites exhibit either negligible or zero coefficient of thermal expansion characteristics. This property makes these materials excellent choices when building structures with exceptional thermal stabilities.

3. ADVANCED COMPOSITES APPLICATIONS IN RCA SATELLITES

The unique properties of advanced fiber reinforced composites have made it possible to utilize these materials in the design of a variety of spacecraft subsystems designed and built at RCA's Space Center (RCA Astro-Electronics). Almost every major subsystem in a satellite can derive special advantages by the application of advanced composites in their design. The satellites of the 1980's are designed to carry increasing numbers and volumes of payloads meant for a variety of missions. The goal is to accommodate a maximum amount of payloads at a minimum overall weight of the satellite system. This means the development of lightweight structures utilizing advanced lightweight materials is required wherever possible. In the following paragraphs, we will review the past and ongoing applications of composite materials to the design of a variety of subsystem components for the various RCA-built satellite systems.

3.1 <u>Advanced Composites are Being Utilized in Lightweight, Over-Lapping, Polarized Antenna Reflector Designs</u>

The application of advanced composites to a major spacecraft subsystem at RCA began with the Kevlar/epoxy sandwich, polarized, parabolic antenna reflector design for Satcom 1 launched in December 1975 (see Fig. 4) and Satcom 2 launched in March 1976. Besides the excellent specific strength and specific stiffness characteristics of Kevlar/epoxy composite, this material was selected for the antenna reflector design for two other reasons. As

mentioned earlier, Kevlar/epoxy
laminates can be designed to have
zero coefficient of expansion.
This property was utilized to provide distortion-free antenna reflectors as the temperature distribution on the reflector ranges
from very cold (approximately
-135°C) in shadowed regions to very
hot (approximately +100°C) in
regions directly exposed to solar
radiation during the satellite's
operational mission. More import-
antly, the combined requirements
of polarization isolation, restricted volume constraint of the
Delta fairing, and alignment
stability resulted in overlapping
reflectors with no deployment.
The two reflecting surfaces themselves consisted of grids of parallel
copper wires arranged orthogonal
to each other. This meant any
structure used to support the
reflecting grid elements should be
transparent to rf signals. Kevlar/
epoxy was found to be such a
material with its low-loss dielectric characteristics. Fig. 5
shows one of the two 50-in focal
length, 70 x 50 in aperture Kevlar/
epoxy composite C-band antenna
reflectors that was designed and
fabricated under an IR&D program to
meet the mission static and dynamic
loads. Each of these reflectors
weighs less than 4.5 lb. These
reflectors have been extensively
tested for their electrical and
mechanical performances for the
purpose of developing antenna

structures that will meet the
future communications payloads
requirements. The knowledge
gained from the design, development,
and testing of these IR&D
reflectors are being utilized to
the design and fabrication of much
larger (94 in x 68 in) antenna
reflectors for Satcom F. These
reflectors combine the unique
properties of Kevlar/epoxy
materials with ingenious structures
designs to meet the more drastic
requirements of ultra lightweight
structures with ultra high
stiffness, natural frequency, tolerance, and stability requirements
of the next generation communications payloads.

3.2 <u>Advanced Composite Feed Tower,
Waveguides, and Feed Horns Take
Advantage of the High Specific
Stiffness and Strength Properties
of Graphite/Epoxy</u>

Also, in the communications system
area, graphite/epoxy designs have
been developed for lightweight
feed tower, waveguides, and feed
horns. Figure 6 shows examples of
straight and ridged waveguides,
E-bends, H-bends, and feed horns
that have been designed and fabricated out of graphite/epoxy
materials. These structures are
coated with special multilayer
metallic films which protect the
graphite/epoxy from deleterious
moisture penetration and provide
rf conductive interior surfaces.
Similar designs were used for the
tower and feed systems of Satcom 1

and Satcom 2 communications satellites shown in Figure 7.

3.3 Lightweight, Thermally Stable Multiplex Microwave Filters Have Been Built Out of Graphite/Epoxy

Invar is the common material used in the design of multiplex microwave filters needed for communications satellites. Although this material meets the frequency stability requirements for the microwave cavities, invar filters result in a heavy weight penalty, even with thin wall designs. The advent of graphite/epoxy technology has met both the lightweight and the thermal stability requirements imposed on multiplex microwave filters for communications satellites. RCA has developed a special forming and plating process for graphite/epoxy filters which provide well adhering and smooth metallic layers ensuring low rf losses. The graphite/epoxy multiplex microwave filters used on Satcom 1 and Satcom 2 are shown in Figure 8.

3.4 Graphite/Epoxy Designs Have Been Developed which Meet the Difficult Weight, Alignment, Stiffness, and Strength Requirements of Precision Mounting Platforms (PMP)

Figure 9 shows an advanced composite design of a precision mounting platform for the Air Force DMSP Weather Satellite. The PMP supports several instruments that must be precisely aligned at assembly and remain aligned in space. The conventional design employs a dip brazed aluminum structure. The aluminum structure needs to be maintained at a constant temperature to avoid any thermal distortion and, hence, misalignment during the spacecraft's mission. Moreover, the aluminum design is heavy and is limited in its stiffness, strength, and natural frequency capabilities. The graphite/epoxy PMP design shown in Figure 9 and its improved versions under development promises to alleviate these drawbacks of aluminum and meet the stringent pointing accuracy requirements of future space missions.

3.5 New Advanced Composites Designs Have Been Developed for Lightweight, Thermally Stable Satellite Structural Subsystems

A spacecraft structural subsystem has several functions. It primarily provides adequate strength and stiffness to support and maintain the relative locations of components. Secondary functions include providing thermal paths for heat conduction, grounding electronic equipment, shielding equipment from space radiation, and protecting against potential thermoelastic and dynamic instabilities. The major constraints on the structure are weight and stability. Historically, aluminum has been the primary material for spacecraft construction. However, the limitations of aluminum in terms of weight, stiffness, strength, and thermal stability have provided us high incentives to design and develop advanced composite, especially graphite/epoxy, structures for structural subsystems. Figure 10 is an example of such an advanced composite structural subsystem development at RCA. The figure illustrates an advanced composite truss structure for the dual launching of two NOVA satellites stacked on a single cradle assembly. The truss structure is designed to meet the very stringent requirements of weight, natural frequency, and minimum distortion under mechanical and thermal loads of the mission environment.

4. COMPOSITE STRUCTURES ARE DESIGNED THROUGH DETAILED MODELING AND STRESS ANALYSIS

Structural components fabricated from composites are designed as multi-layer laminated configurations for most effective use of unidirectional or woven composite materials. Each ply of the laminate has unique orthotropic stress-strain characteristics that contribute to the total stiffness and strength of the multi-ply laminate. Proper design of a composite laminate structure requires detailed modeling and stress analysis of the individual laminae (ply) to completely define the overall behavior of the structure. The properties of a unidirectional ply may be obtained from the properties of the reinforcing fiber and the reinforced matrix using micromechanics techniques. The

most common practice, however, is to measure the stiffness and strength properties of the unidirectional composite by simple tensile, compressive, and shear tests performed by loading in directions parallel and transverse to the fiber. A minimum of four different elastic constants, six different strength constants, and several physical constants (see Table 2) are needed in order to analytically characterize the performance of a multi-ply, multidirectional laminate. Special computer programs have been developed to fully characterize multi-ply laminated composite structures. The analyses use linear theories, but are very complex because the number of independent parameters necessary to specify the complete structural vehavior is many times greater than that needed for the usual homogeneous isotropic analyses of conventional structures.

The failure in a multi-ply laminate is defined as the onset of failure in the weakest ply in the laminate. Failure envelopes, such as the example shown in Figure 11 for a cross-ply laminate, are developed to fully characterize the strength behaviors of a laminated composite structure. Computer programs have been developed to calculate the margins of safety of any laminate orientation under any combined in-plane and out-of-plane mechanical loads as well as environmental

loads such as temperature and
humidity. Also, advanced finite
element modeling (Figure 12) and
analysis are extensively employed
to characterize the response of
complex composite structures under
static, dynamic, and acoustic
loadings.

Specific layup patterns are used
for special purposes. For example,
maximum strength or stiffness for
components such as beams or columns
is obtained from unidirectional
orientation. Quasi-isotropic
properties are obtained with
$(0^{\circ}/+45^{\circ}/-45^{\circ}/90^{\circ})_S$ and
$(0^{\circ}/+60^{\circ}/-60^{\circ})_S$ laminates. Because
of the anisotropic thermal expan-
sion behavior of individual laminae,
it is necessary, in order to
prevent warping due to temperature
changes or during fabrication, to
use laminate orientations that are
symmetric about the mid-plane of
symmetry of the laminate, and are
balanced (equal number of plus and
minus plies). For cylinders or
other surfaces of revolution that
cannot warp, the layup need not be
symmetrical.

    5. PREPREG LAYUP AND FILAMENT
WINDING ARE TWO COMMON TECHNIQUES

    FOR COMPOSITES FABRICATION
Prepregs are purchased in forms of
unidirectional tapes or woven
fabric broadgoods. The most common
practice is to fabricate composite
structures by laying up thin
(approximately 0.005 inch) plies of
prepregs into shapes and filament

orientations as dictated by the
design. Plies, gores, or patterns
are cut from prepreg by use of
templates. The plies are then
carefully stacked so that the
individual laminae are oriented as
required by the design. The next
step is to compact the laminate
thus assembled by simultaneous
application of heat and pressure.
Vacuum bagging is a common technique
for thin laminates and composite
skin honeycomb core sandwich structures which do not require high
compacting pressures. For thick
laminates which require high
pressures for compacting, other
means of pressurizing such as an
autoclave or a hydraulic press is
employed. Figure 13 depicts the
sequence of operations involved in
the fabrication of a communications
antenna reflector.

Composite structures are also
fabricated starting from dry fibers
by means of specialty processes.
For example, structures which may
be described as surfaces of
revolution such as tubes, cylinders,
and pressure vessels are fabricated
by filament winding. In this
process, dry fiber rovings are
saturated by passing them through a
resin bath and are then positioned
over a mandrel in a helical pattern.
The wound structure is then cured
in an oven.

    6. CONCLUSION
The satellite systems of the next
decade demand stringent pointing

accuracies, increased payloads,
compatibility with shuttle launch,
the attendant weight/size constraints, ability to perform under
hostile environments, and hardware
and operational cost efficiencies.
Judicial application of advanced
fiber reinforced composites to
selected satellite structures has
already proven to satisfy some of
these requirements. In the remaining part of the twentieth
century, an expanded and a total
systems approach application of
advanced composites to Astro's
satellite systems is predicted
which will meet all of the above
future mission requirements. The
composites technology is evolving
into a more matured and reliable
technology and it is safe to say
that composites are here to stay.

TABLE 1. ENGINEERING PROPERTIES OF ADVANCED FIBER/EPOXY UNIDIRECTIONAL COMPOSITES COMPARED TO CONVENTIONAL STRUCTURAL MATERIALS

	Glass	Kevlar	Graphite			Boron		
	E Glass/ Epoxy	Kevlar 49/ Epoxy	HS GR/ Epoxy	HM GR/ Epoxy	UHM GR/ Epoxy	Boron/ Epoxy	Aluminum	Titanium
Tensile Strength, KSI	260.	220.	220.	170.	110.	230.	82.	125.
Tensile Modulus, MSI	8.	11.	20.	28.	44.	30.	10.	16.
Density, lb/in^3	0.072	0.05	0.057	0.059	0.061	0.073	0.10	0.16
Coefficient of Thermal Expansion, in/in/$^\circ$F	3.5	-2.2	-0.2	-0.3	-0.65	2.5	13.	5.
Thermal Conductivity, Btu/hr/ft/$^\circ$F	2.	1.	8.	31.	65.	1.	70.	4.
Raw Material Cost, $/lb	10.	15.	50.	75.	500.	500.	2.	30.

TABLE 2. The table shows the stiffness, strength, and physical constants experimentally determined for a unidirectional laminae. Special computer programs are used to calculate laminate modulus and strength properties from the individual ply constants.

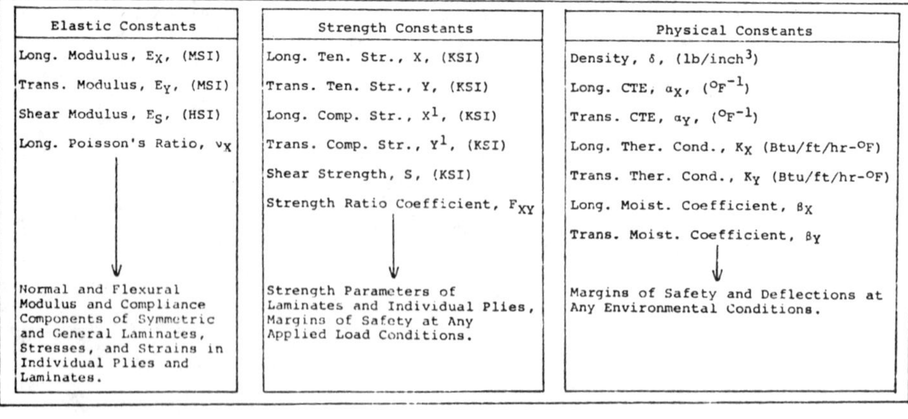

Figure 1.   RCA Satellites and Their Missions

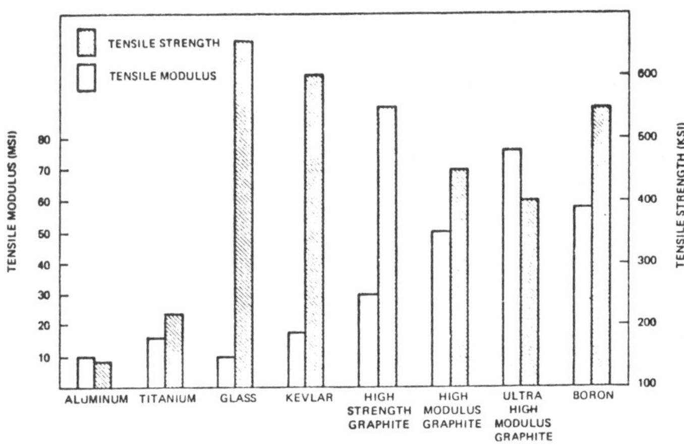

Figure 2.   Tensile Modulus, Strength Characteristics of Advanced Fiber
Reinforcements Used in Composites and Conventional Materials

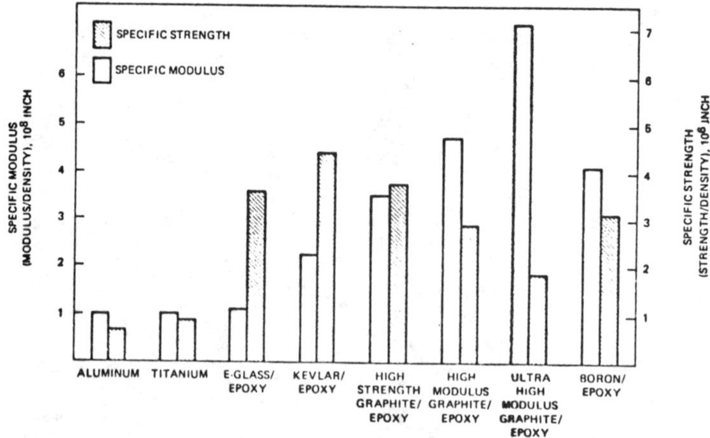

Figure 3.   Specific Modulus (Modulus/Density) and Specific Strength (Strength/Density) of Advanced Composites Compared with Those of Conventional Materials

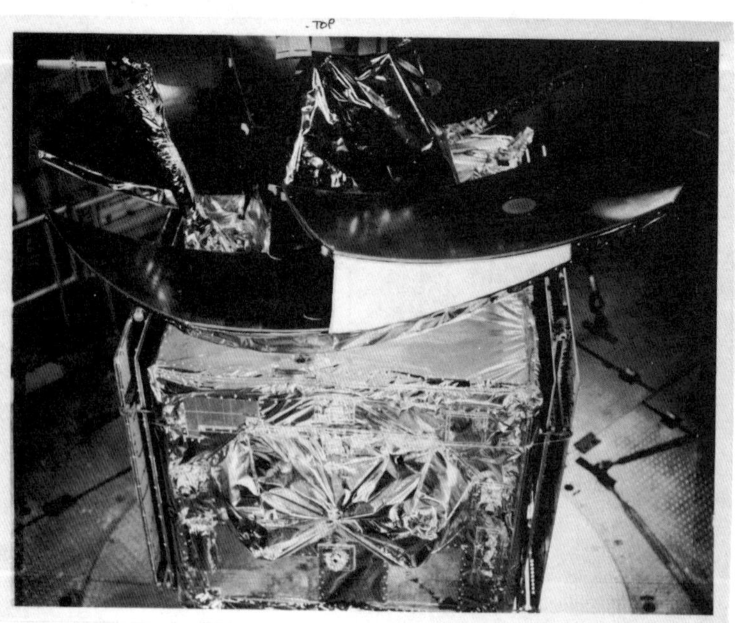

Figure 4. Kevlar/Epoxy Antenna Reflector Assembly of Satcom 1

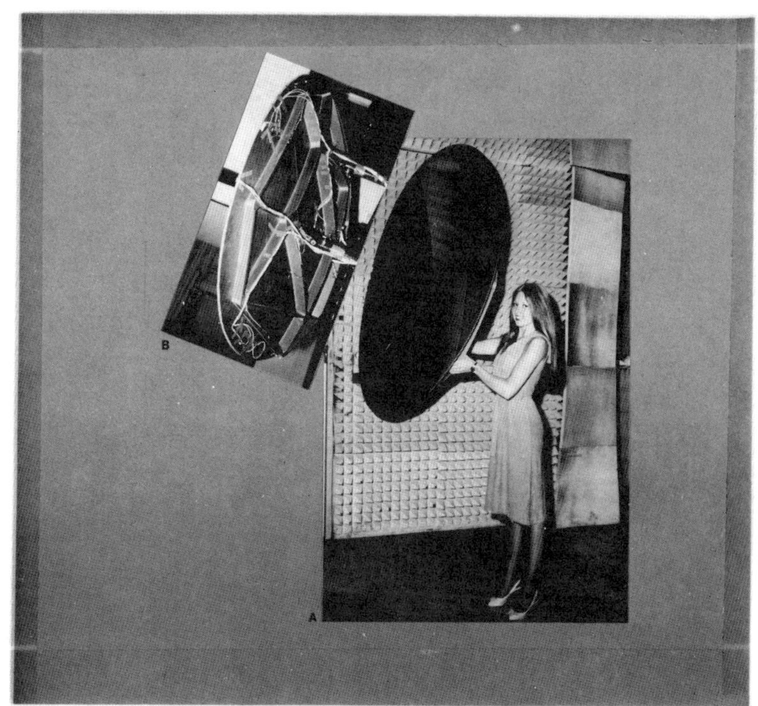

Figure 5. Kevlar/Epoxy Composite IR&D C-Band Antenna Reflector

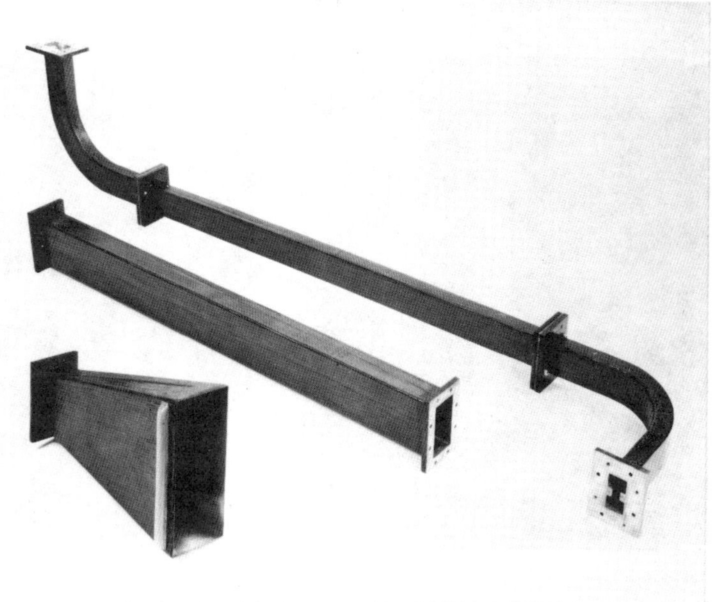

Figure 6. Graphite/Epoxy rf Waveguides, which Include Straight and Ridged
Waveguides, E-Bends, H-Bends, and Feedhorns

227

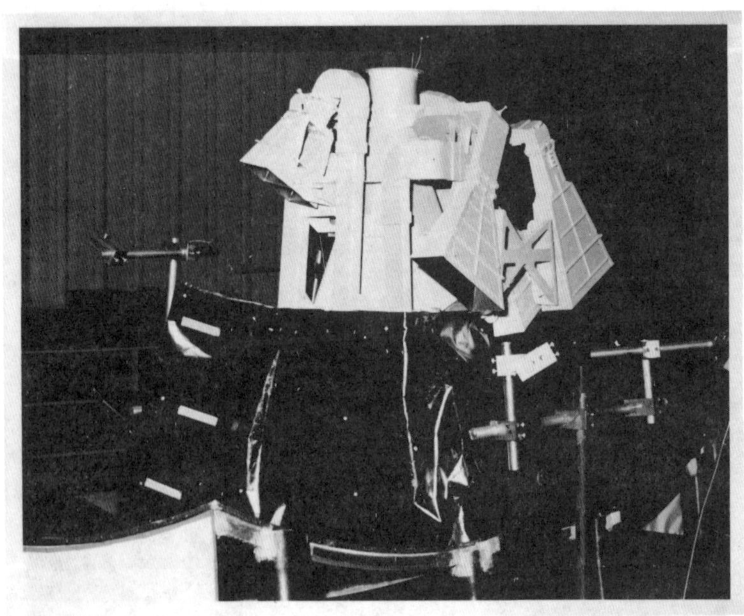

Figure 7. Graphite/Epoxy Tower and Feed System of Satcom Communications Satellite

Figure 8. Graphite/Epoxy Multiplex Microwave Filters Used on Satcom 1

Figure 9.   Advanced Graphite/Epoxy Composite Design of the Precision
Mounting Platform (PMP) Developed for the Air Force DMSP
Weather Satellite

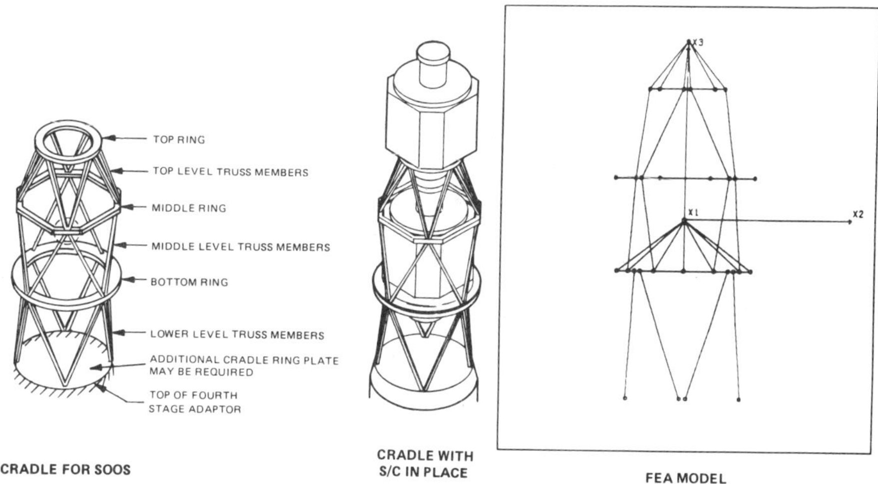

Figure 10.   An Advanced Composite Cradle Design for the Tandem Launching of
Satellites

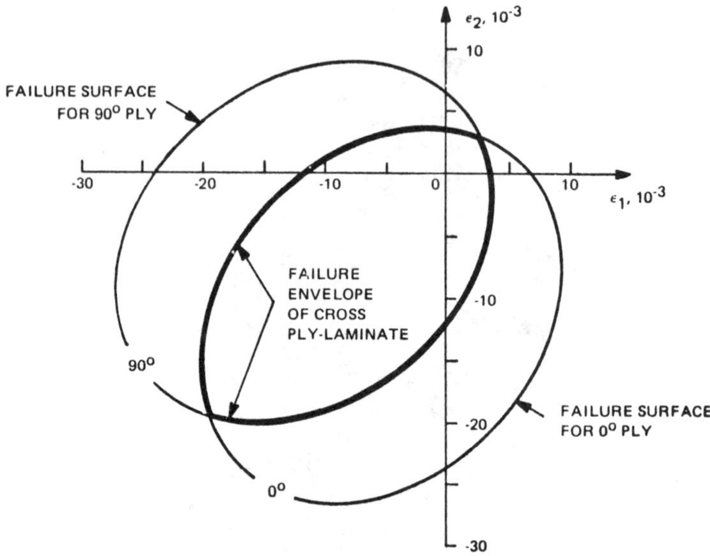

Figure 11. Failure Envelope for a Cross-Plied (0/90) Graphite/Epoxy Laminate in the Strain Space at Zero Applied Shear Strain

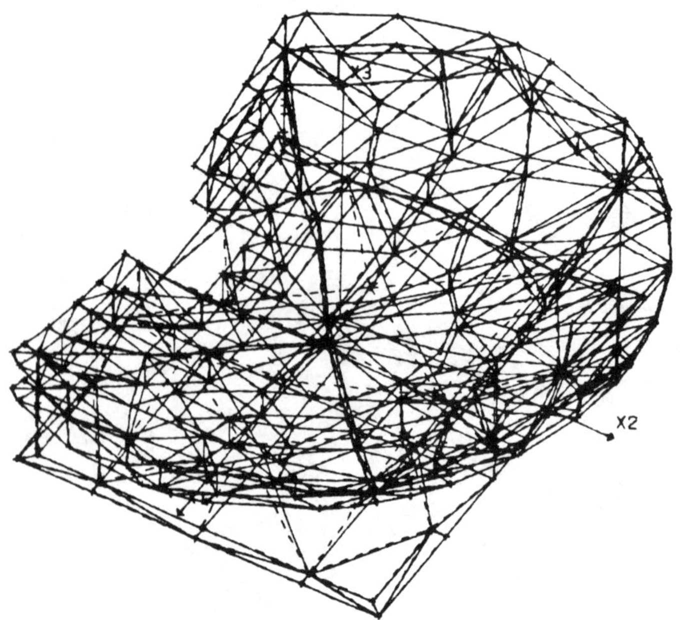

Figure 12. Finite Element Model of a Fully Overlapping Kevlar/Epoxy Antenna Reflector System Showing Distortions Under Mission Thermal Environments

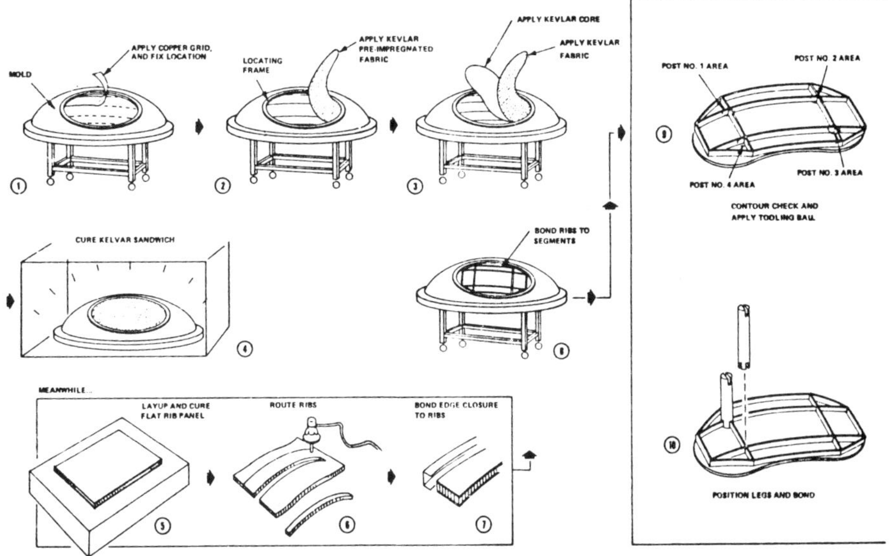

Figure 13. Processing Steps Involved in the Fabrication of a Kevlar/ Epoxy Antenna Reflector

BIOGRAPHY

Raj Gounder, Manager, Composite Materials, RCA Astro-Electronics, is responsible for advanced structures design and development, advanced structures analysis and testing, and advanced materials and fabrication technologies required for future satellite missions. At RCA, Dr. Gounder has led the mechanical design of antenna subsystem for future communications missions and other lightweight structures technology developments including composite solar panel substrates and structural subsystem required for stacked launching of satellites. Before joining RCA in July 1979, he led the application of composite materials to helicopter hardware at Sikorsky Aircraft, Stratford, Connecticut.

26th National SAMPE Symposium
April 28-30, 1981

THE SHUTTLE ORBITER THERMAL PROTECTION SYSTEM: A MATERIAL AND STRUCTURAL OVERVIEW

L. J. Korb, H. M. Clancy
Space Transportation System
Development and Production Division
Rockwell International

Abstract

The Space Shuttle System has been developed to provide an economical method for placing payloads and personnel into low earth orbits. The Shuttle orbiter, the cargo-carrying element of this system, will be launched, orbit, and reenter as a spacecraft. Once into the lower atmosphere, it will be maneuvered to land as an aircraft. One of the critical design problems has been the development of a lightweight, reusable thermal protection system capable of surviving not only the entry heating but also the launch and maneuvering loads. The major portion of this thermal protection system consists of nearly 31,000 pure silica tiles bonded to the vehicle through nylon felt strain isolator pads. The silica tile is an excellent insulator; its glazed surface coating provides high emittance to reject entry heating.

Difficulties have been encountered in both design and manufacture in achieving the aerodynamically smooth, plasma-tight installation required and to ensure that the system has adequate structural integrity. An overview of the thermal protection system design and the methods employed to ensure its structural integrity are described.

Keywords

Space Shuttle, Orbiter, Thermal Protection System, Silica Tiles, Manned Space Program

1. INTRODUCTION

The Space Shuttle System is designed to provide an economical method of transporting men and supplies into low earth orbit. The system (Fig. 1) consists of four major elements: the solid rocket boosters (SRB), the Space Shuttle main engines (SSME), the external tank (ET), and the Space Shuttle orbiter (SSO). The system is capable of launching up to 65,000 pounds of payload into orbit and returning 32,000 pounds of payload from orbit.

Nearly 80 percent of the thrust for launch is provided by the solid rocket boosters. The remaining 20 percent is supplied by the main engines, which burn hydrogen with oxygen supplied by the external tank. The orbiter, launched vertically from a piggyback position

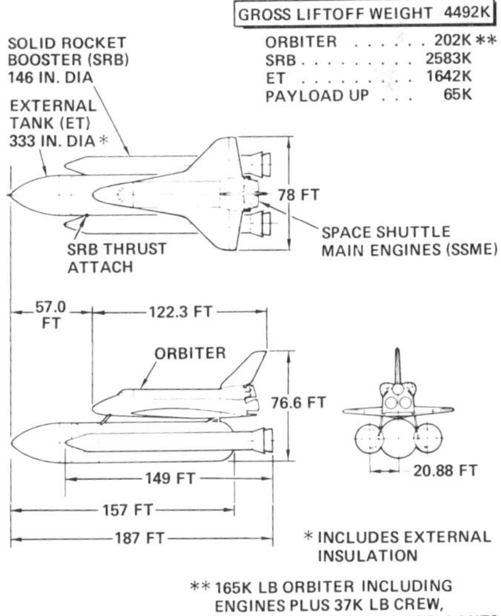

Figure 1. Space Shuttle System

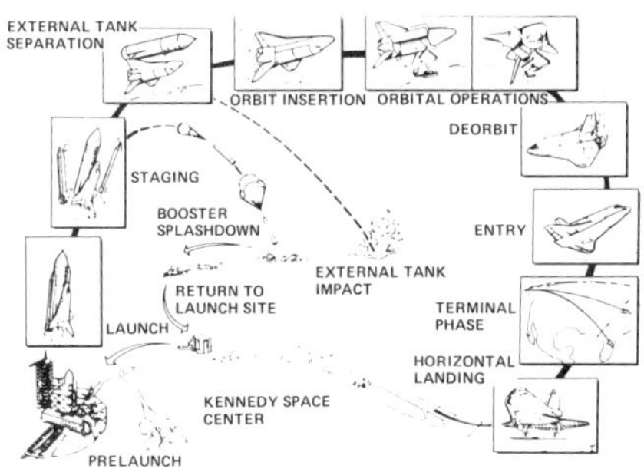

Figure 2. Typical Mission Profile

astride the external tank, contains the astronauts, the main engines, and the payloads.

The orbiter must function as both a spacecraft and an aircraft. During entry from orbit, it must be protected from temperatures exceeding 2300F on its lower fuselage and in excess of 2700F along the leading edges and nose cone. At an altitude of approximately 150,000 feet, the orbiter will slow to about eight times the speed of sound and will pass its maximum heating. At 50,000 feet, the orbiter will enter into a level flight path and will be maneuvered aerodynamically to land as a glider. A typical mission profile is shown in Fig. 2.

The requirement to achieve a minimum weight orbiter (165,000 pounds dry weight) has necessitated use of the most efficient structural materials and processes. The requirement for 100 mission reuse has extended advancements in thermal protection materials well beyond the state of the art existing at the contract inception.

Both weight and cost dictated that the basic orbiter structure be made from aluminum. In many areas, such as the cargo bay doors and orbital maneuvering subsystem (OMS) pods, graphite-epoxy was used to provide the minimum weight structure. Both aluminum and graphite are limited to a maximum of 350F to avoid degradation. Thus the orbiter thermal protection system (TPS) must function within the temperature regime dictated by the temperature limits of the TPS materials on its outer mold line and the temperature constraints of the vehicle structure at the inner mold line.

2. DESIGN REQUIREMENTS

The thermal protection system is designed to perform a variety of missions, each of which modifies the thermal environment of the TPS during its ascent, on-orbit, entry, maneuvering, and landing phases. The inclination of the orbit (e.g., polar vs. east-west), the total time in orbit, the requirements for vehicle altitude with regard to the earth and sun, the payload requirements, and the range and cross range requirements are among some of the mission parameters that must be accommodated. Additional localized heating and impingement from plumes of the solid rocket boosters,

Space Shuttle main engines, orbital maneuvering subsystem, and reaction control subsystem engines (RCS) must also be accommodated. The TPS, under normal mission operations, must be capable of 100 missions with minimum refurbishment and must support a 160-hour vehicle turnaround for relaunch. The TPS must ensure survival in the event of a pad abort or an around-the-earth-once abort (AOA).

One of the keys to achieving a reusable thermal protection system is to minimize the temperatures. To do so, it is necessary to delay the transition from laminar to turbulent flow as long as possible into the entry cycle. This requires that very high levels of aerodynamic smoothness be built into the design.

The TPS not only must ensure that the substructure is kept below 350F but must be able to accommodate the stresses and strains resulting from both thermal, aerodynamic, and structural loads.

Finally, the TPS must be designed in such a manner so as to minimize fluid entrapment or absorption that could occur from a rainy or humid environment. The design must consider the potential for ice formation that could occur during cooling either from tanking (of the hydrogen and oxygen cryogens) or from evaporative cooling of entrapped moisture during launch.

3. SELECTION OF THE TPS CONCEPT

There have been a wide variety of thermal protection system concepts studied in the aerospace industry over the last 20 years. These concepts include ablative materials, hot radiative metallic structures, ceramic insulations, heat sinks, transpiration cooling systems, etc. Only the first three could be considered serious trade-off candidates for a system as large as that on the orbiter (e.g., a heat sink system would be too heavy; transpiration cooling would be too complex.)

Ablatives have been widely used for entry nose cones on military missiles, for linings of rocket engine nozzles, as well as for the primary heat shields for the Mercury, Gemini, and Apollo spacecrafts. In these programs, the ablative material was encapsulated into an open-faced fiberglass honeycomb sandwich to ensure its structural integrity—that is, to prevent ablator loss from cracking at low temperatures and to retain the char surface at elevated temperatures. Although the extensive aerospace experience would support the choice of an ablator, the use of ablators for a major portion of the TPS presented two distinct shortcomings. First, ablators do not retain their aerodynamic surfaces through the entry trajectory. Use of ablators would result in costly refurbishment after each flight and would not readily support either the 100 flight reuse or the 160-hour turnaround requirements. Second, ablator materials such as those used on Apollo had nearly four times the density of the ceramic insulators being considered. In spite of the ability of ablators to operate at higher temperatures to permit faster rates of descent, the authors doubt that ablators were truly competitive from a weight-efficiency standpoint.

Hot radiative structures have been used for many years in aerospace. The X-15 used Inconel alloy X-750 as both the aerodynamic and radiative surfaces to temperatures of approximately 1150F (wing and nose areas included heat sinks; later models also employed a thin ablative layer). The Mercury and Gemini spacecraft used shingles made from Rene 41 on sidewalls to endure entry temperatures up to approximately 1700F. Rocket engine nozzles, such as the columbium nozzles used on Apollo engines, were designed for service up to 2400F. Yet, in spite of this experience base and the excellent development on a columbium metallic heat shield for the Shuttle (funded by NASA Langley Research Center), the use of a hot radiative metallic structure for a major portion of the TPS had several drawbacks. First, the concept was heavy. Second, it had limited overtemperature capability. Its fragile silicide coating is easily damaged. Localized loss of coating could result in rapid oxygen embrittlement at high temperatures and probably, catastrophic ignition above 2730F, the melting temperature of its principle oxide, Cb_2O_5. Perhaps even more significant, however, was the design and manufacturing complexity. The problems of attaching panels of such a system to a typical aircraft substructure, yet maintaining the close step and gap tolerances necessary to avoid excessive heating from early boundary layer transition or plasma ingestion, are considered formidable. The design and thermal stress analyses of even the simplest structural panel is quite time-consuming. The analysis must ensure that panel buckling or excessive deflections would not occur on either initial or subsequent loading cycles. The installation is compounded by the large number of clips, brackets, stand-offs, frames, beams, and fasteners required. The complexities of packaging insulation, of controlling plasma ingestion while allowing for expansion and

contraction, and of limiting fasteners to 1600F (to permit reuse) are difficult challenges.

No single material approach could be used efficiently for the Shuttle orbiter thermal protection system. To a limited extent, both ablators and hot radiative metallic structures are used. The ablator used is the same material system as that used on the Apollo; however, its use is confined to the small area between the elevons where the combination of plasma flow and poor viewing factor (restricting reradiation) results in temperatures well above 3000F. Hot radiative metallic panels of Inconel 625 are used in the engine-mounted heat shields (to approximately 1600F) and for flipper doors and elevon seal panels (to approximately 1400F).

Three material systems are used for the majority of the Shuttle orbiter thermal protection system (Fig. 3). For temperatures in the 350-750F range, felt blankets (FRSI) made from an aromatic nylon (Nomex*) and coated with a room-temperature vulcanizing elastomer are bonded directly to the aluminum substructure. These are located on the upper surfaces of the wings and along the fuselage side walls. (A typical isotherm map of the orbiter is present in Fig. 4.) For the temperature range of 750-2300F, which represents by far the largest area of the orbiter, a coated ceramic tile system, made from amorphous silica, is employed. The wing leading edge and nose cone areas, where temperatures climb, in some locations, above 2700F, use a coated reinforced carbon-carbon (RCC) composite material.

---

*Registered trademark, E. I. du Pont de Nemours and
Company (Inc.).

- TOTAL RSI CERAMIC TILES — 30,759
- REINFORCED CARBON/CARBON (RCC) (44 PANELS/NOSE CAP)
- FELT REUSABLE SURFACE INSULATION (FRSI) (3,581 FT^2)

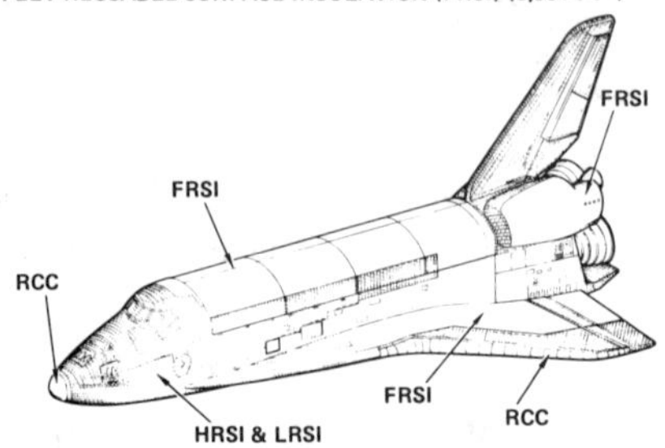

*Figure 3. Thermal Protection
System Materials*

Each material system presents a variety of challenging and intriguing problems. Because the ceramic tile system constitutes the major portion of the Shuttle thermal protection and because this unique system offers such difficult challenges, the authors have chosen to discuss this system in more detail.

4. THE TILE SYSTEM

The basic tile system is composed of three key elements: a ceramic tile, a nylon felt mounting pad, and a room-temperature vulcanizing elastomeric adhesive (RTV). The tile is coated with a high emittance layer of glass

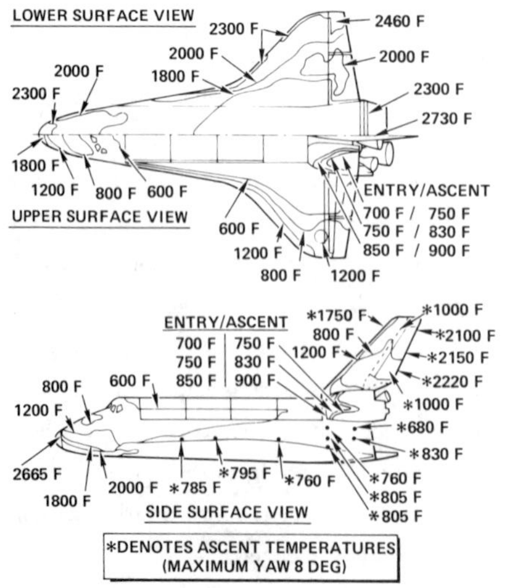

*Figure 4. Orbiter Isotherms —
Typical Trajectory*

and functions as both an insulator and a radiator to limit heating of the structure. The felt mounting pad, called the strain isolator pad (SIP), provides for mechanical isolation of the tile from the vehicle deflections and strains. The RTV bonds the tile to the SIP and the SIP to the structure (see Fig. 5).

4.1 Silica Tiles

Ceramic tiles, made from pure amorphous silica or mullite, have been under laboratory investigation for several years. Prior to the Shuttle proposal, NASA had funded developmental studies on these materials. The system chosen for the Shuttle orbiter was that made from silica. From the beginning, it was clear that a silica tile system offered considerable advantages over the other systems; however, until designs matured in the late 1970's, some of its key design problems were not fully recognized.

The silica tiles are made from a very-high-purity amorphous silica fiber approximately 1.2 to 4 microns in diameter and up to 1/8 inch long. These are felted from a slurry, pressed, and sintered at approximately 2500F into a tile production

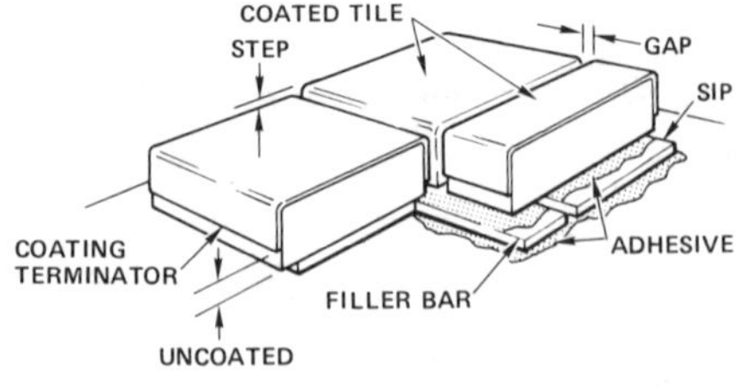

MATERIALS

TILE — 22 LB/FT3 / 9 LB/FT3 PURE SILICA FIBER-FIRED AT ~2500 F

COATING — BOROSILICATE (GLASS) FOR WATERPROOF & THERMAL PROPERTIES—FIRED AT 2200-2300 F

SIP — NOMEX* FELT

FILLER BAR — COATED NOMEX* FELT

ADHESIVE — RTV SILICONE

*REGISTERED TRADEMARK, E. I. du PONT de NEMOURS AND COMPANY, INC.

Figure 5. Silica Tile System Configuration

unit (PU) by Lockheed Missiles & Space Co., Sunnyvale, California. The PU is the starting block of material from which tile banks and shapes are cut. Two densities of tiles are manufactured. Approximately 80 percent of these are made from a nine-pound-per-cubic-foot (PCF) density (LI-900) while areas requiring greater mechanical strength employ the 22 PCF density (LI-2200). The manufacture of these two products is similar, but there are major differences. In the fabrication of LI-900, it is necessary to add a silica binder to improve its strength and adjust its density. The LI-2200 requires impregnation with silicon carbide to reduce radiative heat transfer through the tile.

All tiles have a thin glass coating on five sides to provide the proper thermal properties (a, ϵ, and a/ϵ ratio). In general, tiles on the lower surfaces of the orbiter are coated black for high emittance while those on the upper surfaces and sides have a white coating to limit solar heating. Typically, black-coated tiles are used in areas that exceed 1200F, whereas white-coated tiles are used below this temperature. The tiles are sized in thickness to limit the temperature of the SIP, the adhesive bond line, and the aluminum structure. The planform size is limited to prevent tile cracking from the induced thermal and mechanical stresses of the vehicle structure and adhesive system. The majority of the tiles have a square platform; the black tiles are typically 6 inches by 6 inches and the white tiles are 8 inches by 8 inches. There are, however, many special shapes and sizes (as small as 1.75 inches square) where vehicle geometry dictates. There are 30,759 tiles on the Shuttle orbiter.

The silica tile system has many engineering advantages. The nine-pound-per-cubic-foot density makes the system extremely light in weight. Such a tile is approximately 93 percent void and therefore is an excellent insulator, having through-the-thickness conductivities of 0.01 to 0.03 BTU-ft/hr-ft^{2}F. Although the low density is accompanied by low strength, the tiles are capable of tolerating the high g forces from the severe acoustic levels on the spacecraft, which in some areas approach 170 db and structural responses of 35 g^2/Hertz.

Amorphous silica is an ideal material for the tiles. Its coefficient of expansion is extremely low compared to other ceramics (Fig. 6), resulting in reduced thermal stresses. This is particularly important because the tiles could have a ΔT within the tile of 2300-2400F at the start of entry. The low density of the tile results in a low elastic modulus, also contributing to reduced thermal stresses. The silica chosen has a purity of greater than 99.62 percent to limit devitrification, for the formation of a significant crystalline phase, such as crystobalite, could result in tile cracking due to a large volume change in the 400-500F temperature range (Fig. 6).

Silica has high temperature resistance; it can survive excursions to 2700F. Since silica is already an oxide, it does not require further oxidation protection as do carbon-carbon and columbium materials. The tile coating is a borosilicate glass. The black coating has silicon tetraboride added for increased emittance. Subsequent oxidation of the silicon tetraboride in the glass yields boria and silica, the basic constitutents of the glass itself.

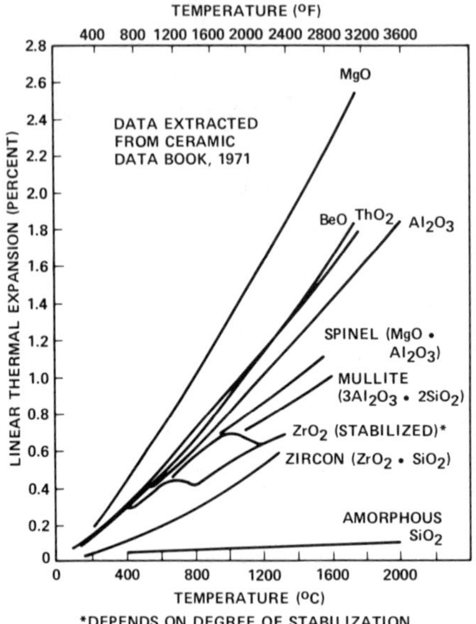

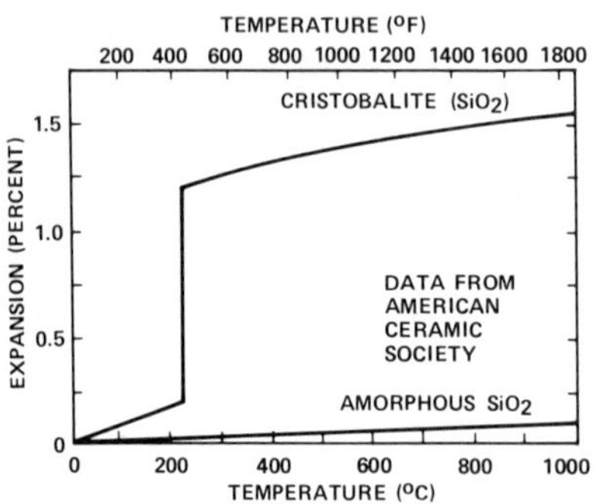

Figure 6. Ceramic Thermal Properties

The coating has an emittance well above 0.8. In addition to providing the desired thermal properties, it is a barrier to rain and atmospheric erosion. It tends to minimize tile-handling damage although it does not eliminate it. The coating thickness ranges from 0.009 to 0.015 inch. The coating terminates above the tile inner mold line (IML) on the sides to limit heat flow toward the IML and to permit the tiles to vent air pressure during both the ascent and descent phases of operation.

While the tiles must "breathe," they must not pick up water, for this could result in a vehicle overweight condition, in the loss of tiles in the vibroacoustic environment, or in coating damage under freezing conditions. Water repellency of the silica tiles is obtained from the vacuum deposition of a silane, Dow Corning Z6070, in a furnace heated to 350F. The silane will burn out that portion of a tile that exceeds 1050F during launch or entry; therefore, the coating terminator is located below the 1050F isotherm of each tile.

4.2 Strain Isolator Pad and Filler Bar

The strain isolator pad is made from Nomex*, an aromatic nylon material that neither supports combustion nor melts. The pad, manufactured by Albany International Research Company, a division of Albany International, is a felt product that is needled to provide strength in the thickness direction.

Two thicknesses of strain isolator pads are used for the majority of the vehicle, each with widely different properties. The 0.160-inch thickness is quite soft (a typical tensile modulus of 26** psi under a load of 5.3 psi), and is generally used under the LI-900 tiles. Where more mounting rigidity (or strength) is required, such as around doors or thermal barriers (discussed later), the 0.090-inch SIP is used. It is roughly ten times as stiff as the 0.160-inch SIP and is generally used under the higher-strength LI-2200 tiles. SIP will retain its room temperature properties, 35-60** psi typical tensile strength, without degradation

---

*Registered trademark, E. I. du Pont de Nemours and Company (Inc.).

**Properties in the thickness dimensions

with exposures to 550F for the equivalent of 100 missions (25 hours). It is capable of supporting limited stresses up to 720F for short times (i.e., on a single-mission basis).

The SIP, like the tile, must also "breathe" to permit equalizing pressures during ascent and descent yet must also avoid water pickup. The latter would result in ice formation during launch, excessive stiffness of the SIP, and overloading of the tile. Therefore SIP is treated with a water-repellent agent, Zepel RN*.

The SIP is generally sized to a footprint an inch smaller than the tile; that is, a 6 by 6 inch tile has a 5 by 5 inch SIP under it, whereas an 8 by 8 inch tile uses a 7 by 7 inch SIP.

Underneath the tile edges, another Nomex* felt product, very similar to SIP, is used. It is called filler bar. The filler bar protects the surface of the aluminum substructure in the tile gap from radiation and plasma heating. The filler bar upper surface is coated with a thin layer of RTV-560. The filler bar is slightly thicker than the SIP and tends to form a mechanical moisture shield against the tile. The filler bar also becomes an attachment point for gap fillers (discussed later in the paper). The filler bar is given a final processing at temperatures of approximately 830F by the manufacturer before the RTV coating is applied. This product is capable of surviving 100 missions of temperatures to 800F (with some RTV damage) and a single mission to nearly 1000F.

4.3 RTV Silicones

The heart of the TPS bonding system is RTV-560, a two-part condensation curing elastomeric sealant/adhesive with a unique combination of properties. Normally the first consideration for a room-temperature curing adhesive system would be an epoxy adhesive. However, the room-temperature curing epoxies are temperature limited. Since the tile bondline sees temperature extremes of -160F to +550F, temperature becomes the driver in selecting an adhesive. Fortunately, exceptional bond strength was not a requirement for the lightweight tile system, allowing consideration of room-temperature vulcanizers. The methyl phenyl RTV silicones have brittle points below -175F, with a high-temperature capability to $>600F$, depending upon exposure conditions. Therefore, the methyl phenyl RTV silicones as a class and specifically RTV-560, was selected early in the Shuttle program as the prime candidate for tile bonding. Early vehicle temperature predictions gave temperatures down to -200F, which is below the brittle point of RTV-560. Therefore, early program tests were concerned with demonstrating that the low-temperature contraction and increase in modulus of the adhesive system would not crack the tile; the RTV-560 was satisfactory while other candidate adhesives were not.

Along with the use of RTV's came the need for a compatible primer, one that developed good adhesion over a variety of surfaces to which tile was to be bonded. The primer

---

*Registered trademark, E. I. du Pont de Nemours and Company (Inc.).

selected was SS-4155, a hydrolyzing titanate coupling agent. This primer was offered the challenge of providing good RTV adhesion on such surfaces as:

- Chromated epoxy primed aluminum
- Bare aluminum
- Titanium
- Inconel
- Graphite epoxy
- Epoxy fiberglass
- Epoxy honeycomb edge filler

Any material that is to be used over the vehicle external surface must be weight-efficient and therefore perform in minimum thicknesses. RTV-560 is used in three bondlines beneath the tile:

- SIP transfer coating, 3 to 7 mils
- SIP to structure, 5 to 9 mils
- SIP transfer coat-to-tile, 4 to 5 mils

The SIP transfer coat was developed to ensure a proper SIP-to-tile bond. Since both the SIP and tile are highly porous, it is not possible to apply a bond to either one of these and ensure consistent RTV penetration into both members at the same time. RTV penetration into the SIP is controlled during the application of the transfer coat in a subassembly operation. Since this penetration affects the SIP modulus, it must be controlled accurately. After the transfer coating has cured, it may then be bonded to the tile.

Other RTV's used in TPS application are RTV-566 and RTV-577. RTV-566 is the low outgassing (but very expensive) version of RTV-560. It is used around windows to reduce contamination from outgassed products. Extensive vacuum tests on RTV-560 showed that the less expensive material could be used without jeopardizing payload experiments. RTV-560 and RTV-566 employ iron oxide as a filler material. RTV-577 has the same starting silicone polymer as RTV-560 and RTV-566, is filled with zinc oxide and calcium carbonate, and has an additive to provide a thixotropic "trowelable" consistency. RTV-577 is used as a "body putty" (screed) to fill in low areas of the vehicle structure and achieve a contour compatible to the TPS installation. The RTV-577 material allows the use of structure with a ± 0.062-inch as-manufactured contour to be faired to the ± 0.032-inch waviness requirements for ceramic tile installation.

5. TPS STRUCTURAL INTEGRITY

It is essential to the function of the orbiter TPS that each tile remain attached to the vehicle through the critical entry heating phase, for the loss of a single tile in many areas could result in severe vehicle damage or even the loss of the orbiter. To ensure the integrity of this system, the loading spectrum to which each tile is subjected must be known, the tile and SIP properties must be adequately defined, and the method of analysis must be verified by laboratory testing. Further, it is important that no significant processing errors be made in the tile installation. To ensure the TPS structural integrity, nearly every tile on the orbiter has been subjected to a proof test.

5.1 Material Strength

There are several unique challenges in designing with tiles and SIP. The first shock to a materials engineer is the extremely low material properties. These are hard to put into perspective with one's aerospace experience base. The tile minimum tensile strength in one critical loading direction (through the thickness) is 13 psi. Compare this with typical airframe structural materials on the orbiter whose tensile strengths are in the range of 10,000 to 290,000 psi. It soon becomes obvious that even the most minor stresses introduced into the system or into laboratory test specimens must be accounted for.

A key property of the SIP is its elastic modulus, for vehicle deflections load the tiles in proportion to this value. One must learn to think in terms of modulus values of 20 psi where one's experience may be with aluminums or steels whose moduli are 500,000 to 1,500,000 times as great.

The second factor that meets the jaundiced eye of the materials engineer is the wide scatter in material properties (Figs. 7 and 8). Tensile test values of coupons cut from LI-900 tiles typically range from 10 to 40 psi; moduli values of 0.160-inch SIP at identical loads often vary from 16 to 40 psi.

The combination of low properties and the wide property scatter impacts the reliability of the design. For example, the minimum tensile strength (99 percent probability, 95 percent confidence) of LI-900 in the thickness direction is 13 psi. Using a design factor of safety of 1.4, the tile stress at limit

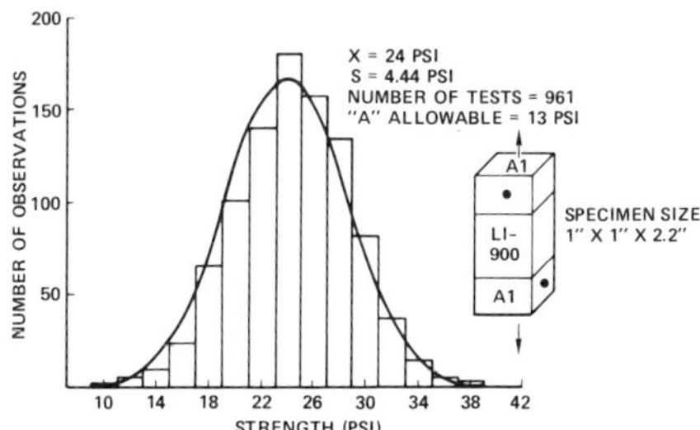

Figure 7. *Tile Coupon Tensile Strength (LI-900)*

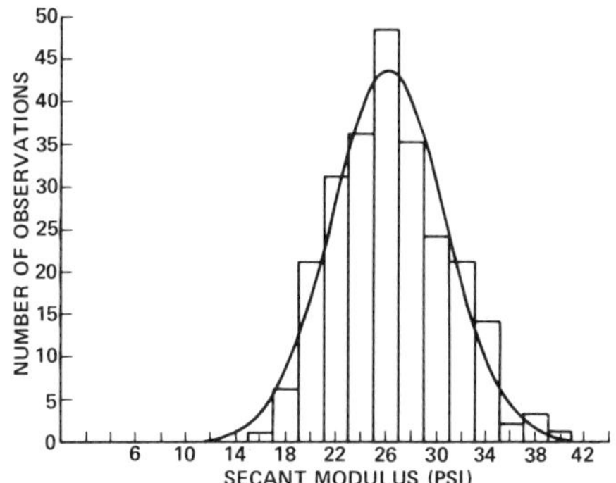

Figure 8. Strain Isolator Pad Secant Modulus at 5.33 PSI (0.160-Inch SIP)

load, 9.3 psi, lies less than one standard deviation below the minimum design tensile strength. In an aluminum design, the stress at limit load typically lies more than seven standard deviations below the design minimum tensile strength.

A third factor that soon becomes obvious is the difficulty of carrying structural loads with brittle ceramic materials. The tiles offer little or no forgiveness for stress concentrations, loading eccentricities, or minor defects in test specimens.

The 0.160-inch SIP is a challenging material to characterize. Its stress-strain behavior is nonlinear, inelastic, and exhibits significant hysteresis changes. The modulus of SIP increases with load. As load is removed, the SIP does not return to its original dimensions but retains some permanent tension or compression set, depending upon which load was applied. Repeated cycling of SIP changes its stress-strain behavior, resulting in a soft zone ("dead band") in which little stress will result in rather large SIP extensions. (Fig. 9.)

As one would expect, the compressive modulus of SIP differs from the tensile modulus at the same stresses. The compressive

241

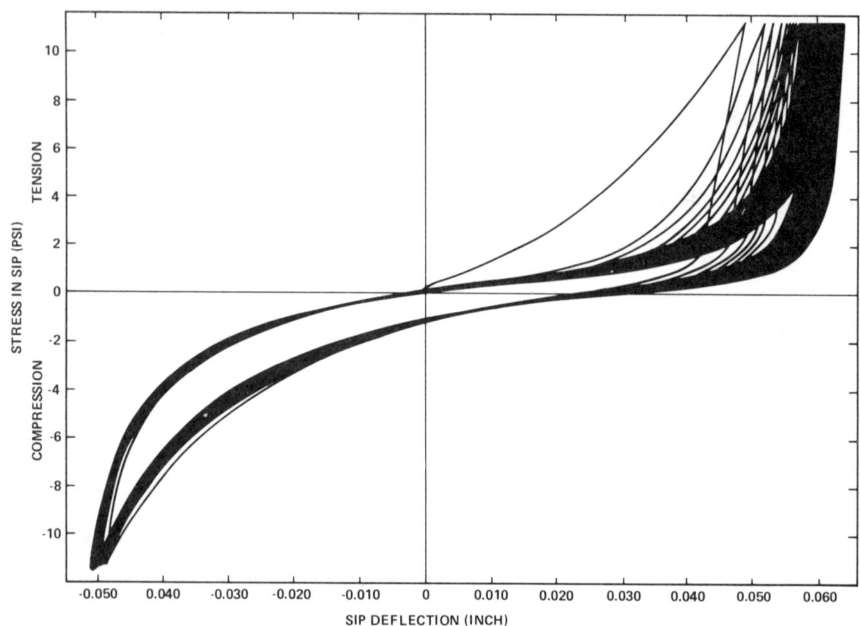

Figure 9. Strain Isolator Pad Hysteresis Behavior (0.160-Inch SIP)

behavior of SIP is typical of that of any batt; as compression increases, fibers are brought into more intimate contact, resulting in increasingly greater resistance to compression. Tensile behavior is controlled by the resistance to extension of bundles of threads needled in the thickness direction of the felt. As further extension occurs, fibers in more bundles are drawn taut, thus increasing its stiffness. The structure of SIP, as seen by the electron microscope, is shown in Fig. 10.

In effect, then, we have a tile system in which both major members have both low properties and wide property scatter. One member of the system, the tile, is brittle and incapable of deformation to reduce localized stresses. The other member, the SIP, is soft and compliant, up to a point, but its elastic properties depend on its load level and its previous loading history.

5.2 Design Loads and Stresses

The tiles experience stresses from a wide variety of loading conditions. For example, the tiles form the outer mold line of the vehicle and must react the aerodynamic maneuvering pressures. They can experience positive or negative pressure loads from ascent or descent where the tile interior cannot equalize the external pressure rapidly enough. Similarly, aeroshock loads, resulting from impinging or dancing aeroshock waves, can add overturning moments and pressure loads to the tiles. Further, the tiles are subjected to high-level vibroacoustic loads from engines firing and air flow.

Pressure differences between the interior and exterior of the vehicle cause out-of-plane vehicle deflections which, in turn, load the tiles through the SIP. These deflections can

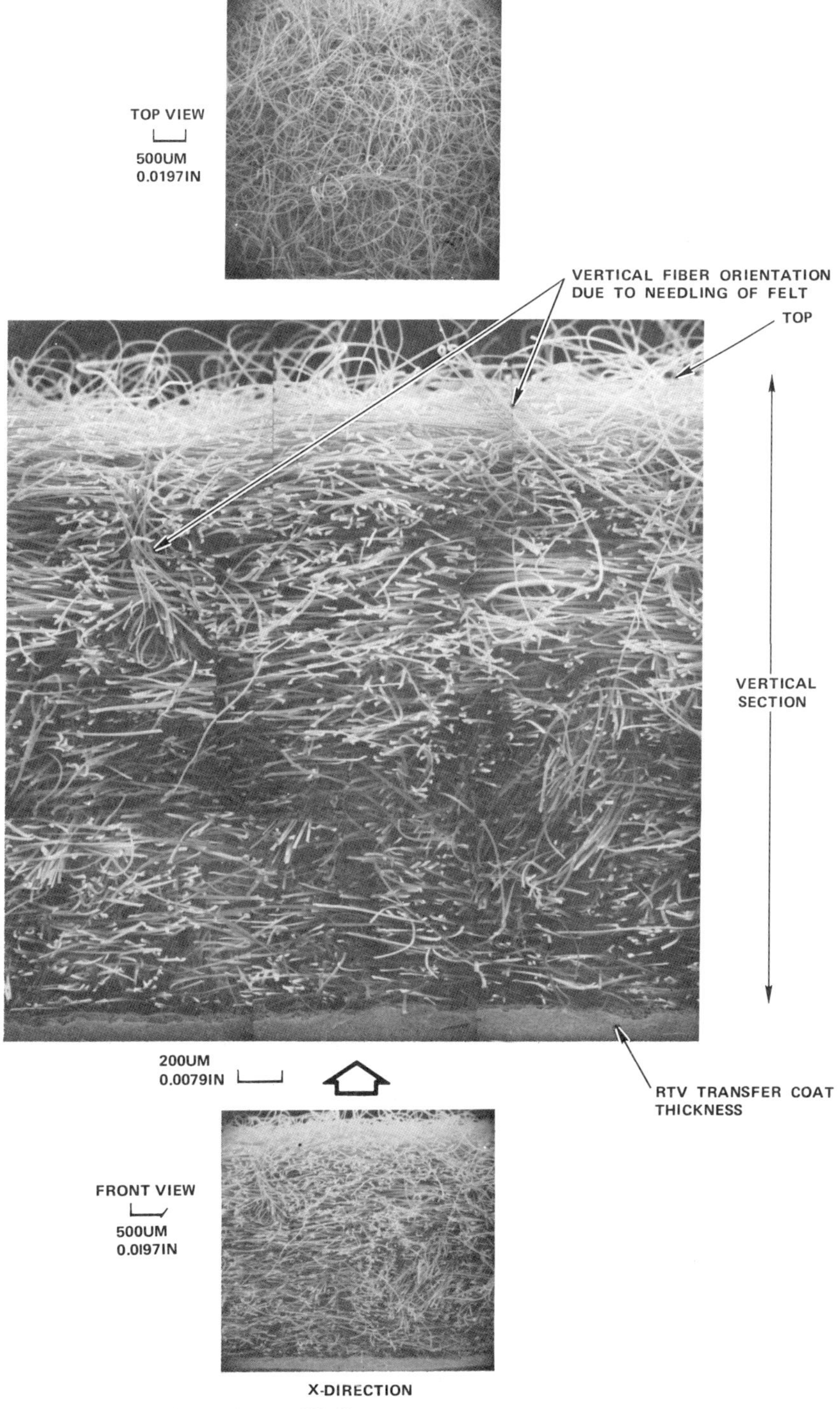

Courtesy Brittle Material Design Group University of Washington
(supported by NASA Grant. NSG 48-002-004)

Figure 10. Strain Isolator Pad Structure, Seen by Electron Microscope

be increased by additional in-plane stresses or thermal stresses in the vehicle substructure.

One potential source of large stresses is that resulting from mismatch during tile installation. Mismatch is the difference in local contour between the tile IML under the SIP and the structure OML to which it is attached. This difference must be accommodated by extension of the SIP, resulting in tile loading. For example, a fit-up difference of 0.020 inch could result in residual stresses exceeding 3 psi in tiles bonded to 0.160-inch SIP, whereas a mismatch as small as 0.005 inch could result in localized residual stresses exceeding 7 psi with 0.090-inch SIP.

Because the tile coating has a different coefficient of expansion than the tile, residual stresses are introduced into the tile upon cooling. The significance of these stresses is quite apparent as coated tiles are machined to a thickness of 0.5 inch or less; for as residual stresses are relieved, the tiles curl slightly, taking somewhat the shape of a potato chip. When such a tile is bonded to the SIP under pressure, the "reflatting" of the tile distributes some residual stress into the tile and introduces stresses into the SIP and its interface.

Finally, mechanical loads imposed on tiles by thermal barriers and gap fillers can result in overturning moments on the tiles, which must be reacted by the SIP and tile. (Thermal barriers consist of ceramic cloth, Nextel 312*; knitted Inconel X750 springs and ceramic batt, Saffil**. They are used to ensure plasma-tight integrity around doors and penetrations. Gap fillers are made from the ceramic cloth and batt and incorporate a piece of Inconel 601 foil to control shape. They are used to restrict plasma flow into tile gaps at pressure gradient areas.)

5.3 Tile System Strength

The strength of the tile system (tile-SIP-RTV) was originally thought to be equal to the strength of its weakest link, the tile. A typical tensile strength of an LI-900 tile averages 24 psi in the thickness direction, whereas the 0.160-inch SIP averages 40 psi and RTV is greater than 250 psi. When test failures occurred, they would occur in the tile just above the RTV layer that bonds the tile to the SIP transfer coat.

Structural element test specimens to confirm the analytical methods used a bending beam configuration. A tile and its SIP were installed on an aluminum plate which, in turn, was bent over various mandrels until failure occurred. The radius of the mandrel at tile failure was observed and stresses in the tile were computed based upon SIP deflections. Although these early tests tended to confirm the analytical methods, large errors could be introduced because the properties of the specific SIP or tile used in the test were unknown.

Subsequent testing later in the program indicated that a flatwise tension test was much more precise in defining the system properties since it eliminated the variability of the SIP stiffness. Flatwise tensile tests of various SIP and tile combinations revealed the average system strength was slightly less than 50 percent of the average tile strength.

*Registered trademark, 3M, St. Paul, Minnesota
**Registered trademark, ICI United States, Inc.

Detailed studies showed that the bundles of fibers that gave SIP its short transverse strength were acting as localized stress concentrations just above the tile bondline, causing early failure. This revelation resulted in two activities: (1) a reappraisal of all tile safety margins on the vehicle; and (2) an effort to develop a higher strength at the tile-SIP bondline.

It was demonstrated that a thin layer of aluminum foil (0.010-inch thick) between the SIP and the tile would bring the tile system strength up to the tile strength. This approach would add approximately 1,000 pounds to the vehicle, yet it could not readily accommodate the vehicle contours machined into the tile IML's nor could it sustain the thermal cycling of the bondline without causing the thermal stress failure in the tile. A thin graphite composite shim was developed that prevented thermal stress failures, but its manufacture to tile contours was prohibitively difficult. The elimination of these detrimental stress concentrations was finally achieved by a process we called densification.

5.4 Densification

The tile is roughly 93 percent void. In the densification process, the voids in the IML surface of the tile are filled with a slurry containing a fine glass slip and a colloidal silica adhesive (Ludox*). This is analogous, in some respects, to putting a sealer coating onto concrete. It is important that the Ludox*-slip mixture be brushed into the tile and not be permitted to build up or the coating would flake off and have limited strength. Such a process increased the outer surface density to approximately 80 pounds per cubic foot and a density gradient penetrated inward approximately one-eighth of an inch. The glass slip is highly abrasive to the tile. Techniques were developed to limit tile IML contour changes to below 0.006 inch during densification. While the densification process was being developed, a second problem was encountered. The Ludox* inactivated the RTV catalyst, preventing its cure against the tile surface. Subsequent investigations revealed the Ludox* effect could be controlled by waterproofing the tile by vacuum vapor deposition of silane after densification in a furnace.

Once the densification process was fully developed, it was necessary to remove critical tiles on the Columbia OV-102, densify them and reinstall them in the same locations. Fig. 11 shows the cross section of a densified tile.

5.5 Proof Loading and Acoustical Emission Testing

While densification was required for a large number of tiles, the minimum system strength of the undensified LI-900 tile with 0.160 SIP, 6 psi, was considered more than adequate in some areas. To verify that these installations were strong enough, each tile was subjected to a proof load on the vehicle to a level 25 percent above the maximum load (limit load) it would see in service, wherever possible. Proof loading of tiles was accomplished by pulling the tile OML using a vacuum chuck. The load applied was limited by the adherence of the vacuum chuck to the tile (theoretically 14.7 psi but approximately

---

*Registered trademark, E. I. du Pont de Nemours & Co. (Inc.)

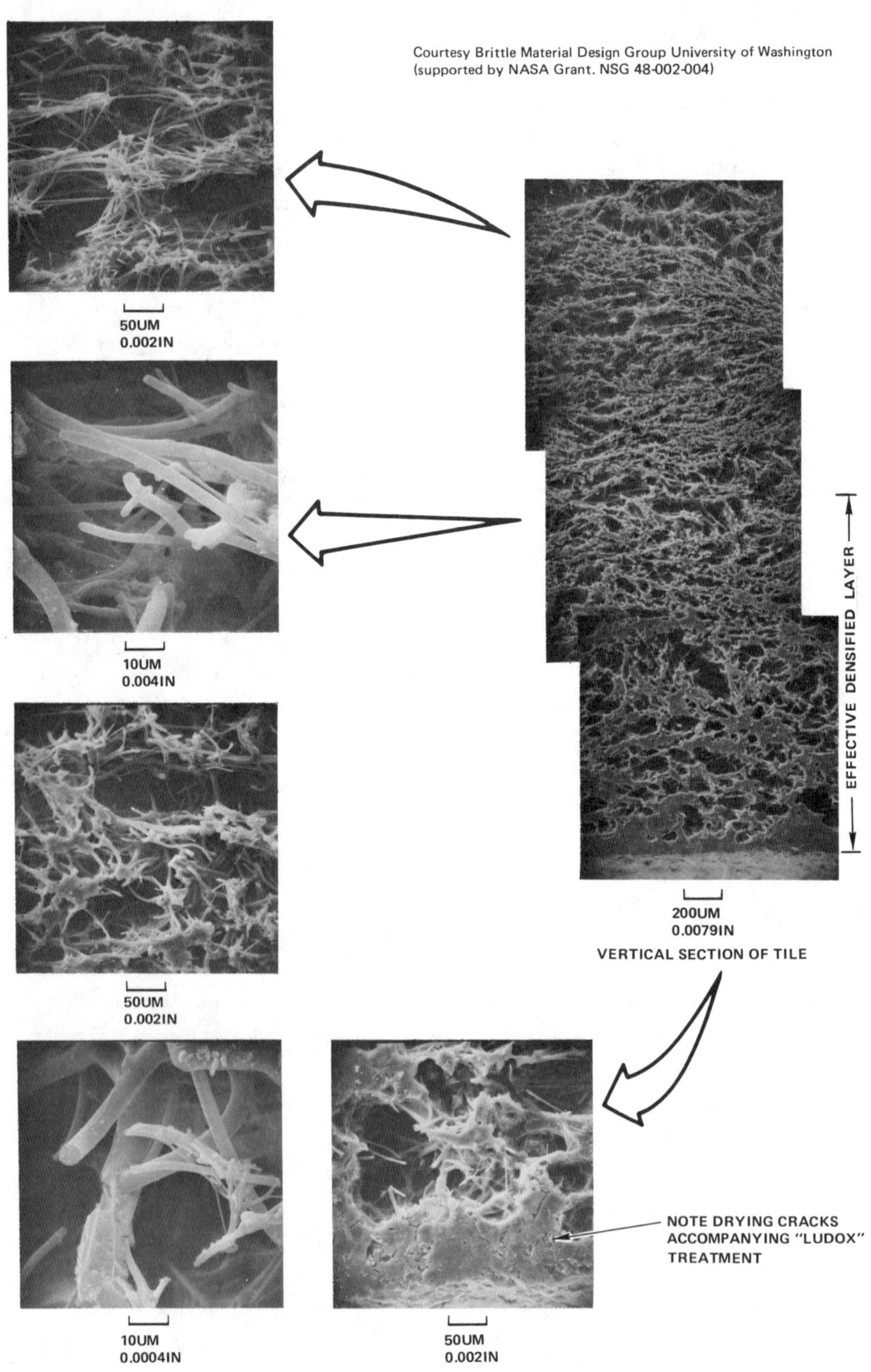

Figure 11. Cross Section of Densified Tile

10-12 psi from a practical standpoint) and the limit to which the substructure could be loaded. Acoustical monitoring transducers were in contact with the tiles during proof loading. Acoustical techniques were developed to ensure no detrimental bond or tile degradation occurred during proof loading. Nearly every tile was proof-loaded; however, acoustic techniques were not required for densified tiles.

5.6 Pulse Velocity Testing

The tile strength used in design and analysis is that minimum strength that 99 percent of the tiles will equal or exceed. With a vehicle covered by more than 30,000 tiles, one could expect more than 300 tiles could be below this minimum strength. Since the loss of a critical tile was so serious, a nondestructive method had to be developed to ferret out the weaker tiles. Such a process was developed based upon two known facts: (1) the speed of sound in a material is related to its elastic modulus; and (2) the strength of low-density ceramics is related to its modulus. All densified tiles were subjected to pulse velocity testing to screen out low-strength tiles. Fig. 12 shows the statistical correlation between strength and velocity of sound in tiles.

5.7 Process Controls

The use of RTV, like any other adhesive system, requires careful attention to process controls to ensure bonding strength requirements are met.

5.7.1 Surface Preparation. Most of the surface area of the Shuttle orbiter is coated with a chromated epoxy polyimide paint applied over the aluminum skin. It is necessary to

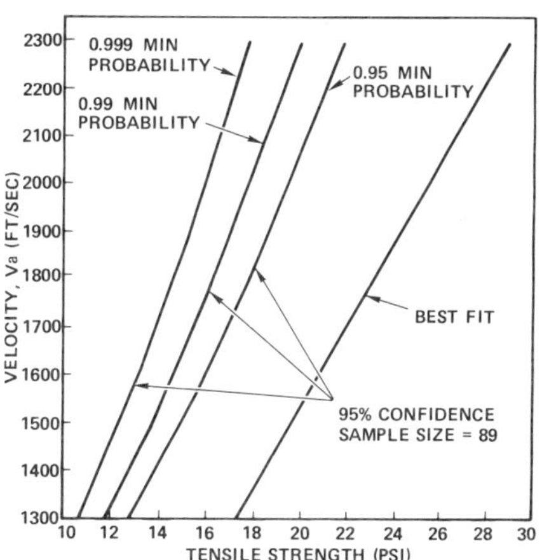

Figure 12. Pulse Velocity vs. Tile Strength (LI-900)

solvent-wipe this surface with methyl ethyl ketone to remove organic contaminants. In addition, sanding of the surface with a fine mesh (400 grit) paper is required to remove the gloss and provide some mechanical attachment. Similarly, the glossy surface of screeded RTV-577 also require roughening to achieve optimum bonding.

Epoxy surfaces are primed with two coats of a silicone primer, SS-4155, in addition to the sanding; RTV surfaces do not require an additional primer.

5.7.2 Pot Life. One of the most important controls applied to the RTV-560 adhesive is the control of pot life (working life). Pot life and curing times of RTV are decreased either by increases in temperature or relative humidity (Fig. 13). In cool dry air, curing may take two to three days or more whereas in warm humid environments, bonding must be accomplished within 15 minutes of the mixing with the catalyst. Each RTV mix has its pot life recorded on the container when mixed, and inspection must verify the bonding

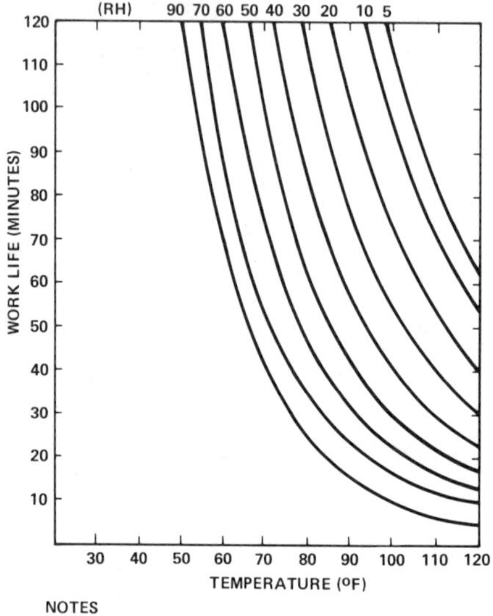

Fig. 13.   RTV Pot Life

is accomplished within the pot life. In addition, each batch is accompanied by a shore hardness coupon to verify proper pot life and cure.

Pot life and shelf life are susceptible to contaminates introduced by the manufacturer of the RTV. For this reason, both the manufacturer and Rockwell verify the pot life and shelf life on each order of RTV-560.

6. SUMMARY

The thermal protection system used on the Space Shuttle orbiter has been selected based upon the requirements for minimum weight, 100-mission reuse capability, and a 160-hour maximum turnaround period for relaunch. The system chosen uses low-density, fired tiles made from fine filaments of pure silica. The tiles exhibit excellent characteristics: a low coefficient of expansion, low thermal conductivity, low density, excellent thermal stability, and compatibility with the environment. The design must accommodate the shortcomings of this material: low strength, wide scatter in mechanical properties, and brittleness. The technique of bonding the tiles to a strain isolation pad and, subsequently, the pad to the structure, using a room-temperature curing elastomer, appears to provide a satisfactory approach. However, because the loss of a single tile could jeopardize the vehicle, extreme care must be taken to fully understand specific tile loads and stresses, material strengths and behavior, and to verify analytical methods. Finally, close process control must be maintained during tile installation. A final process evaluation of each tile bond must be made by proof-testing to ensure the structural integrity of the system.

7. ACKNOWLEDGEMENTS

The authors wish to acknowledge the efforts of J. Rollins in helping with the preparation of this paper.

8. BIBLIOGRAPHY

Black, W. E., et al. Evaluation of Coated Columbium Alloy Heat Shields for Space Shuttle Thermal Protection System Application. NASA CR 11219 (June 1972).

Dotts, Homer. "Materials and Fabrication Methods Used in the Gemini Spacecraft." Metal Progress (March 1963).

Stockett, Stewart J. "Metals and Processing Used in the Mercury." Metal Progress (June 1962).

BIOGRAPHY

L. J. Korb

Mr. Korb is currently the Supervisor of Metallurgy and Ceramics on the Space Shuttle Program at Rockwell and intimately associated with the materials and processes for the orbiter thermal protection system. He has more than 25 years of experience in the materials field, having been Supervisor for Metallurgy in the Apollo program and Supervisor of Advanced Materials at Rockwell. He holds engineering licenses in California in both metallurgy and corrosion and has authored several technical papers. He presently serves on the Handbook Committee of ASM.

BIOGRAPHY

H. M. Clancy

Mr. Clancy is Supervisor for Nonmetallic Materials on the Space Shuttle Program at Rockwell. He has been with Rockwell for 27 years working on programs including the F-86, F-100, X-15, B-70, and Apollo. He has worked in many areas of materials technology, including sealing, elastomers, corrosion, contamination control, flammability, and propellant compatibility. He was instrumental in the development of the present standard test procedures and acceptance criteria for toxicity control of spacecraft cabin materials. He has a BS degree in Chemistry from UCLA.

SURFACE DENSIFICATION OF
HIGH-POROSITY SILICA INSULATION

J. W. Holt and L. W. Smiser
Advanced Manufacturing Technology
STS Development and Production Division
Rockwell International, Downey, California

Abstract

A method of improving the Shuttle orbiter reusable surface insulation bond strength and resistance to surface impact was developed. The low-density insulation tiles are very fragile and susceptible to surface damage. In addition, the tile-to-vehicle attachment bond strength did not meet design requirements.

The method developed consists of applying a silica solution to the silica tile material to increase the material density at the tile surface, thus increasing the tile bond tensile strength to be equivalent or greater than the strength of the parent tile material. This strengthening process, the test program conducted, and the material properties that were identified are outlined.

The process described has been implemented on the current Shuttle orbiter to improve tile attachment bond strength and will be used on subsequent vehicles.

Keywords

Strain isolation pad, Nomex, frit, slurry, densification, glaze, silica, density.

1. INTRODUCTION

The reusable surface insulation (RSI) for the thermal protection of the Shuttle external aluminum skin structure is composed of machined, fused silica tiles, with a thin glazed ceramic coating. The majority of these tiles have a 6- or 8-inch square planform. The remainder are of various shapes and sizes to conform to compound contours and to fit around surface penetrations and openings. The tiles range from 0.2 to 3.5 inches thick.

The density of the fused silica blanks used for the machined tiles is 9 or 22 lb/ft^3. The ceramic coating is a silica glass frit cured by reaction with boron silicide. Use of this low-density silica system for the reusable surface insulation resulted in two basic problems: (1) low adhesion strength at the tile attachment interface due to the tenuous structure of the low-density material and (2) poor surface impact resistance of the tiles due to the low modulus of the system.

To overcome the low adhesion strength problem, a process was developed to increase the density of the silica tile surface. By this treatment, the locus of failure is transferred from the tile attachment surface interface into the tile's interior.

To improve the impact resistance of the tile, the density and modulus of the machined tile surface are increased by the densification process before the glazed ceramic coating is applied. This surface densification is a process whereby a controlled slurry mixture is impregnated into the surface of a machined silica tile. Upon drying, the slurry mixture provides a hard, strong, densified layer or "skin."

2. BACKGROUND

The RSI system design shown in Fig. 1 is composed of the ceramic-coated silica tiles bonded to a strain isolator pad (SIP), which, in turn, is bonded to the aluminum skin structure of the orbiter. The SIP is 0.160 inch thick for the 9 lb/ft^3 tiles, or 0.090 inch thick for the 22 lb/ft^3 tiles.

The 9 lb/ft^3 silica tile base material is formed from high-purity silica fibers (2 to 4 microns in diameter) and a high-purity silica binder. This mixture is formed into loose-knit blocks, then dried and fired to provide a low-strength machinable material with low density, low thermal conductivity, high temperature resistance capability, and a low coefficient of thermal expansion. Fig. 2 is a scanning electron microscopy photograph of this tile material showing the open fibrous structure, and Fig. 3 is a microscopy photograph showing a cross section of the densified layer.

The 22 lb/ft^3 material is similar to the 9 lb/ft^3 material and is manufactured in a like

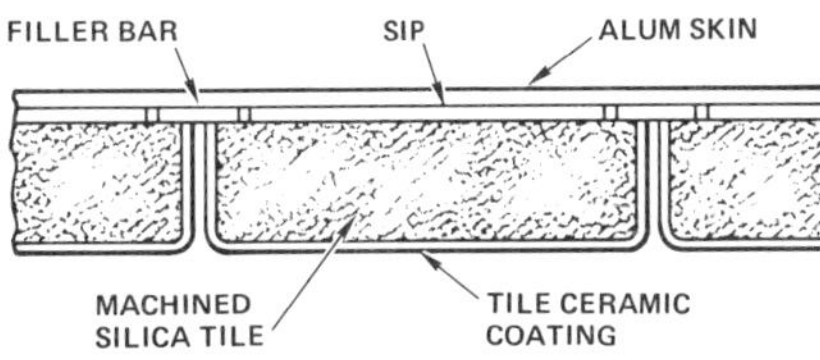

Fig. 1 — Reusable surface insulation (RSI) system design

Fig. 2 — Microscope of silica tile open fiberous structure (2000X)

manner, but with particulate silicon carbide added.

The SIP is fabricated with Aramid fibers from Du Pont Nomex®. These fibers are felted and needled with reverse barb needles into Nomex® felt fabric, in which each fiber element supports the tile relatively independent of the other fibers in the SIP. The silica tile is bonded with RTV-560 silicone adhesive to the Nomex® felt SIP, which in turn is bonded to the aluminum skin of the orbiter with this same adhesive.

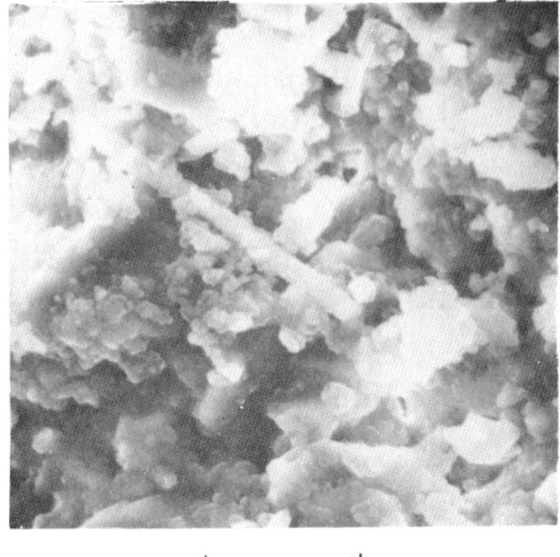

Fig. 3 — Microscopy of densified fiberous structure

Under flatwise tensile stress, the SIP, which is nonuniform due to the method of manufacture (needling), pulls unevenly (responds nonlinearly to stress) at the SIP-tile interface. Thus, the effective area of the interface is reduced and failure occurs consistently at this SIP/tile interface. To prevent this type of failure, the tile bond interface surface must be densified. The surface then becomes a strong link in the system instead of a weak link.

3. SUMMARY

The process of surface densification allows the development of the full tile strength. To demonstrate this, flatwise tensile specimens prepared with undensified 9-lb/ft^3 material failed interfacially with an average of 11 psi, while specimens assembled with the tile surface densified failed in the body of the tile at an average of 22 psi. In the 22-lb/ft^3 material, the flatwise tensile values increased from 30 to 43 psi due to surface densification.

When the machined tile surface is densified before the ceramic coating is applied, the ability to withstand impact damage is substantially improved. A comparison of surface strength with and without densification shows that densification provides improvement of approximately 75 percent.

4. SURFACE DENSIFICATION PROCESS

The densification process consists of impregnating the surface of the silica material with a ceramic slurry in the following manner. The tile is first masked to expose only the surface or surfaces to be densified. Isopropyl alcohol is then applied to the surface to serve as a wetting agent. The amount of alcohol applied depends on the tile thickness. The required amounts for the various tile thicknesses are shown in Table 1. The alcohol is applied with a soft bristle brush over the entire surface.

Immediately after the alcohol is applied, the densification slurry is applied. The method of slurry application to improve bond strength is a little different from that used to improve coating impact strength.

Table 1 − Alcohol required per square inch of tile surface

Tile Thickness (in.)	≥ 0.3	$\leq 0.3\text{-} \geq 0.5$	$\leq 0.5\text{-} \geq 1.0$	< 1.0
Alcohol Required in (g)	1	2	3	4

4.1 Slurry Application

On surfaces to be bonded, 3 g/in.^{2} of densification slurry is applied. The densifying slurry is applied evenly with a soft bristle brush to the entire surface in a "box coat" pattern.

During this process, the slurry awaiting application is continually mixed to keep the fused silica particles in suspension. The object is to impregnate and fill the voids in the low-density porous surface with the fused silica particles without allowing a buildup.
When a buildup begins, the excess slurry is removed by gently wiping the surface with a soft felt or cheese cloth pad. The brushing application process and wiping operations are continued until all of the measured amount of slurry has been applied.

When the surface is densified to improve impact resistance, the tiles are masked in a similar manner, as described, exposing only the surfaces to be densified. The alcohol is also applied as previously described with the same amount per in.^{2} as shown in Table 1. However, the method of slurry application is somewhat different. To minimize erosion of tile corners or edges due to brushing and wiping, the tiles are dipped into the densifying slurry, and all of the exposed surfaces to be densified are submerged in the slurry. The tile remains submerged in the slurry for 3 minutes.

The tile is then removed from the slurry, and the slurry residue remaining on the exposed surfaces is brushed into the surface. Additional slurry is then applied until buildup begins. The excess is then gently removed by wiping, and additional slurry is added by brushing for a total of three brush applications and three wiping operations, the final being the wiping operation.

A basic principle of this process is to use the fused silica particles to act as an aggregate to fill the voids of the exterior tile suface so that, when the material dries, a hard surface is provided. To do so, a final brushing operation is required to work the fused silica aggregate into the surface, filling the voids sufficiently to provide a strong, hard surface.

4.2 Densification Slurry Composition

The slurry used to densify the surface is composed of colloidal silica and fused silica slip, with a trace of boron-silicide (SiB_4) for coloring. The colloidal silica[1] is a Du Pont product known as Ludox® AS, which is composed of deionized water and approximately 30-percent (by weight) silica solids, with a ph in the range of 8 to 9.5. The fused silica slip is a particulate silica with irregularly shaped particles. The fused silica is added to the Ludox® as an aggregate to bring the slurry density to 1.36 to 1.38 (g/ml).

The slurry mix (by weight) contains approximately 55-percent colloidal silica (Ludox®), 45-percent silica slip, and 0.12-percent boron-silicide. This slurry mix is continuously roll-mixed until the time it is applied on the

tile surface to keep the silica particles in suspension.

5. PROCESS VALIDATION

5.1 Density Versus Depth

The densification process does not produce a uniform density in the direction of impregnation, but the density falls off rapidly with depth into the tile, as shown in Fig. 4. At a depth of about 0.1 inch, the density is essentially unaltered for both 9- and 22-lb/ft^3 tiles.

5.2 Flatwise Tensile Testing

The apparent strength of undensified 9-lb/ft^3 tiles bonded to 0.160-inch SIP averaged 11.8 psi, based on a bond area of 25 in.^{2} with failure occurring at the tile/SIP interface. When densified, the average failure was at 22.6 psi, based on a 25-in.^{2} area with failure in the body of the tile—an improvement of about 11 psi.

Similarly, the 22-lb/ft^3 tiles bonded on 0.090-inch-thick SIP failed at an average of 30.2 psi undensified and at 43.6 psi when densified—an increase of about 13 psi.

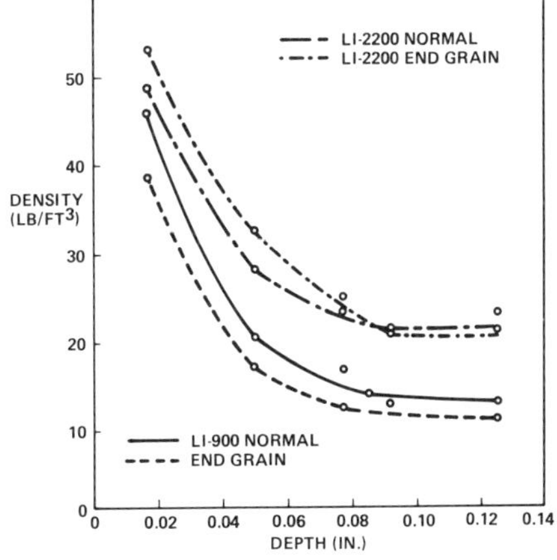

Fig. 4 — Density vs. depth for densified ceramic tiles

This does not mean that the strength of the densified layer was measured but rather that the tile-to-SIP bond was strengthened sufficiently to allow the full tile strength to develop. The bond area under flatwise tensile load was 25 in.^{2} while the tile area was 36 in.^{2}. Thus, the poor bond strength (11.8 psi) was less than the 15 psi average for the 9-lb/ft^3 tile material. Table 2 summarizes the flatwise tensile test results.

Table 2 — Flatwise tensile stress

| Tile Type (lb/ft^3) | Bonded Surface Treatment | | Ultimate Tile Strength (psi) |
	Undensified (psi)	Densified (psi)	
9	Bond[1] 11.8[3]	22.6	
	Tile[2] 8.2	15.6[3]	15
22	Bond[1] 30.2[3]	43.6	
	Tile[2] 21.0	30.6[3]	30

[1] Bond area is 25 in.^{2}

[2] Tile area is 36 in.^{2}

[3] Indicates failure location

5.3 Tensile Modulus of Densified Layer

Table 3 shows that the tensile modulus of the densified tile specimens increased by about 33 ksi in the in-plane direction for both the 9 and 22 lb/ft^3 tiles. It is therefore assumed that the same increase occurs in the thickness direction. This is the basis for calculating the following estimated tile modulus for the densified layer. The last column in Table 3 was calculated by taking a ratio of the square of densities of the tile and the densified layer times the modulus [e.g., $(9/21.5)^2$ x 7 ksi = 40 ksi for the 9 lb/ft^3 tiles]. At 0.05 inch, the average densities from Fig. 4 are 20 to 30 lb/ft^3 for the densified layers of the 9 and 22 lb/ft^3 tiles, respectively.

Table 3 — Tensile modules

| Type of Tile (lb/ft^3) | Bonded Surface Treatment | | Estimated Average Density (lb/ft^3) |
	Undensified (ksi)	Densified (ksi)	
9	7	40	21.5
22	27	60	32.8

5.4 Effect of Sanding the Surface

The tile-to-vehicle surface mismatch must be within 0.010 inch. To achieve this tolerance, it is sometimes necessary to sand the densified surface to the vehicle contour. Therefore, the effect on bond strength of sanding the densified surface was evaluated. The total depth of the densified area is approximately 0.100 inch; however, the highly densified surface (20 to 40-lb/ft^3) thickness is only 0.020 to 0.030 inch deep.

The evaluation showed that bond strength reduction did not occur until 0.015 inch of the densified surface had been removed. Therefore, 0.010-inch surface removal was established as a maximum to maintain a factor of safety.

5.5 Thermal Conductivity

The thermal conductivity (K-factor) of the densified Li-900 (9 lb/ft^3) material in the densified area was increased from 1.2 x 10^{-4} cal/cm/cm^2/sec/°C to 5.0 x 10^{-4} for the outer 0.034-inch-thick surface. The Li-2200 (22 lb/ft^3) material increased in this same densified area from 3.2 x 10^{-4} to 5.7 x 10^{-4}. These values are considered acceptable. For a 2-inch-thick tile, the overall conductivity is increased by about 4 percent by this densified layer for either material.

5.6 Glass Transition Temperature

None was detected below 726°C.

5.7 Environmental Effects

Environmental extremes of heat, vacuum, and humidity were imposed on densified Li-900 material specimens before bonding of flatwise tensile specimens. Tests of these specimens resulted in no degradation due to the exposures. These tests provided assurance that the densified layer would not be degraded by actual orbiter ascent and entry conditions.

5.8 Coating Improvements

The major tile impact damage area on the orbiter is on the tiles where a protruding lip along one or two sides of the tile encapsulates gap fillers installed between the tiles. These tiles, because of the protruding lip, are particularly fragile and are susceptible to impact damage. Therefore, the major effort directed toward improving tile coatings was devoted to strengthening the lip on these tiles. Fig. 5 shows a typical lip tile specimen formed by a 1-inch-radius undercut. The lip and tile surfaces were densified, then tested as a cantilever loaded in a test device shown in Fig. 6. The results of these tests are summarized in Table 4. Both (9- and 22-lb/ft^3) tiles were densified and tested. The amount of densification can be seen in Fig. 4. Similar percentage strength increases were seen in both types of tiles, as shown in the last column of Table 3.

The reaction-cured glass coating also contributes to the tile lip strength. Table 5 shows the result of tests on 9-lb/ft^3 and 22-lb/ft^2 densified and undensified lip tile specimens. The 9 lb/ft^3 shows an approximate linear increase with coating thickness from 0.009 to 0.022 inch. The 22-lb/ft^3 material, however,

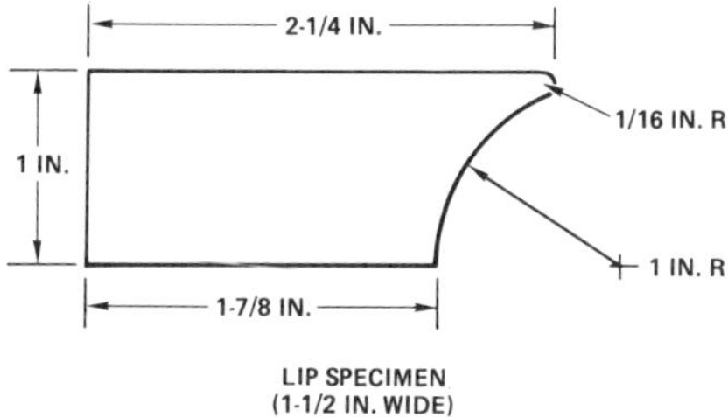

Fig. 5 — Lip tile specimen

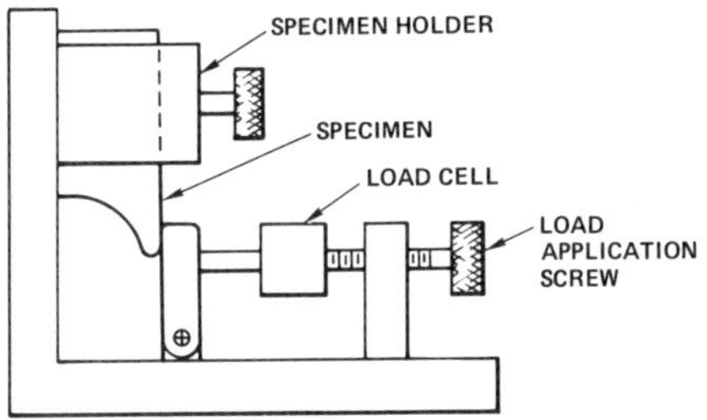

Fig. 6 — Lip tile test device

appears to increase strength rapidly to about 0.011-inch thickness and then level off or increase modestly with coating thickness above 0.025 inch thick.

Table 4 — Lip tile strengthening by densification of surface

Lip Strength (lb/in.)

Tile Type	Undensified	Densified	% Increase
LI-900	13.2	18.7	42
LI-2200	36.7	53.1	45

5.9 Impact Strength of Coated Tiles

Several factors affect the impact resistance of coated Li-900 and Li-2200: residual stress in the coating, coating thickness, elastic modulus or stiffness of the underlying tile, and, strangely enough, the impact energy level.[2] In general, thicker coatings and higher tile modulus (as results from surface densifica-

Table 5 — Effect of coating thickness on lip strength

Failing Load

Tile	Undensified		Densified		% Strength Increase
	Thick. (in.)	Load. (lb/in.)	Thick. (in.)	Load (lb/in.)	
Li-900	$\emptyset$	13.2	$\emptyset$	18.7	41.7
	0.009	27.3			
	0.015	30.7	0.015	40.3	31.3
	0.022	43.3			
	0.024		0.024	79.0	82.4
Li-2200	$\emptyset$	36.7	$\emptyset$	53.1	44.7
	0.011	74.8	0.010	124.3	66.2
			0.015	139.7	
	0.012	64.0	0.015	131.5	105.5
	0.022	67.5	0.025	141.0	108.9

tion); decrease the amount of impact damage to tiles. But this is not a hard and fast rule, since the impact energy magnitude also affects the mode of failure and, hence, the degree of damage. More study is needed before this impact energy magnitude factor is fully understood.

6. REFERENCES

1. *Ludox, Colloidal Silica, Properties, Uses, Storage and Handling,* Du Pont Industrial Chemicals Department Bulletin A85845.

2. Varner, J.R., and Ono, Kanji, *Damage Resistance of Space Shuttle Orbiter Thermal Protection Tiles,* Final Report, Interchange No. NCA2-OR390-901.

7. BIOGRAPHIES

J. W. Holt is a Member of the Technical Staff at the Space Transportation System Development and Production Division, Rockwell International Corporation. Mr. Holt has, as a member of the Advanced Manufacturing Technology Department, been involved for the last five years in the development of processes and equipment for the manufacturing, installation, and repair of the Shuttle orbiter thermal protection system. The principal area of activity has been the reusable surface insulation system and the resolution of the associated problems.

Dr. L. W. Smiser has been employed by Rockwell International in Downey, California, since 1973. Current responsibilities in the Advanced Materials Technology Group include ceramic tooling for composites and development of repair, testing, and processes for the thermal protection system of the Shuttle orbiter. Dr. Smiser received his PhD degree in ceramic science at Rutgers University in 1972 and began working with thermal protection system material at the University of Washington shortly before joining Rockwell International.

26th National SAMPE Symposium
April 28-30, 1981

EVALUATION OF PROOF TESTING AS A MEANS OF ASSURING MISSION SUCCESS FOR THE SPACE SHUTTLE THERMAL PROTECTION SYSTEM

David J. Green, John E. Ritter, Jr.*, and Fred F. Lange
Rockwell International Science Center
Thousand Oaks, California 91360

Abstract

The reliability analysis of a thermal protection system (TPS) incorporating low density, fibrous ceramic materials must take into account the time-dependency and variability of the system's strength. Fracture mechanics concepts can be used to estimate the allowable strength and expected lifetime, as well as to define a proof test scheme for assuring mission success of TPS. The aim of this study was to evaluate the proof testing scheme used on the TPS of the Space Shuttle with these fracture mechanics concepts. The analysis confirmed that proof testing was necessary for the undensified system and that proof testing should lead to adequate reliability with respect to the design stresses. For the undensified system, fracture mechanics predictions were confirmed by measuring the strength of samples that survived proof testing. It was also found that the time-dependent nature of the system's strength was controlled mainly by the ceramic and that this time-strength dependency is similar to that observed for bulk glasses that have compositions similar to the glass fibers present in the ceramic tiles.

"Keywords":
    Thermal Protection System,
    Proof Testing,
    Ceramics,
    Strength Variability,
    Time-delayed Failure,
    Fracture Mechanics

1.   INTRODUCTION

Low density ceramics have a great potential for use in the thermal protection system (TPS) for space vehicles. The current TPS on the Space Shuttle is based on a

---

*On leave from University of Massachusetts, Amherst, MA 01003.

rigidized ceramic tile that has a
network structure of silica glass
fibers. These materials were developed to withstand severe environments, particularly with respect to
reentry and reusability. In addition to the requirements of low density and low thermal diffusivity,
the tiles must also withstand a
variety of thermal and structural
stresses during flight. It is
therefore important to understand
the mechanical behavior of the tiles
in order to design a reliable TPS,
particuarly in situations where the
system strength is controlled by the
ceramic.

Ceramics unfortunately can exhibit
time-delayed failure and a large
variability in strength that com-
plicates the reliability analysis.
It is often necessary to establish
procedures for eliminating weak samples, such as proof testing, where
samples are mechanically tested
before use. Methods of dealing with
design problems involving ceramic
materials have been developed over
the past 10 years through the application of the techniques and
concepts of fracture mech
anics.[1-3] It is believed that
time-dependent failure occurs from
subcritical growth of pre-existing
cracks to dimensions critical for
catastrophic failure. Therefore, by
characterizing the initial flaw
size, the conditions for subcritical

crack growth and final instability,
the strength and lifetime of a ceramic can be determined for any
stress history.

The aim of this paper is to review
briefly the fracture mechanics
theory appropriate to carrying out a
reliability analysis of the TPS; to
present the time-dependent strength
results for the TPS based on LI 900
tiles* and to show how proof testing
can be used to define the strength
allowables necessary for mission
success of the TPS.

2. FRACTURE MECHANICS APPROACH

Fracture mechanics theory is based
on the reasonable assumption that
the time-dependent failure occurs
from a stress-dependent growth of
pre-existing flaws. As the crack
grows to its critical size, the
crack velocity (v) is assumed to be
dependent on the applied stress
intensity factor (K_I) by

$$v = AK_I^N \qquad (1)$$

where A and N are crack growth constants that depend on the environment and material composition. Such
behavior is often observed in
silica-based ceramics and the subcritical crack growth is usually
associated with a stress corrosion
process at ambient temperatures.
From Eq. (1) and the definition

$K_I = \sigma_a Y\sqrt{a}$, it can be shown that there is a relationship between strength (σ_f) and stressing rate $(\dot{\sigma})$,[1] i.e.,

$$\sigma_f^{N+1} = B(N + 1)\, S^{N-2}\, \dot{\sigma} \qquad (2)$$

where $B = 1/(AY^2 (N - 2) K_{IC}^{N-2})$, K_{IC} is the critical stress intensity factor for crack instability, S is the fracture strength in an inert environment where no subcritical crack growth occurs prior to fracture, σ_a is the applied tensile stress, a is the crack size and Y is a constant that depends on the geometry of the specimen and the crack. From its definition, B is a crack propagation constant that depends on the environment and the material composition. From Eq. (2) it can be seen that the strength will be higher the faster the stressing rate, which just reflects a decrease in the amount of time for subcritical crack growth to occur. By making similar assumptions to the above, it is also possible to derive the time-to-failure under constant or cyclic stresses.[4]

Because ceramics exhibit a wide variation in strength, the allowable stress during service is quite low if low failure probabilities are required. Proof testing offers a means for increasing the stress allowables by eliminating weak samples. The value of proof testing is that it characterizes the largest effective flaw possible in a tested component, since any larger flaw would have caused failure during the proof test.

For proof testing, fracture mechanics can be used by considering how the inert strength changes during the proof test, i.e., how much subcritical crack growth has occurred. The inert strength after proof testing (S_a) is related to the initial inert strength (S_i) by[5]

$$S_a^{N_p-2} = S_i^{N_p-2} - D_p/B \qquad (3)$$

where $D_p = \int_0^t \sigma_a(t)^{N_p} dt$ represents the amount of strength degradation during the proof test cycle and N_p is the crack propagation constant for the proof test environment. Equation (3) uses inert strength as the frame of reference, however, it is often more convenient to use the strength as measured in a given environment and at a particular stressing rate $(\dot{\sigma}_s)$, preferably an environment and stressing rate that is similar to that expected in service. For this case and where the proof test is performed in the same environment as that being analyzed for service $(N = N_p)$, Eq. (2) can be substituted in Eq. (3) to give

$$\sigma_{fa}^{N+1} = \sigma_{fi}^{N+1} - (N+1)\dot{\sigma}_s D_p \qquad (4)$$

The probability of failure (F) for a ceramic after proof testing can be obtained by expressing the initial strength in terms of its failure probability distribution. Generally, for ceramics the initial distribution can be approximated by the Weibull relationship

$$\ln \ln \left(\frac{1}{1-F}\right) = m_f \ln \frac{\sigma_{fi}}{\sigma_{of}} \qquad (5)$$

where m_f and σ_{of} are empirical constants evaluated by a fit of the strength data. By substituting Eq. (5) in Eq. (4), the strength after proof testing can be derived to be

$$\sigma_{fa}^{N+1} = (Q_a + Q_p)^{(N+1)/m_f}\sigma_{of}^{(N+1)}$$

$$-(N + 1)\dot{\sigma}_s D_p \qquad (6)$$

where $Q_a = -\ln(1-F_a)$, $Q_p = -\ln(1 - F_p)$, F_a is the failure probability after proof testing and F_p is the failure probability in the proof test. The failure probability in the proof test (F_p) can be determined from the number of samples that break during the test or by setting $\sigma_{fa} = \sigma_{fmin}$, and $Q_a = 0$ in Eq. (6). This procedure gives

$$Q_p^{\frac{N+1}{m_f}} = \frac{(N + 1)\dot{\sigma}_s D_p}{\sigma_{of}^{N+1}} + \left(\frac{\sigma_{fmin}}{\sigma_{of}}\right)^{N+1}$$

$$(7)$$

For a given material σ_{fmin} is determined by the unloading rate ($\dot{\sigma}_u$) and the crack propagation parameters during the proof test. It can be shown from Eq. (2) and the minimum inert strength[5] that

$$\sigma_{fmin}^{N+1} = \dot{\sigma}_s B^{(N+1)/3}[\dot{\sigma}_u(N - 2)]^{(N-2)/3}$$

$$[3^{N-2}/(N + 1)]^{1/(N-3)} \qquad (8)$$

The parameter σ_{fmin} is important in that it represents the lowest possible strength after proof testing, i.e., how the strength distribution is truncated. For situations where the proof testing is not performed in inert environments, σ_{fmin} can be much smaller than the highest proof stress. In these cases the second term on the right hand side of Eq. (7) is negligible, Eq. (6) can be re-written as

$$\left(\frac{\sigma_{fa}}{\sigma_{of}}\right)^{N+1} = (Q_a + Q_p)^{(N+1)/m_f}$$

$$-Q_p^{(N+1)/m_f} \qquad (9)$$

These last two equations are most important to the design engineer.

Equation (8) indicates the conditions for truncation of the strength distribution at the stress allowable; for situations where these conditions are impractical, Eq. (9) allows the stress allowable to be determined at an acceptable failure probability. In the latter equation this is simply accomplished by selecting a particular failure rate during the proof test. This in turn will depend through Eq. (7) on the value of D_p and hence the proof stress cycle. The above equations summarize failure predictions for ceramic materials before and after proof testing. The predictions, as presented, are based on the determination of a strength distribution at a particular stressing rate and the crack propagation parameters, B and N. It was assumed that the proof testing was to be performed in the same environment as that being analyzed for service. There are several techniques for the determination of B and N. For example, using Eq. (1), direct measurement of crack velocity at different stress intensity levels allows A and N to be determined. From the value of A and the determination of K_{IC} it is then possible to determine B. Alternatively, the parameters can be determined, using Eq. (2), from the strengths determined at various stressing rates. It is this technique that was chosen for the TPS study. It has also been shown

that the parameters B and N can be determined from constant or cyclic stress experiments.[1]

3. SYSTEMS AND MATERIALS

The major portion of the TPS for the Space Shuttle is based on glazed silica tiles that are adhesively bonded to a strain isolation pad (SIP), which in turn is bonded to the aluminum surface of the Shuttle. The low thermal conductivity of the tile and the high emittance of the glazed coating thus serve to protect the airframe from the high temperatures (up to ~ 2500°F) associated with re-entry. Since the predominant structural stresses were found to act perpendicular to the aluminum substrate, which is also the weakest direction within the ceramic tile, material testing concentrated on characterizing the system in this direction. It was therefore appropriate to determine the tensile strengths of the system in this direction. This mode of loading was termed flatwise tension.

Experiments and analyses were performed on two systems and on the LI 900 tile material itself. The first system studied is called the undensified LI 900 tile/SIP system and exhibits failure within the tile, but very close to the ceramic tile/SIP interface. It was previously established that failure in this location was the result of an interfacial stress concentration.[6] The second system studied, called the densified LI 900 tile/SIP system, was manufactured with a densified tile surface which eliminated the effect of the stress concentation by strengthening the tile near the interface. This newer system, which comprises the majority of tiles on the TPS of the current Space Shuttle, is approximately twice as strong as the undensified system in flatwise tension testing (FWT). Failure of the densified system still occurs within the tile, but away from the tile/SIP interface.

In order to improve the reliability of the undensified system, a proof testing procedure had been established for these tiles on the Shuttle.[7] For this study therefore, the fracture mechanics approach was used to evaluate this proof test. Both the system characterization and the proof testing were performed in the tensile mode at ambient conditions, so that Eqs. (8) and (9) are appropriate.

4.   EXPERIMENTAL PROCEDURE

The strength of the undensified LI 900/SIP system (25 samples) was determined in FWT for a variety of stressing rates using a hydraulic testing machine. A schematic of the testing configuration is shown in Fig. 1. The tiles were bonded in the same way as on the Space Shuttle but with 1/4 in. thick aluminum plate instead of the airframe surface. A second 1/4 in. thick aluminum plate was glued with epoxy to the upper glazed surface of the tile. The upper loading rod was screwed into this plate while the lower plate was clamped to the hydraulic ram. The system was carefully aligned to minimize any bending stresses. This alignment was facilitated by using universal joints within the load train. The tests were performed in ambient conditions (23°C, 65% relative humidity). The values of B and N were determined using Eq. (2). Statistically, this was performed using a linear regression on the $\ln\bar{\sigma}_f/\ln\dot{\sigma}$ data, where $\bar{\sigma}_f$ is the median strength. The intercept (b) and slope (m) of the regression analysis are related to B and N by[1]

$$m = \frac{1}{N + 1}$$

$$b = \frac{\ln B + \ln(N + 1) + (N - 2)\,\ln\bar{S}_i}{(N + 1)}$$

$$(10)$$

where $\bar{S}_i$ is the median inert strength of samples statistically identical to those used in the experiment. Preliminary strength/ stressing rate data was also available from the STS* for both the undensified and densified systems.[8,9] Comprehensive FWT strength distributions were available for both these systems from STS.[10,11] These distributions were each determined on 60 samples at a displacement rate of 0.03 inch/min, which corresponds to a stressing rate of approximately 0.07 psi/s. Sufficient data was then available to evaluate the proof testing procedure using fracture mechanics.

SC80-11282

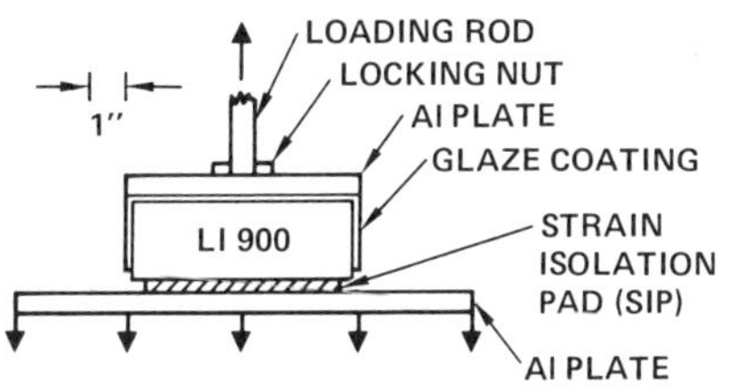

Fig. 1 FWT testing configuration for tile/SIP system.

To confirm these predictions, proof testing was also performed in a laboratory on 27 samples of the undensified system. The proof test cycle used on the Shuttle is shown approximately in Fig. 2. Basically, the stress was raised at a rate of 0.5 psi/s to the proof stress level (σ_p) for 60 seconds but with holds at the $(\sigma_p - 2 \text{ psi})$ and $(\sigma_p - 1 \text{ psi})$ levels for 30 seconds each. The unloading rate was approximately 0.5 psi/s. For this cycle the value of D_p (Eq. (3)) is given by

$$D_p \; [(\text{psi})^N \cdot s] = 30(\sigma_p - 2)^N + 30(\sigma_p - 1)^N + 60\,\sigma_p^N + 4\,\sigma_p^{N+1}/(N + 1)$$

$$(11)$$

SC80-11283

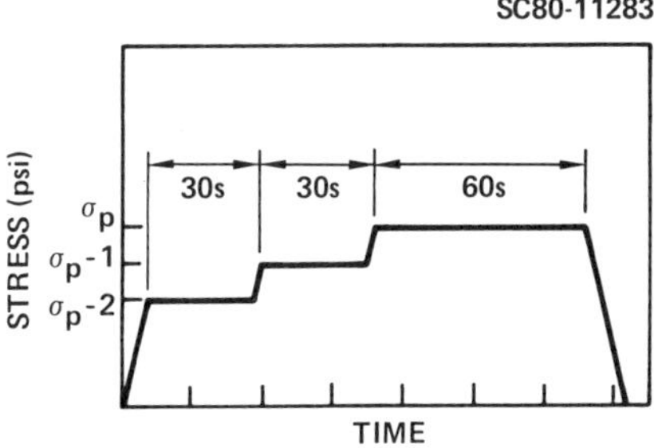

Fig. 2 Schematic of proof test cycle used on Shuttle.

For the laboratory study the proof stress was chosen to be 10 psi. The proof test configuration was the same as that described for the FWT tests. The samples that survived the proof test (23) were then pulled to failure to determine their after-proof test strengths. These results could then be compared directly with the fracture mechanics prediction.

Finally, tensile strengths were also determined for the LI 900 material

---

*Rockwell International, Space Transportation System Development and Production Division, Downey, CA 90241.

263

alone for several stressing rates.
For these tests, 2 in. × 1 in. × 1
in. samples of LI 900 were glued to
aluminum blocks with epoxy to facilitate gripping to the load train.

5. RESULTS AND DISCUSSION

The combined strength/stressing rate
data for the undensified LI 900/SIP
system obtained both at the Science
Center and STS[8] are presented in
Fig. 3 and it was found that the two
data sets correlate very well with
each other. As indicated in the
fracture mechanics theory, the
strength was found to be higher, the
higher the stressing rate. From a
linear regression of this data, it
was found that N = 34.1 and lnB =
-8.87. Therefore, as outlined earlier, once the strength distribution
at a given stressing rate is known,
the influence of a proof test can be
predicted. The predictions are summarized in Fig. 4 for proof tests
where the maximum proof stresses
were 6, 8 and 10 psi. The upper
line in this figure is the FWT
strength distribution determined at
STS.[10] From this data, the
strength distribution was predicted
for a stressing rate of 0.5 psi/s
using Eq. (2), as this was the rate
used in practice to remove tiles
from the Space Shuttle. It is clear
from these distributions, why proof
testing was necessary for the undensified system. For example,

stresses of 4.8 psi were expected
for many tiles during service. For
such stresses, failure probabilities
of about 1/1000 would be expected

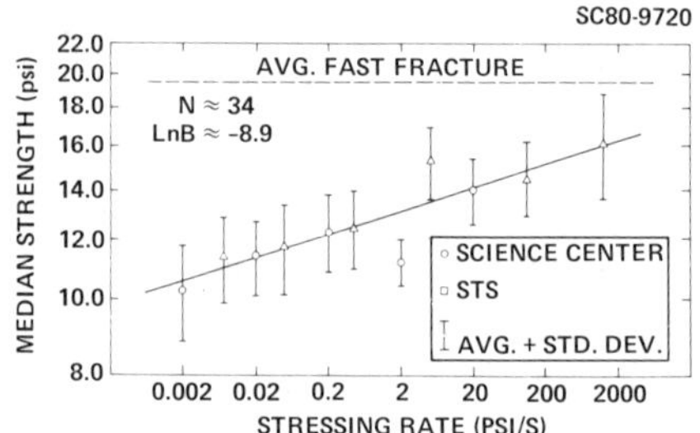

Fig. 3 Strength/stressing rate
data for undensified
system.

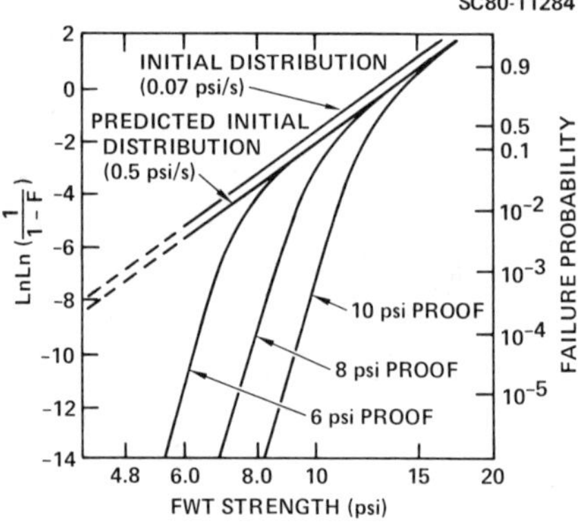

Fig. 4 Influence of various proof
test cycles on initial
strength distribution
(undensified system).

from the predicted initial distribution, which is unacceptable. The
influence of the proof test on the
strength distribution (0.5 psi/s)
was determined from Eq. (9), with Q_p
determined from Eq. (9). It was
established from Eq. (8) that σ_{fmin}
was very small for the proof test
schedule, so that the use of Eq. (9)

was correct. It is apparent from Fig. 4 that all three proof tests increase the system strength especially at low failure probabilities even without truncation. For example, the 6 psi proof test reduces the failure probabilities at 6 psi from about 3×10^{-3} to 10^{-5}, simply by mechanically eliminating about 1% of the tiles. In addition to mechanical failure, tiles on the Shuttle were also rejected on acoustic emission criteria,[7] which should increase the after-proof strengths further.

In order to substantiate the fracture mechanics approach, it was necessary to measure the after-proof test strengths for comparison with the prediction. The results of this comparison is shown in Fig. 5 for a 10 psi proof test. As can be seen the predicted after-proof test strengths agree well with the experimental results particularly in the low failure probability region.

For the densified LI 900/SIP system, failure still occurs within the ceramic so although the densification increases strength it would not necessarily eliminate subcritical crack growth. This was substantiated from the strength/stressing rate data shown in Fig. 6, where although the densified LI 900/SIP system strength was higher, the

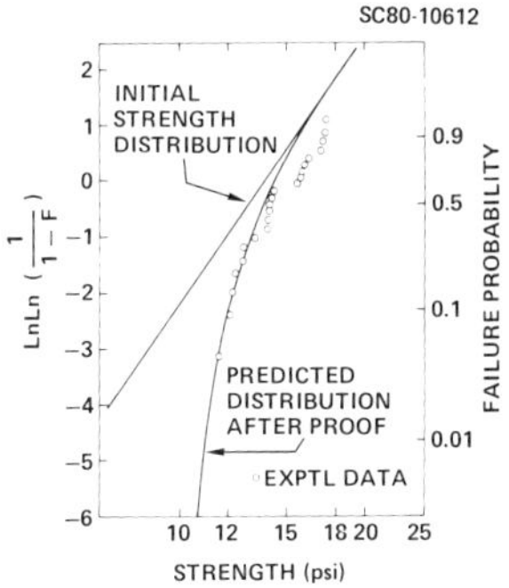

Fig. 5 Comparison of predicted and experimental after proof test strengths (undensified system).

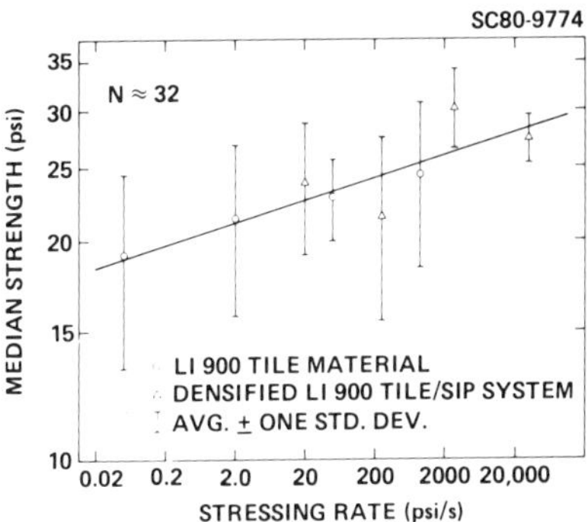

Fig. 6 Strength/stressing rate data for densified system and LI 900 tiles.

value of N (32.0) was approximately the same as that for the undensified system. Indeed, as shown in this figure the strength behavior of the densified system is very similar to that determined for the LI 900 material alone. The strength behavior of the densified system is therefore simply controlled by the ceramic and the interfacial stress concentration no longer plays a role.

It is interesting to note that these
N values correspond well with those
obtained for bulk fused silica (N $\cong$
37) and borosilicate glasses (N $\cong$
31),[12] which are similar in composition to LI 900. Fractographic
analysis of LI 900 indicates that
fracture occurs in the tile by failure and fragmentation of individual
fibers. In essence, the subcritical
crack growth is similar in these
fibrous glassy materials as it is in
the bulk glasses. For bulk glasses,
it has already been established that
the subcritical crack growth is
linked to moisture in the atmosphere, probably through a stress
corrosion reaction.[13] For these
reasons, it is expected that the
amount of subcritical crack growth
in LI 900 tiles will be reduced in
space, once the tiles are de-gassed.
Thus, it is expected that the predictions based on atmosphric testing
will be conservative. Indeed, it
has already been established at the
Science Center that the strength of
LI 900 is higher when tested in dry
nitrogen or at liquid nitrogen
temperatures.

Using the data from Fig. 6 it is
possible to predict the influence of
proof testing on the densified LI
900/SIP system. These are summarized in Fig. 7. The upper line in
the figure is the strength
distribution determined at STS,
which is then used to predict the

strength distribution for a stressing rate of 0.5 psi/s. For this
analysis, a simpler proof test
scheme was used, i.e., the system
would be simply loaded to the proof
stress at 0.5 psi/s, held for 10 s
and unloaded at 0.5 psi/s. It can
again be seen that proof testing
could also be used to increase
reliability of the densified system.
For example, after a 10 psi proof
test, the minimum strength at a
failure probability of 10^{-5} would be
10.1 psi. The number of tiles
expected to fail such a proof test
would be ~ 0.6%.

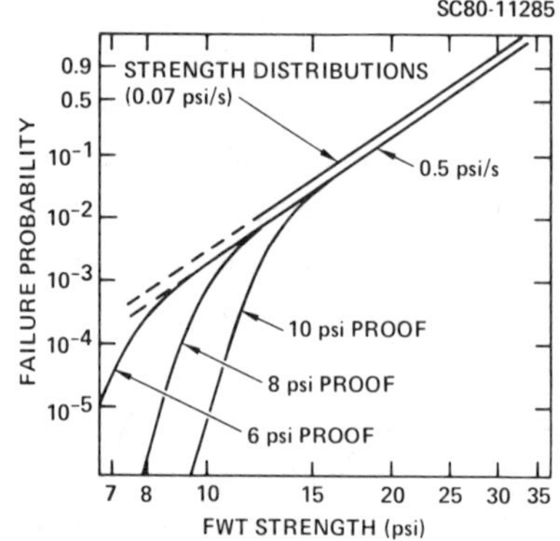

Fig. 7 Influence of various proof
test cycles on initial
strength distribution
(densified system).

Finally, it should be noted that
although fracture mechanics can
successfully be used to predict the
influence of proof testing on
strength, lifetime predictions based
on cyclic loading on the tile during
a mission were not made. It has
been established that cyclic loading

can lead to strength degradation in
the SIP which can then control the
systems strength. The fracture
mechanics approach must therefore be
modified to account for inelastic
behavior of the SIP before TPS lifetimes under cyclic stresses can be
analyzed. The theoretical lifetime
of the ceramic tile by itself, however, could be analyzed for cyclic
loading using the fracture mechanic
concepts outlined in Section 2.

6. CONCLUSIONS

It was shown that fracture mechanics
can be used to predict the beneficial effects of proof testing on the
densified and undensified LI 900/SIP
TPS. Proof test predictions were
confirmed for the undensified system
in the laboratory. For the proof
testing established by STS, it was
shown that the failure probability
associated with the required
strength was reduced to an
acceptable level. It was further
shown how proof testing can be
designed to assure the reliability
of a TPS by choosing a very low
level of failure probability.

It was also found that the strength
behavior of the TPS is controlled
mainly by the ceramic. The subcritical crack growth behavior is
similar for both undensified and
densified systems and for the LI 900
material itself as well as to that
found in bulk glassy materials at
ambient conditions. The absolute
strength of the undensified TPS is
determined by the strength of the
ceramic and the stress concentration
at the tile/SIP interface, whereas,
for the densified system, the
strength is controlled only by the
ceramic.

7. ACKNOWLEDGEMENTS

The authors gratefully acknowledge
the assistance of Arvind Arora,
Milan Metcalf and Chris Hawkins in
this study. The cooperation of many
members of the Space Division was
also invaluable in establishing the
background for the Shuttle developments. The work was conducted in
support of the Rockwell International Space Division, Space Shuttle
Thermal Protection System, NASA
Contract No. NASA 9-14000.

8. REFERENCES

1. J.E. Ritter, Jr., "Engineering
   Design and Fatigure Failure of
   Brittle Materials," pp. 667-686
   in Fracture Mechanics of
   Ceramics 4, ed. by R.C. Bradt,
   D.P.H. Hasselman and F.F. Lange,
   Plenum Press, N.Y. (1978).

2. S.M. Wiederhorn and J.E. Ritter,
   Jr., "Application of Fracture
   Mechanics Concepts to Structural
   Ceramics," pp. 202-214 in
   Fracture Mechanics Applied to

Brittle Materials, ASTM STP 678,
ed. by S.W. Freiman, ASTM
(1979).

3. J.E. Ritter, Jr., S.M.
Wiederhorn, N.J. Tighe and E.R.
Fuller, Jr., "Application of
Fracture Mechanics in Assuring
Against Fatigue Failure of
Ceramic Components," NBSIR 80-
2047 (June 1980).

4. A.G. Evans and E.R. Fuller, Jr.,
"Crack Propagation in Ceramic
Materials Under Cyclic Loading
Conditions," Met Trans. 5, 27-33
(1974).

5. S.M. Wiederhorn, E.R. Fuller,
Jr., J.E. Ritter, Jr., P.B.
Oates, "Proof Testing of
Ceramics: I. Experiment," NBSIR
79-1934; "II. Theory," NBSIR
79-1944 (December 1979).

6. Technical Report, TR S107100,
Space Transportation System
Development and Production
Division, Rockwell
International, Downey,
California, (1980).

7. Quality Control Document MTO
501-538, "Proof Loading TPS
Tiles," ibid, (1980).

8. Technical Report, TR S107075,
ibid, (1980).

9. Technical Report, TR S134381,
ibid, (1980).

10. Technical Report, TR S107067,
ibid, (1979).

11. Technical Report, TR S107073
ibid, (1980).

12. J.E. Ritter, Jr., and C.L.
Sherburne," Dynamic and Static
Fatigue of Silicate Glasses," J.
Am. Ceram. Soc. 54, 601-5
(1971).

13. S.M. Wiederhorn, "Mechanisms of
Subcritical Crack Growth Glass,"
pp. 549-80 in Fracture Mechanics
of Ceramics, 4, ed. by R.C.
Bradt, D.P.H. Hasselman and F.F.
Lange, Plenum Press, N.Y.
(1978).

9. BIOGRAPHIES

Dr. David J. Green: Member of
Technical Staff, Structural Ceramics
Group, Rockwell International
Science Center with 10 years of
experience in ceramic research.
After graduating from the University
of Liverpool, England, he completed
his graduate studies at McMaster
University, Canada. There, he
worked extensively on the microstructural aspects of fracture in
ceramics.

In 1975, Dr. Green joined the
Canadian Government, where his work
was concerned with the preparation
of ultrafine, homogeneous ceramic
powders.

After joining the Science Center in
1979, Dr. Green is continuing to
study the relation between fabrication, microstructure and properties
of ceramics. Dr. Green has

authored, or co-authored 21 technical papers.

Dr. F.F. Lange: Group Manager, Structural Ceramics Group, Rockwell International Science Center. Fifteen years experience in Ceramic Research. Graduated with B.S. Ceramic Science, Rutgers University; Ph.D. Solid State Technology, Pennsylvania State University; post-doctoral capacity at the Atomic Energy Research Est., Harwell, England. Dr. Lange is a co-editor of a four volume book entitled Fracture Mechanics of Ceramics, published by Plenum Press. He helped organize two NATO Advanced Study Institutes on Nitrogen Ceramics. Dr. Lange was chairman of the 1980 Gordon Conference on Ceramics. He is the author and/or co-author of more than 70 technical articles and 7 patents; he is a Fellow of the American Ceramic Society.

Dr. John E. Ritter, Jr.: Professor, Mechanical Engineering, University of Massachusetts, Amherst. Twenty years experience in strength of glass and ceramic materials. Graduated with B.S. in Chemical Engineering, and M.S. in Ceramic Science; Ph.D. in Materials Science, Cornell University. Dr. Ritter joined the faculty at University of Massachusetts in 1965 and on a leave of absence in 1971-2 was a technical product manager at Owens Illinois, Inc. He is the author and/or co-author of more than 70 technical articles.

ADHESIVE EVALUATION
FOR
PRINTED WIRING BOARD BONDING

S. A. Tunick and D. T. Chow
Hughes Aircraft Company
Technology Support Division
Culver City, California

Abstract

Design considerations in selecting adhesives for bonding printed wiring boards to heat sinks are discussed. Several film adhesives that do not require conventional high temperature and pressure laminating processes were evaluated. The use of these materials allows high volume production and easy removal of boards from heat sinks without damage. Performance properties including adhesion, dielectric strength, solvent resistance, and thermal conductivity were determined.

1. INTRODUCTION

Printed Wiring Boards (PWB's) of multilayer construction with high voltage components and high component packing density can generate sufficient heat to damage or reduce the performance of the board material and its components. One method for dissipating heat involves attaching the PWB to a heat sink (Figure 1). Heat then transfers from the board through the attachment interface to the heat sink. The heat sink may be in turn cooled by transforming heat via conduction to chassis, via convection to air, or by circulating cooling air or fluid through or over the heat sink, depending upon the system configuration. Attaching the PWB to a heat sink with mechanical fasteners will not provide optimum heat transfer since the board may bow slightly when attached, or be shimmed off the heat sink by circuit traces or solder pad's on the PWB surface facing the heat sink.

Thus, a thermal transfer medium providing maximum contact between the PWB and heat sink is desired. This requirement can be met by bonding or laminating the PWB to a heat sink. The material and process considerations for adhesive selection are generally as follows (Reference 1):

o Maximum thermal conductivity
o Adequate adhesion
o Balance of resin flow
o Maximum dielectric strength

o   Maximum environmental resistance

o   Resistance to solvent and heat
    used in subsequent processes

o   Rework feasibility

o   Cost effective process

Various thermoset B-stage "pre-preg" films have been used to laminate PWB's to heat sinks. They provide thin, strong bonds, with few voids and controlled resin flow. The film adhesives are much easy to handle in production than liquid or paste adhesives. However, conventional pre-pregs requiring laminating pressures above 100 psi and cure temperatures above 335 $^\circ$F are common. Where heat sinks are not flat plates, special tooling is required to achieve the required laminating pressure. If the heat sink is a honeycomb structure, the high laminating pressure may damage the heat sink. High laminating temperatures may be close to the melting temperature of PWB solder pads creating the potential for scraping parts. In addition, these high temperatures may not be compatible with surrounding assemblies. This becomes especially apparent when a PWB must be removed and either repaired or replaced on a completed assembly. Several commercially available film adhesives that do not require conventional high temperature and pressure laminating processes were evaluated. Performance properties of test specimens bonded with these materials including adhesion, solvent resistance, dielectric strength, and thermal conductivity were determined. The ease of removability of PWB's bonded to heat sinks was also studied.

2. MATERIALS

2.1 Adhesive Systems

2.1.1 System No. 1. System No. 1 consisted of two 5 mil layers of a B-staged, fiberglass cloth supported film adhesive surrounding a single 2 mil layer of cured (C-staged) epoxy/glass laminate. The adhesive was an alumina filled epoxy resin containing a polyether diamine flexibilizer. The cured epoxy glass was used as both a spacer and a barrier against circuit trace burrs that might otherwise penetrate the uncured adhesive layers and short to a heat sink.

2.1.1.1 System No. 2. System No. 2 consisted of two layers of a 7 mil B-staged, fiberglass cloth supported epoxy film adhesive surrounding a 2 mil layer of C-staged epoxy/glass. The adhesive was an alumina filled resin with a nitrile rubber modifier.

2.1.1.1.1 System No. 3. System No. 3 consisted of three layers of a 4 mil B-staged, glass-supported epoxy film adhesive. The adhesive was an unfilled nitrile modified epoxy resin with a polyether amine flexibilizer.

3. SUBSTRATES

Performance properties of the adhesive materials were determined on bonds made between polyimide glass board (MIL-P-13949) and: (1) aluminum, (2) electroless nickel plate, and (3) stainless steel. All bonds were cured at 200 $^{\circ}$F with 10-15 psi pressure.

4. TESTS

4.1 <u>Shear, Peel Strength, and Bond Durability</u>

Lap shear specimens were prepared and tested in accordance with ASTM D1002. Peel strength specimens were prepared as shown in Figure 2 and tested in accordance with MIL-P-13949 at a 90 degree angle. Bond durability specimens made are shown in Figure 3.

Polyimide board, nickel, and stainless steel substrates were prepared by solvent flushing followed by scrubbing with an abrasive cleanser. Aluminum substrates were cleaned using an optimized sodium dichromate/sulfuric acid hot etch.

During the course of PWB fabrication, the board will often see several solvent cleaning cycles to remove dirt, grease, solder flux, etc. Sets of lap shear, peel, and bond durability specimens were exposed to a Freon TE or an isopropyl alcohol immersion cycle followed by an oven bake out at 200 $^{\circ}$F. Lap shear and peel specimens were tested at room temperature and 200 $^{\circ}$F. Bond durability specimens were examined at room temperature. The test results are given in Tables I through IV. Test results in Table I show that System No. 3 has the greatest lap shear and peel strength to aluminum.

Lap shear and peel strength data in Tables II and III show that the solvent exposures did not effect bond strength. However, bond durability data in Table IV did show that bonds made with the System No. 2 adhesive had a significant amount of crack extension. This is an indication that bonds made with this material might degrade after repeated exposures to the test solvents.

4.2 <u>Removal Properties</u>

Polymide circuit boards, 6x9x.075 inches were bonded to aluminum honeycomb heatsinks. The boards and heatsinks were then reheated to 200 $^{\circ}$F. The PWB's were then removed using blunt wooden and thin metal spatulas to avoid damaging either the PWB or heatsink. The System No. 1 material adhered well to both surfaces but the PWB could be removed without damaging either PWB or heatsink surfaces. The cohesive failures observed indicated that the bond to the two faying surfaces (even at 200 $^{\circ}$F) was stronger than the interlaminar adhesive bond. The System No. 2 adhesive specimens were more easily removed, but it was noted that bond failure was adhesive to the polymide surface (no adhesive was left on the PWB). Circuit boards bonded with the System No. 3 adhesive could not be removed at 200 $^{\circ}$F without damaging the board or heatsink. Boards could be removed at 350 $^{\circ}$F. (However, this high temperature could damage transistors, IC's, and other sensitive hardware mounted to the PWB or in areas surrounding the PWB being removed.)

4.3 Thermal Conductivity

Thermal conductivity of the adhesive systems was determined using a Colora thermalconductometer. The test results are given in Table V.

Values of thermal conductivity are based on a unit thickness. Since all three adhesive systems examined were approximately 10 to 12 mils in thickness (before cure), the thermal conductivity values obtained provide a relative comparison of their overall heat transfer ability. The test results showed that the System No. 1 material had the highest thermal conductivity (.37 Btu/ft·hr·$^{\circ}$F) of all the adhesives examined.

4.4 Dielectric Properties

The insulation resistance of the System No. 1 adhesive was determined to be 1.2×10^{14} ohm-cm at room temperature and 1.9×10^{11} ohm-cm at 200 $^{\circ}$F, when tested in accordance with ASTM D257. The dielectric strength of Systems No. 1 and 2 was also determined using the ASTM D257 test method. The results given in Table VI show that the dielectric strength of both systems was greater than 900 volts/mil.

5. DISCUSSION

Commercial film adhesive systems are available for bonding PWB's to heatsinks with a 200 $^{\circ}$F and 15 psi cure schedule. The bonds formed are solvent resistant and will hold PWB's in place at 200 $^{\circ}$F or above. Insulation, heat transfer, and removal properties must also be considered in evaluating and choosing the appropriate adhesive. The test results obtained in this study show that:

(1) The dielectric properties of the epoxy film adhesives studied provide the high breakdown voltages and high insulation resistance needed to eliminate the possibility of PWB to heatsink shorts or leakage currents that can cause excessive noise in analog circuitry.

(2) Thermal conductivity values varied from .37 Btu/ft·hr·$^{\circ}$F for System No. 1 to .15 Btu/ft.hr·$^{\circ}$F (equivalent to approximately a 2 $^{\circ}$C/watt temperature differential across the bond) for System No. 3. All values are considered acceptable for PWB/heatsink bonding.

(3) Removal at 200 $^{\circ}$F of PWB's bonded to heatsinks is possible with some adhesives (Systems No. 1 and 2). However, the adhesive type of failures with System No. 2 indicated that **it** may not be compatible with polyimide substrates. PWB's bonded with System No. 3 could not be removed at 200 $^{\circ}$F, but could be removed at 350 $^{\circ}$F. (Earlier work with the System No. 3 material showed that it is useful in applications where slow cure temperature rises are encountered.)

6. CONCLUSION

It is concluded that the adhesive System No. 1 will provide the optimum combination of good solvent resistance, insulation properties, high heat transfer, and easy removal. It also maintains adequate strength under all service and storage conditions in excess of 200 $^{\circ}$F.

REFERENCES

R. W. Korb et al, Achieving Low Cost and Reliability in Heat Sink Bonding, SAMPE Technical Conference, October, 1980.

BIOGRAPHY

<u>S. A. (Steve) Tunick</u> is Head of Coatings and Finishes Group in the M&P Engineering Department at Hughes Aircraft. He received his M.S. in Chemical Engineering from UCLA in 1973 and his MBA from the University of Santa Clara in 1977. He is currently responsible for circuit board bonding specifications and adhesive selection and evaluation for many structural and electro-optical applications at Hughes. Prior to his present association with Hughes he worked on high temperature vapor **deposition** at Fairchild Semiconductor.

David T. Chow received his B.S.
Degree in Chemical Engineering
from University of Washington in
1965. He is presently Head of
Adhesives and Finishes Section
in the Technology Support Division
of Hughes Aircraft Company. His
previous experience includes
adhesive bonding of aerospace and
space structures, laser systems,
and radar electronics. Before
joining Hughes in 1972, he worked
as a Materials and Processes
Engineer at the Boeing Company and
Rohr Industries.

TABLE I. LAP SHEAR AND PEEL STRENGTH TEST RESULTS

Adhesive System	Test Temp	Lap Shear Strength		Peel Strength		
		Aluminum to Polyimide (psi)	Electroless Nickel to Polyimide (psi)	Aluminum to Polyimide (piw)1/	Electroless Nickel to Polyimide (piw)	Stainless Steel to Polyimide (piw)
No. 1	75 °F 200 °F	1280 337	482 113	15.2 2.1	7.9 1.3	5.0 0.2
No. 2	75 °F 200 °F	680 296	752 152	12.9 3.1	4.7 2.4	1.2 0.7
No. 3	75 °F 200 °F	3230 450	N.T. 2/	17.7 10.0	N.T.	N.T

1/ piw = pounds per inch width peel strength

2/ N. T. = Not Tested

3/ All test values are averages of at least three specimens

276

TABLE II. SOLVENT RESISTANCE OF ADHESIVE BONDS – LAP SHEAR STRENGTH ALUMINUM TO POLYIMIDE GLASS

Adhesive System	Test Temperature	Control (No Solvent Exposure Data From Table I) (psi)	Freon T. E. Cycle (psi)	Isopropyl Alcohol Cycle (psi)
No. 1	R. T. 200 °F	1280 337	1374 393	1370 303
No. 2	R. T. 200 °F	680 296	N.T.	N.T.
No. 3	R. T. 200 °F	3230 450	3160 1360	3265 1286

NOTE: All test values are average of at least three specimens.

N. T. = Not Tested

TABLE III. SOLVENT RESISTANCE OF ADHESIVE BONDS PEEL STRENGTH

Adhesive System	Test Temperature	Control (No Solvent Exposure Data From Table I) (p.i.w.)	Freon T. E. Cycle (p.i.w.)	Isopropyl Alcohol Cycle (p.i.w.)
No. 1	R. T. 200 °F	15.2 3.1	15.3 2.6	14.8 2.7
No. 3	R. T. 200 °F	17.7 10.0	18.0 14.3	15.5 15.3

NOTE: All test values are average of at least three specimens.

P.I.W. = Pounds per inch width.

TABLE IV. SOLVENT RESISTANCE OF ADHESIVE BONDS -
WEDGE DURABILITY TEST

Adhesive System	Crack Extension (Inches)	
	Isopropyl Alcohol Cycle	Freon Cycle
No. 1	0.44	1.04
No. 2	2.27	1.77
No. 3	0.07	0.03

NOTE: All test values are average of at least two specimens.

TABLE V.   THERMAL CONDUCTIVITY

Adhesive System	Thermal Conductivity
No. 1	0.37 BTU/Ft.Hr. $^{\circ}$F
No. 2	0.26 BTU/FT.Hr. $^{\circ}$F
No. 3	0.15 BTU/Ft.Hr. $^{\circ}$F

TABLE VI.   DIELECTRIC STRENGTH (BREAKDOWN
VOLTAGE PER UNIT THICKNESS)
ASTM D 257

Adhesive System	Dielectric Strength (Volts/mil)
No. 1	904
No. 2	920

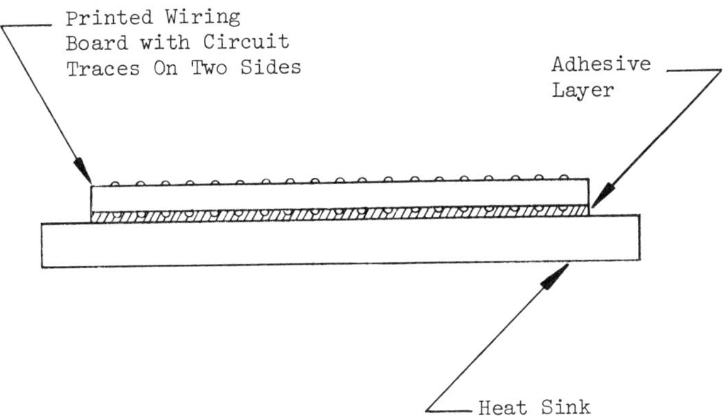

FIGURE 1. SCHEMATIC VIEW OF CIRCUIT
BOARD BONDED TO HEAT SINK

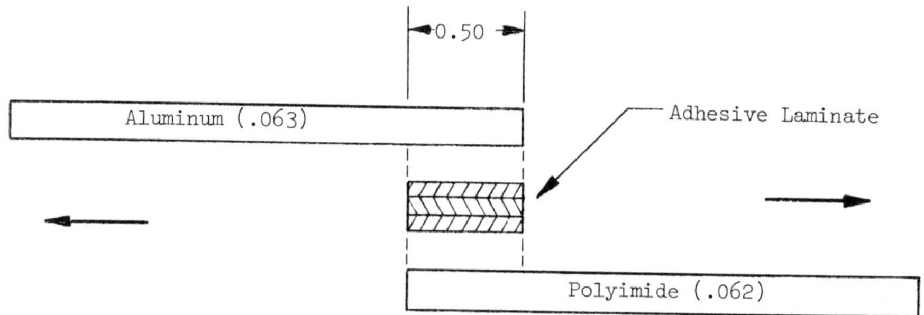

NOTE: All dimensions in inches.

FIGURE 2A.   TENSILE LAP SHEAR SPECIMEN

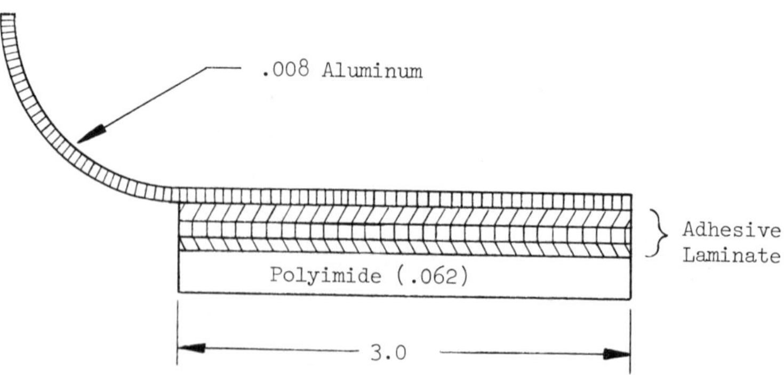

NOTE: All dimensions in inches.

FIGURE 2B. PEEL SPECIMEN

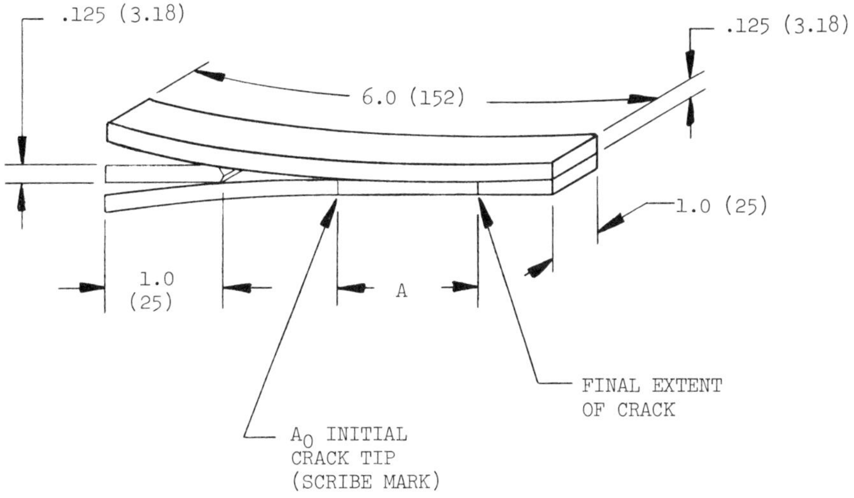

1. Dimensions in inches (millimeters).

FIGURE 3. SOLVENT DURABILITY TEST SPECIMEN CONFIGURATION

26th National SAMPE Symposium
April 28-30, 1981

IMPORTANT ELECTRICAL PROPERTIES OF POLYMERS FOR MODERN ELECTRONIC SYSTEMS

Charles A. Harper
Westinghouse Electric Corporation
Baltimore, Maryland

Abstract

Mission success in modern high performance electrical and electronic systems requires a better understanding than ever before of the critical properties of materials. In the past, for instance, it was often sufficient to know that a polymer was a reasonable electrical insulating material for a given electrical system. Increasingly, however, it is mandatory that electronic systems designers fully understand and intelligently apply the broader range of electrical properties of polymers to assure the required high performance. A fundamental problem exists, however, in what has been called the terminology gap between materials scientists and electrical engineers. The objectives of this paper are (1) to explain, in a manner easily understandable to all technical disciplines, the critical electrical properties of polymers and the major factors which influence those properties and (2) to indicate the practical importance of these critical electrical properties to performance of modern electrical and electronic systems.

To achieve the above objectives, the presentation coverage will include resistance related properties such as volume resistivity, surface resistivity and insulation resistance, along with the major factors which influence those properties and the role of those properties in system performance. Also discussed will be the voltage related properties such as dielectric strength, voltage breakdown, corona, arcing and tracking using the same explanatory approach as with the other electrical properties. In all cases, definitions and terms will be clearly presented and explained so as to readily bridge the terminology gap for all disciplines involved in this important subject. Finally, the increasingly important electrical loss properties such as dissipation factor, loss factor, dielectric constant and other related loss properties will be discussed, along with the major factors which influence those properties and their importance in modern high frequency, high speed, high power electronic systems.

ELECTRICAL PROPERTIES OF POLYMERS

1.1 Resistance Related Properties

Resistance of an insulating material, like that of a conductor, is the resistance offered by the conducting path to passage of electrical current. Resistance is expressed in ohms. Insulating materials are very poor conductors when dry, offering high resistance. For insulating materials, the term volume resistivity is more commonly applied. Volume resistivity is the electrical resistance between opposite faces of a unit cube for a given material and at a given temperature. The relationship between resistance and resistivity is expressed by the equation $\rho = RA/l$, where ρ is the volume resistivity expressed in ohm-cm, R is the resistance in ohms between faces, A is the area of the faces, and l is the distance between faces of the piece on which measurement is made. This is not resistance per unit volume which would be ohm/cm^3, although this term is sometimes erroneously used.

Other terms are sometimes used to describe a specific application or condition. One such term is surface resistivity which is the resistance between two opposite edges of a surface film 1-cm square. Since the length and width of the path are the same, the centimeter terms cancel. Thus, units of surface resistivity are actually ohms. However, to avoid confusion with usual resistance values, surface resistivity is normally given in ohms/sq. Another broadly used term is insulation resistance which, again, is a measurement of ohmic resistance for a given condition, rather than a standardized resistivity test. Insulation resistance values are expressed in ohms, or often megohms, due to the high numerical resistance values of insulating materials. For both surface resistivity and insulation resistance, standardized comparative tests are normally used. Such tests can provide data such as effects of varying environments and ambient conditions on a given insulating material configuration.

Regarding values of resistance and resistivity, the higher the value, the better for a good insulating material. The resistance value for a given material depends upon a number of factors. It varies inversely with temperature, as shown in Figure 1, and is affected by humidity, moisture content of the test part, level of the applied voltage, and time during which the voltage is applied. Some of the differences which exist when different insulations are exposed to

high humidity are illustrated in
Figure 2.

When tests are made on a piece that
has been subjected to moist or humid
conditions, it is important that
measurements be made at controlled
time intervals during or after the
test condition has been applied,
since dry-out and resistance increase occur rapidly. Comparing
or interpreting data is difficult
unless the test period is controlled
and defined.

A fundamental point which is often
overlooked deals with surface wetting and/or contamination. Even the
best high resistivity insulating
materials are susceptible to surface
wetting or contamination, which
greatly reduces the surface resistance properties, and hence the
overall insulation resistance of a
given part. Low surface energy or
non-wetting surfaces are required to
overcome this problem, as shown in
Figure 3.

1.2 <u>Voltage Related Properties</u>

All insulating materials fail at
some level of applied voltage for a
given set of operating conditions.
Dielectric breakdown may result from
an electron avalanche in the insulating material. Each avalanche
results from the acceleration of a
few electrons by the electric field,
which gives rise to secondary
electrons by impact with atoms or
molecules. Each of these produces
other electrons by the same process.

A few free electrons are believed
to be present in all materials. The
dielectric strength is the voltage
an insulating material can withstand
before dielectric breakdown occurs.
Dielectric strength is measured
through the thickness of the material, and is normally expressed in
voltage gradient terms, such as
volts per mil.

Dielectric strength is influenced by
temperature, humidity, voids or
foreign materials in the specimen,
electrode configuration, frequency
and specimen geometry. Because of
this it is often difficult to compare breakdown data from different
sources unless all test conditions
are known. Type of voltage, temperature, and any preconditioning of
the test part must be noted. Also,
thickness of the piece being tested
must be recorded because the voltage
per mil at which breakdown occurs
varies with thickness of test pieces.
Normally, breakdown occurs at a much
higher volt-per-mil value in very
thin test pieces (a few mils thick)
than in thicker sections (1/8 in.
thick for example). This is illustrated in Figure 4. The dielectric
strength term implies a controlled
test. In operating equipment, the
terms voltage breakdown and voltage
endurance are used. These imply the
voltage at which equipment breakdown
occurs and/or the time endurance to
voltage breakdown. Another controlled test term sometimes used is

intrinsic dielectric strength. The intrinsic dielectric strength, which is the maximum voltage gradient a homogeneous substance will withstand in a uniform electric field, is dependent upon the ability of the material to interfere with the acceleration of electrons. It depends greatly on the density of the material, since more closely packed atoms increase the probability of collisions that slow down the free electrons. It also depends on the electron-trapping ability of the atoms and molecules. The intrinsic dielectric strength for polymers ranges from 15,000 to 25,000 volts/mil.

The dielectric strength values obtained with commercial insulating materials in practical tests are very much lower than the intrinsic dielectric strength values, for several reasons: (a) the existence of defects, holes, and conducting and foreign particles introduced by the method of manufacture; (b) the presence of a stress concentration at the electrode edges or the points where the breakdown can be initiated because the electric field is much higher than the average; (c) in ac tests, due to the damaging effect of an electric discharge (corona) during testing; and (d) because of dielectric heating which raises the temperature and lowers the breakdown strength.

The presence of defects lowers the dielectric strength at points and has the effect, in practical tests, of making the test values smaller as the areas under test are increased. The ratio of electric field strength in different insulating materials in series varies inversely as the ratio of their dielectric constants (discussed below). This effect places a higher electric stress on the low-dielectric-constant constituent of a non-homogeneous or multi-material insulating material combination. For example, with oil-impregnated paper, the stress on the oil, which has a dielectric constant of about 2, is approximately three times that on the paper, which has a dielectric constant of about 6. This effect generally lowers the overall dielectric strength of the composite insulation, since, when partial breakdown of one part occurs, the remainder is left under higher electrical stress.

Another failure mode related to voltage stress failure is corona. Corona is ionization, under voltage stress, of air which is inside or at interfaces of insulating materials. Corona is a local breakdown at an edge, point, interface, air void or gap, etc. Corona inception voltage levels are measurable, as are corona extinction voltage levels as voltage stresses are reduced. External corona is also observable. These values are characteristic for given material or part constructions. When corona appears at the surface of insulation under electric stress, it erodes the surface by electron bombardment, associated heat, and sometimes secondary effects due to formation of chemical oxidizing agents such as ozone and oxides of nitrogen. This effect begins immediately and even fractional seconds of exposure at ac voltage near the breakdown voltage lowers the breakdown strength significantly. At lower voltage stresses, with corona still present, breakdown will also occur but the time required for breakdown will be longer. For example, a pair of insulated wires, twisted together and tested in a standard manner, might withstand a 1-min ac test of 6000 V. If 1000 V ac is applied continuously, breakdown will occur in something on the order of 100 hr. For this reason, electrical equipment is designed on the basis of voltage endurance of materials rather than short-time dielectric strength. Short-time dielectric strength is a useful guide for quality control. Voltage endurance is determined by measuring the time required for breakdown of insulated samples of several voltage levels. Accelerated voltage tests can be run in which voltages are applied at higher frequencies, for example, 3000 Hz. Under these conditions, the insulation is subjected to as many cycles in one year as it would normally be subjected to in 50 years at 60 Hz. This method cannot be used with insulation systems that have a high dissipation factor at the test high frequency, since voltage breakdown might result from dielectric heating. Because of the time-dependent limitation, most high-voltage electrical equipment is operated at voltage gradients on the order of 50 V/mil, which is about 10% of the test dielectric strength and less than 1% of the instrinsic dielectric strength.

1.3 Arc and Track Resistance

Arc resistance is a measure of an electrical breakdown condition along an insulating surface, caused by the formation of a conductive path or track on the surface. The track which is created is commonly a carbon path. Being conductive, a once formed track is a continuous and hence catastrophic failure mode. Arc resistance is measured as the time, in seconds, required for breakdown along the surface of the material being measured. Surface breakdown, arcing or electrical tracking along the surface, is also affected by surface cleanliness and dryness. Higher values therefore indicate greater resistance to breakdown along the surface due to arcing or tracking conditions. Reproducible testing is often difficult, and the very erratic surface contamination in actual systems compounds this problem.

Certain plastic materials track easily. For example, polymers based on aromatic rings such as phenolics or polystyrene track readily. Certain other plastics which are based on certain cyclic polymers and polymers containing nitrogen, fluorine or silicon atoms in the structure are more resistant. In general the addition of mineral fillers improves arc and track resistance. Certain hydrated fillers exemplified by hydrated alumina dramatically improve the tracking properties. The waters of hydration have a quenching effect on the arc, thus delaying track formation until the protection so afforded has been exhausted.

1.4 Electrical Loss Properties

1.4.1 Dielectric Constant (Permittivity).

The dielectric constant, or permittivity of an insulating material is the ratio of the capacitance of a capacitor containing that particular material to the capacitance of the same electrode system with air or vacuum replacing the insulation as the dielectric medium. The dielectric constant is also sometimes defined as the property of an insulation which determines the electrostatic energy stored within the solid material. The dielectric constant of electrical insulating materials ranges from a low of about 2 or less for materials with lowest electrical loss characteristics and up to 10 or so for materials with highest electrical losses. Certain types of electrical components and parts require much higher dielectric constants, and this is achieved by adding high dielectric constant fillers such as titanates to the base insulating materials. Aside from such cases, however, low dielectric constant materials are preferred to achieve lowest electrical losses. The dielectric constant for air or vacuum is, of course, approximately 1, depending on the level of contamination or impurities present. The dielectric constant of a given insulating material varies as a function of frequency and temperature. The most drastic change is the increase of dielectric constant with increasing temperature. A rapid increase begins to occur near the glass transition temperature of a given polymer material.

1.4.2 Dissipation Factor, Power Factor and Loss Factor.

Electrical losses are very important, especially in high frequency and power generating electrical systems. These losses, usually released in the form of heat, represent system performance inefficiencies. Highest system efficiency is understandably achieved with lowest electrical losses. As was mentioned above, dielectric constant characteristics play a role in electrical losses. Dissipation factor and related characteristics of electrical insulating materials are also most significant in determining system electrical losses. Dissipation factor is a mathematical relationship of losses in an electric circuit. Dissipation factor is the ratio of the parallel reactance to parallel resistance, i.e., the tangent of the loss angle or the contangent of the phase angle. Dissipation factor ($\tan \delta$), also often expressed simply as $\tan \delta$, is not identical to power factor, another term sometimes used. However, at low values of $\tan \delta$ (less than 0.10), the values for dissipation factor and power factors are nearly identical. The dissipation factor is related to the energy dissipated and hence to the efficiency of the insulation material. Some designers use another term, loss factor, which is the product of the dielectric constant and the dissipation factor. This term is related to the total loss of power (in watts) occuring in the insulating material. As with dielectric constant, the dissipation factor of a given insulating material varies as a function of frequency and temperature. Again, the most drastic change is the increase of dissipation factor with increasing temperature, especially above the glass transition temperature for a given polymer. Some typical examples are shown in Figure 5. Thus, it is obvious that increasing temperatures increase electrical losses and decreases electrical system efficiencies. In fact, above certain temperatures, it is quite possible to obtain a cascading effect, with higher temperatures increasing losses which in turn create higher temperatures. The ultimate result of such cascading is insulating material breakdown or often worse, thermal ignition and fire.

REFERENCES

1. Harper, C. A., Handbook of Plastics and Elastomers, McGraw-Hill Book Co., New York, N.Y.
2. Harper, C. A., Handbook of Materials and Processes for Electronics, McGraw-Hill Book Co., New York, N.Y.
3. Harper, C. A., Handbook of Electronic Packaging, McGraw-Hill Book Co., New York, N.Y.

BIOGRAPHY

Charles A. Harper is a graduate of the Johns Hopkins University, Baltimore, Maryland, holding a degree in Chemical Engineering. As a Technologies Program Manager for Westinghouse Electric Corporation, he has over twenty-five years experience in the field of electronic packaging and materials for electronic systems. He is Editor-in-Chief for a widely known series of McGraw-Hill Handbooks, including those referenced above. He is an active member of many leading professional societies in his field, including SAMPE, and has served as program chairman, session chairman and speaker for dozens of national technical conferences. He also presents a widely known series of seminars in this field.

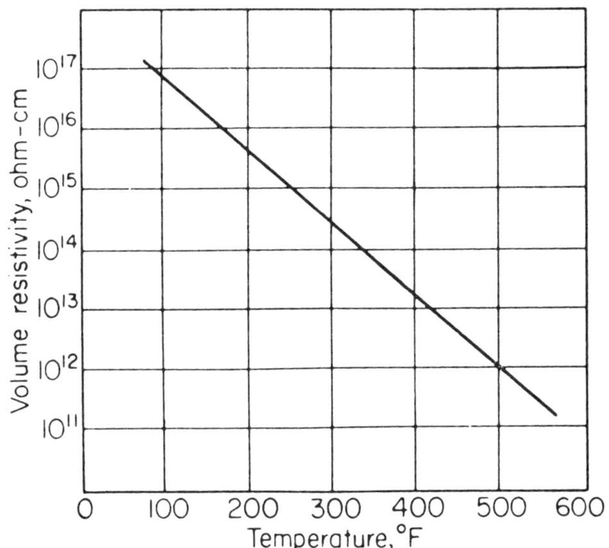

VOLUME RESISTIVITY VS. TEMPERATURE FOR
UNFILLED POLYMIDE
Figure 1

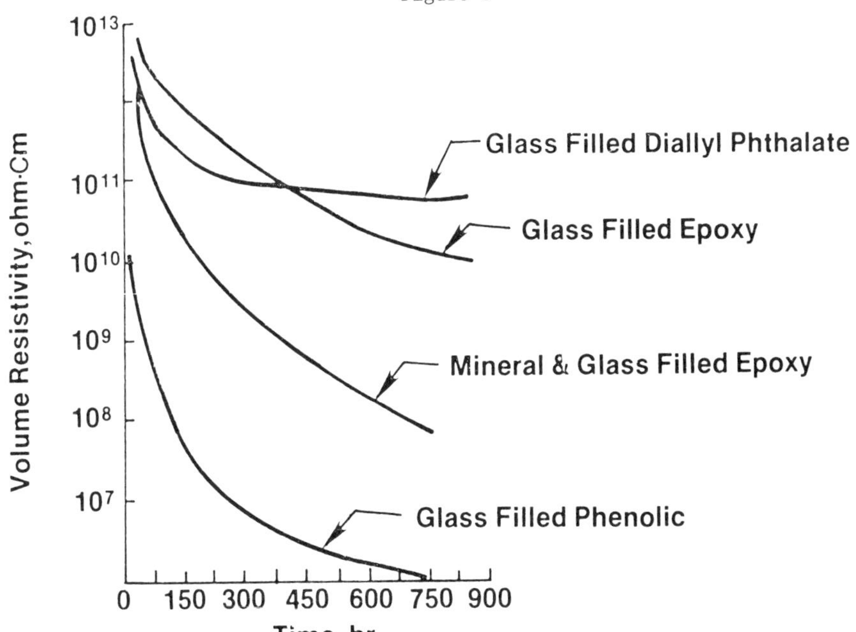

EFFECT OF HUMIDITY ON INSULATION RESISTANCE OF SEVERAL
POLYMER-FILLER SYSTEMS
Figure 2

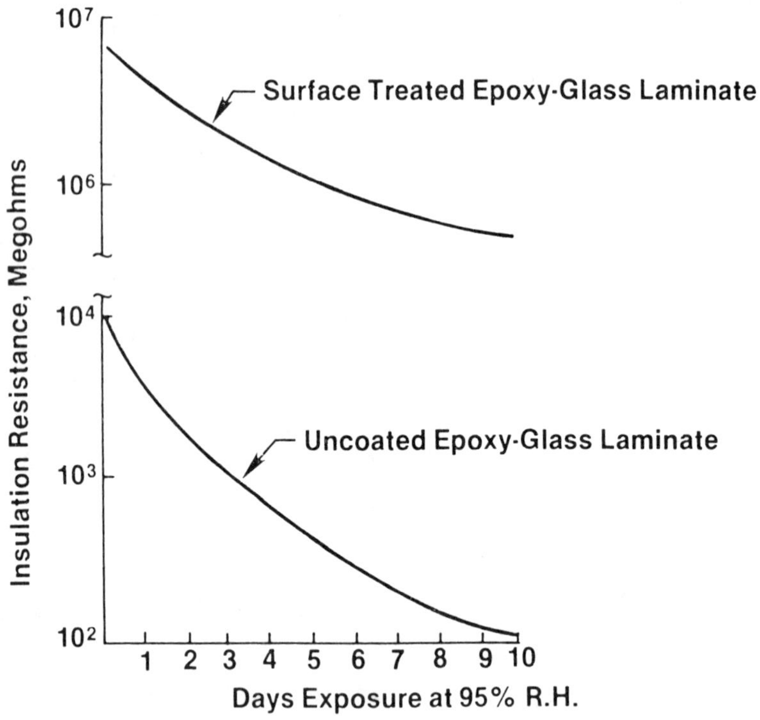

IMPROVEMENT OF PERFORMANCE IN HUMID ENVIRONMENTS THROUGH
APPLICATION OF LOW ENERGY SURFACE TREATMENT
Figure 3

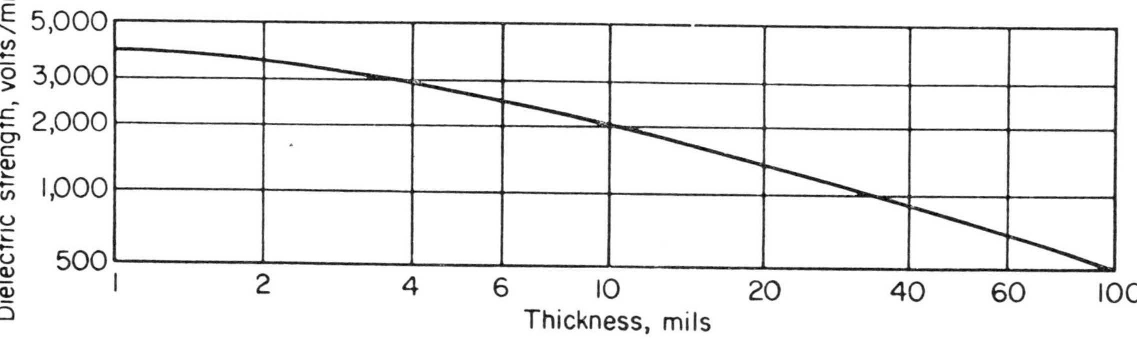

SHORT-TERM DIELECTRIC STRENGTH OF TFE FLUOROCARBON AS A
FUNCTION OF THICKNESS.
Figure 4

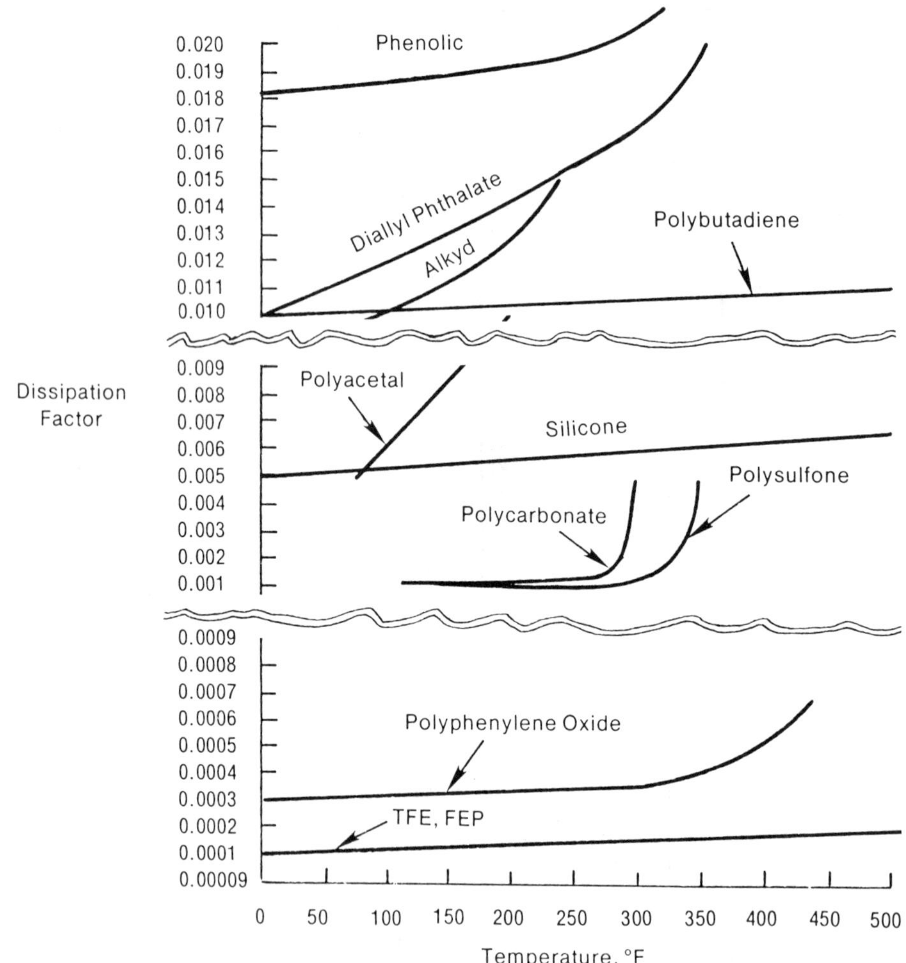

EFFECT OF TEMPERATURE ON DISSIPATION FACTOR FOR SEVERAL POLYMERS
Figure 5

IDENTIFICATION OF NYLONS BY PYROLYSIS GAS CHROMATOGRAPHY

David R. Smith and James R. Myers
Hughes Aircraft Company
Technical Support Division, Tucson, Arizona

Abstract

Hughes Aircraft Company is continually faced with the identification of micro samples of materials employed in missile production programs. A method was developed for determining certain specific nylon polymer types using controlled pyrolysis, followed by gas chromatographic separation of the resulting fractions. The chromatographic "finger-print" was found to be highly repeatable for each material. In addition to providing information about the identity of the polymer, the chromatograph also provided general clues about the structure of the nylon. An example of a typical problem encountered in a Hughes' missile development/production programs is discussed in some detail.

1. INTRODUCTION

Nylon polymers are tough, abrasion resistant thermoplastic polyamides which are not affected by most industrial solvents. These properties, along with excellant electrical resistance and ease of manufacture, have led to their widespread use throughout the aerospace and electronics industries. Nylon 6 and 6,6 are the most widely used nylons, dominating the tire cord and textile industries. However, other nylons show specialized applications. Nylon 9 exhibits significantly higher electrical resistance. Nylon 11 and 12 show less tendency to absorb water providing better dimensional stability. Nylon 6T is highly transparent.

Because of their widespread use in the missile industry these materials are often encountered in hardware failure analysis and contamination evaluation. Identification of polymer type is therefore important in solving these problems and determining the sources of contamination. Infrared Spectroscopy (IR) is the usual method used, however, this technique has several shortcomings. Sample sizes are often so small that they either make IR prohibitive or require the use of a beam condenser. IR instruments can be expensive and therefore, are not found in all laboratories. In addition IR spectra of most nylons show only subtle differences, thus making interpretation difficult.

An alternate method for analysis of nylons was developed using gas chromatography - pyrolysis (GC - pyrolysis) techniques. This technique has been used successfully on other materials in the past, with excellent reviews given by Braner[1], Levey[2] and other[3,4,5,6]. Little effort, however, has been applied to the study of nylons. The principle behind such a method is simple. A small sample of the polymer is pyrolyzed under reproducible conditions. The long polymer chains are thus fragmented into various smaller components. These components are separated and detected using a gas chromatograph (GC) which produces a "finger-print" of the original compound. GC-pyrolysis chromatograms were developed for eight of the most widely used nylons. Using this GC-pyrolysis data base the analysis method was successfully applied to solve a recent engineering problem.

2. EXPERIMENTAL

The GC-pyrolysis technique involves the thermal destruction of the material in an inert gas stream which produces a collection of degradation products. The gas stream carries the degradation products into a chromatographic column where these products are separated into various compounds and are detected by a flame ionization detector. The output of the detector is recorded on a strip-chart recorder which produces a finger-print of the material being analyzed. Commercial instrumentation employed in this study consisted of a Chemical Data Systems Inc. 150 Pyroprobe coupled with a Varian 3700 gas chromatograph.

The pyrolysis temperature was selected by performing thermal gravimetric analysis in a nitrogen atmosphere on a selection of samples to determine the decomposition temperature of the material. GC-pyrolysis chromatograms were then run at ~100 degrees Celsius (^{o}C) intervals above the decomposition temperature. These chromotograms were examined for the most informative finger-print[7,8]. From these data a temperature of 750 ^{o}C was selected as the best pyrolysis temperature for the nylon family. The pyrolysis conditions were then standardized for all identification studies at 750 ^{o}C for 10 sec with a 10 ^{o}C/millisecond linear heating ramp.

In order to select a column for the separation, various liquid phases such as a carbowax 20 M and a SE 30 were evaluated but did not produce acceptable results. Acceptable results were obtained using a six foot long, eighth inch diameter, stainless steel, column packed with 10% carbowax 20 M and 2% KOH on

Chromosorb WHP. This column was selected for its ability to separate organic nitrogen compounds which are produced when polyamides are thermally decomposed.

The chromatographic separation was accomplished using a column temperature program of 100 $^\circ$C for one minute followed by heating to 215 $^\circ$C at 15 $^\circ$C/minute. The flame ionization detector was operated at 300 $^\circ$C and the injection port pyrolysis interface temperature was 200 $^\circ$C. The carrier gas was nitrogen flowing at 30 milliliter/minute. Standard nylon samples were obtained from Aldrich Chemical Company in pellet form. The samples were prepared by slicing off a thin piece of material weighing 0.52 milligram ±10%. Each sample was then placed on the ribbon probe and melted prior to pyrolysis. The melting temperature was varied to accommodate the differing melting points of the various nylons. The melting was accomplished by placing a sample on the probe and heating for 10 seconds at 200 $^\circ$C. Melting of the nylon sample on the ribbon insures good thermal contact during pyrolysis, thus greatly improving the repeatability of the measurement.

3. RESULTS

Figure 1 through 8 show the pyrolysis chromatograms for eight common nylons. These chromatograms were found to be extremely repeatable provided the pyrolysis conditions were held constant. It should be noted here that higher final pyrolysis temperatures produced chromato-grams with more heavily populated light components. This was probably because the higher temperatures were better able to fragment the polymer chain into smaller pieces. Notice that, with the exception of Figure 4 (Nylon 6,T), the first five chromotograms show similar features. These nylons were all made by reacting a diamine with a diacid and therefore have similar backbones. The nylons presented in Figures 6 through 8 are of a different basic type. These were made by polymerizing an amino acid. Figure 8 shows the pyrolysis chromatogram of polycaprolactam (nylon 6). Figure 9 shows the chromatogram for caprolactam which is the starting ingredient for nylon 6 production. These figures clearly show that when nylon 6 is pyrolyzed much of it reverts back to its original, unpolymerized component. Pyrolysis is therefore not only useful as a method for identification of polymer type, but also can be used to provide information about chemical structure.

Recently this method of analysis was successfully applied to an engineering problem on an actuator device. The actuator is used to hold a rocket secure prior to firing. Release of the device at the moment of firing is essential. The part consists of a metal, spring loaded shaft with a heavy nylon coating over the tip. Some of the actuators were found to "stick" in the lock position while others functioned properly. Closer examination showed the problem to involve cold flow of the nylon tip. Infrared analysis was performed on nylon samples taken from both a "good" and "bad" actuator device. The IR spectra were clearly different. The properly behaved nylon was identified as nylon 6,12 whereas, the poorly behaved nylon could not be identified by this method. Extensive thermal analysis was also performed which provided additional insights about the materials. The melting points were measured for both samples and were found to be essentially identical. Thermal mechanical analysis was used to examine the cold flow characteristics of both materials. This measurement clearly demonstrated a difference in the creep behavior of each sample but did not show why such differences existed. Both samples were then analyzed using GC-pyrolysis techniques. The resulting chromatograms are shown in Figures 10 and 11. Figure 10 shows the chromatogram of the properly behaved nylon sample. This sample was found to be nylon 6,12 just as the IR results had shown. Figure 11 shows the poorly behaved nylon material which was identified as nylon 6,9 by comparison with the previously run chromatograms. Thus the engineering problem was due to a difference in the materials used. These results were made available to the vendor with guidelines for corrective action.

A similar set of reference chromatograms have also been made using a variety of Loctites. Loctites are polymethacrylate polymers which cure under anerobic conditions. They are widely used in the aerospace industry to permanently lock metal parts such as nuts, bolts, and screws. Infrared analysis shows all of the various Loctites to be essentially identical. Loctites are color-coded by the manufacturer, however, identification by this method is usually impossible once the material has cured between metal threads. GC-pyrolysis was found to be an easy method for accurate identification of Loctites.

4. CONCLUSIONS

The results presented in this paper demonstrate the power of the GC-pyrolysis method for resolving subtle differences in otherwise chemically similar polymers.

The technique can be implemented using relatively low cost equipment and with a minimal amount of technical expertise. Small amounts of material produce results that allow the identification of a specific polymer. A potential drawback to the technique is the time required per analysis, which is approximately 2 hours.

Other areas of organic and polymer analysis also appear to be potential candidates for the application of this technique. Of particular

interest in the aerospace/electronic
industry is the analysis of soldering fluxes, which are practically
impossible to distinguish by
infrared techniques. A study is
currently being conducted in this
area.

GC-pyrolysis also has the potential
ability to determine small differences between samples of the same
material, but could be particularly
inhanced by the use of pyrolysis-Gas
Chromatography-Mass Spectroscopy.
These techniques are applicable to
the evaluation of other parameters,
such as oxidizer levels, degree of
cure, levels of contamination and
mix of copolymers.

1. G. M. Braner, _J. Polymer Sci._,
    3 (1965)

2. R. L. Levy, 15th Tobacco
    Chemists' Research Conf.,
    October 4-6, 1961.

3. D. F. Nelson, J. Yee, P. L. Kirk,
    Microchem. J. 6, 225 (1962)

4. B. C. Cox and B. Ellis;
    Analytical Chemistry, Vol. 36,
    No. 1 (1964)

5. P. Cukor and C. Persiani;
    J. Macromol Sci.-Chem., A8(1)
    pp. 105-117 (1974)

6. S. G. Perry, _Journal of G. C._,
    (Feb. 1967)

7. W. H. Pariss, P. D. Holland,
    Brit Plastics 33, 372 (1960)

8. J. G. Cobler, E. P. Samsel,
    Soc. Plastics Engrs., Trans 2,
    145 (1962)

BIOGRAPHIES

David R. Smith is head of the Chemical Analysis Group with Hughes
Aircraft Company, where he has been
employed since 1972. He received
his B. Sc (1971) in Chemistry from
the University of Arizona and M. Sc
(1972) in Physical Chemistry from
the University of Oregon. During
his employment at Hughes his responsibilities have included the
analysis of a wide variety of
materials problems involving
polymers, lubricants, contamination
and metals.

James R. Myers is a member of the
technical staff at Hughes Aircraft
Company where he has been employed
since 1978. He received his B. Sc
(1977) in chemistry from the Georgia
Institute of Technology. His responsibilities for the past several
years include contamination analysis, polymeric identification and
selection of materials for use on
new missile development program.

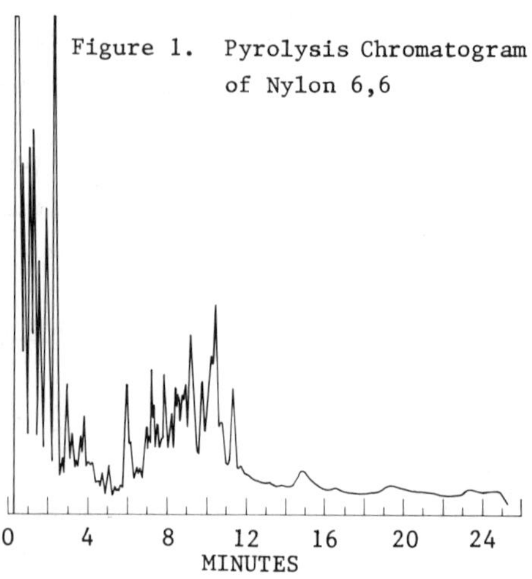

Figure 1.   Pyrolysis Chromatogram of Nylon 6,6

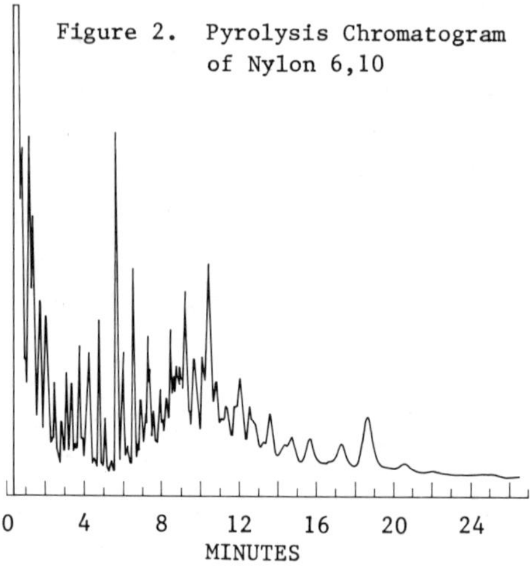

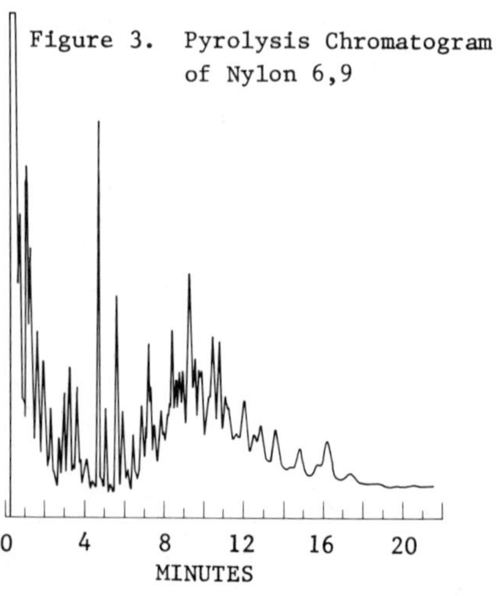

Figure 2.   Pyrolysis Chromatogram of Nylon 6,10

Figure 3.   Pyrolysis Chromatogram of Nylon 6,9

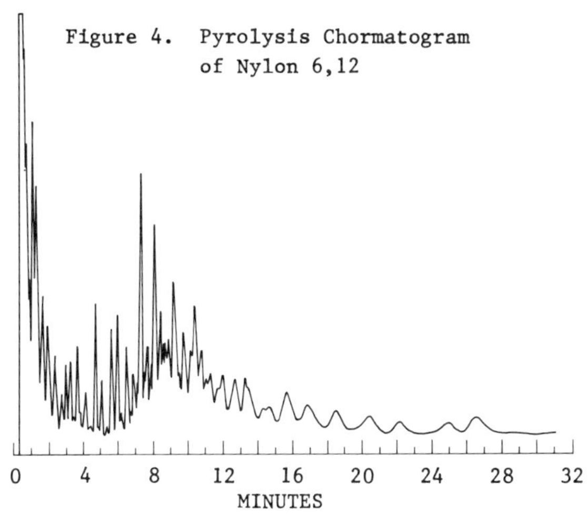

Figure 4. Pyrolysis Chormatogram of Nylon 6,12

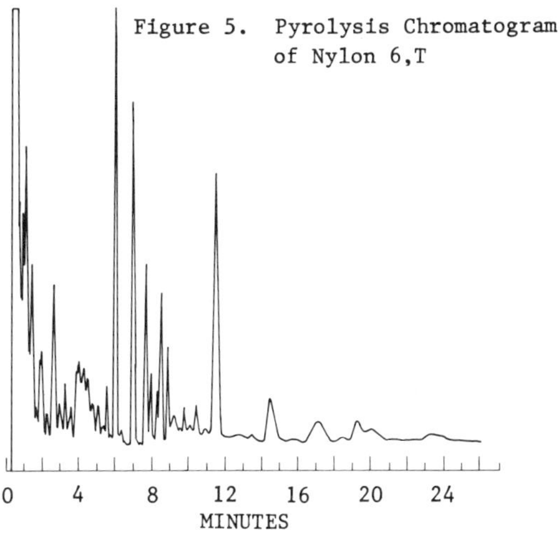

Figure 5. Pyrolysis Chromatogram of Nylon 6,T

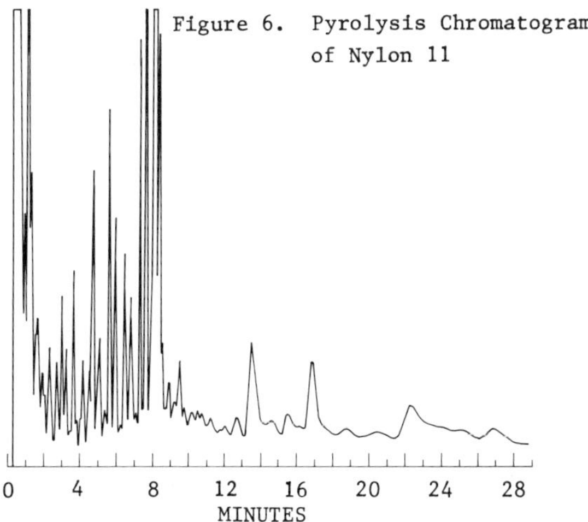

Figure 6. Pyrolysis Chromatogram of Nylon 11

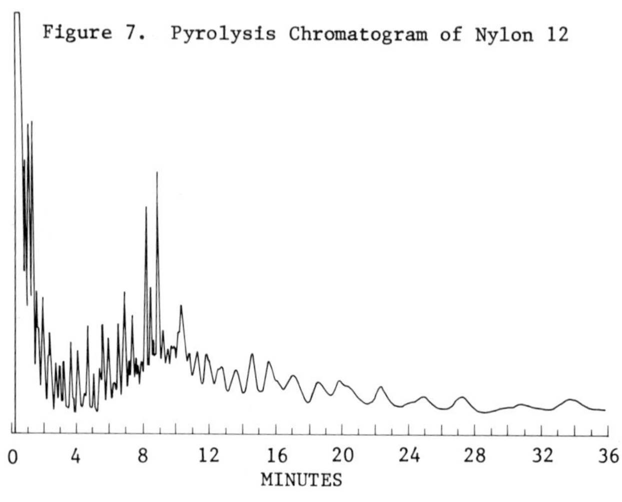

Figure 7. Pyrolysis Chromatogram of Nylon 12

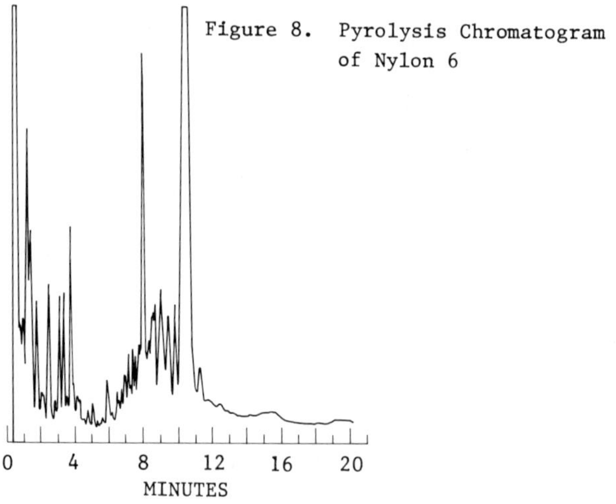

Figure 8. Pyrolysis Chromatogram
of Nylon 6

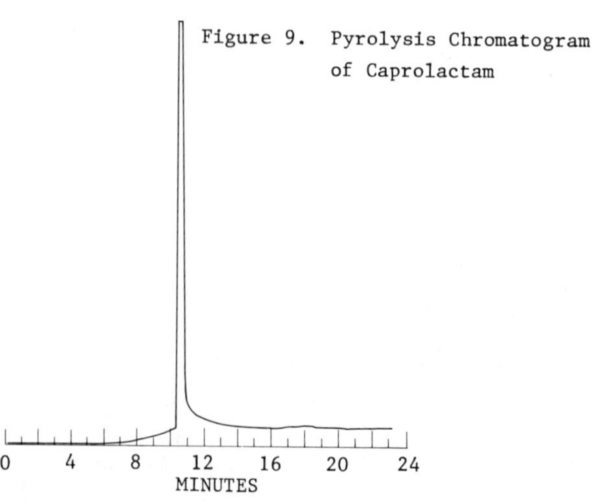

Figure 9. Pyrolysis Chromatogram
of Caprolactam

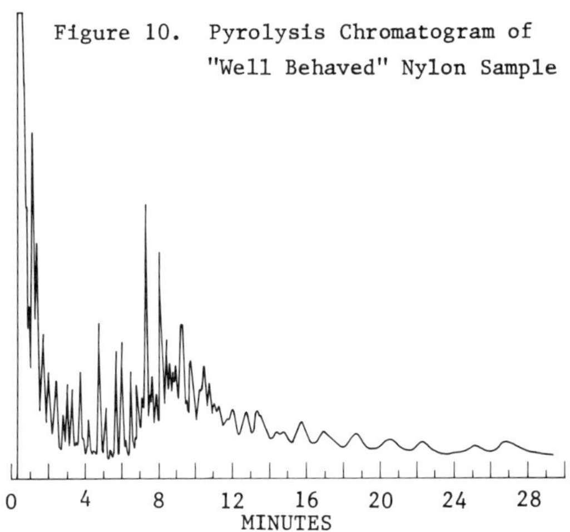

Figure 10.   Pyrolysis Chromatogram of "Well Behaved" Nylon Sample

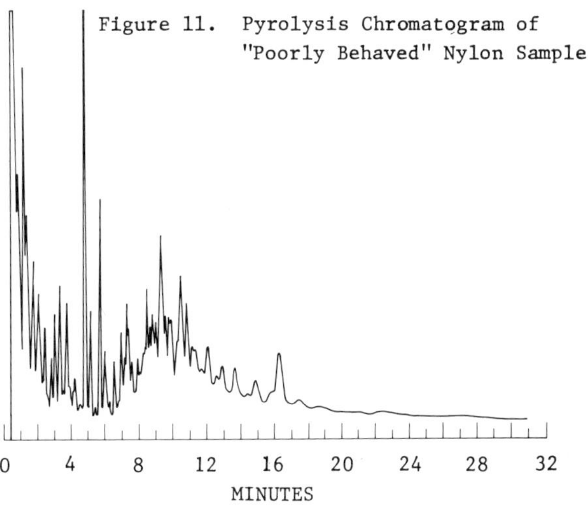

Figure 11.   Pyrolysis Chromatogram of "Poorly Behaved" Nylon Sample

THE APPLICATION OF A REVERSED PHASE LIQUID CHROMATOGRAPHY SEPARATION METHOD TO SELECTED EPOXIES, PLASTICIZERS, AND HALOCARBON FLOTATION FLUIDS.

D.W. Eggimann, J.C. Brand, and C.K. Elliott
Honeywell Inc., Avionics Division
Department of Materials and Process Engineering
13350 U.S. 19 So., Clearwater, FL 33516

Abstract

Selected epoxies, plasticizers, and halocarbon flotation fluids are separated using a single Reversed Phase Liquid Chromatography solvent/column system. Using a μBondapak C18 column and water-tetrahydrofuran-acetonitrile solvents, selectivity is adjusted to provide suitable separations of each material without requiring major time consuming changes in columns and solvents. The versatility of the method makes it attractive for routine quality control analyses.

1. INTRODUCTION

High Performance Liquid Chromatography (HPLC) is a versatile analytical tool. It has been used to separate complex mixtures of organic and inorganic liquids and solids (solvent soluble) that are comprised of compounds spanning over five orders of magnitude in molecular weight.[1] Recent reviews show tremendous growth in HPLC technology.[2-5] The versatility of the method has been well documented in academic and industrial publications.[6-11]

Reversed Phase Liquid Chromatography (RPLC) is a term used to describe the separation of a liquid by a stationary phase that is less polar than the mobile phase. The method has become popular because it is versatile, fast, equilibrates rapidly to new mobile phases, and, in general, is inexpensive to operate. Today, RPLC represents the majority of LC applications[12] and it has been suggested that it should be renamed "normal phase" due to its popularity.[1,9]

A single RPLC solvent-column system was used in this study to analyze three key categories of materials that are important in our industry. The categories included several epoxy adhesives, selected plasticizers, and a halogenated lubricant used as a flotation fluid for guidance sensors.

2. EXPERIMENTAL

A Waters Associates High Performance Liquid Chromatograph with a model 440 UV absorbance detector at a fixed wavelength of 254 nanometers, a model R401 differential refractometer detector, a U6K injector, two model 6000a pumps, and a model 660 solvent programmer was used. Separations were achieved using either one or two (in series) 3.9mm X 30cm μBondapak C18 columns (Waters). These reversed phase columns are packed with octadecylsilane bonded to 10 micron particle size silica. The chromatograms were recorded using a 10 millivolt Honeywell Electronik 194 two pen strip chart recorder at a chart speed of five minutes per inch.

Two mixed solvents were prepared using HPLC grade (MCB R Omnisolv R glass distilled) tetrahydrofuran (THF) and acetonitrile (AN), and distilled-deionized water. The water was filtered through 0.45 micron pore size millipore R membranes. Solvent A was 5 percent (bv) THF in water. Solvent B was 20 percent THF in AN. The mobile phase composition, sample concentrations, injection volumes, and other pertinent conditions are listed with each chromatogram.

3. RESULTS AND DISCUSSION

3.1 Epoxy Resins

Recent advances in Reversed Phase Liquid Chromatography (RPLC) technology have greatly improved selectivity and reproducibility. Only recently have applications of this powerful separation method been reported for epoxy adhesives. Among these, Dark, Conrad, and Crossman[13] published a method in 1974 using water and tetrahydrofuran (THF) solvents to separate epoxy resins on a Bondapak C18/Corasil column. A similar application of RPLC was reported by Hagnauer and Setton[14] in

1978. Using the same solvents, but a smaller particle-sized column (μBondapak C18, 10 micron), they reported excellent separations of epoxies in less than 30 minutes.

Analysts have also begun to turn to more complex solvent mixtures. Studies have shown that control of the mobile phase, using solvent modifiers and gradient elution methods, is a simple and effective means of controlling selectivity.[15]

In addition to using water and tetrahydrofuran solvents, acetonitrile (AN) has become a popular mobile phase.[16] Utilizing this solvent, Dark improved the earlier method[13], by employing AN with a 20 percent THF modifier. This solution was mixed with 40 percent water at injection and programmed to 0 percent water by a 15-minute linear gradient. The method was successfully used to analyze nine common epoxy resins and three prepreg epoxy systems. The method and chromatograms were published in a Waters Associates special publication.[17]

The methods applied in this report are modifications of the Dark[17] method. Modifications were made in the mobile phase composition and in the mobile phase gradients so that it would be applicable to the three categories of materials we tested.

Throughout this report the mobile phase will be referred to as a mixture of the B solvent (20 percent THF in Acetonitrile), and the A solvent (5 percent THF in water). The mixture will be described in terms of the B solvent. For example, 50 percent B means that the solvent is a 50:50 mixture of solvents B and A.

Figure 1 is a chromatogram of epoxy resin DER 332 (Dow Chemical Co.). This diglycidyl ether of bisphenol-A (DGEBA) based resin was separated using two 3.9mm X 30cm μBondapak C18 columns in series. The mobile phase was programmed through a linear gradient from 50 percent to 100 percent B in 20 minutes.

The same analytical conditions were used to separate epoxy resin EPON 828 (Shell Chemical Co.) in Figure 2. The identification of the bisphenol-A peak (BPA) and the monomers of DGEBA (N=0,1,2) were determined by comparison to those of Dark.[17]

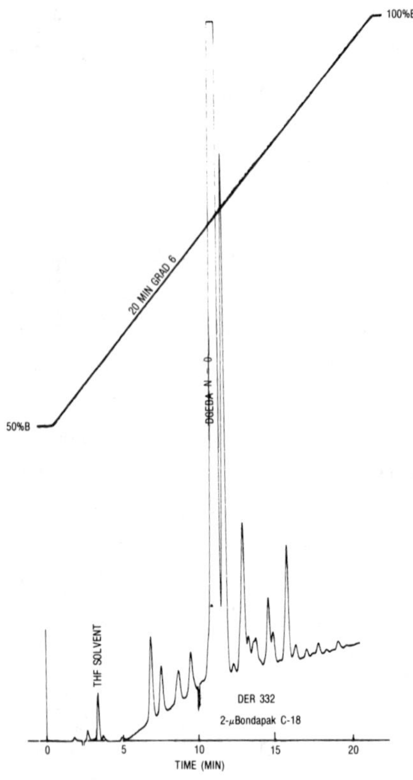

FIGURE 1. EPOXY RESIN DER 332

Conditions: 100mg sample/5 ml THF, 25 microliter injection, 2 μBondapak C18 columns, 20 minute linear gradient - 50 percent B to 100 percent B, solvent A ≤ 5 percent THF in water, solvent B ≤ 20 percent THF in AN, 2 ml/minute, detector: 254 nm UV at 0.2 a.u.f.s.

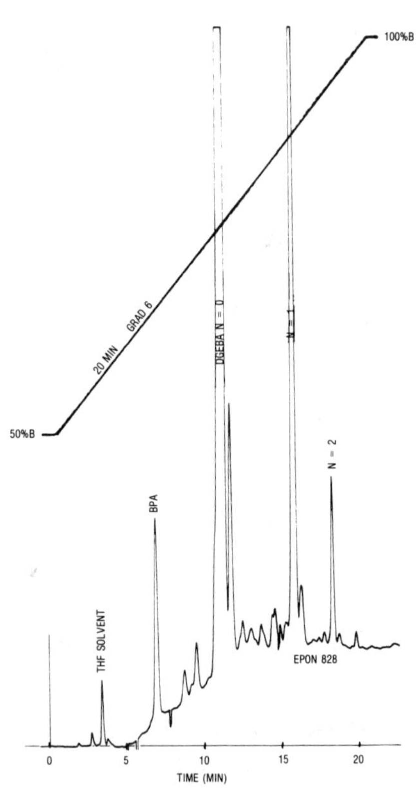

FIGURE 2. EPOXY RESIN EPON 828

Conditions: Same as Figure 1.

In Figure 3, the epoxy resin ERE 1359 (Ciba-Geigy) is separated by both gradient and isocratic elution methods. ERE 1359 is a diglycidyl ether of resorsinol based resin. This resin did not contain as many molecular species as the previous two examples and was easily separated by both elution methods.

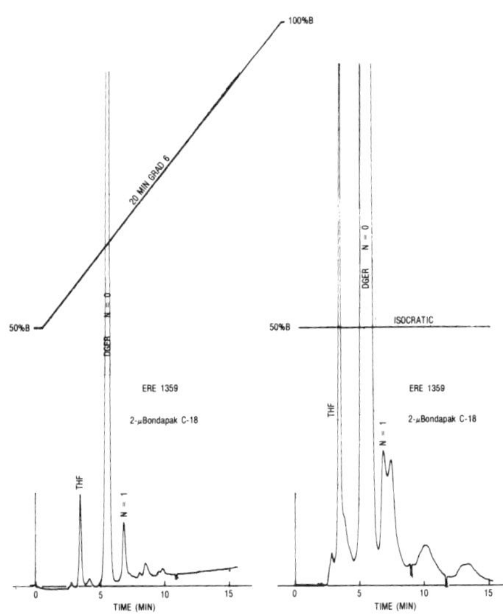

FIGURE 3. EPOXY RESIN ERE 1359

Conditions: Gradient separation - same as Figure 1. Isocratic separation - 100 microliter injection, hold solvent mixture at 50 percent B.

3.2 Epoxy Activators

The analysis of epoxy activators (hardeners) is also important when considering a versatile RPLC method. Activators are often more polar compounds which elute rapidly and appear early in the chromatogram. More polar mobile phases, such as methanol, are often used to force slightly longer retentions on the column. However, since our goal was to utilize only one solvent/column system, the water-THF-AN mobile phase was adjusted toward greater polarity by simply increasing the water content.

The activators (amines) are often single compound materials that elute as a single peak. However, formulators wishing to impart special physical characteristics to the epoxy system will mix additives with the activator. The proportion of these additives in the system can be important. Figure 4 is the chromatogram of an activator (amine) that contains a cure accelerator. The separation is achieved by starting at 40 percent B to elute the accelerator (DMP-30), then rapidly changing the gradient to 100 percent B to elute the activator (menthane diamine).

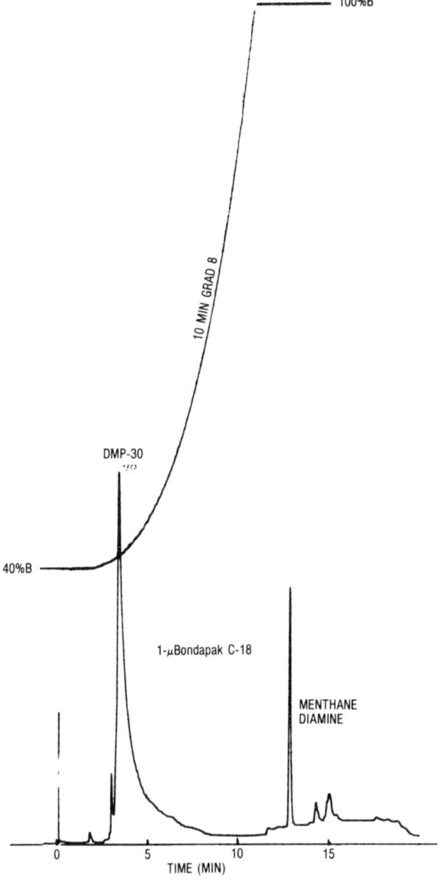

FIGURE 4. EPOXY ACTIVATOR, DMP-30 IN MENTHANE DIAMINE

Conditions: 10 minute curved gradient (8) from 40 percent to 100 percent B, 2.0 a.u.f.s., other conditions same as Figure 1.

3.3 Plasticizers

Phthalate ester plasticizers are added to polymers to impart flexibility. Compatible plasticizers are formulated with specific polymers such that they do not greatly interfere with the polymers configuration. In fact, the ideal plasticizer formulations can establish van der Waal forces with the polymer. They can then impart flexibility and remain compatible.[18]

Plasticizers, in spite of their attractions and compatibility, will leave the polymer system and contaminate the surrounding environment. For this reason their analysis and detection is of interest in our industry.

Liquid chromatography is easily applied to phthalate plasticizers and several methods have recently been published.[19,20] RPLC methods using a water/methanol solvent system provide excellent separations.[20]

The separations of a standard containing mixed dialkyl phthalates (alkyl = C-1,2,3,8,9,10) and a tricresyl phosphate (flame retardant) are in Figure 5. An isocratic elution/separation using the water-THF-AN mobile phase that was used to separate epoxies in the preceding examples, and our application of a water-methanol separation are compared. Although the water-methanol separation is better, the epoxy method provides useful information. In fact, the epoxy method provides better separation for dioctyl-and dinonyl- phthalates, which are important commonly used plasticizers. Although gradient elution was not used in this example, the separation can be improved by using a 15-minute linear gradient to 100 percent B.

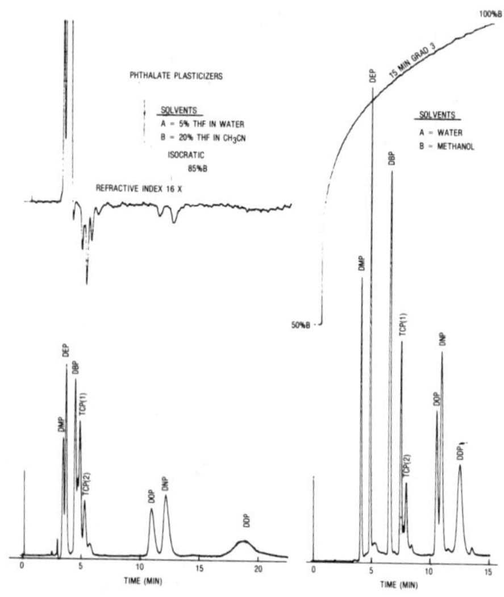

FIGURE 5. MIXED PHTHALATE PLASTICIZER STANDARD

Dimethyl phthalate (DMP), diethyl phthalate (DEP), dibutyl phthalate (DBP), tricresyl phthalate (TCP), dioctyl phthalate (DOP), dinonyl phthalate (DNP), and didecyl phthalate (DDP).

3.4 Halogenated Lubricants

High Performance Liquid Chromatography has been successfully applied to lubricants ranging from petroleum oils to fatty acids and slip agents.[21-26] Although these materials are normally used as lubricants, they often find other specific materials applications.

R.L. Ehrenfeld discussed the use of halofluro lubricants as gyro flotation fluids at the fluorine subdivision of the American Chemical Society Atlantic City meeting in September of 1959.[27] The ensuing development in the application of halogenated lubricants has found widespread application in navigation sensors. The chemical analysis of these proprietary materials has only recently received attention and published reports are rare.

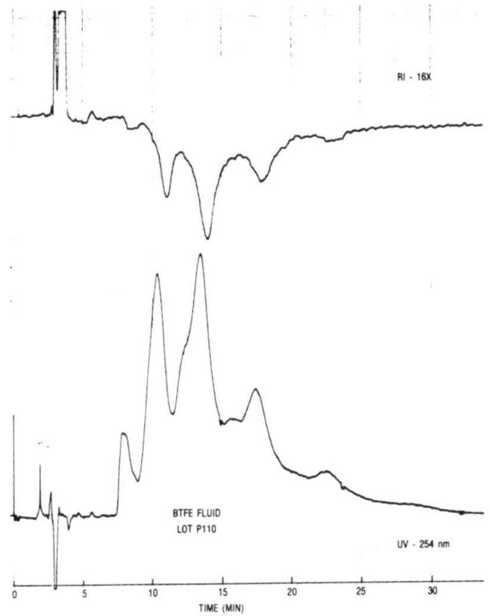

FIGURE 6. BROMO-TRIFLUOROETHYLENE FLOTATION FLUID

Conditions: 500mg sample/1ml freon, 25 microliter injection, 2 μBondapak C18 columns, 85 percent B isocratic, detectors: differential refractometer at 16x, 254 nm UV at 0.05 a.u.f.s.

The method used to separate epoxies and plasticizers above was also applied to two halogenated lubricants. Bromo-trifluoro-ethylene (BTFE) telomers are separated using an isocratic solvent mixture (85 percent B) in Figure 6. Five major components of the fluid are separated. The second fluid, a "monomolecular" fluid made of brominated triazine, was analyzed using a modified method to detect polar contaminants. The chromatogram is in Figure 7. The mobile phase was adjusted to 50 percent B to lengthen retention of the nonpolar contaminant so it could be isolated and collected for analysis. Infrared analysis confirmed that the contaminant was a carbonyl compound, a suspected degradation product of the fluid. After 7.5 minutes a 20 minute gradient was begun to change the solvent to 100 percent B, to elute the triazine peak.

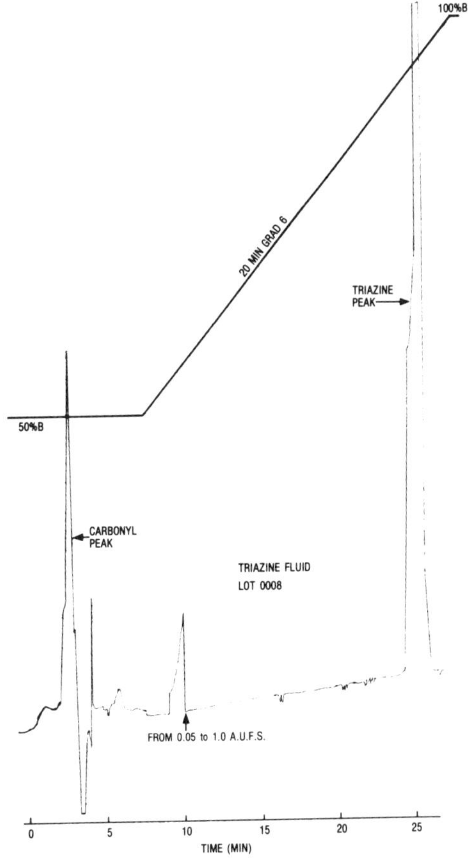

FIGURE 7. TRIAZINE FLOTATION FLUID

Conditions: 100 microliter injections, 50 percent B isocratic for 7.5 minutes, 20 minute linear gradient to 100 percent B, all other conditions same as Figure 6.

4. CONCLUSIONS

The application of a single RPLC solvent-column system to selected epoxies, plasticizers, and halogenated lubricants was adequate for receiving-inspection quality control. The analytical method was simple and easily adapted to the different materials tested. By design, separations were achieved by modifying the mobile phase composition, rather than changing columns and solvents.

The separation of epoxies, for which the original method was used, was excellent. When applied to plasticizers and halogenated lubricants, some compromise in sensitivity and selectivity was needed, as compared to published methods. Such compromises are easily justified for analyses that do not require the "best" method, especially when the time saved by reducing set-up and equilibration procedures is considered.

5. REFERENCES

1. A. P. Graffeo and N.H.C. Cooke (1979). Summary of the U.S./Japan Seminar of Advanced Techniques of Liquid Chromatography. J. Chromatogr. Sci., 17: 202-206.

2. W.A. McKinley, D.J. Popovich, and T. Layne (1980). A review of HPLC detectors. Amer. Lab., Aug: 37-47.

3. R.E. Majors (1980). Recent Advances in HPLC Packings and Columns. J. Chromatogr. Sci., 18: 488-511.

4. F.M. Rabel (1980). Use and Maintenance of Microparticle High Performance Liquid Chromatography Columns. J. Chromatogr. Sci., 18: 394-408.

5. H.F. Walton (1980). Ion Exchange and Liquid Column Chromatography. Anal. Chem., 52: 15R-27R.

6. N.H.C. Cooke and K. Olsen (1980). Some Modern Concepts in Reversed-Phase Liquid Chromatography on Chemically Bonded Alkyl Stationary Phases. J. Chromatogr. Sci., 18: 512-524.

7. J.L. Meek (1980). Nomogram for Adjusting Mobile Phase Composition in Reverse Phase High Pressure Liquid Chromatography. Anal. Chem., 52: 1370-1371.

8. H.M. McNair (1980). Basic Considerations in HPLC. Amer. Lab., May: 33-44.

9. J.J. DeStefano, A.P. Goldberg, J.P. Larmann, and N.A. Parris (1980). Choosing HPLC packings --- practical guidelines for reversed-phase work. Indiv. Resch. and Dev., April: 99-103.

10. I.S. Lurie (1980). Forensic drug analysis by HPLC. Amer. Lab., October: 35-42.

11. S.P. Assenza and P.R. Brown (1980). Evaluation of Reversed-Phase, Radially-Compressed, Flexible-Walled Columns for the Separation of Low Molecular Weight, UV - Absorbing

Compounds in Serum. J. Liq. Chromatogr., 3(1): 41-59.

12. R.E. Majors (1980).
High Performance Liquid Chromatography Columns and Column Technology: A State of the Art Review (Parts I and II) (editorial). J. Chromatogr. Sci., 18: 393.

13. W.A. Dark, E.C. Conrad, and L.W. Crossman, Jr. (1974).
Liquid Chromatographic Analysis of Epoxy Resins. J. of Chromatogr., 91: 247-260.

14. G.L. Hagnauer and I. Setton (1978).
Compositional Analysis of Epoxy Resin Formulations. J. Liq. Chromatogr., 1(1): 55-73.

15. J.E. Twichell, J.Q. Walker, and J.B. Maynard (1979).
An Exploration of Experimental Parameters in the Analysis of Epoxy Resin by Reverse Phase Liquid Chromatography. J. Chromatogr. Sci., 17: 259-263.

16. P. Szap, I. Kesse, and J. Klapp (1978).
The Analysis of Bisphenol A by High Performance Liquid Chromatography. J. Liq. Chromatogr., 1(1): 89-96.

17. W.A. Dark.
Waters Associates Special Publication (WAPP-200) Waters Associates Liquid Chromatographic Technique for Epoxy Prepolymers.

18. Monsanto Special Publication. Plasticizers and Resin Modifiers. Part number IC/PL-361.

19. M.Y. Hellman (1978).
Analysis of Phthalate Plasticizers for PVC by Liquid Chromatography. J. Liq. Chromatogr., 1(4): 491-505.

20. S.C. Amundson (1978).
Determination of Di (2-ethylhexyl) Phthalate, Mono (2-ethylhexyl) Phthalate and Phthalic Acid by High Pressure Liquid Chromatography. J. Chromatogr. Sci., 16: 170-173.

21. H.C. Jordi, U.D. Neue, H.M. Quinn, and C.W. Rausch (1978).
Study of Radially Compressed Columns: Chromatographic Theory and Applications on Selected Lipids. J. Liq. Chromatogr., 1/2: 215-230.

22. R.E. Majors and E.L. Johnson.
Separation of Slip Agents. Varian Instrument Division Publication, 64: 1-5.

23. S.R. Abbott.
Glyceride Separations of Fats and Oils. Varian Instrument Division Publication, 63: 1-6.

24 T. Osburn and G. Mantelli (1980).
Application of HPLC to the Reclamation of Emulsifiable Oils. Amer. Lab., June: 18-30.

25. J.C. Suatoni, H.R. Garber, and B.E. Davis (1975).
Hydrocarbon Group Types in Gasoline-Range Materials by High Performance Liquid Chromatography. J. Chromatogr. Sci., 13: 367-371.

26. J.C. Suatoni and R.E. Swab (1980).
HPLC Preparative Group-Type Separation of Olefins from Synfuels. J. Chromatogr. Sci., 18: 375-378.

27. R.L. Ehrenfeld (1959).
Halofluoro Compounds as Gyro Flotation Fluids. Fluoro-Chem Corporation, New Product Bulletin: 1-7. Paper delivered at the Fluorine Subdivision of the American Chemical Society Atlantic City Meeting (September 17, 1959).

6. BIOGRAPHIES

Donald W. Eggimann is presently employed as a plastics material and process engineer at the Honeywell Inc., Avionics Division, Clearwater, Florida. He is currently engaged in the development of HPLC methods for application to avionics and aerospace materials. Prior to joining Honeywell in 1979, he worked 7 years in basic research as a research associate in the Department of Marine Science, University of South Florida, and published numerous papers in that field. Mr. Eggimann has a BA degree in chemistry and a MS degree in marine chemistry from the University of South Florida.

Jack C. Brand is presently employed as Principal Materials and Process Engineer in charge of polymer testing and applications at Honeywell Inc., Avionics Division, Clearwater, Florida. Over the past 20 years, he has been involved in chemical, spectroscopic, thermal and mechanical testing of aerospace materials as they relate to program application engineering. He has published several papers in the field of materials analysis. Mr. Brand has a BA degree in chemistry and has been engaged in graduate studies in chemisry and materials engineering.

Cynthia K. Elliott is presently employed as a plastics specialist in the Material and Process Laboratory at Honeywell Inc., Avionics Division, Clearwater, Florida. She is currently engaged in thermoanalytical methods of analysis and applications of high performance liquid chromatography. Prior to joining Honeywell in 1980, Mrs. Elliott worked 13 years in Bio-Medical Research. She attended Marshall University and interned at Cabell Huntington Hospital, Huntington, West Virginia.

FRACTURE AND FATIGUE OF
METAL-METAL LAMINATES
L. E. Sloter and D. H. Petersen
Vought Corporation
Advanced Technology Center
Dallas, Texas

Abstract

Improvements in the fracture toughness and fatigue strength of ultrahigh strength materials have been effected through the bonding of thin layers of these materials and separating interleaves to form thick section metal-metal laminates. The results of the fabrication by roll bonding and the tension, fatigue, and fracture toughness testing of these laminates are reported and discussed. All metal laminate panels have been prepared from (1) layers of steel alloy AISI-SAE 4340 interleaved with thin layers of AISI-SAE 1020 carbon steel, (2) steel alloy 300M interleaved with 1020 or 1075 carbon steels or with E52100 ultrahigh carbon steel, (3) Ti-10V-2Fe-3Al titanium alloys interleaved with Ti-15V-3Cr-3Al-3Sn, and (4) Ti-6Al-4V interleaved with Ti-5Al-2.5 Sn. The thin interleaf laminae prevent the major alloy layers from behaving as a homogeneous structure during fast fracture, allowing, therefore, the maintenance of the fracture properties of the individual layers even in thick section through the maintenance of plane or nearly plane stress behavior. Through the proper selection of interleaf alloys and heat treatment, the critical fracture toughness of thick laminate panels has been improved by 100% and more over monolithic or single layer panels of the same layer alloys and equivalent thicknesses. This improvement in fracture toughness further results in an improvement in high cycle tensile fatigue strength in flawed thick laminate panels versus equivalent monolithic panels. The micromechanisms of fracture, which are responsible for the controlled delamination of layers during fast fracture and the concomitant individual layer behavior, are illustrated and discussed, particularly with respect to the 300M/1020 system.

1. INTRODUCTION

Laminate materials, as described herein, are composed of two or more individual material layers which have been joined to form a final product. When all the laminae in the laminate are metals and they are joined by chemical or metallurgical bonds, the material configuration is defined as a metal-metal laminate as shown schematically in Figure 1. The objective of the work reported herein has been the fabrication of high strength metal-metal laminates in which each layer retains its individual fracture properties during fast fracture of the laminate and has avoided the construction of a monolithic system from individual sheets, although such a construction is possible, e.g., diffusion bonded titanium sheets. In this way, the intrinsically higher toughness of thin metal layers may be maintained in thick section - provided, of course, that the layers are not vanishingly thin, i.e., much thinner than the plastic zone size. This is equivalent to saying that during fast fracture the individual metal layers in a thick laminate approach plane stress behavior rather than plane strain. All the laminates discussed herein were roll bonded at elevated temperatures and consisted of major layers of a primary alloy and interleaves (minor layers between the major layers) of a secondary alloy. The primary alloy is the high strength component, and the interleaf prevents fusion of the primary alloy layers into a monolithic structure. The interleaf may be of high or low strength depending on the alloy system and the properties desired. This study compares the mechanical properties of several high strength, high toughness steel laminates with corresponding monolithic steel reference plates and the mechanical properties of several high strength titanium laminates with monolithic titanium. In addition, the fatigue and fracture properties of simulated, flawed structural items fabricated from a steel and a titanium laminate are compared with geometrically identical monolithic items.

Although the unique and desirable properties of adhesively bonded metal laminates; especially their toughness, damage tolerance, and fabricability advantages;[1,2,3,4] have drawn considerable attention to them as potential structural materials, concern about the environmental stability of adhesively bonded laminates has generally limited their use in primary structures.[4] It has been shown[5] that this latter difficulty can be overcome through the substitution of a metal interleaf for the adhesive layer and chemical bonds between the interleaf and layers rather than adhesive bonds. The joining methods which have been investigated[6,7,8,9] have included explosion, diffusion, and roll bonding. In general, however, roll bonding has been considered to be the most cost effective method examined thus far.

2. EXPERIMENTAL PROCEDURE

2.1 Material Selection

The individual metals chosen for evaluation as laminate layers were ultrahigh strength alloys that generally are not used in structures at their highest strength levels because at these strengths they lack sufficient fracture toughness or ductility in thick sections. Other considerations included the potential for joining the alloys by roll bonding, the existence of compatible interleaf alloys, the potential for property control through heat treatment, and commercial availability. Two steel alloys and two titanium alloys were chosen for in-depth study — medium carbon low alloy steels, AISI-SAE 4340[10] and 300M,[11] and the beta and alpha-beta titanium alloys, respectively, Ti-10V-2Fe-3Al[12] and Ti-6Al-4V.[13]

Three steel interleaf alloys were chosen on the basis of carbon content for use with the 300M steel. These were AISI-SAE 1020,[14] SAE 1075,[15] and AISI-SAE E52100.[16] The first two of these alloys are plain carbon steels of low and high carbon content, respectively, and the third is an ultrahigh carbon bearing steel containing 1% carbon. The selection of these various carbon contents permitted the evaluation of a range of interleaf strengths and ductilities while the 300M layer properties remained the same. Only the 1020 alloy was interleaved with 4340 steel layers.

Two titanium alloys were chosen
for evaluation as interleaves.
Ti-15V-3Cr-3Al-3Sn,[17] a beta
alloy, was used in conjunction
with the Ti-10-2-3, and
Ti-5Al-2.5Sn,[18] an alpha alloy, with the Ti-6-4. The
Ti-15-3-3-3 alloy was chosen because it possessed significantly
different aging kinetics than
the Ti-10-2-3 while the Ti-5-2.5
is a non-heat treatable alloy of
moderate strength. Therefore,
in all alloy systems the mechanical properties of the interleaves were significantly different from those of the layers
in the final heat treated laminate.

2.2   Laminate Fabrication

All laminates were fabricated by
roll bonding at temperatures
above the dynamic recrystallization temperatures of their respective alloys. All plates
were prepared for bonding in the
same manner, consisting of grit
blasting of the surfaces to be
bonded and then degreasing. The
steel laminates were surface
welded across the interleaves to
mechanically stabilize the
lay-up and protect the inner
surfaces from oxidation during
heating, and the weldment was
rolled. The titanium laminates
were laid-up and then boxed in a
mild steel container and the
containers were purged with
argon gas until just prior to
rolling. The rolling start temperature for all steel alloys
was 2100°F (1149°C); for the
Ti-10-2-3, 1550°F (843°C);
and the Ti-6-4, 1925°F
(1052°C). In all cases the
precautions and temperatures
were sufficient to assure bonding, the only serious problem
being alligatoring in the case
of two steel laminates for which
the first roll pass reductions
were insufficient. The overall
reductions in area achieved during roll bonding were between
60-70% in all cases.

2.3   Heat Treatment

All alloy systems were heat
treated so as to produce
strength levels which were near
the practical maximum. The details of the various heat treatments used are contained in
Table I.

2.4 Mechanical Testing

All laminate and monolithic
plates were evaluated through
tensile and fracture toughness
testing, and the bond strength
of the laminates was evaluated
through shear tests. The tensile specimens were prepared and
the tests conducted in accordance with American Society for
Testing and Materials (ASTM)
Standard E 8[19] for plate, and
the fracture toughness tests,
with ASTM standard E 399[20] for
compact tension specimens. The
fracture toughness parameter defined as K_C herein was calculated from a measurement of
the maximum load and the corresponding crack length at maximum
load calculated from crack opening displacement. The defining
equation used for critical fracture toughness was

$$K_c = \frac{P_{MAX}}{BW^{1/2}} f(a/W) \qquad 1.0$$

where P_{MAX} = Maximum Load,
     B = Specimen Thickness,
     W = Specimen Width,
     a = Crack Length at Maximum
         Load, and
     f(a/W) = A Polynominal
              Function of a and
              W.
In addition to the above mechanical testing, the fatigue
strength of tension panels fab-
ricted from laminate and monolithic plates of equal thickness
was determined in tension-
tension fatigue with a load
ratio of 0.1. These panels

contained one central fastener
hole with corner flaws.

3. RESULTS AND DISCUSSION

3.1 Micrography

Figure 2 illustrates a typical
bond line between an alloy steel
layer and carbon steel interleaf, in this case 300M interleaved with 1010, in the
quenched and tempered condition. The titanium laminates
Ti-10-2-3/Ti-15-3-3-3 and
Ti-6-4/Ti-5-2.5 are shown in
Figures 3 and 4, respectively,
both in the solution treated and
aged condition. Although some
foreign material may be noted at
the 300M-1020 bond line, the interface is relatively clean and
there are no significant voids
or laminations. This observation is also supported by the
results of ultrasonic evaluation
and shear strength tests, the
latter of which will be reported
below. In all cases the
titanium-titanium bond lines
were exceptionally free of second phase material and alpha due
to contamination, and recrystallization may be seen to have extensively obscured the bond line.

3.2 Mechanical Properties

The tensile properties of the
various alloys and laminates are
listed in Table II. The decreased strengths of the laminates versus their respective
monolithic alloys are due to the
presence of a significant volume
fraction of the weaker interleaf, except in the case of the
300M/E52100 laminate in which
the ultimate properties of the
laminate are controlled by the
fracture strain of the interleaf. With this exception the
measured tensile strengths of
the laminates agree well with
those values based on the tensile strengths of the components
and the rule of mixtures. It is
also of note that there is a
trend towards increased uniform
elongations in the laminates

versus their monolithic components, although the differences
are small. These increases are
probably due to the initial
individual necking behavior of
the laminae, thereby postponing
the development of a single
local neck. The constraint
imposed by the plastic
deformation of the softer
interleaves[21] may also aid in
postponing local necking.
The results of the interlamellar
shear strength tests were as
follows: 300M/1020, Q&T, 77 ksi
(531 MPa); 4340/1020, Q&T, 78
ksi (538 MPa); and
Ti-6-4/Ti-5-2.5, STA 3, 82 ksi
(565 MPa). In all cases the
shear fracture occurred within
the interleaf and the shear
strength correlated well with
the tensile strength of the interleaf alloy, being approximately half its value.

3.3 Fracture Toughness

The principal purpose for
lamination in this study was the
increase in fracture toughness,
especially elastic-plastic
toughness or stable crack growth
behavior, obtainable in thick
section laminates versus equi-
valently thick monolithic materials. The choice of layer
thicknesses for the various
roll-bonded laminates was based
on assuring that the layers
would not fracture under wholly
plane strain conditions if they
fractured independently. Since
toughness intrinsically decreases as the mechanics of
fracture proceed from plane
stress to plane strain until the
plane strain fracture toughness
is reached, and plane strain
conditions are directly related
to section thickness, it was desirable to maintain the toughness inherent in thin section in
thick section by causing crack
growth to occur separately in
the individual layers rather
than singly in the full section
thickness. Nevertheless, prior
to stable crack growth it was
desirable for the laminate to

behave elastically as a monolithic material rather than as a stack of completely uncoupled layers. These two requirements — monolithic behavior initially and individual behavior upon the inception of crack extension — were realized through the use of a ductile interleaf bonded between the alloy layers. The mechanics of layer separation during fracture will be discussed in detail in a subsequent section.

The fracture toughness results obtained for the various roll-bonded laminates are shown and contrasted as a function of tensile strength with the monolithic materials in Figure 5. The critical fracture toughness represents the maximum stress intensity prior to unstable crack growth. For the monolithic materials, however, at these strength levels there was essentially no difference between the critical fracture toughness and the plane strain fracture toughness. For each laminate the heat treatment and layer thickness are noted, and for the monolithic materials the heat treatment and total thickness are listed. In all cases the toughness of the laminates may be seen to be considerably greater than that of the corresponding monolithic alloys. It may also be noted that the toughness increases with decreasing layer thickness. The decrease in strength with increasing toughness that is most pronounced for the steel alloys is a result of the increased volume fraction of mild steel interleaf as the number of layers are increased and their thicknesses decreased. The lower toughness of the bay-quench and tempered material is due to the lowered toughness of the 300M itself and appears to be a result of segregation associated with the very low temperature of the austenite prior to oil quenching[22]. The toughness improvement in the titanium laminates was achieved with very little loss in strength because the tensile properties of the interleaf alloys were much closer to those of the layer alloys than was the case for the steel alloys.

3.4 Fatigue

Table III contains the results of the fatigue and fracture testing of tension panels of 300M, 300M/1020, Ti-10-2-3, and Ti-10-2-3/Ti-15-3-3-3. The steel laminate tested was made up of four layers while the titanium laminate had three layers. The steel panels were cycled with a maximum load of 30 kip (133kN) giving a maximum net section stress of 32 ksi (221 MPa), and the titanium panels were cycled with a maximum load of 15 kip (66.7 kN) giving a maximum net section stress of 20 ksi (138 MPa). As may be noted in the table the 300M laminate evidenced a 53% improvement in fatigue life over the monolithic material, and the titanium laminate, an improvement of 33% over the titanium monolith. In both cases this improvement appears to have resulted from the increased fracture toughness and concomitant critical crack length of the laminates, although there may have been some relative decrease in the fatigue crack propagation rate in the laminates especially at the highest stress intensity factor (longest crack length) levels. Because of the metallurgical integrity of the layer-interleaf bonds, an elastic fatigue crack is not arrested in the crack arrest orientation; however, during fast fracture in these laminate tension fatigue panels, the controlled delamination and consequently the higher fracture toughness do obtain.

3.5 Fracture Mechanisms

Toughness improvement in metal-metal laminates is critically dependent upon delamination during fast fracture and

the concomitant realization of
plane or nearly plane stress
fracture in each individual layer. In general, the titanium
and steel laminates examined in
this program achieved this controlled delamination through
tensile failure between the layer and interleaf materials at
the layer-interleaf bond line.
Macroscopically this delamination manifests itself as shown
in Figure 6 and gives rise to
extensive slant fracture and
shear lip formation for each of
the indivdual layers. The sequence and consequence of this
delamination are illustrated
fractographically in Figures 7
through 10. The stretch zone of
a compact tension specimen fracture surface is shown in Figure
7. On the left of the fractograph may be seen the fatigue
precrack, and it may be noted
that the fatigue crack front is
continuous across the
layer-interleaf bond lines.
Furthermore, there has been no
delamination as a result of fatigue crack growth. Immediately
upon the initiation of stable
crack growth, however, sufficient lateral constraint is generated to cause a tensile delamination of the layers and interleaf. Because of the very large
difference in strength between
the 1075 interleaf and the 300M
layer alloys, the 1075 becomes
plastic at considerably lower
strains than the 300M. This
alone can generate considerably
more lateral contraction in the
interleaf than in the adjacent
layer material, and, therefore,
greatly assists in the establishment of transverse tensile
stresses and tensile delamination. This is conclusively
shown in the compact tension
specimen fracture illustrated in
Figure 8. This specimen was
loaded to the 95% secant load,
unloaded and then cryogenically
sectioned so that the stretch
zone could be examined independently of any fast fracture.
Once again delamination has occured in the stretch zone and

the layers are, therefore, mechanically separate. A detail
of this delamination is shown in
Figure 9. The different fracture morphologies of the 300M
and 1075 are shown clearly in
this fractograph. The 300M
fractures by dimpled rupture under these conditions while the
1075 evidences a quasi-cleavage
and dimpled rupture mixture typical of such a tempered marten-
sitic carbon steel. The delamination may also be seen to be
predominately dimpled, and the
300M, to have already formed a
shear lip. The complete establishment of this shear lip during fast fracture is shown in
Figure 10. Although it is impossible to demonstrate in this
two dimensional figure, the
fracture of the 1075 is slant
fracture during unstable crack
growth.

4. SUMMARY AND CONCLUSIONS

The present investigation
has demonstrated that roll-
bonded metal-metal laminates offer fracture toughness and fatigue strength advantages over
conventional monolithic or homogeneous alloys in thick section
structures. This improvement in
toughness results from the retention of individual layer or
thin section fracture properties
in the laminates through the
operation of controlled delamination during fast fracture. As
a result, laminates possess the
elastic properties and strength
of a monolithic material with
much improved fracture toughness. Since it is fracture
toughness, not ultimate
strength, which determines the
usable strength of any structural item, laminates may be
utilized at higher working
loads, or concomitantly, laminates possess greater damage
tolerance, the ability to withstand a service induced flaw at
equivalent working loads.

The techniques for roll
bonding steel and titanium alloy

laminates developed in the present program have the potential for commercial exploitation and can be especially cost effective if further optimization of the technology is effected. Roll bonding has been demonstrated to be a viable technique for the fabrication of multilayer laminate plate using dissimilar alloy layers within a given alloy system, e.g., alloy steels laminated to other alloy or carbon steels. Furthermore, the roll-bonding procedures developed are applicable to even very reactive materials, such as titanium alloys.

The fabricability of metal-metal laminates coupled with the improved fracture toughness obtainable through lamination make these laminates effective candidates for high strength structural items requiring a high degree of damage tolerance or resistance to fracture.

REFERENCES

1. Kaufman, J. G., J. BASIC ENG., TRANS. ASME, 89, SERIES D(3), 503 (1967).

2. Almond, E. A., J. D. Embury, and E. S. Wright, in INTERFACES IN COMPOSITES, ASTM STP 452, American Society for Testing and Materials, 1969, p. 107.

3. Alic, J. A., INTERNAT. J. FRACTURE, 11 (4), 701 (1975).

4. Ellis, J. R. and G. E. Kuhn, AFML-TR-74-158 and AFFDL-TR-74-105, Air Force Materials and Flight Dynamics Laboratories, 1974.

5. Johnson, R. M. "Stress Corrosion Resistance in Metal/Metal Laminates," ATC B-94400/8TR-1, Vought Corporation, Advanced Technology Center, 1978.

6. Goolsby, R. D., "Fracture and Fatigue of Diffusion, Explosive, and Roll Bonded Al/Al and Ti/Ti Laminates," ATC B-94400/7CR, Naval Air Systems Command, N00019-76-C-0288, 1977.

7. Johnson, R. M., "Fracture and Fatigue of Diffusion, Adhesive, and Roll Bonded Aluminum, Titanium, and Ultrahigh Carbon Steel Lamintes," ATC B-92100/8CR-80, Naval Air Systems Command, N00019-77-C-0287, 1978.

8. Johnson, R. M. and R. D. Goolsby, "Diffusion, Roll and Explosive Bonding of Al/Al, Ti/Al, and Ti/Ti Laminates," in MATERIALS SYNERGISMS, 10th National SAMPE Tech. Conf., Soc. for the Advancement of Mat. and Proc. Eng., 1978, pp. 802-811.

9. Sloter, L. E. and D. H. Petersen, "Fatigue and Fracture of Ultrahigh Strength Steel and Titanium Roll-Bonded and Diffusion-Bonded Laminates," ATC R-92000/0CR-31, Naval Air Systems Command, N00019-78-C-0491, 1980.

10. ALLOY DIGEST, Eng. Alloy Digest, Inc., SA-14.

11. ALLOY DIGEST, Eng. Alloy Digest, Inc., SA-81.

12. Boyer, R. R., J. METALS, 32(3), 61 (March, 1980).

13. ALLOY DIGEST, Eng. Alloy Digest, Inc., Ti-66.

14. ALLOY DIGEST, Eng. Alloy Digest, Inc., CS-39.

15. Metals Handbook, 9th Ed., Am. Soc. for Metals, 1978, p. 125.

16. ALLOY DIGEST, Eng. Alloy Digest, Inc., SA-16.

17. Reinsch, W. A. and H. W. Rosenberg, METAL PROG, 117(4), 64 (March, 1980).

18. ALLOY DIGEST, Eng. Alloy Digest, Inc., Ti-43.

19. "Standard Methods of Tension Testing of Metallic Materials", ASTM E 8-69, American Society for Testing and Materials, Philadelphia, PA, 1975.

20. "Standard Method of Test for Plane-Strain Fracture Toughness of Metallic Materials, ASTM E 399-74, American Society for Testing and Materials, Philadelphia, PA, 1975.

21. Semiatin, S. L. and H. R. Piehler, "Deformation of Sandwich Sheet Materials in Uniaxial Tension", MET. TRANS, 10A, 85-96 (1979).

22. Ritchie, R.O. B. Francis, and W. L. Server, "Evaluation of Toughness in AISI Alloy Steel Austenitized at Low and High Temperatures" <u>MET. TRANS.,7A,</u> 831-838 (1976).

ACKNOWLEDGMENT

This work was performed at the Vought Corporation Advanced Technology Center, and was sponsored by the Department of Navy, Naval Air Systems Command. Mr. W. T. Highberger was the Project Monitor.

BIOGRAPHIES

Dr. Lewis E. Sloter, II is a Metallurgist and Senior Research Scientist in the Advanced Technology Center of Vought Corporation. He holds a B.S. and Ph.D. in Metallurgy from Carnegie-Mellon University and an M.S. in Materials Engineering from Drexel University. After joining Vought, Dr. Sloter served as Principal Investigator for Structures in a Program designed to predict the risk of structural failure for on-condition aircraft maintenance items. This program involved the analysis of failure mechanisms and failure kinetics and the probabilistic determination of safety of flight risk. In addition, he has been Principal Investigator for the Metal Laminate Development for Structures Program. This program has involved the analysis of material, configurational, and thermal-mechanical processing variables necessary for efficient metal-metal laminate fabrication and usage.

Dr. Donald H. Petersen is a Senior Scientist and the Manager of Structures and Materials Research in the Advanced Technology Center of Vought Corporation. He holds a B.S. in Chemistry from the University of Portland and a Ph.D. in Physical Chemistry from the University of Notre Dame. Since joining Vought, Dr. Petersen has been a principal investigator for applied research projects in the areas of bonded structures, adhesives development, metal laminates, non-destructive testing and special graphite filaments. Dr. Petersen also has considerable experience in powder materials and in carbon fiber reinforced composites for aerospace applications, including graphite/epoxy and carbon/carbon composites.

TABLE I: HEAT TREATMENTS

300M/1020, 300M/1075, and 300M/E52100

- QUENCHED AND TEMPERED (Q&T)
  1600°F (871°C),1 hr.
  oil quench
  575°F (302°C),2 hrs.
  air cool, repeat temper

- BAY QUENCHED AND TEMPERED
  (BAY Q&T)
  1600°F (871°C), 1 hr.
  down quench to
  1030°F (544°C), 10 min.
  oil quench
  575°F (302°C), 2 hrs.
  air cool, repeat temper

4340/1020

- QUENCHED AND TEMPERED (Q&T)
  1600°F (871°C), 1 hr.
  oil quench
  450°F (232°C), 2 hrs.
  air cool, repeat temper

Ti-10V-2Fe-3Al/Ti-15V-3Cr-3Al-3Sn

- STA 4
  1450°F (788°C), 1 hr.
  water quench
  950°F (510°C), 4 hrs.

- STA 8
  1400°F (760°C), 1 hr.
  water quench
  900°F (482°C), 8 hrs.

- STA 16
  1400°F (760°C), 1 hr.
  water quench
  900°F (482°C), 16 hrs.

Ti-6Al-4V/Ti-5Al-2.5 Sn

- STA 3
  1700°F (927°C), 1 hr.
  water quench
  1000°F (538°C), 3 hrs.

TABLE II: AVERAGE TENSILE PROPERTIES

ALLOY SYSTEM, HEAT TREATMENT	YIELD STRENGTH, ksi(MPa)	TENSILE STRENGTH, ksi(MPa)	TOTAL ELONGA-TION,%	UNIFORM ELONGA-TION,%
300M, Q&T	250(1724)	295(2034)	9.3	3.6
300M, BAY Q&T	254(1751	299(2062)	14	4.5
300M/1020, Q&T	223(1538)	281(1937)	13	4.7
300M/1020, BAY Q&T	212(1462)	267(1841)	12	4.1
300M/1075, Q&T	237(1634)	287(1979)	11	4.5
300M/1075, BAY Q&T	222(1531)	273(1882)	13	4.6
300M/E52100, Q&T	231(1593)	249(1717)	1.5	1.1
4340, Q&T	222(1531)	258(1779)	14	4.5
4340/1020, Q&T	188(1296)	229(1579)	9.5	4.7
Ti-6-4, STA 3	153(1054)	156(1076)	1.5	1.5
Ti-6-4/Ti-5-2.5, STA 3	150(1034)	160(1103)	3	2.7
Ti-10-2-3, STA 4	195(1344)	196(1351)	1.5	1.5
Ti-10-2-3, STA 8	184(1269)	196(1351)	8.3	3.7
Ti-10-2-3/Ti-15-3-3-3, STA 4	178(1227)	187(1289)	3.9	3.9
Ti-10-2-3/Ti-15-3-3-3, STA 8	161(1110)	185(1276)	4.9	4.0

FIGURE 2: MICROSTRUCTURE OF A QUENCHED AND TEMPERED 300M/1020 STEEL LAMINATE AND BOND LINE. TRANSVERSE SECTION. 2% NITAL ETCHANT.

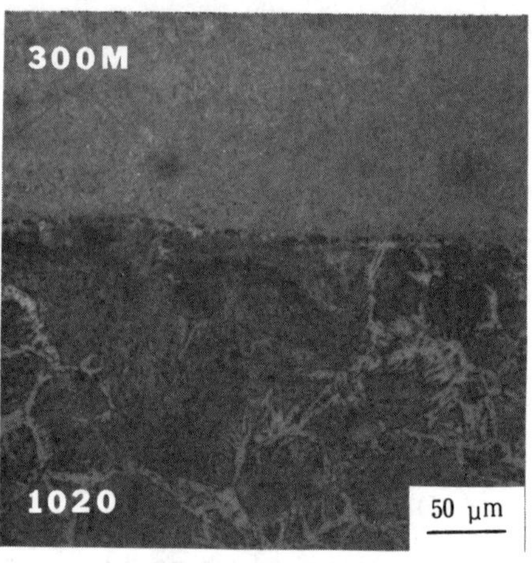

FIGURE 3: MICROSTRUCTURE OF Ti-10V-2Fe-3Al/Ti-15V-3 CR-3Al-3Sn TITANIUM AND BOND LINE. TRANSVERSE SECTION. HF-HNO$_3$ ETCHANT.

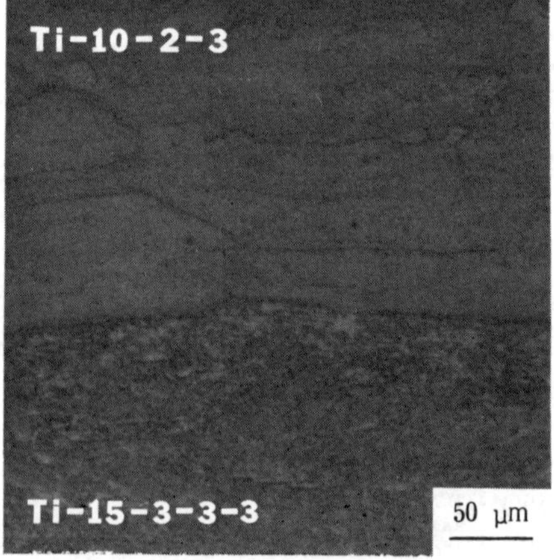

TABLE III: FATIGUE STRENGTH

SPECIMEN, HEAT TREATMENT	CYCLES TO FAILURE
300M, Q&T	118,000
300M/1020, Q&T	180,000
Ti-10 V-2Fe-3Al, STA 4	69,000
Ti-10V-2Fe-3Al/ Ti-15V-3Cr-3Al-3Sn STA 4	92,000

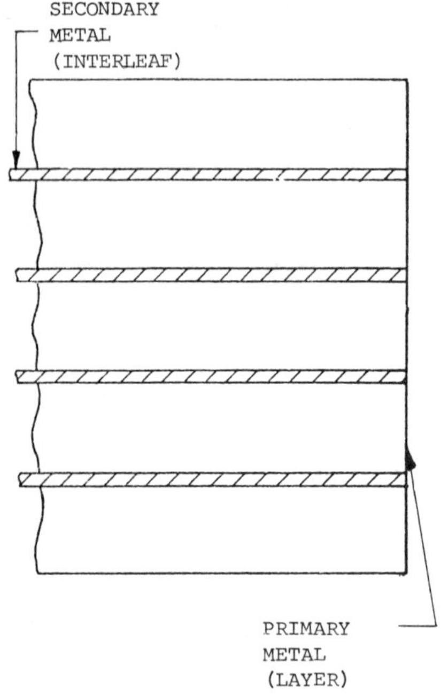

Figure 1: SCHEMATIC METAL-METAL LAMINATE

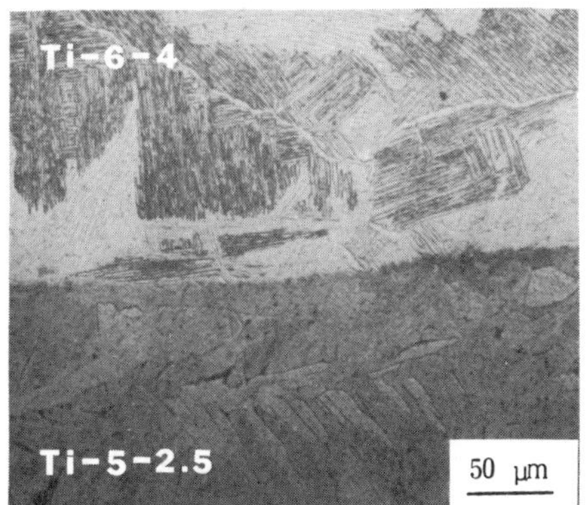

FIGURE 4: MICROSTRUCTURE OF Ti-6Al-4V/Ti-5Al-2.5Sn TITANIUM LAMINATE AND BOND LINE. TRANSVERSE SECTION. HF-HNO$_3$ ETCHANT.

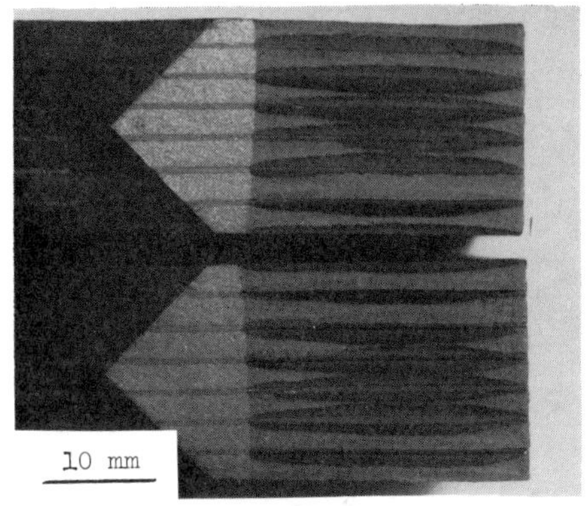

FIGURE 6: 300M/1020 METAL LAMINATE COMPACT TENSION SPECIMEN FRACTURE SURFACES.

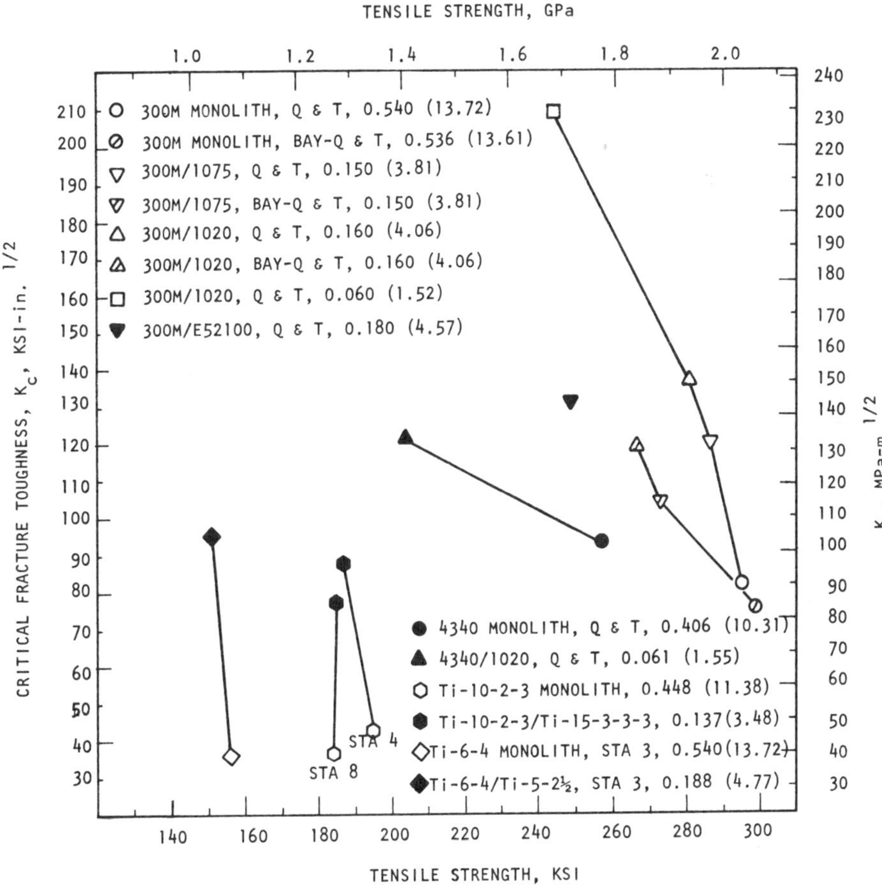

FIGURE 5: TENSILE STRENGTH VERSUS FRACTURE TOUGHNESS FOR THE VARIOUS MONOLITHIC ALLOYS AND LAMINATE SYSTEMS.

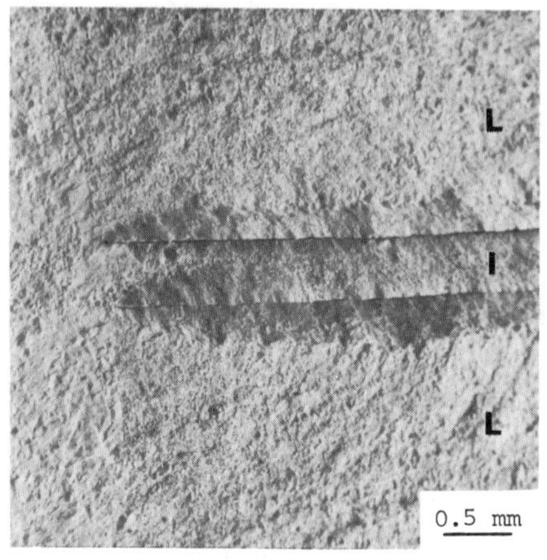

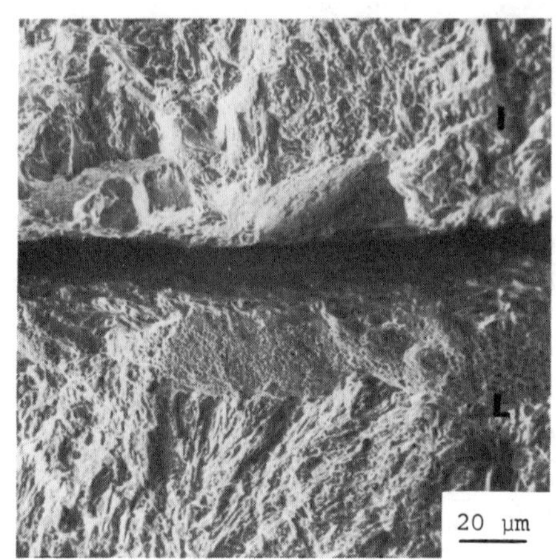

FIGURE 7:   SCANNING ELECTRON
            MICROSCOPE (SEM)
            FRACTORGRAPH OF A
            300M/1075 METAL LAMI-
            NATE. COMPACT TEN-
            SION SPECIMEN FRAC-
            TURE STRETCH ZONE. L
            = LAYER (300M) AND I =
            INTERLEAF (1075).

FIGURE 9:   DETAIL OF THE SPECI-
            MEN OF FIGURE 8. SEM
            FRACTOGRAPH.

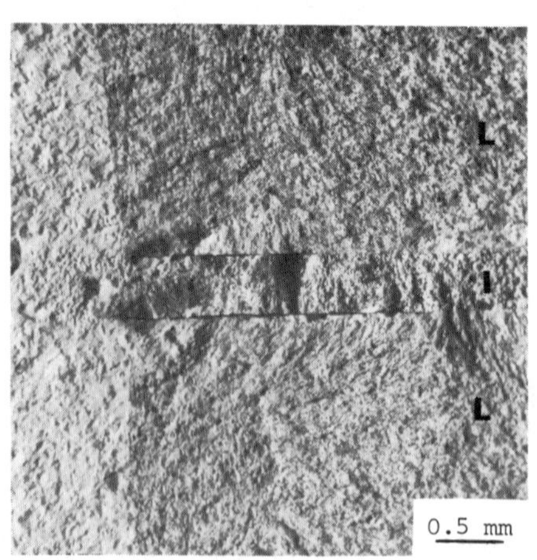

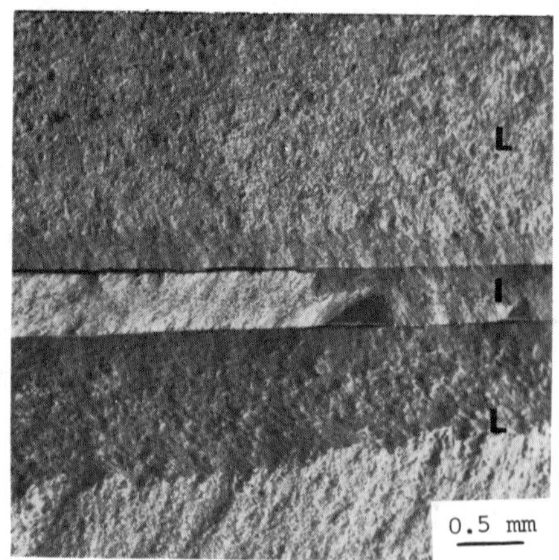

FIGURE 8:   SEM FRACTOGRAPH OF A
            300M/1075 METAL LAMI-
            NATE COMPACT TENSION
            SPECIMEN FRACTURE
            STRETCH ZONE. CRYO-
            GENICALLY SECTIONED.
            L = LAYER (300M) AND I
            = INTERLEAF (1075).

FIGURE 10: SEM FRACTOGRAPH OF A
            300M/1075 METAL LAMI-
            NATE COMPACT TENSION
            SPECIMEN FAST FRAC-
            TURE. L = LAYER
            (300M) AND I = INTER-
            LEAF (1075).

26th National SAMPE Symposium
April 28-30, 1981

COMPARATIVE PERFORMANCE ASSESSMENT
OF GRAPHITE AND KEVLAR MOTOR CASES

Dr. D. C. Sayles
Ballistic Missile Defense Advanced Technology Center
Huntsville, Alabama

Abstract

Several motor cases of two different sizes were fabricated, with and without skirts, from graphite-epoxy, Kevlar-epoxy, and Kevlar-epoxy overwrapped with graphite, and comparatively assessed by hydrotest, burst test, and when loaded with propellant and burst tested. The purpose was to derive data that could be used in the design of future missile motor cases. Modification of the epoxy resins used in the fabrication of motor cases was also undertaken or planned. One of these modifications addressed the reduction or prevention of shrinkage of the epoxy resins by the incorporation of specially synthesized bisspiro-ortho esters. Another area of interest is the replacement of epoxy resins with polyimide resins because these enjoy several attributes that epoxy resins lack, and these materials are now becoming commercially available.

Keywords: graphite filaments, Kevlar filaments, composite materials, composite motor cases, filaments, fibers, epoxy resins, polyimide resins.

1. INTRODUCTION

Composite motor cases that have been fabricated from epoxy resins and a wide variety of filaments are being investigated within the United States for different purposes and for different reasons. The filaments that have been used include fiberglass, graphite, and Kevlar for the motor cases and as overwraps. These are enumerated in Table I.

Table I. Filamentous materials investigated for motor case fabrication within the United States

Filament	Comments
Kevlar	High specific strength High specific modulus High fiber property/density
Kevlar overwrapped with graphite	
Graphite	Thornel, type AS*
Fiberglass	S-2, S-901, X-2285**
Maraged steel overwrapped with graphite	
D6AC steel overwrapped with graphite	
Titanium overwrapped with graphite	

*Specific strength equivalent to fiberglass with greater specific modulus.
**Used on Polaris A3, SPRINT, and Poseidon C3 motor cases. S-2285 glass has increased strength, and the finish of the surface is the same as S-901 glass and thus has the same fiber wetting characteristics.

This manuscript is limited to a discussion of: a) the comparative assessment of motor cases fabricated from graphite, Kevlar, and Kevlar overwrapped with graphite; b) expandable epoxy resins; and c) polyimide resins, which are candidates to replace epoxy resins. The topics that are discussed are enumerated in Table II.

Table II. Topics discussed

```
● Full-scale graphite composite motor cases
● Subscale composite motor cases
        Graphite
        Kevlar
        Kevlar overwrapped with graphite
            Unskirted
            Skirted
● Expandable epoxy polymers
● Polyimide polymers
```

This discussion also considers two different sizes of graphite-composite missile motor cases. The first size was representative of an intermediate test motor case that would be suitable for the evaluation of several hundred pounds of propellant. The second size motor case was more of a pressure vessel. It was designed to be dimensionally one-quarter the size of the other motor case. Its purpose was to provide a means to compare the performance characteristics of newer filamentous materials and resins as these become available and to assess different design features of motor cases.

Several design studies have been carried out within recent years with the objective of assessing the promise of filamentous materials for composite motor case manufacture. Kevlar overwrapped with graphite and graphite alone were assessed as being superior to other fibers for use in hybrid case design. The bases for this assessment are enumerated in Table III. These

Table III. Bases for selection of Kevlar overwrapped with graphite or graphite for motor case design

```
● Higher performance
● Reduced motor case weight
● Lower relative cost
● Superior growth potential (because of
    active ongoing development efforts)
● Reduced sensitivity to nuclear radiation
● Low cycle fatigue
● Higher hoop stiffness than fiberglass (>55%)
    (especially for motors that have large
    propellant web fractions; stiffer cases
    mean reduced propellant bore strains)
```

fibers offer superior performance because of reduced case weight, lower relative cost, reasonable risk in development, greater growth potential, reduced propellant strains resulting from motor pressurization, and lower cycle fatigue. In addition, hoop stiffness of motor cases made from Kevlar-graphite exceeds fiberglass construction of equivalent strength by about 55%.

A comparison of the properties of Kevlar and graphite is contained in Table IV. Values for fiberglass, aluminum, and steel are also included for comparison. These values are representative of those achievable in practice and incorporate manufacturing processes, aging, and test environments. Of particular interest is that Kevlar exhibits the lowest density, the highest design strength, the highest specific strength, and an intermediate tensile modulus and specific modulus. Graphite, on the

Table IV. Comparison of fiber properties

Material	Density (lb/in^3)	Design Tensile Strength (psi)	Specific strength (in-lbf/lbm × 10^{-6})	Tensile Modulus (psi × 10^{-6})	Specific Modulus (in-lbf/lbm × 10^{-6})
Kevlar (12-end roving)	0.0521	480,000	9.2	20.0	384
Kevlar (4-end roving)	0.051	470,000	9.0	19.0	365
Thornel	0.063	275,000	4.4	34.0	540
Graphite AS	0.066	430,000	6.7	32.0	490
S-2 glass	0.092	400,000	4.3	12.6	137
S-901 glass	0.090	550,000	6.1	12.2	136
X-2285 glass	0.088	590,000	6.7	13.4	152
Aluminum	0.100	70,000	0.7	10.0	100
Steel	0.286	180,000	0.6	30.0	105

other hand, offers specific strength equivalent to fiberglass with greater specific modulus.

The effect of materials for motor case fabrication on component weight is depicted in Table V. The

Table V. Effect of motor case material of construction on component weight

Weight Reduction* (%)

Component	Kevlar-Graphite	Graphite
Motor case	39	23
Propellant	9	8
Stage	9	7

*Compared with maraging steel case overwrapped with graphite.

weight reduction obtained through the use of Kevlar overwrapped with graphite or graphite alone, as compared with maraging steel motor cases overwrapped with graphite, is highly significant and responsible for the changeover to Kevlar and graphite filaments for motor case fabrication. Appreciable reductions in weight can be effected in the major components (motor case, propellant, and stage).

The explanation for the reduction in the quantity of propellant that can be loaded into the motor case fabricated from Kevlar or graphite filaments is that the missile's upper stages are volume-limited, and motor case materials of lower specific weight would occupy volume that would otherwise be used for propellant. The stage burnout velocity is not changed appreciably, however. Since the upper stage can be smaller, the overall performance of the missile is improved. The magnitude of the performance improvement is a function of the first-stage size, its performance, and staging ratio.

Typical properties of epoxy resins used in the fabrication of composite motor cases are depicted in Table VI.

The structural chemical formulas for diglycidyl ether of Bisphenol A (EPON 828) and epoxidized dimer of oleic acid (EPON 871) are depicted in Figure 1.

Table VI. Typical properties of epoxy resins used in the fabrication of composite motor cases

Tensile strength	8,477 psi
Tensile modulus	424,000 psi
Percent elongation	3.1%
Compressive strength	16,458 psi
Compressive modulus	560,000 psi
Flexural strength	17,799 psi
Flexural modulus	395,000 psi
Shear strength	7,863 psi
Shear modulus	558,000 psi
Thermal expansion coefficient	28.9×10^{-6} in/in/°F
Density	0.0438 lb/in³
Cure shrinkage	0.067 in/10 in

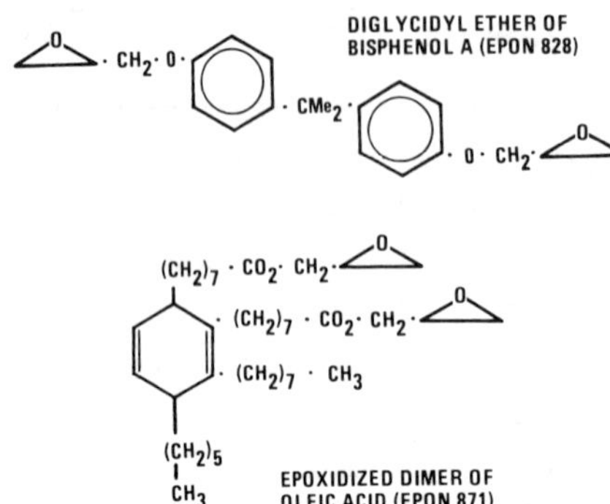

Figure 1. Structural chemical formulas of EPON 828 and EPON 871

2. FULL-SCALE, GRAPHITE-COMPOSITE MOTOR CASE

The preliminary and final designs of the full-scale, graphite-filament-wound composite motor case that had a capacity of approximately 500 lb of propellant are depicted in Figure 2. The motor case and forward skirt were designed to support the combined minimum expected operating pressure of 4,500 psi and 300,000-lb thrust loading. The skirt was designed predicated on a safety factor of 1.5. The design description and design data on the full-scale, graphite-composite motor case are provided in Table VII. The fabrication techniques for this motor case

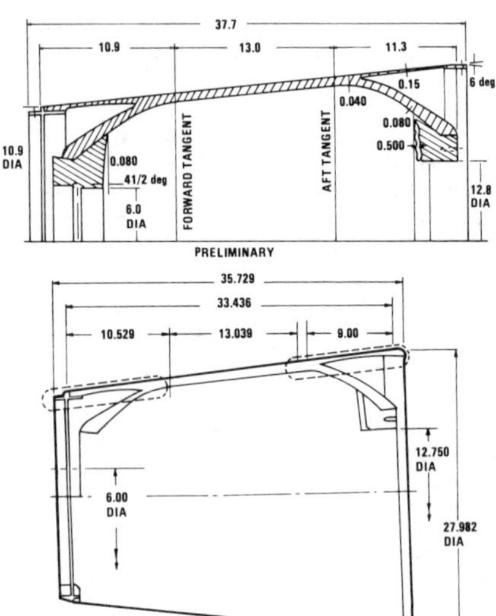

Figure 2. Cross-sectional drawings of the initial and final designs of the full-scale, graphite-composite motor case

are discussed in detail under "Sub-scale Motor Cases."

The forward skirt design consisted of 0- and ±45-deg graphite-epoxy mats and 90-deg hoop layers interspersed throughout the thickness as a nearly balanced and symmetric laminate. This configuration has been assessed to be optimum in both strength and stiffness. The skirt-to-case joint design featured a fiberglass cloth and resin interface between the forward dome and skirt to act as a shear transfer medium between the two structures. Both the forward and aft domes and the polar adapters and insulators were heavyweight and the conical section between the forward and aft tangent points was flightweight in order to simulate the flight strains.

Table VII. Design description and design data on full-scale,
graphite-composite motor case

Geometry

Construction	
Aft polar opening	Deviated winding
Forward polar opening	12.50 in. diameter
Overall length	6.20 in. diameter
Conical section winding angle	35.729 in.
	41.4 deg

Materials

Aft pole piece	2024-T4 aluminum
Forward pole piece	6061-T6 aluminum
Insulation	SBR-asbestos in aft end
	SBR-silica in forward end
Case	ASMS graphite-epoxy blend
Aft splice ring	6061-T4 aluminum
Forward splice ring	2014-T652 aluminum
Forward skirt	ASMS graphite-epoxy blend $(0,90,\pm45)$
Aft skirt	ASMS graphite-epoxy blend $(0,90,\pm45)$

Design Data

Burst pressure = 5,580 psi
Nominal hoop fiber stress = 343 ksi
Nominal helical fiber stress = 294 ksi

Fiber volume = 50%
Helical winding angle: α_1 = 41 deg
Helical winding angle: α_2 = 80 deg

Thicknesses:
Forward tangent = 0.70 in.
Mid-conical   = 0.73 in.
Aft tangent   = 0.85 in.
Forward skirt = 0.34 in.
Aft skirt   = 0.15 in.

The resin composition used in the fabrication of the graphite-composite motor cases is presented in Table VIII.

Table VIII. Resin composition used in the fabrication of graphite-composite motor cases

Ingredient	Composition (parts by weight)
EPON 828*	50
EPON 871**	50
Butanediol diglycidyl ether	25
Tonox 60/40	20

*Bisphenol A-epichlorohydrin.
**Epoxidized trimer acid.
  Mixture of 65% 4,4'-diaminodiphenylmethane,
  10% triamines, and 25% polyamines.

The hydrotest and burst-test results of the full-scale, graphite-composite motor cases are presented in Table IX. The hydrotest procedure consisted of raising the pressure in increments of 200 psig and holding for 30 sec at each step. The skirt on the first case failed around the skirt ring at 2,457 psig and burst at 3,540 psig. The second motor passed the hydroproof and burst-test cycle. It burst at 5,450 psig, and the forward skirt carried the combined burst pressure and the 300,000-lb thrust load that had been applied during the hydrotest. The thrust load had been applied by a piston located in the aft end of the motor case, and the load was reacted by the forward skirt. The third motor case, fabricated to the same design as previously successfully tested, experienced a forward

Table IX. Hydrotest and burst-test results of full-scale,
graphite-composite motor cases

Motor Case	Test	Comments
001	Hydrotest (200 psig/sec to 3,900 psig); held for 30 sec	Forward skirt failed around skirt-ring at 2,457 psig
	Burst test (200 psig/sec to failure)	Motor failed at 3,540 psig
002	Hydrotest (200 psig/sec to 3,000 psig); held for 10 sec	Motor passed satisfactorily
	Burst test (200 psig/sec to failure)	Motor burst at 5,450 psig
003	Hydrotest (200 to 4,500 psig); held for 10 sec	At 6 sec into the hold period, forward skirt failed at the dome interface and skirt separated from the motor
004	Hydrotest (200 psig/sec to 4,000 psig); held for 5 sec	Motor passed
005	Hydrotest (200 to 4,000 psig); held for 5 sec	Motor passed

skirt failure during the hydrotest cycle when the forward skirt was loaded with a 4,500-psi internal pressure and 300,000-lb thrust load. The remaining cases successfully passed hydrotest with the thrust load totally reacted on their forward domes rather than on their skirts.

Because of the skirt failures that were encountered, the remaining cases were hydrotested with the thrust load reacted on the forward domes rather than on the skirts. Both remaining cases successfully passed the hydrotest.

Failure analysis and additional investigation into the cause of the skirt failure and analysis of the case design and fabrication process led to the conclusion that the variation in the strength of the forward skirt could be attributed to problems in the fabrication of the skirt joint. The basic problem with the skirt-to-case joint was identified as being that the inside of the skirt was tapered 6 deg. Windings of 90 deg have been found to slip on a taper angle greater than 4 deg. To compensate for the 6-deg taper angle, the layers within the skirt were shingled horizontally on the 6-deg tapered skirt tooling mandrel. This allowed the 90-deg windings to be wound under tension and the skirt laminate buildup to be consolidated. A blocking dam located at the forward edge of the skirt was used to prevent the entire uncured skirt laminate from slipping down the 6-deg taper. After the skirt-to-case joint had been fabricated, the 0-, 45-, and 90-deg layers were started at the dam and shingled to the joint. The aft skirt was built up in the opposite fashion.

Redesigns for the skirt that were considered were: combinations of unidirectional and ±45-deg graphite fiber-epoxy and S-glass hoop windings as well as all graphite and all fiberglass-epoxy designs. These designs had been found to be superior to all-Kevlar skirts in compression, shear, and bearing-load abilities. These skirts were then bonded to the case with a layer of high-elongation, low-modulus film adhesive and/or an elastic material that would reduce peak shear stresses. The different designs for the joint that were considered are depicted in Table X.

3. SUBSCALE MOTOR CASES

The initial design of the subscale motor cases was arrived at by means of geometric scaling of the full-scale case to one-fourth size. Adjustments were made to satisfy the requirements for integer numbers of layers and minimum manufacturing thicknesses. For example, the aft skirt had to be thicker than one-fourth scale would dictate because the normal thickness of only one layer of material is greater than one-fourth the full-

Table X. Design changes during the development of
skirt-to-case joint

Design No.	Remarks	Joint Schematic	Maximum Shear Stress (psi)	Maximum Meridional Stress (psi)
I	Original design		10,020	174,000
II	Same as I but used graphite skirt		12,500	175,000
III	Partial resin layer Resin modulus $= 5.8 \times 10^5$ psi External hoop windings		13,100	162,000
IV	Same as III, but full resin layer		6,680	162,000
V	Same as IV except removed external hoop windings		6,680	161,000
VI	Added reinforcements in dome; reduced resin-modulus to 10^5 psi	Reinforced Area	5,420	163,000

scale thickness. The contour of the aft dome had to be modified to compensate for the absence of the thrust load present in a static test. Stresses and geometric ratios were used as the basis for scaling. The subscale case was designed to withstand a 5,800-psi burst pressure with the failure mode occurring in the conical section. This failure mode was intended to occur in the identical sector that it would occur in in the full-scale motor.

A cross-sectional drawing of the subscale motor case without skirt is depicted in Figure 3, and the subscale case with both forward and aft skirts is shown in Figure 4.

The initial design of the subscale case was performed using a netting analysis.

For the hybrid case, fabricated from Kevlar helicals and graphite overwrap, a modulus factor was used for the helical layers. This factor was the ratio of the moduli of the two materials at 41 deg. It was intended to adjust the netting equation to account for the lower load-carrying capability in the hoop direction of the Kevlar helicals. Such an adjustment was not necessary for the 41-deg Kevlar

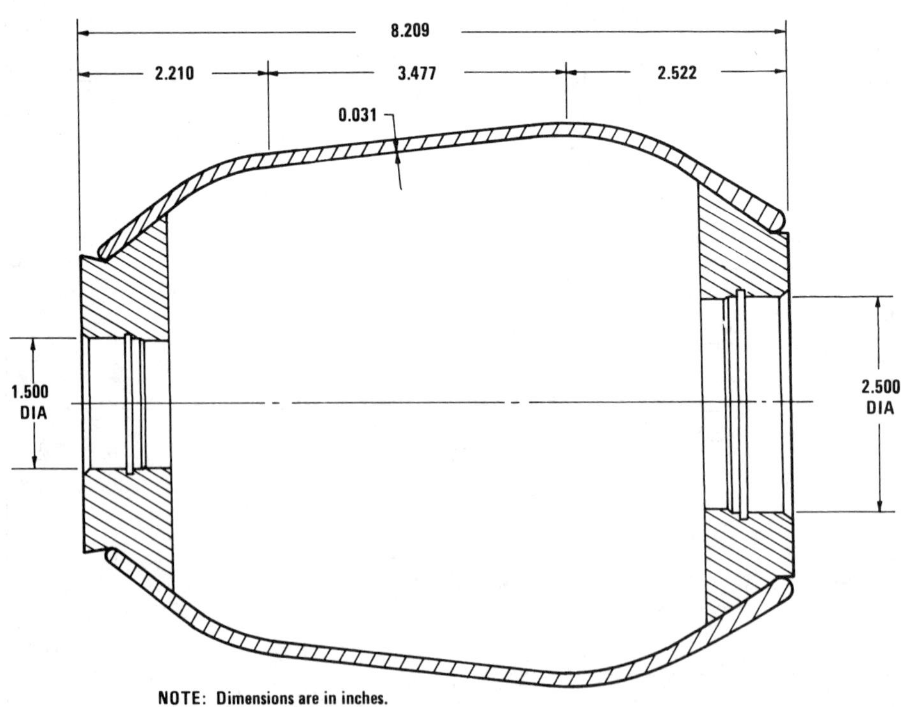

Figure 3. Cross-sectional drawing of subscale composite motor case

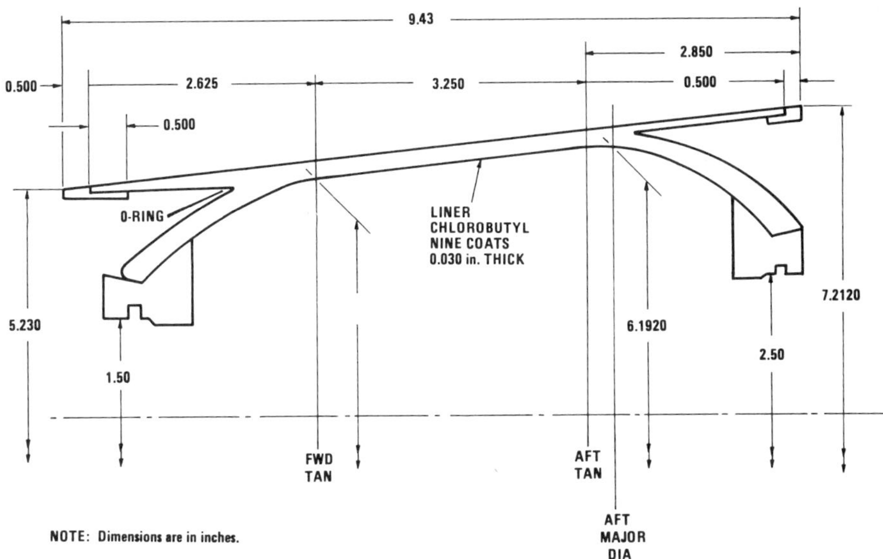

Figure 4. Cross-section drawing of skirted subscale
composite motor case

helicals since this effect was negligible.

The dome contours were calculated based on the geodesic ovaloid and resulted in exactly a one-fourth-scale forward dome. The aft dome was contoured for zero thrust load since the testing and method that was to be used would not involve gas flow. Therefore, the aft dome was not an exact scale of the full-scale because the full-scale was designed for the high thrust of a static firing.

The subscale conical section was designed to deform in a similar manner to the full-scale, although the domes were reinforced. The reinforcing was meant to maintain high margins in the domes and thereby produce a relatively constant stress level in the conical sec-

tion. To accomplish this, a 41-deg helix had to be cut and removed from the forward dome and conical section.

The skirt joints were specifically designed for the subscale case. No attempt was made to scale from the full-scale case because the joint did not perform acceptably in the full-scale case. An improved joint was designed that featured materials and geometries derived from the optimization study with due consideration of layer thickness limitations and small-scale manufacturing tolerances. This design included: a) mating of the forward joint to the dome at an angle of less than 4 deg and b) the use of a film adhesive to control the bond-line thickness and improved shear strength.

The skirts were designed to have a configuration that was common to all three different case designs. The general approach was to have the skirts stiff in the axial direction and soft in the hoop direction. This was done by using about one-third of the buildup as 0-deg graphite layers. One-third was $\pm$45-deg graphite and one-third was glass hoop windings. The inside angle of the forward skirt was set at less than 4 deg to allow compaction of the structure with 90-deg hoop windings and preclude any slippage during manufacture.

The forward skirt was built up outside the desired final contour; the final outside contour was obtained by machining because the outside contour could not be controlled. The adapters for the cases were scaled geometrically. The stress levels were checked to ensure against the occurrence of any problems. Commercially manufactured snap rings were used in the attachment devices for the test closures. The liner was chlorobutyl rubber and consisted of nine layers to produce the 0.030-in. (0.76-cm) thickness.

The winding was done on an Entec winding machine that had been programmed for the 41- and 81.5-deg helix angles. The graphite tow was tensioned to 5 lb (2.27 kg) for all windings. The Kevlar was wound using a glass delivery system and tensioned at 2 to 3 lb (0.91 to 1.36 kg) per roving. The cases were wound by alternating the 41- and 81.5-deg helix angles. There were more 41-deg helix layers, and some overlapping occurred. One layer was cut to reinforce the aft dome, and one was cut to maintain a constant stress in the conical section.

When the winding was complete, the skirt tooling was installed onto the winding shaft. The film adhesive was then placed at the joint area, and the skirts were built up. The buildup consisted of repeating sets of layers of 0-deg graphite, $\pm$45-deg graphite, and glass hoop windings. When the fabrication of the skirt was completed, the cases were cured. After cure, the salt mandrel was washed out with warm water and the skirts machined. The final operation consisted of bonding on the adapter rings.

The first unit of each type of motor case was hydroburst-tested to verify the design before fabrication of the remainder of the units, which were to be delivered to the U.S. Army Missile Command, Redstone Arsenal, Alabama, for special testing involving the use of live propellant as the means of instantaneously pressurizing the case.

The hydrotest procedure that was used differed considerably from that which was generally in use. It consisted of the following: The test closures were installed into the fittings, and approximately 10 strain gages were installed on each unit to monitor the test results. The unit was filled with water, and the air was bled from the system. The unit was proof-tested to the specific operating pressure of approximately 3,500 psi, held for 1 min, and the pressure was released. The water was then drained from the case.

The remaining cases were loaded with a propellant charge; the charge was ignited; and the pressure buildup and the pressure at burst were measured. The burst-test results are presented in Table XI.

Table XI. Burst-test results of subscale composite motor cases

Motor Case	Burst Test	Test Procedure
GR-002	6,500 psi	These motor cases were first hydrotested to 4,000 psi using a pulsating-type pump. The time required to reach 4,000 psi was 2 to 3 min. At that time, the cutoff valve was closed and the pressure was held for a few seconds; then the pressure was bled off.
GR-003	4,900 psi	
GR-005	4,750 psi	
RV-003	4,500 psi	
RV-004	4,500 psi	
GR-GLS-002	5,200 psi	

EXPANDABLE EPOXY POLYMERS

Research activities that have been ongoing at the University of Maryland have resulted in the development of prepolymers capable of undergoing expansion on polymerization rather than shrinkage, which has been found to occur with all polymers. This has been demonstrated in several starting products: namely, spiroortho esters, bicycloketal lactones, and spiroortho carbonates. Figure 5 depicts the structural chemical formulas of these compounds.

The objective of the program that was undertaken in this area was to investigate the effectiveness of certain of these spiro compounds in preventing shrinkage of commercially available epoxy resins. The rationale for investigating expandable epoxy polymers for application to composite motor cases is that they promise to improve adhesion between the resin and the reinforcing agent, thereby reducing cure stresses, cooldown stresses, porosity, and shrinkage.

Figure 6 depicts the synthetic procedure used in the preparation of the simplest expandable polymer units. This was accomplished by reacting butyrolactone with ethylene oxide to produce 1,4,6-trioxaspiro 4.4 nonane. A more practical application is illustrated in the second chemical equation

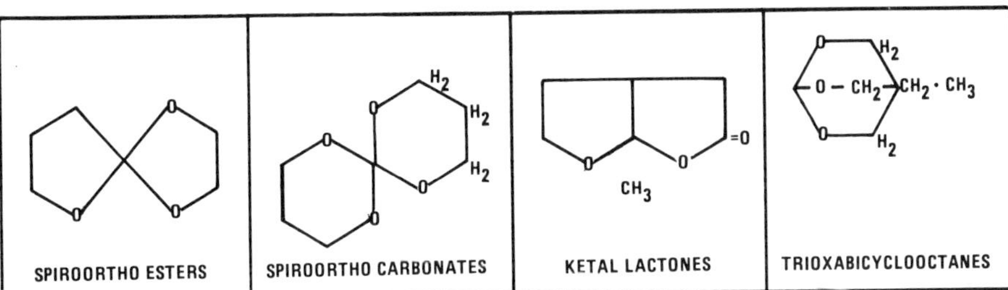

Figure 5. Monomer systems shown to undergo expansion/no shrinkage on polymerization

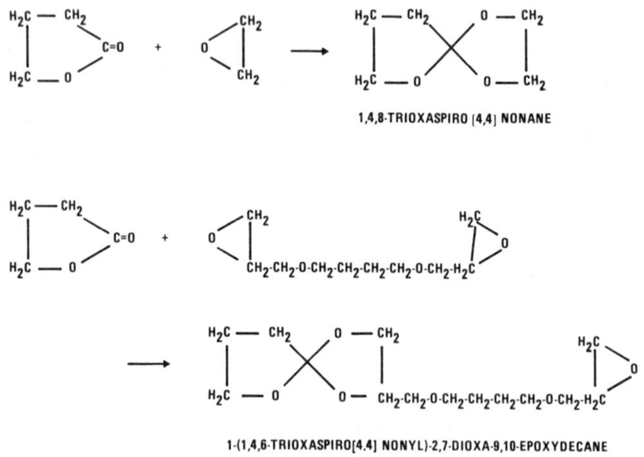

Figure 6. Expandable epoxy polymers for composite motor case fabrication

appearing in Figure 6. This consists of the reaction of RD-2, butanediol diglycidyl ether with one equivalent of butyrolactone to produce 1-(1,4,6-trioxaspiro 4.4 nonyl)-2,7-dioxa-9,10-epoxydecane. The reaction with one of the epoxy functional groups leaves an unreacted epoxy group available for further reaction with the polyamino curatives used in a polyepoxy blend-polyamino system.

Several bisspiroortho esters have already been synthesized from the diepoxides and lactones that are listed in Table XII.

The most recent emphasis in this investigation has been on the spiroortho carbonates that contain two bicycloheptene ring systems, since these were more reactive monomers. The carbonate ring structure was found to be sufficiently reactive to undergo effective copolymerization with epoxi-

Table XII. Bisspiroortho esters that have been synthesized

Diepoxide	Lactone
H_2C-CH-CH-CH_2 (with two epoxide O bridges) 1,2,3,4-diepoxybutane	4-butyrolactone
H_2C-CH-CH_2-O-CH_2-CH-CH_2 (with two epoxide O bridges) bis (2,3-epoxypropyl) ether	5-valerolactone
H_2C-CH_2-CH_2-O-$(CH_2)_4$-O-CH_2-CH-CH_2 (with two epoxide O bridges) 1,4-butanediol diglycidyl ether	6-caprolactone
H_2C-CH-CH_2-O-C_6H_4-O-CH_2-CH-CH_2 (with two epoxide O bridges) 1,4-bis(2,3-epoxypropoxy)benzene	
O=(S ring)—CH-CH_2 (with epoxide O bridge) 4-vinylcyclohexane dioxide	

dized Bisphenol A to produce a product that has essentially no shrinkage but actually underwent a slight increase in volume. At the same time, it produced a matrix that had excellent mechanical properties. To effect polymerization, a complex of boron trifluoride and ortho-phenylenediamine proved to be a very effective catalyst for this system. This reaction is illustrated in Figure 7.

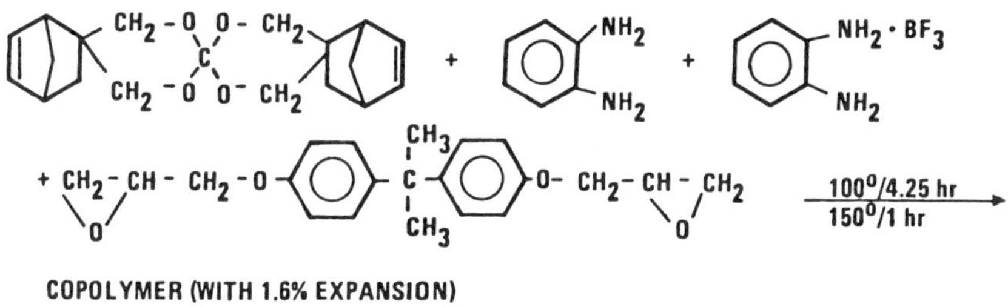

Figure 7. Reaction of spirobicycloheptene carbonate with epoxidized Bisphenol A catalyzed by a complex of boron trifluoride and o-phenylenediamine

POLYIMIDE RESINS

Modified polyimide resins, formed by reaction polymerization with epoxy resins, offer attractive potential for use in thermosetting matrices for filaments in composite motor cases. Several of these polyimide resins have now become commercially available within the United States. Their particularly attractive features are enumerated in Table XIII. Some of these are:

Table XIII. Salient features of flexible polyimide/epoxy resins

Property	Feature
Cure cycle	Cures in 24 to 168 hr in temperature range of 75° to 250°F
Versatility	Cures with aliphatic, cyclo-aliphatic, and aromatic epoxies
Useful property temperature range	-65° to 250°F
Key properties	Toughness, high adhesion, and fuel resistance

They undergo cure without the evolution of gaseous byproducts; they require only a short cure cycle under conditions of low pressure; and the resulting composite materials have excellent retention of their mechanical properties under elevated temperatures and high humidity when compared with epoxy polymers.

A facile method of synthesis of these polyimides involves the reaction of 1,2-bis(maleimido)ethane with a triamino compound, such as triazine, as illustrated in Figure 8. The product is then crosslinked by reacting the pendant amino groups with an epoxy, such as diglycidyl ether of Bisphenol A.

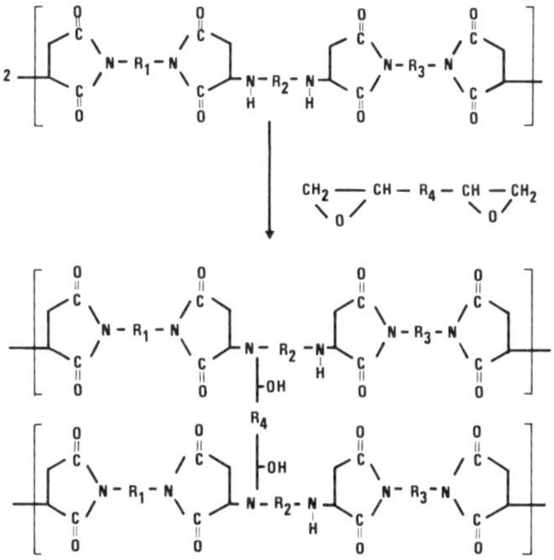

Figure 8. Method of synthesis of epoxy crosslinked polyimide resins

The future plan is to prepare subscale pressure vessels in which the epoxy resin is placed with a modified polyimide resin and their performance compared with their epoxy analogs.

6. SUMMARY

Finally, to recapitulate the subjects that have been discussed: a) Two different sizes of composite motor cases fabricated from either graphite-, Kevlar-, and Kevlar-epoxy overwrapped with graphite, with and without forward skirts, were manufactured and comparatively evaluated. b) Modification of the epoxy resins used in such fabrications by the incorporation of bisspiro compounds to effect expansion or eliminate shrinkage is in the preliminary phase of investigation. c) Replacement of epoxy resins with polyimide resins is planned since the polyimides offer several attractive advantages.

DR. DAVID SAYLES
BIOGRAPHY

Dr. David Cyr Sayles is a native Canadian and became naturalized in 1952. He earned academic degrees at the University of Alberta (B.Sc. in Honors Chemistry), University of Chicago (M.S.), and Purdue University (Ph.D.). He has held positions at Purdue University; Lowe Brothers Company; Wright Air Development Center, Ohio; Air Force Armament Center, Florida; and Army Missile Command, Redstone Arsenal, Alabama. He is currently employed by the Ballistic Missile Defense Advanced Technology Center. He has collaborated in the writing of five technical textbooks; is the author of over one hundred scientific publications; and holds over 200 patents. He has received numerous awards, among these are Sigma Xi, Phi Beta Kappa, Sigma Chi, Tau Beta Pi, and Phi Lambda Upsilon. In 1980, he lectured in Greece on an all-expense trip paid for by the Greek government.

26th National SAMPE Symposium
April 28-30, 1981

THE CHALLENGES OF MANUFACTURING GRAPHITE-EPOXY STRUCTURAL COLUMNS FOR SPACE PLATFORMS

R. L. Vaughn, Manager
Manufacturing Engineering
and
C. A. Friend, Supervisor
Advanced Manufacturing Technology
Lockheed Missiles and Space Company
Space Systems Division, Sunnyvale, California

Abstract

The manufacturing aspects of supplying graphite-epoxy structural tubes in large quantities to support one concept of NASA's program for large space platforms is described. Under this concept, platforms will be built as components on the earth's surface, delivered by Space Shuttle to the assembly site in space, and erected by astronaut-assisted assembly machines. Thousands of tubular components will be required for each platform. Lockheed's approach to tube manufacture is presented in depth. The Lockheed developed equipment, tooling and manufacturing system utilizes a unique method of dry fiber placement which holds the dry fibers in exact location, a resin injection process which forces the epoxy resin into the tooling, and a non-autoclave curing method.

The means of controlling dimensional tolerances, differential thermal expansion between metal tooling and graphite tubes, resin shrinkage during cure, hot pressurized resins in close tolerance tooling, and tool wear are examined.

Lockheed has demonstrated by successful development and pilot production programs that these manufacturing challenges can be met. During these demonstration programs, 172 graphite-epoxy tubes were fabricated and delivered to NASA for testing.

Figure I

1. INTRODUCTION

Numerous manufacturing challenges
are encountered during the development of graphite-epoxy structural
components. The degree of difficulty is proportional to the geometric complexity of the components and the dimensional precision and performance requirements
of the finished articles.
Components for use on space craft
are, in general, high precision,
high performance items with a
multiplicity of secondary requirements including minimum weight and
controlled thermal expansion characteristics.

One of several NASA concepts for
large space platforms calls for the
manufacture of tubular graphite-
epoxy columns on the earth's surface, and the subsequent delivery
of these structural columns by
Space Shuttle to the erection site
in space where they will be assembled by astronaut assisted machines.
Columns designed to meet the performance requirements, the weight
and packing efficiency restrictions,
and the ease of assembly requirements in space present a combination of manufacturing challenges
which are both interesting and
demanding of Advanced Manufacturing
Technology Engineers, Materials
Engineers, Processing Methods Engineers and Tooling Engineers.

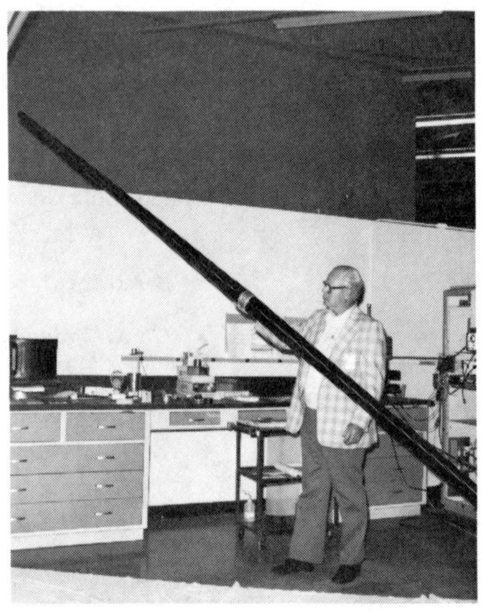

Figure II

2. APPROACH

The tubes are 108" in length with a
constant taper from 2" at the small
end to 4" at the large end. The
wall thickness of 0.025" is made up
of an inner layer of 90° (hoop)
fibers, a layer of 0° (longitudinal)
fibers, and an outer layer of 90°

Figure III

fibers. Pre-machined, integral
aluminum end fittings are cocured
in place during tube manufacture.
The manufacturing approach used by
Lockheed Space System Division to
fabricate these tubes consists of
placing and holding the dry graphite fiber and the aluminum end
fittings on a heated steel mandrel,
covering the mandrel with a closed
caul and injecting the resin under
pressure into the closed cavity.

quired. The tubes are ready for
packaging and shipment with no further machining, alignment, or finishing necessary.

3. THE MANUFACTURING SYSTEM

The Lockheed manufacturing system
developed for tube fabrication places the dry fibers and metal fittings
in close tolerance tooling,evacuates
the tool cavity to remove any moisture and air, and then injects the
resin under pressure into the tool
cavity.

Figure IV

Figure V

These tubes are cured in the steam-
heated pressurized tool; no autoclave is required. The tubes are
removed from the tools while hot to
avoid thermal stresses due to differential thermal expansion/contrac-
tion. No additional work is re-

The heated tooling is maintained at
elevated temperature during the
total process to negate thermal expansion and shrinkage. The tooling
serves as the alignment fixture,
the mold, and since the tooling is

integrally heated and pressurized, it replaces the autoclave. Adequate insulation is provided to control heat loss. The result is a very energy efficient, non-polluting system. Resin injection into the closed tooling is performed while the tooling is hot. This pressure injection places the resin in close contact with the fibers, and with the bonding surfaces of the end fittings to provide strong, void free bonds.

Figure VI

A pressure head is applied to the resin injection ports at both ends of the tooling to provide "make-up" resin as shrinkage occurs during cure. This prevents shrinkage crack and "orange-peel" surface blemishes on the tubes. Non-autoclave curing of the composite and the adhesive bond between the tube end fitting is achieved by using closed, close-tolerance, O-ring sealed, steam heated tooling. The compact, well insulated tooling requires minimum energy to maintain temperature. No large vessels to heat and cool; no large volumes to pressurize; i.e., energy is conserved compared to auto-clave processing, and pollution of the environment (air, water) cannot occur as only closed systems are employed to heat the tooling used by the Lockheed tube manufacturing system.

4. PROBLEMS - CHALLENGES
Numerous interesting challenges pre-sented themselves during the tube development effort. Meeting and solving any one of these on an indi-vidual basis would have been rela-tively straight forward, but due to the combination of the requirements the solutions were obtained only with extensive consideration and evalua-tion. Only the more interesting and critical challenges are briefly dis-cussed here.

Differential thermal expansion of the graphite-epoxy composite, the metal (aluminum) end fittings, and the metal (steel) tooling could cause dislocation of components and/or part breakage if the tooling were heated or cooled after the materials (resin, fiber, fittings) were appli-ed to the mandrel. To accommodate this problem, the fibers were applied to a hot mandrel and the resins were injected into hot mold cavity.

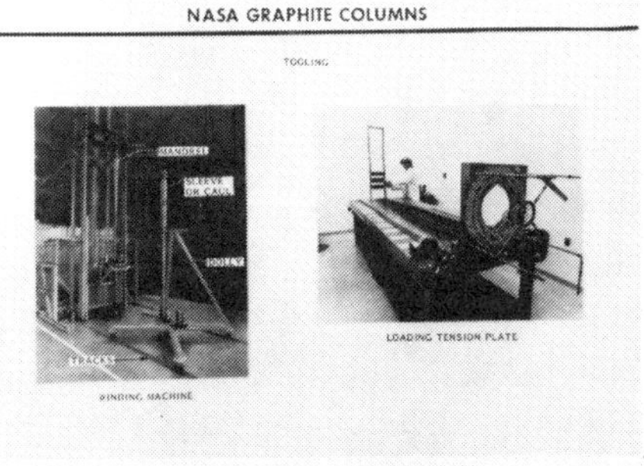

Figure VII

Finally, the cured tubes were removed from the hot tooling without cooldown. In other words, the entire process was carried out using heated tools maintained at temperature. This introduces secondary problems; the people performing the fabrication operations had to work with hot tooling with the attendant risk of burns. By using efficient insulation blankets and planning the process for machine handling, burns were not a problem. Resin injection into heated tools did present problems, however, and will be discussed in later paragraphs.

Holding the graphite fibers in a position during resin injection, and during the mold handling operations, required the application of several special procedures and the control of operation sequences. The tapered mandrel also presented a share of concern for maintaining the hoop (90°) fibers in position. By main-

taining a constant temperature hot mandrel, the application of fiber and the retention of fiber was made possible by vertically winding an inner 90° layer of fibers, then hanging the 0° longitudinal fibers from the top of the mandrel and overwinding the 0° fibers with a layer of 90° fibers which form the outer layer of the tube.

Figure VIII

The 0° fibers were tensioned by passing them through ceramic guides which applied a mild drag of a few ounces on each of the several hundred 0° graphite strands. Several types of graphite-fiber were successfully handled by this method, including the 100 million psi modulus pitch fiber which was the most common fiber used for 0° layers in this program. Lower modulus, less expensive T-300 fiber was the most common fiber used for the 90° layers; since the extreme stiffness properties of pitch was not needed, the T-300 was more readily available, and costs much less. By overwinding the 0° fibers with a layer of 90° fiber, the three fiber layers served to hold each other in place on the mandrel during tooling transfer operations and during resin injection.

also to remove air and any moisture that might have been present on the fiber surface.

4.1 MAINTAINING CONSTANT WALL THICKNESS

A constant 0.025 inch wall thickness was achieved on the tapered tube by maintaining a relatively constant 135 fiber strands per inch of circumference. This was done by "dropping" 0° fibers as the tube tapered down from 4" diameter at the top (large end) to 2" diameter at the small end. The 0° fibers were hand loaded into a tensioning device which kept a constant tension on the 0° fiber strands until they were pressed against the mandrel by the outer layer of 90° (hoop) wound fibers.

Figure IX

The 90° outer layer was tied off at the upper and lower ends of the mandrel, and the winding transferred into the caul in preparation for resin injection. After the wound mandrel was placed in the mold and closed, a vacuum was drawn on the mold cavity to assure that the O-ring seals were functioning, and

Figure X

The filament winding of hoop fibers was performed by the winding machine. Accuracy and efficiency were achieved. The 0° fibers were loaded into the tension plate by hand, a costly, time consuming process.

Approximately half the labor cost involved in manufacturing a tube was due to the hand loading of 0° fiber into the tension plates.

Figure XI

A planned 1981 R & D program will automate this operation and make possible "continuous draw" machine controlled fiber loading for the first time. This key development is expected to cut labor cost about 50%.

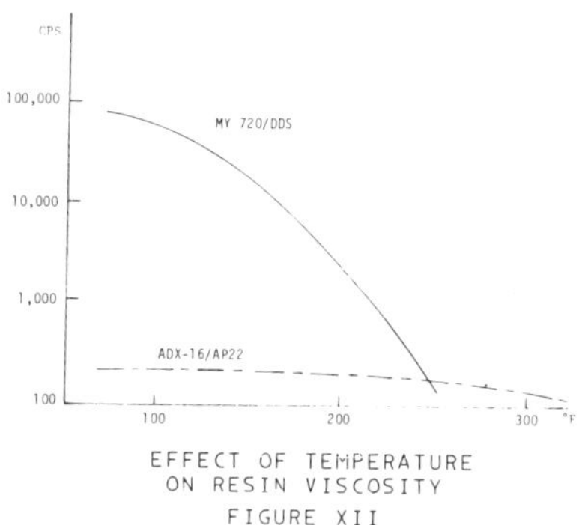

EFFECT OF TEMPERATURE
ON RESIN VISCOSITY
FIGURE XII

4.2 MAINTAINING RESIN VISCOSITY DURING INJECTION

The resin was injected under pres-sure into a hot closed mold which had already been filled by graphite fiber. The resin followed a path inside the tooling which was in excess of 108 inches. To flow under these conditions and to wet the surface of each individual fiber without damaging or moving the fibers (fiber wash) required a low viscosity resin. The viscosity must remain low during the injection period, normally about 20 minutes. The resin should then cure to completion to permit recycling the tool to fabricate the next tube. This ideal resin system may not exist, but a satisfactory alternative was developed by Lockheed which serves in the present application. The resin used by Lockheed has a low viscosity at room temperature and therefore requires no heating prior to injection. Viscosity was further reduced as the resin made contact with the heated tooling and flowed through the mold cavity which contained the dry graphite fiber.

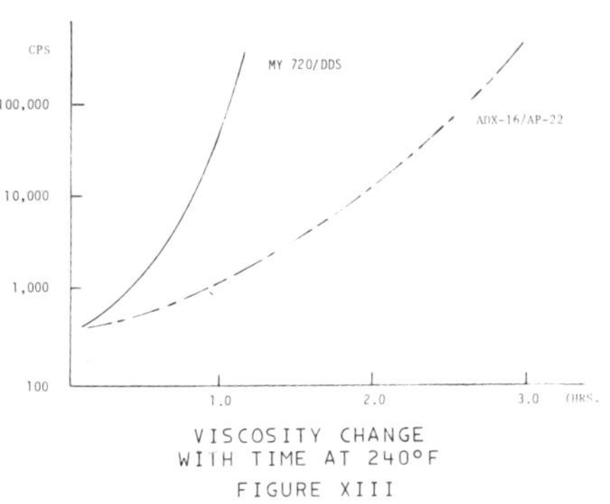

VISCOSITY CHANGE
WITH TIME AT 240°F
FIGURE XIII

Following injection, the resin was held under elevated temperatures (240°F) and pressure (100 psig) conditions for a minimum of four hours to permit the resin to cure to a hard gel state. The tools were opened and the tube removed hot. A thermal blanket was used to cover the tube and control the cooldown rate to prevent resin cracks due to thermal shrinkage stresses.

4.3 SEALING

A major sealing challenge was effectively met only after careful evaluation of tooling materials and design. The tooling was integrally heated by steam. A steam path through the mandrel enters and exits at the large dia. 4" end. The mandrel was made of segments approximately 18" long, each of which was threaded to its neighbor and sealed by an O-ring seal to prevent steam leaks. The mandrel was placed into the caul and sealed prior to injecting resin. O-ring seals between the mandrel outside diameter (O.D.) and caul inside diameter (I.D.) prevents leakage between these major tooling items. These latter seals serve as vacuum seals during mold evacuation prior to resin injection, and then as pressure seals during resin injection and resin cure. These seals must withstand considerable abuse during mold closing at which time they are subject to the effects of hot epoxy resin, fluid pressure, and the hot mold surfaces and the scuffing effects of the sliding mold surfaces as the mandrel is forced into the caul. In the closed configuration, the tooling has steam inside the full length of the mandrel, resin in the full length of the mold cavity, and steam inside the full length of the caul. The assembly therefore consists of numerous seals, both static and dynamic, which separate steam from resin steam from ambient air, and resin from ambient air. The sealing of the mandrel-to-caul interfaces required additional attention to the sealing surfaces to prevent wear and to prevent scuffing in this area which metal-to-metal contact was made as the tools were forced closed and later forced open. During early tool trials, the mild steel mandrel and caul seized during opening and closing operations. This led to high mold opening and closing forces and rapid wear of the affected metal surfaces. To overcome this problem, the contacting metal surfaces of the tooling were ground and plated with nickel, then reground to a smooth finish and precision dimension. No further difficulty was encountered in this area, and after 172 tubes, the tool sealing surfaces remained smooth with no appreciable wear.

4.3 MAINTAIN MOLD RELEASE COATING ON TOOL SURFACES

The need for effective mold release in a mold of the dimensions and precision of the tube tooling is obvious, particularly when there is no cooling cycle to shrink the tool away from the molded part. During the fabrication operation, there were two events which tended to remove or erode the mold release materials applied to the mandrel and caul. The first of these was the scraping or scuffing action of the graphite fiber as the fiber was applied to the mandrel and again as the mandrel was installed in the caul. The second and more severe action affecting the mold release material was the resin injection operation. The hot resin, injected under pressure and forced to flow the full length of the tool set, tended to wash or erode all but the most tenacious mold release materials. Several brands of liquid mold release, waxes, powders, and teflon coating were tried with varying degrees of success. Mold parts coated with teflon released well during the first cycle, but as the coating separated from the metal in local spots, the parts adhered to the tool, causing local sticking and high mold opening forces. The liquid, wax, and powder mold releases, with one notable exception, were subject to erosion by the hot resin on each cycle. The application of heavy coatings or multiple coating of release proved to be counterproductive as the composite matrix resin would adhere to the heavy release layer. Under these conditions, sticking was severe enough to break the tube as excessive mold opening force was required to free the mandrel from the caul. During early mold release trials, some of the tubes were totally destroyed during removal operations. The effectiveness of the mold release is dependent on the type of resin in the composite. For the ADX-16/AP-22 resin used for this program; it was found that Frekote 33, a product of Frekote, Inc., Boca Raton, Florida, was the most effective mold release product for this application. A thin, uniform coating of the mold release over the tooling surfaces, followed by drying, baking, and polishing with a soft cloth, resulted in trouble-free releases for three or more tubes. Repeat applications were made when the mold opening forces showed the first indication of rising. This was monitored by observing the hydraulic pressure required to operate the ram which forced the mandrel out of the caul after the tube was cured. Effective mold release - a mundane but critical requirement for efficiently molding precision, quality, composite parts.

4.4 RESIN SHRINKAGE DURING CURE
The epoxy resin used in the tubes
shrinks during cure. Even though
the shrinkage was limited, effect on
the thin-wall tube is profound.
Cracks in the resin and "orange-
peel" appearance were observed, as
well as surface pits or craters.
Solutions of various kinds were
tried without success, including
mandrel position adjustments in the
caul, adjusting the percentage fiber
in the composite, and various injection rates, controlled by the
resin pressure. Complete elimination of the flaws caused by shrinkage was achieved by putting a resin
supply, under pressure, at the resin
entry and exit ports located at the
ends of the tooling. This pressurized resin supply provided "makeup"
resin to replace shrinkage losses.
This process eliminated the cracks,
orange-peel, and surface pits, giving a tube which needed no further
finishing prior to shipment.

4.5 MAINTAINING TUBE STRAIGHTNESS
    DURING FABRICATION OPERATIONS
Dimensional precision and alignment
of features of the tube are requirements stemming from the need to fit
up during platform assembly and also
from the need for each column to
carry the loads in a symmetric fashion. Any distortion in the tube
lessens the load carrying capacity
and the efficiency of the platform
structure. To assure absolute
straightness, the fabrication pro-

cess was designed around eliminating forces which would distort the
finished tubes. All operations were
carried out vertically: filament
winding, resin injection, resin curing, resin postcuring. This
eliminated any bending from gravity effects which would occur if
the columns were fabricated in a
horizontal attitude. Other features
were also designed for symmetry including the mandrel, cauls, steam
heaters and insulation layers applied over the outside of the tooling for temperature control and
personnel safety. All lifting, all
mandrel removal forces were applied
axially to prevent distortion of
tooling or product. The filament
winding machine used was designed
and fabricated by Lockheed Space
Systems Division personnel.

5. CONCLUSIONS

The manufacturing system defined
herein has demonstrated that this
method can produce high performance,
close tolerance Gr/Epoxy structural
columns without the need for an
autoclave. Successful development
and pilot production programs were
completed by Lockheed Space Systems
Division during which 172 columns
with integral metal end-fittings
were manufactured and delivered to
NASA for structural test and assembly. This program demonstrated that
complex manufacturing problems
associated with precision high per-

formance composite structural hardware for use in space applications can be effectively solved.

Figure XIV

The Lockheed developed and manufactured structural columns were 102 inches long, had a wall thickness of 0.025 inch, with a constant taper from the large end diameter of 4 inches to a small end diameter of 2 inches, were aligned along a longitudinal axis to very close tolerance to permit stacking in "dixie-cup" fashion to assure high density packaging for space shuttle delivery to space, and had an individual tube weight of 1.8 pounds of which more than half (0.93 lbs.) was aluminum end-fitting weight. The length, diameter, and wall thicknesses were determined by the tooling. Diligent attention was exerted by Lockheed to insure that the columns were both structurally and cosmetically of high quality which the customer stated was acceptable.

6. ACKNOWLEDGEMENTS

The authors wish to extend their appreciation to the many persons in the aerospace industry who have in so many ways contributed to this program. Government agencies, industry, universities, and above all, individual contributions by many people in many diverse disciplines made it possible for Lockheed Space Systems Division, in support of NASA, to be in a position to perform the developments which made this program a success. Credit is granted accordingly.

The authors wish to identify a number of contributors who had special influence: Harold Bush of NASA, who held the vision which kept the team on the true path to success in spite of the challenges and occasional setbacks encountered; Hank Cohan and Bob Johnson of Lockheed's Space Technology Organization who served as Program Office and Program Manager, respectively, and provided encouragement to the team to strive for excellence, and to Fred Keck, John Fritzen, Bill Reighard, Ray Bluck, Ken Metherell, Dennis Froidevaux and Bob Fahrni who spent years developing the many technology areas required for this program.

26th National SAMPE Symposium
April 28-30, 1981

DIMENSIONALLY STABLE MESH FOR
SPACECRAFT LARGE ANTENNAS

Donald J. Levy and Charles R. Arnold
Lockheed Research Laboratory
Palo Alto, California 94304

David H. Ma and William D. Wade
Lockheed Missiles & Space Co., Inc.
Sunnyvale, California 94088

Abstract

There is need in the near future for highly-accurate space-erectable antenna reflectors in the size range of 100 to 1000 m diameter. The development of such large apertures is dependent, to a large extent, on the availability of dimensionally stable rf-reflective mesh materials. Two concepts for preparing an ultra-low-expansion antenna mesh were experimentally evaluated and a mesh with CTE of $0.4 \times 10^{-6}/^{\circ}C$ was successfully developed. This material is composed of metallized, woven, continuous-filament quartz. The mesh exhibits light weight, excellent electrical characteristics, good strength and good durability.

1. INTRODUCTION

Predictions of performance of an antenna system are based not only on the knowledge of the absolute value of the various material properties but also on the degree of uncertainty of that knowledge. To illustrate the requirement for accurate knowledge of material properties, a design of a relatively small 30 m wrap-rib antenna was reviewed.[1] In this case, the reflector surface error was developed for assumed errors in structural element material properties. Two error levels were assumed which typically bound the accuracy with which properties are presently deter-mined. The 30 m aperture was then scaled to 150 m to trace the effects of the error contribution as a function of aperture size. These results are presented in Figure 1. The effect of these errors on aperture performance is most dramatic when presented in terms of antenna efficiency as shown in Figure 2. This simple analysis readily points out the need for significant improvement in the ability to manufacture and test structures if we are going to be able to provide the structures for the more advanced missions. It further illustrates the requirement for being able

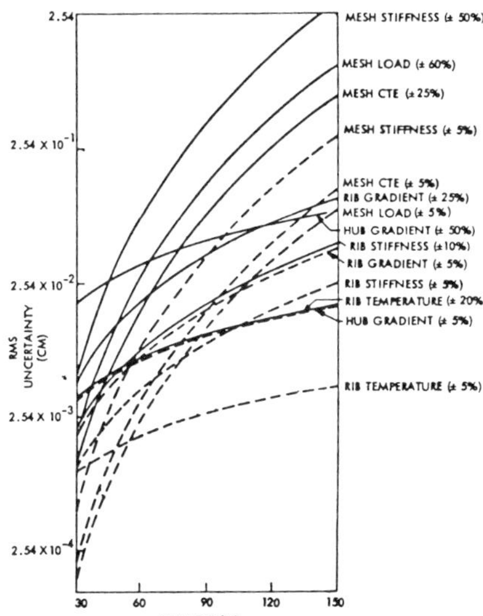

Fig. 1   Surface Figure Error Due
to Uncertainties (Ref. 1)

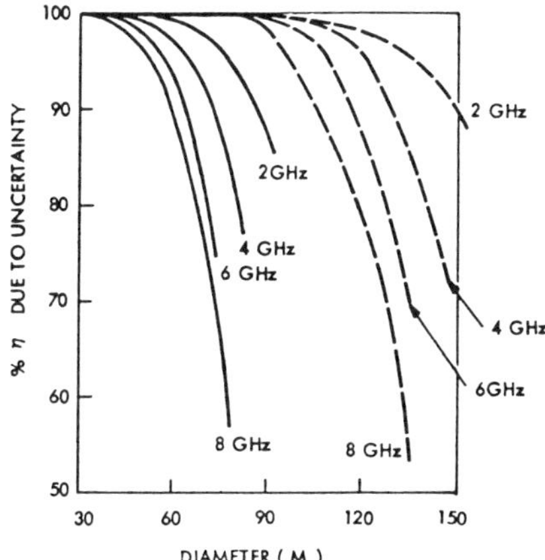

Fig. 2   Surface Figure Uncertainty
Effect on Performance
Efficiency (Ref. 1)

to build large structures with almost
no batch-to-batch variation in material characteristics. Considering
the two major structural components,
the rib and mesh, the data shows
that performance degradation occurs
as the fourth power of mesh errors
and the second power of rib errors.

The unfortunate result is that
mesh effects dominate with a relatively small growth in diameter.
The mesh is typically a knitted wire
material with nonlinear stiffness.
Previous material characteristics
were dictated by the knitting process and procedures and identified
through biaxial stiffness tests
performed in comparatively small
samples. With a dimensionally
stable mesh, improvement of mesh
manufacturing controls and properties measurement to the 5% uncertainty level should be possible.

The development of an ultra-low expansion antenna mesh is described.
Two concepts were considered and
experimentally evaluated. One
approach was to coat a negative-CTE
(coefficient of thermal expansion)
substrate yarn with a suitable thickness of positive-CTE metal. The
other approach was to bond a thin
coating of rf-reflective metal to a
substrate yarn with very low CTE.
Antenna mesh materials were prepared, characterized for physical
properties, and evaluated for durability.

2.   POSITIVE-CTE METAL ON NEGATIVE-CTE MESH

2.1   Principle of Zero-Expansion Yarn

The concept of producing an ultra-low
expansion yarn involves selecting a
high-strength negative-CTE fiber and
coating it with a sufficient thickness of a positive-CTE metal to balance the substrate contractive forces

as temperature is increased. Feasibility of the concept depends upon a strong interfacial bond between the nonmetallic fiber and the deposited metal coating.

The required thickness of a metal coating for a zero-expansion composite yarn can be calculated from Schapery's Equation when α_c, CTE of the composite, is set equal to zero.

$$\alpha_c = \frac{E_m \alpha_m V_m + E_f \alpha_f V_f}{E_m V_m + E_f V_f}$$

where E = Young's modulus

    α = CTE

    V = volume fraction

c, m, f = composite, matrix, and fiber properties

A number of approximations and assumptions are necessary to compute the metal thickness. It was assumed that α and E are constant throughout the range of temperature measurement; average properties of the metals were used.

2.2 Materials and Apparatus

GY-70 graphite fiber (Celanese Corp.) with a negative α was selected as a candidate substrate for study. It has one of the highest moduli commercially available and, in the case of graphite, the higher the modulus the more negative the α. The fiber has a dog-bone cross-section with a nominal 8-μm diameter, porous structure, 74.7×10^6 psi tensile modulus, 99.9% carbon, and zero moisture regain (21°C, 56% R.H.). The yarn has 400 fibers and 0.5 twist per inch. Reported α is $-1.1 \times 10^{-6}/^{\circ}$C.

Kevlar-49 polybenzamide fiber (Du Pont) was the second candidate substrate selected. It has a negative α, oriented structure, high strength, 19×10^6 psi tensile modulus and a round cross-section of 12-μm diameter. Moisture regain is 3.5% (22°C). The yarn has 134 fibers and no twist. The literature values for α are $-2 \times 10^{-6}/^{\circ}$C($0$-$100^{\circ}$C) and $-4 \times 10^{-6}/^{\circ}$C ($100$-$200^{\circ}$C).

Two classes of metals were selected for coatings. Nickel represents a strong, corrosion-resistant metal suitable for structural purposes. A metal exhibiting high RF-reflectivity was also desired. While copper is the logical candidate for practical use, gold was selected to avoid any possibility of oxidation or corrosion effects in the experimental work.

Modulus for gold and nickel was assumed to be 12 and 20×10^6 psi respectively. α for gold and nickel was assumed to be 14.2 and 13.0 x $10^6/^{\circ}$C respectively. "Electroless" autocatalytic deposition and electrodeposition of the metal coatings were used because of thickness uniformity, yarn bundle penetration, and cost considerations.

2.3 Properties of Bare Yarn

Scanning electron microscopy shows that the Kevlar surface is relatively smooth with some swarf attached. The GY-70 surface is rough at high magnification, which should improve the metal-graphite bonding.

The Kevlar is a uniform material exhibiting a consistent α. An α of $-3.72 \pm 0.39 \times 10^{-6}/^{\circ}C$ was found for the temperature range $-150^{\circ}C$ to $+150^{\circ}C$.

CTE of the graphite, $-1.07 \pm 0.73 \times 10^{-6}/^{\circ}C$, showed large devaiations even between adjacent samples from the same yarn. Part of the problem is due to the twist in the yarn. It is also possible that properties of the GY-70 may vary.

Preloads of about 1.5 g were used with the Kevlar and GY-70 yarns. Plots of specimen length versus temperature were linear throughout the entire temperature range from $-150^{\circ}C$ to $+150^{\circ}C$ for both yarns. A computer was used for data reduction to obtain a best fit of the data.

2.4 Characterization of Low-Expansion Yarns

Kevlar yarn has no twist and therefore the loose filaments are each readily coated. Electrodeposition is not possible for coating these nonconductive fibers, so other techniques were used. Electroless nickel (phosphide type) achieved a smooth, uniformly thick coating on the fibers.

Initial metallization with gold was achieved by a variation of the LocksprayGold process, after which the yarn was electroplated with a high-purity gold. The electrodeposited gold was quite nodular compared with the electroless nickel coating.

GY-70 graphite has a tighter bundle because of the twist. Nevertheless the yarn bundle is penetrated and the individual filaments are each coated by using electroless nickel. The yarns are satisfactorily metallized with gold by using Lockspray Gold followed by electroless gold, but there is little penetration and the metal is deposited predominantly on the outside of the yarn bundle. The electroless gold appears quite nodular at high magnification although it is not as nodular as the electrodeposited gold. Direct electrodeposition of gold and nickel on the conductive graphite is feasible but was not used. Less-uniform coating thickness would be expected.

The preload required to straighten single yarns for CTE measurements was determined in the usual manner and found to be about 0.5 - 2 g. This is less than 0.1% of the ultimate tensile strength of the yarns.

Starting with bare yarn and a negative α, the α becomes more positive in direct proportion to the amount of metallization. Efforts were made to achieve near-zero α with each metal on each yarn. Figure 3 shows the the α as a function of the weight of metal coating in the region near zero α.

These data indicate that yarns with an α of less than $0.5 \times 10^{-6}/^{\circ}C$ can be prepared. The precision in measurement of α is estimated to be about $0.2 \times 10^{-6}/^{\circ}C$ and the standard devia-

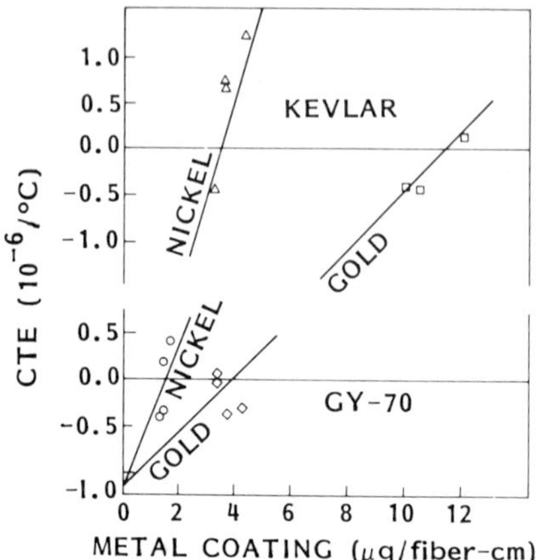

Fig. 3 CTE of Metallized Yarns

tion between α measurements on identical yarns is approximately the same. Most of the error is in α measurements; the metallization goals were achieved within ±5%.

The α of all yarns was measured from -150°C to +150°C and the α remained fairly constant throughout the temperature range. α was the same whether measurements were made during heating or cooling, and very little hysteresis in length was exhibited during a complate cycle of heating and cooling.

The metallization shown earlier was expressed in terms of a weight per unit length for convenience experimentally and in comparing the several fibers and metals. The amount of nickel required for a zero-expansion composite fiber is about 1/4 by volume and 2/3 by weight. Experimental measurements of bare yarn strength were 8-10% below literature values. Metallization contributed little, if any, to the yarn strength and the process of depositing

nickel on Kevlar apparently caused a small degradation in strength. The data indicated a weak interfacial bond as expected when no special effort at improvement was made.

2.5 Durability of Nickel/Kevlar Yarn

At this point, the graphite substrate was eliminated from further consideration because of its brittleness and further studies were conducted with Ni/Kevlar yarn. Several techniques were used to improve the metal-nonmetal bonding and the metallized yarn was evaluated for durability.

Creep measurements show negligible creep of bare Kevlar-49 (139 fibers) or Ni/Kevlar yarn after 700 days at ambient. The data in Table 1 indicate an initial relaxation on the order of 0.03 percent during the first few hours after the specimens are loaded, followed by a change of no more than 0.02 percent during the next 700 days. The small changes observed are believed to be close to the experimental error of the optical cathetometer method and are not considered significant.

Table 1. Yarn Creep at 14 and 28% UTS Loads

	Creep (%)				
	Kevlar		Ni/Kevlar		
Days	28	28	14	14	28
0.01	0.000	0.000	0.000	0.000	0.000
1	.014	.067	.022	.015	.041
10	.034	.084	.024	.015	.050
100	.053	.079	.022	.010	.053
700	.030	.061	.018	.035	.051

The minimum bend radius of Ni/Kevlar
yarn increases with the magnitude of
working load as shown in Fig. 4. It
is well known that knot strength is
considerably less than yarn tensile
strength because of the very small
bending radius of the fiber. The
bending radii shown are satisfactory
for dynamic loads.

Yarns in use will be subjected to
thermal cycling which will create
differential stresses at the metal-
fiber interface. The composite yarn
must be durable and largely unaffected by these stresses. Figure 5
shows that significant interfacial
stresses are created by thermal cycling over the range of -180°C to
+60°C and that these stresses cause
partial failure of the interfacial
bond and degradation of the CTE.

3. LOW-CTE SUBSTRATE YARN

3.1 Properties of Quartz Yarn and Mesh

Continuous-filament quartz yarn and
mesh were obtained from J. P. Stevens
& Co., N.Y., under the trade name
Astroquartz. Quartz yarn is a very
attractive substrate material for
antenna mesh because of its low
and resistance to moisture, radiation, and elevated temperature.
This study phase included compiling
the physical properties of quartz
filaments and yarn, selection of a
mesh substrate, and measurement of
the yarn CTE and creep.

Filament properties are: density 2.20

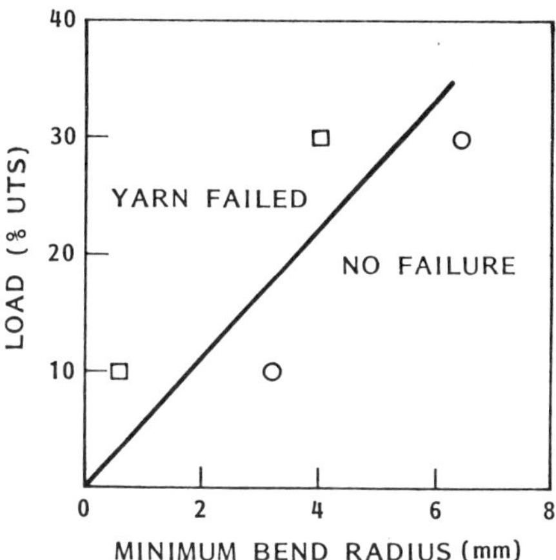

Fig. 4 Minimum Bend Radius of
Ni/Kevlar Yarn Increases
With Load

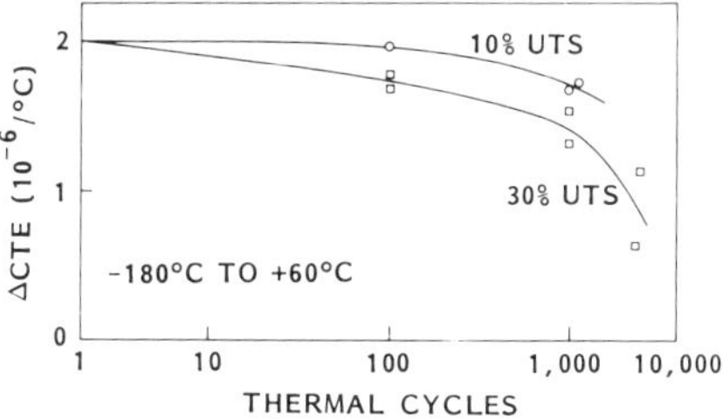

Fig. 5 Effect of Thermal Cycling on
CTE of Nickel/Kevlar

g/cm^3, diameter 0.010 mm, tensile
strength at 25°C 900 MPa (0.13 x 10^6
psi), elongation at break 1%, devitrification temperature 867°C, Youngs
modulus 70 GPa (10 x 10^6 psi).

Yarn in the selected mesh was type
300-2/2 which is comprised of 480
filaments. This yarn has an average
0.30 mm diameter, 1.4 kg tensile
strength, 65 Tx and yield of 15,100
m/kg. Yarns are available with 240
and 120 filaments.

The mesh selected was a relatively
open leno wave, Style 594/38

355

(Fig. 6) with a weight of 82 g/m^2, thickness 0.20 mm, thread count/in. of 20 warp and 10 fill, and tensile strength (lb/in) of 50 warp and 25 fill.

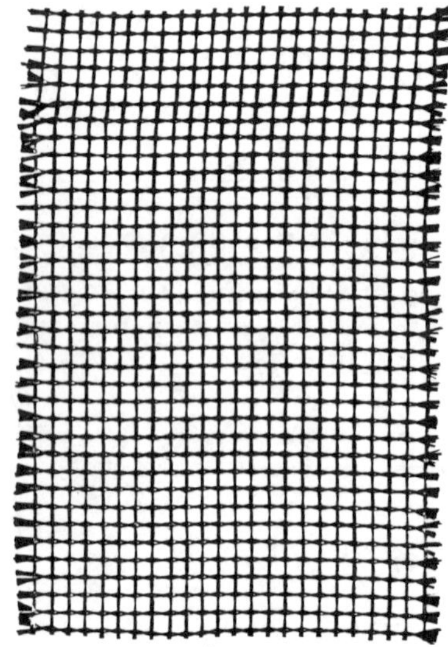

Fig. 6   Leno Weave Quartz Mesh (75% Actual Size)

The characteristic of greatest importance of the quartz yarn is its low CTE. A standard reference length of solid fused silica obtained from the National Bureau of Standards had an average CTE of 0.15 x 10^{-6}/^{o}C for the temperature range of -180°C to 60°C which is the approximate temperature range of interest for antenna mesh. However, it should be pointed out that the low average CTE is composed of a low negative CTE (-0.29 x 10^{-6}/^{o}C) for the range of -180°C to -100°C and a low positive CTE (0.34 x 10^{-6}/^{o}C) for the range -100°C to 60°C.

The dimensional change of 300-2/2 quartz yarn with temperature is similar to that of solid fused silica (Fig. 7) except for a slight displacement of temperature. The temperature displacement may be experimental error in attempting to estimate the temperature of the quartz yarn by means of small thermcouples placed near it.

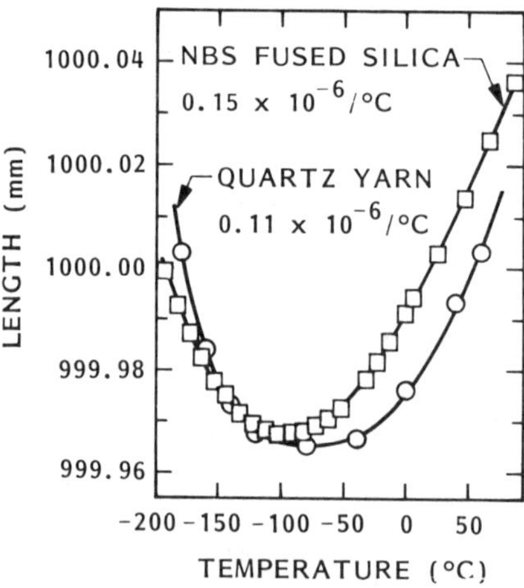

Fig. 7   Dimensional Change of Fused Silica and Quartz Yarn with Temperature (-180 to 60°C)

CTE measurements were also made of quartz yarn after 100 thermal cycles (-180°C to 60°C) which caused no deterioration, as expected. In fact, the measured CTE was slightly lower after cycling:

Material	TCE (10^{-6}/^{o}C)
Fused silica	0.15
Quartz yarn	0.11
Quartz yarn (100 T/C)	0.04

Creep measurements were made on the 300-2/2 yarn electronically (DCDT)

and with a cathetometer. The results (Fig. 8) show an apparent
stretch of the twisted yarns in the
amount of 0.006-0.008% occurring
during the initial 10 days under load
after which there is no evidence of
creep, as expected.

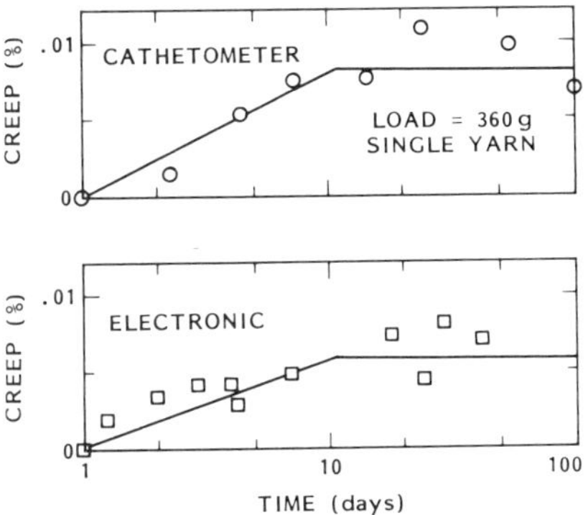

Fig. 8 Creep Test of Quartz Yarn

3.2 RF-Reflective Coatings

The best metals for RF reflectivity
are those with a high electrical conductivity: gold, silver, copper, and
aluminum. Past experience has shown
that rf reflectivity is excellent
when surface electrical resistivity
of the mesh is less than 1.0 ohm.

Gold coatings were found to be very
satisfactory; the metallized quartz
mesh remains soft and flexible. Approximately 18-20% gold, based on
quartz weight, is required to
achieve the desired electrical conductivity.

It is technically feasible to apply
a silver coating. Past work has
shown that silver is an effective
and durable coating on mesh for rf
reflection and its use would reduce

both cost and weight compared to
gold.

Copper metallization on polyester
mesh has been used very satisfactorily on 10-m spacecraft antennas in
the past and some quartz mesh was
prepared with copper. The 18% copper
used was very satisfactory and durability compared favorably with the
copper/polyester mesh. When the cop-
per-coated mesh is crushed to simulate the stowed condition, copper on
the mesh is cracked and air oxidation of the surface within the crack
causes an increase in electrical resistivity. It was found that the
oxidation rate of the copper after
crushing is comparable to and slightly less than that of flight-proven
copper/polyester mesh (Fig. 9) used
on NASA Application Technology Satellites F and G.

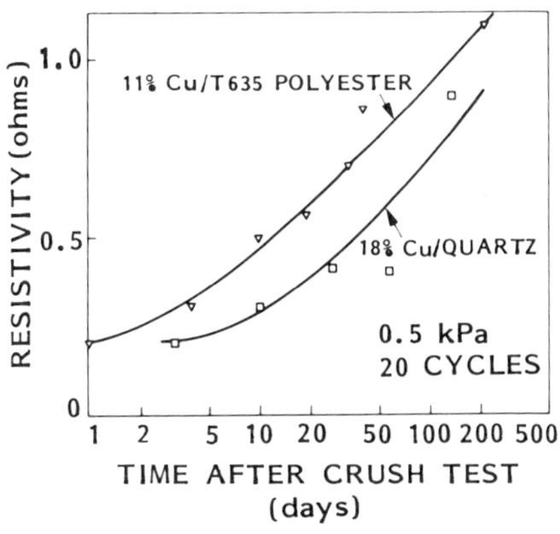

**Fig. 9 Durability of Copper/
Quartz Mesh**

Aluminum is also a candidate coating
for rf-reflection. Since the metal
tends to bond well to quartz, it was
deposited directly on clean, bare

mesh by sputtering. It was found that surface resistivity of a 2.8% Al coating was infinity, but that of a 5.5% Al coating exhibited 1.0 and 0.7 ohms in the warp and fill directions, respectively. Perhaps 6-8% metallization would provide adequate rf properties.

3.3 Mesh Lubricant

It is essential that yarns made of glass or quartz filaments be lubricated in order to possess good strength and handling properties. Otherwise, the brittle filaments rub against one another and fracture easily. The poor tensile strength of the quartz mesh without lubrication can be illustrated by comparing it with the strength of the mesh as supplied by the mill with a proprietary "sizing."

	Mesh UTS N/mm (lb/in)	Strength Retent.,%
Bare mesh from mill	6.1 (35)	100
Cleaned mesh	2.3 (13)	38

Lockheed has a stringent specification for lubricants used on spacecraft in order to prevent these materials from outgassing in the space vacuum and condensing on critical optical surfaces. The standard test consists of heating the material at 126°C for 24 hours in a high vacuum. Qualified materials must have less than 1% weight loss and less than 0.1% VCM (vacuum condensible material on a cold plate at 25°C). One such qualified oil was selected for lubricating the quartz mesh. Although the techniques for using this lubricant have not been optimized, the results are quite good:

	Mesh UTS N/mm (lb/in)	Strength Retent.,%
Clean mesh + lube	5.0 (29)	83

3.4 Mesh Properties

The leno weave causes CTE to be much higher in the warp (double-yarn) direction than in the fill (single-yarn) direction. This problem could be overcome by using a plain weave. Test data for the 20% gold/quartz mesh over the temperature range of -170°C to 60°C showed an excellent CTE in the fill direction.

	$CTE(10^{-6}/°C)$
fill	0.41 ± 0.03
Warp	8.5 ± 0.3

Modulus measurements were made on 24% gold/quartz mesh containing an efficient (95% strength retention) but nonqualified lubricant. Figure 10 shows the stress-strain relationship in the fill direction.

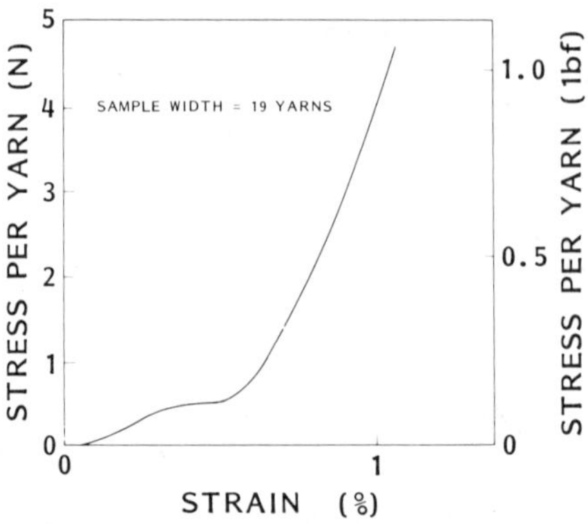

Fig. 10 Modulus of Quartz Mesh (Fill)

<u>Crush testing</u> is used to simulate effects of furl/deployment cycles on the mesh. The Lockheed-developed[2] crush tester,with an anvil pressure of 2 KPa (0.29 psi), was used to determine whether crushing causes any strength degradation in the quartz mesh. While low cycle testing is normally adequate to simulate the maximum of 10 cycles that an antenna mesh would encounter, high cycle testing was conducted on the mill-supplied bare mesh to learn more about the nature of this particular test. It was found that this test tended to repetitively cause sharp bends (and strength degradation) at specific areas in the mesh rather than abuse the mesh all over in a random fashion. Our antenna experts believe that this is a valid test because localized abuse occurs in actual use.

Short cycle testing (up to 30 cycles) shows an average and maximum degradation of 2.7% and 6.6%, respectively. The fill direction was measured because it shows the greatest degradation due to crushing.

Several 10-cycle crush tests of gold/ quartz containing the efficient, non-qualified lubricant showed that the adherence and durability of the gold coating is excellent; there is little, if any, degradation of the electrical resistivity. Strength retention of the mesh was 80% in the fill direction. Resistivity of test panels was the same in both warp and fill directions and ranged from 0.05 to 0.3 ohms before testing. Crush tests caused no change.

4. SUMMARY AND CONCLUSIONS

It was experimentally demonstrated that an ultra-low-expansion yarn could be produced by coating a negative-CTE substrate yarn with a positive-CTE metal. Graphite yarns exhibit the requisite negative CTE but they appear to be too brittle for use in antenna mesh. A nickel/Kevlar yarn was produced with CTE of 0.5 x $10^{-6}/^{\circ}C$ but the interfacial bonding was weak and thermal cycling caused an excessive degradation of CTE. It was found that several thousand thermal cycles at an axial load of 30% UTS causes the CTE to decrease toward that of the substrate Kevlar by about $1 \times 10^{-6}/^{\circ}C$. While bonding was the main cause for rejection of this concept, the amount of metal required to counteract the large negative CTE of Kevlar was also excessive. The composite Ni/Kevlar yarn was 67 wt.% Ni and an rf-reflective metal would also have been quite heavy. Successful use of this concept requires a yarn with only a slightly negative CTE and good metal-nonmetal bonding.

The goal of developing dimensionally stable, durable mesh for spacecraft large antennas was successfully achieved with a substrate mesh of woven, continuous-filament quartz yarns. Flight-quality meshes were produced with gold and copper metallization.

A gold/quartz antenna mesh was prepared which exhibited excellent metal adherence, excellent electrical
resistivity, 80% retention of tensile strength compared to the original bare mesh, and a low CTE of 0.4
x 10^{-6}/^{o}C (single strand, leno
weave). Crush tests to simulate
furl/deployment cycles showed no
significant degradation of electrical resistivity. Gold represented
about 20% of the bare mesh weight
(82 g/m^{2}).

A copper/quartz antenna mesh was prepared and evaluated. Crush tests
showed that the durability of this
mesh was equal to or better than
that of a space-proven copper/poly-
ester antenna mesh used on the NASA
Applications Technology Satellite.
Because of the lower density of copper compared to gold, satisfactory
electrical properties are achieved
when the copper represents about 17%
of the bare quartz weight.

Ceramic filaments require some lubrication to prevent direct rubbing contact which causes filament breakage
and reduced tensile strength. Techniques were developed to enable use
of a space-qualified, low-outgassing
liquid as the lubricant for the antenna mesh and retain 83% of the
bare mesh strength.

Continuous-filament quartz yarns provide an excellent substrate for producing stable antenna mesh. Besides
the low CTE, quartz offers resistance
to moisture, radiation, and elevated
temperatures. Tests confirmed that
there is no creep after an initial
0.007% stretch of the twisted yarns.
It is also technically feasible to
use silver or possibly aluminum for
the rf-reflective coating.

5. ACKNOWLEDGEMENT

This work was funded entirely by the
Lockheed Independent Research Program. Permission to publish is
gratefully acknowledged.

6. REFERENCES

1. Lt.Col. H. L. Staubs, G. G. Chad-
wick and A. A. Woods, Jr., Advanced
Antenna System Requirements and Im-
plications on Material Characteristics, SAMPE National Symposium, San
Francisco, California, 24,320, May
1979.

2. D. J. Levy and W. R. Momyer,
"Metallic Meshes for Deployable
Spacecraft Antennas, Part II,"
SAMPE J., 9, (3), 12, Jul/Aug 1973.

BIOGRAPHIES

Donald J. Levy, Senior Staff Scientist in the Chemistry Research Laboratory, graduated from Washington
University in St. Louis with a degree in Chemical Engineering.
Since joining the Lockheed Research
Laboratory in 1960, he has been
active in the development of antenna mesh and other aerospace materials. At present, he is responsible for chemical process R&D. His
previous experience was in the magnesium and copper industries.

Charles Robbins Arnold, Research
Scientist in the Chemistry Laboratory, received his MA degree in
Chemistry from the University of
Illinois in 1961. He joined the
Lockheed Research Laboratory in
1962, and since that time has been
active in various phases of aerospace materials research, including
a number of years' experience with
the coating and testing of antenna
mesh materials.

David H. Ma is a Senior Process
Engineer with Lockheed Missiles &
Space Company, Inc. He was involved earlier in R&D work in metal
finishing and in studies of interfacial bonding between metallic
and nonmetallic materials at the
Lockheed Palo Alto Research Laboratory. He joined Lockheed in 1974
after receiving his BS degree in
Chemistry from the University of
California at Berkeley.

William D. Wade, Design Specialist.
Mr. Wade received his BSME degree
from Utah State University in 1966.
He has been at Lockheed Missiles &
Space Company, Inc., since that
time, 12 years of which have been
with the Antenna Systems group. He
was design leader and assistant to
the program manager on the ATS-6
reflector program at LMSC. He is
presently the project leader on an
independent program to develop
large reflector antennas.

DEVELOPMENT AND HARDWARE VERIFICATION OF AN AEROELASTICALLY TAILORED FORWARD SWEPT COMPOSITE WING

Joseph W. Ellis
Rockwell International/North American Aircraft Division

Abstract

With the development and verification of new design procedures, aeroelastic tailoring of advanced composite aerodynamic surfaces appears destined to become the accepted approach to composite design. An understanding of this emerging technology is of importance to all who are engaged in the design and manufacture of composite aircraft.

In a recent Defense Advanced Research Projects Agency (DARPA)-sponsored program, application of this technology to a forward swept fighter wing was shown to effectively eliminate flutter and the problem of aeroelastic divergence, a characteristic long associated with forward swept wing (FSW) planforms. This paper describes the development, fabrication, and testing of a divergence and flutter-resistant subscale wing structure representing a high-performance forward swept fighter wing. Development of the unorthodox unbalanced ply wing skin design providing wing special elastic characteristics is described. Verification of the design was provided through the construction of a 0.6-scale structural replica wing scaled for wind tunnel testing. The laboratory and wind tunnel verification testing of this wing is described and results shown.

Key words: forward swept wing, composites, divergence, aeroelastic tailoring

1. INTRODUCTION

Although aircraft wings and empennage surfaces have been tailored to stiffness requirements for many years to avoid flutter and control effectivenesss problems, only recently has the full potential of design for aeroeleastic benefits been recognized. In the composite structure, not only the arrangement and material distribution but the material itself can be tailored to produce controlled flexural characteristics providing substantial benefits in aerodynamic efficiency, dynamic behavior, control effectiveness, and aircraft configuration design freedom. It is becoming apparent that aeroelastically tailored structures will soon be the normal mode of design for advanced composite aerodynamic surfaces. Not only the design engineer, but also those those responsible for specification and control of the fabrication processes should understand fundamental factors underlying aeroelastically, tailored composite structures.

Structures may be tailored to many different criteria. The most common is aerodynamic flutter. In addition to the high stiffness obtainable from advanced composites, flutter-

prone structures may also benefit
from altered ratios of bending-to-
torsional stiffness. Tailoring of
wing cover composite layups can vary
this ratio to a degree impossible
with metallic structures. Control
effectiveness, often a problem on
high-speed aircraft, especially
those with trailing edge control surfaces, can also be significantly
enhanced through aeroelastic tailoring of the aerodynamic surfaces.

Recently, two new applications of
aeroelastic tailoring to fighter
wing structures have been developed.
Although diverse in nature, both of
these applications were of fundamental importance to the aircraft
configuration design upon which they
were used. The first was the con-
trolled-twist wing employed on the
highly maneuverable aircraft technology (HiMAT) research vehicle
shown in figure 1. This remotely

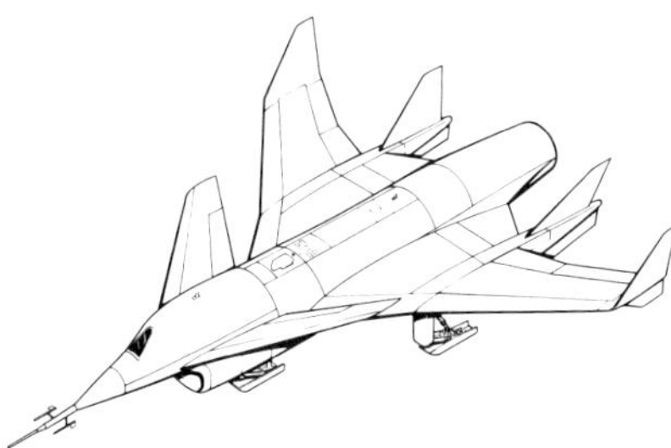

Figure 1. HiMAT Research Vehicle

piloted subscale aircraft developed
by Rockwell International for the
National Aeronautics and Space
Administration incorporates a number
of advanced technologies related to
improved maneuvering efficiency for
fighter aircraft. The HiMAT composite wing incorporates unbalanced
wing skin layups to impart a larger
than normal wingtip washout twist
under maneuvering loads. In this
design, optimum maneuver twist is
achieved at maneuvering load levels,
while a different twist can be maintained under cruise conditions,
improving aerodynamic efficiency in
both modes.

In a second recent development, sponsored by DARPA, application of aeroelastically tailored composite
structures to the forward swept
fighter configuration shown in figure 2 was investigated. Although the

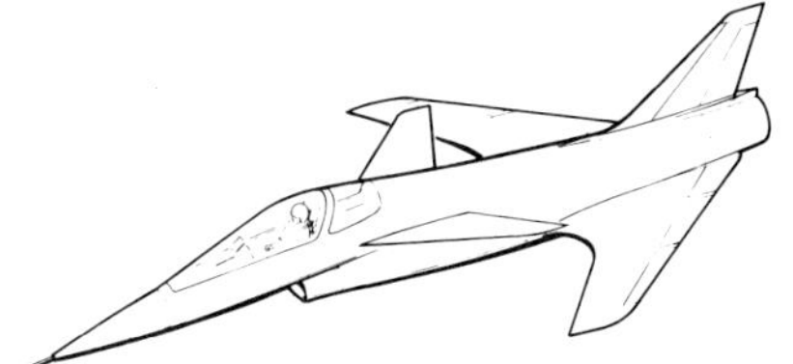

Figure 2. Forward Swept Wing Fighter
Demonstrator

advantages in aerodynamic drag,
usable lift, and controllability at
high angle of attack of FSW's have
been known previously, the use of
forward swept planforms on high-performance aircraft has been curtailed
by the phenomenon of aeroelastic
divergence. Prevention of this aero-
structural instability under fighter
flight conditions results in a prohibitive weight penalty to meet
stiffness requirements when conventional structures are employed. The
use of aeroelastically tailored composite structures promises to reduce
the weight increment to a small,
acceptable level. This paper describes the development, fabrication,
and verification testing of a subscale forward swept fighter wing
which was performed under DARPA sponsorship by Rockwell International.
The program was under the administration and technical direction of the
Air Force Flight Dynamics Laboratory.

2. DIVERGENCE AND TAILORING

The phenomenon of divergence is
little known outside of aeroelastic-
ity analysis groups because it does
not normally occur in aft swept wing
designs and infrequently in straight
wings. It is characteristic, however, of forward swept planforms.
As diagrammed in figure 3, divergence in an FSW can result from the

instability of two opposing sets of
forces: the aerodynamic lift forces
generated as a wing deflects upward
under load and the structural restoring forces which inherently

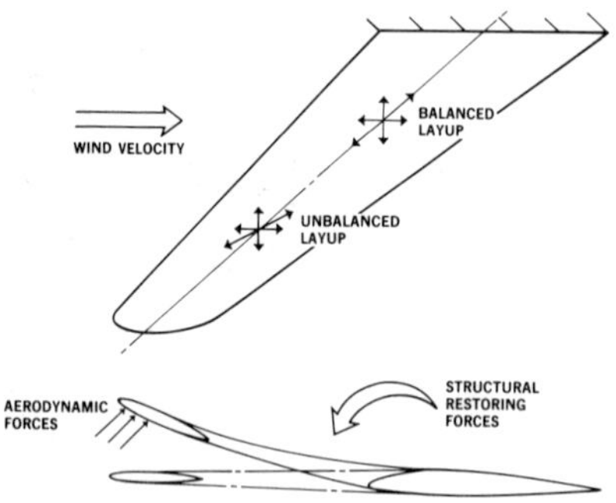

Figure 3. Divergence Prevention
Forward Swept Wing

resist deformation. As shown in the
diagram, the forward swept geometry
tends to produce increasing angles
of attack in the outer portions of
the wing as the airloads bend the
wing upward. If the dynamic pressure due to the flight velocity is
sufficiently high, the aerodynamic
forces will rise so rapidly as the
wing deflects that the more slowly
developing structural restoring
forces will be overcome, and the
wing will experience an unstable
divergence capable of catastrophically destroying it.

Divergence in FSW's can be delayed in
two ways. If the wing stiffness is
increased, either by adding more
material or by increasing its thickness, the magnitude of the structural forces will be larger, and divergence will be delayed until a higher
flight velocity is reached. Another
means of delaying divergence (used
in the DARPA program) employs a design option known as bend-twist
coupling. Through the use of unbalanced ply layups, a warping of the
wing (twist) under the influence of
bending deflections can be induced.
If this induced twist is in the
direction to reduce the angle of
attack of the tip as the wing is

deflected upward, the divergence
characteristics of the wing will be
improved.

Figure 3 shows examples of conventional balanced ply orientations and
an unbalanced layup such as previously mentioned. In the balanced layup
(usually 0/ $\pm$45 degrees), the ply
angles and material distribution are
symmetrical about the wing axis (in
addition to the usual symmetry about
the center plane of the laminate).
In the unbalanced example shown, the
spanwise fibers have been reoriented
off the wing axis. Unbalance is
also produced if the diagonal plies
($\pm$45 degrees) are rotated or the
quantity of fibers is distributed
unequally. The effect of unbalance
in a wing skin is to induce a shear
deformation in the skin when it is
loaded in compression or tension.
This, in turn, produces a coupling
of bending and torsional deflections
of the complete wing box structure,
an effect which can be beneficially
used in aeroelastically tailored
designs.

3. WING DESIGN

A full-scale fighter wing was designed to fighter requirements in the
program, and later a subscale (0.6-
scale) test wing was derived from
the full-scale version. The structural arrangement of the fighter
wing is shown in figure 4. The for-

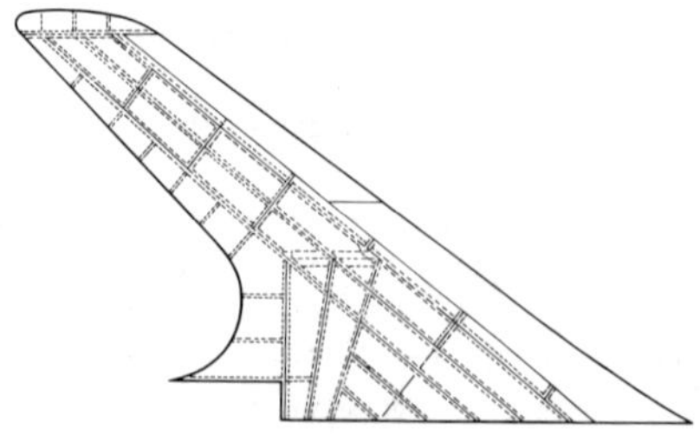

Figure 4. Forward Swept Wing
Structural Arrangement

ward sweep of the primary wing box
breaks well outboard in the wing to
permit the wing center section box

to cross the fuselage forward of the
engine face, where adequate space is
available under the inlet duct.
This arrangement permits a center
section structure integral with the
wing and presents minimum frontal
area. All wing bending loads are
carried across the aircraft centerline in the center section box.
Fuselage attachments are made at the
front spar and at the rear flap support spar.

Preliminary design and optimization
of the wing were carried out with
the Tailoring and Structural Optimization (TSO) program. This computer
program, developed under the auspic-
es of the Air Force Flight Dynamics
Laboratory, is a composite design
synthesis and optimization program
capable of optimizing the structural
covers of a wing box to meet simultaneous requirements for strength,
flutter, and divergence. Ply orientations, as well as material distributions, are included in the optimization, making possible full use of
the bend-twist coupling principle.

Results of some exploratory TSO runs,
illustrating the effect of ply orientation on the weight and divergence
dynamic pressure (q_d) of the wing,
are shown in figure 5. In candidate
1, the diagonally oriented plies
were fixed at angles as close as
were considered practical to the
spanwise plies to obtain a bending
stiffness contribution from their
presence. The optimized cover
design using this orientation
scheme (0/$\pm$30 degrees) produced
weight and divergence characteristics superior to conventional
balanced 0/$\pm$45-degree layups (not
shown), but the fiber directions
would not work well in the inboard
wing, and only minimal bend-twist
coupling is possible with this
fiber orientation.

In candidate 2, the computer was
free to optimize the direction of the
spanwise ply. The optimized design
rotated the spanwise fibers 9 degrees
forward of the box axis to give increased beneficial coupling. A
reduction in weight and increase in
divergence speed was indicated.

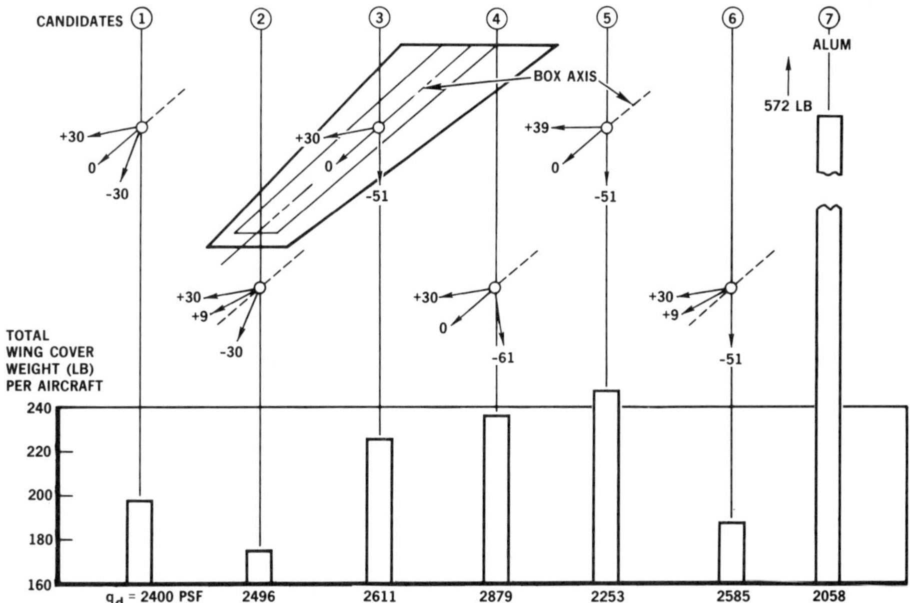

Figure 5. TSO Tailoring Exploration

Candidates 3, 4, and 5 explored sensitivity to orientations of the diagonal plies. It was desired to find a set of orientations which would be satisfactory for both inboard and outboard wing segments. A one-piece wing skin was desired from root to tip, and if a compatible set of orientations could be found, many of the plies could be run through the structural sweep-break area without interruption, eliminating a cover splice, reducing load transfer problems, and minimizing thickness buildup at this point. Adverse effects on weight were noted for all of these variations.

In the configuration finally chosen (candidate 6), the diagonals were fixed in the directions of candidate 3, which had proven the most efficient of the candidates with inboard wing compatibility. The spanwise fiber orientation was left free for optimization by the computer. Again, TSO found a small angle (9 degrees) forward of the box axis was the optimum orientation, producing a very satisfactory result nearly as light as candidate 1, with improved divergence speed and fiber directions compatible with the inboard wing.

An aluminum box was run for comparison using the same design conditions. As shown, the weight was over three times that of the composite design, and the divergence q only marginally exceeded the required 1,980 psf.

Using the TSO definition of the outer panel as a guide, a complete design definition for the wing was developed. Because of the TSO limitations to trapezoidal structures only, the entire wing could not be included in the TSO optimization process. The

inboard area and sweep break transition were designed primarily for strength using conventional composite design procedures. Detailed analysis, used in several iterations to achieve the final design, was performed with the NASTRAN structural analysis program using the finite-element model shown in figure 6. Flutter and divergence analyses were performed with a combination of NASTRAN, FASTOP, and Rockwell proprietary programs, employing established analytical technology.

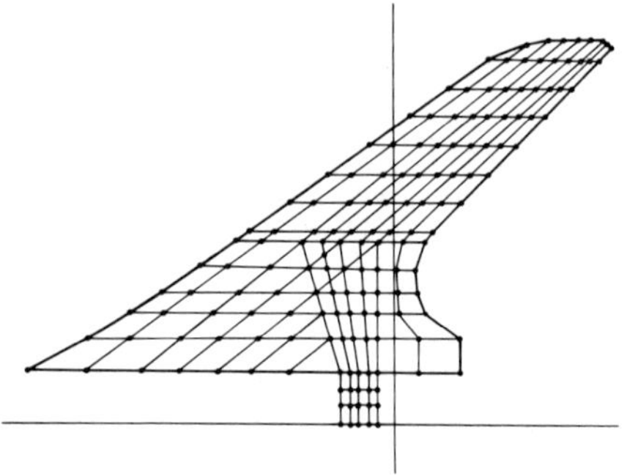

Figure 6. Fighter Wing Finite-Element Model

The results of a deflection analysis shown in figure 7 reveal the effectiveness of the unbalanced layup design. Although the positive streamwise aerodynamic twist under load is not entirely negated, the presence of a substantial beneficial torsional structural twist in the outer panel is evident when it is viewed along its axis. The effectiveness of the unbalanced layup in producing wing twist is apparent when it is realized that this leading-edge-down twist is in opposition to the applied torsional moments, as is evidenced by the relative chordwise positions of the wing box axis and the locus of aerodynamic lift forces.

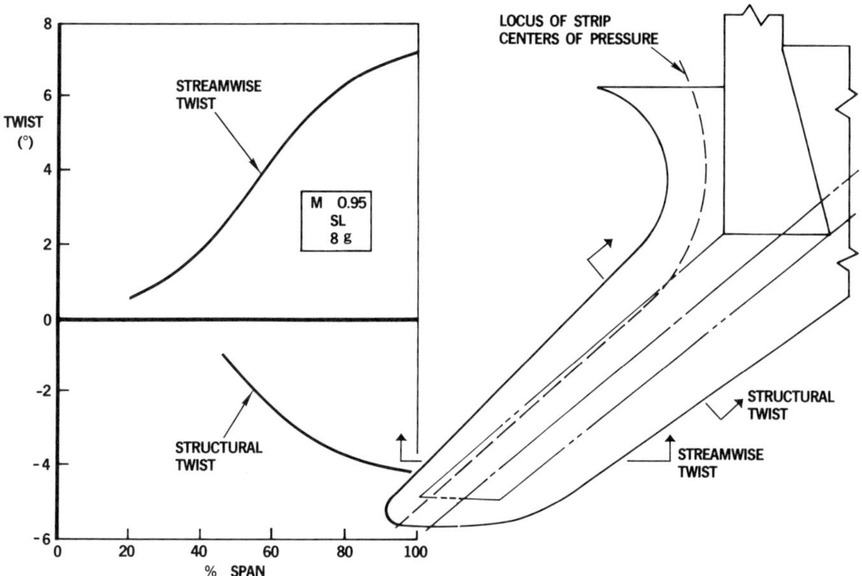

Figure 7. Forward Swept Wing Twist

4. SUBSCALE WING DESIGN

A 0.60-scale structural replica wing was designed, fabricated, and tested to verify the technology employed in the fighter wing design. Because this wing was to be wind tunnel tested to verify its aeroelastic performance, it was scaled to conditions available in the 16-foot Transonic Dynamics Tunnel at the NASA Langley Research Center. This meant that although the geometrical scale was 0.60 of full scale, the design loads and, consequently, the structural gages were substantially less (0.071 of full scale).

The subscale wing was a structural replica in that all of the major structural elements; i.e., spars, ribs, etc, were represented, and the primary box skins reproduced the materials and ply definition of the original wing. A urethane foam core was introduced in place of the inter-mediate spars and ribs, serving the multiple purposes of skin stabiliz-ation, shear load path, and substruc-ture layup mandrel and to establish the contour of the wing. The leading and trailing edge structures were foam cores covered with fiberglass fabric. All elements of the struc-ture were sized to represent the structural behavior of the full-sized counterpart.

The subscale wing was designed for wind tunnel mounting as shown in fig-ure 8. The wing panel was supported from a tunnel wall turntable by means of a flexible steel mounting beam simulating wing center section and fuselage flexibilities. An aerody-namic reflection plane and boundary-layer splitter plate was supported on the tunnel wall in close proximity to, but not touching, the wing root. The angle of attack of the wing was varied during testing by rotating the tunnel wall turntable.

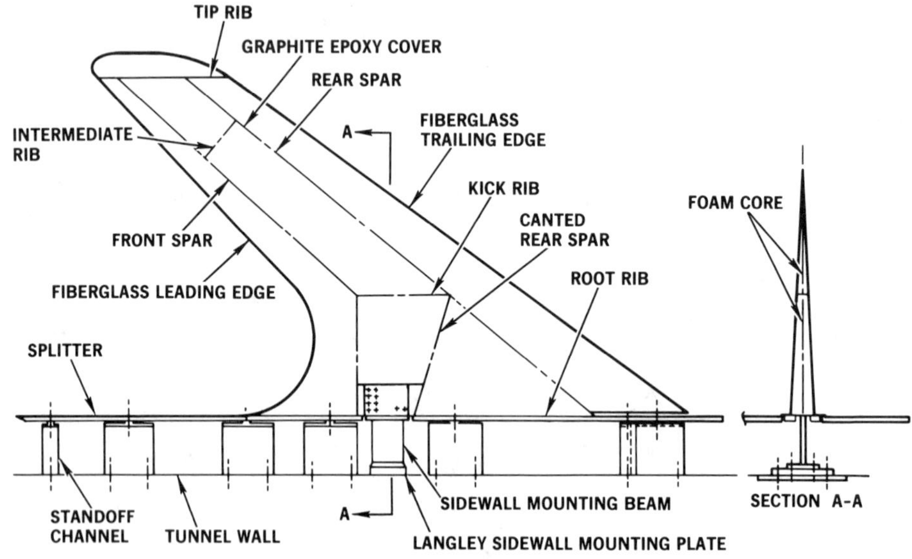

Figure 8. Subscale Verification Model Configuration

5.   SUBSCALE WING FABRICATION

Primary requirements for fabrication of the subscale wing were accuracy, because of its function as a verification wind tunnel model and low cost, to meet very limited program budget allocations. These two seemingly contradictory requirements were met using an unusual fabrication approach using a numerically controlled (NC) machine program to produce and dimensionally control both tooling and wing hardware. Through this approach, skin layup dies, substructure fabrication, and urethane foam fabrication tooling were all dimensionally coordinated through a master NC tape.

As shown in figure 9, the wing assembly was comprised of a primary structural box substructure assembly, leading and trailing edge and wingtip foam cores, metal root fittings, and graphite/epoxy and fiberglass skins. The primary box substructure assembly consisted of a contoured urethane foam core in which the major spars and ribs were embedded.

A master NC computer program was generated, defining the outer mold line surface of the wing. From this definition, a second NC program defining the inner mold line was generated by incorporating skin thickness offsets. Using these tapes, inner mold line and outer mold line lower surface female work holder/assembly jigs were numerically machined of 15 pcf polyurethane tooling foam as in-

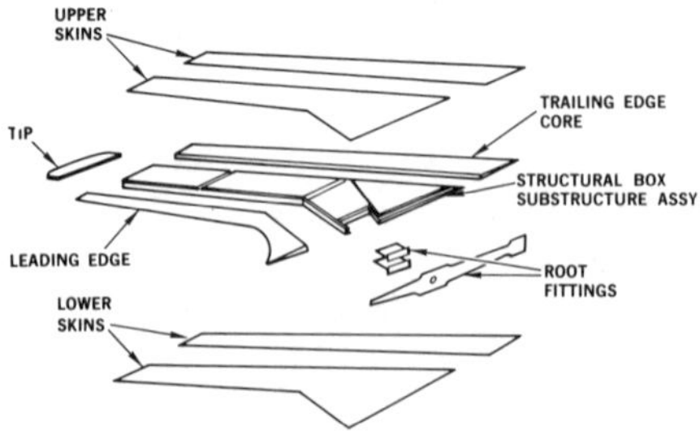

Figure 9.   Exploded View - Subscale Wing

dicated in figure 10. These tools
were used throughout the wing fabrication process as assembly alignment
tooling and as machining work supports.

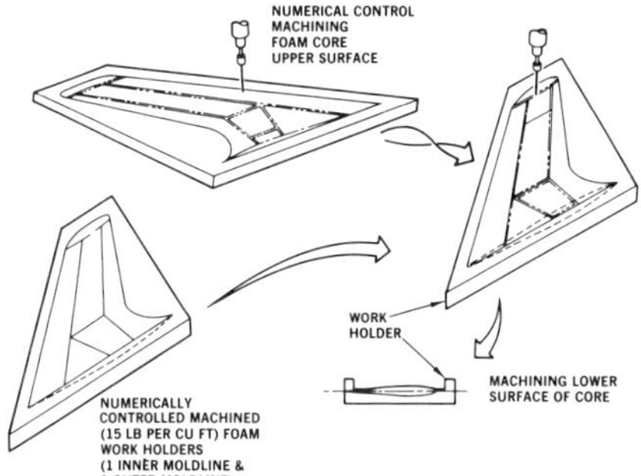

Figure 10.   Foam Tooling and
Cone Concept

A complete wing foam core was next NC
machined from a built-up foam blank
incorporating 4, 10, and 15 pcf
closed cell, rigid urethane foam as
specified on the engineering drawings. The inner mold line NC program,
with signs reversed to create male
contours, was used for core machining.

The upper surface was machined with
the blank supported on the milling
machine table, then the blank was
inverted onto the foam work holder,
indexed, and machined on the reverse
side. In addition to the inner mold
line definition, the computer program
also provided shallow recesses in
the core surface to accommodate the
flanges of the spars and ribs which
would be added later. The foremost
consideration during this operation
was accurate indexing of the work
for machining of the second surface
to assure registration of the upper
and lower contours and accurate
thickness.

In preparation for the addition of
the spar and rib structures, the
core was cut into segments along the
spar and rib heel lines with a band
saw, as shown in figure 11. The
work holder used to support the core
during sawing was also cut in this

operation. After trueing the sawed
surfaces and rounding corners, the
spar and rib structures consisting of
three plies of 0.009-inch MB0135-009
fiberglass fabric were laid up directly on the core segments and room-
temperature cured. All bonding and
laminating in conjunction with the
foam core was done at room temperature because of the temperature limitation of the urethane foam. Foam
surfaces were sealed with a light
brush coat of Epocast 2 resin prior
to layup of the fiberglass.

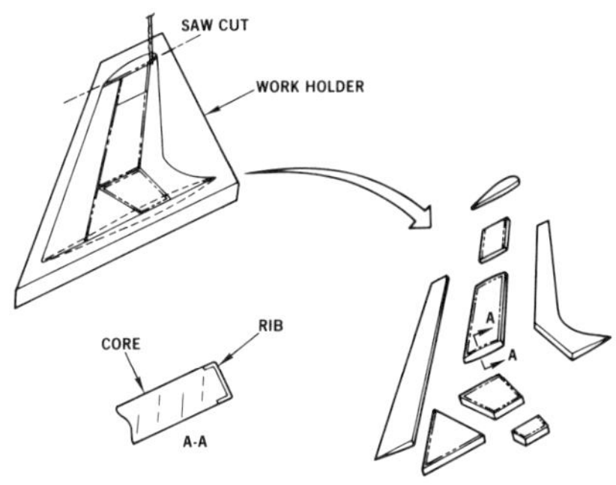

Figure 11.   Substructure Fabrication

After curing, the fiberglass elements
were hand dressed to remove excess
resin and projections. The core segments with their integral spar and
rib elements were then reunited and
bonded together with Epibond 1210
resin in the work holder which had
been reassembled on a plane table.
The result was an accurately aligned,
bonded assembly of foam core and embedded substructure elements.

The primary structural box skins of
0.0025-inch-thick T-300/934 graphite/
epoxy were laid up on aluminum female
dies machined from the master NC
tapes. The special-thickness composite tape was necessary to permit a
sufficient number of plies for representation of the full-scale design
laminate layup. The skins were
vacuum bagged and autoclave cured at
elevated temperatures. Considerable
curling of the skins was experienced
upon removal from the dies. This is
attributed to the unavoidable lack

of symmetry about the laminate center
plane due to the necessarily small
number of plies in some areas of the
layup to meet scaling requirements.
The unorthodox unbalanced ply orientation is not considered to be the
source of warpage because cooling
distortion from this source should be
in-plane and of secondary magnitude.
Successful experience with the highly
unbalanced HiMAT wing skins appears
to bear this out.

The completed primary box skins were
bonded to the structural box substructure assembly in a single operation using vacuum bagging in the
outer mold line work holder fixture
as indicated in figure 12. Bonding
was performed at room temperature
using Epibond 1210, a thixotropic
bulk adhesive.

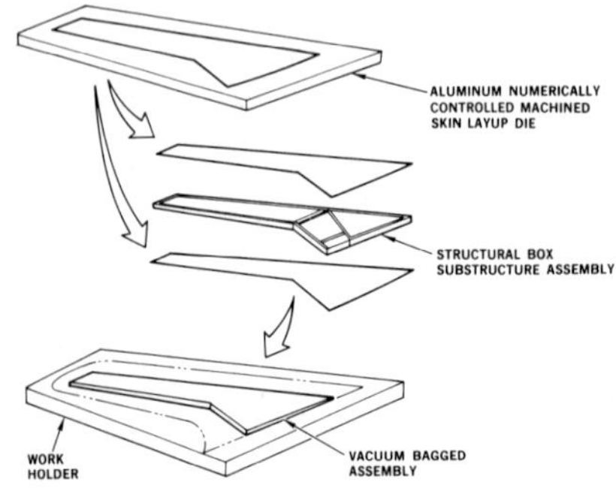

Figure 12. Structural Box
Fabrication

As shown in figure 13, final assembly
performed in the outer mold line work
holder consisted of bonding leading
and trailing edge cores to the primary box assembly and then bonding
precured flat fiberglass skins to the
trailing edge in a vacuum bag operation. Leading edge and wingtip
cores were later fiberglassed using
wet layup procedures.

The fabrication process described
worked well and produced a satisfactory test wing with good structural
integrity. Problems were encountered, however, which affected the
dimensional accuracy and dynamic mass

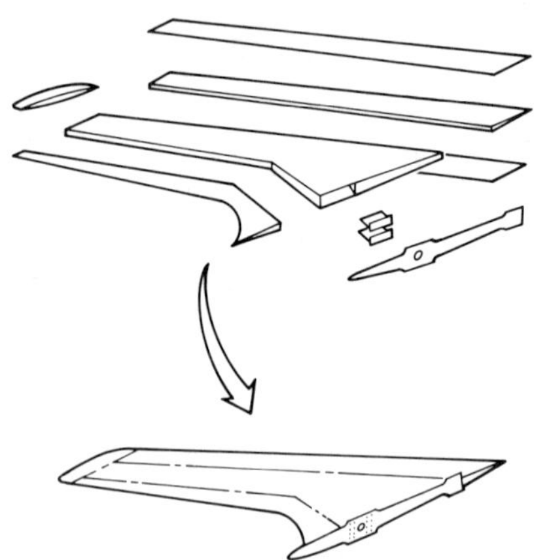

Figure 13. Final Assembly

representation of the wing to some
degree. These discrepancies did not
appreciably compromise the test results and could be avoided in future
construction. The primary problem
was that of bond-line thickness control. The overzealous application of
the bulk adhesive, with the intent to
assure a void-free bond line, resulted in a bond-line thickness ranging
from 0.020 inch to as high as 0.035
inch, instead of the specified 0.010
inch. The resulting wing exceeded
its specified thickness and weight as
the result of this problem. A similar overdimension, overweight condition and loss of some aerodynamic
smoothness occurred on the leading
edge as the result of poor control
of resin content during the wet lay-
up of the fiberglass covers. Both
of these discrepancies could be overcome by more careful control of manual procedures.

6. TESTING

6.1 Laboratory Testing

Laboratory testing was performed on
the subscale wing to determine its
characteristics before wind tunnel
testing. Structural integrity verification required static proof testing to 100 percent of design limit
load. Distributed loads were applied

incrementally up to the required
levels by means of a system of
weights and loading saddles. The
loads were carried without sign of
breakdown, and wingtip deflection
measurements showed completely linear
load/deflection behavior, indicating
the linear nature of the skin ply
design and the adequacy of the foam
core and substructure.

Vibration testing was performed to
verify predictions of vibrational
frequencies and flutter speed.
Results agreed reasonably well with
predictions, and results of all three
wings were very close, indicating
repeatability.

Structural influence coefficients
were measured at 36 points to establish the flexibility characteristics
of the wing. When these measured
flexibility data were used in a rean-
alysis of the divergence speed of the
wing, the result was within 4.4-per-
cent of the originally calculated
speed, indicating a good analytical
capability to predict wing characteristics.

6.2 Wind Tunnel Testing

Final verification of the effectiveness of the aeroelastic design and
fabrication approach used in the program was obtained in wind tunnel
tests of the 0.6-scale wing. Figure
14 shows the wing mounted in the
Langley 16-foot Transonic Dynamics
Tunnel. Because divergence is catastrophic to a wing structure and
gives very little advance notice,
divergence testing is performed at
below divergence critical speeds.
The dynamic pressure for divergence
is predicted from wind tunnel data by
extrapolation of measured wing load
trends as angle of attack and wind
tunnel dynamic pressure are varied.

Figure 14. Subscale Wing in NASA
Langley 16-Foot Transonic
Dynamics Tunnel

Figure 15 shows data derived from a
typical tunnel run in which dynamic
pressure (q) and angle of attack (α)
were varied at a constant mach
number. By extrapolation of the
slopes of these wing bending moment
curves, it is possible to predict the
q at which the slope of the curve
will be infinite; i.e., the divergence q. A somewhat unexpected
result of the testing was a slight
nonlinearity of the load curves
(figure 15). The result of extrapolation of the nonlinear curves is a
small dependence of predicted divergence q on wing loading level.

The results of the wind tunnel data
reduction are compared with the analytical predictions based on measured
wing flexibility data in figure 16.
As noted previously, nonlinearity of
the data produces slightly different
predicted divergence q's for full-
load (100 percent) and unloaded
conditions.

The close correlation of analysis and
test results obtained in these tests
is considered conclusive verification
of the principles and methods used in
the development of this aeroelastically tailored structure.

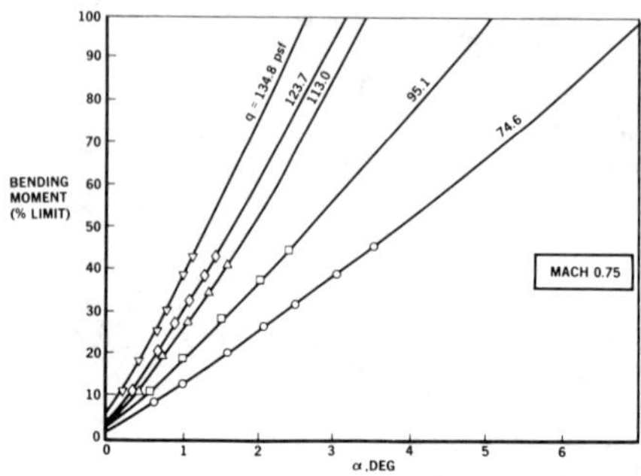

Figure 15.   Typical Divergence
Testing Wind Tunnel Data

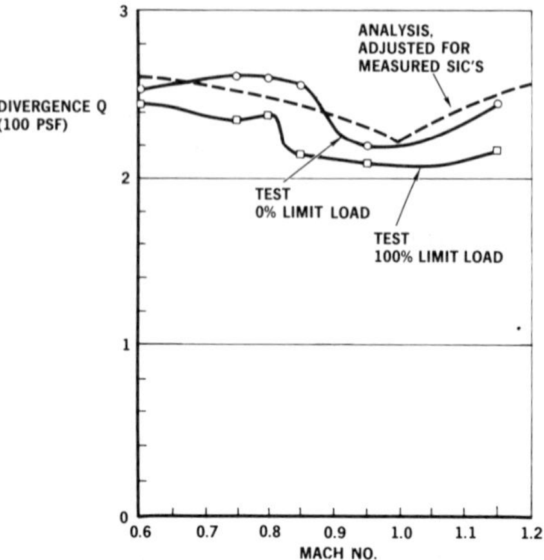

Figure 16.   Analytical Predictions
and Wind Tunnel Results

7.   AEROELASTIC TAILORING IMPACT ON COMPOSITE MATERIALS AND PROCESSES

The results of recent technology development efforts, such as the one described, have demonstrated the viability of special approaches to aeroelastic tailoring available with advanced composites. Because of the substantial demonstrated advantages of aeroelastically tailored composite structures, this new design dimension will in all likelihood accelerate the overall usage of advanced composites on future aircraft. It is becoming apparent that aeroelastic tailoring will become the accepted approach to future composite design. Because of this, composites materials and processes engineers and those who have the responsibility for manufacturing composite structures must be alert to the potential impact of this new technology.

Aeroelastic tailoring of wings and empennage surfaces may create a demand for materials with properties differing from those available today. Special modulus composite fibers to further enhance the tailored structure characteristics are likely candidates for material development. Structures such as the HiMAT wing tend to benefit from lower modulus fibers to provide structures of greater flexibility. In many applications, involving flutter, flight control, and special structures such as the FSW, higher fiber modulus will be beneficial.

In conjunction with structures designed for high flexibility, such as HiMAT, high strain resin systems would benefit structural performance because this type of structure is matrix critical, limited by the maximum strains permissible in the resin system.

When structures are designed to carefully specified elastic criteria, modulus control in the material and later in component fabrication is important. To this point, fiber modulus has been secondary to strength and has been a relatively loosely controlled property. Closer control of material modulus in manufacture, inspection, batch selection, and component fabrication will be in order for aeroelastically tailored structures.

Unfamiliar layup patterns will often be used in aeroelastically tailored structures. This need not cause manufacturing problems if basic composite design rules are followed. The success in manufacture of the highly unbalanced HiMAT wing skins (+35/±50 degrees) illustrates the fabrication feasibility of such unconventional laminates.

Hybrid laminates using materials of
different properties in a single lay-
up may also be expected to provide
improvement of some tailored structures. Thus, combinations of graphite fibers of different elastic
moduli or such diverse materials at
Kevlar, glass, and graphite might be
specified in a single laminate.

In summary, aeroelastically tailored
composite structures offer significant benefits on a variety of modern
aircraft. The methodology required
for incorporation of this new technology is rapidly being assembled
and will be ready for the next generation of aircraft. The exploitation
of the benefits of aeroelastic tailoring may induce the development of
new materials, but the basic nature
of the composite systems will remain
unaltered. The new materials and
processes required, if recognized
and considered in a timely manner,
should be readily assimilated into
the state-of-the-art.

REFERENCES

J. W. Ellis, S. K. Dobbs and G. D.
Miller, Structural Design and Wind
Tunnel Testing of a Forward Swept
Fighter Wing, AFWAL-TR-80-3073, Air
Force Wright Aeronautical Labora-
tories.

This final report of the FSW verification program will be published
in the latter part of 1981.

BIOGRAPHY

Joe W. Ellis is program manager of
the DARPA Forward Swept Wing Verification Program at the North American
Aircraft Division of Rockwell
International. Mr. Ellis has
specialized for many years in technical development and management
of advanced structures and systems
for advanced military and civil
aircraft. He was recently design
project engineer on the HiMAT
research vehicle and previously
supervisor of the Advanced Design
Group at Rockwell. There, he was
responsible for structures and
subsystems definition on advanced
projects. Mr. Ellis received a BS
and MS from the California Institute of Technology.

SPRAY METAL COMPOSITE TOOLING

Merle L. Thorpe
TAFA Metallisation Inc.
Concord, N.H.
Joseph W. Minge
Boeing Wichita
Div. of Boeing Company
Wichita, K.S.

Abstract

Large sprayed metal tooling fabrication concepts have evolved during the past three years to the point where the process is beginning to be used for large aircraft and industrial autoclave and layup tools. The technology is still on a steep learning curve; however, examples of sucessful production tools are given. The arcspray metal tooling concept is described in detail and the various components of the system including patterns, parting agents, sprayed metal shell and backup procedures are discussed. Two other applications of the technology are also mentioned: aircraft skin repair, and buildup and shaping of contoured tools formed from aluminum plate. "Keywords": sprayed metal, mold, tooling, composite, tool repair, autoclave.

1. BACKGROUND

During the past 20 years there have been many attempts to make sprayed metal tools via the flame spray process. All were relatively unsuccessful and the process was never accepted because of excessive heat buildup in the pattern, associated release agent deterioration, and mold distortion or destruction. These problems were caused by the nature of the flame spray process--its very low thermal efficiency. To spray zinc at 20 lbs/hr. via flame spray requires 17 kW with 2 kW entering the wire and the remainder blowing on the surface in the form of high temperature billowing flame. The Arcspray by comparison requires only 2 kW. Once the molten material is atomized, (see Figure 1) the only energy impinging on the surface is that which remains in the molten particles. This results in a pattern temperature never rising above 150°F. The coolness of the spray can be demonstrated by the fact that an operator may spray his hand without becoming uncomfortable.

In 1972, TAFA began to apply the Arcspray technique to sprayed metal mold manufacture. Many problems had to be solved. A miniaturized hand-held spray torch was required to make the process practical since the complex shapes and low volume nature of the tooling industry dictate that the spray operation be manual. In addition, it was found early in the development that the sprayed metal particle velocities from conventional guns were not adequate to produce dense surface coats or acceptable surface reproduction. Similarly, release-parting agents had to be located which would not destroy the surface detail of the master, resist the impact of the hot sprayed metal particles, withstand the errosive effect of the high velocity air stream, and promote adhesion of the sprayed metal. In addition, proper wire metallurgy and physical properties had to be developed which would produce the detail required and feed reliably in the spray gun. All of these problems were solved during the 1972-1975 period and many units were delivered to mold makers for use on simple low temperature and pressure molding applications. During that period, the primary market centered around the shoe industry,

since either urethanes were processed at near atmospheric pressure and 150°F or soft rubber materials were used which again have limited pressure and temperature requirements. These relatively simple applications gave needed experience and application development time to perfect the process. Since 1977 application to other molding and tooling applications have been more aggressively pursued and the current state-of-the-art can be classified as fairly well developed with thousands of production tools operating.

2. INTRODUCTION:

The moldmaking technique described here is a process to reproduce shapes very accurately. Given any model or pattern made of materials--such as metal, wood, combinations of wood, wax, cloth or plaster--a metal shell can be built up around it to desired thickness. This is an exact copy not only of every surface detail but also of precise form and dimensions. With the cool spray technique there is no problem of thermal distortion and minimal shrinkage.

Reproduction is so true that the metal shell is comparable to an electroform in its precision. However, the process is much quicker. For instance, it takes only ten minutes to spray one square foot to a thickness of one-sixteenth of an inch.

The process is not size limited and it has been used to reproduce pieces as small as a coin and as large as 120 square feet. The process is simple and straightforward and can be learned by virtually anyone after a brief hands-on training session.

Although a number of organizations have tried over the years to develop metal spray moldmaking methods, all have had some fundamental deficiencies such as model overheating, distortion, weak, soft or low-melting mold surfaces, poor release properties and inadequate backup technology. Consequently, until recently, plastics engineers have not been able to consider metal spray methods as practical alternatives to machining, cast metal, electroforms or epoxy tooling.

The techniques described here have surmounted these shortcomings through the perfection of a total moldmaking system including a unique, high velocity, electric arc metal spray-generating system, parting agents, backup materials and fabricating procedures.

The spray itself is fine and thoroughly atomized. The compressed air/arc power/metal feed ratios are engineered to produce a cool, forceful spray and avoid model heating and distortion--substrate temperatures do not exceed 150°F.

The fine atomization and force of the spray also ensure--in addition to faithful reprodution of details-- buildup of an unusually dense, hard, metal shell. The spray gun can be operated continuously so that the results produced are always consistent, controlled by a simple on-off trigger.

The process is now being used or evaluated for most tooling and plastic molding processes. For some high pressure/high temperature or abrasive plastics molding applications, part numbers are limited and the process can only be classified as a prototype system capable of 25-1,000 parts. In other cases, where less severe point pressures and temperatures are involved, medium run sizes are possible in the range of 10,000-100,000 parts. Typical performance of such tools are shown in Table I.

3. TOOL FABRICATION

One of the appealing aspects of this tool making technique is that actual surfaces do not need to be generated with the machine tool and an experienced machine tool operator is not required. The process is fast, and within three to four hours of training, an operator can be reasonably competent in producing small uncomplicated tools. Normal tool making knowledge is required in the area of tool design and in adapting the tool to the specific molding processes. The steps which must be followed in producing a sprayed metal tool are relatively simple. The basic procedure varies little from application to application. Obviously, as complexity of the tool changes and the pressures and temperatures to be sustained increase, more care must be taken relative to backup, chasing, and mounting. Basic steps then include: 1. Preparing a suitable pattern, model or part for which a faithful reproduction tool is needed. 2. Mounting the pattern on a base plate utilizing conventional mold making knowledge to achieve appropriate parting lines. 3. Cleaning the pattern surface thoroughly of all wax, oils, fingerprints, and other foreign material. (Defects on the surface will be reproduced in the final mold. Contamination in many cases will prevent the sprayed metal from bonding in its early stages of spraying). 4. Applying a release/bond agent to the surface. This is required to assure adhesion of the first coat of sprayed metal and to facilitate removal of the sprayed metal tool after completion. 5. Spraying metal until 0.060-0.090" thickness is achieved. 6. After the metal shell is complete, it is usually backed up with a high temperature, fast hardening epoxy system.

3.1 Composite Tool Fabrication

The following sections discuss each of the steps mentioned in the previous paragraph as they apply to large composite tool manufacture. Other types of molds follow similar procedures and are discussed in an SPI preprint[1].

3.1.1 Patterns and Masters

Plaster patterns which are quite commonly used for large composite work characteristically contain moisture. If the surface is not sealed properly, this moisture can form in unpredictable pockets on the surface of the master between the sprayed metal and the plaster. When this occurs, steam pressures build up during curing and this can deform the surface of the metal. Such bubbles can be less than an inch in diameter to many inches in diameter, with a thickness of a few mils to 1/8 inch. The ideal solution is to use a pattern material other than plaster or a plastic faced plaster. When this is not possible, the surface must be thoroughly sealed.

3.1.2 Release of Sprayed Shell

A special parting film (release agent) has been developed to permit separation of the sprayed metal shell from the pattern. It is specially formulated PVA to achieve good metal spray adherence to the pattern, remain stable at higher temperatures and give superior separating characteristics when applied as directed. In a true sense it acts not only as a bonding agent but as a water assisted parting film.

3.1.3 <u>Sprayed Metal</u>

Essentially any metal or alloy which can be made into wire can be sprayed by the cool spray process, including all steels, nickel, copper and aluminum. However, from a practical standpoint, the current spray metal tooling process utilizes a kirksite type metal. Materials with melting points higher than this (805°F) overheat most conventional patterns and distort, crack or warp upon application. Kirksite type metals, on the other hand, when applied by the cool spray process described here, have unique characteristics which reproduce the pattern surface exactly, have low shrinkage or warpage, maintain a pattern temperature warm to the touch, and can be applied at rates up to five square feet of pattern per hour. Most users find the sprayed metal kirksite surface surprisingly hard. A 10 RMS finish can be reproduced. If the lower melting point materials like tin/zinc are used, a softer surface is produced (melting point 350 – 400°F) which is too low for composite tool applications. Aluminum is not a good candidate for sprayed metal tooling in that it has a high shrink rate, a higher melting point and a softer surface.

Normally the metal surface of a sprayed tool weighs only 3-5 lbs/ sq. ft. Thus, the actual weight of the sprayed metal material is usually not a significant factor. The properties of the sprayed metal are given in Table II. Attempts have been made to fabricate tools of higher temperature materials like steels, stainless and nickel; however, it has been found that the metal particles are so hot when they hit the surface that they overheat the model, destroy the mold release agent, or have too much shrinkage to produce a distortion-free, crack-free surface. Some simple shapes have been fabricated with these metals and one might envision with further R & D that higher temperature tools can be made by this concept; however, as of this writing one should consider only zinc and zinc alloys as practical materials. Sprayed metal tools can be electroplated with nickel and other nickel coating techniques are being evaluated.

3.1.4 <u>Spray System</u>

The Arcspray gun is similar in size and appearance to those used for paint spraying. Wires are fed automatically. Although metal particles as they impact the surface are in a molten state, a significant characteristic is that the substrate undergoes only a small temperature rise. The system consumes 2 kW of power (100 amperes at 20 volts d.c.) and 35 CFM of 80 psi air.

Compressed air is supplied to preset automatic controls which eliminate operator judgment relative to spraying parameters. Once the electric system is energized only one trigger is pulled to start and stop the preset spray rate (wire feed). The power supply automatically maintains the proper wire intersection geometry by maintaining arc voltage constant to give a consistent, controllable spray jet with a cone diameter of approximately three inches at an eight-inch standoff.

Patience must be exercised during the spray application. The gun has the potential of spraying at rates far higher than a pattern can receive the metal. The general procedure is that of a good spray painter--flowing motions over a long area with overlapping coats building up a uniform coating over the entire surface rather than having a thick deposit in one area and a thin one in another.

3.1.5 Backup

With small tools, cast epoxy/aluminum granule backup has been used successfully; however, with large area composite tools this approach has proved unsatisfactory because of weight and the problems associated with mass casting of thick epoxy sections. To circumvent this problem glass fiber/epoxy layups on sprayed metal were tried but shrinkage of the layup distorted the tool. Conventional glass laminating pastes were tried but

these, in addition to shrinkage, had inadequate strength.

This experience pointed out the need for a new, lightweight backup material which would have adequate physicals, match expansion of the sprayed metal, and be easy to apply. Such a material, designated TAFITE 4360 (properties given in Table II), was developed and is now in use. This material, a laminating paste, has been used to back up all the tools discussed in this paper.

The laminating paste must be applied to the clean, uncontaminated spray surface, preferably not more than one day old. First, a compatible viscous liquid "surface coat" is applied to the back of the sprayed surface 30 mils thick to assure a sealed surface and promote adhesion of the backup. The laminating paste is applied after the surface coat becomes tacky, working into the surface coat using conventional grooved laminating rollers. A heat gun may be used to increase resin flow and facilitate air removal. The paste is covered (while laminating) with one layer of glass cloth (1500 open mesh) which lets the air out and holds excess resin. Once the paste is applied it is allowed to room set until hard (2-8 hours). After the initial curing of the first layer, an additional layer is applied with glass cloth to reach required tool thickness (total laminate thickness depends on tool size – for 3 ft^2-1/4 inch is adequate; over 3 ft^2-3/8 inch). An optional perforated teflon or peel ply cloth can be used while rolling the final surface if a more cosmetically appealing surface is desired.

3.1.6   Curing & Frame Attachment

The finished tool is then allowed to set overnight before it is cured (without removing from pattern) in an oven – first for two hours at 150°F then for two hours at 200°F.

After the first cure, and while the tool is still attached to the master, a support frame is attached. Two approaches have been used here. First, phenolic or fiberboard ribs have been shaped and then attached directly to the part with the same 4360 paste (using fillets) to create an egg crate support. A second approach involves the spray attachment of bolts to the sprayed metal shell before the laminate is applied. In this case a metal egg crate frame is then attached to the bolts.

Once the backing frame is attached, the frame and tool are removed from the pattern (plaster will calcine above 200°F) and then post cured as follows: two hours at 250°F, one hour at 300°F and two hours at 350°F. Note that vacuum bagging was not used during the entire fabrication procedure.

4. SPECIFIC APPLICATIONS

4.1 Tail Cone

A replacement was required for the fiberglass tool used to manufacture the tail cone of the Boeing 737. The old tool leaked and release characteristics were poor especially from the deepest part of the tool. It was decided to remake the tool with sprayed metal. After a plaster pattern was produced and sealed (Figure 2) it was sprayed with kirksite type (204) wire (Figure 3) and backed with 4360 laminating paste (Figure 4). After room curing for four hours phenolic ribs were then cut to shape, fitted to the back of the tool and attached with fillets of 4360 paste (Figure 5). The tool was then given two cure cycles as outlined previously. The completed tool on the left side of Figure 2 shows surface detail. Once the pattern was ready, two days were required to fabricate the tool. This unit had been in production use for eight months as of December 1980 and had made over 50 fiberglass parts (Figure 6). The surface finish as sprayed met requirements and surprisingly, parts continue to drop out using a release agent but without extraction force. Location holes are drilled in the part before it is removed to permit mounting on a truing fixture with similarily positioned locating pins.

4.2 Inlet Duct

A second tool was required for a complicated inlet duct lip to be used on the Boeing 767. This tool was originally fabricated using epoxy and glass cloth but a new tool was required because of the high (25 percent) rejection rate on parts produced by this tool. Figure 7 is the plastic faced plaster mandrel, Figures 8 and 9 are exterior and interior view of the finished sprayed metal tool, and Figure 10 a finished Kevlar graphite part ready for trimming. This tool was sprayed in two hours and backed up in two more. Due to the configuration of the tool no ribs were needed. Part rejection rate with this unit was reduced to zero.

Because of the cool nature of this spray process (minimal substrate heating) it has also been used to repair, rebuild and fill in metal surfaces. Two aircraft related applications include the cosmetic repair of aircraft skins and the buildup to dimension of tools formed from aluminum plate.

4.3 Skin Repair

In the case of aircraft skin repair the process is limited to scratches and gouges. It involves masking undamaged adjacent areas and contouring and roughening the defect. The prepared surface is then arcsprayed with excess aluminum, ground and polished. The portability of the apparatus permits the repair to be done on the flight line. Figure 11 shows a repair in progress just aft of the main entry door on the 737. This process is now used as a standard cosmetic repair technique on selected applications at Boeing. Plasma spraying has been evaluated for this application but the excessive heat and reduced coating density were found detrimental. Bond strengths in the range of 4200 psi are obtained with arcspray. A variation of this concept is also being used to repair damaged tooling surfaces as well.

4.4 Tool Repair

Another application involved the buildup of certain sections of a engine cowling metal bonding fixture used to form aluminum skin. The main portion of the die is made from aluminum plate which was formed to shape and bolted to backup ribs for rigidity. The ribs were attached by countersinking bolts through the forming surface; the countersink areas were then filled with epoxy. These epoxy patches caused markoffs on the formed skins. The problem was solved by spraying aluminum into the countersink holes and finishing conventionally-this eliminated the markoffs. Similarly the opposite half of the tool, (Figure 12) was built up in low areas after initial forming was completed to achieve desired contour. This figure shows the tool marked for build up in one area and sprayed in another. Thicknesses is excessive of 1/8 inch have been applied to large

large surfaces after gritblasting and arcspraying a special 5 mil nickel aluminum bond coat. No distortion of the tool occurred and skin quality, parts were produced using .062 inch aluminum.

5. TIME & COST SAVINGS

Significant time and cost savings have been demonstrated using this process. Comparisons with other moldmaking processes are shown in Figures 13 and 14. On large parts the shortage of large electroform tanks and related technology and the ability to fabricate tools in-house are added benefits.

6. CONCLUSIONS

It appears that sprayed metal composite tooling when produced in a methodical controlled fashion can be made to withstand the rigors of the autoclave tooling process. The current learning curve in this technology is steep. Much still has to be learned and new shapes sometimes require experimentation before fabricating can proceed. It is not the intent of this paper to leave the reader with the impression that all problems have been solved. However, results to date has been quite promising and exciting.

7. REFERENCES

[1] Merle L. Thorpe: Progress Report: Sprayed Metal Faced Plastic Tooling (Preprint) Paper No. 24-A, Presented, 35th Annual Conference Reinforced Plastics/Composites, Feb. 4-8, 1980, New Orleans, LA.

8. BIOGRAPHIES

Joseph W. Minge is a Manufacturing Development Engineer in the Welding and Brazing Group of the Airframe Manufacturing Technology Section of Boeing Military Airplane Company. He has been with The Boeing Company thirty-four years-- nineteen of which have been in the Engineering Materials & Processes Group involved in metal spraying. During the past two years he has been instrumental in developing a method to repair scratches in clad aluminum aircraft skins using metal spray techniques and involved in developing economical high temperature autoclave tooling processes.

Merle L. Thorpe has been an innovator in high temperature and thermal spray technology for the past 25 years. Since 1970, he has been active in the sprayed metal tooling field. He holds over 30 patents, is a member of the American Welding Society, Thermal Spray Committee; he also is a member of numerous technical societies including the SPE. Mr. Thorpe holds an A.B., M.S. in Engineering Science from Dartmouth College and has been a Professor and Assistant Dean at the Thayer School, Dartmouth.

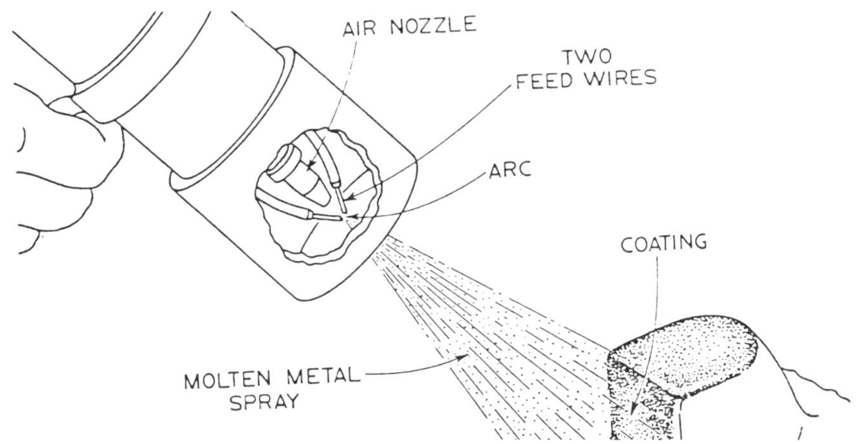

Figure 1
How Arcspray Works

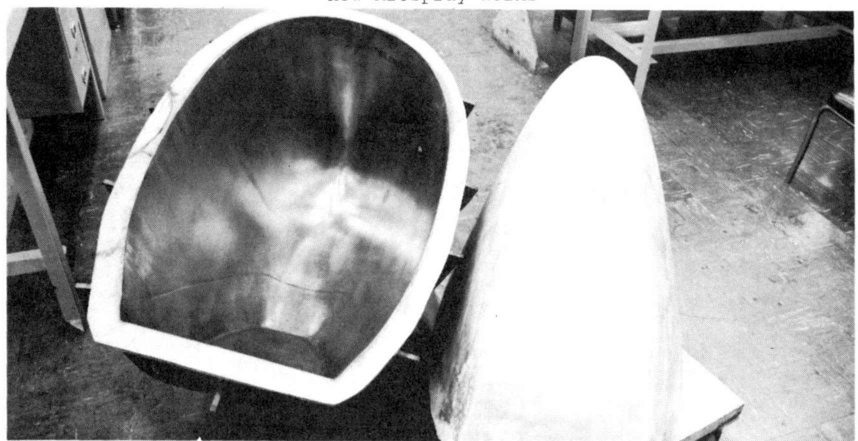

Figure 2
Finished Tool & Plaster Pattern

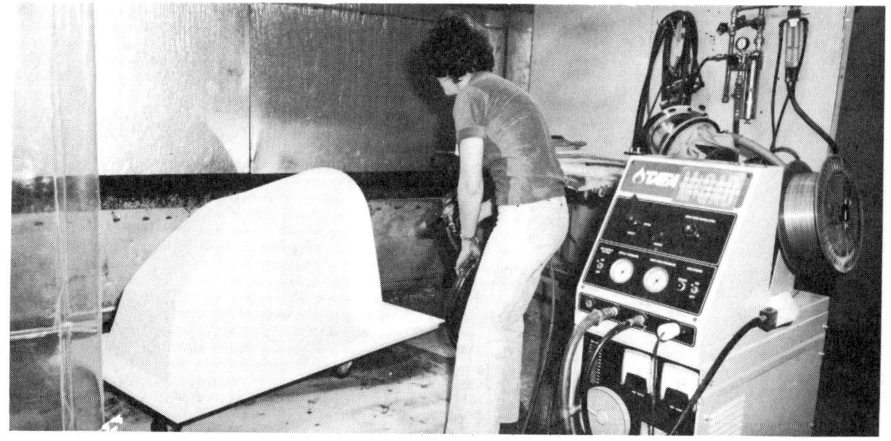

Figure 3
Pattern Mounted on Plate and Sprayed with .060" Metal

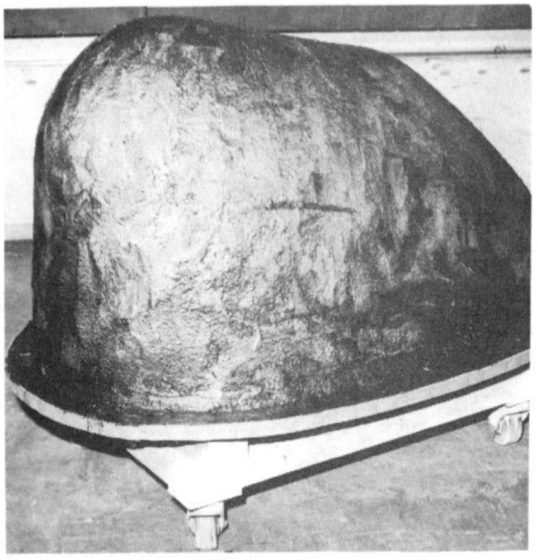

Figure 4
Tool Backed with Laminating Paste

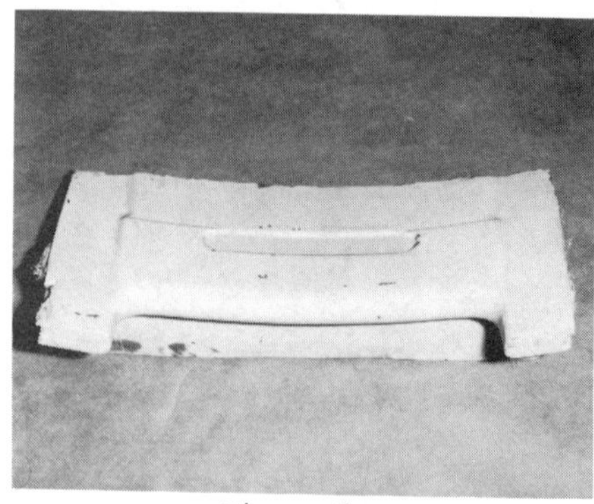

Figure 7
Plastic Faced Plaster
Mandrel 767 Part

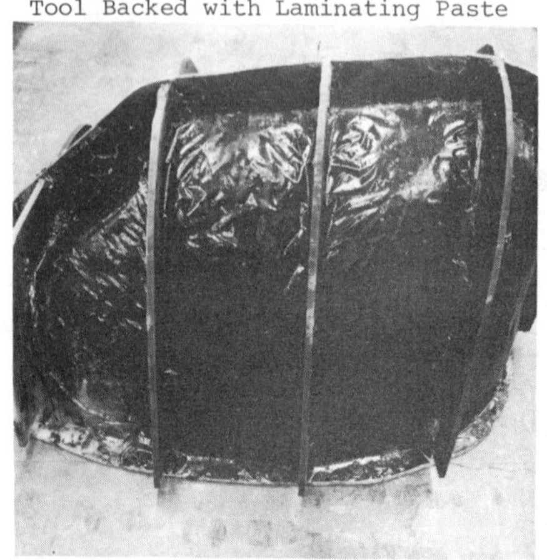

Figure 5
Phenolic Ribs
Attached to Back of Tool

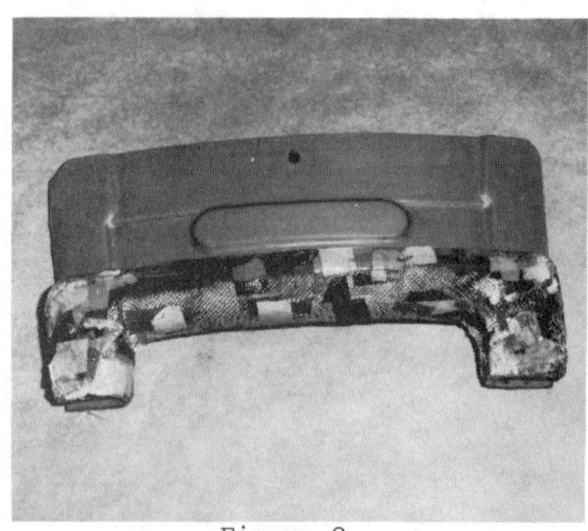

Figure 8
Exterior View
Completed 767 Tool

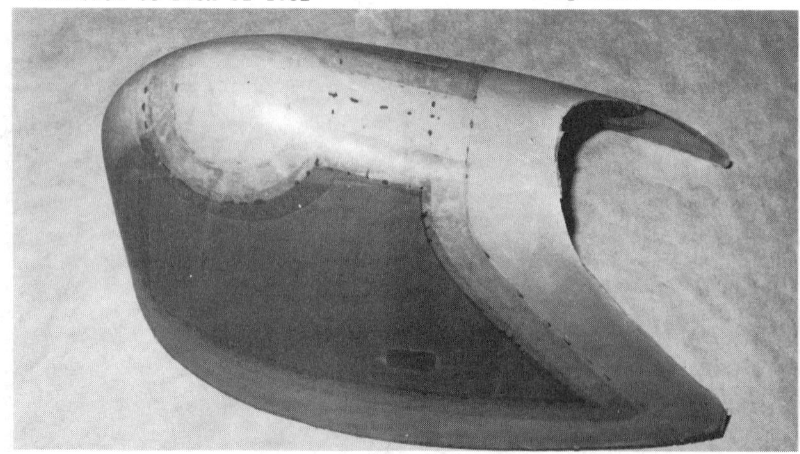

Figure 6
Finished Fiberglass Part

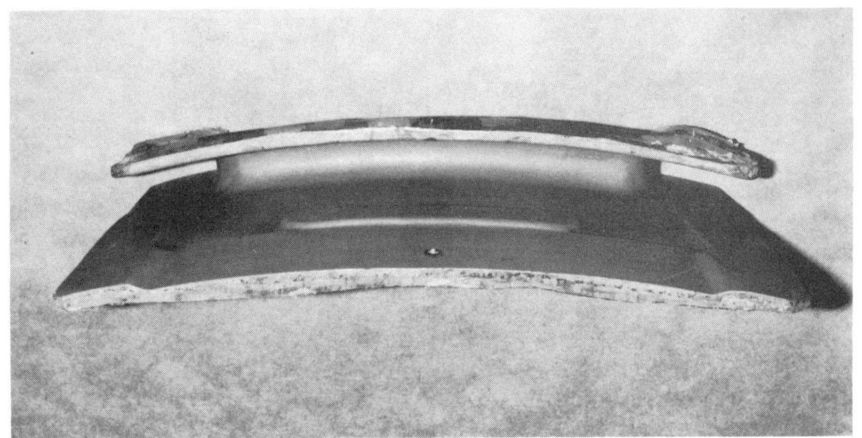

Figure 9
Interior View Completed 767 Tool

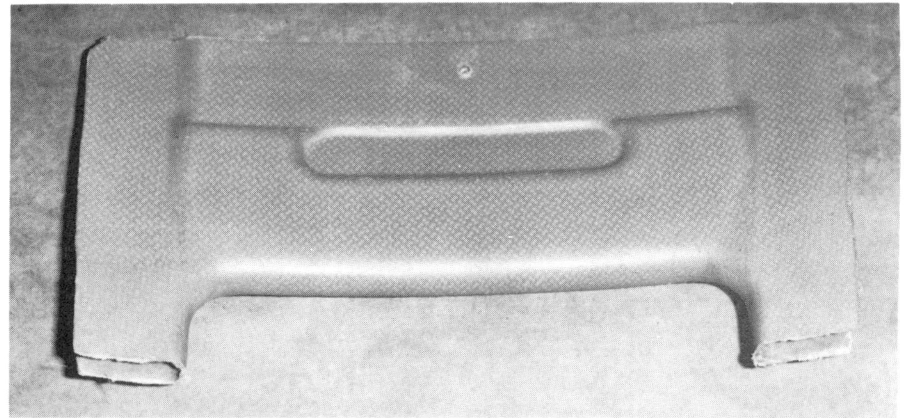

Figure 10
Finished Kevlar Graphite 767 Part Ready for Trimming

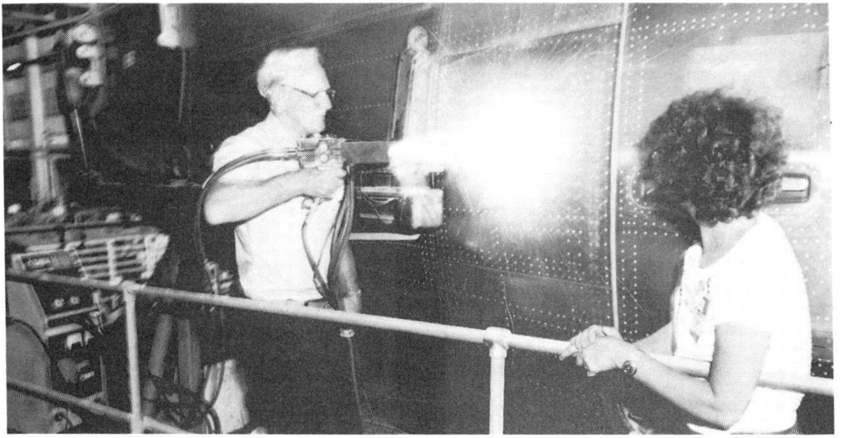

Figure 11
Aircraft Skin Being Repaired by Arcsprayed Aluminum

Figure 12
Aluminum Bonding Tool Showing One Area Built Up,
Second Area Marked for Buildup

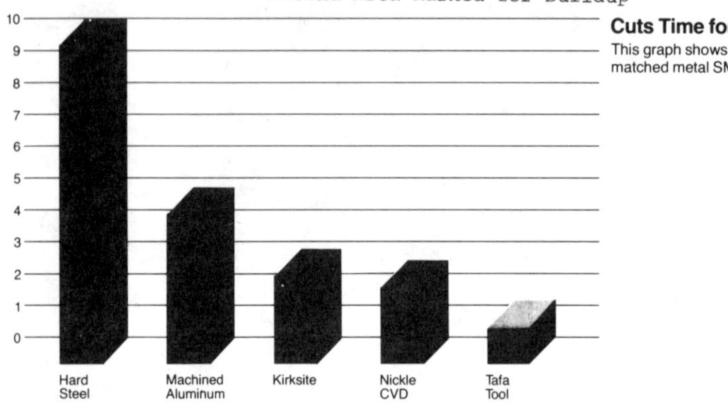

Figure 13
Time Savings - Sprayed Metal Tooling

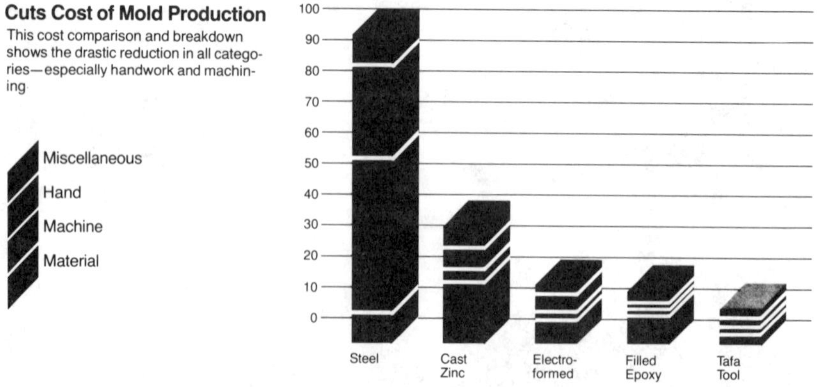

Figure 14

Cost Savings - Sprayed Metal Tooling

Table I
Molding Experience

Process	Pressure psi	Parts
SMC	1-250	25-200
Structural Foam	200-500	4000+
RIM	10-100	5000+
Roto	0	8 mos.[a]
Blow	5-60	6 mos.[a]
Vacuum Form ABS	0	18 mos.[a]
Composite Laminates	0	8 mos.[a]
Cast Urethane	0	5,000-40,000
Injection	15,000+	50-1000

a- Negligible wear

Table II
Physical Properties, Laminating Paste
and Sprayed Metal

	4360 Laminating Paste	204 Kirksite Type Sprayed Metal
Mixing Ratio, wt	100A/25B	none
Pot Life	6-10 hours	infinite
Specific Gravity	1.20	6.36
Volumetric Weight	0.043 lbs/in^3	0.230 lbs/in^3
Tool Weight	3.9 $lbs/ft^2/1/2''$	2.3 $lbs/ft^2/.070''$
Hardness	85 Shore D	52 Rh
Shrinkage	<0.001 in/in.	0.001 in/in
Heat Deflection 264 psi	385°F	700°F
Melting Point	–	805°F
Ultimate Tensile Strength, psi	70°F 8500 300°F 5400	18,500 18,500
Ultimate Compression Strength, psi	70°F 23,400 300°F 18,400	28,400 28,400
Ultimate Flexural Strength, psi	70°F 17,500 300°F 13,000	– –
Flexural Modulus	70°F $.95 \times 10^6$ 300°F $.58 \times 10^6$	– –

MODEL 206L COMPOSITE VERTICAL FIN

James Howard Harse
Research Project Engineer
Bell Helicopter Textron
Fort Worth, Texas

Abstract

This paper describes a program to design and manufacture a composite vertical fin for the Model 206L helicopter. The composite vertical fin is of sandwich construction with precured facesheets of graphite/epoxy unidirectional tape adhesively bonded to Fibertruss core of varying density. A layer of aluminum wire screen is adhesively bonded to the outer surface of each facesheet for lightning protection. A separate precured leading edge of Kevlar 49 is bonded to the main fin structure. A filament-wound, tapered S-glass tail skid is mounted by two injection-molded fittings bonded into a graphite/epoxy closure at the base of the fin. A two-part closed cavity tool is used with each facesheet being laid up and cured on its half of the tool. The leading edge and lower closure are cured on separate tools. The entire assembly is bonded together in the cavity tool. This paper discusses in detail the design requirements, manufacturing process development, tooling approach, structural tests and follow-on production for the NASA/Army/BHT Flight Service Evaluation Program.

INTRODUCTION

In 1977, Bell Helicopter Textron (BHT) began a research project to design and fabricate a composite vertical fin that would be structurally equivalent to the existing metal vertical fin. The composite vertical fin was designed to meet the strength and stiffness requirements specified in the design criteria for the original metal vertical fin. Other considerations were lightning protection, fatigue strength, environmental durability, and ease of manufacturing. The objectives of this project were to develop a composite vertical fin as a potential replacement for the metal vertical fin that would have lower weight, lower cost, and the same aerodynamic contour and interface with the tailboom.

The composite vertical fin program evolved from an earlier BHT research program that investigated structures made from advanced composite materials for the Model 206L helicopter. The vertical fin was one of the components selected for further development because it would demonstrate a variety of materials, construction techniques, and tooling concepts. The other components selected for further development were the baggage door, litter door, forward cowling, and horizontal stabilizer. These components were to be used in the NASA/Army/BHT Flight Service Evaluation Program (FSE). The FSE program involved producing 45 shipsets of the selected composite components for field service experience on BHT customers' Model 206L helicopters. However, in early 1978, before the contract had gotten underway, NASA was directed to prohibit the use of graphite fibers in new flight programs until the potential electrical hazards of graphite were assessed. This resulted in deletion of the vertical fin and the horizontal stabilizer from the FSE program because they were to be made with graphite. BHT decided to continue the development of the composite vertical fin through FAA Certification. In early 1980, the Government concluded its assessment of graphite and agreed to produce 45 vertical fins and place them into service along with the other components under the FSE program.

Some of the basic design considerations in developing the composite vertical fin will be discussed followed by the material selection rationale, manufacturing process, and tooling approach. The FAA Certification test results are presented along with a detailed weight comparison of the composite fin relative to the metal fin.

BASELINE METAL VERTICAL FIN

The vertical fin for the Model 206L helicopter, shown in Figure 1, is 78 inches high. The leading edge sweep of the lower 33 inches is 27 degrees, and the sweep of the upper 45 inches is 32 degrees. The chord at the apex is 21 inches and tapers to 16 inches at the lower end and 13 inches at the upper end. The vertical fin is attached to the right side of the tailboom by four potted inserts. A former with seven nutplates is riveted and bonded to the left side of the fin (Figure 2). A fairing that shields the fin-to-tailboom attachment area is bolted to the nutplates.

Figure 1. Model 206L Helicopter

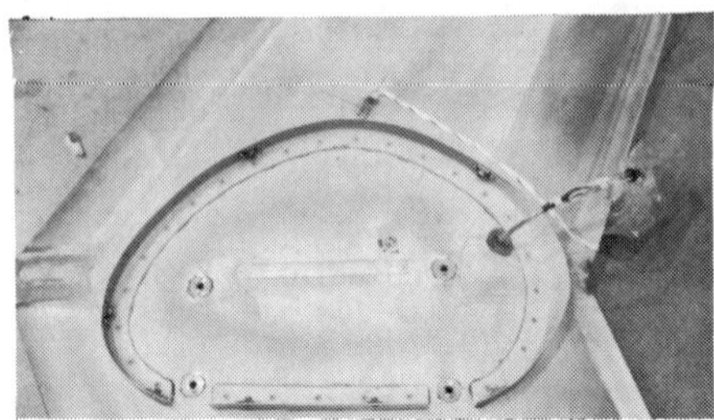

Figure 2. Former and Attachment
Area of Baseline Fin

The production vertical fin is a bonded and riveted sandwich construction assembly with 1.25-inch thick aluminum honeycomb core of varying density. The facesheets are made of 2024-T3 aluminum alloy sheet metal that is chemically milled to taper in thickness from 0.040 inch on the right side and 0.063 inch on the left side at the attachment area to 0.015 inch at the fin tips. The leading and trailing edges are 0.020 and 0.016-inch aluminum sheet, respectively. They are riveted into position and bonded to the basic sandwich portion of the fin. A steel tubular tail skid is cantilevered from a fitting bonded to the lower fin tip (Figure 3). A fairing that houses the navigation beacon is riveted and bonded to the upper fin tip.

Figure 3. Steel Tail Skid Assembly

DESIGN REQUIREMENTS

The geometric requirements of the composite vertical fin were to maintain the same planform area, aerodynamic contour, leading edge sweep, attachment to the tailboom, and attachment to the fairing. The static strength requirements are an aerodynamic side loading based on rolling-pullout maneuvers, and a tail skid loading based on tail-down landings. The calculated aerodynamic side loadings are uniform 0.50 psig limit load and 0.75 psig ultimate load. The surface area of the fin is 1387 square inches resulting in a total limit load of 693 pounds and a total ultimate load of 1040 pounds. The calculated tail skid vertical loads are 273 pounds at the end of the tail skid for the limit loading (based on a drop of 4.0 feet per second) and 315 pounds at the end of the tail skid for the reserve energy loading (based on a drop of 4.9 feet per second). In addition, a side load equal to one-half of the vertical load is applied simultaneously at the end of the tail skid for the limit and reserve energy conditions.

The composite vertical fin must withstand the oscillatory loads induced from the main rotor plus have sufficient stiffness to assure no significant dynamic magnification of the oscillatory loads or induce any vibration problems.

The remaining consideration is lightning protection. Since graphite/epoxy was selected for the facesheets of the composite vertical fin, a means of protecting the graphite and dissipating P-static electrical charges from lightning strike was required.

MATERIAL SELECTION

In order to meet the geometric design requirements, trade-offs were made on the construction of the composite vertical fin. Sandwich construction with 1.25-inch honeycomb core was selected to assure the same aerodynamic contour as the metal vertical fin. The subsequent material selections were made based on this design concept.

Facesheets

Since one of the main objectives was to reduce weight while maintaining high stiffness, graphite/epoxy was selected for the facesheets. The requirement for the facesheet laminate was to have high spanwise and torsional stiffness along the swept axes with high stiffness and strength in the attachment area. Graphite fabric was ruled out because once adequate spanwise and torsional stiffness was achieved, there was excessive chordwise stiffness, overall thickness, and weight. Unidirectional graphite tape of 0°, +45°, -45°, and 0° orientation along the upper and lower span with overlap in the attachment area was selected. The

chordwise components of the +45° and -45° plies provide sufficient chordwise stiffness without any 90° plies. Graphite prepreg unidirectional tape 0.004-inch thick is used because it produces the lightest possible laminate and a desirable thickness distribution. The resulting facesheet laminate is stepped in thickness from 0.016 inch at the fin tips (4 plies of 0°, +45°, -45°, and 0°) to 0.064 inch in the attachment area (4 sets of plies of 0°, +45°, -45°, and 0°). The set of plies that form the upper and lower fin tips also form the trailing edge. A layer of 200 x 200 aluminum wire screen is bonded with FM1000 film adhesive to the outer surface of each facesheet for lightning protection. The FM1000 insulates the aluminum from the graphite to prevent galvanic corrosion. The wire screen also provides small-particle impingement protection for the graphite.

Leading Edge

Kevlar fabric was chosen for the leading edge because of its low weight and its small-particle impingement resistance. Two plies of 0.010-inch thick fabric orientated at ±45° form the upper and lower leading edges. The leading edges overlap at the apex.

Core

The requirements for the core are to redistribute loads from the tail skid into the fin and aerodynamic loads from the fin into the tail-boom. Nomex core was eliminated from consideration because of its low shear modulus (7800 - 16,000 psi in the W and L directions respectively) and static shear strength (190 - 335 psi) for 8 pounds-per-cubic-foot density core. Aluminum core of 8.1 pounds per cubic foot has a shear modulus of 54,000 - 135,000 psi and a static shear strength of 400 - 670 psi; however, the aluminum would be subject to galvanic corrosion when in contact with the graphite facesheets. This would require sealing the aluminum core from the graphite, which was undesirable. Fibertruss core, made by Hexcell, was selected because for 8 pounds-per-cubic-foot density nonmetallic core, it offers the highest shear modulus (24,000 - 48,000 psi) and highest static shear strength (300 - 490 psi). Also, Fibertruss core will not corrode, has low moisture pickup, good elevated temperature properties to 350°F, and is available in the desired densities applicable for the fin. The Fibertruss is made from ±45° bias weave fiberglass that gives it the high shear modulus and strength. The high density (8 pounds per cubic foot) Fibertruss core is used in the attachment area and at the lower portion of the fin where the tail skid attaches. Low density (3 pounds per cubic foot) Fibertruss core is used for the remainder of the sandwich structure.

The leading and trailing edge areas are devoid of honeycomb core.

During the design support testing for the composite vertical fin, the static strength of the core proved adequate. However, the high density Fibertruss core failed near the potted inserts in shear during fatigue testing of the attachment area. The calculated oscillatory shear was 87 psi and failure occured at 890,000 cycles. Since this core offers the highest shear strength of a nonmetallic core, and aluminum core had been ruled out, graphite doublers were incorporated between the facesheets and the core in the area of the inserts on both sides of the fin. The doublers are made of graphite tape oriented at 0°, +45°, -45°, and 90° and are 0.13 inch thick. They reduced the oscillatory shear in the core to 40 psi. This design was verified by 2-bolt specimen tests (Figure 4). The graphite doublers are 2 inches by 6.5 inches and tie the upper inserts with the lower inserts (Figure 5). Steel inserts are used instead of the standard aluminum inserts to reduce the effects of galvanic corrosion.

Tail Skid

The design of the tail skid housing was changed from the cantilevered fitting to a lower closure with two bearings that position the tail skid. The lower closure is made

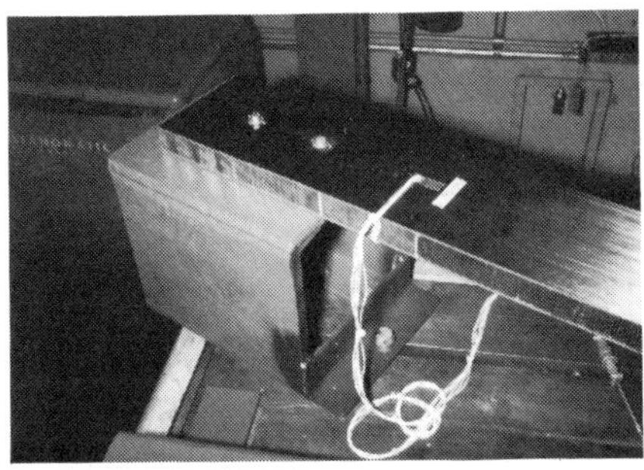

Figure 4. Two-Bolt Insert Fatigue
Test Specimen

Figure 5. Graphite Doublers
Positioned in Core

with the same graphite tape as the
facesheets with plies oriented at
+45°, -45°, and 90°. The outer
surface has a ply of fiberglass
cloth to provide a better finish for
assembly bonding and painting.
Injection-molded bearings are bonded
into the closure during its cure
cycle. The tail skid is a filament-
wound, tapered, S-glass tube with
roving oriented at ±15°. A short
length of steel tubing is bonded at
the end of the tail skid for abrasion protection. A lower closure
with the tail skid is shown in
Figure 6.

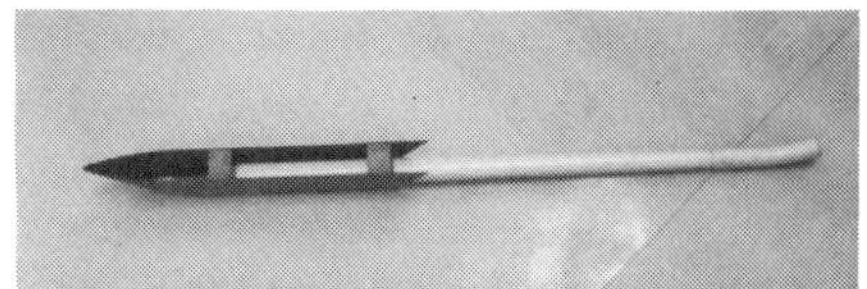

Figure 6. Lower Closure with
Glass-Wound Tail Skid

MANUFACTURING

<u>Tooling</u>

The major tool for this program is
the main assembly cavity mold. The
tool was designed to serve in two
capacities: fabrication of precured
facesheets, and to position and
secure all details during the final
assembly bonding operation.

Several alternative materials were
evaluated for the main cavity mold.
The logical choice for thermal
expansion compatibility appeared to
be graphite epoxy. However, this
approach would require the development of patterns and increase
tooling costs. It was also considered that in a production environment the fins should be assembled in
a bond press. Given these manufacturing parameters, the use of steel
became more appropriate because of
its durability.

Lines for the tapered tool surfaces
to mold the facesheet exteriors were
created by splining dimensional
elements at the edge and height of
each ply drop-off. These values
were incorporated into a program for

sculpturing the molds with numerically controlled equipment (Figure
7).

Figure 7. Sculpturing of
Steel Mold

Closed cavity or match mold assembly
tools have been proven effective for
this type composite structure,
particularly when Nomex-type honeycomb core is used. The closed mold
represents dimensional surfaces and
volume that must be totally occupied
by the component to be bonded if the
finished part is to be void free.
Compensation for the thickness
variations due to dimensional tolerances was planned to be accomplished
by producing the honeycomb core
slightly oversize. With bonding
temperature and pressure, Nomex
exhibits some thermoforming characteristics that enable it to conform to a shorter dimension in the
cell direction.

Fibertruss core has negligible heat
forming capability and presented
problems in the assembly bonding.
In order to restore a compensatory

element in the dimensional stack up,
a thin silicone rubber liner was
added to one face of the assembly
cavity mold and acceptable bonding
was achieved.

A two-piece mold consisting of a
machined steel lay up mandrel and a
fiberglass epoxy overpress (Figure
8) is used for curing the Kevlar
leading edge.

Figure 8. Tool and Overpress for
Leading Edge

Mold tooling similar to the leading
edge was fabricated for the lower
closure (Figure 9). Graphite is
used for the overpress to avoid
thermal lock up of the nested contour after cure. The tool also
positions the injection-molded
bearings.

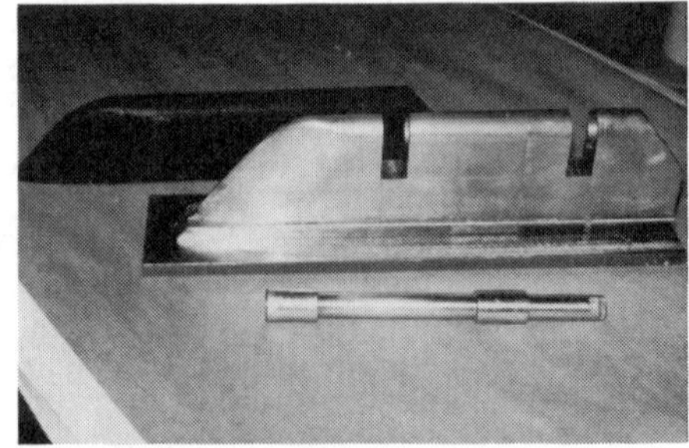

Figure 9. Tool and Overpress
for Lower Closure

For the tapered tail skid, an internal cavity tool is used to control the contour of the outside surface. The tool was made from two aluminum halves with a cavity cast from a high temperature tooling resin, Furane 31-D.

Fabrication

Design features of the fin were finalized with careful consideration for manufacturability.

Since the graphite facesheets represent the major structural elements, maximum laminate properties were essential. Cocuring, as a manufacturing method for the facesheets, was considered unsuitable due to the laminating pressure limitations imposed by the honeycomb core. The decision was made to produce the facesheets as details on each half of the tool precured at 80 psi.

The facesheets are ply oriented and laid up by hand. The first ply placed in the mold is the FM1000 film adhesive followed by the aluminum wire screen. The remaining graphite plies are trimmed using mylar pattern templates overlaying a 48-inch wide roll of the graphite tape (Figure 10). Once all of the graphite plies are layed into position, the facesheets are vacuum bagged to the steel mold and autoclave cured at 350°F and 80 psi.

Metal templates are used to mark the periphery trim of each core detail.

Figure 10. Mylar Templates Overlayed on Graphite Tape

In order to avoid the cost of sculpturing, the core was designed with flat and parallel sides. This arrangement allowed the tapered side of the facesheets to be manufactured as external surfaces.

The V-shaped leading edge is constructed from two plies of Kevlar/epoxy fabric. The two plies are laid up on a male punch type tool, then an overpress is positioned over the layup, vacuum bagged, and autoclave cured at 270°F.

The lower closure is layed up with the injection molded bearings in place. FM1000 film adhesive is placed around the bearings, then the seven plies of graphite tape followed by the fiberglass cloth are layed up. The lower closure is then vacuumed bagged to the tool and autoclave cured at 80 psi and 350°F. The tail skid is filament wound by machine on a tapered steel mandrel (Figure 11). A series of seven layers of roving oriented at ±15° are used to make the tail skid. Three of the layers are shortened and staggered to maintain a uniform wall thickness at the tapered end of the tail skid. After the winding is complete, it is placed in a cavity mold with an internal pressure bladder to ensure proper debulking and cured at 250°F.

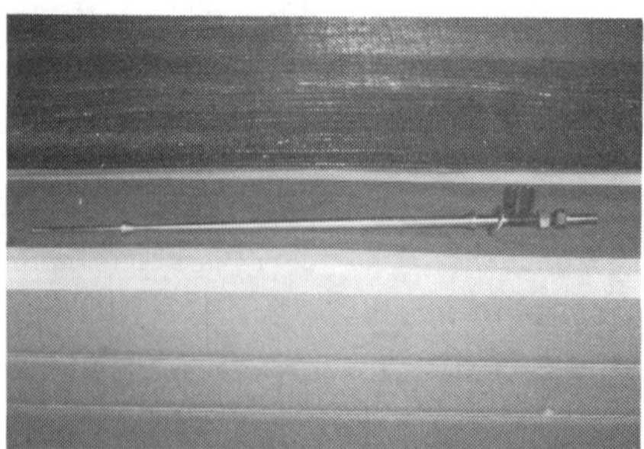

Figure 11. Steel Mandrel for Glass-Wound Tail Skid

A facesheet with adhesive film (FM-53) applied is placed in the outboard half of the cavity mold to begin final assembly (Figure 12). The individual core details are placed in position with an expandable film adhesive (AF 3020) interposed between each core splice. The remaining facesheet, leading edge, and lower closure assembly with adhesive applied are added to layup (Figure 13). The bond mold is then closed, vacuum bagged, and placed in the autoclave for 90 minutes at 30 psi pressure and 270°F.

When cure and bond quality inspection operations are complete, the assembly (Figure 14) is trimmed and the four steel inserts are installed. The final operation consists of hardware installation such as the

former with the nutplates, tail skid, and fairing with the navigation beacon. The fin is then ready for fill and fair, painting, and installation.

Figure 12. Facesheet with Adhesive Film in Mold

Figure 13. Core and Leading Edge Position in Mold

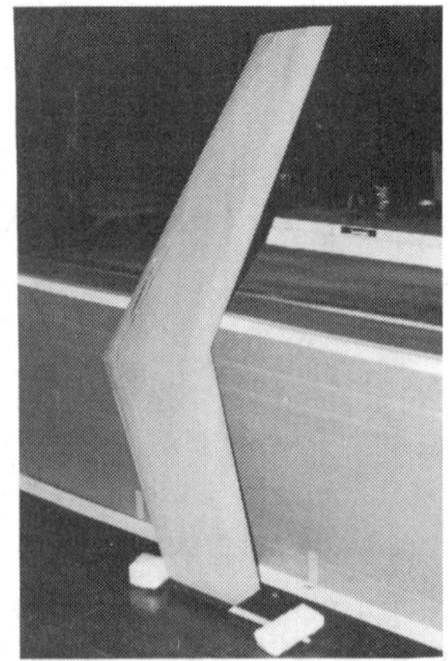

Figure 14. Composite Vertical Fin After Assembly Bonding

The FAA Certification portion of this program involved demonstrating the structural quality of the composite vertical fin by both static and fatigue testing. The design requirements, mentioned earlier, were met with the effects of moisture and temperature taken into account by either environmentally soaking the test specimens prior to testing or elevating the applied loads by an amount determined from coupon tests on the materials that are used.

Static Tests

The static testing was accomplished with two fins that had passed FAA conformity inspection. One fin was subjected to the aerodynamic side loading and the other was tested for the tail skid loading condition. The effects of moisture and temperature were addressed by elevating the applied loads by the highest "knock-down-factor" that was determined during the graphite and adhesive qualification programs (40 percent). This dictated that each fin had to withstand 140 percent of the limit loads without permanent deformation that would adversely affect further use, and each fin had to withstand 140 percent of the ultimate loads (or reserve energy for the tail skid loading) without failure.

Figure 15 shows the fin that was tested' for the aerodynamic side loading after failure. Failure occurred with 2025 pounds, or 195 percent of the design ultimate load, uniformly distributed over the fin. The facesheet failed in compression in the area where the graphite plies begin to drop off.

Figure 15. Composite Fin After Failure with the Aerodynamic Side Loading

Figure 16 shows the fin that was tested for the tail skid loading after failure. Failure occurred with 927 pounds, or 263 percent of the reserve energy load, applied to the end of the tail skid at 26.5 degrees to provide a side load equal to one-half the vertical bond. The facesheet failed by buckling in the area of the core splice between the 8 and 3 pounds-per-cubic-foot Fibertruss core.

Fatigue Tests

Fatigue tests of structural elements were conducted to determine the structural integrity of the fin-to-tailboom attachment area. The structural elements have four potted inserts in the same orientation relative to the load application as the fin (Figure 17). The facesheet laminate and doublers, as well as the core densities, ribbon directions, and splices are duplicated in the elements. Four structural elements were tested at 125 percent of maximum measured steady and oscillatory level flight loads. Each element was environmentally preconditioned at 115 to 125°F and 90 to 100 percent relative humidity for 42 days prior to testing.

Figure 16. Composite Fin After Failure With the Tail Skid Loading

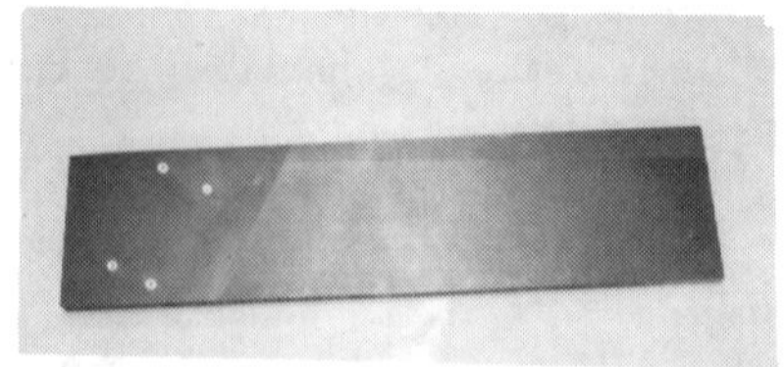

Figure 17. Structural Element for Fatigue Testing

All of the elements accumulated in excess of 10 million cycles with no failures. After the 10 million cycles were applied to the fourth specimen, the loads were evaluated to 185 percent of the maximum measured flight loads and additional 34 million cycles were applied without failure.

In order to assure that there will be no significant dynamic magnification of the oscillatory loads or any vibration problems, a conformed composite vertical fin was installed on a Model 206L helicopter and shake tested to determine its natural frequencies and their separation from main rotor two-per-rev. These results showed no significant changes for the in-flight response relative to the metal fin.

Lightning Protection

The worst possible lightning strike that the composite vertical fin must survive is 200,000 amperes. The layer of 200 x 200 aluminum wire screen was selected for dissipating the p-static electrical charges into the tailboom through the steel inserts. Studies by BHT and Lightning Transients Research Institute (LTRI) have concluded that should the composite vertical fin become the attachment point of a lightning strike, the wire screen will cause the current path to flash over the outer surface and attach to the metal fuselage. With a lightning attachment at another part of the airframe, the current will sweep across the airframe and exit at one or more extremities, usually as streamers, and the current sweeping to the composite vertical fin will flash over the fin exiting at the extremities.

The aftereffects of a lightning strike could result in some minor degree of composite fin surface damage, which will depend on the amperage of the strike and whether the attachment point is the fin.

It was concluded that the composite fin should provide lightning protection equivalent to the metal fin.

All of the test results were submitted to the FAA and certification was granted in December, 1980.

WEIGHT

One of the main objectives of the composite vertical fin program was to reduce the weight significantly over that of the metal vertical fin. The areas that offered the best potential for weight savings are the facesheets and tail skid assembly, and the rivets, adhesive, and sealant that are used. The fairing with the navigation beacon and the former with the nutplates are the same for the composite vertical fin. The Fibertruss core resulted in an increase in weight over the metal core used in the metal vertical fin. A breakdown of actual weights of a small sample of fins is shown in the following table.

WEIGHT COMPARISON

Item	Composite Fin	Metal Fin
Facesheets	5.3	6.2
Leading Edge	0.7	0.75
Trailing Edge	-	0.85
Tail Skid Assembly	0.8	1.6
Core	2.6	1.8
Former with Nutplates	0.2	0.2
Fairing with Beacon	0.8	0.8
Doublers	.35	-
Miscellaneous	1.88	3.1
TOTAL	12.63	15.3

The weight for the facesheets of the composite fin include the graphite, wire screen, FM1000 adhesive, and the trailing edge. The weight for the tail skid assembly includes the glass-wound tail skid, graphite end cap, and the injection-molded bearings for the composite fin. The steel tubular tail skid, fitting, and rubber bumper are included for the metal fin.

The core for the composite fin is heavier because of the 8 pounds-per-cubic-foot Fibertruss core needed at the lower portion of the fin and 3 pounds-per-cubic-foot Fibertruss core used in the remainder of the fin. The metal fin uses 2.5 pounds-per-cubic-foot aluminum core in these same areas. Both fins use 8 pounds-per-cubic-foot density core in the attachment area. The miscellaneous weight includes film adhesive, steel plugs and sleeves; splice foam adhesive and potting for the composite fin; and film adhesive, aluminum plugs and sleeves, splice foam adhesive, potting, sealant, and rivets for the metal fin. Fill-and-fair materials and paint are not included in the weight comparison.

The composite fin shows a weight savings of 2.67 pounds or 17 percent over the metal fin.

FSE PROGRAM

The FSE program involves producing 45 shipsets of the selected composite components for field service experience on BHT customers' Model 206L helicopters. The purpose of the FSE program is to determine the serviceability, durability, and environmental effects of composite components after up to five years in service. Concurrently, exposure specimens manufactured from the same material batches will be mounted on exposure racks. Components and exposure specimens will be removed at appropriate intervals for post-service test and evaluation.

Another aspect of the FSE program is to document the cost of each composite component and establish learning curves for the factory labor. At this writing, approximately half of the 45 composite vertical fins have been fabricated. Therefore, a final cost per fin can be made only as a projected estimate. Costs are anticipated to be twice that of the metal unit. It should be noted that direct cost comparisons for the 45 fins are not conclusive for two reasons. First, the metal fin is an efficient, simplistic design with a minimum of manufacturing tasks. Second, over 8000 metal fins of this type have been produced, providing ample opportunity for implementation of cost-reducing tooling and methods. Cost projections for a lot of 8000 composite fins show the cost can approach that of the metal fin.

SUMMARY AND CONCLUSIONS

The objectives of the composite vertical fin program were to design a composite vertical fin that would be structurally equivalent to the existing production metal vertical fin and have lower weight and cost. Other considerations were lightning protection and structural tests for FAA Certification. The resulting composite fin is significantly lighter than the metal fin, and design support tests have verified its structural quality.

The manufacturing process used in this program produces very reliable parts with maximum properties realized in the graphite/epoxy facesheets. The criticality of the shear fatigue strength of the nonmetallic honeycomb cores became evident and graphite doublers had to be added in the attachment area of the fin to reduce the shear in the core. The design of the tail skid assembly showed a significant reduction in weight over the metal tail skid.

In the final analysis, the composite fin will be slightly more costly to produce than the metal fin, even in large quantities. However, ultimate evaluations will include the value of a significantly extended service life. With these evaluation criterion, the fin is a cost-effective application of composite materials and technology.

BIOGRAPHY

The author graduated from the University of Texas at Austin in 1971 with a Bachelor of Science in Engineering. He went to Martin Marietta Orlando Division and worked in the aerodynamic group for the SAM-D Missile. In 1973, he came to Bell Helicopter Textron as a Research Analysis Engineer working on vibration and loads characteristics of new rotor system concepts. Currently working as a Research Project Engineer on programs involved with the application of composites on helicopter airframes and rotor systems.

FEASIBILITY OF DIRECT/BOND CERAMIC THERMAL PROTECTION SYSTEM TILE ON GRAPHITE/POLYIMIDE STRUCTURE

W. H. Morita and S. R. Graves
Space Transportation System Development and Production Division
Rockwell International
Downey, California

Abstract

A new structure thermal protection system concept has recently been demonstrated for advanced Space Shuttle orbiter applications at Rockwell using advanced state-of-the-art technology. The new concept employs direct-bond fibrous refractory composite insulation (FRCI) tiles (developed as reusable surface insulation and improved by NASA's Ames Research Center) on graphite/polyimide (GR/PI) structure. The high-temperature GR/PI composite has adequate structural properties at 600°F as compared to 350°F for aluminum structure. The high-temperature GR/PI allows for significant reduction in orbiter thermal protection system (TPS) weight which, in turn, would increase the Space Shuttle mission payload capability. The direct-bond tile concept eliminates the strain isolation pad (SIP), which is indispensable with the aluminum structure/TPS technology in the present orbiter design. The paper will present the concepts, analysis, and test results of a GR/PI sandwich deflection panel with direct-bond FRCI tile, representative of the Shuttle orbiter body flap. A comparison of the analysis and test data is presented for the deflection panel, which was tested under orbiter ultimate pressure loads at room temperature and 500°F. The feasibility of direct-bond FRCI on GR/PI structure is the result of structure/TPS thermal compatibility (similarity of thermal expansion coefficients for small thermal stresses) and sufficient structural stiffness to preclude TPS tile fracture caused by structural deflection. The large stiffness/weight ratio of GR/PI composite allows design of a lightweight structure with the required stiffness. Tests were constrained to a 500°F temperature by the RTV-560 adhesive. However, for application on the Shuttle orbiter where GR/PI skin temperatures will reach 600°F, an advanced technology (650°F) adhesive (modified RTV silicone) is currently under development.

1. INTRODUCTION

A new structure thermal protection system (TPS) concept has recently been experimentally demonstrated for advanced Space Shuttle orbiter applications at Rockwell using advanced state-of-the-art technology. The new concept employs direct-bond fibrous refractory composite insulation (FRCI)-

improved NASA Ames Research Center reusable surface insulation tiles*) on graphite/polyimide (GR/PI) structure. The direct-bond tile concept shown in Fig. 1 eliminates the strain isolation pad (SIP), which is indispensable with the aluminum structure TPS technology in the present orbiter design. The high-temperature GR/PI composite structure has adequate structural properties at 600°F as compared to 350°F for aluminum structure. The high-temperature GR/PI allows for significant reduction in orbiter structure and TPS weight which, in turn, would increase the Space Shuttle mission payload capability. The feasibility of direct-bond FRCI TPS tile on GR/PI structure is the result of structure/TPS thermal compatibility (similarity of thermal expansion coefficients for small thermal stresses), and sufficient structural stiffness to preclude reusable surface insulation (RSI) tile fracture caused by structural deflection. The large stiffness to weight ratio of GR/PI composite allows design of a lightweight structure with the required stiffness.

Direct-bond also holds the promise for cost savings, through decreased RSI tile installation and repair times. The reduction in tile-to-tile steps achieved by direct bond tiles is a result of the improved dimensional control of the single bondline versus the tolerances from multiple bonds and the strain isolator on the baseline configuration. The reduction of gaps between tiles is primarily a function of the reduced thermal expansion of the substrate GR/PI versus aluminum on the baseline, although reduced structural deflections and elimination of the dynamic vibrator excursion allowance for tile mounted on a strain isolator pad contribute to this gap reduction. Direct-bond also eliminates stress concentrations associated with SIP, which greatly reduce the load-carrying capability of the tile. The predicted weight savings for FRCI on GR/PI structure over the baseline design is shown in Fig. 2; a summary of the advantages is given below:

- Direct-bond tiles on GR/PI can reduce gaps and steps to improve the aerodynamic surface

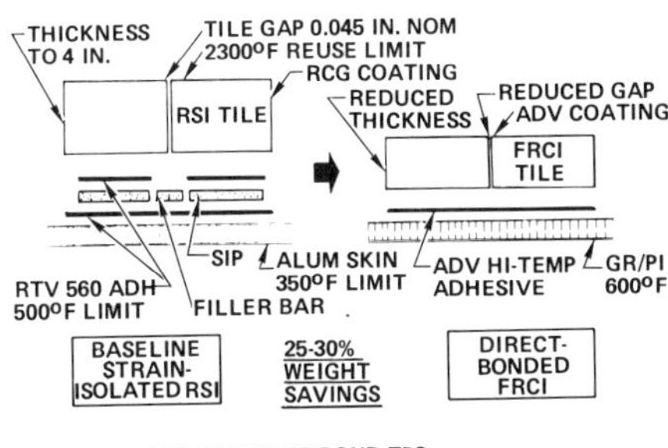

Fig. 1. *Direct-Bond FRCI Concept*

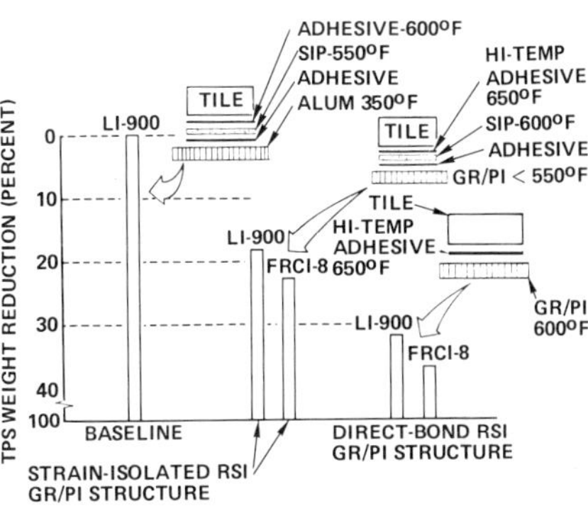

Fig. 2. *Advantage of Direct-Bond TPS Weight Reduction; TPS Concept Comparison*

*The authors acknowledge H. Goldstein of NASA Ames Research Center for data and FRCI tiles received.

- Reduces aeroheating by delaying transition from laminar to turbulent boundary layer
- Results in thinner tiles and additional weight savings
- Direct-bond FRCI allows more precise maintenance of outer mold line
  - Stack tolerances are reduced
  - Flexibility of SIP is eliminated
- Direct-bond tiles can eliminate most gap fillers
  - Coefficient of thermal expansion of tile similar to GR/PI
  - No SIP deflection under aeroacoustic loads
  - Manufacturing tolerances tighter; eliminates SIP and one bondline
- Direct-bond eliminates tile stress concentractions due to SIP

2. DIRECT-BOND FRCI DEFLECTION PANEL TEST

The orbiter body flap has been identified as a candidate for structural weight reduction using GR/PI composite, and was selected by NASA as a technology demonstration component for the Composites for Advanced Space Transportation Systems (CASTS) program.* The body flap structural concept is illustrated in Fig. 3 with a weight comparison between the baseline design (aluminum/SIP/LI-900) and the advanced composite design (FRCI-8/direct-bond/GR-PI).

A GR/PI sandwich deflection panel with direct-bond FRCI tile, representative of a portion of the Shuttle orbiter body flap as

CONFIGURATION		BODY FLAP WEIGHT (LB)		
		STRUCT	TPS	TOTAL
TILE ADHESIVE -600°F SIP-550°F ADHESIVE ALUM 350°F	BASELINE — LI-900/ SIP/ ALUMINUM	450	872	1322
TILE HI-TEMP ADHESIVE 650°F GR/PI 600°F	FRCI-8/DIR BOND/GR-PI DELTA	357 -93	646 -226	1003 -319

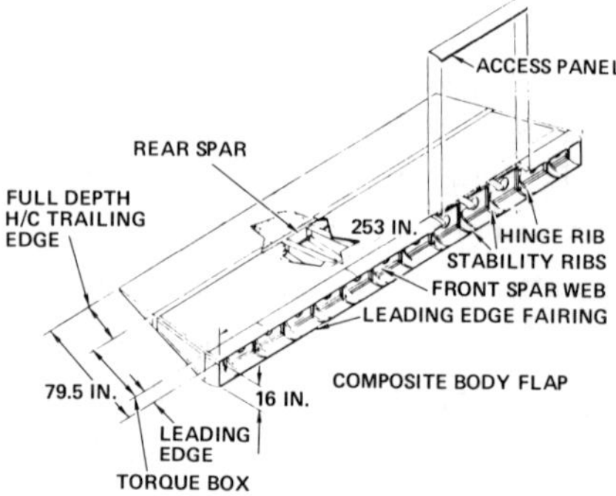

Fig. 3. GR/PI Body Flap Potential Weight Savings

illustrated in Fig. 4, was successfully constructed and tested under a Rockwell-sponsored program.** The FRCI tiles were bonded to the GR/PI panel with RTV-560, and the test panel was subjected to orbiter body flap ultimate pressure loads at room temperature and at 500°F. The feasibility of this advanced structure TPS concept was further demonstrated when the test article was twice cycled from room temperature to -160°F with no damage to either structure, RTV bond, or TPS.

The deflection test set-up is depicted in Figs. 5 and 6. The tests were performed using tiles that were bonded to the panel with RTV-

*Program managed by Langley Research Center under Dr. John G. Davis, Jr.
**IR&D Project 202 Orbiter Subsystem Design Improvements, S/A 20200, Structural Weight Reduction.

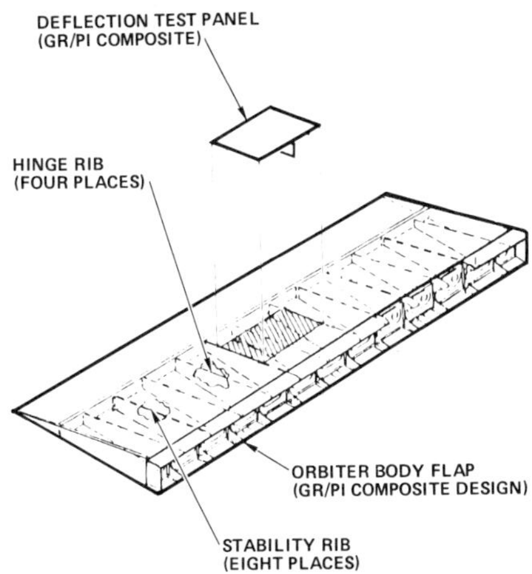

Fig. 4. Deflection Test Panel Location

560, a 500°F allowable adhesive. However, for application on the Shuttle orbiter where GR/PI skin temperatures will reach 600°F, an advanced-technology (650°F) adhesive (modified RTV silicone) is currently under development.*

The direct-bond FRCI deflection panel test objectives were: (1) to determine the TPS/ structure response to airloads for a typical rib-to-cover panel configuration, (2) to verify the direct-bond RSI tile concept under static

Fig. 5. Direct-Bond FRCI Test Set-Up

Fig. 6. Direct-Bond FRCI Test Set-Up

loading conditions at temperatures up to 500°F, (3) to determine the panel stiffening contribution of the tile, (4) to verify the structure/TPS thermal compatibility in a simulated space environment, and (5) to verify the analytical model of the test panel. The panel radius of curvature effect on the direct-bond tile is depicted along with the test objectives in Fig. 7.

The panel and test fixture are shown in Fig. 8. The test panel with an active area 40 inches

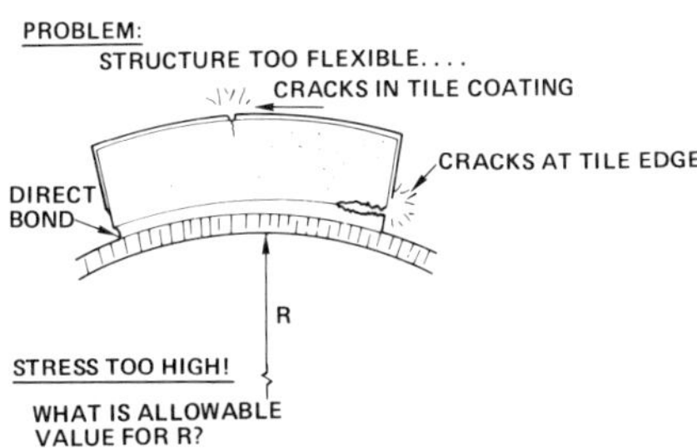

*Fig. 7. Test Objectives: Direct-Bond
FRCI on GR/PI Structure*

Initial development was conducted in 1977 under NASA LaRC Contract NAS1-15152, "High Temperature Adhesive for Direct Tile Bonding" with subsequent development being conducted under Rockwell-sponsored IR&D Project 276, Advanced Materials Technology.

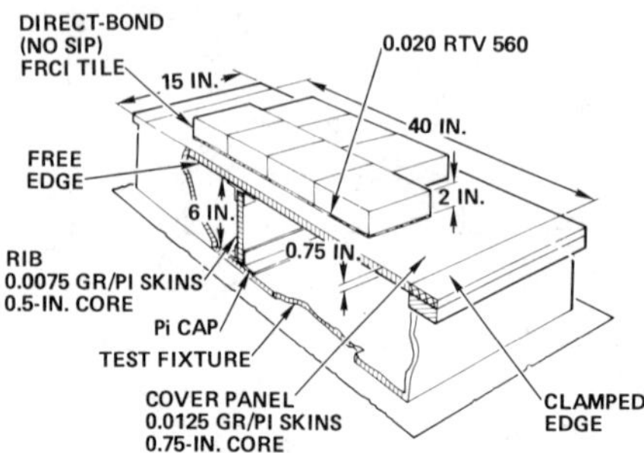

*Fig. 8. Direct-Bond FRCI Deflection Panel
and Test Set-Up*

long by 15 inches wide represents a portion of the lower cover of the orbiter composite body flap spanning two 20-inch bays centered about a typical rib (see Fig. 4). The panel is constructed of Celion-6000/LARC-160 GR/PI with a unidirectional modulus of 20.4 Msi. The honeycomb core is a glass/polyimide material (HRH-327-3/16-3.0) with a nominal density of 3.0 pcf. The short sides of the panel are clamped in the test fixture, representing the condition to be found where the cover panel passes over adjacent ribs. The long sides are left free to deflect, as they effectively would in the central portion of the body flap.

A seven-tile array was used to provide as much information as possible with a limited number of tiles. The 8-pcf NASA Ames FRCI tiles were 2.0 inches thick, and were bonded to the cover panel with a nominal 0.020-inch-thick RTV-560 adhesive layer. The test panel was instrumented with 40 data channels of high-temperature strain gauge instrumentation and six deflectometers. The gauge pressure in the test fixture was monitored via a digital pressure transducer. The instrumentation data was automatically read out by a computer-controlled data acquisition system.

3. DIRECT-BOND FRCI ON GR/PI TEST PROCEDURES & RESULTS

Tests were performed first without, then with RSI tiles directly bonded to the GR/PI cover panels to determine the additional panel stiffening provided by the tiles. During the room-temperature test without tiles, the panel was loaded to 2.2 psi in 0.4-psi increments by drawing a partial vacuum inside the test fixture. Data readings were taken at each increment, and the procedure was repeated to remove settling effects. Next, FRCI-20-8 tiles were direct-bonded to the cover panels, and the panel was loaded to 3.4 psi following the above procedure. This load exceeds the 1.8-psi ultimate body flap load defined by design condition EA106 (4.4 reentry symmetric maneuver), and corresponds to the body flap minimum cover panel radius of curvature of 300 inches. At this radius of curvature, the predicted tile stress is 37 psi. The tile stress analysis is discussed in detail in Section 5. The test load rationale is outlined in Fig. 9, and the actual applied test loads with the corresponding panel minimum radius of curvature are given in Table 1. The tiles were found to reduce the panel deflections and increase the panel curvature by 18.6 percent, as illustrated in Table 2.

The panel was then loaded to 1.8 psi at 200°F, 300°F, and 400°F. At 500°F, the panel was loaded incrementally to 2.8 psi with no tile failure. The panel response to mechanical loads at 500°F was nearly identical to that at room temperature; when the thermal strains are subtracted, the mechanical strains are very similar (Fig. 10).

The GR/PI composite structure/direct-bond TPS thermal compatibility has also been

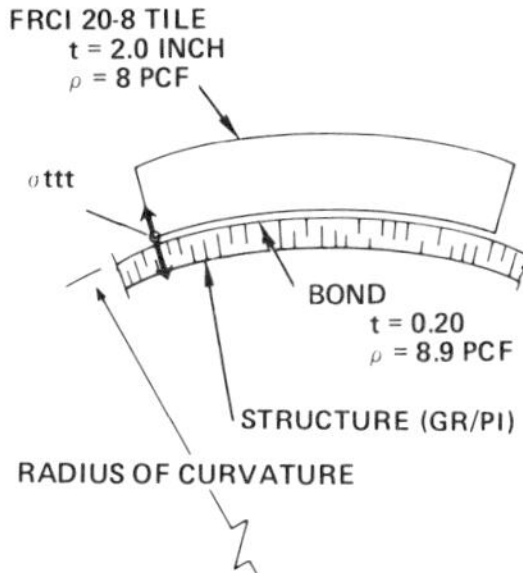

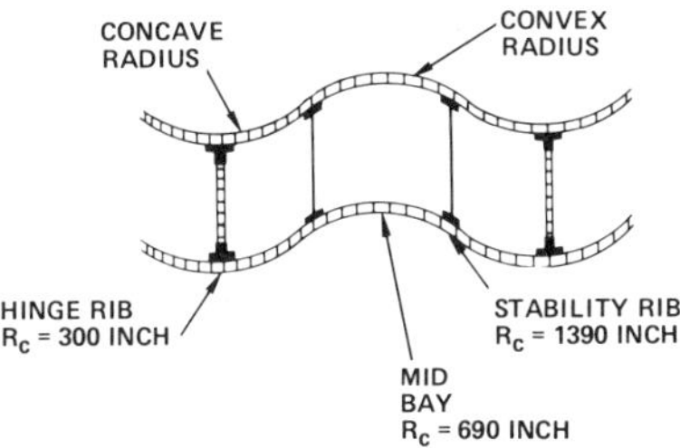

Fig. 9. Test Load Rationale: Direct-Bond FRCI on GR/PI Structure

verified at cryogenic temperatures. As discussed previously, the structure TPS elevated temperature thermal compatibility was verified when the test article was loaded to 2.8 psi at 500°F with no tile failure. However,

Table 1. Test Procedure: Direct-Bond FRCI on GR/PI Structure; Test Panel Applied Load Levels

Condition	Panel Load (psi)	Radius of Curvature (inches)
RT test without tile	2.2	420
RT test with tile	3.4	300
Elevated temp. (200, 300, & 400°F)	1.8	600
Elevated temp (500°F)	2.8	380
Notes: Loads applied by drawing a partial vacuum inside the test fixture Tests performed, first without, then with RSI tiles to determine additional panel stiffening provided by the tiles		

Table 2. Room-Temperature Test Results; Panel Stiffening Due to Direct-Bond FRCI Tiles

Condition	1.8 psi		3.4 psi	
Item	Without Tile	With Tile	Without Tile	With Tile
Maximum panel deflection	0.043	0.035	0.080*	0.065
Minimum radius of curvature *Calculated	514	608	253*	300
Note: Direct-bond RSI tile (2-inch thick 8 pcf FRCI) adds 18.6 percent to cover panel stiffness				

material properties change with temperature, especially the RTV tile bonding adhesive, which becomes less elastic with lower temperatures and provides little strain relief at cryogenic temperatures. Because of these changing material properties, a test was required to evaluate the direct-bond FRCI on GR/PI concept in a simulated space environment. Because no aerodynamic loads are applied in space, the test panel was removed from the metal test fixture and placed in an environmental chamber (Fig. 11). The test

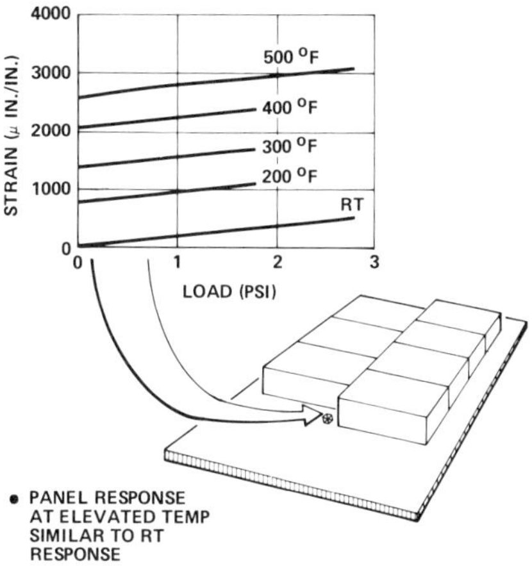

Fig. 10. Elevated Temperature Test Results: Typical Strain Gauge Response at Various

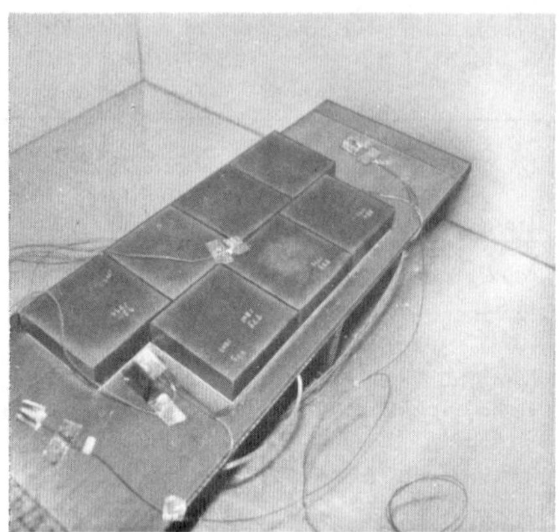

Fig. 11. Direct-Bond FRCI Deflection
Panel

article was twice cooled to -160°F with no failures, thus verifying the structure/TPS thermal compatibility at cryogenic temperatures.

4. TEST PANEL DEFLECTION ANALYSIS

The test panel was modeled as a statically indeterminate beam under distributed load having three simple supports. The middle support is allowed to deflect to model the settling and compression of the middle rib. The model is shown in Fig. 12 with the calculated deflected shape, and the test data for the panel at room temperature under 1.8-psi simulated airload. The panel stiffening of the tiles is demonstrated by the decrease in panel deflections. Prior to the conducting of the deflection panel test, the amount of panel stiffening associated with the direct-bond FRCI tile was unknown, and the stiffening effect was estimated by models that bounded the solution. The upper bound was the model without tile, and the lower bound was a model that assumed the tile would pick up load in proportion to the tile stiffness and moment of inertia. The lower bound model predicted that the tile would reduce the panel deflections by 70 percent, but this model neglected the strain relief of the RTV adhesive and the shear lag of the tile. The test determined an actual decrease in panel deflections of 18.6 percent. The strain relief of the RTV is illustrated in Fig. 13, which shows the relative deflections of the tile and cover panel.

The solution for the deflected shape of the statically indeterminate beam in Fig. 12 was obtained by replacing the center support with a concentrated load. The problem is then solved by superposition of the deflected shape of the beam subject to a distributed load on the deflected shape of the beam subject to a concentrated load acting in the opposite direction. Because the panel is a honeycomb sandwich structure, shear deformations are included in the model. Fig. 14 presents a comparison of the analytical model to the test

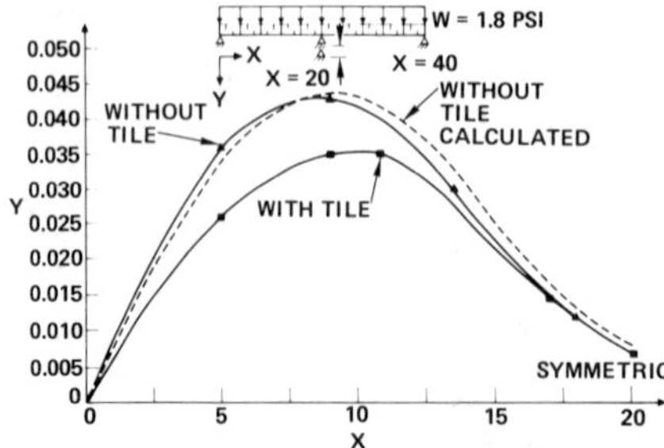

Fig. 12. Analytical Model and Deflected
Shape; Direct-Bond FRCI on
GR/PI Structure

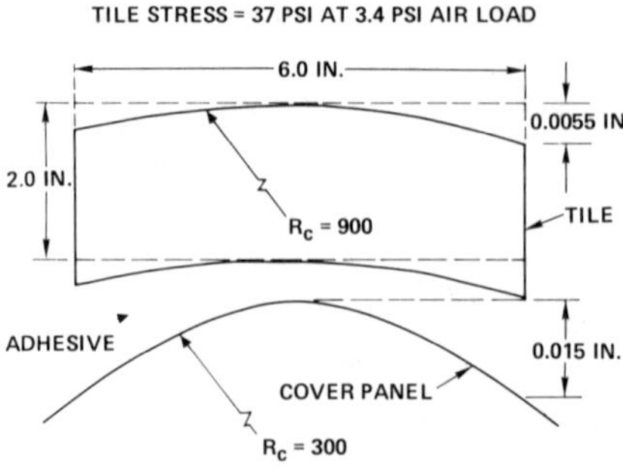

Fig. 13. Cover Panel/Tile Relative
Deflected Shape

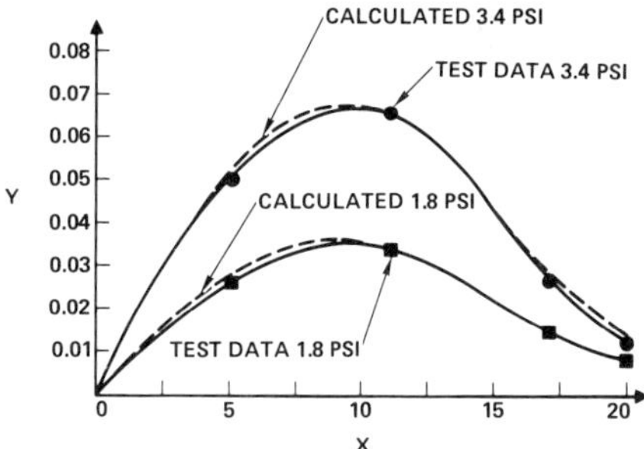

Fig. 14. Analytical Model/Test Data Correlation – With Tile

data for the test panel with tiles under 1.8 and 3.4 psi simulated air load.

An emphasis was placed on an accurate analytical model of the panel-deflected shape because the model is used to calculate the panel radius of curvature. The panel radius of curvature is used to determine the direct-bond tile stress state as discussed in the following section.

5. DIRECT-BOND FRCI TILE STRESS ANALYSIS

A dual analysis approach was used to predict the tile stress response as a function of substructure radii of curvature. A closed-form solution for a beam-on-an-elastic-foundation model[1] was developed and programmed on a Hewlett-Packard 9825 calculator in order to generate large quantities of parametric data relating tile stresses to variations in tile geometry and material, adhesives, bondline thickness, etc. The results of the trade study are given in detail in Ref. 2. For a more detailed stress analysis, including the effects of tile coating, a detailed two-dimensional finite element model was developed for use with the ASKA* computer program. The results of these analyses were compared with RSI test data obtained from NASA Ames and good correlation was found.[3]

Results of the trade study indicate the feasibility of direct-bond RSI tile on GR/PI substructure, which is designed by thermal and stiffness considerations, for orbiter components such as the body flap, rudder, or elevons. The direct-bond tile concept feasibility is predicated on the use of the higher-strength FRCI tile on a 600°F GR/PI substrate. Design options include the use of FRCI-20-8 tile on locally reinforced composite structure or hi-modulus GR/PI structure, and the use of higher strength FRCI-20-12 tiles on GR/PI panels in areas having large out-of-plane deflections.

For a brittle ceramic tile bonded to a substrate with low in-plane mechanical and thermal deformations such as the GR/PI honeycomb body flap skin panels, the primary mechanism for inducing a stress field into the tile is the out-of-plane deflection of the substrate. The deflected shape of the substrate over the short span covered by a single tile can be most conveniently approximated by using a radius of curvature as shown in Fig. 15. As shown, the substrate deflection introduces tension and compression loads into the tile at the bond interface. The areas of maximum stress on the tile are, first, an in-plane tension due to bending in the coating and outer fibers (σ_X), which predominates in thin tiles approximately 1 inch or less thick, and secondly, a through-tile-tensile stress (σ_{ttt}) at the tile/bond interface, which predominates in thicker tiles.

ASKA – Automated Systems for Kinematic Analysis – a general-purpose finite element modeling program developed at the University of Stuttgart, Germany.

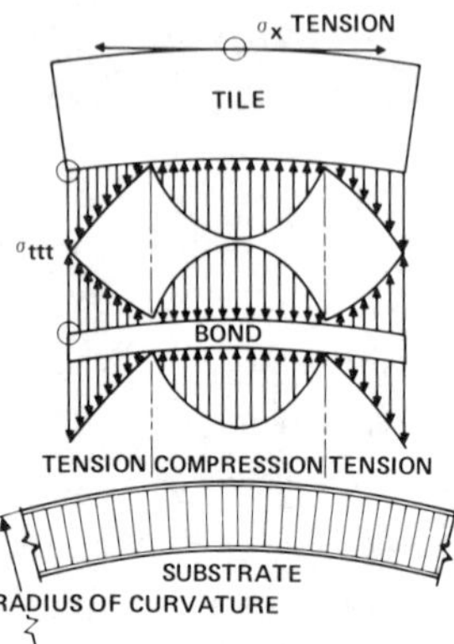

- BASED ON BEAM-ON-ELASTIC-FOUNDATION SOLUTIONS, CASE OF FINITE BEAM DEFLECTED BY CONCENTRATED END MOMENTS
- DEVELOPED FOR HP 9825A EQUIPMENT – ALLOWS VARIATION OF REQUIRED PARAMETERS
  - TILE GEOMETRY & PROPERTIES
  - BOND CHARACTERISTICS
- ASSUMPTIONS
  - ORTHOGONAL TILE PATTERN
  - SHEAR LAG IN TILE OR TILE/BOND SHEAR COUPLING MINIMAL
  - STRESS DUE TO MECHANICAL LOADS ONLY
- PROGRAM IMPOSES A DEFLECTION ON SUBSTRATE & SOLVES FOR TILE STRESS AT TWO LOCATIONS
  - THROUGH TILE TENSION AT CORNER
  - TENSION DUE TO BENDING AT OUTER FIBER

Fig. 15. Direct-Bond RSI Tiles; Analytical Approach – HP-9825A Program

The use of the substructure radius of curvature provided an efficient, easily understood boundary condition to correlate the structure and the tile analysis. The tile analysis was centered on the geometry and material constraints shown in Fig. 15.

In order to perform a detailed tile stress analysis, a finite element model was developed to be used with the ASKA computer program. The two-dimensional model was based on an existing model developed on the main line Shuttle program, and comprises one half of a slice through a 6-inch-by-2.2-inch coated tile (Fig. 16). The tile and bondline are modeled with nine-node biquadratic membrane elements while the 0.009-inch ceramic coating is modeled as three-node quad-

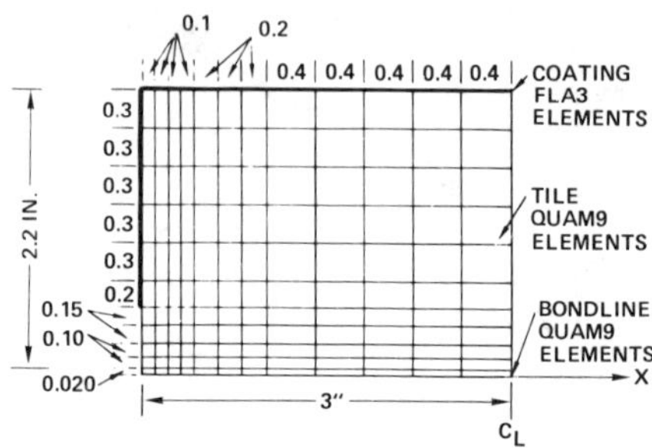

Fig. 16. ASKA Finite Element Tile Model

ratic bar elements. Displacement symmetric boundary conditions are maintained along the tile centerline. The model was modified to a finer grid at the corner, which is the most critical area in the tile. The coating of the tile was also uncoupled from the RTV 560 bondline, which gives a more realistic appraisal of the state of the stress in the directly bonded tile.

The HP9825A program was developed as a low-cost tool to provide a large volume of parametric data in order to spot trends, conduct trade-offs, and make an initial assessment of the feasibility of direct-bond RSI. This program is a closed-form analytical solution derived from the treatment of beams-on-elastic-foundations.[1] The closed-form, beams-on-elastic-solution mathematics is based on classical beam theory for homogeneous isotropic materials. The program is designed to accept modifications that will enhance its accuracy. These are intended to be in the form of empirical correction factors developed from test data, finite element analysis, or a combination of both.

The inputs to the program are the geometry and mechanical properties of the tile and bond material in question, and the range of radii of curvature of the substructure in question. This allows the variation of the material length, width and thickness, as well as the variation of bond properties and bondline thickness. The substructure radius of curvature can be varied between limits of 0 to 99,999 inches, which bounds all cases of practical interest.

The assumptions in the program are basically those of classical beam theory, but, in addition, an orthogonal tile pattern is assumed, and stress is assumed to be generated by out-of-plane deflections of the substructure only.

The program imposes a deflected shape on the tile, derived from the defined radius of curvature, and then solves the beam-on-elastic-foundation equations for the through-tile-tension stress at the corners (σ_{ttt}) and the in-plane tension stress on the outer fiber (σ_x), which are the critical cases. Full details of the program are given in Ref. 2.

Figs. 17, 18, and 19 present a summary of the results to date concerning direct-bond RSI tile, based on the parametric studies conducted. Fig. 17 shows that direct-bond TPS would not be practical on any GR/PI body flap configuration that used the present orbiter baseline TPS tile as the RSI material. This is basically because of the very low tensile strength of the LI-900 silica material $(\sigma_{ttt\ ult} = 13\ psi)$. This reconfirms the deci-

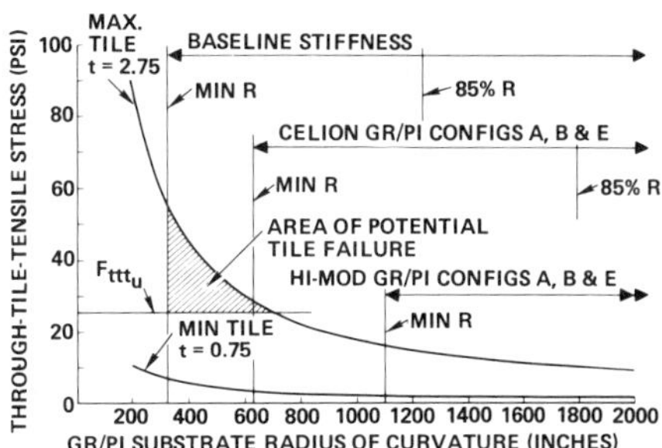

Fig. 18. Potential Structure/TPS Options; GR/PI With Direct-Bond FRCI-20-8 Tile

sion of main line orbiter to go to strain-isolated installations for the LI-900 tile.

However, if the newly developed NASA Ames FRCI material is chosen as the body-flap RSI material, the picture changes. As shown on Fig. 18, using the 8-pcf density FRCI-20-8, GR/PI body flaps could be designed to take advantage of direct-bond TPS with little or no penalty. Fig. 19 shows the same information with 12-pcf density FRCI-20-12 tile. In this case, there is adequate margin on all the configurations studied. This improvement in performance is basically due to the dramatic increase in tile allowables in the FRCI-20 series of materials.

The analytical model used to calculate the deflected shape of the test panel (Fig. 12) was

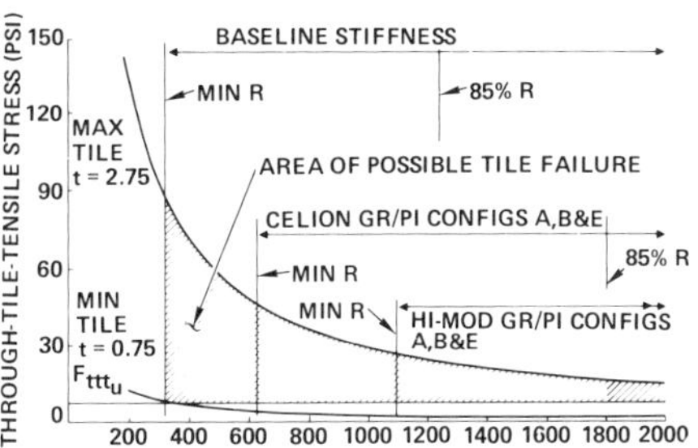

Fig. 17. Potential Structure/TPS Options; GR/PI With Direct-Bond LI-900 Tile

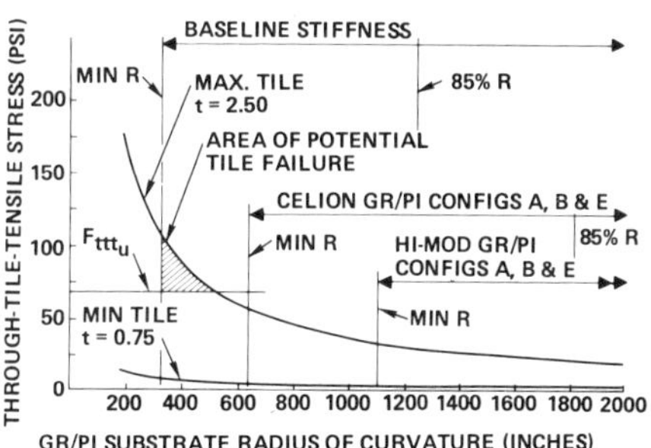

Fig. 19. Potential Structure/TPS Options; GR/PI With Direct-Bond FRCI-20-12 Tile

also used to calculate the panel radii of curvatures. The deflections were converted to panel radii of curvature using a three-point approximation. Additional information is available in Refs. 4 and 5.

The direct-bond FRCI on GR/PI test panel radius of curvature (without and with tile) as a function of air load is shown in Fig. 20. The figure includes the various body flap curvatures and the material allowables of the 8-pcf tile. The panel stiffening effect of the tile is demonstrated by the rightward shifting of the curve with tile.

The predicted through-tile tensile stress as a function of the substrate radius of curvature is presented in Fig. 21. The critical tile stress location and the panel radius of curvature concept are depicted in Fig. 9. A dual analysis approach (an ASKA finite element model and a closed-form beam-on-elastic-foundation model) was used to predict the tile stress. The tile material properties used in the analysis are presented in Table 3. Referring to Fig. 21,

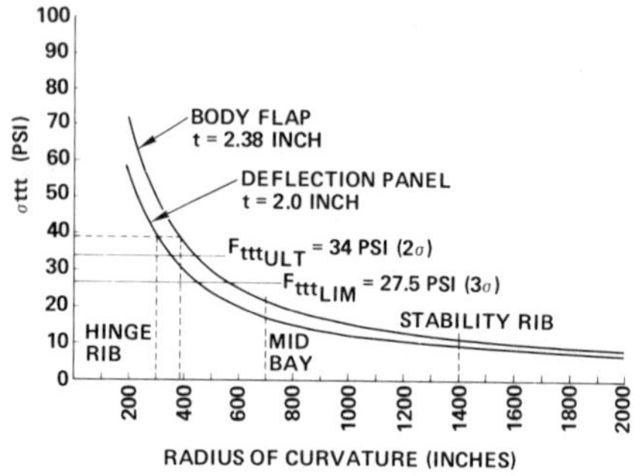

- MINIMUM BODY FLAP (WITH TILE) RADIUS UNDER ULTIMATE LOAD IS 375 INCHES OVER HINGE RIBS
  - BODY FLAP TILE IS 2.38 INCH THICK
  - DEFLECTION PANEL TILE IS 2.0 INCH THICK
- DEFLECTION PANEL TILES SUBJECTED TO 300-INCH RADIUS FOR EQUIVALENT BODY FLAP TILE STRESS
- BODY FLAP WILL REQUIRE LOCAL REINFORCEMENT TO REDUCE TILE STRESS OVER HINGE RIBS

Fig. 21. FRCI-20-8 Tile Stress vs. Body Flap Radius of Curvature

the minimum body flap radius of curvature is 375 inches over the hinge ribs resulting in a tile stress of 37 psi for the 2.38-inch-thick body flap tile. To subject the deflection panel 2.0-inch-thick tile to an equivalent stress level, the tile was subjected to a 300-inch panel radius of curvature. The body flap will require local reinforcement or the use of 12-pcf tile over the hinge ribs to reduce the tile stress below the 2σ (34 psi) ultimate design value.

7. SUMMARY AND CONCLUSIONS

The direct-bond FRCI on GR/PI structure deflection panel was successfully tested under Shuttle orbiter ultimate pressure loads at room temperature and 500°F. The structure/TPS thermal compatibility was further demonstrated when the test article was cooled to -160°F with no tile failure. All initial test goals and objectives were met or exceeded. The direct-bonded RSI tiles will add 18.6 percent to the composite body flap cover panel bending stiffness.

The primary conclusion that can be drawn from this effort is that direct-bond RSI is

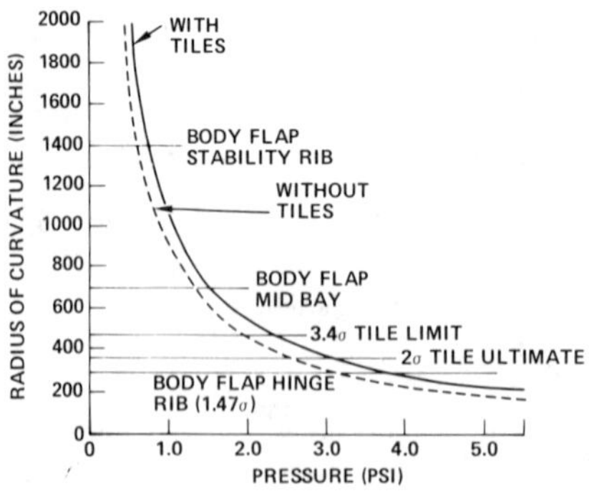

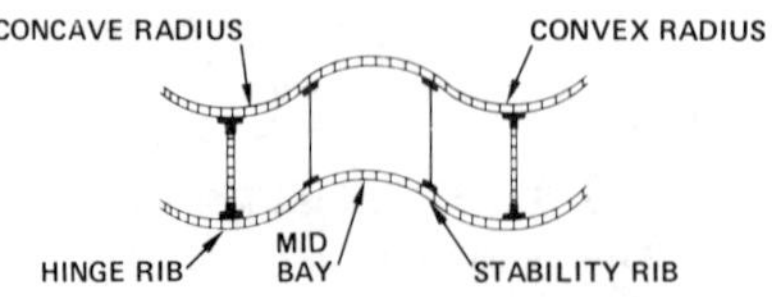

Fig. 20. Panel Radius of Curvature vs. Pressure Load, Direct-Bond FRCI on GR/PI Structure

Table 3. Mechanical Properties of FRCI Tile and FRCI/Baseline Strength Comparison

FRCI Mechanical Properties*

Property		FRCI-20-8			FRCI-20-12		
		Average	2σ	3σ	Average	2σ	3σ
Tensile Strength (psi)	Transverse	46.1	33.7	27.5	122.6	91.9	76.6
	In-plane	97.9	72.5	59.8	230.9	182.9	159.0
Modulus of Elasticity (psi)	Transverse	4700	–	–	10910	–	–
	In-plane	18900	–	–	51440	–	–
Modulus of Rupture (psi)	Transverse	76.5	–	–	189	–	–
	In-plane	137.8	–	–	367	–	–

FRCI/Baseline Tile Strength Comparison*

Property		LI-900	FRCI-8	LI-2200	FRCI-12
Transverse Tensile Strength (psi)	Average	24	46	70	123
	"A" allowable	13	28	35	77

*Data from Refs. 6-8

feasible on properly designed GR/PI substrate. The direct-bond concept feasibility is predicated on the use of 600°F GR/PI, designed with stiffness criteria, and the use of new "high-strength" FRCI tile and the availability of a 650°F flexible adhesive.

Several design options are possible for an integrated direct-bond TPS/GR/PI body flap system. These options are: first: use an 8-pcf FRCI tile on Celion/PI structure, with local reinforcement to limit the minimum radius of curvature of the structure; second, use 8-pcf tile over most of a Celion/PI structure, and 12 pcf tile in areas of maximum deflection; and third, use 8 pcf tile on hi-modulus GR/PI structure.

REFERENCES

1. Heteny, M., *Beams on Elastic Foundation.* Ann Arbor: The University of Michigan Press (1958).

2. Morita, W. H., et. al., *Advanced Composites Application To Orbiter IR&D Study CFY 1979.* Rockwell International Corporation, SOD 79-0245 (December 1979).

3. Pitoniak, F. J., *Preliminary Data on Direct-Bonded RSI Experiments.* NASA Ames Research Center (January 1979).

4. Morita, W. H., et. al., *Advanced Composites Application to Orbiter IR&D Study, FY 1980.* Rockwell International Corporation, SOD 80-0355 (October 1980).

5. Morita, W. H. and Graves, S. R., *Deflection Test, Direct-Bond FRCI on GR/PI Structure.* Rockwell International Corporation, SSV 80-64 (October 1980).

6. Vaughn, M. L., *Inplane and Transverse Tensile and Compressive Strengths of FRCI-8 Material From Six Blocks.* Lockheed Missiles and Space Co., Inc., RK1.6.8-S3 (February 1980).

7. Vaughn, M. L., *FRCI-12 Inplane and Transverse Tension, Compression Test Results.* Lockheed Missiles and Space Co., Inc., RK1.6.8-S2 (January 1980).

8. Goldstein, H. E., *Summary of FRCI Data.* In-House NASA Document, NASA Ames Research Center (October 1979).

BIOGRAPHIES

W. H. Morita is principal investigator on Orbiter Subsystem Design Improvements, IR&D, and is Study Manager of Composite Structures for Advanced Systems at Rockwell's Space Transportation System Development & Production Division. He was a Study Manager of the recent NASA MSFC Shuttle Growth Study and the NASA LaRC contract for the Design and Test Requirements for the Application of Composites to Space Shuttle Orbiter. Prior to this, he was engaged in systems analysis and conceptual design of Space Shuttle systems, encompassing the Space Shuttle Phase A study and Phase C/D program. Earlier project assignments include study manager of the Mars/Venus Flyby S-IIB Injection Stage Study; program manager of the Titanium Tankage Program (USAF); the Utilization of Spent Launch Vehicle Stages (NASA) contract; project engineer on the Apollo-LOR contract study, S-II Structural Weight Savings; and the USAF Cost Optimized Booster study contract. Mr. Morita has a Bachelor of Science degree in mechanical engineering from the University of Nebraska (1944) and a Master of Science degree in mechanical engineering from USC (1954).

S. R. Graves has 1½ years of experience in the design and analysis of composite structures in Advanced Systems at Rockwell's Space Transportation System Development and Production Division. He is currently involved in the application of advanced composites to the Space Shuttle orbiter for structural weight reduction. He has been responsible for stress analysis and computer modeling in support of several NASA contracts and IR&D projects, including GR/PI LaRC-160 Fabrication Development and the design and development of the GR/PI Technology Demonstration Segment. Mr. Graves graduated with a Master of Science degree in mechanical engineering in 1979 from the University of Wyoming where he spent one year in the Advanced Composite Material research group.

UNIQUE SOFTGOODS APPLICATIONS
FOR THE
SPACE SHUTTLE

Ed Yung, P.E.
ILC Space Systems
Houston, Texas   77058

Abstract

High and low temperature softgoods
are vital to the Space Shuttle.
The feasibility, merits, and problems of softgoods in dynamic and
high temperature environments is
discussed, including how they
compare with competitive materials.
Both fatigue experienced under
dynamic loading conditions and
strength loss resulting from extremely high temperatures may be overcome by proper techniques.

1.   INTRODUCTION

It is pretty difficult to find an
industry, product, business, or
individual not relying on fabrics,
and the Space Shuttle Program is no
exception.

When you think of softgoods in
aerospace you probably think of
"rags and bags": Space Suits
(Figure 1), Clothing (Figure 2),
Straps, Harnesses, Flotation Gear,
Parachutes, Bags to Contain Equipment and Personal Items (Figures
3 and 4), Hoses, and Upholstery.

Famous Apollo Moon Photo and Shuttle
Space Suit. Both suits were manufactured by ILC Dover.

FIGURE 1

Some Typical Shuttle Clothing

FIGURE 2

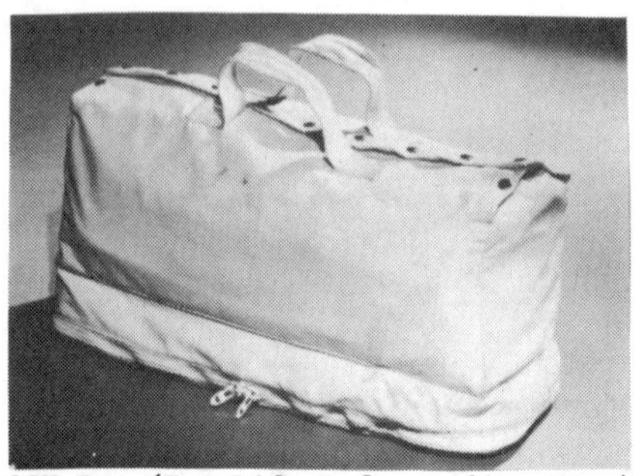

OSK Bag (For Life Raft, Radio, Etc.)

FIGURE 3

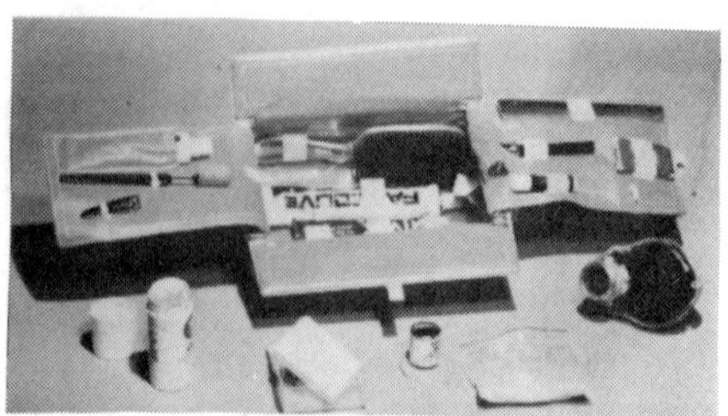

A Typical Personal Hygiene Kit

FIGURE 4

But shades of World War I, who would ever expect a ragwing spacecraft, in the vernacular of private aviation? It's difficult to imagine the fabric cover of a Jenny or a 1911 Cessna (Figure 5) modernized sufficiently to suffice for the Space Shuttle. As shown by the solid white areas in Figure 6, a significant portion of the Space Shuttle is now covered with nomex, a DuPont Aramid fiber that looks and feels much like untreated canvas. Nomex does not melt and has extremely low flammability, but at temperatures above 700°F it degrades to a friable char at a rate proportional to the intensity of the heat source. In the event that it does burn, it is self-extinguishing when the flame source is withdrawn.

FIGURE 5

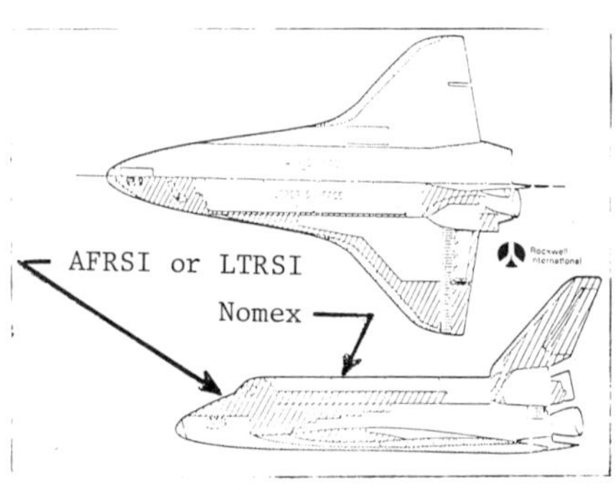

Outer Cover Material Location

FIGURE 6

Modern quartz cloth sewn over micro-quartz insulation in a quilted configuration may also be used to cover a large portion of the Shuttle exterior (per the cross hatched areas of Figure 6). This application is, in fact, the primary emphasis of this technical paper.

This quartz composite quilt will experience temperatures up to 1200° F from aerodynamic heating if it is used as a replacement for the white tile (the Low Temperature Reusable Surface Insulation, LTRSI). It is called the AFRSI (Advanced Flexible Reusable Surface Insulation). Similar quilts are used to insulate critical components inside the conical nose cavity from 2700°F carbon-carbon nose cone re-entry temperatures. On the opposite end of the system, similar quilts protect the rocket engine directional control hydraulic system from the high temperature rocket exhaust.

ILC Space Systems, a division of the International Latex Corporation, a subsidiary of the Rapid American Corporation, is involved in all of these products and projects. ILC Space Systems is currently under contract to NASA, Rockwell International and United Space Boosters, Inc., for most of the items mentioned.

2. SOFTGOODS UNDER SEVERE ENVIRONMENTS, INCLUDING HIGH TEMPERATURES

2.1 Rocket Exhaust Thermal Barrier

The USBI thermal shields being made by ILC protect the hydraulic system which controls the gimbled rocket engine steering system from hot rocket exhaust gasses. This environment, however, includes shock, as over pressure snaps the blankets upward, and then back as a low pressure area develops. Although quartz and other glass cloth and threads have high strength, they have poor fatigue characteristics and abrasion resistance.

The nature of the machine sewn products (and most hand sewn applications) include the sharp 180° bend of thread. This loop, or sharp bend, is highly susceptible to

fatigue if movement results in shifting (bend moving along the thread).

For the USBI thermal shields, this problem is solved by covering the high temperature blankets with a heavy, high strength ballistic nylon blanket to support it during the initial severe shock.

The nylon burns away immediately, but only after it has protected the high temperature blankets from the initial shock. Thereafter, the "ceramic" insulators perform their function to protect vital spacecraft systems and components from high temperatures.

The USBI blankets vary appreciably, but a typical one is ½" thick, tapered from 2 to 2½' wide by 5' long. They use quartz cloth not drastically different than that used in building fiberglass boats as an OML (outer material layer) and glass cloth as IML (inner material layer). Between two layers of this cloth is placed a ½" thick layer of fiberfrax, a high temperature ceramic insulation material somewhat like residential fiberglass insulation. The resulting quilt is secured by Q-18, a teflon coated quartz thread, sewn on 2" centers in a grid pattern quilt. It is sewn on a large commercial sewing machine not drastically different from your wife's Singer. The thread, although quartz, looks much like conventional nylon or cotton threads.

2.2   <u>Orbiter Nose Cone Insulation</u>

The carbon-carbon nose cone sees the highest temperatures on the vehicle; 2700°F. Immediately behind the nose material are a myriad of critical items, including, of course, the crew.

Last year, ILC Space Systems constructed a set of nose cone blankets (totaling 3" thickness) which employed lacing to secure it to the vehicle and adjacent layers. See Figure 7.

High Temperature Quilts

FIGURE 7

More recently, a pressure transducer system was added for use on later flights. This called for a new nose cone blanket with the addition of protection for portions of the pressure transducer system.

Although there are no aerodynamic forces involved, and accelerations are not severe, this was an extremely difficult task because of the 2700°F temperature anticipated in some areas.

Some ceramics retain their shape at this temperature; quartz and other glasses flow. Unfortunately, all of these materials lose strength at elevated temperatures. Ceramics are not available in machine sewing fibers, although great effort to develop them is being expended. An excellent approach to the problem seems to be to place iridium or platinum wire inside a woven ceramic tubing to secure the quilts around the manifolds of the pressure transducer system, or to secure the large conical blankets to the polyimid cone which they protect. The use of these high temperatures wires (4449°F and 3217°F melting points, respectively) inside of ceramic tubing provides several protective services. The taut wire will pull tightly against the inside of the toroid formed by the ceramic tubing. This will form a thermally protective space between the hotter outer portion of the toroid and the wire. This is dramatically illustrated by Santa Fe Textiles' advertising brochure photograph shown in Figure 8. The lady's finger is protected by the space within a 3M Nextel 312 ceramic tubing. The ceramic tubing alone may not retain sufficient strength at 2700°F to carry the necessary loads.

FIGURE 8

2.3 Orbiter Outer Cover

The white tile on the Orbiter will experience temperatures as high as 1200°F. Unfortunately, rigid, fragile ceramic tile are not compatible with the typical flexible aircraft type aluminum structure of the Space Shuttle; thus, the well publicized "tile problem".

A logical substitute would be a material which is more flexible. Perhaps the best suited "material" is a flexible high temperature composite of quartz or glass cloth and similar insulation sewn into a quilt with quartz thread.

It goes without saying that people are not accustomed to thinking in terms of softgoods on the exterior of spacecraft - or even jet aircraft. Experience with more conventional softgoods justifies the skepticism. Even the exotic high temperature quartz and carbon cloths and threads have limitations in a re-entry situation.

Some obvious questions which arise relate to:

2.3.1 Length, width and thickness (including pillowing effect). tolerances.

2.3.2 Dimensional stability (all three axes).

2.3.3 Insulation shift within quilts; thus increasing dimensional instability.

2.3.4 Strength at elevated temperatures.

2.3.5 Strength under aerodynamic stresses.

2.3.6 Fatigue resistance.

2.3.7 Quality assurance uncertainties (how may interior be proven suitable; especially after several missions).

2.3.8 Weight limitations.

2.3.9 Thermal insulation characteristics and values.

2.3.10 Moisture adsorption and retention (especially in driving rain).

Some concerns which these questions generate include: (1) the possibility of destruction by boundary layer reattachment after surface irregularity induced separation; (2) fatigue from shift of 180° bends in threads during aerodynamic turbulence; (3) non-uniform structural distortion. Such non-uniform heating would occur from a mismatch of insulation K valve gradient from the design goal.

Conversely, softgoods do not present tolerance build-up problems common to rigid items. Each blanket may be stretched or compressed slightly to achieve nominal dimensions. Conformance to compound curved surfaces is simplified, and a major advantage. Vehicle structural flexing does not damage AFRSI.

2.4 Addressing the above listed questions and concerns we see that softgoods may indeed again be a better approach:

2.4.1 Thickness tolerances may be held to suitable levels by stacking various thicknesses of insulation and by varying thread tension and

sewing machine foot pressure. This
is not to say that tolerances
common to metal working are possible, but tolerances acceptable for
this application do seem feasible.
Length, width and thickness tolerances are all affected by the
quilting or pillowing effect. The
greater the thread tension and the
more resilient the insulation, the
greater the surface irregularity
caused by the stitch rows pulling
the OML and IML closer together,
resulting in larger bulges between
stitch rows. This pillowing effect
also changes final length and width
of the quilts by decreasing the
size covered by a given size material. This variation is best
handled by manufacturing oversized
quilts and trimming after sewing to
more precise dimensions. Stitches
cut during this trimming operation
may be locked by a stitch around
the final periphery either before
or after trimming. Again, for
large area coverage by multiple
quilts, the tolerances do not build
up, as they do with rigid materials, because of the flexibility
of softgoods.

2.4.2   Dimensional stability under
aerodynamic turbulence forces is
obviously not as great as with
ceramics or other rigid materials.
It seems to be, however, adequate
for the portions of the orbiter
proposed for this softgood material. Conversely, the flexibility
may permit conformance to the
natural bounday layer and actually
reduce cavitation damage hazards.

2.4.3   Insulation shift within the
quilt might result in pooling of
insulation within the individual
quilt squares. Given sufficiently
small quilt pattern (e.g., ½" or 1"
square), this pooling is not expected to occur, and if it does, the
small spacing will prevent serious
thickness change.

2.4.4   Strength at elevated temperatures is not a problem in the
AFRSI application, since the quartz
and ceramic materials considered
have excellent strength characteristics at the temperatures anticipated. In the nose cone application
it is definitely a problem, but it
appears to be a solvable one.

2.4.5   Strength under aerodynamic
stresses is a major factor for the
AFRSI tile replacement. Aerodynamic
testing was performed for Rockwell
by Ames Research Lab. Several
combinations of materials and manufacturing methods were tested.
Variations included stitch type
(Figure 9) and spacing, edge close-
out methods, insulation thickness,
OML and IML material type and
weight, and thread (material, type
and thickness. Figures 10 and 11
illustrate some manufacturing sample
quilts.

Fatigue resistance is only a problem
where severe aerodynamic turbulence
exists. Testing has demonstrated
the importance of minimizing abrasion of one fiber or thread against
another; especially where 180° bends
occur and it is possible for one

thread to abrade against another by
shifting the locations of the 180°
bends along each thread. This type
of fatigue can be prevented by
anchoring the loops with RTV in
cases like stitch type 301 modified
and 401 (chain stitch) shown in
Figure 9.

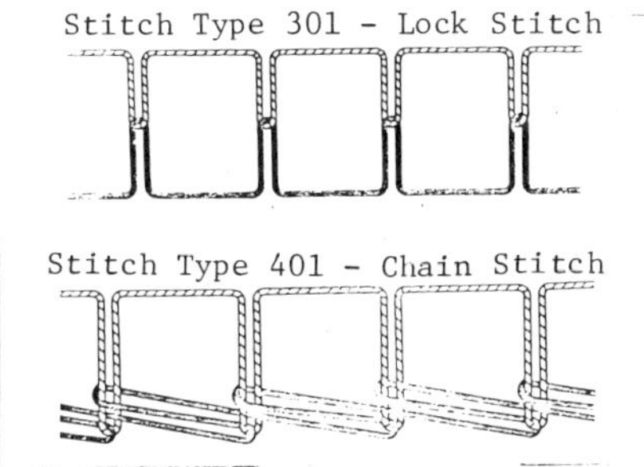

Stitch Type 301 - Lock Stitch

Stitch Type 401 - Chain Stitch

(Modified Type 301 places loop at
bottom of material)

FIGURE 9

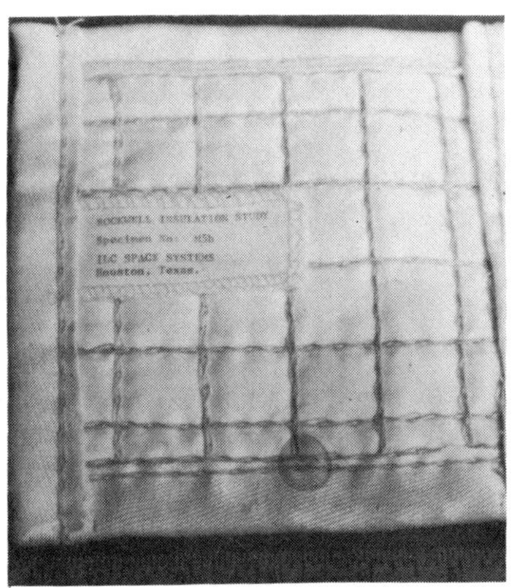

Sample Quilt, Orbiter Side, 1"
Square Quilt (Grid) Q-18 Thread,
Chain Stitch
FIGURE 10

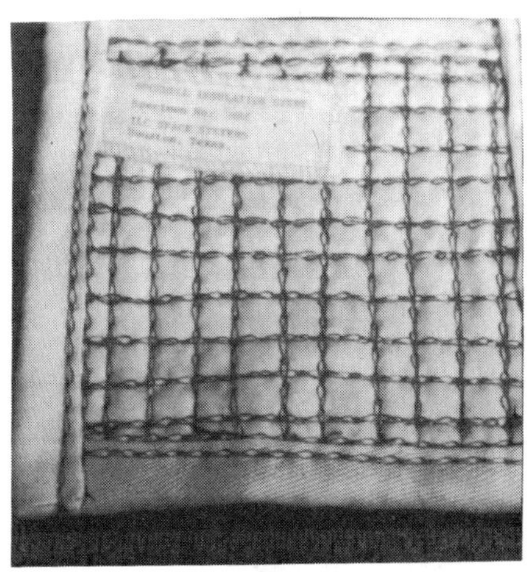

Sample Quilt, Orbiter Side, 1/2"
Quilt, 5 Mil SS Thread, Chain Stitch
FIGURE 11

Unmodified stitch type 301 (lock
stitch) is not practical because it
is not possible to anchor the 180°
bends (loops) with RTV. Figures 12
and 13 are enlargements of loops to
illustrate the ease with which loops
might be anchored with RTV. Note in
Figure 12 that 5 mil SS (stainless
steel) thread involves much looser,
larger loops because of the extremely low tension required.

2.4.7   Quality Assurance uncertainties may be less difficult than
with ceramic tile. Working in areas
most likely to have experienced
dangerously high loads or fatigue
environments, individual threads
could be carefully tested using
round pointed picks and tensiometers.
Microscopic and/or macroscopic
surface (OML) yarn examination would
also be valuable in determining
both OML and thread condition after
each mission.

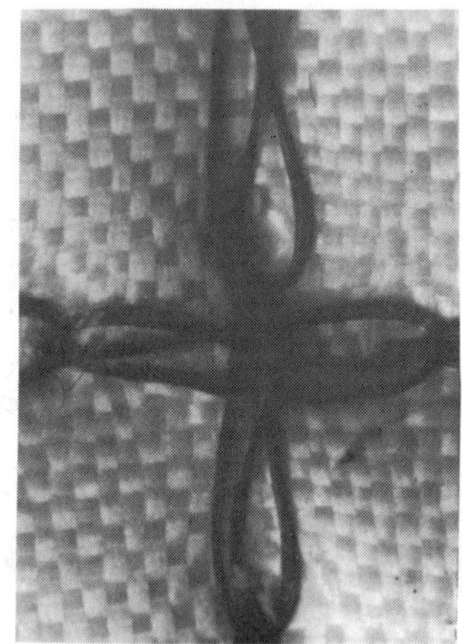

Enlarged 5 Mil SS Chain Stitches
FIGURE 12

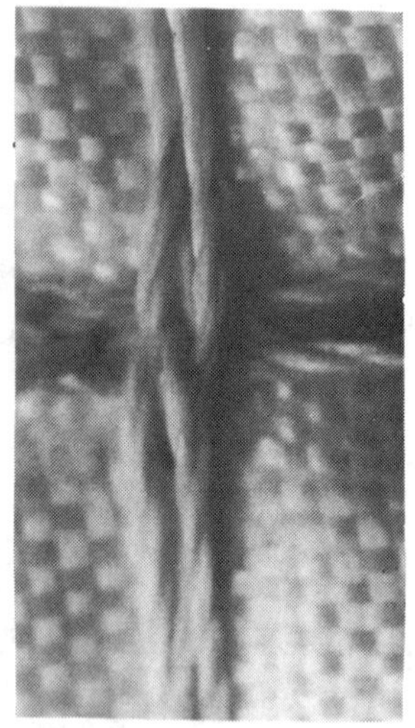

Enlarged Q-18 Chain Stitches
FIGURE 13

2.4.8 Weight limitations are always a problem with spacecraft. Although lightweight is often a major advantage for softgoods, this is not necessarily true in the AFRSI application. The reason for this is that the microquartz insulation used in the AFRSI softgoods is precisely the same material as that used in the hard ceramic tile. That used in the tile is simply fused to bond the loose fibers together to form a solid which looks and feels much like the common styrofoam coffee cup material. The OML, IML and thread typically weigh slightly more than the thin, hard ceramic shell which has been formed over the fused ceramic fiber insulation.

2.4.9 Thermal insulation characteristics and valves are very similar to those of hard tile, since the major material is the same, and other materials are so similar.

2.4.10 Moisture adsorption and retention may be a serious problem, although several possible solutions exist. Microquartz does not absorb moisture, but it does adsorb it. Fog and gentle rains would be unlikely to present problems, but a driving rain might force sufficient moisture through an unprotected OML to saturate the microquartz. In this event, pooling of microquartz might occur, since weight of the water would tend to cause it to collapse to the bottom of each quilted square. The resulting surface irregularity and/or weight increase might be unacceptable. OML sealing methods which would prevent moisture penetration appear to afford a practical solution to this problem. The difficulty is in finding a sealant that is suitable in all respects and under all conditions. A simple acrylic spray for example, would provide adequate prelaunch protection against moisture and would probably survive cloud penetrations during launch. Re-entry heating would, however, burn the sealer away so that immediate post-flight protection would be lost. Similar protection would be afforded by a material such as polyethylene taped over the quilts. Neither would be likely to present problems during their natural removal during launch or re-entry. Sealing methods allowing breathing during pressure changes were also considered. Weight penalty of sealing methods is again a factor. The most desirable sealer would, of course, last many missions, pass air but not water, and have minimal weight.

2.5   Study and Testing RE AFRSI Tile Replacement

A study and associated testing was conducted by ILC, under contract to Rockwell International to optimize the design of softgoods to replace white tiles.   These included sewing of numerous 1/4 to 1 1/3 square foot samples employing a variety of materials, quilt patterns, stitch types, edge binding methods, and combinations of same.

Some of these samples were evaluated analytically, others by wind tunnel test, plasma arc, guarded hot plate, tensile tests, and a driving rain test facility.

2.5.1   Several types and weights of glass and quartz cloth were used as OML and IML (outer and inner material layers).   Variations in OML and IML weight and material made only minor differences in the manufacturing process and the test results.

Various methods of reducing insulation compaction at and between seams for the purposes of thickness control and weight reduction met with limited success.

2.5.2   Several types of threads were employed.   Several types of each of three materials (glass, quartz, and stainless steel) of threads were tested and used in the manufacturing process.   These presented excellent illustrations of the complexity of the softgoods approach.   The success of the machine sewing of a thread varies with several critical machine adjustments, the physical characteristics of the thread, the materials being sewn, and even the speed of the sewing operation.   A typical speed of 1,000 stitches/minute, for example, is not at all practical with sewing with 5 mil SS thread which consists of 60 strands of extremely fine oxidized stainless filament.   This thread must be sewn with a very slow, uniform speed and the lightest possible tension on both top and bottom threads.   Balling, or tangling occurs when even minor tension is applied in pulling this thread over the many moderate to sharp edges in the thread feed, looping and needling paths.   As a result, it is a very poor machine sewing thread.   It obviously also slows the manufacturing process.

A simple illustration of the sensitivity of this thread to the separation that leads to balling and tangling is achieved with one's thumbnail. Simply grip a short length of 5 mil SS thread between thumb and index finger of each hand. Bend one thumb so that the thumbnail presses the thread approximately 1/32" into the finger and pull steadily. When you release the tension the portion which passed over the thumbnail will enlarge into a series of mini coils and have a fuzzy appearance. The diameter of these coils will be greater than 1/16". When, during the sewing process, this occurs and some of these fine filaments break, they tend to gather other filaments and break them also, so that a ball is soon formed. The ball, of course, results in breaking of the remaining filaments.

Threads are made like rope, typically. A common method is accomplished by twisting a definite (usually relatively small) number of staple or filament fibers into yarns, and then combining several of these yarns, by twisting, into a larger yarn. A staple fiber is a short one that requires twisting to achieve the necessary friction and self tightening effect to obtain tensile strength. Staple fibers are short enough to be measured in inches. Filament fibers may be continuous, but are at least long enough to be measured in yards. Twisting may not actually be necessary in the case of filament fibers; e.g., monofilament fishing line.

Some threads tested during the AFRSI panel manufacturing process had insufficient degree of twist or staple length. Insufficient degree or twist (too few twists per inch) and/or staple length result in insufficient strength to withstand the flexing and tension of the machine sewing operation.

Thread binders, such as sintered and unsintered teflon, also influence the flexibility, strength and lubrication of sewing threads. Silicon oil and similar lubricants are often used to improve the machine sewability of threads.

Although the sewing of high temperature materials and indeed softgoods in general is often held in awe as something approaching black magic, it is actually quite scientific. The textiles, sewing machines and manufacturing methods are all quite modern and highly technical. Like any technical process, it does appear mysterious at first glance, and may appear mysterious after years of study. But given sufficient time to analyze the materials, machines and processes, an adequate technical background and sufficient empirical data, the process is quite predictable. At the time of this writing, there are still some mysteries, or unsolved problems, to be more exact.

2.6   <u>Tensile Tests</u>

The tensile load capacities of the threads are an unknown that must be evaluated to determine the capability of the AFRSI quilts to withstand re-entry aerodynamic forces. One method selected for this type test was to enclose a bladder within the quilt. Several rubber-like materials were selected for tests. The bladders will be sealed or clamped around the microquartz insulation before the OML and IML are sewn in place. These bladders will then be inflated in an attempt to determine the tension capacity of the quilting stitches.

2.7   Edge Binding Techniques

Numerous possible methods of securing the edges of adjacent panels were evaluated analytically. Detailed design concepts were completed for the more promising approaches. These methods are being studied at the time of this writing. It is obvious, however, that the edges of adjacent quilts must be firmly secured to the orbiter structure and/or one another. Smoothness of the transition is also a major consideration in edge joining technique studies.

2.8   Repair Techniques

Advanced studies are scheduled for work under an existing contract to consider repair techniques. The merits of repair vs. total replacement of damaged AFRSI quilts seem to justify an extensive effort. Such repairs must obviously result in a product equal to the original. Likewise, quality determinations must be feasible.

3.   RESULTS AND CONCLUSIONS

Softgoods are indeed far more vital and versatile than most would imagine, even in applications where stresses, fatigue and high temperatures are severe.

Recent technological softgoods advances have permitted breakthroughs such as manpowered aircraft, as well as numerous other high stress and high temperature applications which would not otherwise have been possible.

The Space Shuttle is literally filled and covered with softgoods. Some of these items are nearly as conventional as the clothing you wear; others revolutionary and exotic.

The AFRSI quilts for white tile replacement have both drawbacks and merits. For a reusable space vehicle requiring high reliability and safety, it appears that softgoods may represent the best solution.

ION PLATING - A UNIQUE COATING PROCESS
R. A. Whitbeck
Materials and Process Laboratory
Autonetics Strategic Systems Division
Rockwell International
Anaheim, California

Abstract

Ion plating is a vacuum coating process which offers much promise for high reliability and performance. In the process, source material is evaporated by an electron gun, ionized in a plasma and accelerated to the substrate under an electrical bias of up to 5000 volts. The effects of process variables such as pressure and substrate orientation are discussed. Benefits to aerospace hardware accrue from (1) a pollution free process carried out in an ultra clean environment, (2) plating rates for some applications which are comparable to electroplating, (3) good adhesion to ceramic substrates without the use of a primer, and (4) deposition of the refractory metals such as tantalum, tungsten, osmium, and the metals which are not electroplatable such as aluminum and titanium. Examples of these will be cited.

Keywords: Vapor plating, metal films, wear resistance, cleaning, refractory metals, ceramic finishing.

1. INTRODUCTION

Ceramic and glass parts are used in navigation instruments. These materials often require metallization to impart surface conductivity, wear resistance or corrosion resistance. Furthermore, it is difficult to achieve good adhesion of metals to ceramics. The processes used presently are complex entailing many separate steps. Two instrument components are considered for the application of ion plating. Figure 1 shows an electromagnetic accelerometer proof mass. A conductive pattern is deposited by sputtering a layer of chromium to impart adhesion to the fused quartz

Figure 1. Accelerometer Proof Mass

which is followed by a layer of
gold. The goal is to provide an
adherent deposit of gold without
the use of chromium. This is
desirable for two reasons:

(1) The process would be
    simplified.

(2) The chromium layer is be-
    lieved to cause instrument
    drift.

Figure 2 shows two electrostatic
gyro cavities machined from beryllium oxide. The electrode material
is vacuum deposited chromium and
gold which is overplated with electroless nickel. In operation, a
1 centimeter diameter beryllium ball
is electrostatically suspended in
the spherical cavity formed by two
halves. The ball is electronically
spun up to approximately 2500 revo-
lutions per second with a 300 microinch gap between ball and cavity.
Thus, if the spinning ball touches
the cavity walls for any reason,
the failure is generally catastrophic. Therefore, the goal is
to deposit a harder, more wear resistant coating system to improve
instrument life.

The ion plating process was explored to obtain improved coating
processes and materials for inertial instruments. Ion plating concepts have appeared in literature
for about 15 years.[1] It has been
developed as a pollution free process which is carried out in an
ultra clean environment. Plating
rates, throwing power and adhesion
can be compared favorably with
other vacuum processes and with
electroplating.[2][3]

2. MATERIALS AND METHODS

2.1 Plating Equipment

The ion plating equipment shown in
Figure 3 utilizes an electron beam
and a single crucible source. The
electron beam is powered by an
8 kVA supply. A 2 kVA dc supply is
used for plasma generation and for
biasing substrates up to 5000 volts.
A separate RF powered plasma etch
chamber was used for precleaning
the ceramic substrates. Since the
ion plating process, its mechanism
and variations have been well explained in literature[1][4], a
brief description will suffice.

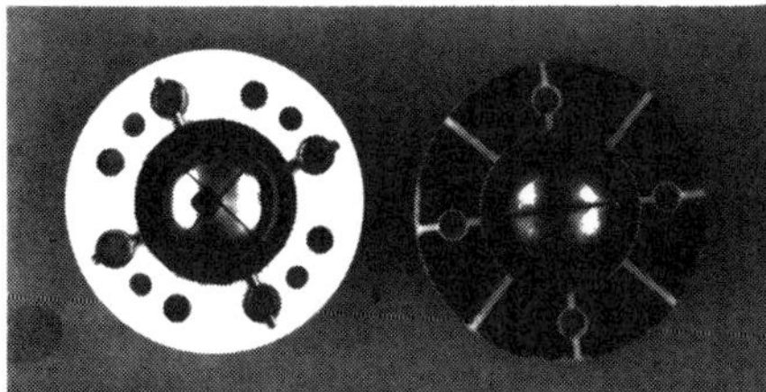

Figure 2.   Electrostatic Gyro
            Cavity Halves

Figure 3.   Ion Plater

Ion plating, shown schematically in Figure 4, is an atomistic deposition in which the substrate is negatively charged and bathed in an argon plasma. Coating material is evaporated by the electron beam, partially ionized in the plasma and accelerated to the negatively charged substrate. The substrate is glow discharge and sputter cleaned before and during deposition.

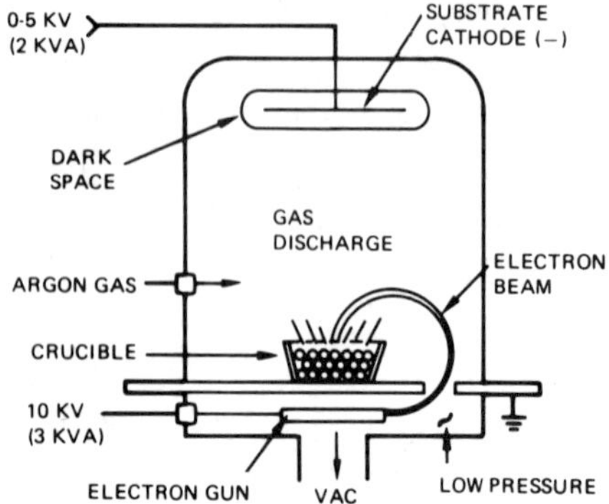

Figure 4. Ion Plating System Schematic

Substrates were held stationary at various levels above the hearth but their temperatures were not monitored. Substrate heating was accomplished by glow discharge or sputtering and by radiant heating from exposure to the hearth at a power input below the evaporation level.

2.2 <u>Wear Testing</u>

Wear testing was done with a disc-on-disc tester. One disc is spun in the horizontal plane against a stationary disc. The stationary disc is held in a flexible diaphragm and is loaded with successively more weight. Discs were made of Inconel overplated with electroless nickel and of beryllium. The discs were precision ground to slight concavity to assure contact near the periphery.

2.3 <u>Adhesion</u>

Deposit adhesion was measured by various means to obtain tensile and shear strengths. Most used was the Sebastian Adhesion Tester in which a small mandrel is cemented to the specimen and tension pulled.

2.4 <u>Materials</u>

Metals of particular interest for the work here were gold for the electromagnetic accelerometer, and osmium and rhenium for the electrostatic gyro electrodes. Table 1 shows the important properties for the two platinum group metals. Data

Table 1. Candidate Gyro Electrode Materials

	Osmium	Rhenium	Electroless Nickel
Density	22.5	20.5	7.8
Melting Point (°C)	2850	3180	900
Hardness (Kg/MM^2)	1500	1700	800
Thermal Conductance (Cal/cm-sec-°C)	0.21	0.17	0.011
Coefficient Thermal Expansion (10^{-6}/°C)	6.1	6.6	13.5
Electric Resistivity ($\mu\Omega$-cm)	8.1	19.1	60.0

for electroless nickel is shown for comparison since this material is used presently in electrostatic gyros. Properties of major importance are hardness (for wear resistance), thermal conductivity and high melting point (for dissipation of heat), and thermal expansion coefficient (to match beryllia).

3. RESULTS AND DISCUSSION

Preliminary experiments were conducted in plating a variety of coatings on several substrates. Information was obtained on the deposition rate, the degree of adhesion, and the spatial distribution of deposits for various coating materials and process parameters.

3.1 Rates of Metal Deposition

Figure 5 shows some experimental plating rates. The rates are a function of the equipment and parameter selection and not necessarily of the ion plating process. For tungsten and tantalum (rates not shown), the full output of the electron beam was applied to the two inch crucible. Osmium and rhenium were evaporated from an insert crucible of 3/4 inch diam-

eter after melting down their powder to crucible shape in the plating equipment. The crucible insert insulated the melt from the water cooled main crucible, thus allowing a greater heat density at lower power settings. The smaller crucible also restricts the evaporating pool area and thereby limits the plating rate.

The source to substrate distance was generally 7 to 9 inches. Figure 6 shows the thickness of nickel for 15 minute plating periods and various conditions. At the lower pressure shown by curves A and B, there is little throwing power, virtually all coating material going to the front face. "B" shows an exponential relationship of rate to distance. It is not the

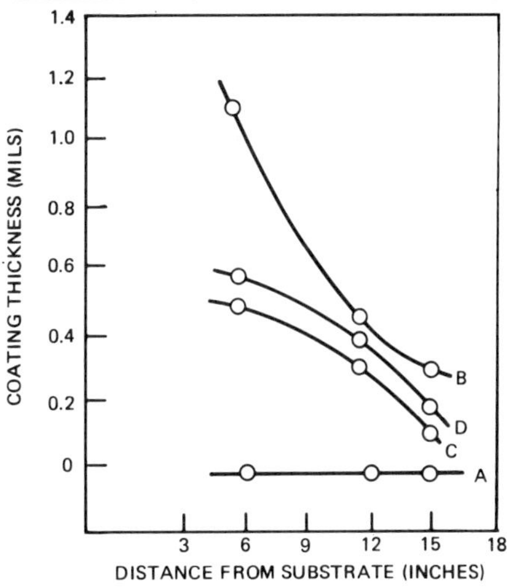

SUBSTRATE BIAS (KV)	2		0	
ARGON PRESSURE (TORR)	8×10^{-6}		6×10^{-3}	
ORIENTATION (TO SOURCE)	FRONT	BACK	FRONT	BACK
DESIGNATION	B	A	C	D

Figure 6. Coating Thickness vs Distance of Substrate from Source for Ion Plated on Aluminum

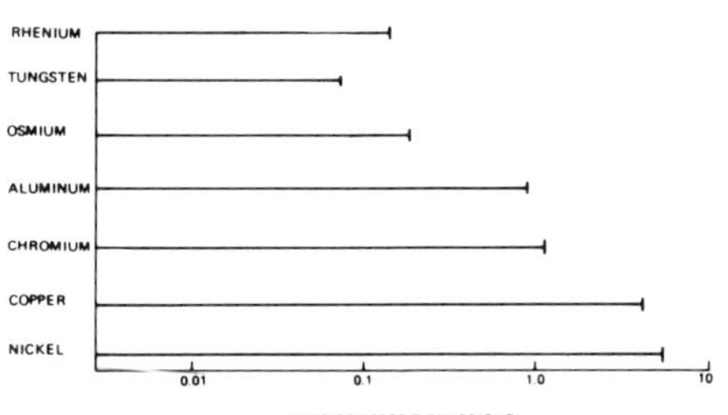

Figure 5. Ion Plating Rates for Some Elements

distance squared relationship which
is typical of vacuum evaporation.
At 6×10^{-3} Torr, curves C and D
show nearly uniform front/back
coverage and a nearly linear rate-
distance relationship.

3.2 Plating Distribution

The plating rate, as shown also in
Figure 6, and the spatial distribution of plated deposits are both
strongly dependent on pressure.[5]
Figure 7 illustrates the effect of
impact scattering for a flat specimen with one side facing the source.
However, the total metal deposited
on a given flat specimen remains
the same as the evaporant is distributed to the back side. It is
further seen that substrate bias
has no measurable effect on
distribution.

For fabrication of gyro electrodes,
the desired approach was to totally
replace the chrome-gold-electroless
nickel with osmium or rhenium.
Plate distribution readings were
taken of osmium deposits on cavity
surfaces. The cavity was held
stationary facing the source crucible and plated for 10 minutes in an
argon pressure of 3 milliTorr and
a substrate-source distance of
7 inches. The distribution shown
in Figure 8 is not good since finish machining calls for 1 microinch
(total indicated runout). However,
it is better than ordinary evaporation and sputtering methods which
yield nearly zero deposit under
the cavity lip at the 5 degree location. Cavity rotation coupled
with a small pressure increase can
be expected to generate the desired
uniformity of deposit.

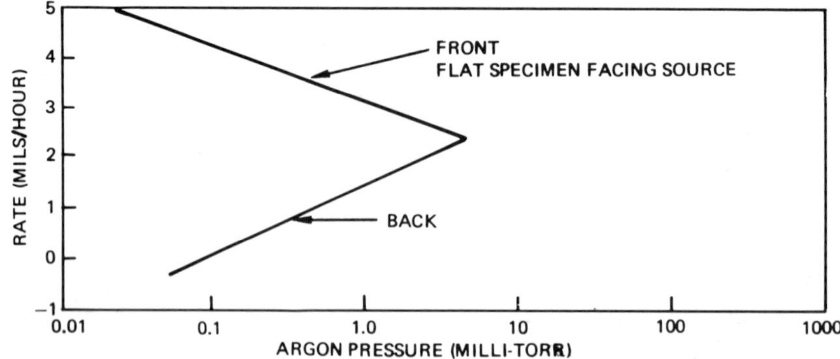

Figure 7. Ion Plating Distribution

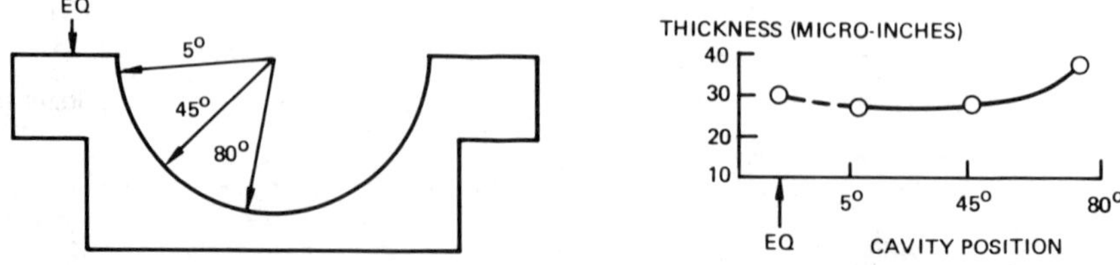

Figure 8. Distribution of Osmium Plate in ESG Cavity

3.3 Adhesion of Plated Materials

Adhesion test results shown in Table 2 did not, in general, yield a value for interfacial strength. If the tape test passed, the capacity of the next test was usually exceeded. Substrate fracture occurred in a few cases. The basic processes used are indicated by letter in Table 2 and cated by letter in Table 2 and shown in Table 3. To improve impact energy of the evaporant upon dielectric substrates, a fine wire charged grid was placed in front of the silica wafers. No improvement was realized, however. D.M. Mattox faced the problem of depositing gold on silica in 1965.[7] He found that gold could be sputtered to an adherent deposit on fuzed

Table 2. Adhesion Values for Ion Plated Metals over Various Substrates.

Plate	Substrate	Process	Adhesion Test	Strength (psi)	Remarks
Au	Silica	A	Tape	–	Failed
Au	Silica	B	Shear, Peel, Die Bond Tester	620	No bond failure (NF)
Au	Silica	B	Tensile, Soldered Wire	950	NF, SiO_2 fractured
Au	Silica	C	Tensile, Soldered Wire	1100	NF, Wire pulled out of solder joint
Os	Beryllia	B	Tensile, Sebastian Tester	8700	NF
Os	E'less Ni over BeO	D	Sebastian	8700	NF
Os	Glass	B	Sebastian	7600	NF Glass fractured
Re	E'less Ni	D	Sebastian	8700	NF
Cr	Beryllium	D	Tensile, Instron	10100	NF, 1" dia mandrel

Table 3. Cleaning and Plating Processes Referenced in Table 2

	A	B	C	D
PRECLEAN	a. Boiling Aqua Regia (AR) 1 minute	a. Boiling AR, 1 minute b. RF Plasma etch, air, 3 minutes	a. Boiling AR, 1 minute b. RF Plasma etch, air, 3 minutes	a. Acetone Wipe
ION PLATE	a. Glow discharge -Argon - 10 milliTorr -2 Kv/300 mA -1 minute b. Plate 3 minutes -E beam 1400 Watts -bias - 1-1/2 Kv -Argon - 6 milliTorr	a. Glow discharge -Argon - 10 milliTorr -1-1/2 Kv/200 mA -5 minutes b. Preheat substrate c. Plate 3 minutes -E beam 1400 Watts -bias - 1-1/2 Kv -Argon - 6 milliTorr d. Post bake 800 F/20 minutes (optional)	a. Glow discharge -Argon - 10 milliTorr -1-1/2 FV/200 mA -5 minutes b. Preheat substrate c. Switch to O_2 plasma Plate 3 minutes -E beam 1400 Watts -bias 1-1/2 Kv -O_2 5x10-4 Torr	a. Sputter clean -Argon - 6 milliTorr -2-1/2 Kv/100 mA -10 minutes b. Preheat substrate c. Plate 15 minutes -E beam 3500 Watts -bias 1-1/2 Kv -Argon - 6 milliTorr

silica in a partial pressure of oxygen. We tried air without success. Pure oxygen as shown in process C worked well. Again, much has been written about adhesion and ion plating.[1][8] The literature states that the graded interface is the important aspect of adhesion. It is said to be caused by (1) high impact energy resulting in ion implantation to depths of about 50 Å, (2) high surface temperatures causing diffusion, (3) bombardment induced effects, and (4) pre-deposition sputtering of substrate. All of these four are at work on a metal substrate in a dc biased ion plating set up. The first three must do the job on dielectric substrates. Auger spectrometry of an osmium deposit over electroless nickel indicates a diffused interface of a little over 1000 Å as shown in Figure 9. A similar Auger analysis for gold over silica shows about a 500 Å graded region. Baking appeared to aid the gold-silica adhesion as did the act of soldering for the tensile test. The extent of the improvement is not defined in these tests nor is the contribution of precleaning by plasma etch. However, the process combinations found in B, C and D, Table 3, resulted in functional satisfaction.

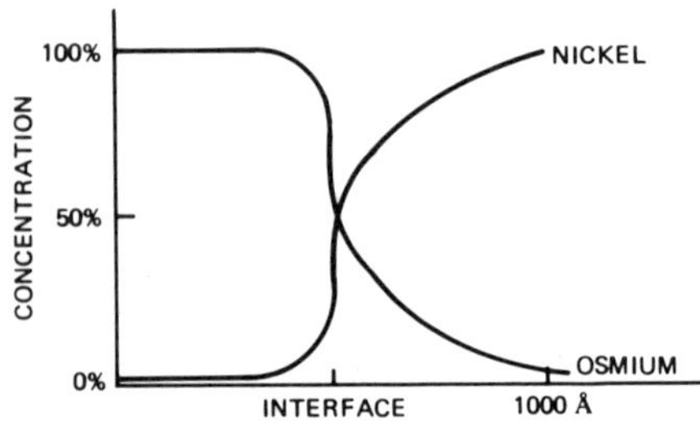

Figure 9. Auger Spectrometer Ion-Plated Osmium of Electroless Nickel

3.4 Wear Resistance of Plated Metals

Beryllium surfaces against nickel in this test self-destruct quickly (see Table 4). STP applied as a boundary lubricant or molybdenum

Table 4. Dynamic Wear Test: Disc-on-Disc for Ion Plated Specimens

Test No.	Rotating Disc Base Coat Lube			Static Disc Base Coat Lube			Load (gm)	Start Torque (gm-cm)	Run Torque (gm-cm)	Time (min)	Remarks
1	Be	–	–	Ni	–	–	165	75	–	<2	Vibration: stick-slip galled with Ni trans. to Be. Be torn.
2	Be	–	STP	Ni	–	STP	165	30	–	<5	
3	Be	–	MOS_2	Ni	–	MOS_2	165	50	–	<5	Stick-slip vibration.
4	Be	–	–	Ni	Os	–	165	–	–	–	Stick-slip vibration.
5	Be	–	STP	Ni	Os	STP	165	28	30	12	Smooth: Vibration toward end.
6	Be	Os	MOS_2	Ni	–	MOS_2	165	32	22	15	No start.
							212	–	–	–	
7	Be	Os	MOS_2	Ni	Os	MOS_2	165	45	40	15	Smooth
							212	60	38	15	
							260	60	40	15	Slight vibration at 260.
							307	101	71	15	
							352	123	120	2	Os gone from Be and Ni.
8	Be	Os	STP	Ni	Os	STP	165	20	22	15	Run OK.
							212	25	33	15	
							260	33	51	15	Slight vibration.
							307	60	59	15	
							352	75	66	15	Os intact on Be. Work off Ni.

disulfide sputtered on in the range
of 100 Å improves wear resistance.
The best results for the series,
and considerably better than
beryllium against nickel, was
osmium plate on both discs with the
oil additive STP acting as a
boundary lube. Plating the gyro
rotor appears out of the question
at this time because of extremely
small tolerances for weight, balance and dimensions. Additionally,
anything that may outgas or generate particulates is disqualified
since this gyro is electrostatically levitated in high
vacuum.

In summary, osmium plated smoothly.
A 15 microinch thick plate of osmium
on glass is shown in Figure 10 in
fracture cross section. The highest
plating rate obtained in our equipment for osmium was 180 microinches/
hour. The thickest specimen of plate
was 80 microinches with a hardness
estimated at about 1000 to 1200
Kg/mm^2. X-ray diffraction showed a
very fine polycrystalline structure.

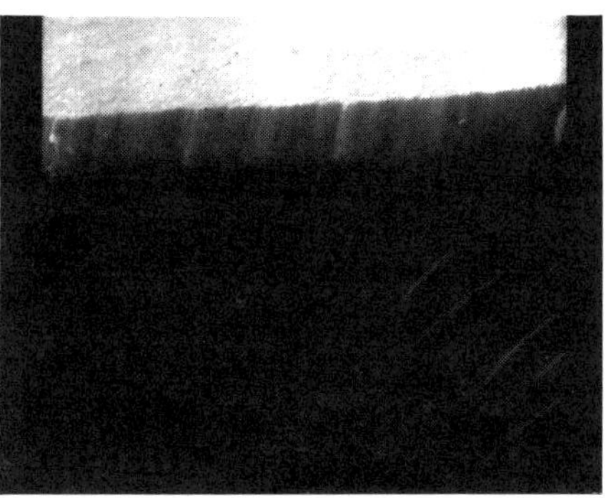

Figure 10. Scanning Electron
Micrograph Fracture
Cross Section of Ion
Plated Osmium

4. SUMMARY

Significant advances were made in
advancing ion plating technology in
this study program. First, it was
shown that plating can be achieved
with high boiling metals on ceramic
substrates. Excellent adhesion and
wear properties were obtained.
Secondly, the process variables of
temperature, pressure, electrode
spacing, geometry and applied voltage were pinned down for different
metals and applications. Thirdly,
the plating rates of metals were
shown to be a function of the equipment rather than inherent to the
metal properties. The spatial distribution of ion-plated deposits was
shown to be uniform rather than line-
of-sight so the process is ideally
suited for many applications. The
unique advantage of ion-plating is
that cleanliness can be maintained
and the bonding strengths of plated
deposits to a variety of substrates
are exceptionally good.

Ion plating, in the several forms in
which it is developing today, appears
to have great versatility and applicability to aerospace hardware.

5. REFERENCES

1. D. M. Mattox, Bibliography, Ion
    Plating and Related Ion Bombard-
    ment Effects; Prepared by Sandia
    Laboratories for United States
    Energy R&D Administration

2. D. L. Chambers and D. C.
    Carmichael, "Electron Beam
    Techniques for Ion Plating,"
    R/D Magazine 22 No. 5, 32
    (1971)

3. R. Carpenter, "Ion Plating:
    Increasing Adhesion of
    Deposited Films" EPP Inter-
    national 7, 5 (1971)

4. R. F. Bunshah, "Hard Coatings
    for Wear Resistance by Physical
    Vapor Deposition Processes"
    Sampe Quarterly, 12 No. 1, 1980

5. Chambers and Carmichael,
    Research and Development
    Magazine, 22 32, 1971

6. J. A. Thornton, "High Rate
    Thick Film Growth" Ann. Rev.
    Material, Sci. 7, 239, 1977

7. D. M. Mattox, "Influence of
    Oxygen on Adherence of Gold
    Films to Oxide Substrates"
    J. of Applied Physics, 37,
    9, August 1966

8. D. G. Teer, "The Ion Plating
    Technique" Proceedings of the
    Conference on Ion Plating and
    Allied Techniques, Edinburgh,
    June 1977

BIOGRAPHY

RICHARD A. WHITBECK

Mr. Whitbeck is a member of the
Materials and Processes Staff at
Rockwell International, Anaheim,
California. He has been responsible
for projects in metal finishing and
related processes for the past
13 years at Rockwell International.
He has 30 years experience in this
field.

Mr. Whitbeck received the BS degree
in chemistry from the University of
California at Berkeley. He is a
contributor to the Metals Handbook,
published by the American Society
for Metals.

RIGID-FLEXIBLE MULTILAYER INTERCONNECTS

J. M. Shaheen
Materials and Process Laboratories,
Autonetics Strategic Systems Division,
Anaheim, California

Abstract

Rigid-flexible multilayer interconnects are finding new applications in aerospace designs and
products. They are hybrid structures and combine the advantages of
rigid circuit boards (for easy
assembly of components) with those
of flexible cables (which eliminate
the need for wires and ribbon
tapes). Although these parts are
used by many leading companies,
improvements are needed. Several
factors must be better understood
in order to minimize their cost and
maximize their reliability. This
discussion centers on one factor
which determines performance,
namely the thermal expansion of
adhesively bonded laminates and
the effect which this has on the
through-hole reliability. It addresses the reasons why these
laminates expand differently from
other interconnects and demonstrates
methods for correcting this
difficiency.

Keywords: laminated plastics,
circuit interconnects, thermal
expansion, electronic reliability,
boards, adhesives, polyimides,
acrylics.

1. INTRODUCTION

Currently, the rigid multilayer
interconnection boards used in
electronic hardware systems are
interconnected with wires or tape
ribbon cables. There is a connector
pair at each transition from the
boards to the flexible portion.
This type of construction has many
disadvantages including weight and
volume penalties, reliability and
cost factors which increase with
every solder joint.

New interconnect technology is needed to achieve better performance and
lower costs for future programs.
The solution to the problem is to
design and fabricate interconnect
systems with fewer solder connections. Rigid-flexible multilayer
interconnects (abbreviated to R/F
MIBs) are especially suited for low
cost mass terminations, high packaging density, and low weight
applications. They accomplish this
by combining rigid and flexible circuit boards into laminated composites. These are then drilled
and plated so that the interconnections can all be accomplished
simultaneously. Figure 1 shows a
typical rigid-flexible multilayer

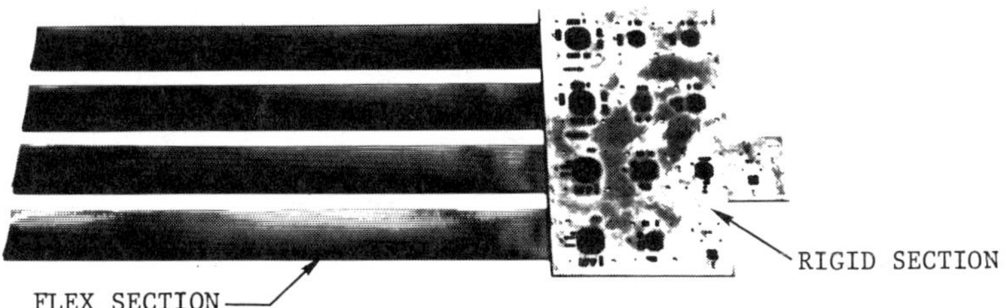

Figure 1. R/F Multilayer Interconnect

interconnect which was designed and fabricated for Autonetics Marine Systems Division.

During the past year, Rockwell International conducted an investigation of R/F MIBs and succeeded in characterizing them for performance and reliability. Both materials and combinations were considered from the standpoint of producibility and integrity of the finished product. The work reported deals with one phase of the investigation which considered the expansion characteristics of the composite laminate.

2. MATERIALS AND METHODS

2.1 Types of Construction

From a survey conducted with several manufacturers, it was revealed that R/F MIBs exist in a variety of constructions using different materials. Table 1 lists the materials which are most commonly used.

Table 1. Materials for R/F MIBs

	Rigid	Flexible	Adhesive
Epoxy-Glass	X		
Polyimide-Glass	X		
Kapton/Acrylic		X	
Epoxy-Glass Prepreg			X
Acrylic			X

These materials are combined together as laminated composites and generally rolled annealed copper is used as the conductor on the flexible portion of the laminate while electrodeposited copper is used on the rigid portion. R/F MIBs can be constructed as follows:

(a) With all flexible layers as shown in Figure 2

(b) With rigid outside layers and flexible internal layers as shown in Figure 3

(c) With alternating layers of flexible and rigid as shown in Figure 4.

2.2 Test Samples

Several R/F MIBs were obtained from the different manufacturers as well as our internal prototype facility. These structures represented a selection of R/F MIBs presently available. Table 2 is a characterization of these sample parts.

A variety of constructions are represented. Some use epoxy-glass as the rigid portion of the boards whereas others use polyimide-glass. Some use epoxy-glass prepreg as the adhesive. However, acrylic is commonly used as the adhesive. Kapton/acrylic is used as the flexible portion of the laminate.

Figure 5 illustrates Sample No. 2 of Table 2 which uses epoxy-glass as a stiffener, and the rigid board was not involved in the plated-through-hole interconnect. All the other boards involve the rigid section as part of the plated-through-hole.

Sample No. 1 did not use plasma for etchback, the Kapton film was pre-punched in the pad areas and these areas were filled with epoxy resin during the lamination cycle. Epoxy-glass prepreg was used as the adhesive. Cleaning of the drilled hole was chemical and etchback was insignificant. This is shown in Figure 6.

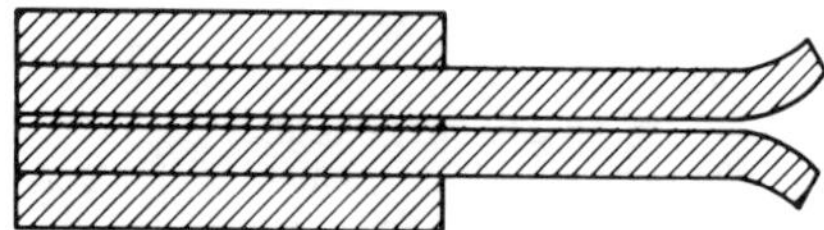

Figure 2. All Flexible Layers

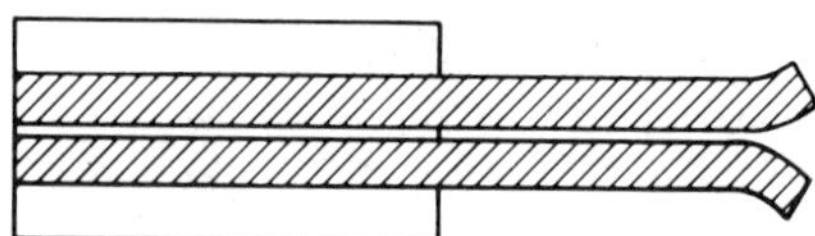

Figure 3. Rigid Layer Outside, Flexible Layer Inside

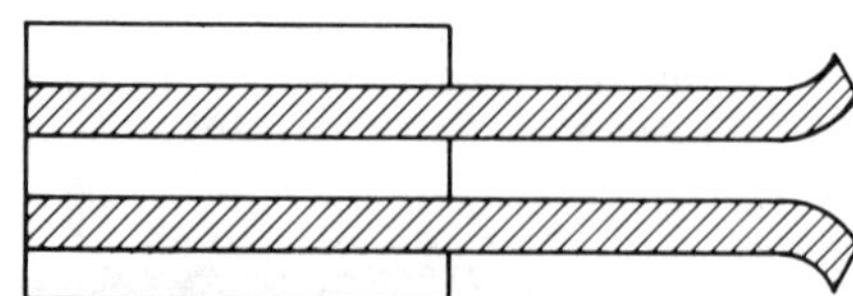

Figure 4. Alternating Layers of Flexible and Rigid

Table 2. R/F Multilayer-Interconnections Tested

Characteristic	Manufacturers' Parts						In-House Parts		
	1	2	3	4	5	6	7	8	9
Rigid Material	←— EG —→						←— PIG —→		
Flexible Material	←————————— KA —————————→								
Adhesive	EG	←——————— A ———————→							
Construction	RFR	FR	←——— RFR ———→				F	RFR	RFRFR
Foil Thickness (oz)	2	←— 1 —→					←— 2 —→		
Laminate Thickness (inch)	0.074	0.064	0.057	0.025	0.057	0.036	0.050	0.052	0.055
Plated Hole Wall Thickness (inch)	0.0014	0.001	0.0014	0.0012	0.0014	0.0012	0.001	0.001	0.001

EG = Epoxy-Glass   RFR = Rigid outside layers, flexible internal layers

PIG = Polyimide-Glass   RF = Flexible circuit mounted on rigid stiffener

K/A = Kapton/Acrylic   F = All flexible layers

A = Acrylic   RFRFR = Alternating layers of R/F

Figure 5. R/F Board No. 2 Table 1
Use of Epoxy-Glass as a
Stiffener

Figure 6. R/F Board No. 1 Table 1
Use of Epoxy-Glass as
Adhesive

The remaining samples with the exception of No. 3 used a plasma to clean the drilled hole. Figure 7 shows a typical plasma cleaned hole wall. It is positively etched back. However a nonuniform hole wall results due to the different etch rates of the materials involved.

Sample No. 3 has a hole wall with negative etched back, (see Figure 8). The internal copper circuit layer has been chemically removed some distance. It is not known what was used as a hole cleaner in this sample.

Samples No. 7 and 9 were fabricated at Rockwell International and provided structures which supplemented those from the various manufacturers. One board used all flexible circuit layers and the other board used alternating layers of rigid and flexible circuits. All such boards were plated with acid copper with 10-20 percent elongation.

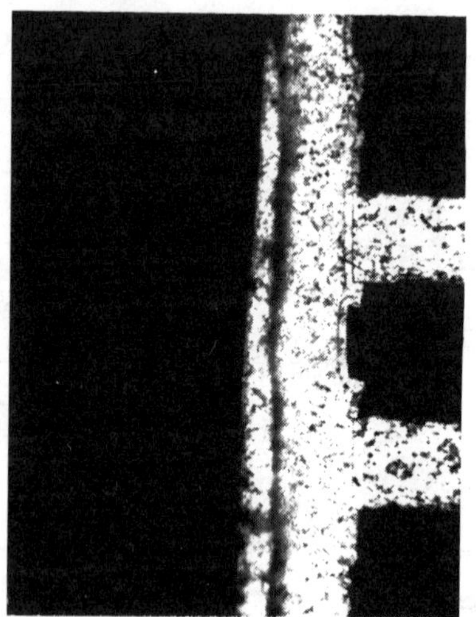

Figure 7. R/F Board No. 4 Table 1
Plasma Cleaned Hole
Positive Etchback

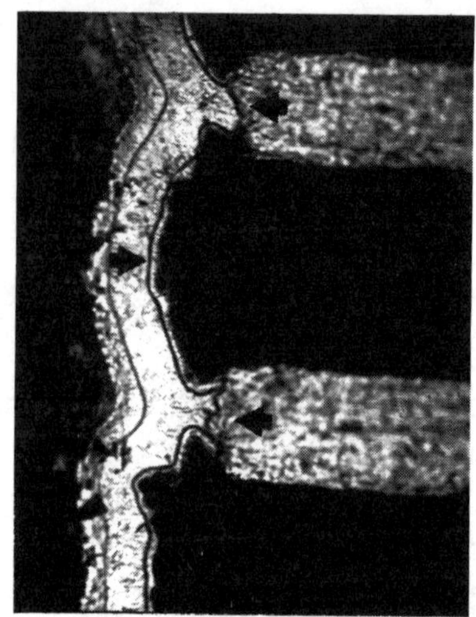

Figure 8. R/F Board No. 3 Table 1
Negative Etchback

2.3 Stress Testing

The parts were stress tested by solder float per MIL-P-55110. This test is used as a screening procedure for multilayer interconnection boards. Smith, Zimmerman, and Strohmer[1] state that process problems are the major causes of copper cracks and this solder float test screen them out. In addition, a step stress test was conducted using 10 cycles respectively at three different temperatures -65 to 125°C; -65 to 150°C; and -65 to 175°C. The interconnects were continuously monitored electrically. The test used a convection oven and a freezer to provide the temperature extremes. Finally, a cryogenic shock of 10 cycles between room and liquid nitrogen temperatures was conducted on the Rockwell International samples. The failed parts were then cross sectioned and examined to determine cause of failure. Table 3 presents the results.

Table 3. Thermal Tests on R/F Multilayer Interconnects

		Sample No.								
	Test	1	2	3	4	5	6	7	8	9
1.	Solder Float 10 seconds 550F(MIL-P-55110)	F	P	F	F	F	F	P	F	F
	Type of Failure	D CC	–	D TJ	D BC	CC	D CC	–	D	D CC
2.	Step Stress 10 cycles @									
	−65 – 125°C	P	P	P	P	P	P	–	–	–
	−65 – 150°C	P	P	P	P	P	P	–	–	–
	−65 – 175°C	P	P	P	P	P	P	–	–	–
3.	Cyrogenic Shock 10 cycles, Room temperature to liquid nitrogen	–	–	–	–	–	–	P	P	P

F	= Failed	BC =	Plated through Hole Wall Crack
P	= Passed	CC =	Plated through Hole Corner Crack
D	= Delamination	TJ =	T-Joint Separation

3. RESULTS AND DISCUSSION

3.1 Thermal Resistance

The data in Table 3 show that R/F MIBs passed exposure to cryogenic temperatures and step stress testing.

Samples No. 2 and 7 passed the solder float test whereas all other samples failed. The through-hole-interconnection in both samples which passed were made in flexible circuit layers and the softness of these materials could be forgiving thereby not stressing the inter-connect. A problem associated with multilayer interconnects using all flexible materials is resolder-ability. Pads have a tendency to move during part replacement.

A preponderance of failures were identified as corner cracks. This type of failure is generally assoc-iated with high "Z" direction expansion. The differential expan-sion between copper and the laminate material applies stress to the copper barrel during thermal cycling and cracked copper results. To reinforce this assessment, a more detailed analysis of the various samples was made. The materials present in each laminate composite were identified and the volume percent of each was calcu-lated. Table 4 shows the results of this analysis.

A major component of most laminates is acrylic adhesive and for remain-ing laminates it is epoxy-glass. A determination of the relationship of these materials to the expansion characteristics of the laminates was then made.

3.2 Coefficients of Expansion

Using a Perkin-Elmer's Thermal Mechanical Analyzer, measurements of both the glass transition temperature (Tg) and the coef-ficient of expansion in the "Z" direction was then conducted. These are reported in Table 5 in comparison with standard multi-layer board materials. A temperature range of 0 to 150°C was used.

The R/F laminate composites exhibit a very low glass transition tempera-ture compared to those for epoxy-glass or polyimide glass laminates. In addition, above the Tg a rela-tively high coefficient of expan-sion is demonstrated. Thus, the R/F laminates start a high expan-sion rate at around 25°C. This is significant since the use tempera-tures of multilayer boards generally extend up to 100°C. Mechanical exercising of the copper barrel (through-holes) at operating temperatures is potentially a serious problem. If the expansion is large enough at storage and/or operating conditions, then work hardening of the copper barrel results.

Table 4. Volume Percent of Materials in R/F Multilayer Interconnects

Materials	Volume (Percent)								
	1	2	3	4	5	6	7	8	9
Copper	13	12.4	14.1	21.0	22.3	21.0	28.0	21.3	20.3
Epoxy-glass	75	72.0	59.9	33.3	-	-	-	-	-
Polyimide-glass	-	-	-	-	21.6	22.8	-	19.3	27.2
Kapton	7.5	7.5	4.3	12.3	9.1	7.5	32.0	23.0	18.1
Acrylic	4.5	8.1	21.7	33.4	47.0	48.7	40.0	36.4	34.4
Total	100	100	100	100	100	100	100	100	100

Table 5. Glass Transition Temperature and Coefficient of Linear Expansion of R/F Composites

Sample No.	Sample Thickness (inch)	Tg_1 (°C)	Tg_2 (°C)	Coefficient of Linear Expansion x 10^{-5} in/in/°C "Z" Direction			
				Below Tg_1	Above Tg_1	Between Rg_1-Tg_2	Above Tg_2
2	0.064	20	111	50.1	–	147	291
4	0.025	21	119	69.1	–	235	376
5	0.057	17	–	96.7	–	259	–
6	0.036	24	–	97.8	–	290	–
7	0.050	21	–	76	–	249	–
8	0.052	22	–	97	–	222	–
9	0.055	27	–	91	–	258	–
Epoxy-glass MLB	0.062	109	–	38	460	–	–
Polyimide-glass MLB	0.062	282	–	47	298	–	–

Furthermore, those laminates using epoxy-glass as the rigid portion exhibit two Tg's. The second Tg is that of epoxy-glass. After the second Tg is reached, expansion can be quite high, e.g. 376 microinches/in/°C for Sample No. 4 in Table 5. This fact argues toward the elimination of epoxy-glass from R/F laminates to minimize expansion.

Finally, the acrylic type of adhesive contributes to high "Z" direction expansion as well as exhibiting a very low glass transition temperature (Tg). The relationship between acrylic expansion and standard multilayer board materials is demonstrated in Figure 9.

Thus acrylic adhesive has approximately three times the expansion coefficient of epoxy-glass between room temperature and 110°C, and eight times that of polyimide-glass. The high "Z" expansion of acrylic material contributes to the change in the expansion of the laminate. It is shown in Table 5 that different coefficients of expansion exist for each composite. This difference is dependent upon the percent acrylic present in the laminates. This is unusual in MIB technology, since the insulating material of standard MIBs is both the laminate resin and adhesive combined with the copper conductor. The resultant effect was that a single expansion coefficient

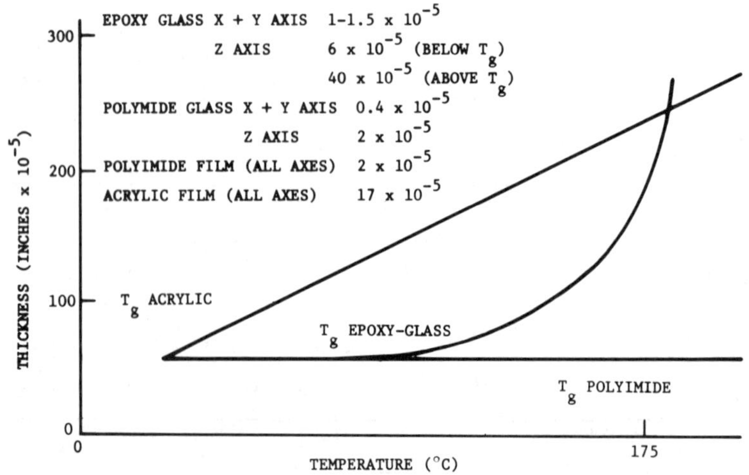

Figure 9. Expansion Characteristics of MLB Materials

existed for MIBs made with epoxy-glass or with polyimide-glass. This inconsistency in the R/F laminates indicates a need to evaluate what effect high expansion will have on these R/F interconnects and what maximum level of expansion should be stipulated.

3.3 Interconnection Integrity

Studies made by Geshner and Jolly[2] indicated that conditions leading to plated-through-hole failure will be the differential expansion of the materials, the use of low ductility copper and existence of rough hole walls. The latter condition contributes to the reliability problem by forming stress risers and/or thin spots in the deposits which can cause hole wall failure during thermal excursions.

The thin areas in the deposits also are indicative of a change in current density and thereby can effectively produce a reduction in elongation of the deposit.

The potential for having a hole wall condition which will cause stress risers and/or thin spots in the copper deposit appears to be present in R/F MIBs. The R/F MIBs, being composed of both hard and soft materials are subject to having tear outs during the drilling operations. Secondly, the composite nature of the board results in differing etch rates during hole cleaning processes thereby providing hole wall surface irregularities. This condition in conjunction with the high expansion in the laminate sets the stage for plated-through-hole wall integrity problems.

3.4 Thermal Expansion of R/F MIBs

An analysis of the thermal expansion properties of the R/F MIBs was made in an attempt to determine whether the elongation could produce copper barrel failure. Yuan[3] provided a method for analyzing thermal expansion properties of multilayer laminates to determine whether a condition for failure exists. He indicates that the thermal expansion properties of the laminate should be minimized to prevent cracks in plated-through-holes. He arrives at an expression which takes the expansion of the laminate between two temperatures and subtracts the expansion and elongation of the copper barrel. If the resultant value is greater than "0" a copper cracking condition exists; if less than "0" no cracking will occur. Table 6 calculates the amount of expansion which some of the laminates we tested would experience in the 550°F test.

It is readily seen that these laminates expand upwards 6 to 8 percent. Assuming a copper deposit of 7 percent was provided in the holes, and using Yuan's method of analysis on the No. 5 sample, the laminate will expand 0.0039 inch; minus the expansion of the copper (0.0002 inch); minus the elongation of the copper (0.0034 inch); which is greater than "0". This condition can be considered a potential failure.

The R/F MIBs tested and reported in Table 3 exhibited a high plated hole corner crack failure mode and therefore we can conclude that the expansion of those laminates was excessive.

Table 6. Expansion of R/F Laminates at 282°C (550°F)

Sample Number	Sample Thickness (Inch)	Changes in °C			Coefficient of Expansion in/in/°C			Expansion	
		Above T_{g1}	Between $T_{g1}-T_{g2}$	Between $T_{g2}-282°C$	Above T_{g1}	Between $T_{g1}-T_{g2}$	Between $T_{g2}-282°C$	(Inch)	%
2	0.064	–	91	171	–	147	291	0.0044	6.8
4	0.025	–	98	163	–	235	276	0.0021	8.4
5	0.057	265	–	–	259	–	–	0.00369	6.9
6	0.036	258	–	–	290	–	–	0.0026	7.2

3.5 Analysis of Laminate Expansion

An analysis was made using a range of expansion coefficients to determine approximately what value should be sought when fabricating R/F MIBs. A board thickness of 0.062 inch was used; copper elongation of 7 percent was assumed; and the coefficient of linear expansion in the "Z" direction was ranged between 150-400 x 10^{-6} in/in/°C. A temperature range between 25-280° was used and comparison to standard multilayer board materials was made. Table 7 reports and shows that R/F MIBs with linear coefficient of expansion above 250 x 10^{-6} in/in/°C have thermal expansion characteristics which are greater than "0" and therefore can be considered as potential problems. The data also show that the epoxy-glass MIB materials also have a thermal expansion greater than "0". The ability to control the irregularities of an epoxy-glass hole wall are better than the composite structure of R/F MIBs and therefore the production of stress and/or reduction in the elongation of plated copper would be considerably less. Finally the data show that the PI glass MIB materials show very low expansion. This was also pointed out by Quintana[4], who recommended use of polyimide-glass MIBs as a method to prevent plated-through-hole cracking.

3.6 Control of Expansion in R/F MIBs

Methods to control the coefficient of expansion of R/F MIBs were considered. Formulations of composite laminates were made such that the materials exhibiting low expansion were increased in volume percent. The coefficient of expansion of these composites was determined. One approach used an increased thickness of the PI-glass rigid boards. A second approach changed the thickness of adhesive between layers. Table 8 reports the formulation differences and the test results.

By increasing the thickness of the rigid portion of the board and maintaining all other materials constant, a change in coefficient from 292 microinches/inch/°C to 157 microinches/inch/°C is effected. Compare boards No. 3, 4, and 1 of Table 8. The polyimide glass thickness was changed from 0.010 inch to 0.020 inch and 0.040 inch respectively. By increasing the adhesive thickness between layers and maintaining all other materials constant a change in the coefficient from 224 microinches/inch/°C to 275 microinches/inch/°C is effected. Compare boards No. 2 and 4 where the adhesive was reduced from 0.024 inch to 0.017 inch respectively.

Table 7. Thermal Expansion Properties of Multilayer Laminates Using a Range of Expansion Coefficients

"Z" Direction Linear Coefficient of Expansion x 10^{-6} in/in/°C	Laminate Thickness (inch)	Temperature Change 25-280° (°C)	Expansion of Laminate (inch)	Expansion of Copper Barrel (inch)	Elongation of Copper @ 7% (inch)	Relation to "0"
400	0.062	255	0.0063	0.0002	0.0043	>0
350	0.062	255	0.0054	0.0002	0.0043	>0
300	0.062	255	0.0047	0.0002	0.0043	>0
250	0.062	255	0.0040	0.0002	0.0043	<0
200	0.062	255	0.0032	0.0002	0.0043	<0
150	0.062	255	0.0024	0.0002	0.0043	<0
Epoxy-Glass 38 x 10^{-6} Below Tg 460 x 10^{-6} Above Tg	0.062	85 170	0.0050	0.0002	0.0043	>0
Polyimide-Glass 47 x 10^{-6} Below Tg	0.062	255	0.0007	0.0002	0.0043	<0

Table 8. Expansion of Rigid Flexible MIB

Material	Board No.			
	1	2	3	4
Total Thickness (inch)	0.0802	0.0672	0.0502	0.0602
Copper Thickness (inch)	0.0112	0.0112	0.0112	0.0112
% Copper	14	16.7	7.5	18.6 –
Polyimide-Glass Thickness (inch)	0.040	0.020	0.010	0.020
% PI Glass	50	29.8	19.9	33.2 –
Kapton Thickness (inch)	0.012	0.012	0.012	0.012
% Kapton	15	17.8	23.9	20 –
Acrylic Thickness (inch)	0.017	0.024	0.017	0.017
% Acrylic	21	35.7	34	28.2
Linear Coefficient of Expansion $\times 10^{-6}$ in/in/°C "Z" Direction	157	275	292	224

Another method which would produce the same result is to increase the thickness of Kapton at each layer. This could be achieved simply by substituting laminate materials using 2 oz copper on 3 mil Kapton in place of materials using 2 oz copper on 2 mil Kapton.

Alternate adhesive materials could also be considered as an additional method; however, this would require effort by the material manufacturer.

4. SUMMARY

In summary, we have identified that R/F laminate materials have coefficients of expansion which differ one from another depending upon the composition. When epoxy-glass laminate is used as the rigid material in the R/F MIBs, there are two glass transition temperatures and thermal expansion is large. Expansion of R/F laminates is proportional to the volume percent of the acrylic adhesive used.

We have also shown that the R/F MIBs have through holes with irregular surfaces. These can result in low ductility and/or thickness of copper deposits. Such conditions often lead to failure under thermal stress.

Finally, we have defined a method for controlling the expansion in these composites by increasing the thickness of the more stable materials and by insuring that acrylic is maintained at a minimum percent by volume.

REFERENCES

(1) G. Smith, B. Zimmerman, and J. Strohmer, "Multilayer Board Plated-Through-Hole Barrel and Foil Cracks," Department of Defense, IPC Technical Review, March/April 1978.

(2) R. Geshner and S. Jolly, "Preventive Medicine for PTH Failure in MLB's," Circuits Manufacturing, August 1970.

(3) E. Y. Yuan, of DuPont de Nemours, Co., "Interrelations of Laminate Thermal Properties and Multilayer Circuit Board Processibility."

(4) J. Quintana, "Production Controls to Prevent Plated-Through-Hole Cracking." Contract No. DAAK 40-78-C-0308 for USA MICOM. Redstone Arsenal.

BIOGRAPHY

JOSEPH M. SHAHEEN

Mr. Shaheen received a B.S. in chemistry from Boston College. He is a member of the Technical Staff, Chemical and Process Unit of the Materials and Process Laboratories. Mr. Shaheen has a total of 28 years experience and 24 years directly related to electronic packaging developments. He has developed multilayer printed circuit boards, flexible circuitry, rigid-flexible multilayer boards, core memory and plated wire memory packaging. He holds more than 19 patents in these fields.

Mr. Shaheen is presently the responsible M&P Engineer for the development of advanced rigid-flexible multilayer boards and leadless chip packaging.

IN SITU GROWN FIBER REINFORCEMENT FOR
IMPROVED HIGH VOLTAGE COMPONENT ENCAPSULANTS

R. E. Kelchner and L. B. Keller
Technology Support Division
Hughes Aircraft Co.
Culver City, California

Abstract

A feasibility study directed to the development of composite, high voltage encapsulants using fiber reinforcement grown by a new solution agitation process is described. Initially, the fiber forming capabilities of eight polymers, linear polyethylene (PE), isotactic polypropylene (PP), isotactic poly-1-butene (PBu), poly-4-methyl-1-pentene (PMP), polyvinylidene fluoride (PVF_2), polyacrylonitrile (PAN), poly(hexamethyleneadipamide) and polyvinylchloride (PVC) were investigated. All of these polymers formed fibers or fibrous structures by the agitation process with the exception of PVC. The polymers PE, PP, PBu, PMP, and PVF_2 uniformly fiberized tine-like geometries by the process.

Fibers formed from PE and PBu exhibited preferred molecular orientation by x-ray diffraction. Scanning electron micrographs of PE, PP, PBu, and PMP revealed fibers with the shish kebab morphology. In addition, the PE, PP, PBu, PMP, and PVF_2 fibers exhibited melting behavior characteristic of the shish kebab morphology. Tensile strengths up to 140,000 psi were measured for the PE and PP fibers and 18,000 psi for PBu fibers.

Simulated high voltage devices reinforced with PE, PP, PBu, PMP, and PVF_2 fibers and encapsulated with Epon 825/HV resisted corona discharge under high voltage conditions after thermal cycling to liquid nitrogen temperatures.

"Keywords"

In situ grown polymer fibers, solution agitation process, preferred molecular orientation, shish kebab morphology, fiber reinforcement, composite encapsulants, high voltage component encapsulant.

1. INTRODUCTION

The current trend in airborne power supplies is to use solid polymeric encapsulation in place of liquid dielectrics to insulate high voltage electronic components. The change to solid encapsulation offers significant advantages for aero-

space applications such as lightweight construction, high reliability, decreased maintenance cost,
and simplified spares needs.

Solid encapsulated high voltage
systems are particularly vulner-
able to electrical failures when
subjected to thermal cycling. Such
failures are almost always traced
to corona or partial discharge
degradation in minute voids that
have developed because of mechanical overstressing. Contemporary
encapsulating materials often lack
the intrinsic strength to sustain
the thermomechanical stresses that
result from large temperature excursions realized in military applications.

A potential solution to this problem is to use composite encapsulants that have high strength
derived from fiber reinforcement.
Unfortunately, encapsulation with
fiber reinforced composites is
virtually impossible because of
processing difficulties. Electrical components are commonly characterized by small interstices,
windings, and complex geometries.
Impregnation of these small spaces
with resin-fiber mixtures results
in a filtration of the resin. In-
evitably, those areas where the
thermomechanical stresses and electrical field strengths are potentially highest remain unreinforced
with fiber.

1.1 Background

Some years ago a study was conducted to investigate the in situ
formation of polymer fibers from
solution by agitation-induced
crystallization. This work was
motivated by the concept that low
viscosity solutions could easily
permeate the complex spaces of electronic components and, if induced
to precipitate fibers by a simple
agitation process could ultimately
produce a uniformly, fiber reinforced encapsulant. Pennings and
Kiel[1] had demonstrated that polyethylene fibers with unique molecular morphology and extraordinary
strength could be crystallized from
xylene solutions under conditions of
isothermal stirring. To be determined was whether or not a reciprocating hydrodynamics produced by a
solution agitation process would be
capable of effecting the molecular
uncoiling and preferred orientation
required to produce high strength
fibers.

Initial experiments wherein 7% w/v
isotactic polypropylene/xylene
solutions were simultaneously cooled and agitated indeed demonstrated
that fibrous masses could be formed
by such a process[2,3]. The microstructure of such masses typically
consisted of interconnected, three
dimensional fibrous networks as
shown in Figure 1. It was further
observed that wire windings could
be fiberized by agitation in dilute
polyethylene/xylene and isotactic
polypropylene/xylene solutions,1-2%
w/v, under isothermal conditions.

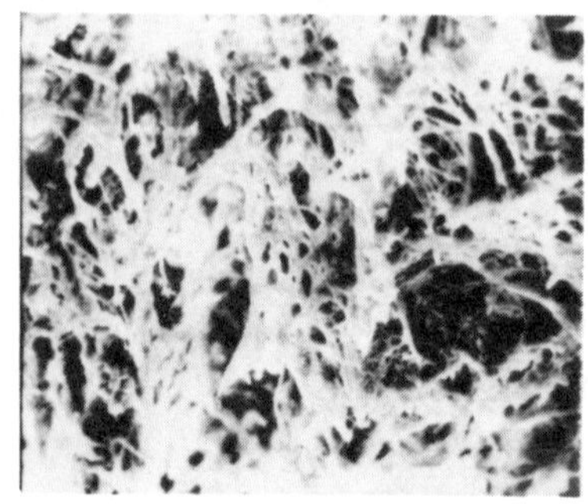

1000X

Figure 1. Scanning electron micrograph of isotactic polypropylene
fibrous masses formed by solution
agitation.

Again interconnected fibrous structures, quite similar to those obtained by solution agitation, formed between and around the wire
windings at temperatures upwards to
$10^{o}C$ above the supercooled temperatures. Scanning electron micrographs of polyethylene fibers crystallized at $104^{o}C$ are shown in Figures 2a and 2b. The fibers are
observed to exhibit the "shish
kebab" morphology obtained by the
stir-induced crystallization process[1,4]. Examination of the
micrograph in Figure 2b at 100,000X
reveals the characteristic smooth
central core along which disk-like
lamellae overgrowths form a row
structure. Significantly, an x-ray
diffraction pattern of these fibers
in Figure 3 illustrates the high
degree of preferred molecular orientation which is attained.

a) 9400X

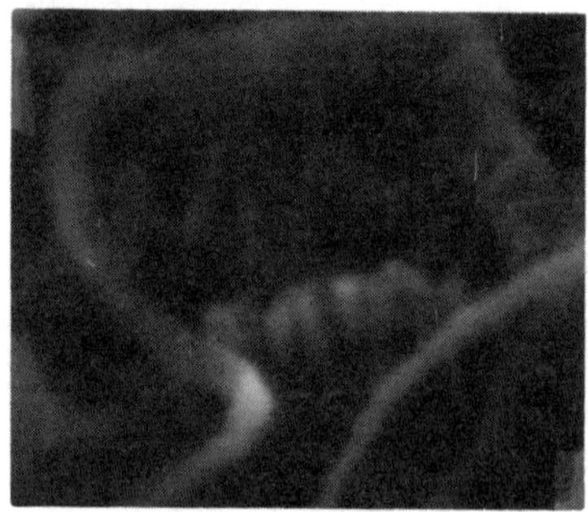

b) 100,000X

Figure 2. Scanning electron micrographs of polyethylene formed by
agitation induced crystallization
from dilute solution at $T = 104^{o}C$.

Figure 3. X-ray diffraction pattern of polyethylene formed by
agitation induced crystallization
from dilute solution at T = 104ºC.

The in situ formation of polymer fibers by the agitation of their solutions appeared to be not only a feasible but highly promising process for the reinforcement of high voltage component encapsulants. Reported here are the results of an investigation directed to the development of in situ grown fibers suitable for high voltage applications and the evaluation of simulated devices, fiberized by the agitation process and encapsulated, under conditions of high voltage as a function of thermal cycling.

2. EXPERIMENTAL PROCEDURES

2.1 Materials

Eight linear polymers were selected for this study based primarily on 1) crystallinity and 2) demonstrated fiber forming capability by flow-induced crystallization. These polymers are listed in Table 1 along with their sources, reported molecular weights (mol. wt.), and melt temperatures (T_m). In some cases the molecular weights were determined in this laboratory from intrinsic viscosity measurements.

2.2 Fiber Formation and Characterization

As a first step, attempts were made to find solvents conducive to fiber formation for the polymers of interest. These efforts were conducted in an exploratory fashion wherein solubility, critical miscibility temperatures, and fiber forming tendency were evaluated. Solvents were selected in terms of their ability to allow recrystallization of the polymers at elevated temperatures. This was motivated by the desire to conduct isothermal fiber formations carefully controlled by a thermal bath. Approximate critical miscibility temperatures were determined simply by observation at concentrations of interest, <2% w/v. The fiber forming tendency of promising polymer/solvent systems were evaluated by manual agitation of test tubes containing the solutions simultaneous with cooling. The occurrence of fibrous structure was evaluated by observing the precipitate between polarized lenses.

The actual fiber formation experiments were conducted with those systems showing promise by agitating fixtures with tine geometries in the solutions. The objective was to fiberize the fixtures. An MB Electronics Exciter/Dynaco Amplifier/Hewlett Packard oscillator combination was used to drive and control the agitation. The apparatus was equipped with an Endevco accelerometer, piezoelectric transducer, for direct monitoring of the acceleration and/or displacement of agitation. The tines were agitated in a vessel which could be continuously replenished with fresh solution through a tube attached to an adjoining reservoir. These containers, the fiber formation vessel and reservoir, were mounted in an oil bath controlled to $\pm$0.2°C. The set-up is shown in Figure 4.

Figure 4. Apparatus used for agitation-induced crystallizations. Shown are the MB Exciter and thermal control bath containing the fiber formation vessel and the reservoir.

molecular morphologies were characterized by scanning electron microscopy (SEM) and x-ray diffraction. The scanning electron microscopy studies were conducted with a Kent Cambridge Scientific Equipment Inc. microscope. The x-ray diffraction characterizations were conducted by Dr. Christian N. J. Wagner of the Materials Department, University of California, Los Angeles, using a micro-diffraction camera (Philips) with Cu or Co Kα radiation. A primary beam size of 50 μm was used to focus on the small fibers. This x-ray work was performed with the primary objective to elucidate preferred molecular orientation. The

Table 1. Polymers investigated, their sources, reported molecular weights and melt temperatures.

Polymer	Sources	Mol. Wt.	T_m, °C
Polyethylene	Union Carbide Corp.	2.7×10^5 (Mw)	136
Isotactic Polypropylene	Exxon	1.9×10^5 (Mv)	171
Isotactic Poly-1-butene	Scientific Polymer Products, Inc.	9.8×10^5 (Mv)	122-142
Isotactic Poly-4-methyl-1-pentene	Polysciences, Inc.		235
Polyvinylidene Fluoride	Polysciences, Inc.	4.55×10^5 (Mw)	
Polyacrylonitrile	Polysciences, Inc.	7.2×10^4 (Mv)	319
Polyvinylchloride	Univ. of Mass.	1.1×10^5 (Mw)	212
Poly(hexamethylene adipamide),Nylon 6-6	Polysciences, Inc.		265

Attempts were made to collect sufficient fibrous product from the fixtures for characterization and evaluation in terms of morphological, thermal, and mechanical strength properties. Fiber and

melting behavior of the fibers was determined by Dr. J. J. Aklonis of the Chemistry Department, University of Southern California, using a Perkin-Elmer differential scanning calorimeter. Small amounts of samples, several milligrams, necessitated the use of 20°C/min heating rates to obtain suitable responses.

Tensile strengths were measured where fibers large enough for handling could be selected from fiber masses. A wire testing instrument capable of measuring gram loads over ranges of 0-5, 0-25, and 0-50 grams was used. The tensile strength values were based on cross sectional areas of the fractured ends determined by SEM.

2.3 Simulated Device Fiberization and Evaluation

Simulated devices were chosen for evaluation of the in situ formed, fiber reinforced encapsulants which offered the following qualities:

o Geometries similar to that of the tines to allow evaluation of fibers formed under known optimum conditions,

o Configurations offering definable field strengths for accurate corona inception voltage evaluation.

The design consisted of two identical sets of tine-like electrodes in each part (Figure 5). Initially, these devices were fiberized by agitation in a fashion which caused the solution to flow through the tines, i.e., the tines were aligned in a perpendicular manner to the reciprocating motion. They were submerged in the solutions by attachment to an

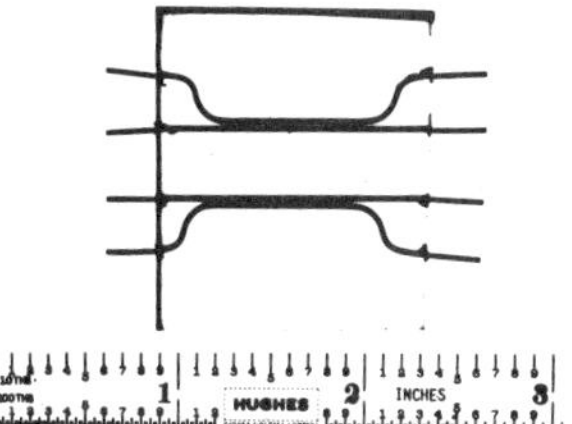

Figure 5. Epoxy encapsulated simulated high voltage device.

extension rod connected to the MB Exciter. Following fiberization, the products were washed by agitating the tines in the corresponding solvents at temperatures a few degrees higher than the crystallization temperatures. Subsequently, the fiberized tines were solvent extracted, dried, and encapsulated with an Epon 825/HV epoxy to produce devices for evaluation. The 825/HV system is widely specified in high reliability, high voltage composite encapsulations at Hughes Aircraft Co.

The evaluation procedure consisted of cycling fiberized devices along with controls (no fiber reinforcement) from ambient up to the matrix resin glass transition temperature and then sequentially down to lower and lower temperatures starting at -10°C. Corona inception voltages (CIV) were measured, when they existed, over a limited test voltage range at 60 Hz prior to and after each cycling exposure. A modified

449

Biddle corona testing apparatus
capable of measuring partial dis-
charges in subpicocoulomb ranges
was used.

In addition, stress patterns developing in the devices as a result
of the thermal treatments were
monitored with the aid of cross
polarized optical transmission
photos. This offered qualitative
analyses of the response of reinforced devices to thermomechanical
stressing relative to control
devices.

3. FIBER FORMATION RESULTS

3.1 Polyethylene (PE) and Iso-
tactic Polypropylene (PP)

Preliminary experiments were conducted with these known fiber formers to determine the optimum solution concentration, temperature,
agitation parameters and tine
separation conditions required
for preferred molecular orientation
crystallization. The PE fibers
which exhibited the most uniform
shish kebab structure by SEM and
preferred orientation by x-ray
diffraction were formed on tines
agitated in 1% w/v xylene solutions
at 103°C. An example of a fiberized tine is shown in Figure 6
along with a SEM of the fibers.
The best quality PP fibers in
terms of structure uniformity were
formed from 0.5% w/v xylene solutions at 100°C. A tine fiberized
with these fibers and a corresponding SEM are shown in Figure 7. The
PP fibers were observed to form

with a more bifurcated structure
and a less distinct shish kebab
morphology than that observed for
the PE fibers. In both cases, tine
spacings of 0.007 cm. were coupled
with agitation parameters of 1.78 x
10^2 cm/sec. maximum velocity and
42 Hz frequency to yield the best
results.

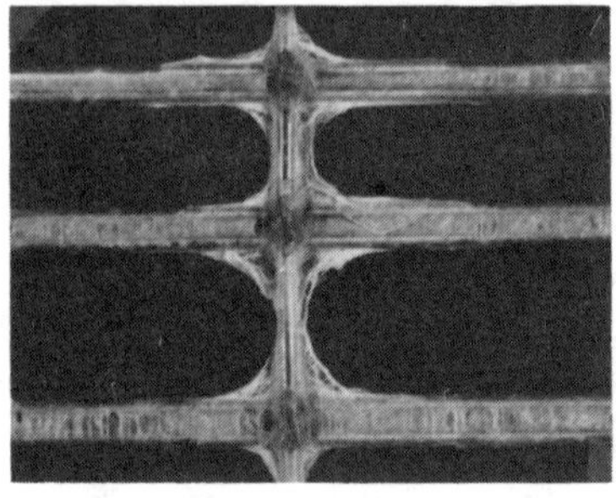

a) 5X

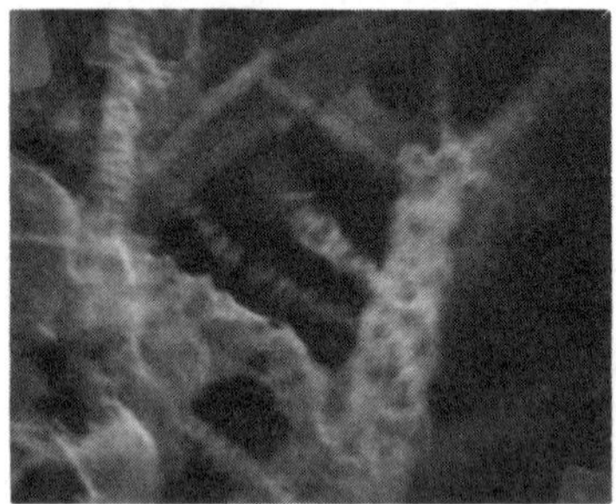

b) 15,200X

Figure 6. Polyethylene fibers
a) Fixture showing fibers formed
from 1% w/v xylene solution at
103°C and b) Scanning electron
micrograph of fibers.

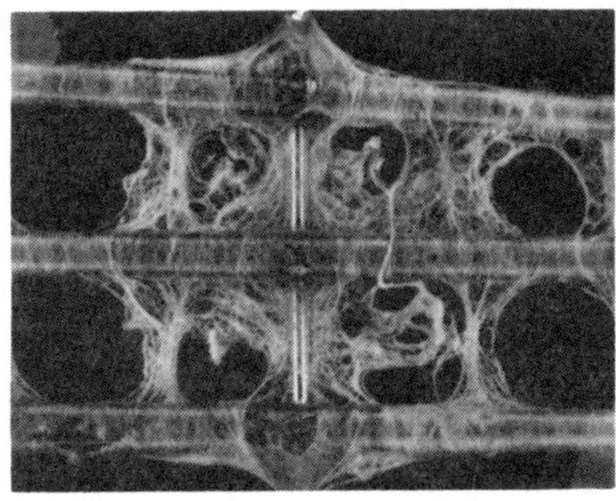

a) 5X

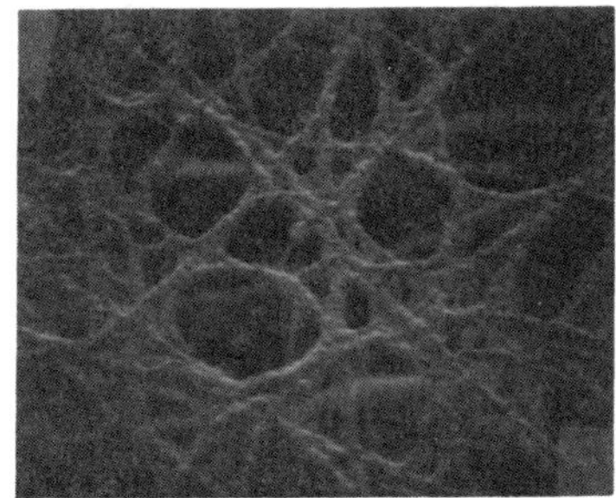

b) 16,400X

Figure 7. Isotactic polypropylene fibers a) Fixture showing fibers formed from 0.5% w/v xylene solution at 100°C and b) Scanning electron micrograph of fibers.

It was discovered during the course of these experiments that a pre-coating of the tines with the respective fiber forming polymer greatly enhanced fiberization in terms of fiber density and process time. The coating procedure simply consisted of dipping the tines into a dilute solution of the polymer and drying. This technique was used for all of the polymers studied.

The task of collecting fibers from the tines for tensile strength measurements was a tedious process because of their size (∼3 mm lengths) and bifurcated structures. Fibers of uniform girth and shape over suitable portions of their lengths to allow reasonable tensile pulls were selected. In addition, calculation of the tensile strengths was complicated by the fact that the fibers were actually bundles of fibrils. It was necessary to approximate the cross-sectional areas of the bundles from SEM photos of the fractured ends. In such fashion, strength values of 1.4×10^5 psi were determined in one case for both PE and PP fibers listed in Table 2.

Table 2. Tensile strengths measured for polyethylene, isotactic polypropylene, and isotactic poly-1-butene fibers.

Fibers	Sample #	Tensile Strengths $\times 10^{-4}$ PSI
Polyethylene	1	9.5
	2	3.8
	3	14.
	4	3.8
Isotactic Polypropylene	1	2.9
	3	1.8
	4	14.
	5	8.1
	6	4.2
	7	1.4
Isotactic Poly-1-Butene	1	0.45
	2	0.67
	3	0.45
	4	1.8
	5	0.55

The melting behavior of the PE and PP fibers are compared with controls by the DSC thermograms in Figures 8 and 9 respectively. The PE fibers, curve a in Figure 8,

yield a large endotherm at ∿137°C
and a smaller shoulder at ∿127°C,
whereas the control or native material, curve c, yields a single
endotherm at ∿130°C. Curve b demonstrates that the shoulder observed for the fibers disappears
to yield a single endotherm at
∿130°C when the sample is cooled
and the run repeated. The PP
curves in Figure 9 show similar
behavior where again a first run
and repeat run are given for the
fibers along with that of the
native material. The double endotherms exhibited by the fibers are
typical of the behavior observed
for shish kebab fibers formed by
stir-induced crystallization[5].

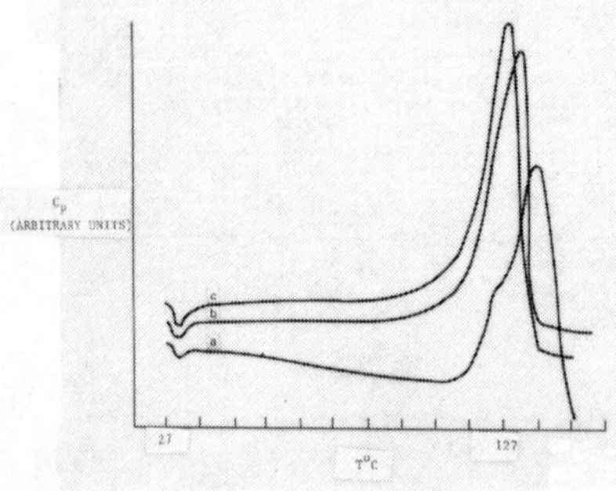

Figure 8. Melting behavior of
polyethylene; a) first run,
fibers, b) repeat run, fibers,
and c) native material.

3.2 Isotactic Poly-2-butene (PBu)
During the course of conducting
viscosity - molecular weight experiments, heptane was observed to
be a potential fiber forming solvent for this polymer. Initially,
the polymer was found to crystal-

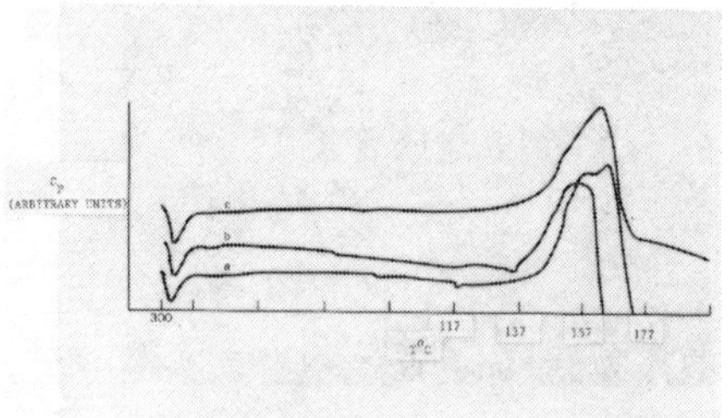

Figure 9. Melting behavior of
isotactic polypropylene; a)
first run, fibers, b) repeat
run, fibers, and c) native
material.

lize readily from even dilute solutions, ∿0.05 to 0.25% w/v. Subsequently, manual solution agitation
tests conducted simultaneously with
cooling were found to produce fibrous material when temperatures
around ambient were reached.

Thus, actual fiber formation experiments were conducted by agitating
tines in 1% w/v PBu-heptane solutions at various temperatures
above ambient, the apparent region
of the supercooled temperature.
The highest temperature at which
fibers were observed to form was
58°C. A SEM and x-ray diffraction
pattern illustrate the structure
and preferred orientation achieved
under these conditions in Figure 10.

The PBu fibers formed at 58°C were
evaluated in terms of tensile
strength and melting behavior.

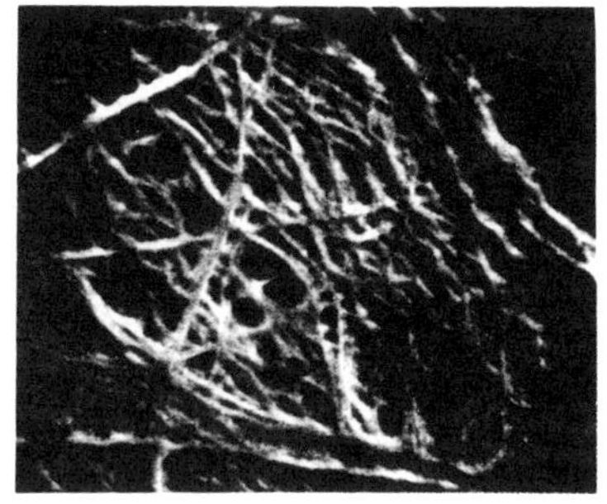

a)   4500X

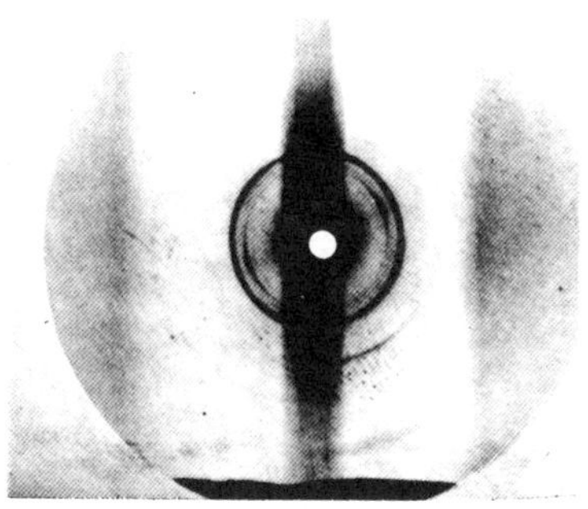

b)

Figure 10.   a) Scanning electron
micrograph and b) x-ray diffraction
pattern of poly-1-butene fibers
formed from 1% w/v heptane solution at 58°C.

The highest tensile strength value
determined for these fibers, listed
in Table 2, was 1.8×10^4 psi. The
distinctive structure which characterizes these fibers is suggested
by their melting behavior shown in
Figure 11. Again, curve a exhibits
the first run behavior of the fibers, curve b repeat run of the
fibers, and curve c that of the
native material. Curve a shows the
double peak which was exhibited by

the melting behavior of the PE and
PP fibers.

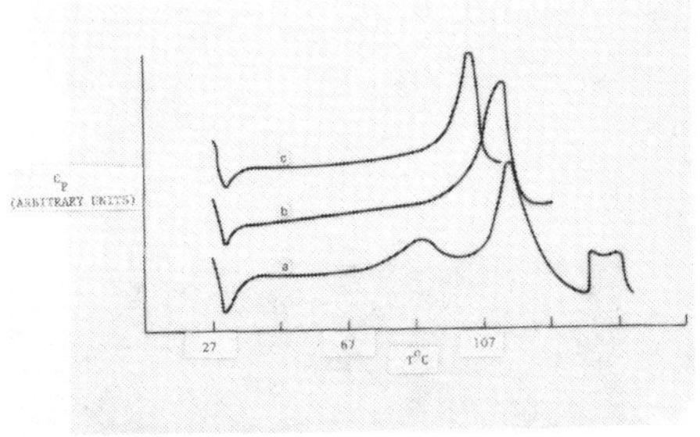

Figure 11. Melting behavior of
isotactic poly-1-butene; a) first
run, fibers, b) repeat run, fibers, and c) native material.

3.3   Isotactic Poly-4-methyl-1-pentene (PMP)

Attempts to determine the viscosity
average molecular weight of PMP
from its intrinsic viscosity in
diisobutylene were not successful.
Difficulty was encountered in retaining complete dissolution at a
$20°C$ temperature for which Mark-
Houwink constants are reported[6].

The solubility and fiber forming
tendency of PMP was evaluated with
the solvents diisobutylene, deca-
hydronaphthalene, heptane, o, m,
p-xylene mixture (xylenes) and p-
xylene. The latter two were the
only solvents which showed promise.

Tine agitation experiments were conducted with 1% w/v solutions of the
PMP in the xylene solvents at temperatures of $79°C$ and higher. The
starting temperature of $79°C$ was
based on approximate miscibility

temperatures observed over a range
of 75° to 78°C. Fiber crystallizations were observed only for the
p-xylene solutions up to a temperature of 82.5°C. A SEM and x-ray
diffraction pattern of these fibers
are shown in Figure 12. A distinct

a)   9,500X

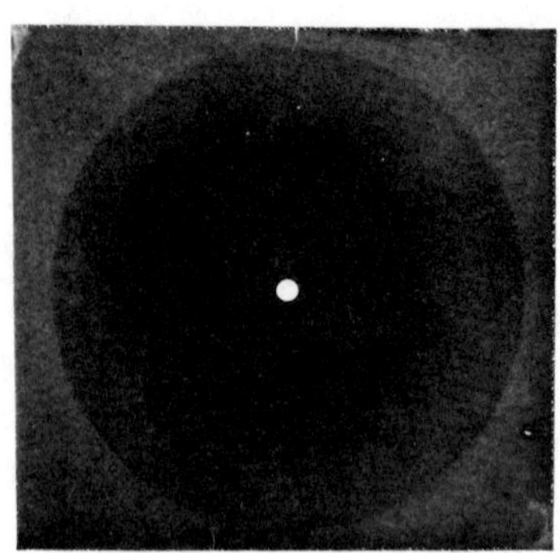

b)

Figure 12. a) Scanning electron
micrograph and b) x-ray diffraction
pattern of poly-4-methyl-1-pentene
fibers formed from a 1% w/v p-
xylene solution at 82.5°C.

shish kebab structure is evident
from the SEM photo. The sharp rings
of the x-ray diffraction indicate a
highly crystalline material, but pre-

ferred molecular orientation is not
apparent. The preferred molecular
orientation may be masked in this
case by a random orientation of fibers within the mass through which
diffraction occurred.

The melting behavior of the PMP fibers was quite unusual in that a
large exotherm was observed to occur
in the region of the melting point
as shown by curve a in Figure 13.
The exotherm was observed to dis-
appear with repeat runs of the fibers as illustrated by curve b. No
exotherm was exhibited by the native
material, curve c in Figure 13.
Tensile strength measurements were
not obtainable because the material
was too brittle to allow specimen
preparation.

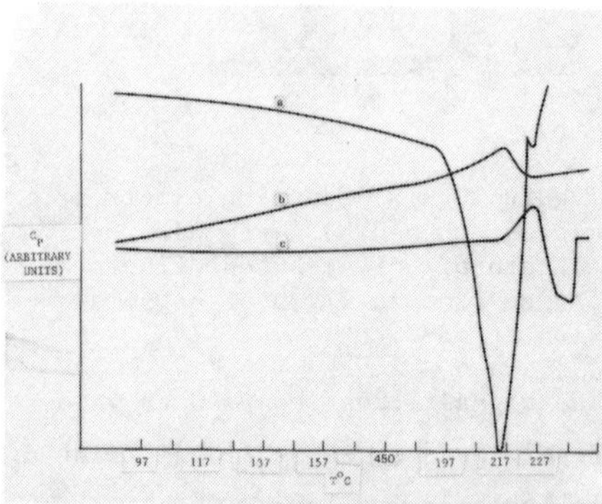

Figure 13. Melting behavior of poly-
4-methyl-1-pentene; a) first run,
fibers, b) repeat run, fibers,
c) native material.

3.4 <u>Polyvinylidene Fluoride (PVF_2)</u>
Numerous solvents and solvent pairs
were investigated in an effort to
find a medium conducive to the fiber
formation of PVF_2. This polymer

has a tendency to separate by gelation rather than crystallization.
Mixtures of methyl isobutyl ketone
(MIBK) and butyl acetate (BA) showed the most promise in that fib-
rous-like structures could be produced by manual solution agitations.

The best results were achieved with
1% w/v MIBK-BA (50/50) solutions.
Structured PVF_2 was observed to
form not only between the tines but
uniformly over their entire surfaces. Shown in Figure 14 is a
photograph of a fixture liberally
coated with fibrous PVF_2 formed at
$64^{\circ}C$.

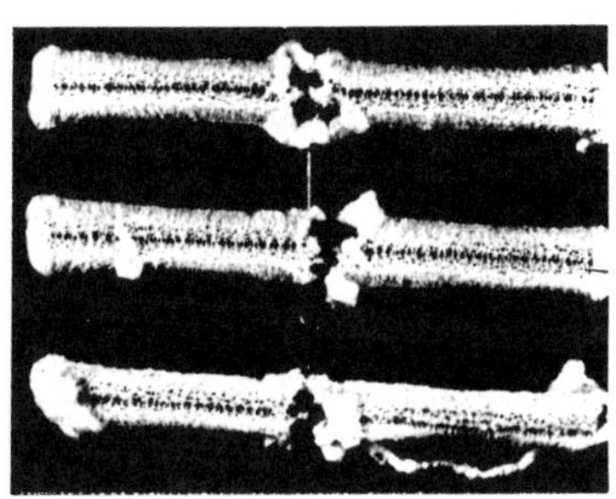

5X

Figure 14. Fixture with polyvinylidene fluoride fiber structure
formed from 1% w/v MIBK/BA (50/50)
solution at T = $64^{\circ}C$.

The x-ray diffraction pattern of
this material revealed little with
regards to preferred molecular orientation. The melting behavior
of the PVF_2 fibrous material, however, exhibited uniqueness analogous to the previously discussed
fibers. In Figure 15 curve a shows

the double endotherm for the fibrous material which upon rerunning
disappears, curve b. As before,
the native material, curve c, exhibits a single endotherm.

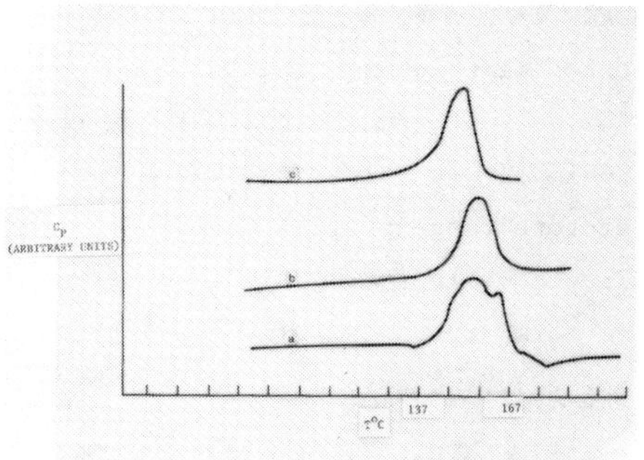

Figure 15. Melting behavior of
polyvinylidene fluoride; a) first
run, fibers, b) repeat run, fibers,
and c) native material.

3.5 Polyacrylonitrile (PAN),
     Polyvinylchloride (PVC) and
     Poly(hexamethylene adipamide)
     (Nylon 6-6)

The fiber formation of PAN was investigated using a number of solvents. The solutions worthy of note
consisted of 2% w/v PAN in dimethyl-
formamide (DMF), DMF/heptane (80/20)
and dimethyl sulfoxide/dioxane
(50/50). Fibers were produced
from these solutions by tine agitation but in each case they dispersed freely throughout the medium
without adhering to the tines.

PVC fiber formation experiments
were conducted using a toluene/p-
chlorotoluene (84/16) solvent mixture after exploring a number of
other solvents. The experiments
were conducted with 1% w/v con-

centrations at various temperatures from 70° to 75°C. In all cases, the results were the same in that PVC showed a predominant tendency for film formation.

The fiber formation of a Nylon 6-6 was limited to a few tine agitation experiments using benzyl alcohol as the solvent. The tines were agitated in 1% w/v solutions over a temperature range of 145° to 150°C. Traces of long fibers were observed to form around the tines at 146°C but only in very small quantities.

4. SIMULATED DEVICE FIBERIZATION, ENCAPSULATION, AND EVALUATION

4.1 Introduction

Based on the results of the fiber formation work, the five polymers PE, PP, PBu, PMP and PVF_2 were selected for the simulated device investigation. First, tine fixtures were fiberized under conditions found optimum in the previous study. The fiber formations were conducted to an extent which, by observation under an optical microscope, yielded uniformly fiberized parts. Second, the fiberized tines along with unfiberized controls were encapsulated with the Epon 825/HV resin to produce devices for evaluation. Finally, the encapsulated devices were subjected to thermal cycling and evaluated in terms of their resistance to thermomechanical failure. Essentially, the occurrence of corona under high voltage conditions was taken as an indicator of void or crack formation.

4.2 Fiberization

The fiber formations were conducted using tine separations of $\sim$0.007 cm. and the optimum concentration, temperature, and velocity-frequency conditions determined earlier. The precoating process was used again in all cases.

Once fiberized, the tine devices were washed by gentle agitation in pure solvent for a period of one hour at 2°C above the fiber crystallization temperature. Subsequently, the tines were removed and solvent extracted by submersion in acetone for one hour. The tines were then dried under room conditions overnight, followed by vacuum drying in an oven at a temperature about 10°C below the boiling point of acetone. The fiberized tines and unfiberized controls were encapsulated while mounted in plastic containers which served to produce the rectangular shaped devices, as shown in Figure 5. The parts were vacuum impregnated with the Epon 825/HV resin and then precured for 16 hours under 5.5 atmospheres of dry N_2 pressure at 80°C. They were given a final cure for 24 hours in air at 100°C.

4.3 Evaluation

4.3.1 Prior to Thermal Cycling.

As a preliminary step, fiberized devices with no encapsulation were characterized in terms of their resistance to high voltage corona and/or breakdown while submerged in Freon. These tests, as well as those discussed later, involved the application of 60 Hz voltages across the tine-like electrodes in an increasing amount until an electrical response occurred or a maximum voltage was reached. The Biddle Instruments Corona Tester used for these evaluations recorded the root mean square applied voltages at which corona or breakdown took place. These recorded voltages were converted to peak, intrinsic field strength voltages using a general formula for parallel cylinders[7]. The results are given in Table 3 as peak voltages per mil for the CIV and breakdown voltages.

Table 3. Corona inception voltages and breakdown voltages in terms of peak, intrinsic field strength voltages for fiberized devices with no encapsulation submerged in Freon.

Polymer	CIV V_{peak}/mil	Breakdown V_{peak}/mil
Polyethylene	---	1043
	---	1068
Polypropylene	---	390
	---	469
Poly-1-Butene	---	829
	---	1144
Poly-4-Methyl-1-Pentene	466	744
	---	700
Polyvinylidene Fluoride	---	549
	---	589

The incidence of corona was observed only for the PMP fiberized devices at a CIV of 466 V_p/mil. All the other fiberized devices exhibited no CIV but avalanched to yield breakdown potentials from 390 V_p/mil in the case of the PP fibers to 1144 V_p/mil for the PBu fibers. The breakdown strength of Freon is reported to be on the order of 400 to 500 V_p/mil[8]. These results were considered to indicate that the fibers do not contain significant amounts of impermeable voids within which premature failure can occur.

In addition, epoxy encapsulated devices, one fiberized with PE, another with PP, and an unfiberized control, were characterized with respect to CIV and breakdown strength prior to thermal cycling evaluations. No corona was observed in any case. The breakdown voltages measured were at levels greater than 1000 V_p/mil, which are comparable to the intrinsic strengths of the materials. Of particular significance, there appears to be no problem in thoroughly encapsulating the fibers which are fibrillated in structure.

4.3.2 <u>Thermal Cycling Results</u>. The thermal cycling consisted of heating the devices to the epoxy glass transition temperature, measured as 110°C, then cooling to a temperature below ambient (1st cycle to -10°C) and finally returning to ambient temperature. The devices were evaluated with respect to corona after each cycle.

Initially, the PE and PP fiberized devices were tested for corona up to a test voltage limit of 600 V_p/mil at ambient temperature. A limit to the best voltage was used in an effort to avoid premature failures resulting from intrinsic breakdown. However, the 600 V_p/mil limit was

Table 4. Low extreme cycling temperature after
which corona was first observed and CIV
values for PE, PP, PBu, PMP and PVF_2
fiberized devices and epoxy controls.
Cold temperature CIV tests.

Fiber Reinforcement	Sample	Low Extreme Cycling Temperature, °C	CIV V_p/mil
Polyethylene	63B	-150°C and liq. N_2 shock	None
	65B	-150°C and liq. N_2 shock	639
Polypropylene	70B	-150°C and liq. N_2 shock	None
	71B	-150°C and liq. N_2 shock	None
Poly-1-butene	7A	-80	366
	7B	-80	315
	8A	-150°C and liq. N_2 shock	None
	8B	-150°C and liq. N_2 shock	None
Poly-4-Methyl-1-pentene	11A	-150°C and liq. N_2 shock	None
	11B	-150°C and liq. N_2 shock	None
	12A	-150°C and liq. N_2 shock	None
	12B	-150°C and liq. N_2 shock	None
Polyvinylidene fluoride	3A	-150°C and liq. N_2 shock	None
	3B	-150°C and liq. N_2 shock	None
	4A	-150°C and liq. N_2 shock	None
	4B	-150°C and liq. N_2 shock	None
None (controls)	1A	-80	473
	1B	-80	251
	2A	-70	600
	2B	-80	178
	3B	-150°C and liq. N_2 shock	197
	4B	-150°C and liq. N_2 shock	657
	14A	-150°C and liq. N_2 shock	None
	14B	-150°C and liq. N_2 shock	263

increased to 800 V_p/mil when cycles down to -60°C had been reached and no corona was observed for any of the devices. In spite of this increase in the test voltage limit, no corona was observed for the devices after undergoing cycles down to -150°C in addition to liquid nitrogen thermal shocks. The evaluation of these thermally cycled devices was continued by conducting corona tests up to the 800 V_p/mil limit again but this time at a test temperature of -70°C. Acetone dry ice baths were used to cool the devices prior to testing.

The devices fiberized with PBu, PMP and PVF_2 were tested for corona at ambient temperature up to the 800 V_p/mil test voltage limit after each thermal cycle until a lower temperature of -60°C was reached. Thereafter, corona tests were conducted at -70°C after thermal cycling reached this temperature and below.

The CIV results for all the fiberized devices and unfiberized controls are listed in Table 4. Corona was observed for only two of the fiberized devices, one PE and one PBu. All the unfiberized controls, on the other hand, exhibited corona with the exception of one device.

Optical birefringence photos were used to record changes in stress patterns within the devices after various thermal cycles. White light transmitted through the devices in a manner perpendicular to the plane of the tines and through 0-90 degree polarized plates produced the patterns which were photographed.

Birefringence photos of an unfiberized control are shown in Figure 16 after a) thermal cycling to -10°C, and b) thermal cycling to -100°C.

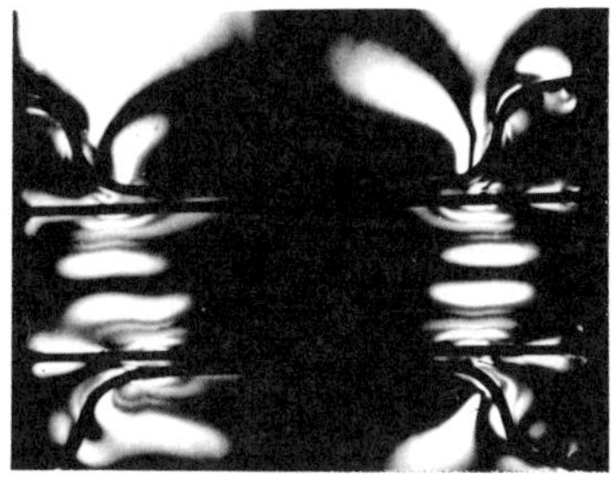

a)   1.5X

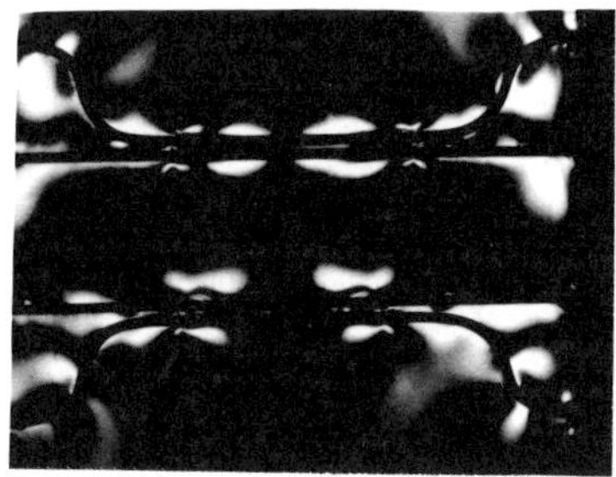

b)   1.5X

Figure 16. Birefringence photos of an unfiberized control after a) thermal cycling to -10°C, and b) thermal cycling to -100°C.

The light regions represent areas of thermomechanical stress concentration whereas the dark regions represent areas of relatively uniform stress.

The first photo in Figure 16 reveals the stresses to be concentrated at the ends of the tines after cycling to -10°C. Clearly, the epoxy matrix, having a coefficient of thermal expansion approximately ten times that of the metal wires, has attempted to contract upon cooling an amount greater than that of the tines. Interfacial bonding, however, has restrained this differential contraction of the matrix and thus, the reason for the stress concentration at the ends. In contrast, the second photo shows the stresses to be concentrated along the central portions of the tines after cycling to -100°C. In this case, debonding has occurred at the matrix-wire interface allowing the matrix to partially contract. The points of stress concentration have moved inward along the tines with the partial contraction.

Birefringence photos of a device fiberized with PP are shown in Figure 17. In comparing these photos with those of the controls it is important to examine the regions between the tines, i.e., the location of fiber reinforcement. Such comparisons reveal that, in general, the fiberized devices have experienced less debonding than the controls. Similar behavior was observed for the PE, PBu, PMP, and PVF$_2$ devices.

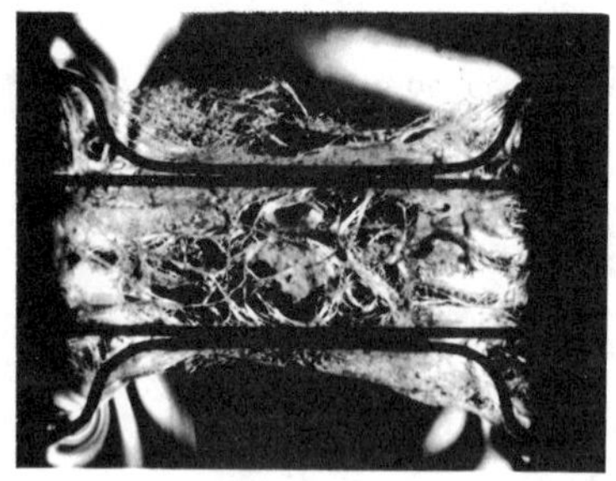

a) 1.5X

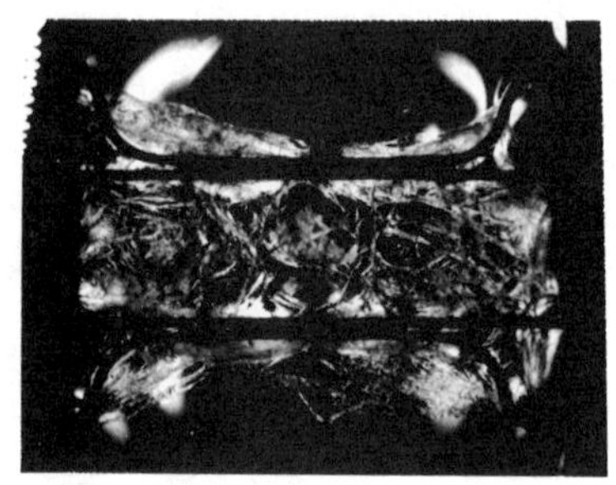

b) 1.5X

Fgiure 17. Birefringence photos of a PP fiberized device after a) thermal cycling to -10°C, and b) thermal cycling to -100°C.

5. CONCLUSIONS

Wire tine fixtures can be fiberized around and throughout their spaces with PE, PP, PBu, PMP or PVF$_2$ by isothermal agitation in solutions of these polymers. Fibers with unusual strength, structure, and thermal properties are produced. Tensile strength measurements up to 140,000 psi for PE and PP fibers formed by agitation demonstrate the potential for producing extraordinary reinforcement by the process. X-ray diffraction patterns of the PE and PBu fibers illustrate the preferred molecular orientation which can be obtained with fiber crystallized from a reciprocating flow field. PE, PP, PBu, PMP, and PVF$_2$ fiber structures similar to stir-induced crystallization products are obtained as illustrated by observed shish kebab morphologies and/or melt temperature endotherms with a characteristic shoulder. The agitation process is not particularly effective in producing PVC, PAN or Nylon 6-6 fibers using the solvents investigated in this study.

Fiberized, simulated devices having the tine geometry exhibit resistance to high voltage corona under conditions of liquid dielectric impregnation and epoxy encapsulation. The epoxy encapsulated devices fiber reinforced by the agitation process demonstrate superior resistance to thermomechanical failures and high voltage corona when compared to devices without the reinforcement.

6. REFERENCES

1. Pennings, A. J. and A. M. Kiel, Kolloid-Z., <u>205</u>, 160 (1965).

2. L. B. Keller and R. K. Jenkins, Nature, <u>278</u>(5703), 439 (1979).

3. L. B. Keller and R. K. Jenkins,
U. S. Patent Nos. 4,127,624,
4,037,010, and 4,198,010.

4. Pennings, A. J., J.M.A.A.
van der Mark and A. M. Kiel,
Kolloid-Z., 237, 336 (1970).

5. Keller, A., and F. M. Willmonth,
J. Macromol. Sci., Phys., B6 (3),
493-538 (1972).

6. "Polymer Handbook", ed. by
J. Brandrup and E. H. Immergut,
Interscience, New York, 1966.

7. Dunbar, W. G. and J. W. Seabrook,
"High Voltage Design Guide for Airborne Equipment", AFAPL-TR-76-41,
1976 .

8. Clark, Frank M., "Insulating
Materials for Design and Engineer-
ing Practice," John Wiley & Sons
Inc., N. Y., 1962 .

7. ACKNOWLEDGEMENTS

A major portion of the work reported here was supported by the Air
Force Materials Laboratory, Wright-
Patterson AFB, Dayton, Ohio under
the project monitorship of Mr.
Walter H. Gloor. Mr. D. J. Mueller
performed almost all of the fiberization experiments. We are grate-
ful to Dr. J. J. Aklonis, University
of Southern California and Dr.
Christian N. J. Wagner, University
of California, Los Angeles, for
their consultation and experimental
assistance.

8. BIOGRAPHIES

Ray Kelchner is Head of the Material
Sciences Section in the Materials
and Processes Laboratory, Technology Support Division, of Hughes
Aircraft Company. Generally, he
has worked on the application of
polymer physics to the characterization, development and processing
of polymer systems since joining
Hughes in 1974. Dr. Kelchner
received his B.S. degree in Chemistry from the University of
California at Los Angeles in 1966
and his Ph.D. degree from the
University of Southern California
in 1970.

Brian Keller is Assistant Division
Manager of the Technology Support
Division of Hughes Aircraft Company. He has approximately thirty
years experience in polymer and
composite material development and
numerous patents, one of which
describes the agitation-induced
fiber crystallization process.
Mr. Keller received his B.A.
degree in Chemistry from Columbia
College of Columbia University in
1948.

26th National SAMPE Symposium
April 28-30, 1981

INJECTION MOLDED CERAMIC ROCKET ENGINE COMPONENTS

Gary D. Schnittgrund
Rockwell International/Rocketdyne Division
Canoga Park, California

Abstract

An injection molding and sintering process has been developed to make near net-shape ceramics. These materials exhibit an inherently high Weibull modulus due to the absence of machining damage. High-strength, high-density silicon nitrides intended for use in rocket engines and other high-temperature applications have proved to be stable in simulated rocket engine combustion products. The net-shape sintered parts have excellent surface finish and superior dimensional control. The elevated temperature flexural strength (45 ksi at 2500 F) is equivalent to that of hot-pressed silicon nitride.

Keywords: silicon nitride, ceramic, rocket engine components, high-temperature turbine

1. INTRODUCTION

For several years the Rocketdyne Division of Rockwell International has actively worked under company-funded programs to define the characteristics of ceramics and to determine if they are capable of functioning in rocket engine structural, thermal, and chemical environments. Utilization of ceramic components in liquid rocket engine turbines is one approach currently being evaluated to increase turbine inlet temperature (TIT) from the 1600 F maximum of the Space Shuttle Main Engine (SSME) to 2000 F or greater.

Like any heat engine, the turbine of a pump-fed rocket engine achieves greater efficiency as working fluid temperature increases. Evolution of pump-fed liquid rocket engines is linked to increased turbine operating temperatures which in turn are limited by the properties of the high-temperature materials of construction.

Today's rocket engine designs are limited by the properties of nickel base superalloys. One way to operate with an increased TIT is to cool the superalloy component with propellant. However, unlike air-breathing turbines that use externally available air as both oxidizer and coolant, rocket engines are cooled by propellants transported as part of the vehicle. Therefore, there is a strong drive to develop structural ceramics, which by virtue of their high-temperature strength, do not require cooling. This paper is a discussion of the ceramic materials selection criteria and candidate ceramics plus an assessment of ceramic applicability in rocket engine turbine components.

2. CERAMIC MATERIALS SELECTION CRITERIA

The lure of ceramics is easy to understand and is based on:

- Its high-temperature capability

- Its potential for low cost manufacture

- The abundance of raw materials

Rocket engine turbine combustion environments usually involve the reaction products of oxygen with either hydrogen or an organic fuel such as methane or kerosene. Proposed turbine inlet temperatures are greater than 2000 F. Because of the very rapid heating and cooling rates encountered during start/stop transients in rocket engine turbines, thermal shock resistance is a necessary characteristic in any candidate material. For a given thermal shock characterized by a boundary layer heat transfer coefficient (h) and temperature change (ΔT), the transient thermal stresses (σ_t) can be minimized by choosing a material with small values of thermal expansion (α), elastic modulus (E), Poisson's ratio (μ), and a large value of thermal conductivity (K), as shown by:

$$\sigma_t = \frac{\alpha E \Delta T}{1-\mu}\ f\left(\frac{h}{K}\right)$$

where, by definition, $f(h/K) \rightarrow 1$ as $h \rightarrow \infty$. It can be shown that for large bodies subjected to a very high heat flux (large h), the thermal expansion properties of the body have a more pronounced effect on thermal stress development than does thermal conductivity. Thus, ceramics having a low coefficient of thermal expansion and high strength are the most desirable candidates for withstanding transient thermal stresses.

A comparison of properties favorable to thermal shock resistance (Table 1) reveals that the combination of high strength, moderate elastic modulus, low thermal expansion, and good thermal conductivity make silicon nitride a favored candidate over other ceramic materials.

Si_3N_4 and SiC best meet these requirements and, as shown by the critical property data in Fig. 1 and 2, Si_3N_4 is superior to SiC at high

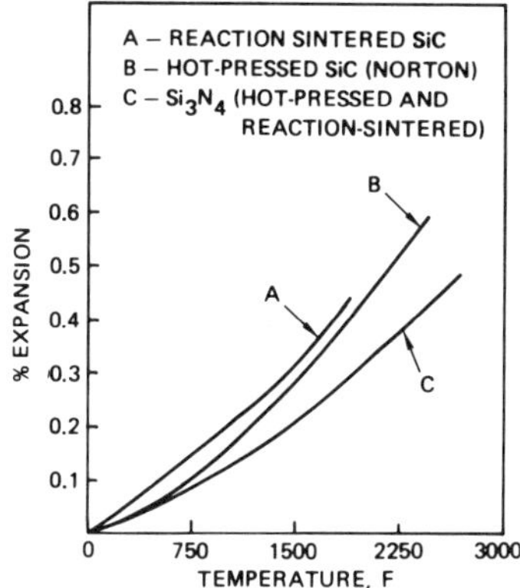

Fig. 1. Thermal Expansion of Si_3N_4 and SiC Materials

Table 1. Properties of Several Ceramics

	Modulus of Elasticity (in tension) 10^6 psi (GPa)		Flexural Strength ksi (MPa)		Coefficient of Thermal Expansion $10^{-6} \cdot C^{-1}$	Thermal Conductivity W/m·K	
	25 C	1000 C	25 C	1000 C		25 C	1000 C
Si_3N_4-Hot Pressed	44 (303)	40 (275)	108 (750)	108 (750)	2.9 (25 C)	30	17
α-SiC-Sintered	59 (406)	55 (378)	67 (459)	64 (442)	4.0 (25-700 C) 5.3 (700-2000 C)	87	67
Al_2O_3-Polycrystal	50 (345)	41 (323)	48 (331)	30 (207)	8.0 (25 C)	30	6
SiO_2-Fused	11 (75)		16 (110)		1.0 (25-1260 C)	2	3
ZrO_2-Stabilized	22 (150)		12 (83)		6.0 (25-1200 C)	2	2
BeO	50 (345)		35 (240)		6.7 (25-125 C) 10 (25-1500 C)	260	20
Mullite ($3Al_2O_3 \cdot 2SiO_2$)	10 (70)		10 (70)		2 (25 C)	6	4
Graphite	1.3 (9)				1.1 (longitudinal) 4.6 (transverse)	180	63
TiC	45 (310)		160 (1130)		7.4 (25-1000 C)	25	6
ThO_2					9.5 (25-1500 C)	10	3
MgO	31 (215)		15 (103)		14 (25-1500 C)	38	7

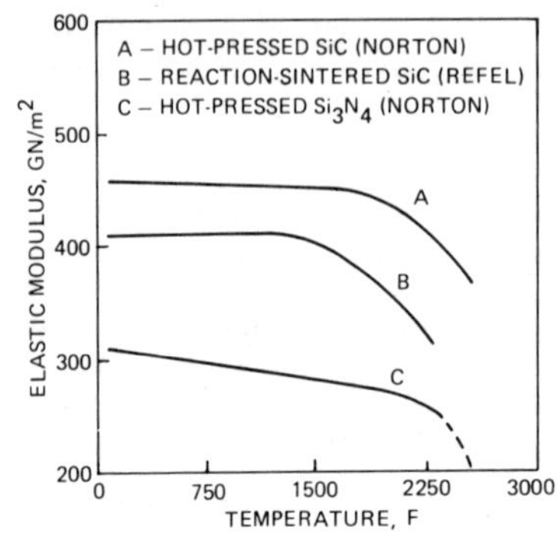

Fig. 2.   Elastic Modulus of Si_3N_4 and SiC

surface-heat-transfer rates. For conditions involving high gas flow velocities at high-temperatures, Si_3N_4 is thus the better choice to minimize thermal stresses and has emerged as the leading candidate for high-temperature structural applications.

3.   PROCESSING SILICON NITRIDE

Silicon nitride products are usually manufactured by variations on three basic processes: hot pressing, reaction bonding, or sintering. The strength of a ceramic and its resultant performance depends strongly on the method of manufacture.

Hot-pressed products require the addition of a sintering aid (e.g., MgO). Although quite dense, they are not readily made to net shape, and thus must be diamond-ground to final shape at a significant increase in cost. Reaction-bonded products require no sintering aid, but are generally not fully dense and, thus, lack strength.

Rocketdyne has developed a process in which net-shape, high-density

products are made by <u>injection molding</u> and <u>sintering</u>, thus avoiding the penalties of high cost or excessive porosity. This process, which also requires a sintering aid, provides a cost-effective route to produce net-shape Si_3N_4 components. Densities 95% of theoretical are routinely attained with strengths as indicated in Table 2.

Rocketdyne injection molded/sintered SN-104 Si_3N_4 is prepared from a commercially available, high-purity precursor Si_3N_4 powder. The sintered material has a bulk density of 3.3 Mg/m^3 and 4-point modulus of rupture (MOR) values of 55 to 65 ksi at room temperature. MOR values at 2000 F obtained in air for specimens from this dense material are 50 to 60 ksi. The primary attributes of the injection molding (IM) process are:

1.   It is an efficient method for mass producing large number of parts

2.   The unit cost is very low for large numbers of units. Depending on the size and complexity of the part, IM parts can be produced as much as 2 orders of magnitude lower than parts made by alternate processes.

3.   The final dimensions of hardware made by the IM process can usually be controlled so that additional finishing is not required.

4.   Complex ceramic shapes with irregular surfaces, channel, and holes can be made easily and economically.

5.   The process is adaptable to numerous materials and alloy compositions.

Because of these unique attributes, Rocketdyne has developed IM products for advanced high-performance

Table 2.   SN-104 Silicon Nitride Flexural Strength Summary

Sample Condition	Test Temperature, F	Weibull Distribution		Normal Distribution	
		Characteristic Strength, ksi	P_{10}*, ksi	Mean Strength, ksi	P_{10}*, ksi
As-sintered	Ambient	52.2	37.0	48.4	36.9
	2500	67.6	49.7	63.5	51.0
Oxidized	Ambient	59.5	43.2	55.7	46.1
	2500	46.2	36.6	44.0	37.6
*P_{10} is that value below which 10% of the population lies.					

applications in lasers, heat exchangers, and rocket engines. Since 1973, Rocketdyne has had on-going IR&D programs to investigate a variety of metallic and ceramic materials, including nickel, columbium, tungsten, iron, copper, aluminum oxide, silicon nitride, and silicon carbide.

The overall IM process is shown schematically in Fig. 3. Fine ceramic powder, together with sintering aids, is mixed with a proprietary plasticizer binder system. this mixture has rheological properties that allow it to flow uniformly into a complex die without segregation. This plasticized material can be used with conventional IM equipment similar to that of the plastics industry. After the as-molded part is ejected from the die, the plastic binder is extracted and the molded compact is sintered to full density.

3.1 Component Fabrication

The IM process offers a unique approach for fabricating intricately shaped hardware to net dimensions. Nearly any shape that can be produced by conventional molding also can be produced in a variety of ceramic materials. Under IR&D programs, Rocketdyne has developed the technology to IM/sinter precision

Si_3N_4 shapes such as the turbine blades and vanes shown in Fig. 4 and 5. Figure 6 shows the nozzle vanes stacked together. These vanes have a measured density greater than 95% of theoretical when sintered. Compansion test bars exhibited a characteristic strength in 4-point loading of 60,000 psi.

Fig. 4. As-Molded and Sintered Si_3N_4 Turbine Blades

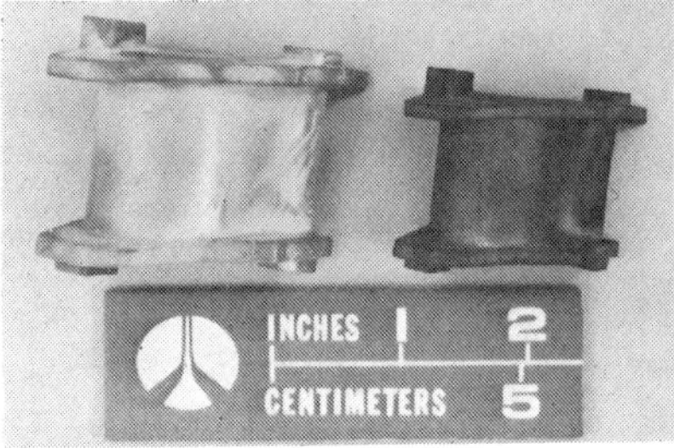

Fig. 5. As-Molded and Sintered Si3N4 Nozzle Vanes

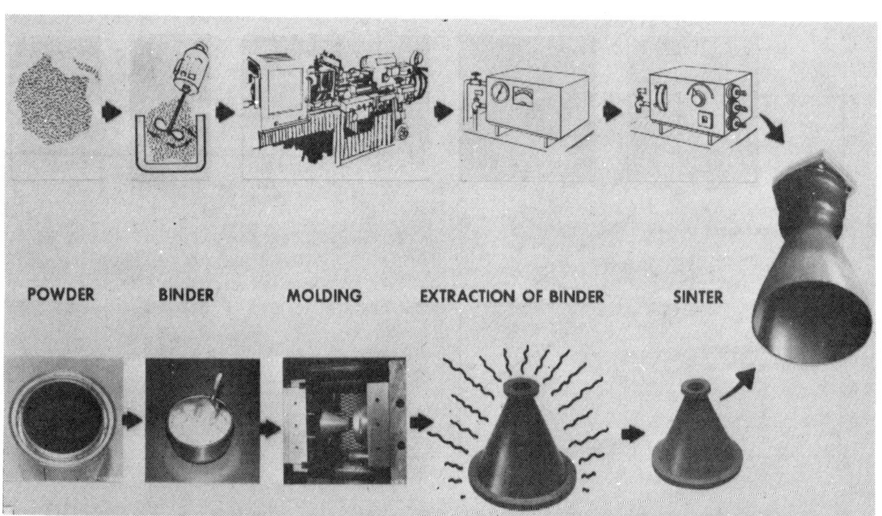

Fig. 3. Injection Molding Process

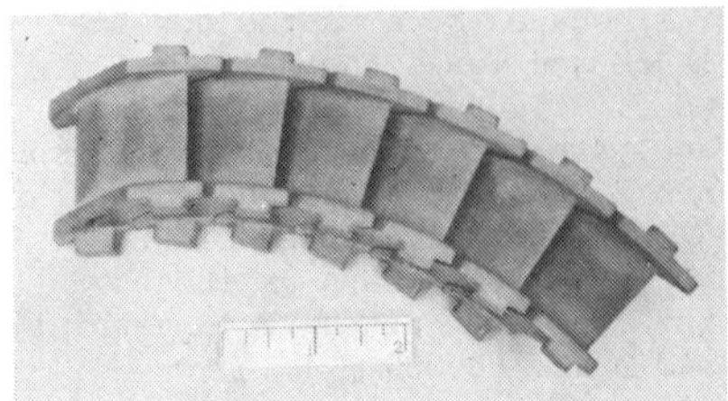

Fig. 6. Silicon Nitride Nozzle
         Vanes

3.2 Mechanical Properties

Figure 7 is a plot of the flexural
strength of SN-104, a Rocketdyne
designation for an advanced $Si_3N_4/Y_2O_3/SiO_2$ alloy. These values are
compared with those obtained from
the literature for other Si_3N_4 materials (Fig. 8). As can be seen in
Fig. 8, the SN-104 strength at temperatures above 2000 F is superior
or competitive to that of other
candidates.

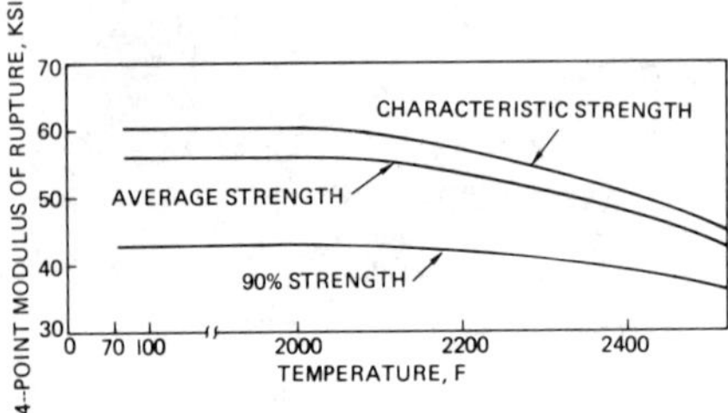

Fig. 7. SN-104 Silicon Nitride
         Strengths (oxidized
         2 hours at 2000 F)

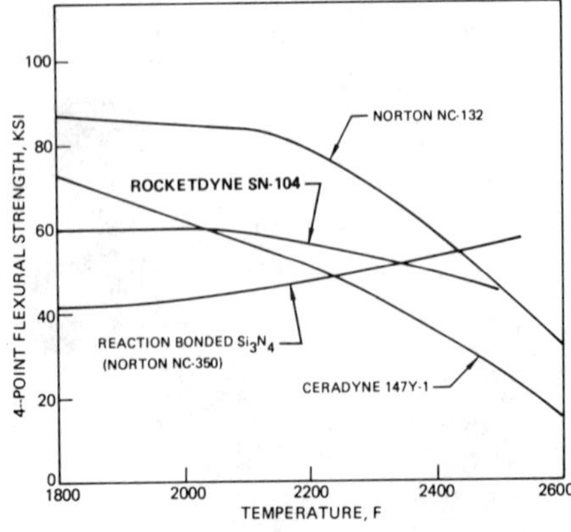

Fig. 8. Strength of Several
         Silicon Nitride Materials

4. ENVIRONMENTAL COMPATIBILITY

The chemical formula Si_3N_4 is usually
used to represent a family of different materials where each family member is fabricated by a different
method or to a different composition.
Each of these variations can affect
the microstructure, the mechanical
and physical properties, and chemical
stability of the resultant product.

Environmental effects on mechanical
behavior and performance of IM/
sintered silicon nitride have been
of importance to Rocketdyne since the
inception of the injection molding
programs. For this reason, Si_3N_4 specimens modified with sintering aids
such as CeO_2, MgO, Y_2O_3/SiO_2, and
Y_2O_3/Al_2O_3 were first exposed at elevated temperatures to the reaction
products of O_2/CH_4 and O_2/H_2 and then
subjected to flexural strength tests
to assess the effect upon their mechanical properties. These tests showed
no significant effect on strength of
$Si_3N_4/Y_2O_3/SiO_2$ and $Si_3N_4/Y_2O_3/Al_2O_3$
(Fig. 9). Si_3N_4 modified with CeO_2
or CeO_2/SiO_2 demonstrated excessive
strength degradation after exposure
to the rocket engine environments,
as can be seen from Table 3.

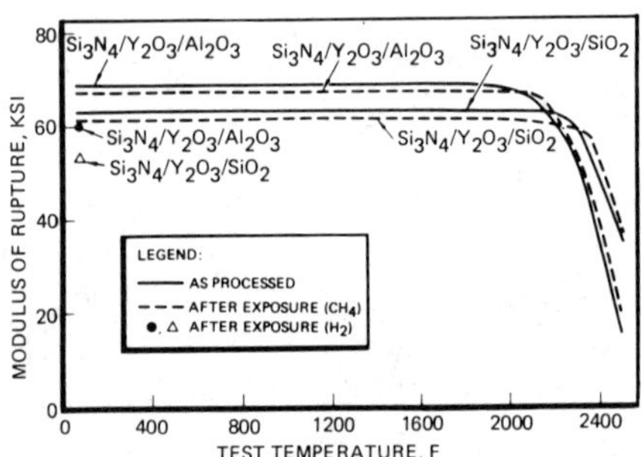

Fig. 9. Test Results for Rocketdyne
         IM/Sintered Silicon Nitride
         Environmental Exposure at
         2500 R (2040 F)

5. DESIGN CONSIDERATIONS

Because of the brittle nature of ceramics, a major concern is their survival during the thermal shock of
rocket engine turbine start and shutdown transients. A rocket engine

Table 3. Silicon Nitride Strength After Rocket Engine Exposure

| Composition | Flexural Strength, ksi | | | | Fraction Mean Strength Retained, % |
| | As-Sintered | | After Exposure | | |
	Mean	Standard Deviation	Mean	Standard Deviation	
Sintered					
Si_3N_4/MgO	55.5	13.0	47.1	20.8	85
Si_3N_4/CeO	56.4	15.2	21.2	21.6	38
$Si_3N_4/Y_2O_3/Al_2O_3$	64.3	11.5	59.9	13.9	93
$Si_3N_4/Y_2O_3/SiO_2$	61.7	6.9	52.4	4.7	85
$Si_3N_4/CeO_2/SiO_2$	38.8	10.0	6.1	4.4	16
Hot-Pressed					
Si_3N_4/MgO	81.4	9.7	96.1	9.3	118
Si_3N_4/CeO_2	84.9	10.6	30.0	-	35

turbine nozzle vane will illustrate the factors considered during design with structural ceramics.

Both mechanical and thermal stress have been analyzed using the SSME, high pressure fuel turbopump (HPFTP) first-stage turbine nozzle as a reference design. Figure 10 is a cross section of the HPFTP turbine showing the first-stage nozzle. Several nozzle vane configurations and environments were analyzed in an attempt to lower transient thermal stress.

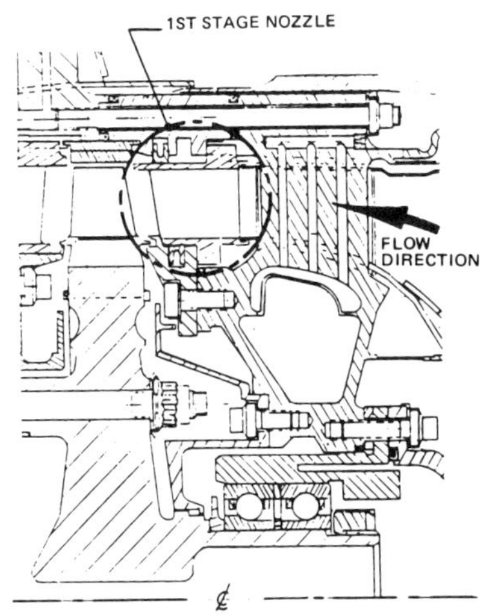

Fig. 10. SSME HPFTP Turbine Cross Section

Given that the allowable stress for SN-104 is 43 ksi (90th percentile expected minimum), it is shown that silicon nitride nozzle vanes can survive the rocket engine shutdown

provided that thermal gradients are minimized and engine shutdown is sufficiently extended to moderate thermal transients.

For mechanical loads, the vane was treated as a beam with uniformly distributed pressure. Three different end conditions were examined: simply supported, cantilevered, and fixed guided. As can be seen in Fig. 11, the simply supported vane has the lowest tensile stress (7.2 ksi). The maximum tensile stress for the cantilevered and fixed-guided supported vanes are 39.6 and 26.0 ksi, respectively. For this reason, a simply supported mounting is preferred.

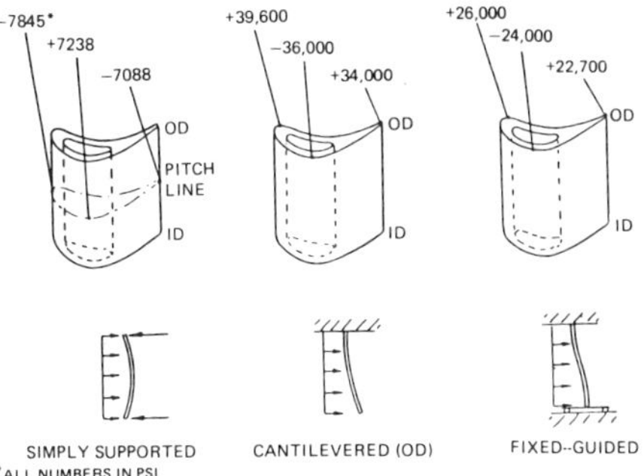

Fig. 11. Peak Steady-State Mechanical Stresses for the Three Candidate Vane Support Schemes

The nozzle vane is subjected to thermal loading during the start/stop of the engine cycle. During start, the vane is suddenly subjected to a high-temperature, high-velocity gas stream. The outer surfaces are put in compression while the cool center of the vane is in tension. Since ceramics are 10 to 20 times stronger in compression than in tension, however, startup is not considered to be as critical as shutdown.

During shutdown, the hot vane is subjected to a liquid hydrogen quench. The outer surfaces, and especially the leading and trailing edges respond quickly, become cold, and try to shrink. This results in high-tensile stresses on the outer surfaces, which is considered the limiting stress situation for ceramic rocket engine components.

To analyze the stresses, a 2-D plane
strain finite element model was generated for both solid and hollow
vanes, and the peak tensile stress
was calculated as a function of
time. This resulted in a "best
available" stress, which the ceramic
must survive.

Several alternatives were considered
that would reduce the thermal
stresses:

   1.   Hollow airfoil

   2.   Thermal conditioning (i.e.,
passing a fixed flow of combustion
gas products through the center of
the hollow airfoil)

   3.   Extension of the transient
time span

Figure 12 is a plot of thermal
stresses versus time for the hollow
airfoil with 0, 2, and 5% conditioning flow of the external gas through
the hollow center. This way, duct-
ing flow results in a lower ΔT because of simultaneously cooling
outer and inner surfaces of the vane
during shutdown. Conditioning
markedly reduced the stress levels
from 61 ksi for the solid vane to
37 ksi for 5% conditioning flow.

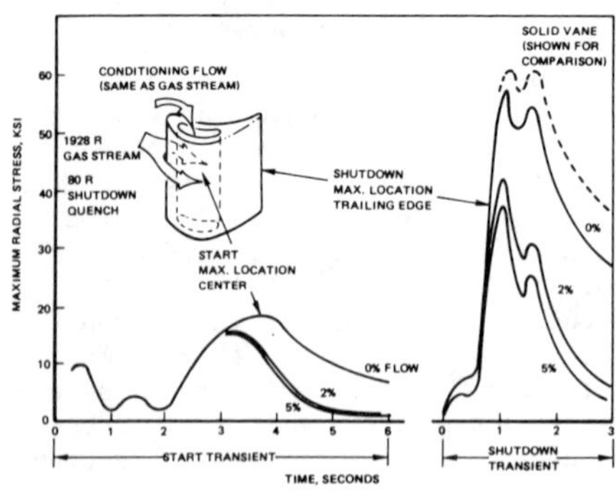

Fig. 12.   Plot of Maximum Stress
of a Hollow Vane With
Conditioning Flow

Extension of the shutdown transient
was examined in conjunction with
various percentages of conditioning
flow. Figure 13 is a plot of peak
thermal stress versus percent

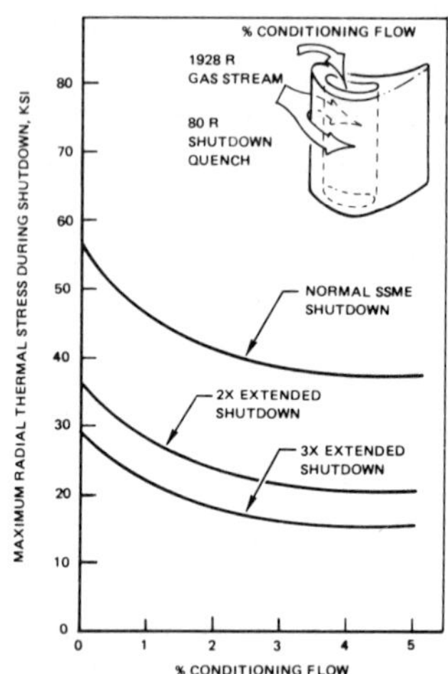

Fig. 13.   Peak Stress During Shutdown vs % Conditioning
Flow (TIT = 1928 R)

conditioning flow for 1X, 2X, and 3X
shutdown extension, using the SSME
as a baseline. The 3X shutdown extension plus 5% conditioning flow
peak tensile stress is 15 ksi.

The hollow vane with extended shutdown and various percentages of conditioning flow also were analyzed at
high turbine inlet temperatures.
Figure 14 plots peak stresses versus
percent conditioning flow for normal

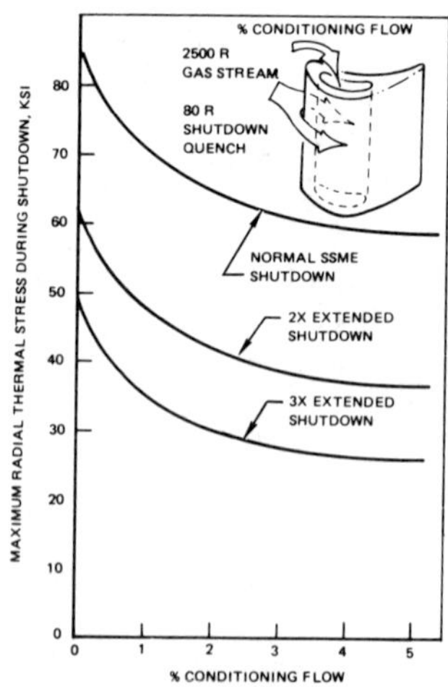

Fig. 14.   Peak Stress During Shutdown vs % Conditioning
Flow (TIT = 2500 R)

and extended shutdowns with an inlet temperature of 2500 R (2040 F). The 3X shutdown extension plus 5% conditioning flow stress is 26 ksi.

It should be noted that the maximum mechanical gas bending stresses and thermal transient stresses are not additive, since gas flow is not maximized when the start or shutdown stresses occur. There are mechanical stresses acting in the vane in addition to the thermal stresses, but they are less than the maximums considered. In addition, effects on turbine efficiency of the conditioning flow must be balanced against extension of the shutdown. Stress can be reduced by a factor of 2 by conditioning and again by a factor of 2 by extending the cutoff transient time duration.

6. SUMMARY

Silicon nitride and silicon carbide ceramics are candidates as uncooled, hot-gas path component materials in a rocket engine turbine with a turbine inlet temperature of 2000 F or above. Silicon nitride has been identified as the most promising ceramic candidate based on superior thermal shock resistance. It can be fabricated by IM and sintering to produce complex component designs economically.

The major concern for ceramic components is their ability to survive thermal transients during turbine start and shutdown. Their most practical near-term application is for nozzle vanes or similar stationary components, which are not subject to centrifugal or high cyclic loads.

It has been shown analytically that silicon nitride nozzle vanes can survive typical (i.e., SSME) turbine shutdown stresses provided the shutdown is sufficiently extended and the vanes are "conditioned" with internal gas flow. Conditioning reduces the thermal gradient by cooling the interior of the airfoil at the same rate as the exterior.

Silicon nitride samples have successfully passed preliminary thermal shock and environmental stability tests at Rocketdyne, giving an excellent indication that we are on the right track. In future programs, IM silicon nitride nozzle vanes will be tested to verify these analytical results.

7. BIOGRAPHY

Gary D. Schnittgrund is a Member of the Technical Staff specializing in high temperature ceramic materials for the Rocketdyne Division of Rockwell International. He earned a B.S. in 1970, an M.S. in 1971, and a Ph.D. in 1975 from the University of Illinois; all of his degrees are in Ceramic Engineering. Dr. Schnittgrund has worked at Rocketdyne for 2 years, where he has been involved in development of injection molded and sintered silicon nitride components for high temperature turbines. Prior to joining Rocketdyne he was with Consolidated Aluminum and the Aluminum Company of America where he conducted high temperature materials research. He has published several papers and is a Member of the American Ceramic Society. His technical interest emphasize the interrelation between fabrication, microstructure, and properties.

26th National SAMPE Symposium
April 28-30, 1981

POLYIMIDE APPLICATIONS IN ADVANCED MISSILES

R. J. Jones, R. W. Vaughan and A. E. Samsonov
TRW Defense and Space Systems Group
One Space Park
Redondo Beach, California 90278

Abstract

The resins employed in many missile systems today were selected and qualified prior to 1965. Hence, advanced aromatic/heterocyclic polymers such as polyimides were not considered at that time due to their fledgling state of development in the early to mid 1960's. However, significant property and processability improvement combined with commercial development of polyimides has occurred in recent years making these polymers attractive candidates for many missile applications.

Selected polyimides must be viewed as an available polymer technology suitable for solving currently employed resin disadvantages in missile components. For instance, room temperature curing and hot melt processable flexible polyimides are available for filament winding fabrication of tough, structurally integral motor cases or moisture resistant liners. Linear and crosslinkable aromatic polyimides are available that should offer significant improvement in fabrication reproducibility of carbon-carbon structures at a potential cost benefit over phenolic and pitches. A number of applicable polyimide polymer types and technical and processing data will be presented and discussed.

1. INTRODUCTION

Since the inception of the United States commitment to intercontinental and shorter range ballistic missiles in the 1950's, key factors such as range/delivery capability, component reliability and long-term storability have been key drivers. In the missile development programs, it soon became apparent that plastics and elastomers offered a significant weight savings and reliability advantages over many metallic structural and non-structural components.

The incorporation of plastics and

elastomers into missiles have seen
increasing use from the early Titan
and Minuteman Systems up through the
current MX, Cruise and advanced
Pershing prototype systems. Because
of the nature and significance of
our network of missile systems as a
prime line of defense, it has been
necessary to use a great deal of
conservatism and caution in selection of plastics and elastomers for
incorporation into primary and
secondary component hardware. The
basic epoxy, phenolic, urethane
and butyl rubber components that
have demonstrated acceptable long-
term storability and performance
characteristics for the Titan and
Minuteman, continue to be specified
for many advanced missile system
uses. Very few new polymers have
been investigated with the objective
of improving missile performance
from the mid-1960's until the present. During this time significant
advances have been made in engineering plastics. These plastics could
offer significant benefit if used
in advanced missiles.

One particular generic resin family
whose time is thought to have arrived
for advanced missile applications is
the polyimide resins. The early
aromatic condensation class of
polyimides discovered in the 1960's
suffered from problems of poor processability and high product void
contents; however, significant improvements have been made which make
polyimides superior alternatives to

many resins used today in production missile hardware.

A list of projected advantages of
selected types of polyimides as a
class over several commonly used
and qualified resins for missile
applications are as follows:

- Flexible Polyimides - Higher
  toughness and/or high temperature stability than epoxy
  matrices used to fabricate or
  repair filament wound motor
  cases.

- Addition-curing Polyimides -
  Higher hydrolytic stability
  and/or high temperature stability
  than polyurethanes in linear
  propellant binder applications

- Addition-and Condensation Type
  Polyimides - Greater char yield
  and reproducibility than phenolics employed as precursors to
  fabricate carbon-carbon components.

The remainder of this paper will
cite a few selected polyimide resin
systems which are thought to be
directly applicable to the fabrication of improved missile component
hardware. The technologies cited
are only intended to be representative of the types or classes of
polyimides, and not the only particular resin formulation(s) or
polymer structures considered as
viable replacements for the currently used resins to which they
are compared.

    2. TECHNICAL PROGRESS

Beginning in approximately 1960, the

trade began experimental studies in a new area of polymer research which has led to many new generic resin types known as aromatic/heterocylic systems. Such polymers normally contain aromatic rings most commonly joined through nitrogen containing linkages resulting from the reaction of carboxylic acid or carbonyl containing monomers with di-, tri-and tetramines. This research work had led to the reduction to practice of many new polymer types, the prime feature of each being high temperature stability (normally > 200°C) and high char yields (normally > 50%) due to the high aromatic/heterocylic character. A number of these new polymers have received considerable development work sponsored by government and industry. Several new resin types which still are held to have high promise for future use or are in limited to full-scale production today are reviewed in Table 1. It is TRW's opinion that the greatest innovation and technical progress have been made since 1965 in developing polyimides. The polymers have been introduced successfully in many product areas, and appear suited for use in advanced missile components.

Perhaps the most widely used polyimide is Kapton film from DuPont (Reference 1). This product, utilizing high molecular weight, condensation-type polyimides is a qualified material for an ever increasing number of electrical applications as wire wrap, capacitor film, etc.

The success of Kapton-type, high molecular weight polyimides as high temperature resistant polymers has spurred the government and industry to seek a wide variety of other product applications for polyimides. The high molecular weight, condensation-type polyimides are unsuitable for autoclave fabrication of complex structures of simple injection moldings of repetitive, large volume production parts. Consequently, an area that has received much attention during the period of 1965 to 1975 is that of addition curable mechanisms for transforming easily processable low molecular weight oligomers into thermosetting resin matrices.

The significant contributions in the area of thermosetting polyimides are the PMR polyimides developed by NASA/Lewis Research Center (Reference 2) and the Thermid resins developed by Hughes (Reference 3). These resins, particularly the PMR polyimides, are finding rapidly increasing use in autoclaved or filament wound complex structures for airframe and engine applications. However, the commercially available addition-type polyimides require a $\geq$ 550°F cure temperature which restricts general use to those few companies possessing high temperature autoclave

processing hardware. Also, this
high cure temperature excludes these
resins from many structural or bonding applications where aluminum is
part of the prime structure or is
the adherend. Thus, addition-type
polyimides which cure at $\leq$ 400°F,
and offer thermal resistance at
500°F and above should be of significant future utility.
Also, because polyimides generally
possess higher thermal and hydrolytic stability than polyurethanes,
flexible polyimides should be of
significant utility as a replacement for the latter. Also, elas-
tomeric-type or tough polyimides
could go far to overcome the inherent brittleness problems frequently encountered in the use of
epoxies as composite matrices in
autoclave and filament wound composite structures.
For the reasons stated above, TRW
during the 1970's concentrated on
providing innovative approaches to
improve the processability and/or
thermo mechanical properties of
polyimide systems available at the
time. Key contributions from this
work resulted in the development
of three new polymer technologies
as follows:
- Polyimides that are highly
  flexible
- Polyimides that are autoclavable
  at $\leq$ 400°F
- Polyimides that are compression
  moldable and possess a Tg > 750°F.

A chemical and property overview
of the new TRW polyimides is presented in the remainder of this
paper.
Over the past decade, TRW has developed two types of a new generic
polyimide family designated flexible polyimides. These resins are
prepared from selected aromatic and
hydrocarbon ether ingredients which
allow extreme versatility to vary
the tensile elongation to break
from 10% to 200% depending on the
particular product requirements.
One group of flexible polyimides is
tailored for facile cure by commercially available epoxies in the
temperature range of 50°F to 150°F
and is suitable for use in the temperature range of -65°F to 250°F.
The other flexible polyimide family
is hot melt processable in the temperature range of 200°F to 300°F,
which upon cure at 300°F to 400°F,
is suitable for use in the temperature range of -65°F to 350°F.
Each of these resin systems is
described separately below.
Flexible Polyimides - The flexible
polyimide formulations which could
be cured at or near room temperature
was performed for USAF on Contract
F33615-C-5134 and is reported in
AFML Report Number TR-79-5054
(Reference 4).
The chemistry of the flexible poly-
imide/epoxy resins is shown in
Equation 1. A number of available
epoxies recommended for use with

the polyimide polymers are shown in Table 2. Highlight properties of the resins determined during Contract F33615-77-C-1534 are given in Table 3.

The work to date on the flexible polyimide/epoxy resins has resulted in the allowance of three United States patents which are expected to issue in 1981.

A key projected use for these resins is the field repair of filament wound motor cases. A flexible polyimide sample was recently supplied to Army Ballistic Missile Defense Advanced Technical Center and is undergoing tests for this purpose.

A second, flexible polyimide family which was developed by TRW possess the novel feature of hot melt thermo-plastic behavior in the temperature range of 200°F to 300°F. The linear polymers then cure *in situ* by an addition reaction at temperatures above 300°F into tough, structurally integral resin matrices demonstrating excellent properties. To date, these resins have been evaluated as wire coatings, adhesives and barrier sealants by hot melt extrusion processes.

The chemistry of the hot melt pro-cessable flexible polyimides is given in Equation 2. The properties determined to date on the resins are highlighted in Table 4. Because of the excellent processability and cured resin properties highlighted in Table 4, superior motor cases

may be filament wound from these resins. Also, it should be fea-sible to melt cast motor case liners from the polymers.

The work performed to date on these polyimide polymers has resulted in the issue to TRW of United States Patent Numbers 4,116,937 and 4,179,551 (References 5 and 6).

Autoclave Processing of Polyimides

During the last decade, TRW con-ceived and successfully demon-strated the potential of polyimide resins prepared by a novel poly(Diels-Alder) polymerization mechanism employing selected bis(dienophile) and bis(diene) reactants. The novel feature of the poly(Diels-Alder) approach is that autoclave (or filament winding) processing is easily achieved at ≤ 400°F. Upon postcure, struc-tures fabricated from polyimide systems demonstrate excellent thermo-mechanical structural properties in the range of 500°F to 600°F.

The chemistry of the poly(Diels-Alder) apporach is shown in Equation 3. The properties obtained on graphite reinforced composites are highlighted in Table 5. A graphic representation of a proto-type YF-12 composite wing panel autoclave fabricated from the poly(Diels-Alder) polyimide resin employing high modulus graphite skins, co-cured to a polyimide honeycomb core is shown in Figure 1. The work conducted to date on the poly(Diels-Alder) polyimides has resulted in the allowance to TRW of United States Patent Numbers 3,927,027,; 3,951,902 and 3,975,363 (References 7, 8 and 9). Key autoclave fabrication studies and prototype part production are described in NASA final report document CR-134900 (Reference 10).

Polyimides that are Stable at > 700°F

Perhaps the TRW polyimide family with the greatest future potential to yield extremely high performance properties are the partially fluorinated polyimide polymers. Work conducted to date in 1980 at TRW and GE has given strong evidence that these polymers can be fabricated into jet engine (or rocket engine) parts suitable to maintain excellent thermo-mechanical properties at temperatures $\geq$ 700°F.

The chemistry of the partially fluorinated polyimides is shown in Equation 4. The high performance properties obtained to date on the resins are given in Table 6.

The work conducted to date on the partially fluorinated polyimide resins has resulted in the allowance to TRW of United States Patent Numbers 4,111,906; 4,196,27 and 4,203,922 (References 11, 12 and 13).

The preceding examples of novel polyimide types of technology are thought to be available for development and qualification as advanced missile structural and sub-system components. It should not be interpreted that the TRW polyimide resins constitute the only approach to improved polymer matrices. For instance, a proprietary, flexible or tough polyimide has recently been introduced as V-378A by the U. S. Polymeric Division of Hitco, Inc. (Reference 14). Also, a high temperature, partially fluorinated polyimide family is available from DuPont as the NR-150 family (Reference 15). It is anticipated that several other new polyimides emphasizing novel combinations of processability and thermo-mechanical properties will be introduced in the 1980's.

3. CONCLUSIONS

Based upon the current state-of-the-art involving polymeric material uses in deployed and in-production missiles, TRW offers the following conclusions regarding qualified resins and available alternative technology:

- Currently qualified and utilized materials (e.g., epoxies, phenolics and polyurethanes) have performed very well in the past in a number of key missile applications

- New polymer technologies and types exist today which may offer significant performance benefits over presently qualified polymers

- Polyimides as a general class offer significant potential benefits over presently qualified polymers.

4. ACKNOWLEDGEMENT

The authors wish to express their appreciation to Dr. David C. Sayles, Army Ballistic Missile Defense Advanced Technical Center, for his interest and guidance in the preparation of this presentation.

REFERENCES

1. Lee, H., *et al*, New Linear Polymers, Mc Graw-Hill Co., New York, 1967

2. Serafini, T. T., Delvigs, P., and Lightsey, G. R., J. Appl. Polym. Sci., 16, 905 (1972).

3. Landis, A. and Andres, R., "High Temperature Laminating Resins," Summary Report, AFML-TR-70-250, December 1970

3. Jones, R. J., Hom, J. M. and Markles, O. F., Jr., "Development of High Adhesion Sealants," Final Report, Contract F33615-77-C-5134, AFML-TR-79-4054, May, 1979.

5. Jones, R. J., O'Rell, M. K., and Sutherland, J. D., U. S. Patent 4,116,937 (1978).

6. Jones, R. J., Green, H. E. and Quinlivan, S. C., U. S. Patent 4,179,551 (1979).

7. Jones, R. J. and Green, H. E., U. S. Patent 3,927,027 (1975).

8. Jones, R. J., Cassey, H. N. and Green, H. E., U. S. Patent 3,951,902 (1976).

9. Jones, R. J., U. S. Patent 3,975,363 (1976).

10. O'Rell, M. K. and Zakrezewski, G. A., "Development of Auto-clavable Polyimides," Final Report, Contract NAS 3-17726, NASA CR-134900, dated 1976.

11. Jones, R. J., O'Rell, M. K. and Hom, J. M., U. S. Patent 4,111,906 (1978).

12. Jones, R. J. and O'Rell, M. K., U. S. Patent 4,196,277 (1980).

13. Jones, R. J., O'Rell, M. K. and Hom, J. M., U. S. Patent 4,203,922 (1980).

14. Street, S. W., "V-378 A, A New Modified Bis-Maleimide Matrix Resin," SAMPE, vol. 25, May 1980.

15. Gibbs, H. H. and Breder, C. V., "High Temperature Laminating Resins Based on Melt Fusible Polyimides," Adv. in Chemistry Series No. 142, Copolymers, Polyblends, and Composites, American Chemical Society, Washington, D. C., 1975.

BIOGRAPHIES

R. J. Jones

Dr. Jones is currently Manager, Chemical Research and Applications Department. He is responsible for planning, coordination, solicitation and implementation of government contracts related to new and improved polymer/plastics R&D. He serves as a consultant to other major groups of TRW in plastic products. During his 13 years at TRW he has been a major contributor or sole inventor of several unique polymer systems, including flexible polyimides, poly(Diels-Alder)

476

polyimides and partially fluorinated
polyimides. Dr. Jones received his
B.S. degree from the University of
Missouri (Columbia) in 1961 and his
Ph.D. in 1965 from the same
institution. Prior to joining TRW,
Dr. Jones worked for the DuPont
Company for three years in Nylon
Research.

R. W. VAUGHAN

Mr. Vaughan is currently Program
Manager of the Air Force Elastomeric
Diaphragms manufacturing technology
program and the NASA Improved Tack
PMR Resin program. During his
fourteen years at TRW he has
previously served as Program
Manager on a variety of programs
of different sizes for NASA and
the Air Force involving many
technologies including composite
matrix resins with improved toughness, moisture resistance, and
high temperature servicability and
manufacturing processes for weldbonding, filament winding and
elastomeric device fabrication.
He also serves as Head, Elastomer
R&D Section. Mr. Vaughan has
previous experience with EFMC,
Canadair, Ltd., and De Havilland,
Great Britain. Mr. Vaughan
received his B.S. in Chemical
Engineering in London, England.

ALEXANDER E. SAMSONOV

Mr. Samsonov is Head, Engineering
Applications Section in Chemical
Engineering Department. At present,
provides technical support to Ogden
ALC on the Chemical Engineering

Tasks for the Minuteman LRSLA
(Long Range Service Life Analysis)
Program and to BMO on the MX Missile
Development Program. Also provided
technical support to SAMSO on
Minuteman Aging and Surveillance
Programs covering three solid
propellant boosters, liquid pro-
pellant (N_2O_4 & MMH), PBPS and Pen
Aids. Also involved in systems
support to MX production program.
Specialization in Non-Metallic
materials, propellant and ordnance
properties and service life analysis.
Mr. Samsonov received his Master of
Chemical Engineering at Polytechnic
Institute of Brooklyn and his
Bachelor of Chemical Engineering
from City College of New York.

TABLE 1.

REVIEW OF PROMISING AROMATIC/HETEROCYCLIC RESINS

POLYMER TYPE	GENERAL STRUCTURE	CHIEF ATTRIBUTES
AROMATIC NYLONS OR ARAMIDS	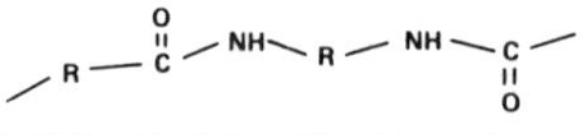	• UP TO 400°F THERMAL STABILITY • FORM TOUGH, HIGH MODULUS ORGANIC FIBERS
POLYIMIDES	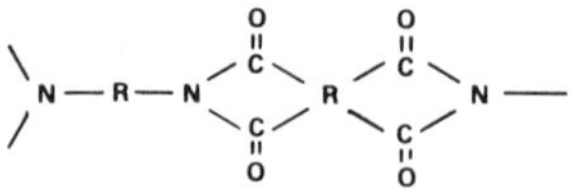	• UP TO 700°F THERMAL STABILITY • USEFUL IN THERMOPLASTIC OR THERMOSET FORM • NORMALLY RESISTANT TO HYDROLYSIS AND SOLVENT ATTACK • PRODUCT APPLICATIONS ARE MANY
POLYBENZIMIDAZOLES	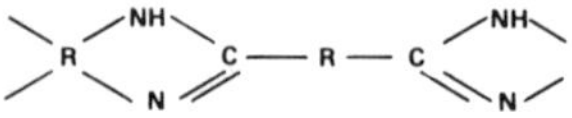	• UP TO 600°F THERMAL STABILITY • FORM TOUGH, HIGH MODULUS ORGANIC FIBERS
POLYPHENYLQUINOXALINES	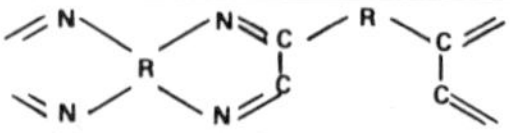	• UP TO 600°F THERMAL STABILITY • FORM EXCELLENT HIGH PERFORMANCE BONDED JOINTS

TABLE 2.

REPRESENTATIVE PROPERTIES
OF FLEXIBLE POLYIMIDE/EPOXY RESINS

Property Measurement	Property Value[a]
1. Specific Gravity	1.1 to 1.2
2. Cure Time	48 Hours to 168 Hours
3. Initial Thermo-Oxidative Stability (TGA)	460°F to 500°F (Nitrogen) 430°F to 460°F (Air)
4. Low Temperature Flexibility	-50°F to -65°F (Static Tests)
5. Weight Pick-Up in Jet Reference Fuel for 720 Hours at 140°F	5% to 8%
6. Tensile Elongation at RT	20% to 200%
7. 180° Peel Strength	20 pli to 60 pli
8. Shore A Hardness	60 to > 100

[a]This table gives the normal range of properties obtained
when curing with aromatic epoxies (e.g., DEN 431) and
aliphatic ether epoxies (e.g., DER 736).

TABLE 3.

REPRESENTATIVE EPOXY RESINS SUGGESTED FOR
USE WITH FLEXIBLE POLYIMIDES

NAME OF EPOXY RESIN	MANUFACTURER	STRUCTURE
MY-720	CIBA-GEIGY	Tetraglycidyl diaminodiphenylmethane: each of two para-phenylene rings bears an $N(CH_2\text{-}CHCH_2\text{ (epoxide)})_2$ group, the rings joined by $\text{-}CH_2\text{-}$.
DEN-431	DOW	Novolac-type epoxy: $O\text{-}CH_2\text{-}CH\text{-}CH_2$ (epoxide) glycidyl ethers on cyclohexyl rings linked by $\text{-}CH_2\text{-}$ units, $[\;]_n$.
DER-732	DOW	$CH_2\text{-}CH\text{-}CH_2\text{-}O\text{-}[CH_2\text{-}CH\text{-}O]_n\text{-}CH_2\text{-}CH\text{-}O\text{-}CH_2\text{-}CH\text{-}CH_2$, with pendant R groups, $n = 3$.

TABLE 4.

REPRESENTATIVE PROPERTIES OF
HOT MELT PROCESSABLE FLEXIBLE POLYIMIDES

Property Management	Property Value
1. Specific Gravity (Neat Resin)	~ 1.2
2. 180° T-Peel Strength	15 lbs/in to 20 lbs/in on urethane primed substrates as fabricated; > 30 lbs/in after exposure at 300°F.
3. Lap Shear Strengths	• C-1010 Steel 1200 psi[c] • T-2040 Aluminum 1500 psi[c]
4. Adhesive Failure Mode on Test	Cohesive
5. Fuel Resistance	
• JP-4[a]	No degradation observed on aging for 500 hrs at 300°F
• JRF[b]	Improvement in 180° peel strengths to 20 lbs/in to 30 lbs/in after aging for 168 hours at 140°F
6. Thermo-oxidative Stability[a]	High property retention obtained on exposure to flowing air for 168 hrs at 300°F
7. Low Temperature Behavior[a]	Film samples remained tough and flexible at -60°F
8. Hydrolytic Stability[b]	No degradation observed in properties after 1000-hour aging at 160°F and 95% R.H.
9. Extrusion Behavior (Hand tool and screw-type extruder)	Excellent in temperature range of 300°F to 390°F under low pressure (< 100 psig)

[a] Thin ($\sim$ 5-mil) films employed for test; mechanical property measurements made at 72°F.

[b] 180° peel specimens employed for test; mechanical property measurements made at 72°F

[c] Neat resin values obtained at 72°F on surfaces prepared by methyl-ethyl ketone wipe only.

TABLE 5.

REPRESENTATIVE PROPERTIES
OF POLY(DIELS-ALDER) POLYIMIDES

Property Measurement	Property Value
1. Specific Gravity (neat resin)	1.28-1.32
2. Resin Cure Temperature (Autoclave)	350°F to 400°F
3. Normal Resin Postcure Temperature Range (unrestrained)	400°F to 600°F
4. Representative Composite Properties Employing A-S Fiber	
• Weight loss after 1000 hours aging in air at 500°F	1.6%
• Weight loss after 1000 hours aging in air at 550°F	6.9%
• Interlaminar shear strengths at RT, 500°F and 550°F prior to aging (Ksi)	14.6, 7.1, 6.2
• Interlaminar shear strengths after aging at 500°F and 550°F for 1000 hours in air (Ksi)	• Aged at 500°F: 7.4 RT Test 6.2 500°F Test • Aged at 550°F: 7.2 RT Test 3.8 550°F Test
• Void volumes measured (% v/v)	0.0-1.5

TABLE 6.

REPRESENTATIVE PARTIALLY FLUORINATED
POLYIMIDE PROPERTIES

Property Measurement	Property Value[a]
1. Specific Gravity	$\sim$ 1.4[a]
2. Glass Transition Temperature Range (T_g) by Thermal Mechanical Analysis (TMA)	290°C to 390°C ($\sim$ 580°F to 740°F)[a,b]
3. Thermo-Oxidative Stability	
• Initial by Thermalgravimetric Analysis (TGA)	$\sim$ 500°C[c]
• Weight Loss on Aging at 675°F for 100 Hours in 5 Atmospheres Air	10%[b]
4. Hydrolytic Stability by Two Hour Water Boil	1.5% Weight Gain (No Discernible Change in Resin/Fiber Surface Characterization)[b]
5. Solvent Resistance to Methylethyl Ketone and Jet Fuel	2.0% to 5.0% Weight Gain (No Discernible Change in Film)[c]
6. Monomer and Polymer Preparation Yield	98% to 100% (Determined in up to 8 lbs./Batch Scale)
7. Resin/Fiber Wetting by Photomicroscopy	Excellent (No Voids Detected)[b]
8. Composite Machinability	Excellent (Stator Bushing Components)[b]

[a]A range is obtained depending upon resin molecular formulary.

[b]Determined on 50% resin/50% chopped GY-70 fiber (by weight) compression molded composites.

[c]Determined on thin cast film samples of $\sim$ 5-mil thickness.

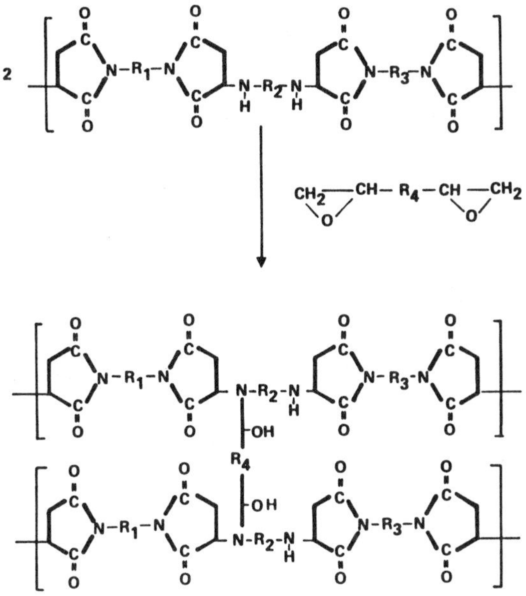

Equation 1. Chemistry of Flexible Polyimide/Epoxy Resins

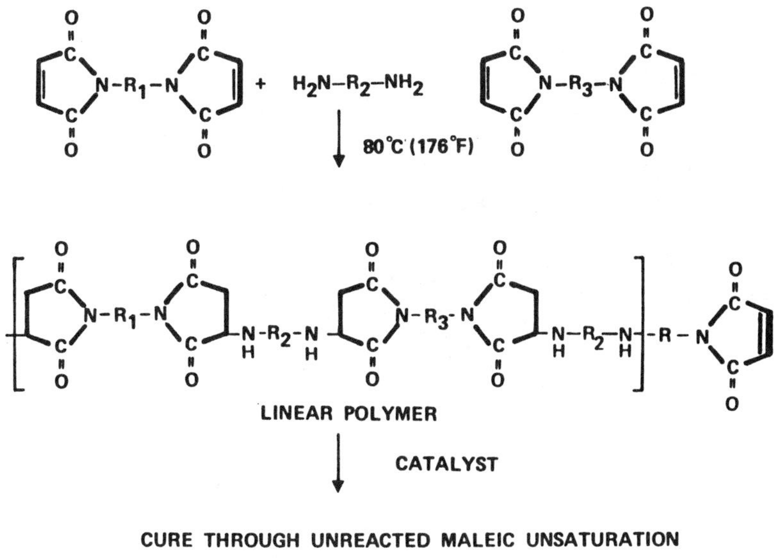
80°C (176°F)
LINEAR POLYMER
CATALYST
CURE THROUGH UNREACTED MALEIC UNSATURATION

Equation 2. Chemistry of Hot Melt Processable Flexible Polyimides

400°F

400°F TO 600°F

Equation 3. Chemistry of Poly(Diels-Alder) Polyimide Resins.

POLYIMIDE FROM BFDA DIANHYDRIDE

WHERE R = AROMATIC RING STRUCTURE

POLYIMIDE FROM BDAF DIAMINE

WHERE R = AROMATIC RING STRUCTURE

Equation 4. Chemistry of Partially Fluorinated Polyimide Resi

Figure 1.   Prototype YF-12 Wing Panel Fabricated
from a Poly(Diels-Alder) Polyimide.

MISSILE CONTAMINATION PROBLEMS
W. B. Tolley
Hughes Aircraft Company
Technical Support Division Tucson, Arizona

Abstract

Six contamination related problems which have occurred in Tucson Missile Production Programs are discussed. All problems require solution. However, problems involving loss of aircraft or life obviously require major action as compared to cosmetic problems.

Some problems that cause high cost, risk and require mandatory and absolute solutions are not in the norm of what most of us associate with contamination. Examples follows:

o Stress corrosion of launch hooks - caused by salt environment access to high strength metal from poor or cracked platings - can cause loss of aircraft.

o Adhesive debond in missile cold-walls - due to introduction of minor amounts of moisture into the system while being handled and checked out on aircraft carriers - would cause loss of missile.

o Stress corrosion of a missile guidance wire-due to miscompounding of an adhesive system — would cause loss of missile.

In general, contamination problems encompass a wide range from the expected to the totally unexpected. Minimizing such problems requires:

o Good design

o Sound manufacturing practice

o Quick reaction and problem identification

o Reasonable and cost effective engineering action.

1. INTRODUCTION

Contamination is usually viewed as the introduction of an easily recognized unwanted material into a system. However, in failure analysis efforts over a period of years at the Hughes Aircraft Missile Production Plant in Tucson, Arizona, a number of contamination problems have surfaced which were completely unexpected and in which the contaminant was not readily recognized as a contaminating material. In some cases, given the knowledge that the specific material would be present during system life, one would be hard pressed to recognize that it might react to cause problems. We generally recognize acids, bases, chlorides, etc, as materials one would not want in most electrical-mechanical systems as even minute impurities. However, other contaminants such as, moisture, alcohols, oils might not be recognized as problem materials, especially when introduced into a system which already is known to contain similar materials.

In the following text six (6) contamination problems are presented, some fit the norm and once recognized would immediately be suspect, others are not so easily identified as being detrimental to the hardware.

2. DISCUSSION

2.1 Effects of Contamination

Many operations and hardware can be affected by minor amounts of a contaminating substance. Table I identifies operations and processes in which contaminants can be a degrading factor.

TABLE I
MISSILE CONTAMINATION

1. Wide Range of Hardware Affected

    o Optical Systems
    o Electrical Circuitry
    o Bearings - Sliding Parts
    o Mechanical Structural Elements

2. Covers a Broad Spectrum

    o Coatings - Paints/Adhesion
    o Potting - Encapsulation/Sealing
    o Bonding - Staking/Adhesion
    o Corrosion - Stress Corrosion
        - Hydrogen Embrittlement

Yearly costs attributed to contamination induced problems runs into the billions of dollars, affects the reliability of many critical systems and in cases can cause loss of vehicle/aircraft and life.

2.2 Control of Contamination

The control of contamination problems is a responsibility of both the design functions as well as the manufacturing operations. In many cases good design can minimize susceptibility to contamination induced problems. However, until the problem occurs, it may be very difficult to recognize the subtle deficiencies in the design which can lead to a problem. In fact, problem identification requires rather sophisticated detective work prior to being able to effect corrective action.

In general, contamination problems range from the easily recognized to the totally unexpected. All problems require corrective action. However, problems involving loss of aircraft or life require major action as compared to cosmetic problems. It is the responsibility of both the design and manufacturing organizations to put these problems in perspective in defining the effort to be expended and corrective action to be taken. There is a wide difference in impact between cosmetic paint blistering and a fracture of a missile launcher hook. Both

require solution but the latter
obviously requires mandatory solution with engineering control of
affected hardware. Contamination
problems and associated loss of
hardware, cost of problem identification and costs of corrective
action can be minimized through:

o Good Design
o Sound Manufacturing Practice
o Quick Reaction and Problem
   Identification
o Reasonable Cost Effective
   Engineering Action

Reasonable engineering action
requires putting the problem in
perspective and providing a solution commensurate with the magnitude of the problem. We have
often seen minor problems blown out
of perspective causing huge costs
that might have been minimized.

3. MISSILE CONTAMINATION PROBLEMS
Six (6) missile problems are
identified in Table II. Four*
(4) of these are discussed in some
detail

TABLE II
VARIABILITY OF CONTAMINATION PROBLEMS

1. Missile Relay Malfunction

    o In-House Assembly Con-
       tamination

2. Missile Launcher Transformer
    Solder Problem

    o Vendor Potting Contamination

3. TOW Guidance Wire-Stress Corro-
    sion *

    o Vendor Adhesive Formulation
       Error

4. TOW Missile Fusible Resistor
    Failure *

    o Vendor Contamination

5. Phoenix Missile Coldwall Ad-
    hesive Debond *

    o User Moisture Contamination

TABLE II. - continued

6. TOW Wire Spool Paint Adhesion *

    o Anodize Rinse Water Con-
       tamination

3.1 TOW Missile Wire Fracture
The TOW Missile is a wire guided
system which contains two (2) wire
bobbins in the aft end of the
missile dispensing two (2) strands
of wire during flight. Control
signals are automatically programmed through this command link
to provide guidance to the missile.
Reliability has been greater than 95
percent but does require wire
integrity at transonic flight velocities. The guidance wire is a
0.0050 inch diameter 500 KSI steel
strand insulated with a polyester
coating to a nominal diameter of
0.0058 inches. Approximately two
miles of guidance wire is precision
wound on a tapered aluminum spool
to form a bobbin. Photo I shows
the threaded spool, a wound bobbin
and the completed wire dispenser
with protective aluminum shroud.

Photo II shows a typical fracture
of wire with greater than 30 percent reduction in area, indicating
that the wire did achieve full
strength of approximately 10 pounds
at failure.

Photo III shows a non-typical
brittle fracture associated with
problem wire. Generally such failure can not be detected in normal
tensile testing as full strength is
achieved. However, in bending the

PHOTO I
TOW WIRE DISPENSER

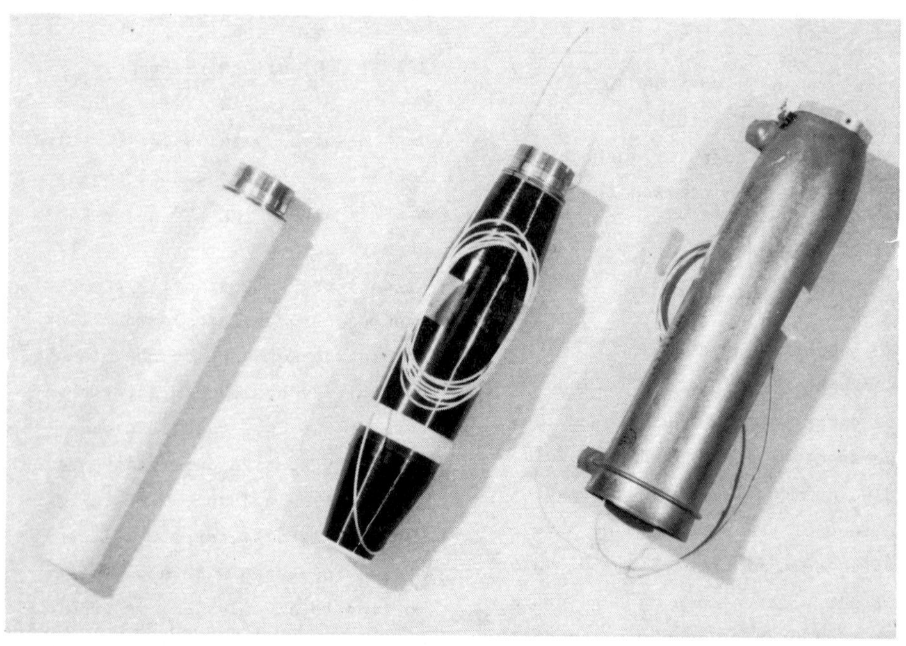

Spool
Threaded

Bobbin
with Leader

Dispenser
with Protective Shroud

PHOTO II
DUCTILE WIRE FRACTURE
∿450 X

PHOTO III
BRITTLE WIRE FRACTURES
∿450 X

>30% RA – Tensile Fracture

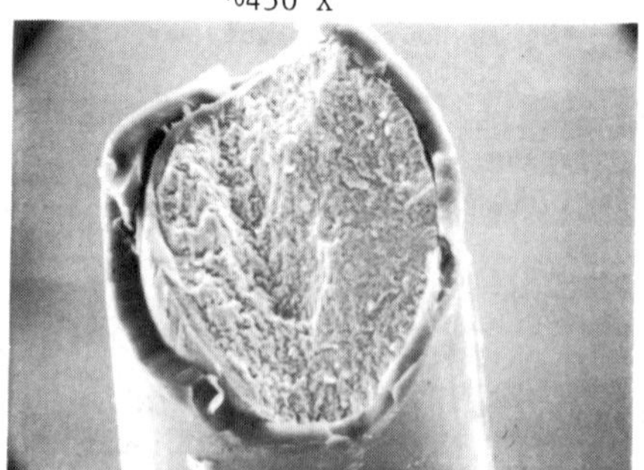

No Ductility – Static Failure

wire strength is severely degraded. During dispense from the bobbin at high velocity, the wire is subjected to severe bending and tensile loads.

During bobbin winding the wire is adhesively staked at each winding layer termination where a transition occurs and wind direction changes to start the next wire layer. This staking is critical to maintain wire pack stability during the bobbin life. An acrylic adhesive with rapid drying properties, good staking adhesion to insulation and very minimum resistance to wire peel from the adhesive during flight dispense, is employed for the staking of each layer transition zone. After winding bobbins and prior to assembly into missiles, opens were identified on some bobbins only a few weeks old. Further investigations revealed brittle fractures through varied stages of fabrication including recently wound bobbins, assembled dispensers and missiles ready for delivery. The magnitude of the problem was extensive and required positive/rapid identification of the failure mechanism, location and control of hardware that could have been affected followed by corrective action to prevent further wire failure. Table III identifies wire fracture properties.

TABLE III
TOW BOBBIN WIRE FRACTURES

1. Low Ductile Fracture
2. Delayed Failure
3. Recognized Corrosion of Faces
4. Fractures Associated with
    Insulation Damage

After detailed evaluations, it became evident that a hydrogen embrittlement/stress corrosion mechanism was active.

Finally we identified that several hundred gallons of acrylic staking adhesive had been miscompounded by a vendor. Cellosolve acetate had accidently been replaced with cellosolve, a four carbon alcohol. Although given the information that cellosolve was present, one would very likely not have identified that a problem could exist. However, cellosolve does react with the wire insulation to cause embrittlement. The exact corrosion/embrittlement mechanism was not fully identified but failures were reproduced in the laboratory and cellosolve was positively identified as the problem material.

The control of affected hardware was a major undertaking. Scrap of bobbin systems in the winding area, removal of units from missiles ready for delivery as well as recovery of units delivered was accomplished. Specifications on the acrylic adhesive were revised and mandatory in-house chemical analysis of each adhesive lot was implemented.

3.2 TOW Battery Fusible Resistor

The TOW Missile battery contains
a fusible resistor link consisting
of a fine manganin wire which is 83
percent copper, 13 percent manganese, 4 percent nickel. Open
circuits caused by fractures of the
manganin resistance wire were found
with no indication of improper
handling or damage to the link.
Table IV identifies wire fracture
properties.

TABLE IV
FUSIBLE RESISTOR FRACTURES

1. Mid-Wire Fractures – Open
    Circuits

2. Fractures Due to Stress
    Corrosion Cracking (SCC)

3. SCC Is Intergranular

4. SCC Is Faster in Coarse Grain
    Material

When identified, the failures were
low ductile breaks in a battery
system that had been supplied for
years with no apparent changes in
the hardware. Fractures were
identified as stress corrosion, an
intergranular failure mode. Requirements for stress corrosion consist
of a corrodant material and an
electrolyte. Analysis showed that
the manganin wire supplied by the
vendor contained minute amounts of
chloride contamination. Literature
indicated that at high relative
humidity (80-89 percent) copper
alloys will form and maintain a film
of liquid moisture on their surface.
The ingredients for stress corrosion
were then available. During failure
analysis studies, it was noted that
two different grain sizes existed
in the manganin wire links. One
had a minimum grain size of 0.002
mm compared to 0.02 mm for the
coarse grained material. Failures
had been found only in the coarse
grained material which had recently
been supplied by the wire vendor.
This change had not been recognized
as a problem and in fact grain size
was not controlled. Failure in
the coarse grained material is
attributed to the fact that stress
corrosion cracking (SCC) is faster
in coarse grained materials (grain
boundary corrosion path is shorter).
In summary, a product used for
years suddenly showed a high
failure rate. Minor chloride contamination was found along with a
change in material grain size.
Corrective action consisted of required cleaning of the wire after
drawing, control of grain size and
inspection testing to verify on a
lot basis that these requirements
were being met. Suspect batteries
were purged from the system and
the wire link replaced.

3.3 Missile Coldwall Failures

The Phoenix Missile, a long range
air to air system, contains coldwalls employed to maintain acceptable operating temperature for
electronic components. The coldwalls consist of 6061-T6 aluminum
plates bonded to 3003 aluminum fin
stock to form a sandwich channel
through which an ester base coolant
fluid is pumped. Surfaces were
MIL-C-5541 treated and the assembly

488

bonded using EC 2219 adhesive.
Seven (7) missiles on the aircraft
carrier Constellation were found to
have the ester base coolant fluid
leaking into the missile. Failure
analysis revealed that for unknown
reasons the adhesive bond was failing in an adhesive mode. The
bonded coldwalls were unzipping in
use after more than two year in the
field. Coolant fluid from eight
(8) missiles on the affected carrier
revealed moisture concentration in
the coolant fluid to vary from 198
to 595 parts per million. Allowable limits were 150 ppm. With the
indication that moisture was a
contributing factor, production
coldwalls were subjected to 100 percent humidity at varied temperatures
and pressures. Test results are
presented in Table V.

average time to failure of 6.4
hours. At 100 percent RH, 75 PSI
and 75°F failure occurred in 48
hours with average time to failure
of 63.4 hours. A second series of
tests employed production coldwalls
and coolant fluid contaminated with
moisture to specific levels. Table
VI shows the results of these tests
which were surprising in that
failures could not be produced when
known moisture contamination up to
600 PPM had been added to the coolant fluid. A sample of coolant
fluid from the aircraft carrier
Constellation was obtained which
contained 443 PPM moisture by
analysis but contained beaded water
indicating moisture saturation.
The failure mode of adhesive debond
was reproduced after testing for
only 147 hours at 150 PSI and 150°F.

TABLE V

FAILURE MODE REPRODUCTION

Coldwall No.	Moisture	Press PSI	Temp $^{\circ}$F	Hours to Failure
1	100% RH	150	150	3.9
2	100% RH	150	150	0.66
3	100% RH	150	150	5.1
4	100% RH	150	150	16.0
5	100% RH	75	75	99.
6	100% RH	75	75	55.5
7	100% RH	75	75	48
8	100% RH	75	75	55
9	100% RH	75	75	52
10	100% RH	75	75	71

RH = relative humidity

As seen in Table V at 100 percent,
RH, 150 PSI and 150°F failure
occurred in only 0.66 hours with

Although the failure mechanism had
been reproduced in 100 percent
humidity and when using moisture

TABLE VI

PRODUCTION COLDWALLS MOISTURE/
PRESSURE DATA AT 75%

Specimen #	Moisture PPM	Static Pressure PSI	Time Hours No Failure
1	150	80	1250
2	150	150	1176
3	150	80	2250
4	150	150	1176
5	300	80	2100
6	300	80	2100
7	600	150	1750
8	600	150	1726
9	600	80	1840
10	600	80	1840

contaminated coolant fluid from the Constellation it was difficult to explain why freshly contaminated coolant fluid did not cause bond failure after several thousand hours testing. Further work identified that the moisture and coolant fluid tended to react reducing the effective moisture in the system. Only when moisture had been introduced over a long time period were we able to maintain a high concentration of moisture in the fluid. Freshly introduced moisture to 600 PPM would be reduced significantly if analyzed after several days. Debond occurs rapidly when the moisture in the coolant fluid reaches the saturation level (100 percent RH). Below saturation bond degradation is extremely slow and could not be reproduced in laboratory testing after several thousand hours. A cooperative effort between HAC and Navy personnel identified the mechanism on shipboard that allows slow build up of water into the coolant system during missile checkout. In the carrier environment with up and down loading of the missiles on the coolant checkout stand moisture was easily introduced over a long time period from condensation or spray collection in connectors coupling missile to test stations and would eventually reach a saturation level in the fluid. Corrective action consisted of maintenance instruction to the field crews to eliminate contaminating the coolant systems with moisture and the development of two alternate coldwall designs. A brazed coldwall was designed and qualified and an alternate adhesive system was developed and tested with excellent results.

The alternate system consisted of FM-73 adhesive, BR-127 primer and a phosphoric acid anodize of the mating aluminum surfaces. Humidity (100 percent RH) exposure tests were performed on the redesigned coldwalls with no failures in several thousands of hours of exposure as compared to as low as 0.66 hour to failure in the old system.

3.4 TOW Spool Paint Adhesion

A paint layer is applied to the TOW missile wire spool and a fine thread impressed into the partially cured paint which is then fully cured. The insulated guidance wire is wound into this thread which provides a controlled pitch spacing and aids in mechanical stabilization of the fully wound bobbin. The production line for painting the spools is a high rate automated facility. Over six hundred spools are painted at one time in a fully automated operation. A problem occurred where painted spools were found to contain blisters and craters which precluded being able to produce acceptable painted base lay threads for wire bobbin winding. Although the spools could be stripped and repainted the high reject volume caused missile line stoppages. Failure analysis by surface analysis techniques revealed that the anodized, unsealed aluminum spool surface was contaminated with minor amounts of orgainc materials. Ultrasonic cleaning of the spool surfaces provided an adequate substrate for painting as an interim measure to maintain production. Further efforts identified that an anodize rinse tank was contaminated with minor amounts of organic oils which caused the blister problem. Controls were implemented to maintain purity of the processing tanks and for handling spools prior to painting.

4. CONCLUSIONS

Controlling and minimizing contamination becomes especially important in a high rate production operation. Control of contamination is a responsibility of the design organization as well as the manufacturing operations. We in engineering often tend to overlook the fact that we can minimize contamination induced problems through careful design.

To minimize the impact of such problems when they occur requires quick reaction and problem identification. Finally, problem correction requires reasonable and cost effective action. Cost effective action means that the corrective action should fit the criticallity of the problem. Costs soar when minor problems are blown out of proportion.

5. BIOGRAPHY

Mr. W. B. Tolley is manager of the Materials and Processes Analysis Support function for the Hughes Aircraft Tucson Plant Site. Mr. Tolley received his B.S. and M.S. degrees in Chemistry from the University of Arizona and is a Regestered Professional

Metallurgical Engineer. He has
two patents and has authored
numerous reports in the fields of
nuclear chemistry - metallurgy and
materials and processes. Mr. Tolley
has been active in ACS, SPE SAMPE
and is post Tucson Chapter Chairman
of ASM.

RESIN IMPREGNATED CERAMIC
FIBER REPLACEMENT FOR ASBESTOS FIBER
INSULATION SYSTEMS

J. F. Binnie, Jr.; R. C. Johnson; and W. B. Tolley
Hughes Aircraft Company
Technical Support Division-Tucson, Arizona

Abstract

Two insulation systems were developed and qualified utilizing a common ceramic fiber felt consisting of aluminum-silicon oxides impregnated with either a high temperature epoxy or phenolic resin. Specifications were written to cover both insulation systems and the epoxy resin for which there was no existing specification.

1. INTRODUCTION

The initial ablative insulation material for the TOW wire guided missile flight motor (Figure 1) was a bonded aluminum silicate covered by NAVORD OS 10152. However, since production began in the late sixties, essentially all of the flight motors produced have used an asbestos fiber felt impregnated with phenolic resin. This material, specified by Army Missile Interim Specification, MIS 10049, was superior to the silicate in its bonding to the motor housing.

In 1979, a shortage of the required long fiber asbestos occurred primarily as a consequence of its limited availability from only a few mine locations throughout the world. In addition, the Army is responding to a directive from President Carter which calls for a phasing out of the use of all asbestos materials over the next two years. This directive has resulted from an increased awareness of the occupational health risks associated with exposure to asbestos. These two events have provided the impetus to develop non-asbestos containing systems.

2. EXPERIMENTAL

2.1 Materials

Ceramic paper is a generic name applied to a manufactured refractory felt composed almost totally of synthetic inorganic fibers. Material covered in NAVORD WS 8437 specifies a fiber composition which is essentially 50-50 aluminum oxide (Al_2O_3) and

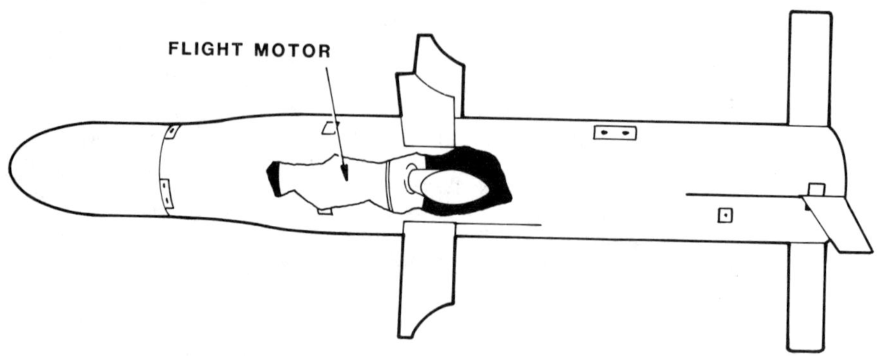

FIGURE 1. TOW MISSILE FLIGHT MOTOR LOCATION

silicon dioxide (SiO_2). Type I material under this specification may contain a maximum of about 5% organic binder. The Type II material is devoid of any organic content. Ceramic paper composed of aluminum and silicon oxides has excellent refractory properties comparable with those of asbestos. For example, chrysotile, the actual type of asbestos used, fuses at 2770° F. Aluminum-silicon oxide composed ceramic paper melts somewhat higher at 3260° F. For purposes of impregnation discussed below, we found the Type I material containing binder desirable over Type II which tended to flake and crumble easily during handling.

The ceramic paper used in this study was obtained from Carbonundum as Fiberfrax 970J. The presently used asbestos prepreg RPD 41 is available from Raybestos Manhattan. The phenolic resin RPD 984 per MIL-R-9299C, used in the RPD-41 prepreg, was employed to fabricate the ceramic fiber prepreg. The epoxy resin was purchased from Fiberite Corp. (7714) and is now covered by MIS-34312.

2.2 <u>Impregnation and Lay-up Process</u>

For over ten years, the process for applying insulation on the interior of the TOW flight motor has consisted of laying up preform patterns of asbestos felt impregnated with phenolic resin (RPD-41) and subsequently molding these in place with the application of heat and pressure. The resin content of the impregnated felt has been 40-45%. The key quality criterion has been the thickness of the cured insulation subsequent to the application of 300 psi at 300°F for 1 hour. When this thickness is within the range 0.025/0.040 inches, all qualification tests to be

494

discussed have been met. Insulation thickness is determined primarily by two production parameters: number of plies of material and pressure. Bond strength between insulation and motor case is determined by temperature, pressure and time of cure.

The asbestos-phenolic ablative insulation was highly effective. As mentioned, the impetus to find alternatives did not derive from quality failures but rather from supply problems and health considerations.

Fiberfrax ceramic paper is available in fractional thicknesses beginning at 1/32 inches. Judging from the thickness of the asbestos-phenolic lay-up prior to cure, it appeared that 1/8 inch ceramic paper would be a suitable starting point for development of a ceramic-phenolic system. Initial impregnation was done manually using a dip technique. Commercial phenolic resin was diluted with approximately three volumes of an appropriate solvent such as isopropyl alcohol or methylethyl ketone. Using this dilution ratio, it was found that the paper completely absorbed a pre-weighed quantity of the resin to give a homogeneously impregnated felt. Excess solvent was allowed to evaporate and the product was oven dried by heating at 150°F for 1 to $1\frac{1}{2}$ hours in a forced air oven. It

was stored at $35\text{-}40^{\circ}$F prior to use.

Generally, lay up of material inside the rocket case followed the same process procedure developed for the asbestos-phenolic system. However, since the 1/8 inch thick material was being used, the ceramic-phenolic system was somewhat simpler. While the asbestos-phenolic system uses up to 8 plies (ply thickness varies somewhat from lot to lot), the ceramic-phenolic uses only one. Preforms are cut for the circular end portions of both forward and aft sections (Photos I and II). Rectangular strips were used for the sidewall. The preforms are cut slightly oversized so that there is a small overlap joint. A fifth preform covers a circular grain support mounted within the interior of the motor.

The resin content of the asbestos-phenolic system is in the range of 40-45%. Within this range, the product insulation has the specified thickness and shows excellent bonding qualities. The ceramic-phenolic system on the other hand requires a slightly higher resin content (50-55%) in order to achieve comparable quality. When the resin content is somewhat less than 50%, there is a tendency to fail the magnetic forming operation (see 2.3).

The ceramic-epoxy system was

developed concurrently with the ceramic-phenolic system. Epoxy was considered as a candidate along with the phenolic because of its excellent thermal properties. The epoxy resin content requirement tended to be slightly higher (53% -57%) than in the case of the phenolic. Resin content in this range insured against magnetic forming failure (see 2.3) and damage during firing tests.

Physical data on both 1/16 and 1/8 inch ceramic-phenolic insulation are shown in Table I. This data was provided to us courtesy of Fiberite Corporation, which was contracted to produce pilot quantities of the prepreg material. The ceramic-phenolic prepreg is the least expensive of the two ceramic systems and the one which would be picked initially for production purposes. For this reason, physical data (Table I) has been accumulated for the ceramic-phenolic system first and at this time no similar data exists for the ceramic-epoxy system.

2.3 Qualification Tests

A formal test plan was followed in order to qualify the two ceramic insulation materials for use in the TOW flight motor case. Twenty complete motors were insulated with each system. In order to simulate production conditions, a pilot quantity of each of the insulation materials was prepared by Fiberite using the Fiberfrax 970J ceramic paper impregnated with phenolic RPD 984 and epoxy 7714 resins. The test sequence as shown in Figure 2 consists of temperature shock (MIL-STD-810B, method 503), magnetic forming shock and firing tests at -25 and +145°F. The insulated cases and grain supports were inspected for insulation thickness, resin content and workmanship prior to testing. The fired cases and grain supports were inspected for hot spots, internal damage and insulation integrity.

The TOW forward fuselage and the TOW center structure are joined to

TABLE I
Physical Properties of Ceramic-Phenolic Insulation

	1/16 inch Thickness	1/8 inch Thickness
Tensile Strength (psi)	8,350	8,820
Tensile Modulus (X10^6 psi)	1.16	1.11
Tensile Elongation (%)	0.81	0.93
Flexural Strength (psi)	12,370	10,090
Flexural Modulus (X10^6 psi)	1.18	1.04
Compression Strength (psi)	18,830	18,830
Specific Gravity	1.35	1.50
Shear Strength (psi)	3,710	--
Residual Volatile (%)	1.51	1.50
Cured Resin Solids (%)	48.4	49.2
Barcol Hardness	29	46
Cured Ply Thickness (In)	0.015	0.026

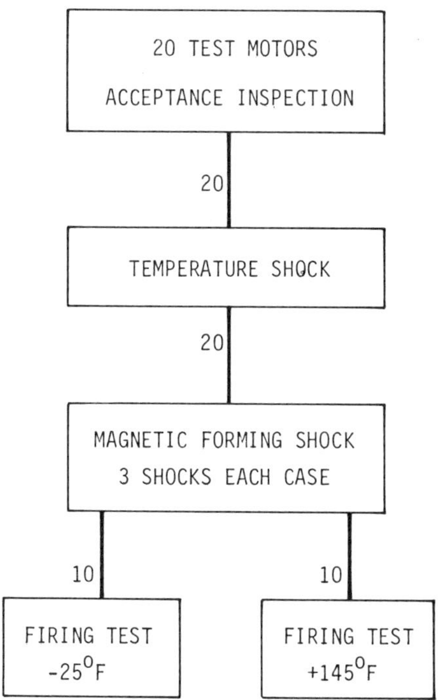

FIGURE 2.   TEST SEQUENCE

the forward and aft sections of the flight motor respectively by means of a magnetic forming process which applies a severe shock to the case. During this process, any weakness in the insulation may show up as cracks or delaminations. It is imperative that the insulation withstand this production operation.

During firing tests, thermocouples were installed at three positions on the external surface of the motor (Figure 3). In addition, flight motor pressure and thrust were monitored and visual observation of smoke was compared with the present configuration.

3.   RESULTS

Test results for the ceramic-phenolic and ceramic-epoxy are described separately below.

3.1 Ceramic Systems Motor Testing

Insulation met all thickness requirements and all insulated motor cases met the minimum grain clearance dimensions. This is important since ignition begins on the surface of the grain thereby exposing the insulated motor surface to the direct heat of combustion throughout the propellant burn. It should be noted that the nature of the grain support insulation process required that an extra 1/16 inch of insulation material (in addition to the basic 1/8 inch thickness used for all parts) be used in order to meet minimum thickness in some areas. There were no visible insulation defects. There was no loosening of the insulation or visible damage resulting from temperature shock. There was no cracking, delamination or other visible damage as a result of magnetic forming shocks.

Typical firing test results are shown in Figures 4 and 5 for the ceramic-phenolic system and in Figure 6 and 7 for the ceramic-epoxy system in which temperature versus time is plotted for each of the three external motor case positions as noted in Figure 3. Results are shown for firings at both -25°F and +145°F. Test requirements specify a maximum temperature at each location 30 seconds after firing. Table II shows that actual temperatures were well within their requirements and compare favorably with the asbestos-phenolic system. In fact the temperature results for the ceramic-phenolic and ceramic-epoxy were generally lower than for the presently used asbestos phenolic system. Other firing test parameters such as pressure, impulse and ignition

TABLE II

Flight Motor Case Temperatures

System	Test Temperature $^{\circ}$F	Fwd Case Dome Temp Max Thru 30 Sec	Fwd Case Sidewall Temp Max Thru 30 Sec	Aft Case Dome Temp Max Thru 30 Sec
Requirement	Cold -25°F Hot $+145^{\circ}$F	600°F Max	600°F Max	800°F Max
Ceramic-Phenolic	-25°F $+145^{\circ}$F	485 473	337 427	457 403
Ceramic-Epoxy	-25°F $+145^{\circ}$F	344 492	262 310	347 452
Asbestos-Phenolic	-25°F $+145^{\circ}$F	381 538	388 452	457 685

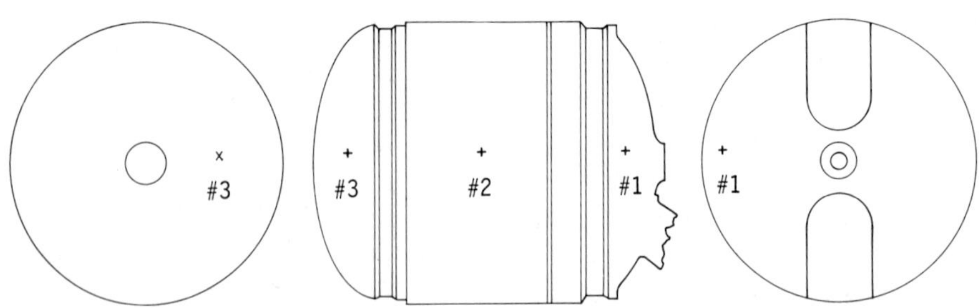

LEGEND

#1 Aft Dome
#2 Center Sidewall
#3 Forward Dome

FIGURE 3.   THERMOCOUPLE LOCATIONS
ON EXTERIOR OF MOTOR CASE

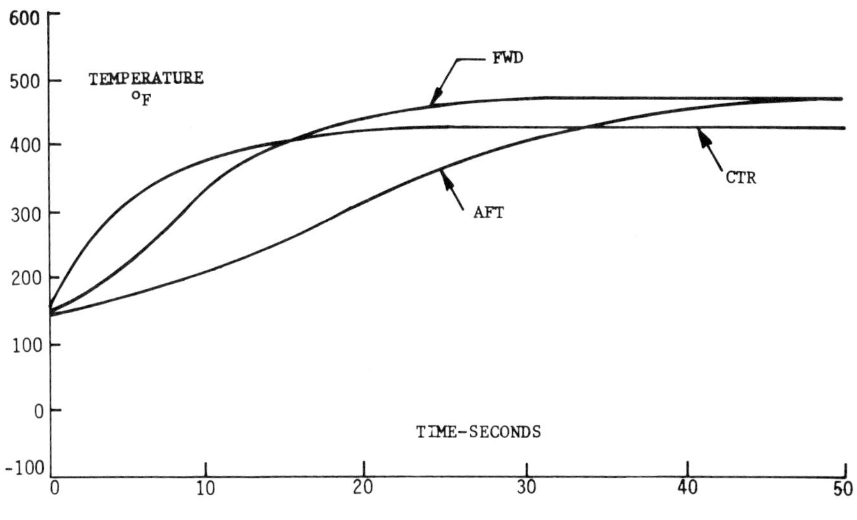

FIGURE 4.   CERAMIC-PHENOLIC SYSTEM CASE TEMPERATURE
CURVES AT +145° TEST TEMPERATURE

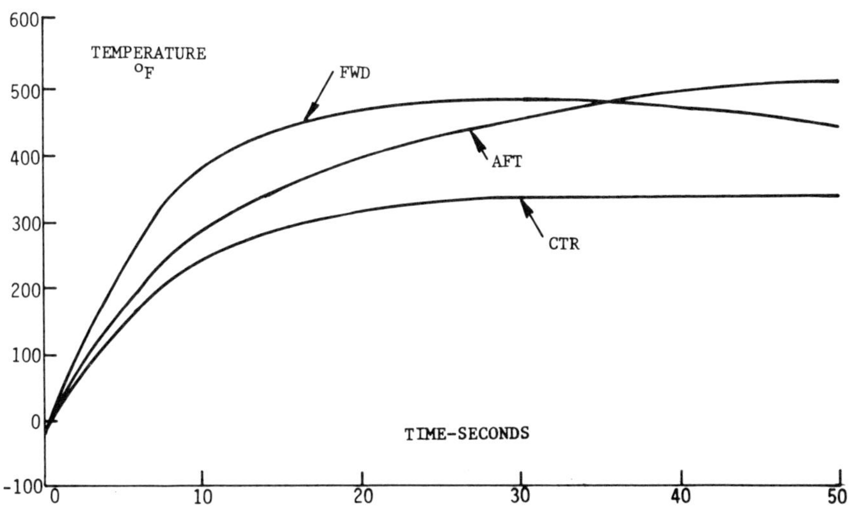

FIGURE 5.   CERAMIC-PHENOLIC SYSTEM CASE TEMPERATURE
CURVES AT -25° TEST TEMPERATURE

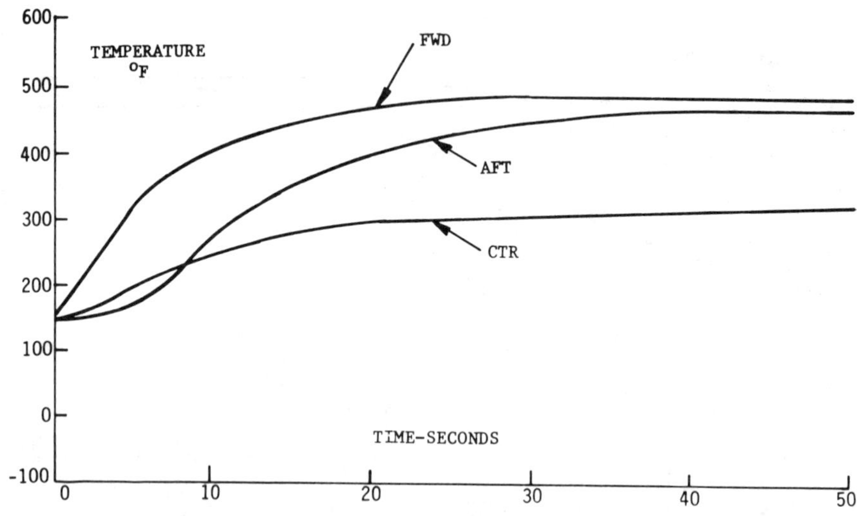

FIGURE 6.   CERAMIC-EPOXY SYSTEM CASE TEMPERATURE
CURVES AT +145° TEST TEMPERATURE

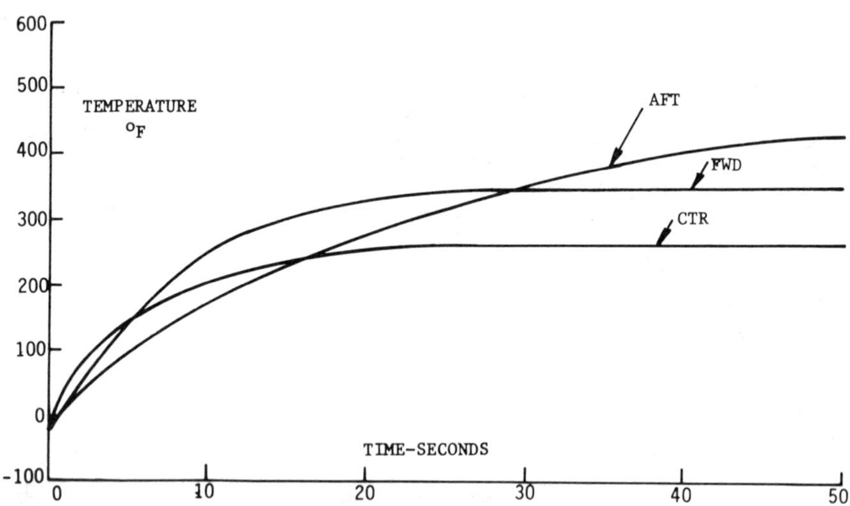

FIGURE 7.   CERAMIC-EPOXY SYSTEM CASE TEMPERATURE
CURVES AT -25° TEST TEMPERATURE

500

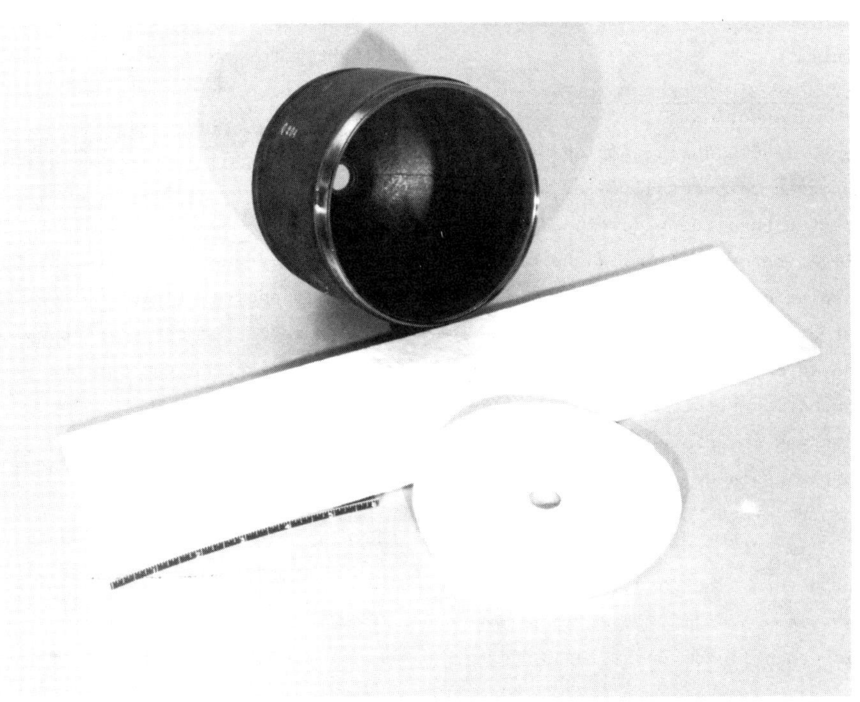

PHOTO I. CERAMIC-PHENOLIC SYSTEM PREFORMS FOR FORWARD MOTOR CASE-
TOW MISSILE. BACKGROUND SHOWS FORWARD CASE AFTER CURE

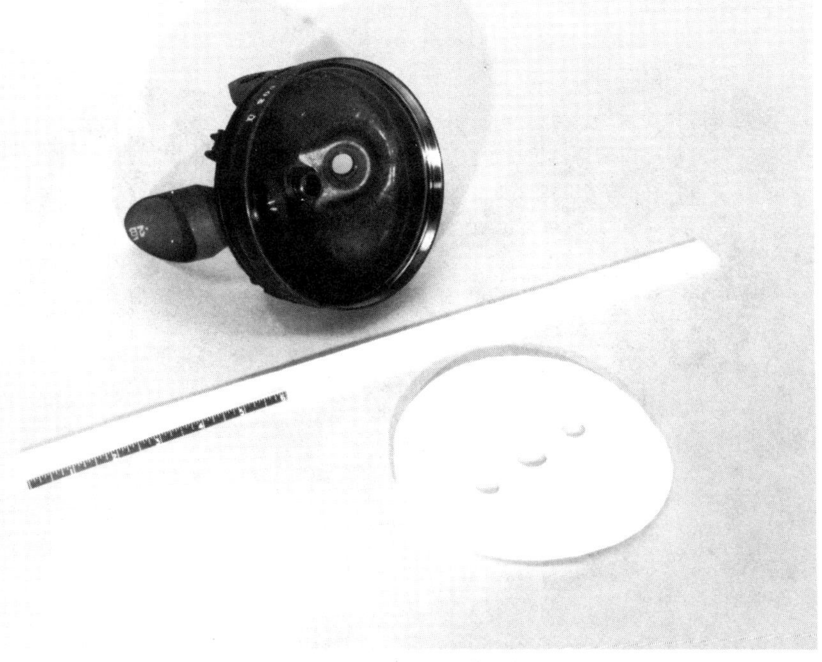

PHOTO II. CERAMIC-PHENOLIC SYSTEM PERFORMS FOR AFT MOTOR CASE-
TOW MISSILE. BACKGROUND SHOWS AFT CASE AFTER CURE

time were also well within requirements.

3.2 Ceramic System Post Fire Inspections

Smoke levels during firing were observed to be equal to or slightly greater than those developed when flight motors utilizing the present insulation were fired. This was not considered to be significant and in anyway detrimental. Visual inspection showed that the insulation was intact and that no damage to the cases or grain supports had occurred (see Photo III).

4. SUMMARY

Two alternative insulation materials, ceramic paper impregnated with phenolic or epoxy resin, have been developed for use as the flight motor ablative lining in the TOW wire guided missile. Each of these materials has met the requirements of a formal qualification test.

The ceramic-phenolic system is the least expensive of the two systems and therefore is the prime candidate to succeed the current system.

The new Army specifications have been written to cover materials developed in the study. The epoxy resin is now covered by MIS-34312. The resin impregnated ceramic paper is specified by MIS-31396. Two

PHOTO III. CERAMIC-PHENOLIC SYSTEM CASES AFTER FIRING

types of material are specified.
Type I refers to the phenolic
prepreg and Type II refers to the
epoxy prepreg system.

5. BIOGRAPHIES

Mr. Binnie is currently head of the
Process Analysis Group at Hughes
Aircraft Company. He is responsible
for directing investigations in the
areas of joining and fabrication
processes for a wide range of
materials for missile applications.
Mr. Binnie received his B.A.
degree in Chemistry from the
University of Arizona in 1965 and
has been employed by Hughes Aircraft Company since 1966.

Richard C. Johnson has a B.A. in
chemistry from the University of
Idaho. He has had 9 years
experience in industry most of it
in organic synthesis and process
development. He is currently
materials and applications engineer
at Hughes Aircraft Company in
Tucson.

Mr. W. B. Tolley is manager of the
Materials and Processes Analysis
Support function for the Hughes
Aircraft Tucson Plant site.
Mr. Tolley received his B.S. and
M.S. degrees in Chemistry from
the University of Arizona and
is a Registered Professional
Metallurgical Engineer. He has
two patents and has authored
numerous reports in the fields of
nuclear chemistry-metallurgy and

materials and processes. Mr. Tolley
has been active in ACS, SPE SAMPE
and is post Tucson Chapter
Chairman of ASM.

26th National SAMPE Symposium
April 28-30, 1981

PAN-BASED CARBON FABRICS
FOR THERMAL PROTECTION COMPOSITES
Donald L. Schmidt
AFWAL/Materials Laboratory/MLBE
Wright-Patterson Air Force Base, Ohio

Abstract

Rayon-based carbon fabrics are widely used in ablative and insulative composites. Due to declining commercial markets and profitability, fibrous carbon products derived from continuous filament rayon may eventually be unobtainable or in limited supply. An investigation was therefore conducted to obtain suitable polyacrylonitrile (PAN)-based carbon fabrics, measure their properties, and fabricate tapewrapped phenolic composites. Commercially available PAN-based carbon fabrics exhibited high strength and uniformity, but they were deficient in insulative and chemical purity characteristics. Special PAN-based carbon fabrics were then obtained at lower (less than 1,350°C) processing temperatures and utilized in 20° tapewrapped phenolic composites. Ablative, thermophysical, mechanical and chemical properties of the composites were measured. The developmental composites were found to be suitable replacements for rayon-based carbon fabric/ phenolics, but further improvements in fiber-to-matrix bonding and fiber purity are desired.

Keywords: Carbon Fabrics, Thermal Protection, Composites, Alternate Materials, Ablation, Fibers.

1. INTRODUCTION

Carbon fibers are widely used in thermal protection composites because of their low thermal conductivity, dimensional stability at high temperatures, low density, nonflammability, low cost and availability in a variety of physical forms. These fibers are typically woven into various fabric constructions and then combined with a char-forming resin to yield the desired protective composite. [1]

For almost two decades, rayon fiber has been the preferred precursor for the manufacture of carbon fibers. Interest in rayon precursor yarns has been declining, however, because of the loss of suitable domestic sources. At present, only one U.S. source exists for high purity, continuous filament rayon. The continued availability of filament rayon for carbon fiber manufacture is thus suscept because only a small volume is required for defense needs and those material requirements are highly cyclic. [2-5]

Polyacrylonitrile (PAN) fibers have also been available for many years. Most of the world's interest has centered on PAN fibers because of their wider availability and suitable carbon fiber properties. The objective of this materials program was to determine the suitability of commercial and developmental fibers for missile heat-shielding uses.

2.   PRECURSOR PAN FIBERS

PAN fibers are produced from high molecular weight, linear organic compounds. They are either homo-polymers, terpolymers, or graft polymers. Copolymeric and ter-polymeric fibers have been found to be the most suitable for producing fibrous carbons.

PAN fibers are available in the U.S. from about three commercial sources and ten foreign sources. Fibers used in carbonaceous products are largely imported from Great Britain and Japan. Tow, staple yarn and continuous yarns are used, but for economic reasons, large diameter tows such as 10,000 to 160,000 filaments are most common.

PAN fibers are excellent precursors for various carbonaceous materials. They have a high carbon content, low in costs, available in numerous filament count yarns and tows, stable commercial market, and multiple producers. There are two obvious limitations of most PAN yarns in considering defense requirements; namely, few sources for small diameter yarns or tows and the presence of alkali metals. One thousand filament yarns are needed to produce thin fabrics and contaminant levels of 500 parts per million alkali metals or less are required for specialized heat-shielding applications.

3. CARBON FIBER FORMATION

PAN fibers are converted to carbon fibers by pyrolysis in an inert gaseous atmosphere. Typical carbonization temperatures range from 800° to about 1,400°C (graphite fibers are produced at about 2,000°C or higher). The structure and properties of carbon fibers are controlled by the choice of raw material, heat treating cycle, use of chemical treatments during carbonization, and application of stress on the fibers during processing. Fig. 1 illustrates the manufacturing process for fibrous carbons. The process is invariably a multiple step, batch, or combined batch and continuous method. (6)

Carbon fibers derived from PAN retain their general morphology, although they shrink biaxially during pyrolysis. Carbon fibers manufactured from wet spun PAN have characteristic round cross-sections and striated surfaces, as shown in Fig. 2.

4. DEVELOPMENTAL CARBON TOWS

Commercially available carbon fibers derived from PAN precursors have been optimized for structural composites and sporting goods. Their high strength and high modulus characteristics are not needed in thermal protective composites,

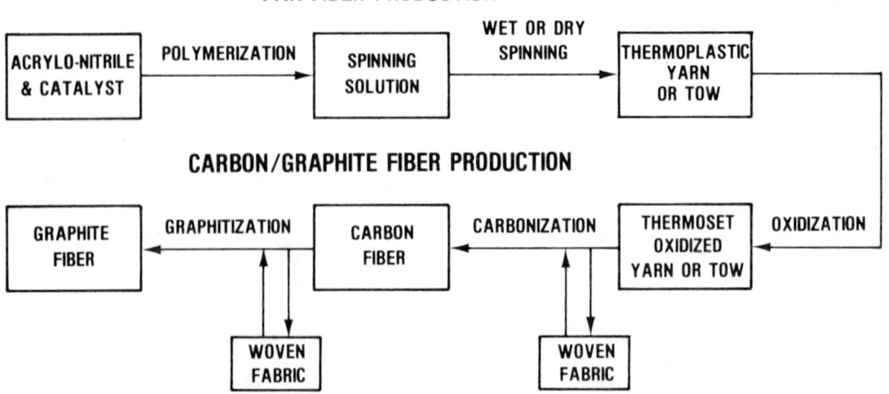

FIG. 1. MANUFACTURING PROCESS FOR PAN-BASED FIBROUS CARBON PRODUCTS.

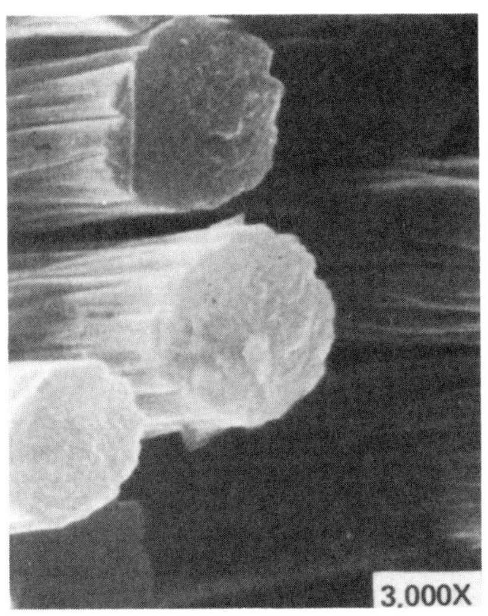

FIG. 2. MORPHOLOGY OF WET-SPUN PAN-BASED CARBON FIBERS.

because the structural load is usually carried by a metallic substructure. Some sacrifice in fiber strength and modulus could thus be tolerated to achieve improved insulative, ablative and high purity fiber properties.

A systematic investigation was conducted on the relationship between carbon fiber properties and the maximum pyrolysis temperature used in their manufacture. As shown in Table 1, a typical processing temperature of 1,350°C yielded carbon fibers with the highest: carbon content, strength, modulus, ash content and trace metals. Lower processing temperatures (900° to 1,125°C produced carbon fibers with substantially lower thermal conductivity. These specialty tows had desirable tensile strengths in excess of 220,000 psi, which is about double the value for rayon-based carbon fibers. Note, however, that the low temperature fired tows had a low carbon content (about 80%) and a high nitrogen content (17%). A standard, commercially available carbon tow and four developmental grades of carbon tow were selected for further investigation.

5.   WOVEN FABRICS

PAN-based carbon fabrics are manufactured by two different processes. Thermoset PAN yarns are woven into the desired fabric construction and then pyrolyzed, or the fabric is woven directly from carbon yarn or tow. Most PAN-based carbon fabrics manufactured today are obtained by weaving yarn or tow. This method permits a wide range of woven products, excellent structural properties and good handleability. Developmental and commercially available 3,000 filament tows were woven into 8-harness satin fabrics by a commercial weaver. The 24 inch

507

TABLE 1

PROPERTIES OF COMMERCIAL AND DEVELOPMENTAL PAN-BASED FIBERS AND TOWS*

TOW DESIGNATION	HTA-7	DG-125	DG-112	DG-114	DG-110	DG-129
MAX. PROCESS TEMP, $^{\circ}$C	1,350	1,125	1,075	1,000	950	900
FIBER PROPERTIES						
DENSITY, gm/cc	1.76	1.74	1.80	1.80	1.78	1.76
TENSILE STRENGTH, Kpsi	400	375	348	291	278	220
TENSILE MODULUS, Mpsi	34.0	28.9	25.8	22.9	19.8	17.0
CARBON CONTENT, %	96	91	87	84	81	70
ASH CONTENT, WT. %	0.21	–	0.16	–	0.088	–
TRACE METALS, ppm	499	366	352	340	–	294
THERMAL STABILITY, % LOSS 1 hr in 500°C AIR	4	41	37	55	60	77
TOW PROPERTIES						
DENIER, gm/9,000 m	1,824	1,873	2,080	2,180	2,260	2,280
ELECTRICAL RESISTIVITY, micro ohm-cm	1,500	3,228	6,610	19,153	57,219	114,400
THERMAL CONDUCTIVITY, cal/cm sec-$^{\circ}$C (Est.)	0.032	0.010	0.009	0.007	0.006	0.005

* CELANESE CORP. 3,000 CONTINUOUS FILAMENT TOWS

wide products were of excellent quality and uniformity.

Certain ablative applications require the use of very pure carbon fabrics. Less then several hundred parts per million of alkali metals are required because these metals can be ionized at relatively low temperature and greatly increase the electron concentration in ablation products-air boundary layer surrounding a re-entry body. (7-8)

The developmental PAN-based carbon fabrics had alkali metal contents of only 495 parts per million or less. Such products were obtained by using high purity, foreign source PAN fibers and clean weaving conditions.

Woven carbon fabrics are relatively low cost products. Standard rayon-based and PAN-based carbon fabrics are on the order of $50,00 per pound. Special carbon fabrics woven from developmental carbon tows ranged in costs from $150 to $200 per pound. With increased usage, however, the costs of these specialty fabrics would be significantly reduced.

6. PHENOLIC PREPREG

Standard phenolic resin impregnation techniques were used to obtain prepregged fabrics. Compared to rayon-based carbon fabrics, the newly available PAN precursor carbon fabrics exhibited lower resin wettability and slightly lower resin contents. Suitable prepregs were obtained, however, with resin contents on the order of 32 to 36 weight percent, five to seven percent volatiles and four to six percent flow.

7. CYLINDRICAL COMPOSITES

The tapewrapping process is widely used to obtain cylindrical and frustra configurations. [9] Hollow cylinders were therefore fabricated to gain experience with the developmental prepregs, obtain baseline composite properties, and to identify candidate materials for follow-on evaluation efforts. The developmental, non-surface treated carbon fabrics were prepregged with an unfilled phenolic resin, cut into tapes at 45° to the warp direction, sewn together into continuous strips, and slit into the desired tape width. The tape was then wrapped into a hollow cylindrical part using standard fabrication procedures. The cylindrical specimens were nine inches in length, nine inches in outside dimension and one-half inch in thickness. The fabric angle was 20° to the axial direction.

7.1 <u>Ablative Characteristics</u>. The recession characteristics of various PAN-based carbon fabric/phenolic composites were determined in high temperature, arc-heated air. Flat plate specimens were immersed in turbulent, supersonic flow environments using the Avco Corporation 10MW Air Arc Facility. The commercially available (high temperature fired) PAN-based carbon fabric reinforced phenolic composites exhibited a substantially higher (34%) linear ablative recession compared to rayon-based carbon/phenolics, as shown in Fig. 3. This be-

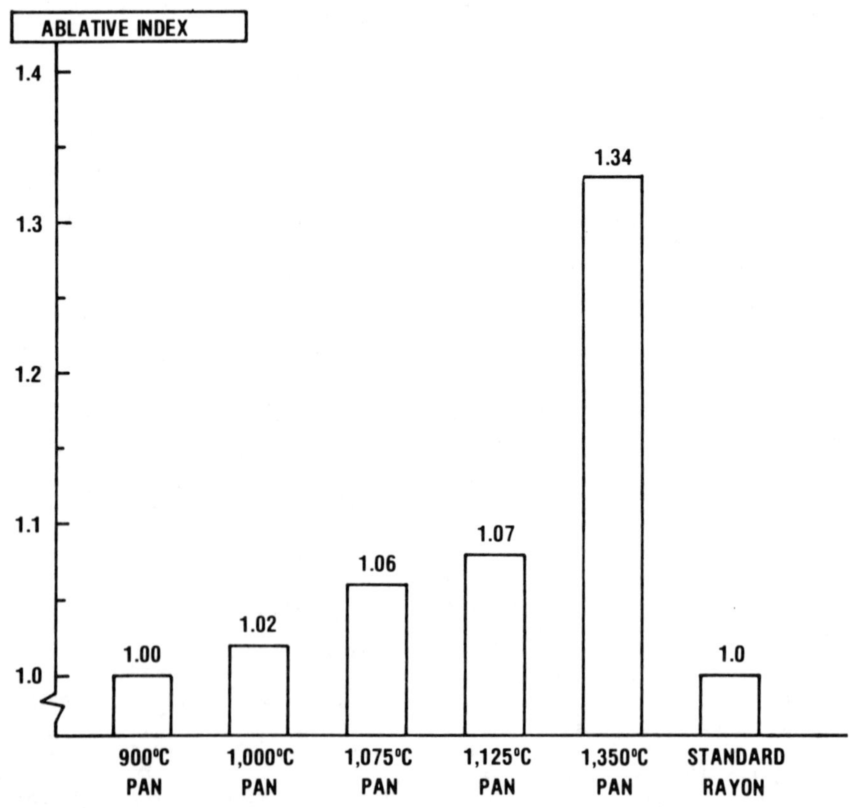

FIG. 3. RELATIVE LINEAR ABLATION RATES OF PAN-BASED AND RAYON-BASED CARBON FABRIC/PHENOLIC COMPOSITES.

havior was contrary to expectations because the PAN-based carbon fibers were of higher density. Composites with lower temperature fired carbon fabrics had comparable ablative recession rates.

7.2 Thermophysical Properties.

7.2.1 Density. Composites containing PAN-based carbon fabrics were typically higher in density compared to rayon-based analogues. A representative PAN-based carbon fabric/phenolic had a density of 1.58 gm/cc whereas a similar rayon-based carbon/phenolic had a density of 1.46 gm/cc.

7.2.2 Thermal Conductivity. Carbon fabrics woven from high temperature fired (1,350°C) carbon tows are quite crystalline in structure, and consequently have high thermal conductivities. Tapewrapped carbon fabric/phenolic composites, therefore, were expected to be less insulative than rayon-based carbon fabric/phenolics. Such is the case

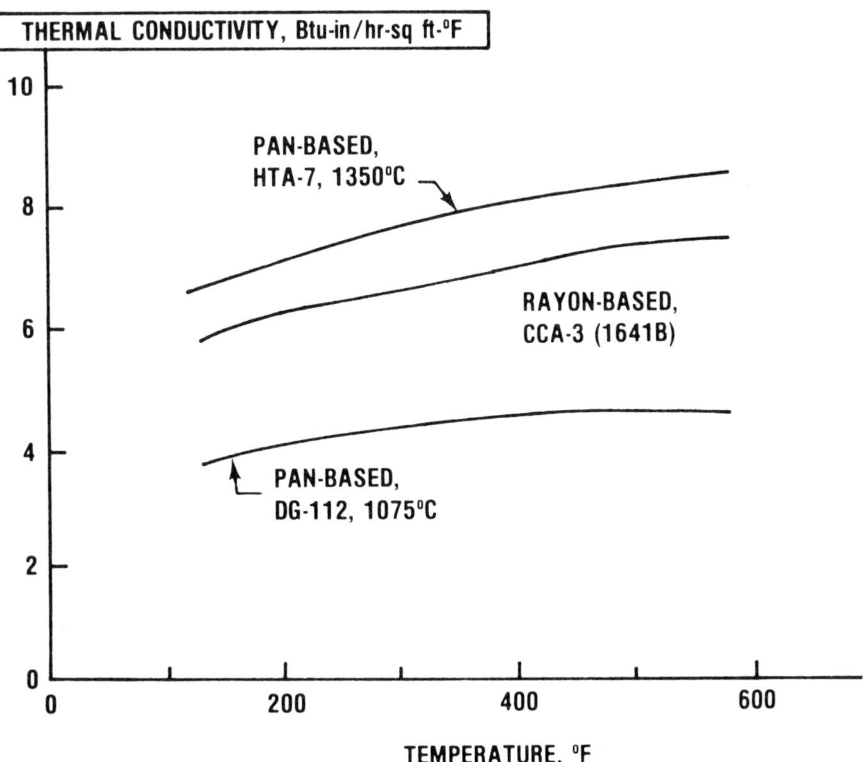

FIG. 4. RADIAL THERMAL CONDUCTIVITIES OF 20° TAPEWRAPPED CARBON FABRIC/ PHENOLIC COMPOSITES AT ELEVATED TEMPERATURES.

as shown in Fig. 4. On the other
hand, low temperature processed PAN-
based carbon fabric composites had
lower radial composite thermal conductivity at all measurement temperatures.

7.2.3 <u>Thermal Expansion</u>. Resinous
composites typically have high
thermal expansion coefficients because of the matrix. Carbon fabrics
significantly restrain the
expansion and contraction of
matrices, depending upon the fiber
properties and the direction of
reinforcement. Phenolic composites
reinforced with low modulus, rayon-
based carbon fabrics exhibit significant expansion with temperature
as shown in Fig. 5. The more
thermally stable PAN-based carbon
fabric reinforced composites exhibited little change with temperature, but the lower temperature
processed carbon fabric/phenolics
underwent some shrinkage with
temperature.

7.3 <u>Mechanical Properties</u>. PAN-
based carbon fabric reinforced composites exhibited very high mechanical properties when stressed

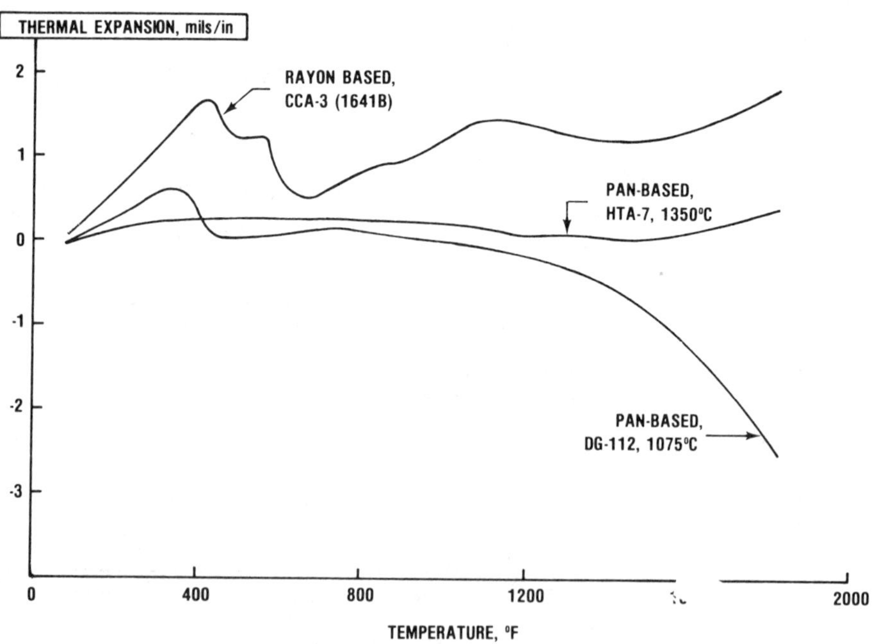

FIG. 5. HOOP THERMAL EXPANSION OF 20° TAPEWRAPPED CARBON FABRIC/PHENOLIC
COMPOSITES AT ELEVATED TEMPERATURES.

parallel to the direction of the reinforcement. At other angles of reinforcement, the mechanical properties were greatly reduced. Typical hoop mechanical properties for the optimized 1,075°C processed carbon fabric/phenolic included a tensile strength of 15,600 psi, a tensile modulus of 2.97 million psi, a tensile strain-to-failure of 11.5 mils per inch, a compressive strength of 18,200 psi, and a compressive modulus of 2.17 million psi. Although these structural properties are lower than rayon-based carbon fabric analogues, they appear to be adequate for most ablative heat-shielding applications.

The interlaminar shear strength of composites is typically the lowest mechanical property, and much attention has been devoted to increasing this design-limiting property. (10) The nontreated 1,075°C carbon fabric reinforced phenolic composite had an interlaminar shear strength at room temperature of 3,300 psi, whereas the rayon-based carbon fabric/ phenolic composite had a shear strength of 6,570 psi. At 500°F,

the shear strength of the PAN-based composite was reduced to 1,600 psi whereas the rayon-based composite had a shear strength of only 3,000 psi.

7.4 <u>Chemical Purity</u>. The PAN-based carbon fabric/phenolics were analyzed for alkali metal contents to determine their potential for missile heatshielding. Phenolic composites containing the 1,075°C processed PAN-based carbon fabric had a total alkali metal concentration of 425 parts per million. Similar contaminant levels were also found for other developmental grades of PAN-based carbon/phenolics. These impurity levels are about an order of magnitude higher than desired.

8.   CONCLUSIONS

Thermal shielding composites were fabricated with commercially available and developmental PAN-based carbon fabrics. The latter reinforcements were shown to have excellent insulative, ablative and mechanical properties, but some improvement is required in chemical purity and fiber-to-matrix bonding.

9.   ACKNOWLEDGEMENT

The author gratefully acknowledges the contributions of various personnel in the Avco Corporation, Celanese Corporation, Textile Products, Inc., U.S. Polymeric, Inc., HITCO Corporation, and the Southern Research Corporation.

10.   REFERENCES

1.   D. L. Schmidt, "Hypersonic Atmospheric Flight," Environmental Effects on Polymeric Materials, John Wiley & Sons, New York City, 1967, pp. 487-587.

2.   R. B. Millington, "Carbon and Graphite Ablative Reinforcements," 8th National SAMPE Synposium Preprints, Section 1, San Francisco, CA, 25-28 May 1965.

3.   J. C. Bowman and J. H. Brannon, "Graphite Reinforcement for Plastic Composites," Chapter II, Handbook of Fiberglass and Advanced Plastic Composites, G. Lubin, Ed., Van Nostrand Reinhold Co., New York City, 1969. pp. 237-254.

4.   D. L. Schmidt and W. C. Jones, "Carbon-Based Fiber Reinforced Plastics," Chem. Engr. Prog. 58(10) 42-50, October 1962.

5.   D. L. Schmidt, "Alternate Carbon Fabrics," SAMPE Quarterly 8 (4) 48-54, July 1977.

6.   W. Watt and J. Green, "The Pyrolysis of Polyacrylonitrile," Carbon Fibers: Their Composites and Applications, Plastics Institute, London, England, 1971, pp. 23-31.

7.   R. A. Bonsall, "Design and Weaving of Experimental Carbon Fabrics," AFWAL-TR-80-4054, April 1980.   Available Nat'l Tech. Infor. Service, Arlington, VA.

8.   S. L. Channon and W. T. Barry, "Status of Reentry Vehicle Heatshields," Proceedings of the AIAA/ASME 8th Structures, Structural Dynamics and Materials Conference, Palm Springs, CA, March 29-31, 1967, pp. 236-240.

9.   W. C. Jones and D. C. Siverts, "Optimization of Reinforced Plastics in Ablative Rocket Nozzle and Re-Entry Body Applications," SAMPE J. 1(4) Aug/Sept 1965, pp. 20-27.

10.   E. Fitzer, et. al., "Surface Chemistry of C-Fibers and Its Influence on Mechanical Properties of Phenolic Based Composites," Proceedings of the Fifth London Int'l

<u>Carbon and Graphite Conf</u>., Vol. I,
London, England, Sept 18-22, 1978,
pp. 405-418.

10.   BIOGRAPHY

Donald L. Schmidt is the Technical
Manager for Thermal Protection
Materials, AFWAL/Materials Labora-
tory, Wright-Patterson AFB, Ohio.
He has authored over 100 publications
on high temperature materials. Mr
Schmidt received a B.S. in Science
from Wisconsin State University in
1952 and a M.S. in Chemistry from
Oklahoma State University in 1954.
He is a member of the SAMPE Quarterly
Review Board, and has served on
numerous government and professional
society technical committees. He is
the recipient of the Air Force
Meritorious Service Award, the
Wisconsin State University Dis-
tinguished Alumnus Award and other
professional awards.

DUCTILITY INFLUENCING VARIABLES FOR A357-T6 ALUMINUM CASTINGS

D.L. McLellan
The Boeing Commercial Airplane Company
Seattle, Wa.

Abstract

Investigation of ductility influencing variables for A357-T6 castings shows that three features of the microstructure are primarily responsible and measurable using automated metallographic techniques. The number of cells per unit area provides a basic ductility potential. The amount of porosity must be used to reduce that potential to an effective value. Another consideration must be made for the amount of modification of silicon resulting from solution treatment. A model containing these three factors is being developed. At the same time individual influences of selected alloying elements and each stage of the total heat treatment process are being evaluated for their contributions to improved ductility.

1. INTRODUCTION

Utilization of high-strength aluminum alloy castings as primary structure in aerospace applications requires product assurances equivalent to replaced products. Methods of manufacture of wrought products are relatively standardized compared to castings. Furthermore, destructive sampling plans currently used for wrought products require only small portions for testing leaving the majority of a lot for use as structure. These concepts cannot be applied directly to castings as the methods of manufacture vary from part to part and between producers. In addition, each large casting may have a unique chemistry and heat treatment. To further complicate matters, the properties within a single casting are known to cover significant ranges. Because such castings cannot be both destructively tested and used as structure, nondestructive inspection methods are required to establish their structural integrities. In some cases, proof testing may prove satisfactory. But in general, this is both costly and time consuming. The key to product assurance for high-strength aluminum alloy castings is the development of reliable nondestructive measurements of the physical characteristics which control mechanical performance.

Efforts to identify and characterize the physical properties of aluminum castings are leading to new guidelines from which tensile properties can be accurately predicted. Of primary importance is ductility, or total elongation. For the A357 alloy investigated, ductilities in a single casting have been observed to vary from less than 1 percent to more than 10 percent. These zonal characteristics appear to be mainly attributable to size variations of the cellular structure which results from different solidification rates. In addition to predicting a ductility potential from the number of cells per unit area or average cell size, it is also believed that considerations must be made for (a) chemistry, (b) porosity, (c) silicon particle sizes and disperson patterns, and (d) quench and artificial aging influences. This presentation reviews recent developments and current work to establish reliable mechanical property prediction criteria for A357-T6 castings.

2. BACKGROUND

The Air Force/Boeing CAST program[1] provided an initial opportunity to develop an extensive data base from which static property design criteria were established. As a result, tensile properties of approximately 550 specimens excised from 18 bulkhead and segment castings were obtained and thoroughly evalulated. Factors evaluated included foundry variables (distances from gates, risers and chills), geometric variables (casting zone thicknesses and specimen configurations), and physical characteristics. The physical parameters measured included Dendrite Arm Spacing (DAS) and ASTM E155 Soundness Grades A, B, C and D taken from zones adjacent to the fracture surfaces of all specimens.

Correlation analyses showed that the best relation for ductility was provided with DAS when data were partitioned into the various soundness grades. Although data from all 18 castings provided the same general trend, scatter among results was excessive. Fig. 1 shows a normal probability plot for 157 ductility results. In Phase II, 14 bulkhead segments were examined and in Phase V, 4 complete bulkheads were cut up, tested, and results analyzed. These results are indicative of the ductility variability observed for a very small range of DAS from .0013 to .0018 inch and soundness grade A. The coefficient of variation is about 50 percent. This amount of dispersion is too large for either product assurance or efficient design use.

Following the CAST program, these same data and specimens were re-analyzed. Other methods of characterizing the microstructure were evaluated. Fig. 2 shows 100X magnifications of the A357 microstructure in the as-cast and -T6 heat treated conditions. The major difference is the appearance of silicon which forms cell boundaries. In the as-cast condition, light-colored aluminum cells are clearly defined by dark-colored silicon boundaries. In this condition, silicon appears as long, needle-like slivers. Solution heat treatment transforms silicon into rounded individual particles without noticeable changes to the sizes or shapes of aluminum cells. Following quench and natural aging, artificial aging causes Mg_2 Si to precipitate within the aluminum cells but has no

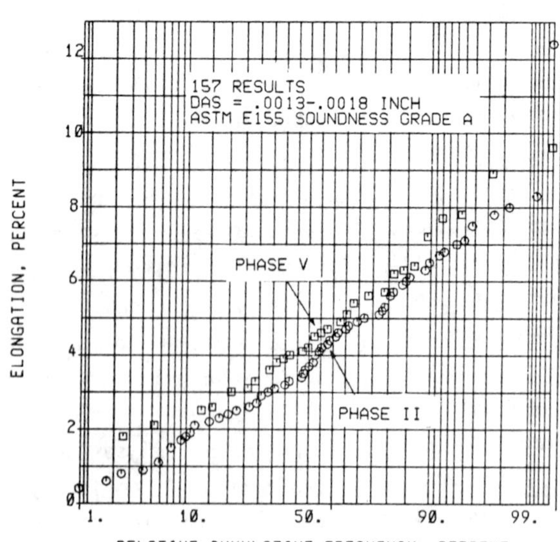

FIGURE 1   A357-T6 DUCTILITY SCATTER

apparent effect on either the aluminium cell or silicon particle geometries. In the fully heat treated condition, cell boundaries are less definite but are still outlined by silicon particles.

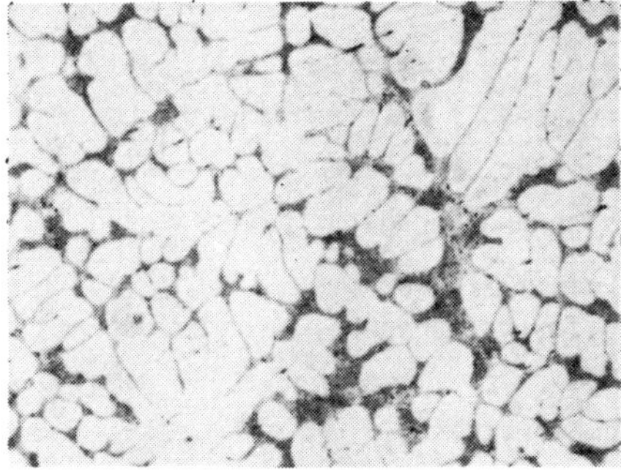

(A) AS-CAST

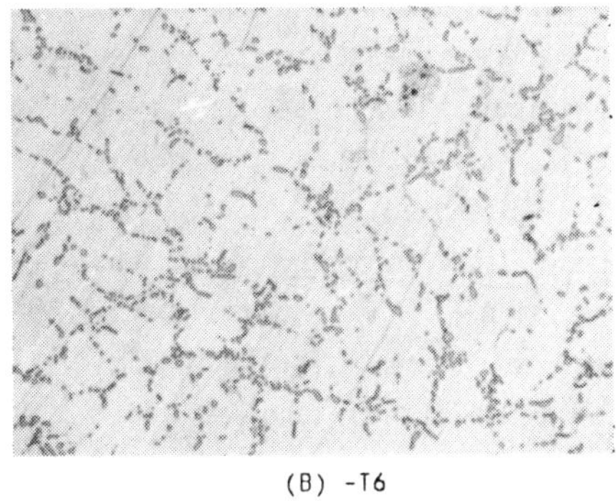

(B) -T6

FIGURE 2   A357 MICROSTRUCTURES

Development of an improved concept for predicting ductility is based on the fundamental deformation model for the alloy in question. Both translational and rotational flow potentials of aluminum cells are functions of cell sizes with silicon particles acting as primary restrictions. Formation of this type of

microstructure[2] and its reaction to loading including fracture characteristics[3, 4] have been reported by others. Using these criteria, a model was developed which relates the number of aluminum cells per unit area to the ductility potential of the matrix. This concept is illustrated in Fig. 3. Seven results from the CAST program all having a DAS of .0013 inch and soundness grade A create a range of ductilities from about 2.4 to 12.4 percent strain. The number of cells counted from zones of .0001 square inch using 100X magnification photographs ranges from 18 to 47. This shows that DAS is not representative of the overall matrix cellular structure contributing to the deformation process. A degree of concept verification was provided by

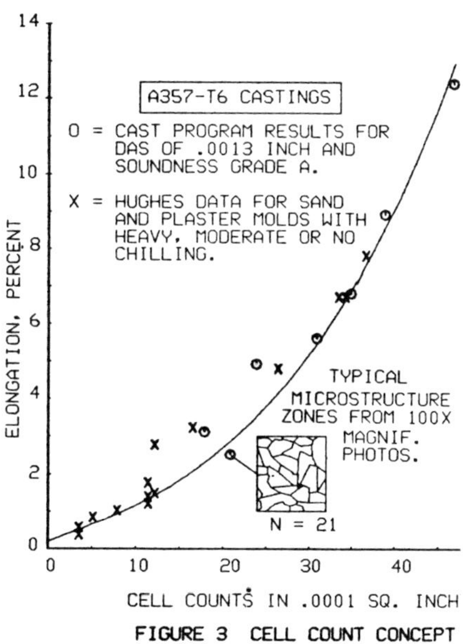

FIGURE 3 CELL COUNT CONCEPT

analyzing independently generated data[5] and microstructure records. These results were reported with Dendrite Cell Size (DCS) measurements, but also exhibited scatter in the ductility-DCS trend. In Fig. 3, these results follow the same ductility-cell count curve used for CAST results. These results are indicative of properties for both plaster and sand mold A357-T6 castings with no, moderate and heavy chilling. Insofar that the elongation-cell count trend applies equally to both sets of data, it can be considered as a tentatively general relationship. Obviously, more data from different producers must be evaluated. Influences of chemistry variations within MIL-A-21180C[6] specification limits, gas and shrinkage porosities and heat treatment must be explained to promote the utility of this concept.

3. CURRENT ACTIVITIES

In March, 1980, the Air Force awarded Boeing a manufacturing development contract to determine A357-T6 ductility influencing variables, establish the manufacturing functions which influence these variables, and develop the necessary manufacturing controls to obtain consistent and improved ductility. This program is currently in progress as a four-phase effort, and is referred to as the Casting Ductility program. It is further identified as Air Force Contract F33615-80-C-3209. Progress in this program has been reported quarterly (7, 8).

Phase I has been essentially completed. It consisted of the following measurements and analyses on CAST program specimens and ductility data: Electrical conductivity, hardness, cell counts, porosity, silicon particles and chemistry

Neither electrical conductivity nor hardness measurements showed significant correlations with ductility. The chemistry investigation showed that zinc, manganese and chromium did not vary significantly among specimens to account for ductility scatter. Titanium and beryllium contents appeared to benefit ductility while silicon, copper and magnesium seemed to reduce ductilities. These results are inconclusive insofar that CAST program castings were produced with nominal levels of the elements described in MIL-A-21180C. As a result, multiple element variations occurred between castings. At best, the above results obtained from multiple regression statistical analyses were considered as basic indicators of relative effects on ductility.

It was also possible during Phase I to establish the basic validity of the original manually-made cell count measurements using automated metallographic techniques. This method involves a computerized analysis of selected constituents in the micro-structure. In our case, aluminum cells, porosity and silicon particles are represented by different gray-scale values. The video portion of a Leitz Texture Analysis System (TAS) observes a portion of a polished and chemically etched metallurgical fracture zone mount through a selected magnification objective lens. This image is then stored in the core of the LSI-11 computer for processing. The image is also displayed on a video screen to allow the operator to select, focus, stage and measure constituents of interest. The main interest was devoted to counts of aluminum cells including primary and secondary arms as well as those resulting from coarsening and multiplication. Typical results from this re-examination are shown in Fig. 4. Two sets of symbols are shown representing original (manual) and current (machine) cell count measurements plotted against elongation data obtained from full-range stress-strain curves. Machine measurements are not only less subjective and faster than those obtained manually, they also scanned much larger areas. All of the machine-measured cell counts were made from 15 field scans in which each field was approximately .0005 square inch.

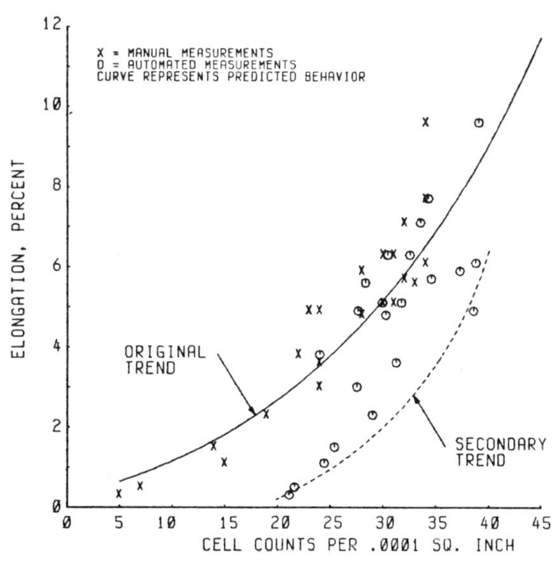

FIGURE 4 AUTOMATED CELL COUNTS

521

Calibration of the TAS showed that between 5 and 10 field scans would provide 90 percent or more repeatibility. See Fig. 5. Repetitive measurements for the numbers of cells were made using 1, 3, 5 and 10 field scans. As expected, scatter among repetitive measurements decreases as the number of fields scanned is increased.

concentrations of those chemical elements deleterious to ductility. In the case of these data, such effects could be combinations of each of these three factors. Such effects have not been completely determined. However, several initial attempts to evaluate porosity and silicon particle size effects on ductility have been made.

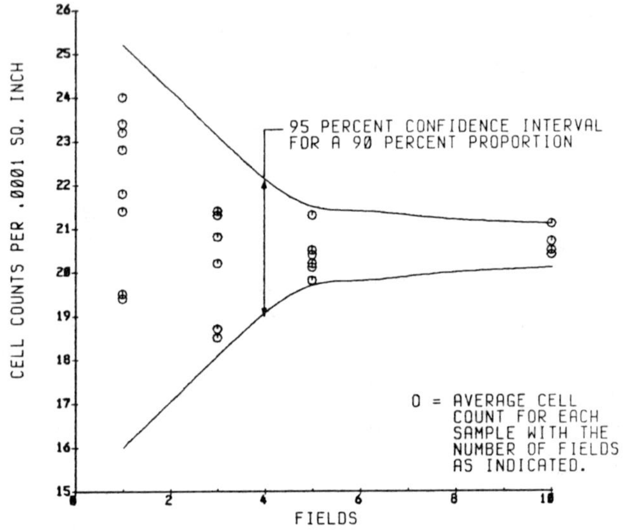

FIGURE 5   CELL COUNT MEASUREMENT CALIBRATION

Considering the TAS cell count measurements in Fig. 4 to be accurate, it is necessary to conclude that there exists more than a unique cell count-elongation relationship as originally developed and shown as the solid curve. Most likely, there are a family of curves, one of which is shown as the dashed curve representing higher porosity, less modified silicon, or

Porosity decreases the ductility of A357-T6. Fig. 6 shows TAS measurements for porosity on selected CAST program specimens. Initial measurements used 5 field scans and in a few cases single field scans. These results show that as the percentage of porosity in the matrix increases, deviations increase between measured and predicted values of ductility. The ordinate of Fig. 6

is the percentage of predicted elongation actually achieved in tensile tests. More recently, calibration efforts have shown that 20 field scans are required to obtain accurate porosity measurements. Even though these porosity data have questionable accuracies, the trend is clear.

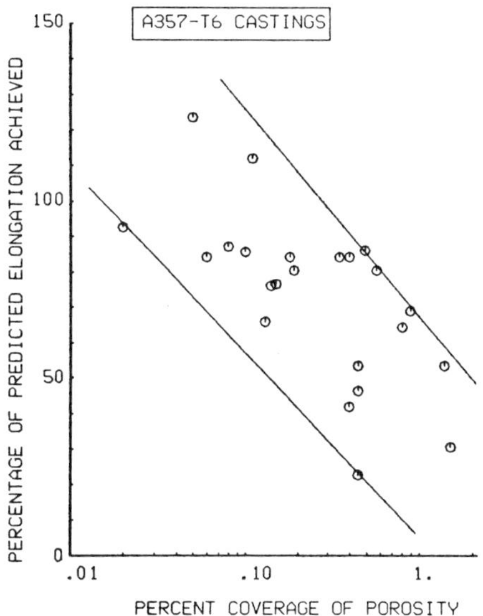

FIGURE 6 EFFECT OF POROSITY ON DUCTILITY

Future efforts will provide improved measurements from which a suitable modification parameter will be developed for the basic elongation-cell count model. A limited amount of TAS measurements for silicon particle lengths shows a ductility decreasing trend similar to that for porosity. Between 500 and 2000 silicon particles were measured from various specimens to determine the mean particle dimension. These results indicate that measured ductilities approach cell count predictions as silicon particle sizes become smaller. Efforts are currently in progress to develop a silicon particle length correction term for use in conjunction with a porosity parameter.

As a means of comparison, another independent investigation [9] has also recently attempted to apply quantitative metallography to predict mechanical properties from microstructural features of four Al-Si-Mg casting alloys. Porosity information were obtained from extremely precise density measurements of samples weighed in air and water. Counts of 20 to 30 silicon particles from small areas by eye through a microscope were used to extrapolate a particle count in a larger area, preferably 1000 square micrometers. These data were used to represent the degree of silicon modification. The basic microstructural feature was taken as dendrite arm spacing. With the exception of DAS, this work seems to contain the relevant features for predicting tensile properties. But the methods utilized for measurements are extremely difficult, time consuming, and results are subjective. Automated metallographic methods of measurement as used in the

4. FUTURE EFFORTS

Phase II of the Casting Ductility program is currently in progress, to be completed in May, 1981. Specific controlled chemistries and heat treatments are to be evaluated to develop an economical, optimum ductility A357 product. Present work is being done on 4.5 inch by 7 inch by 1 inch cast test plates as shown in Fig. 7. All 30 test plates are being produced by Teledyne Cast Products for Boeing under the same conditions of pouring and chilling to obtain the same microstructural conditions. In the first task of this effort, 14 plates will provide information regarding specific chemistry effects on ductility. All of these plates will be heat treated by Boeing to obtain a nominal -T6 condition. Following tensile tests and NDI examinations, a second task involving 12 plates will be conducted to evaluate specific influences of heat treatment processing on ductility. A preferred chemistry will be established and used in both unmodified and sodium-modified conditions. Artificial aging and quench rate will be examined for a selected solution heat treatment.

In the third task, the optimum ductility chemistry and heat treatment producing combination will be established. Four more

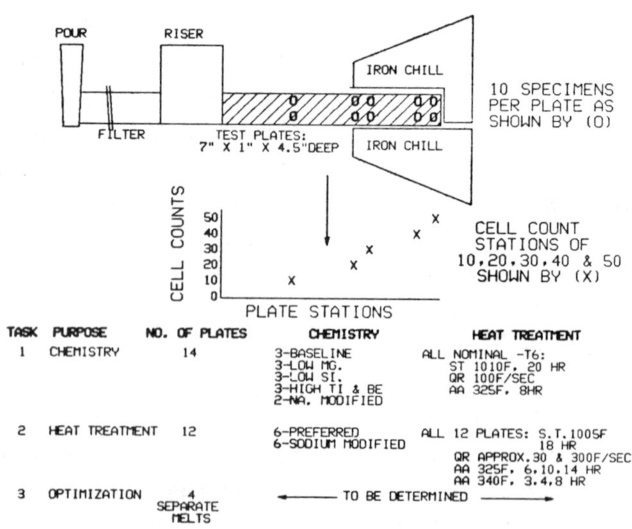

TASK	PURPOSE	NO. OF PLATES	CHEMISTRY	HEAT TREATMENT
1	CHEMISTRY	14	3-BASELINE 3-LOW MG. 3-LOW SI. 3-HIGH TI & BE 2-NA. MODIFIED	ALL NOMINAL -T6: ST 1010F. 20 HR QR 100F/SEC AA 325F. 8HR
2	HEAT TREATMENT	12	6-PREFERRED 6-SODIUM MODIFIED	ALL 12 PLATES: S.T.1005F 18 HR QR APPROX.30 & 300F/SEC AA 325F. 6.10.14 HR AA 340F. 3.4,8 HR
3	OPTIMIZATION	4 SEPARATE MELTS	← TO BE DETERMINED →	

FIGURE 7 CHEMISTRY AND HEAT TREATMENT EVALUATION

plates will be poured under these conditions to evaluate process repeatibility. Each plate will be poured from a separate melt.

Subsequent optional phases of this program are designed to provide verifications of A357 ductility prediction criteria and improved ductility manufacturing methods. In Phase III, five A357-T6 structural castings will be obtained from various foundries to assess ductility predictions. In Phase IV, Teledyne will produce 10 castings of a Boeing Air Launched Cruise Missile tank section using the chemistry and heat treatment criteria from Phase II which provide improved and consistent ductility. Results of these program phases will be reported in a future paper.

5. CONCLUSIONS

The acceptance of high-strength aluminum alloy castings for primary structure requires that such products be economical and reliable. Development of product assurance requires identification of those physical characteristics that dictate mechanical behavior and development of suitable nondestructive methods for inspection. Ideally, these criteria should be independent of foundry and specific casting variables. Under the sponsorship of the Air Force, these criteria are being developed. Physical characteristics of the microstructure are being examined using automated metallographic methods and these information appear to explain ductility variations. Establishment of these criteria will eventually benefit both producers and customers by achieving improved products and reliability.

6. REFERENCES

1. McLellan, D.L., "Tensile Properties of A357-T6 Aluminum Castings," Journal of Testing and Evaluation, JTEVA, Vol. 8, No. 4, July 1980, pp. 170-176.

2. Flemings, M.C., "The Solidification of Castings," Sci. Amer., Vol. 231, No. 6, p. 88, 1974.

3. Frederick, S.F. and Bailey, W.A., "The Relation of Ductility to Dendrite Cell Size in a Cast Al-Si-Mg Alloy," Trans. of the Metallurgical Society of AIME, Vol. 242, p. 2063, Oct. 1968.

4. Armstrong, G.R. and Jones, H., "Effect of Decreased Section Thickness on the Formation, Structure and Properties of a Chill-Cast

Aluminum-Silicon Alloy," The Metals Society, p. 454, 1971.

5.   Binder, H.F., French, J.C., Lake, W.W., and Thorp, J.M., "Program for Evaluation of Effect of Dendrite Cell Size on Mechanical Properties of Aluminum Alloy Castings, "Hughes Helicopter Report HH75-173, report to U.S. Army Aviation Systems Command, St. Louis, Mo., Contract DAAJO1-74-C-0282, July 1975.

6.   MIL-A-21180C, Military Specification, Aluminum-Alloy Castings, High Strength.

7.   McLellan, D.L., "Manufacturing Methodology Improvement for Casting Ductility," First Technical Report, 3 March - 3 June 1980, Air Force Contract F33615-80-C-3209, Boeing Report D180-26028-1, dated 3 June 1980.

8.   McLellan, D.L., "Manufacturing Methodology Improvement for Casting Ductility, "Second Technical Report, 3 June - 3 Sept. 1980, Air Force Contract F33615-80-C-3209, Boeing Report D180-26028-2, dated 3 Sept. 1980.

9.   Arbenz, H., "The Use of Structural Features to Determine the Quality of Aluminum Castings," Giesserei 66 (1979) Nr. 19-17 September, printed in both German and English.

ACKNOWLEDGEMENTS

Results reported in this paper were obtained under Air Force Contract F33615-80-C-3209, under the direction of the Air Force Flight Dynamics Laboratory, Wright-Patterson Air Force Base, Ohio.

Dale L. McLellan is a Senior Specialist Engineer in the Structural Methods and Allowables Group, Structures Research Unit, Boeing Commercial Airplane Company, Seattle, Washington. During the Air Force development program Cast Aluminum Structures Technology (CAST), he was responsible for developing a physical basis format to classify static design properties. He is the principal investigator on the current Air Force contract. His previous work at Boeing has included analysis of manufacturing and environmental effects on static design properties of both metallic and non-metallic materials. Mr. McLellan graduated from Oregon State University in 1959 with a Master of Science degree in Mechanical Engineering.

POLYMER QUENCHING OF ALUMINUM CASTINGS

Tom Croucher and Denny Butler
Metallurgical Consultants
Tom Croucher and Associates
Culver City, California

Abstract

A program is being conducted in an effort to gain an understanding of the fundamental aspects of the quenching process used for heat treating premium quality aluminum alloy castings. A major portion of this effort is devoted to comparing the quenching characteristics of polymer quenching versus hot water quenching for premium cast parts. This paper presents that portion of the effort involving alloy A356. Cooling rate determinations were made on different candidate polymer and hot water quenchants. Mechanical property tests were made on different casting heats using polymer quenchants and a quench rate sensivity curve was developed which showed the effect of different quench rates on the tensile and yield strengths of the A356 casting alloy. A series of production parts were polymer quenched and test bars were excised from the parts to demonstrate the benefits of using polymer quenchants over hot water quenchants for some situations.

1. INTRODUCTION

The heat treatment of high strength aluminum alloy castings involves a quenching operation from the solution heat treating temperature in order to retain sufficient alloying elements in solution so that subsequent age hardening will achieve the desire level of mechanical properties. As in most aluminum alloys, water has historically been used as the quenching medium because it possesses sufficient quenching power, is readily available and the cost is low. When quenching production castings, many considerations must be taken into account in order to select the proper quench water temperature. Water temperatures from ice water to boiling are commonly used. As most castings are net or near net shape, distortion of the castings during the quenching operation is a common occurrence. Generally the casting heat treater will use hot or boiling water in an attempt to eliminate the distortion and to reduce the subsequent check and straightening operations. Quenching in cold water can also result in a high level of residual stress being imparted to the casting. Consequently, most military, society and company specifications recommend that hot or boiling water be used.

For many foundry applications, hot water quenching has proved to be an effective quenchant. However, for premium quality castings where high strengths are required, the quenching power of hot water has proven to be inadequate in many instances to achieve a quench rate necessary to develop the higher level of machanical properties. Thus, the casting heat treater has been forced to use colder water temperatures and accept the consequences of high levels of distortion and residual stress.

In recent years, the use of polymer quenchants has shown widespread acceptance within the aluminum industry. As in the case of sheet metal heat treating, many foundryman have found that water soluble polymers can have a beneficial effect in controlling the distortion of aluminum castings. Croucher reported in Reference (1) that high mechanical properties could be achieved when aluminum castings were quenched in polymers. Poirier (Reference 2) presented date regarding the quenching of 356 in different media, and demonstrated that quench rates do have some effect on mechanical properties. He did not present any data on the higher purity "A" grade material. In the Boeing CAST program (Reference 3) Christner and Goehler investigated the use of polymers for quenching high strength A357 premium quality castings and concluded that polymers resulted in less distortion than 160F water but resulted in a slight reduction in mechanical properties. No test results were presented for a higher water temperature.

To date in the aluminum casting industry, polymer quenching parameters have been mostly determined on a trial and error basis. A fundamental understanding regarding the application of polymer quenchants has not been achieved so that proper specification criteria can be developed. Also, the relative effects of the cooling power of the polymers versus water quenching at different temperatures for casting applications and the quench rate effects on mechanical properties (for both water and polymer quenchants) have not been accurately determined.

It was the purpose of this program to study some of the fundamental aspects of quenching aluminum alloy castings, particularly the effects of quench rates on mechanical properties. The portion of the study reported in this paper involves the alloy A356 and further studies are planned for alloys A357 and A201. The overall objective of the program is to determine comparative quenching characteristics of both water and polymers for casting applications and to determine the relative effects of different quenchants on distortion and mechanical properties.

2. EXPERIMENTAL

The first task of this program was to determine the quenching characteristics for the different candidate quenchants through the use of cooling curves. It was initially decided to perform the cooling rate tests using cast test specimens in which a thermocouple was cast into a 1/2 inch A356 cast test plate. Inspection of the test plate after fabrication revealed however, that the tip of the thermocouple in the center of the plate did not maintain intimate contact with the cast material after the metal had solidified. Previous tests had shown that in order to obtain reliably accurate cooling rate data, continuous contact of the thermocouple with the test material was an absolute necessity,

It was thus decided that for the purpose of this program, cooling rate data would be obtained on 1/2 inch wrought alloy plates using the technique described in Reference (1). The data was then correlated to round bar test

TABLE I

QUENCHANTS EVALUATED

WATER	POLYMER
	10%
90F	20%
130F	30%
200F	40F
	60%

specimens using the data shown in Figure 1 adapted from Willey's tests described in Reference (4) Cooling curves were then performed on the candidate quenchants shown in Table 1.

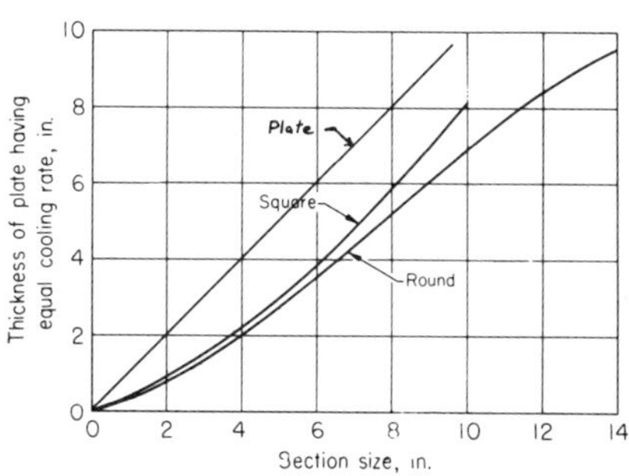

FIGURE 1 GEOMETRY EFFECTS ON QUENCH RATE

A356 test bars were then procurred from two different foundries. Attempts were made to closely follow premium quality foundry practice used by most aluminum foundries in achieving high strength, premium quality castings

529

according to Mil-A-21180. The
magnesium contents were kept as
high as possible within the
specification tolerances and the
iron content was kept as close to
0.100% as possible in order to
achieve maximum heat treat
response. The chemistry of the two
heats used is shown in Table 2.
The polymer used was Jo-Quench 50,
a single type polymer qunchant
manufactured by Jonell Chemical
Company and obtained to AMS
specification 3025 from Maxx
Products, Torrence, California.

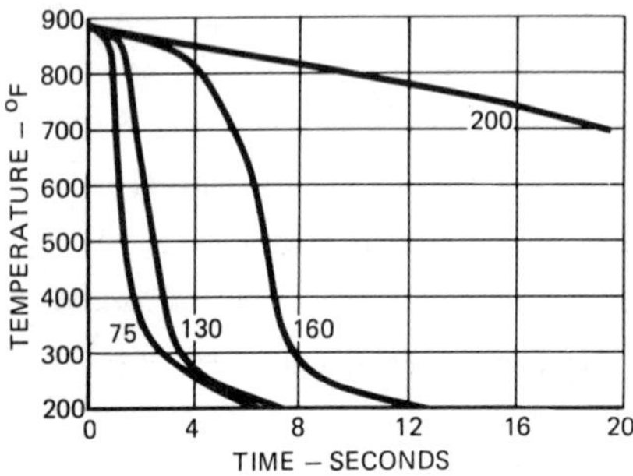

FIGURE 2   WATER COOLING CURVES
FOR 1/2 INCH PLATE

The test bars were then heat
treated according to Mil-H-6088 at
1000F for ten hours and quenched
in the appropriate quenchant. All

TABLE II

CHEMICAL ANALYSIS OF TWO HEATS

ELEMENT	#1	#2
Si	6.80	6.98
Mg	.38	.40
Zn	.05	.02
Mn	.01	.01
Fe	.11	.09
Ti	.08	.10
Al	Rem	Rem

quenching was performed in an
unagitated twenty gallon tank. All
test bars from a given heat of
material were heat treated
together and a load thermocouple
was used in the furnace zone to
ensure close temperature control
All bars were quenched in
duplicate. To ensure that all
specimens were aged to the T6
condition in an identical manner
and to minimize the possibility of
regression or prior natural aging
introducing a variable in the test
results, the test bars were stored

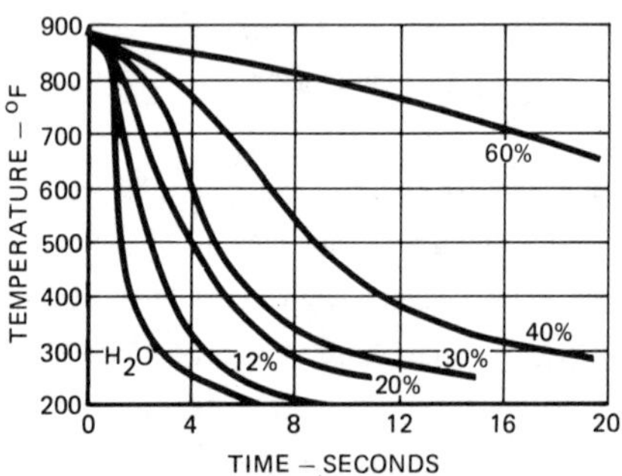

FIGURE 3   POLYMER COOLING CURVES
FOR 1/2 INCH PLATE

in an ice box at zero degrees
until just before artificial
aging.

In current aluminum foundry
practice, the achievement of T6
properties in the A356 and A357
alloys involves a delicate balance
between foundry control and heat
treat practice. There are many
different time-temperature aging
cycles used in the industry. The
foundry heat treater will select
his particular aging treatment
based on many considerations.

The aging time-temperature cycles selected for this program were chosen to be generally representative of premium quality practice. Two different aging cycles were used. These were

310F (+-10) for 18 hours
350F (+-10) for 3 hours

The tensile bars were tested and the tensile strength, yield strength and percent elongation were plotted as a function of the quench rate for each condition and conclusions were drawn from the data.

The final effort of the program involved the heat treatment of production castings to confirm the conclusions regarding the relative effects of polymers versus hot water quenching.

3 DISCUSSION OF THE RESULTS

3.1. Cooling curves

The cooling curves obtained for the different variables are shown in Figures 2 and 3. It is noted that completely different effects are achieved when an attempt is made to slow down the cooling rate from that achieved in cold water by using a polymer or hot water quench. Progressively heating the water results in a reduction of the 750-550F cooling rate, but also significantly extends the "A"

phase of cooling - i.e. that cooling which takes place by a boiling action at the surface immediately after immersion. Increasing the concentration of the polymer quench from that of 100% water also decreases the 750-550F cooling rate but does not significantly change the "A" phase of quench. Thus at approximately the same (750-550F) quench rate, the hot water quench gives a completely different quench path from the solution heat treating temperature to room temperature. The time spent in a specific temperature range will also vary and thus the tendency towards diffusion and precipitation during the quench in this range will differ. As a consequence, mechanical properties can be expected to differ significantly due to the dissimilar quench paths and not just the quench rate through the generally recognized critical range which only represents a portion of the total quenching period.

It was also noted that the quench rates achieved in the water quench from room temperature to about 160F declined only slightly, however, there was a significant drop in the cooling rate as the water temperature was increased above 160F. This effect would indicate that polymer quenching should give more consistant results than a 160-212F water quench tank. Due to variations in agitation or a high part mass/tank volume ratio in some locations within the tank, large variations in the cooling rate may occur which would cause differences in mechanical properties. Because the polymers are normally used at room temperature and are not as susceptible to bath temperature changes as the hot water quench, a more uniform quench would be achieved throughout the tank.

3.2 Quench Rate Sensitivity of Mechanical Properties

The effects of quench rate on the mechanical properties of A356 aluminum alloy castings is shown in Figure 4 for two different aging treatments. It can be seen that the rate of quench has a significant effect on the final tensile and yield strengths of A356 castings and as expected, the fastest quench achieved the highest properties. Little effect was noted on the ductility. It should also be noted that although the lowest mechanical properties observed in this program were

obtained through the use of an air quench, the quench rate achieved in these bars (i.e. 3 degrees per second) can be experienced in production parts when thicker castings are quenched in hot water or a high concentration of polymer. For reference purposes Figure 5 is included and shows the quench rate/thickness relationship for some water and polymer quenchants.

In order to achieve the required mechanical properties in premium quality castings through the use of polymer quenchants, the same balance between chemistry, foundry practice and heat treat procedures apply as in the case of water quenching. However the polymers appeared to offer the feasibility of achieving higher quench rates with less tendency for distortion.

From the cooling rate data, it appeared that polymer quenching would be extremely effective in achieving better distortion

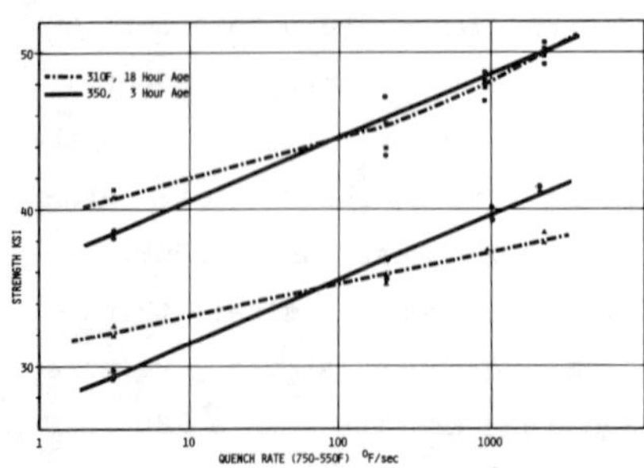

FIGURE 4   QUENCH RATE SENSITIVITY FOR A356-T6

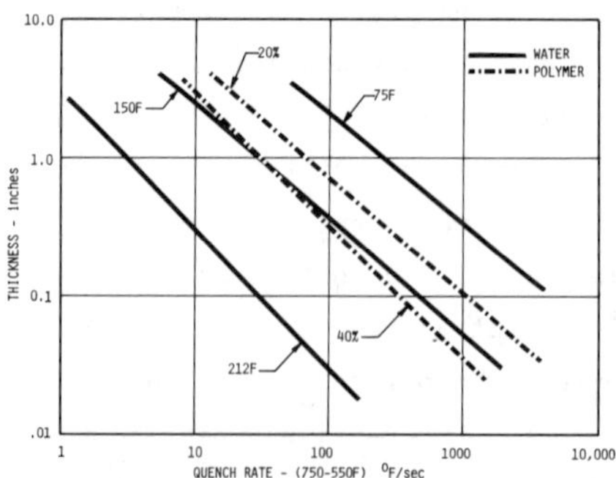

FIGURE 5   QUENCH RATE/THICKNESS RELATIONSHIP FOR VARIOUS QUENCHANTS

control than hot water in some distortion prone parts. The selection of the particular polymer concentration for a particular part would be dependent upon foundry practice, the properties required, the thickness of the part and the prior quench water temperature. To determine the applicability of applying polymers to premium quality castings, a number of production parts were quenched in various polymer quenchants and the resultant mechanical properties measured in the castings

3.3   Polymer Quenching of
      Production Castings

Various castings from several suppliers were quenched in different polymers depending upon (1) the foundry source and (2) the properties desired. Two of these parts are shown in Figures 6 and 7. All test bars excised from these castings passed the minimum mechanical properties specified.

The part shown in Figure 6 was A357-T6 which had to meet the minimum requirements of 50,000 psi tensile, 40,000 psi yield and 5% elongation. The part shown in Figure 7 was A356-T6 with the requirements of 38,000 psi tensile, 28,000 psi yield and 3% elongation.

4. SUMMARY

This program quantitatively showed the effect of the quench rate sensitivity on the mechanical properties of the A356 aluminum premium quality casting alloy. It showed that at equivalent quench rates, aluminum casting alloys quenched from solution heat treating temperature to room temperature in polymer quenchants are being quenched more rapidly but still more gently than when quenched in extremely hot water. This condition is due to the presence of the extended "A" phase in the water quench which is not present in the polymers.

FIGURE 6   PREMIUM QUALITY A357-T6
CASTING

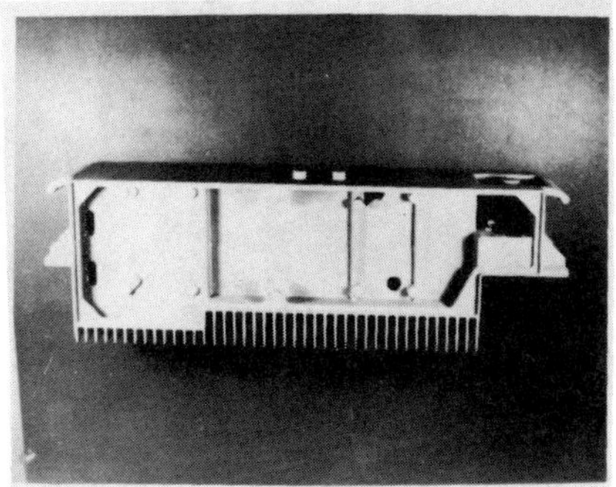

FIGURE 7   PREMIUM QUALITY A356-T6
CASTING

The results also showed that a
significant decrease in cooling
rate occurs when water
temperatures above 160F are used
for the quenching of castings and
that these reduced cooling rates
could result in lowered mechanical
properties. This conclusion indicates that the quench temperature
tolerances currently recommended
in some specification criteria
(i.e. 150-212F or 130-200F) are
too large and that a more closely
controlled temperature range would
be necessary in many cases. It
also indicates that the polymer
quenchants are a viable
alternative to water quenching for
premium quality castings and that
more work should be performed to
characterize other alloys to
enable the proper determination of
specification criteria. The
program also demonstrated that the
polymer quenchants can be
extremely effective in achieving a
high level of mechanical properties
in distortion prone premium
quality castings.

5. REFERENCES

5.1. "Applying Synthetic Quenchants
to High Strength Alloy Heat
Treatment" by Thomas R. Croucher,
Metals Engineering Quarterly, May
1971.

5.2. "Characterizing the Quenching
Of Aluminum Alloys in QA-Water
Solutions, By D.R. Poirier and
R.W. Heins, University of
Wisconsin, 1968.

5.3. "Cast Aluminum Structures
Technology (CAST) Manufacturing
Methods, Phase II," by Richard G.
Christner and Donald D. Goehler,
The Boeing Co., AFFDL Report
TR-78-62, 1978.

5.4. ALUMINUM, VOLUME I, PROPERTIES
PHYSICAL METALLURGY AND PHASE
DIAGRAMS, PUBLISHED BY THE
American Society for Metals, 1967.

6. Biographies

Tom Croucher is a Consulting
Engineer with Tom Croucher and
Associates, Culver City,
California. He received his
degree in Metallurgical Engineering
at the University of Michigan and
did graduate studies at the
University of Southern California.
He previously served as President
and General Manager of Progressive
Metallurgical Industries and was a
Program Manager and Lead
Metallurgical Engineer with the
Manufacturing Technology and
Materials and Process Departments
at the Northorp Corporation. He
also has previous experience as a
Development Engineer with the Great
Lakes Steel Corporation and as a
Metallurgical Laboratory Manager
for the U.S. Air Force.

Denny Butler is a Metallurgical
Consultant with Tom Croucher and
Associates, Culver City
California. He previously was
Assistant General Manager with
Southern California Aluminum
Treating, a Part Owner of S&H
Heat Treating and was Vice President
and Production Manager of Progres-
sive Metallurgical Industries.

PROGRESS REPORT: THE EFFECT OF SPECIMEN SIZE ON SHORT ROD
FRACTURE TOUGHNESS MEASUREMENTS OF ALUMINUM
R. V. Guest
Terra Tek, Inc.
Salt Lake City, Utah

Abstract

Interest in the short rod method of fracture toughness testing is continuing to grow. As the experience in this method increases, knowledge is gained concerning the necessary criteria for valid test results. One criterion is the minimum specimen size constraint. Early testing experience indicated that for valid fracture toughness measurements, the specimen diameter (B) should satisfy:

$$B \geq \left(\frac{K_Q}{\sigma_{ys}}\right)^2 \tag{1}$$

where K_Q is the measured fracture toughness, and σ_{ys} is the tensile yield stress of the material. The purpose of the on-going study reported here is to confirm the minimum specimen size constraint by testing a wide range of specimen sizes in two aluminum alloys. Tests were performed on specimens with 1/4 inch, 1/2 inch, 1 inch, and 2 inch diameters. The results were inconclusive due to insufficient data. Further testing will be required before a firm minimum specimen size constraint can be set. Information was gained, however, about the short rod chevron notch-bottom requirements.

Keywords: fracture mechanics, size effects, short rods, aluminum.

1. INTRODUCTION

The short rod technique[1,2] is a relatively new method of fracture toughness measurement, which may be used with specimens small enough to exhibit some elastic-plastic behavior. Advantages of the short rod method include a low testing cost[3] and very small specimens[2,3]. Work to date on the short rod testing method includes laboratory compliance calibrations[4], emperical calibrations[5], numerical analysis calibrations[6], and studies of dependence on specimen geometry[7]. Currently underway are further specimen geometry dependence studies and specimen size effect studies. The goal of the specimen size effect study is to find the minimum specimen diameter consistent with valid fracture toughness measurements. The current size effect study involves tests of several steel alloys, several aluminum alloys, and two other non-ferrous alloys. This paper presents partial results for that section of the size effect study involving the testing of aluminum alloys. The findings in this interim report are preliminary, as the study at the time of this writing is still in progress.

2. PROCEDURES

Samples of two aluminum alloys were procured for purposes of this study: 6061-T651 in 2.5 inch plate and 7475-T351 in 3 inch plate. Short rod specimens (Figure 1) were machined from the plates in the L-T orientation, so that the specimen crack-opening load is applied parallel to the rolling direction and the direction of crack growth is in the long transverse direction. Specimens were machined in four sizes: 1/4 inch, 1/2 inch, 1 inch, and 2 inch diameters. Grip grooves and thin chevron notches were machined in the specimens. Because of machining limitations, the chevron notches in the 1/2 inch and 2 inch diameter specimens were relatively wider than the slots in the 1/4 inch and 1 inch diameter specimens. All notch bottoms were rounded, except for the 2 inch diameter specimens, which had V-bottomed notches. The purpose of the V-bottom in the 2 inch diameter chevron notches was to enhance the crack front flank-end side restraint, providing for more nearly plane strain conditions at the crack front.

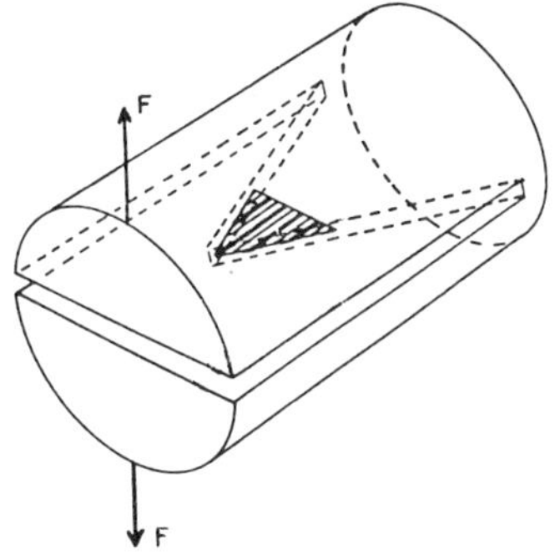

Figure 1.   Short Rod Specimen

The specimens were tested using equipment in the Terra Tek line of fracture toughness test machines. Test records were made, and the records were analyzed in accordance with standard short rod technique[8] to determine K_Q values, candidate fracture toughness measurements. The specimen crack surfaces were examined for anomalies indicating situations which may invalidate the tests. The test records were also examined for such situations. The yield strengths of the samples, from manufacturer's data[9], were compared to K_Q values which were not invalidated in the crack surface examination or the test record examination, in an effort to determine the minimum specimen size requirement.

3. RESULTS

The results of the 6061-T651 and of the 7475-T351 materials tests are shown in Table I. In the 6061-T651

Table I

Candidate Fracture Toughness (K_Q) and Yield Strength for Two Aluminum Alloys

Material		6061-T651	7475-T351
K_Q (ksi$\sqrt{\text{in}}$)	Specimen Diameter		
	1/4 inch	--	--
	1/2 inch	33.7	62.6
	1 inch	28.1	50.3
	2 inch	31.8	49.2
Yield Strength (ksi)		40	59

tests, the 1/4 inch diameter specimen simply bent open, and yielded no value of K_Q. The test records of the 1/2 inch and 2 inch specimens showed considerable crack flank-end non-plane strain effects as described by Barker[8]. These effects are a result of the chevron notch geometry and of other conditions not well understood at present. Since the K_Q values for the 6061-T651 material in 1/2 inch and 2 inch diameter specimens show variations not caused only by specimen size variations, they are of little use in the determination of specimen size constraints. The 1 inch diameter 6061-T651 specimens showed only elastic-plastic effects, and not crack flank-end non-plane strain effects, and gave a valid toughness measurement. In the 7475-T351 tests, the 1/4 inch specimens bent open. Of the 1/2 inch, 1 inch, and 2 inch diameter 7475-T351 specimens, none showed significant crack flank-end non-plane strain effects. These three specimen sizes showed decreasing elastic-plastic behavior with increasing specimen diameter. These two conditions make the 7475-T351 specimen tests useful for the purposes of this study.

4. DISCUSSION

The usual method of specifying a minimum specimen size constraint is to require that the specimen diameter (or breadth), B, be equal to or greater than some multiple of the square of the ratio of the fracture toughness to the yield strength:

$$B_{min} = X \left(\frac{K_Q}{\sigma_{ys}} \right)^2 \qquad (2)$$

Where K_Q and σ_{ys} are expressed in consistent units, the squared ratio above has units of length. For example, for ASTM standard specimens[10], the X parameter is equal to 2.5. The X parameter in the above relation is found emperically. The tests of the 7475-T351 material, while not sufficiently broad to allow a firm determination of the X parameter, still allow a tentative evaluation. Figure 2 shows a plot of $(K_Q/\sigma_{ys})^2$ versus specimen diameter B for the 7475-

538

T351 tests. The curve joining the points does not necessarily reflect the true relation; it is just the power curve containing the three points. The power curve suffices for the present, although of course the testing now underway will aid in defining the true shape of the curve. The slope of a line on the $(K_Q/\sigma_{ys})^2$ versus B curve is the inverse of the X parameter. On the graph, there are three lines drawn. The line with unity slope represents the currently used X parameter. It seems likely that the best values for the X parameter lies between .75 and 1.5, represented by the other two lines shown on the plot of Figure 2.

Whereas the 7475-T351 tests gave an indication of the minimum specimen size constraint, the tests of the 6061-T651 material did not. The fact that two of the three specimen sizes which yielded K_Q values exhibited obvious crack flank-end non-plane strain effects causes some concern. For a valid K_{IcSR} value, it is necessary that a crack exist in a specimen such that a state of plane strain prevail over a significant portion of the crack front. Now, a plane strain condition can exist only in a tri-axial stress state described by[10]:

$$\sigma_z = v(\sigma_x + \sigma_y) , \qquad (3)$$

where, for the case of a crack growing in a specimen, the x-axis

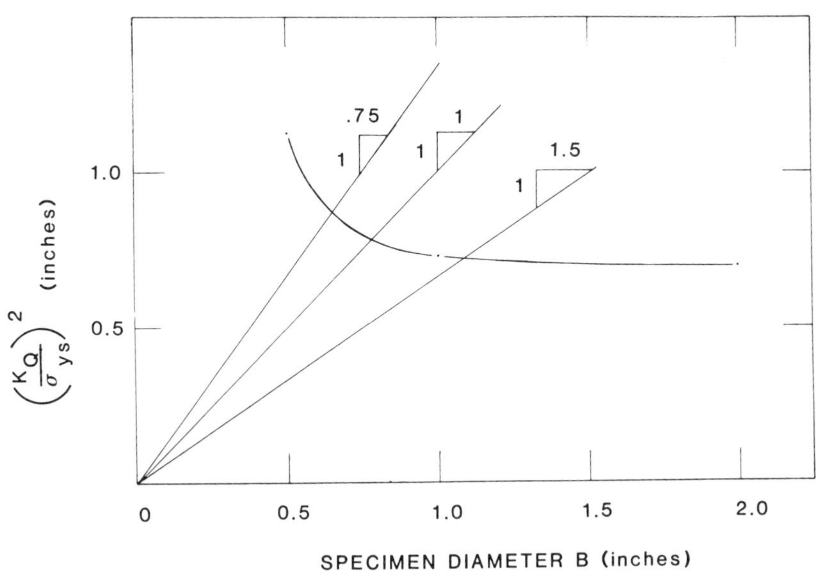

Figure 2. $(K_Q/\sigma_{ys})^2$ versus Specimen Diameter. The three data points are joined by a power curve to show the trend only. The lines through the origin represent possible values for the X parameter in the minimum specimen size relationship.

is the direction of crack propagation, the y-axis is the direction of crack opening stress, and the z-axis is parallel to the crack front. For the case where a crack exits perpendicularly to a surface (as at the flank-ends of a crack in a compact toughness specimen) it is clear that the above expression is not satisfied (since σ_z must be zero at the surface) and therefore that these flank-ends of the crack are not in a plane-strain condition. For this case, the non-plane strain zone size is virtually independent of the specimen size. Therefore, to avoid the effects of the region, it is necessary to use a specimen large enough that the non-plane strain region at the flank-ends of the crack is of negligible size relative to the specimen width. The short rod method uses side chevron notches in an attempt to decrease the actual size of the non-plane strain region at the flank-ends of a crack, allowing smaller specimens to be used. This classifies the short rod as a side-grooved specimen. The value of side grooving for constraining a crack in a crack-line-loaded specimen (such as the short rod) has been recognized[1215]. The thickness of the chevron notches is normally kept very small, to limit the amount of material around the flank-end of the crack which cannot supply a σ_z stress. As an alternative to very thin chevron notches, the use of moderately thin rounded or V-bottom chevron notches[15] has been tried. The 6061-T651 1/2 inch and 2 inch diameter specimens had rounded and 90° V-bottom chevron notches respectively. It is apparent from the test record and from examination of the 2 inch diameter specimen that the notches used were too thick (about 3% of specimen diameter), and that the rounding, in the 1/2 inch diameter case, and the 90° V, in the 2 inch diameter case, were not severe enough[8] to supply the necessary z-axis constraint. The non-plane strain regions at the crack flank-ends was too large for these specimens, making the K_Q values useless for the purposes of this study. However, valuable information about undesirable notch-bottom geometry has been gained which should improve the probability of obtaining good data on the remainder of the test series.

5. CONCLUSIONS AND RECOMMENDATIONS

1. It appears that the minimum specimen size is within the following range:

$$.75 \left(\frac{K_Q}{\sigma_{ys}}\right)^2 < B_{min} < 1.5 \left(\frac{K_Q}{\sigma_{ys}}\right)^2$$

2. Additional testing should be done to determine the shape of the $\left(\dfrac{K_Q}{\sigma_{ys}}\right)^2$ versus specimen diameter curve.

3. For future tests, the chevron notches should be thinner than 3% of the specimen diameter,

or use a V-notch bottom more acute than 90° should be used. It is recommended that for the next tests, 2% of the specimen diameter be used for the notch thickness, or 60° be used for the V-notch bottom.

6. REFERENCES

1. L.M. Barker, "A Simplified Method for Measuring Plane Strain Fracture Toughness," Engineering Fracture Mechanics, Vol. 9, p. 361 (1977).

2. L.M. Barker, "Theory for Determining K_{Ic} from Small, Non-LEFM Specimens, Supported by Experiments on Aluminum," International Journal of Fracture, Vol. 15, No. 6, pp. 515-536 (Dec., 1979).

3. R.V. Guest, "Fracture Toughness Testing Using Short Rod Specimens, Illustrated by Controlled Temperature Tests on M-2 High Speed Tool Steel," presented at the Symposium on New Developments in the Processing and Properties of High Speed Tool Steels, AIME Annual Meeting, Las Vegas, Nevada, February, 1980.

4. L.M. Barker and R.V. Guest, "Compliance Calibration of the Short Rod Fracture Toughness Specimen," Terra Tek Report TR 78-20 (April, 1978).

5. L.M. Barker and F.I. Baratta, "Comparisons of K_{Ic} Measurements by the Short Rod and ASTM E 399 Methods," ASTM Journal of Testing and Evaluation, Vol. 8, No. 3, pp. 97-102, May, 1980.

6. J.F. Beech and A.R. Ingraffea, "Three-Dimensional Finite Element Calibration of the Short Rod Specimen," Geotechnical Engineering Report 80-3, Cornell University, Ithaca, NY (1980).

7. L.M. Barker, "Evaluation of a Simple Method for Measuring Fracture Toughness in Both Brittle and Ductile Materials," Proceedings of the ICM-II Conference, Boston, MA, p. 1547 (1976).

8. L.M. Barker, "Short Rod and Short Bar Fracture Toughness Specimen Geometries and Test Methods for Metallic Materials," Terra Tek Report TR 80-11, presented at the 13th National Symposium on Fracture Mechanics, Philadelphia, PA, June 16-18, 1980.

9. Verbal Communication with Dr. Kevin Brown of Kaiser Aluminum, January, 1981.

10. ASTM E 399-74 "Standard Method of Test for Plane-Strain Fracture Toughness of Metallic Materials."

11. R.W. Hertzberg, <u>Deformation and Fracture-Mechanics of Engineering Materials</u>, John Wiley & Sons, New York, p. 265.

12. Mostovoy, S., Crosley, P.B., and Ripling, E.J., "Use of Crack-Line-Loaded Specimens for Measuring Plane-Strain Fracture Toughness," Journal of Materials, Vol. 2, pp. 661-681 (1967).

13. Marcus, H.L. and Shih, G.C., "A Crack-Line-Loaded Edge-Crack Stress Corrosion Specimen", Eng. Frac. Mech., Vol. 3, pp. 453 (1971).

14. Shih, C.F., Andres, W.R., DeLorenzi, H.G., Van Stone, R.H., Yukowa, S., and Wilkinson, J.P.D., "Crack Initiation and Growth Under Fully Plastic Conditions: A Methodology for Plastic Fracture," Report No. 78CRD038, General Electric Coporate Research and Development, Schenectady, New York (1978).

15. Freed, C.N. and Kraft, J.M.,
"Effect of Side Grooving on
Measurements of Plane-Strain
Fracture Toughness", Journal
of Materials, Vol. 11, p. 770
(1966).

7. BIBLIOGRAPHY

542

Randall V. Guest is a Research Engineer working for Terra Tek, Inc. He received a B.S. degree in Mechanical Engineering from Brigham Young University in 1977. He is 27 years old and lives in Salt Lake City, Utah with his wife and four children.

SPACE TRANSPORTATION SYSTEM PAYLOAD CARRIERS:
AN OVERVIEW

Nancy G. Williamson, Administrator, Space Transportation System (STS) User Service Center

Space Operations and Satellite Systems Division, Rockwell International

Downey, California

Abstract

This paper describes Space Shuttle payload bay capabilities and presents an overview of various payload carriers currently planned for use with payloads on the Space Shuttle. These carriers range from the small, self-contained payload canister to upper stages, the Long-Duration Exposure Facility, Spacelab, and the Multimission Modular Spacecraft. Of primary interest are the STS carriers relating to materials and processes research.

Key words: payload carriers, research in space, Space Shuttle, materials processing in space

1. INTRODUCTION

The Space Transportation System (STS) consists of many elements, such as the Shuttle orbiter, external tank (ET), and solid rocket boosters (SRB's). Launch and landing facilities, control centers, mission planning, and ground support are also key elements of the Space Transportation System. Collectively, another important element is the family of STS carriers available to support or assist payloads placed aboard Space Shuttle.

The Space Transportation System provides its users with a wide variety of facilities in space. Once a particular project has been determined, whether it is a deployable satellite, orbiting research facility, or an on-board research and manufacturing facility, an appropriate payload carrier must be considered. The capabilities of the Shuttle and the characteristics of its payload carriers can accommodate a broad spectrum of projects.

Materials processing research in space, for instance, involves processing materials under low-gravity conditions, commonly referred to as weightlessness. This research is expected to contribute significantly to the nation's technology base and ultimately lead to the development of new materials and processes with potential commercial applications. Research studies in this field emphasize materials and processes that will best explain the limitations of gravity and demonstrate the enhanced sensitivity of control processes made possible by the weightless environment of space. Primary fields of investigation include contained solidification and crystal growth, containerless processing, fluid and chemical processing, and vacuum processing. These experiments will require such hardware as furnaces, acoustic wavelength devices for containerless processing, quenching tanks, etc.

2. THE SPACE SHUTTLE ORBITER AND CARGO BAY

The Space Shuttle orbiter (Fig. 1) was designed to satisfy many payload and mission requirements, such as the following:

- Allow for payload maximum size of 4.8 m (15 ft) in diameter by 18.3 m (60 ft) in length.
- Carry a maximum of 29,500 kg (65,000 lb) into low earth orbit.
- Return up to 14,500 kg (32,000 lb) to earth.
- Provide protection, support, and handling of payloads.
- Limit maximum acceleration to 3 g's.
- Provide a nominal crew of two to four on standard missions (up to ten in an emergency).
- Provide for mission duration of 7 days (capability of 30 days with additional consumables, etc.).
- Land like an airplane.
- Return payloads to users (as applicable).

The cargo bay (or payload bay) of the Shuttle orbiter is designated to carry multiple payloads to and from earth orbit (Fig. 2).

To date, documented results of experiments performed on Apollo, Skylab, and Apollo-Soyuz can be obtained from the National Aeronautics and Space Administration Head-

Fig. 1 – Space Shuttle orbiter

Fig. 2 – Payload bay

quarters in Washington, D.C. The information is in the public domain for projects funded by the government. Experiments flown on Space Shuttle can be completely proprietary as long as they are privately funded. There are other options, however, for flying a Shuttle experiment in which the government shares in the research and the cost. These options are negotiable in accordance with the risks, involvements, and investments of the parties.

The following are examples of previously flown experiments:

- Electrophoresis
- Surface-tension-induced convection
- Monotectic and synthetic alloys
- Interface marking of crystals
- Zero-g processing of magnets
- Metals melting
- Exothermic brazing
- Immiscible alloy composition
- Whisker-reinforced composites

3. PAYLOAD CARRIERS

Payload carriers are items of hardware used to secure a payload properly within the cargo bay of the Shuttle orbiter. In very few cases does a payload attach directly to the cargo bay. The current family of STS carriers is depicted in Fig. 3. Specific carriers addressed in this overview are listed below:

- Small self-contained payload (SSCP), also known as "Getaway Special" (GAS)
- Inertial Upper Stage (IUS)

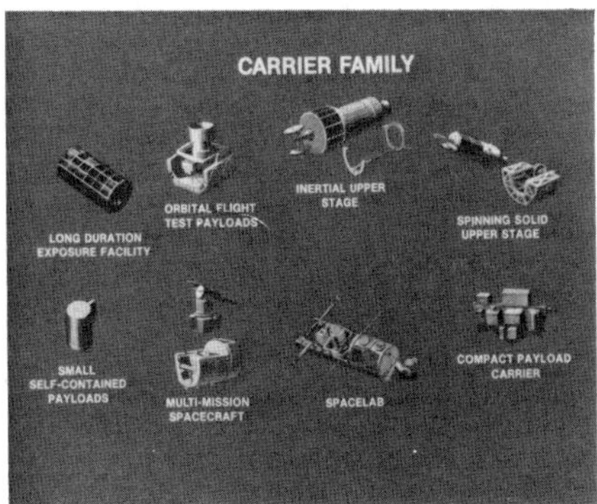

Fig. 3 — Carrier family

- Payload Assist Module (PAM), also called spinning solid upper stage (SSUS), Class A and Class D
- Multimission Modular Spacecraft (MMS)
- Long-Duration Exposure Facility (LDEF)
- Spacelab

3.1 Small Self-Contained Payload Carrier ("Getaway Special")

The "Getaway Special" (GAS), as it is nicknamed by NASA, is the smallest, least expensive carrier available to a user. The GAS canister is 1.5 m (5 ft) in diameter and approximately 0.9 m (3 ft) in length (Fig. 4). It can carry multiple experiments weighing up to 91 kg (200 lb).

The experiments must be self-contained, i.e., they will not be permitted to draw on any Shuttle services beyond the three on/off switches provided by NASA and operated by

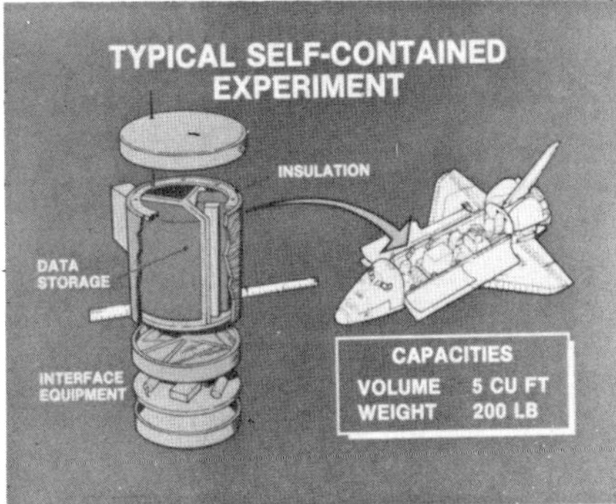

Fig. 4 — Typical self-contained experiment

one of the Shuttle crew members. Each experimenter is responsible for providing his own electrical power, heating, cooling, and data-handling facilities. After completing a flight aboard Shuttle, the experiment is returned to the user; the canister is maintained by NASA.

This particular project carrier was designed to accommodate a wide variety of users from government, military, industry, private citizens, and the academic communities. Experiments will be flown on a first-come, first-served basis; however, they will also be considered as standby payloads and will be manifested when space is available. Current plans are to launch four Getaway Specials on a flight.

Examples of experiments planned for the small self-contained flights are listed below:

- Threshold-of-acceleration perception by higher plants as measured from seedling geotropism
- Development of permanent magnets with high coercion strength, energy product, and directional solidification in a low-gravity environment
- Formation of amorphous alloys
- Effects of thermal gradient on levitated droplets
- Lizard tail regrowth
- Solar-activity-induced drag on inflatable structures
- Biology of human blood
- Alloy homogeneity studies
- Polymer formation quality versus that on earth
- Materials processing of selenium
- Inorganic chemistry
- Soldering instrument tests
- Thin-film semiconductor processes

This is an exciting and relatively inexpensive way to be a part of the space program and test project ideas in weightlessness. Any safe R&D project is acceptable. However, such ideas as sending coins, stamps, or souvenirs are prohibited. To date, over 300 reservations for Getaway Specials have been received by NASA.

3.2 Long-Duration Exposure Facility (LDEF)

The Long-Duration Exposure Facility can carry numerous experiments or payloads simultaneously. The LDEF is deployed by Shuttle in low earth orbit, where it remains from six months to one year. The LDEF is then retrieved by Shuttle and all experiments are returned to the user (Fig. 5).

The LDEF is basically an aluminum canister 4.3 m (14 ft) in diameter and 9 m (30 ft) long. Its outer surfaces are divided into bays that hold shallow trays (72) around the periphery and 2 on each end. The trays are 25.4 to 76.2 mm (1 to 3 in.) deep (Fig. 6). Some are open and some are closed. Acceptable experiments will come from scientists, engineers, universities, industry (foreign and domestic), and government agencies (U.S. and foreign).

This free-flying spacecraft has no maneuvering ability, attitude controls, power supply, data-collecting equipment, communication, or

Fig. 5 – Long-Duration Exposure Facility

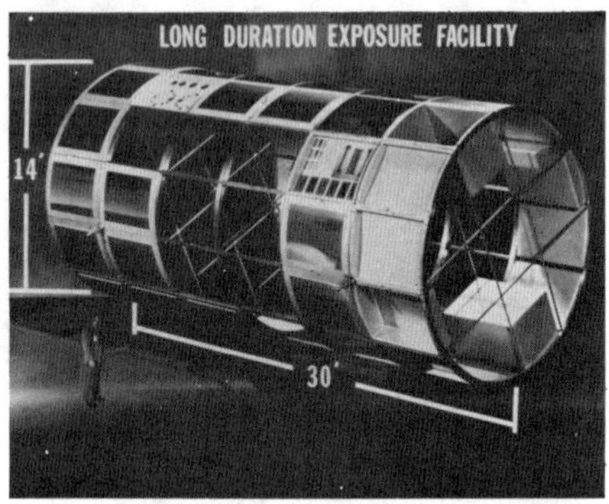

Fig. 6 – Long-Duration Exposure Facility dimensions

instruments of its own. It will be deployed into a circular orbit at an altitude of approximately 556 km (300 nmi) with an inclination to the equatorial plane between 28.5 deg and 57 deg.

The Shuttle's cargo bay contains a remote manipulator system (RMS) to deploy and retrieve payloads. The RMS will remove the LDEF from the cargo bay. After proper orientation, the LDEF will begin its six-month to one-year orbit.

Experiments on LDEF may be passive or active. For passive experiments, data measurements will be made in the ground-based laboratory before and after the mission. Such active systems as power, data storage, etc., must be provided by the user and be an integral part of his experiment assembly.

Experimenters whose projects require long exposure (six months or more) to space may want to consider LDEF (especially if they wish to have the experiment returned). The LDEF is being developed by the government at the Langley Research Center in Hampton, Virginia.

The following are examples of experiments planned on the Long-Duration Exposure Facility:

- Influence of extended exposure in space on mechanical properties of high-toughness graphite-epoxy composite material
- Atomic oxygen stimulated outgassing
- Space exposure of composite materials for large space structures
- Collection of dust debris with stacked detectors
- Exposure to space radiation of high-performance infrared multilayer filters
- Fiber optics data transmission
- Micrometeoroids chemistry
- Advanced photovoltaic experiment
- Space plasma – high-voltage drainage

3.3 Upper Stages

Free-flying satellites use a variety of carriers whose propulsive stages assist in attaining orbital altitudes beyond the reach of Shuttle, such as geosynchronous and synchronous orbits. These carriers are referred to as upper stages.

One such upper stage is the Payload Assist Module (PAM) built by McDonnell Douglas Corporation. The PAM is a spinning solid upper stage and is sometimes referred to as SSUS by NASA. The SSUS, or PAM, that assists Atlas-Centaur-type payloads weighing up to 1,997.2 kg (4,400 lb) is referred to as PAM-A. The PAM that assists the Delta-launch-vehicle-size payload weighing up to 1,248 kg (2,750 lb) is referred to as PAM-D (Fig. 7).

PAM-D's will first be flown on STS Flight 6 to launch a Telesat and SBS satellite. In addition, a PAM-A will also be used to launch an Intelsat on Flight 6.

Another propulsive carrier called the Inertial Upper Stage (IUS) is being developed for the Air Force and NASA by the Boeing Company

Fig. 7 – Payload Assist Module

for use with both military and civilian spacecraft. The IUS can place up to 2,270 kg (5,000 lb) into geosynchronous orbit. It can also propel spacecraft into planetary trajectories (Fig. 8).

IUS functions include stage structure, solid rocket motors, reaction control, thermal control, and avionics. The avionics functions include guidance, navigation, and control; telemetry, tracking, and command; communications; instrumentation; data management; and electrical power and distribution.

Two solid rocket motor (SRM) combinations of different sizes provide the two IUS vehicles with the range of performance and interfacing requirements for IUS missions.

The IUS will place the first Tracking and Data Relay Satellite (TDRS) into orbit on the first operational flight of the Shuttle. (The TDRS

Fig. 8 – Inertial Upper Stage

will handle communications among all elements of the STS.)

3.4 Spacelab

Spacelab is a versatile laboratory for manned and automated research in the low-g, high-vacuum environment of space. Spacelab's main elements are the pressurized manned laboratory and an instrument-carrying platform called the pallet, which exposes experiments and materials directly to space. The manned module and pallet can be flown together or separately (Fig. 9). When pallets alone are used, equipment that provides essential services for the experiment, such as a power distribution box and computers, is protected in a small, pressurized, temperature-controlled housing called an igloo (Fig. 10).

Spacelab is carried in the cargo bay of the Shuttle orbiter and remains attached to the orbiter through the mission. Its modular design also allows for off-line integration of experiments (Fig. 11). The lab module can best accommodate three people regularly and a fourth briefly, e.g., to change work shifts (Fig. 12).

Fig. 9 – Spacelab manned module and pallet

The purpose of Spacelab is to provide ready access to space for a broad spectrum of experiments in many fields and from many nations. Some of Spacelab's advantages are listed below:

- Reusability, allowing for improvement of the experiment
- Simultaneous observation of complementary experiments
- Recovery of experiments
- Manned or unmanned experiment capability

The Space Shuttle orbiter supplies Spacelab with such basic resources as electrical power,

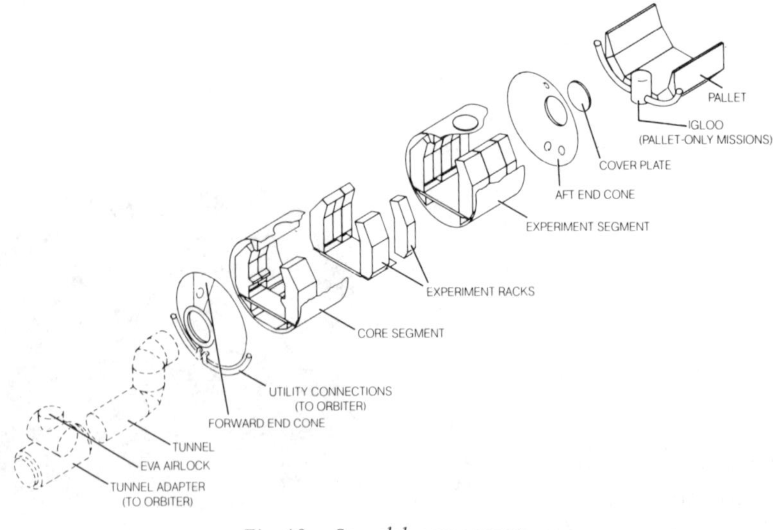

Fig. 10 – Spacelab components

Fig. 11 — Spacelab modular design features

heat rejection, and telecommunication. Standard services are furnished by Spacelab's own subsystems. A primary requirement in Spacelab subsystems design has been to provide standard interfaces for experiments with well-defined characteristics.

Spacelab is being built by a consortium of 11 European countries forming the European Space Agency (ESA): England, Spain, France, Denmark, Netherlands, Belgium, Switzerland, Italy, Germany, Sweden, and Austria.

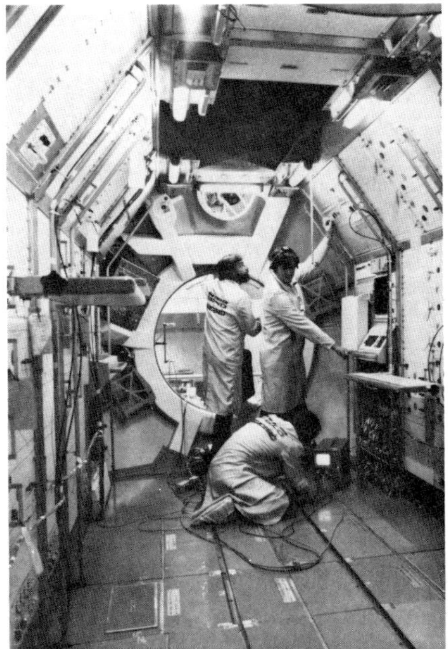

Fig. 12 — Spacelab's shirtsleeve interior

Spacelab is designed to offer the user full laboratory support and to place a minimum of constraints on experiment equipment design. Equipment should be designed for reuse whenever possible. The use of commercially available equipment which meets safety requirements is encouraged.

Examples of experiments planned for Spacelab flights are as follows:

- Solidification of immiscible alloys
- Emulsion and dispersion alloys
- Adhesion of metals in ultra high vacuum chamber
- Growth of lead telluride
- Thermodiffusion in tin alloys
- Unidirectional solidification of cast iron
- Production of ultrapure metals in space
- Firing of glass in space
- Containerless preparation of advanced optical glass
- Bubble-reinforced materials
- Solid electrolytes containing dispersed particles
- Semiconductor materials growth in low-g environment
- Ultrahigh vacuum semiconductor thin-film technology
- Kinetics of spreading of liquids on solids
- Floating-zone stability in zero g
- Surface-tension-driven convection phenomena
- Organic crystal growth

3.5 Multimission Modular Spacecraft (MMS)

The Multimission Modular Spacecraft (MMS) is an unmanned spacecraft being built at NASA's Goddard Space Flight Center in Maryland. It will be carried in the Shuttle cargo bay and is designed to take advantage of Shuttle's unique capabilities (Fig. 13).

Fig. 13 – Multimission Modular Spacecraft

The spacecraft has its own propulsion stabilization and guidance equipment (Fig. 14). Once deployed into earth orbit, the MMS can rendezvous with the Shuttle for servicing in orbit or for return to earth for overhaul. It is capable of taking on a variety of mission assignments directed toward the sun, earth, and stars.

The Solar Maximum Mission, using the MMS, is currently on orbit (Fig. 15). The MMS was

Fig. 15 – Multimission Modular Spacecraft with Solar Maximum Mission

launched by an expendable system and will be retrieved by Shuttle's flight support system developed by Rockwell International.

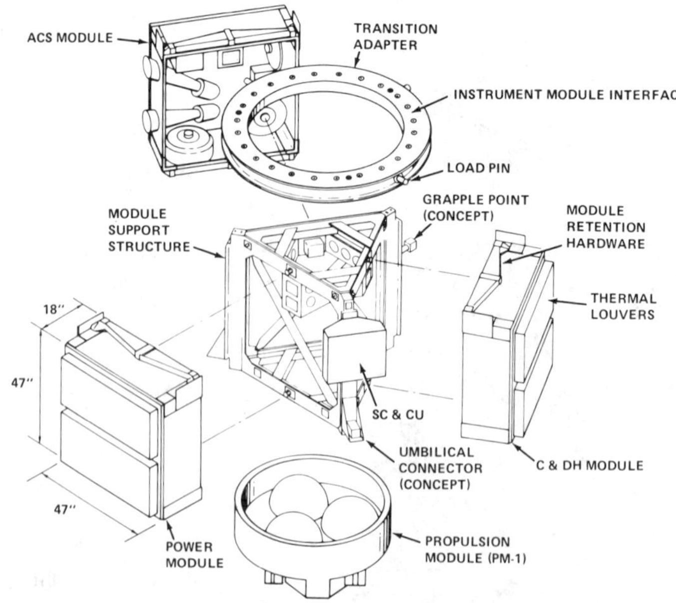

Fig. 14 – Multimission Modular Spacecraft components.

4. BIOGRAPHY

As administrator of the Rockwell International Space Transportation System User Service Center, Nancy Williamson is responsible for communicating with current and potential Space Shuttle users. Specific areas include experiment opportunities in space, pricing of Shuttle payload, flight costs, availability of flights, and the best flight accommodations available for particular project ideas. Ms. Williamson meets with organizations interested in opportunities for and the values of research and development in the weightless environment.

Ms. Williamson has also served in other key positions with Rockwell International Space Operations regarding public and customer information programs. She has traveled extensively throughout the nation describing the benefits of America's space programs before academic, civic, public, industry, and government groups.

MATERIALS PROCESSING IN SPACE
Neville J. Barter
Project Manager
Materials Processing in Space
Attached Shuttle Payloads Organization
TRW Defense and Space Systems Group
Redondo Beach, California 90278

Abstract

This paper discusses the Materials Processing in Space program from two aspects - the technical descriptions and expectations of the first generation space processing payloads for Space Shuttle/Spacelab missions in the mid-1980's and the planning actions underway for second generation payloads to be flown on the Space Shuttle/Spacelab and aboard automated carrier systems in the late 1980's.

The earlier payload systems were directed at the science issues of materials processing in microgravity and the development of processing apparatus. The payload systems, and their host vehicles, to follow promise advanced development and commercial applications.

The anticipated benefits are presented for both the early year payloads and the follow-on capability for advanced space processing.

1. INTRODUCTION

If we could develop materials to their full physical, thermal, chemical, optical, and electric theoretical potential we would make major breakthroughs in medicine, electronics, energy, instrumentation, and construction. In many respects, the quality of life depends directly on the quality of materials.

Our Earth-bound environment limits certain materials processes. All our experience on Earth is constrained by the force of gravity - the mutual attraction that the Earth and every body near it exert on each other. The weight of the objects, their falling if unsupported, and the pressure they exert on their supports and manifestations of gravity. In almost all human activities and industrial processes, gravity is a dominant force we must contend with: it constrains our motion, produces friction which must be overcome, and separates some fluids. It causes layering in alloys, contamination in glasses, separation in chemicals, and defects in electronic materials.

Space flight offers science and industry a new environment in which the effects of gravity are essentially absent. We have only begun to understand or take advantage of this new phenomenon but it holds much promise.

Currently, the United States National Aeronautics and Space Administration (NASA) is sponsoring a Materials Processing in Space (MPS) Program. The program involves both **ground- and space-based research** and looks to frequent and cost effective access to the space environment for necessary progress.

The first generation payloads for research are under active design and development. They will be hosted by the Space Shuttle/Spacelab on earth orbital flights in the early 1980's. The technical definition of several of these payload is presented. Specific experiments are briefly summarized and their payload apparatus described for Space Shuttle/Spacelab missions in 1984. Focus for these missions is on: acquisition of materials behavior research data, the potential enhancement of earth-based technology, and the implementation of space-based processing for spe-

cialized, high-value materials. Some materials may, under study in these payloads, provide future breakthroughs for stronger alloys, ultrapure glasses, superior electronic components and new or better chemicals.

2. THE ROLE OF GRAVITY IN MATERIALS PROCESSING

Space processing near-term research objectives are directed to the use of near zero gravity conditions which will prevail on Shuttle flights. The absence of significant gravitational effects over extended periods of time can give new insights into a number of fluid and metallurgical processes and can offer a degree of process control not now feasible. We have achieved promising preliminary results of sounding rocket experiments which can provide a zero-g environment of five or six minutes; in drop towers which provide these conditions for a few seconds; and in experiments that were conducted during the Skylab program. The absence of significant gravitational acceleration results in a suppression of settling sedimentation and buoyancy-driven or natural thermal convection which allows the enhanced control of temperature fields. This permits us to produce crystals of increased compositional uniformity and to process materials in a containerless fashion. These techniques show great promise for exploitation of selected materials (Figure 1).

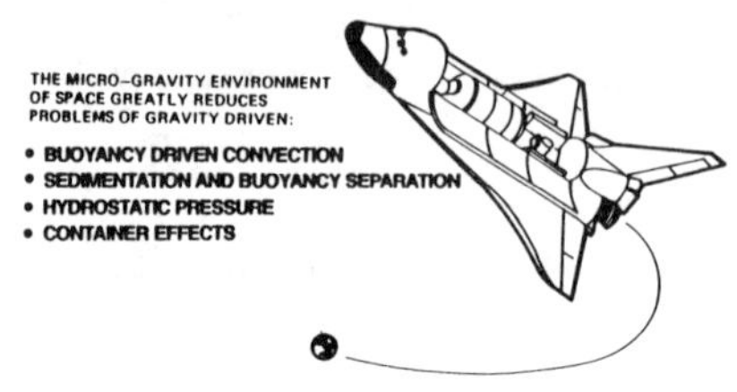

Figure 1. Why Space?

Major tasks are to develop a thorough understanding of the advantages to be gained by processing materials in space and to determine what phenomena are important in controlling low-g processes. Previous space experiments on Apollo, Skylab, and SPAR proved that natural convection arising from either thermal or concentration gradients could be adequately suppressed by going into space. Such flows are important because they give rise to nonuniform diffusion boundary layers as well as transient segregation due to temperature and velocity fluctuations during processes such as crystal growth (Figure 2). It was also demonstrated that crystals with fewer defects and with uniform composition both on a micro as well as a macro scale could be grown by taking advantage of the quiescent growth conditions in space. There are, however, other nongravitational flows, such as those induced by surface-tension gradients, volume change, or spacecraft (laboratory) motion, that operate in space as well as on earth. Such flows are often masked by gravity-driven convection and are therefore difficult to study. Experiments in space provide a means for separating gravity-driven from nongravity-driven flows and studying them separately. This information is fundamental to the design of space experiments and processes. Additionally, many terrestrial processes may be improved by a better understanding of flows from which better control strategies can be devised.

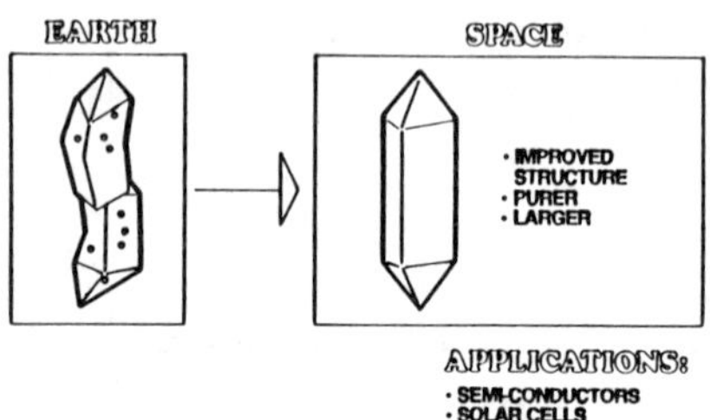

Figure 2. Crystal Formation

The ability to handle liquids and melts in a containerless mode offers unique opportunities to perform such scientific experiments as determining thermodynamic properties of chemically active materials at high temperatures; studying solidification at extreme undercooling; preparing ultrapure samples of material; and avoiding container-induced nucleation of such difficult-to-prepare amorphous solids as bulk metallic glasses and a variety of exotic glasses, particularly the refractory oxide glasses which tend to devitrify because of heterogenous nucleation. Figure 3 shows the benefits and applications of space processing to glasses and ceramics.

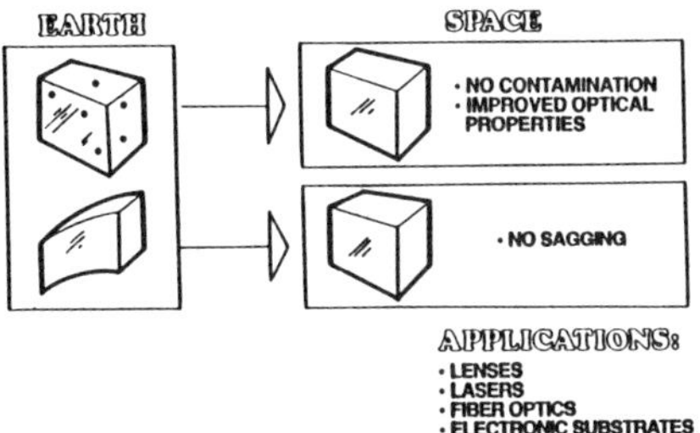

Figure 3. Glass/Ceramic Processing

The elimination of sedimentation and Stokes flow in low gravity allows the study of a number of phenomena that cannot be adequately studied terrestrially: bubble dissolution by chemical fining agents in glass, bubble centering mechanisms in thin glass shells, and bubble deformation and motion in a thermal gradient.

Figure 4 illustrates the improved structure and homogenous mixing in the space processing of metals and alloys. Figure 5 depicts the advantages of reduced gravity in the processing of organics and biological materials.

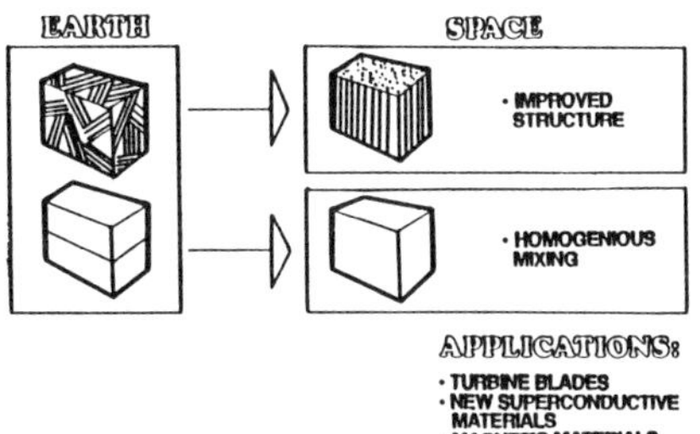

Figure 4. Metal/Alloy Processing

3. MPS PROGRAM EVOLUTION

The Materials Processing in Space (MPS) Program has conducted approximately 70 flight experiments in space on Apollo, Skylab, ASTP, and Space Processing Applications Rockets (SPAR) flights (Figure 6). These experiments demonstrated that

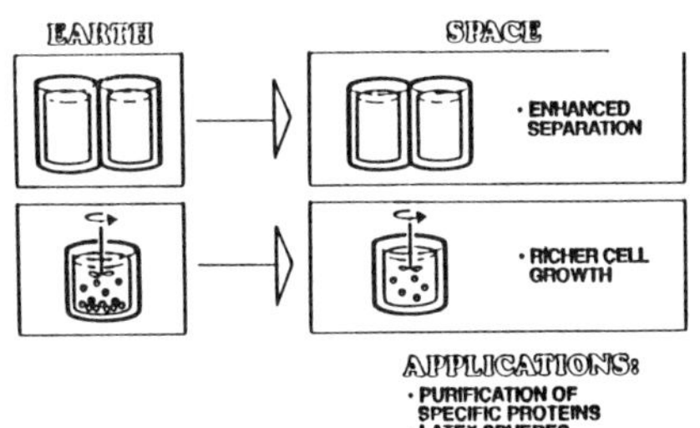

Figure 5. Organics/Biological
           Processing

The weightlessness of the space environment dramatically affects key phenomena involved in technologically important processes. This experimental work will be conducted on Space Shuttle flights starting as early as 1984 through the integration of SPAR-type experiment apparatus in the Materials Experiment Assembly (MEA).

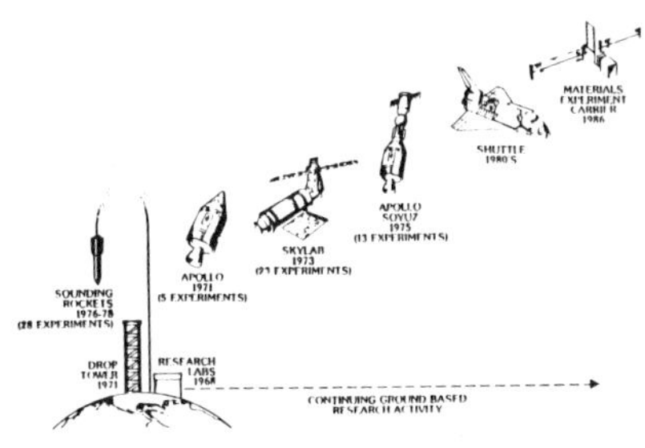

Figure 6. MPS Evolution

The MPS Program has initiated the MPS Spacelab Payload Project. Three MPS payloads are currently being developed for flight on Space Shuttle missions in 1984. These payloads will be accommodated in the Spacelab module or on the Spacelab pallet.

The MPS Spacelab module payloads will accommodate a comprehensive variety of sophisticated research in fluid mechanics and crystal growth.

Currently in the definition
stage are the Materials Experiment
Carrier (MEC) and its MPS payloads.
The MEC is intended to fly attached
to the NASA 25 kW Power System on
long-duration missions starting in
the late 1980's.

4. INITIAL SHUTTLE/SPACELAB FLIGHTS - FIRST GENERATION PAYLOADS

NASA is currently sponsoring
development of experimental payload
hardware for the first generation
payload capability to conduct Space
Shuttle/Spacelab-hosted research to
begin in the mid-1980's. The space
experimentation will be carried out
in the hardware designed to be
carried to Earth orbit by NASA's
Space Shuttle -- the recoverable and
reusable Space Transportation System of the 1980's and beyond (Figure 7). Payload hardware is currently being designed to conduct
solidification experiments using
high temperature furnaces. To take
advantantage of the high power
available due to the absence of the
pressurized Spacelab module, this

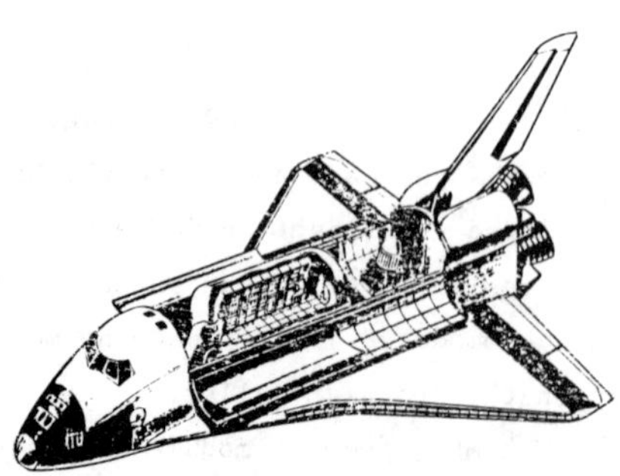

Figure 7. Materials Processing
in the Space Shuttle

payload will be located on the pallet in the Shuttle bay.

A fluids experiment system and
a vapor crystal growth experiment
are also being developed to conduct
crystal growth studies under weightless conditions. These experiments
will be conducted with the aid of
the payload specialist in the habitable environment of the Spacelab
module.

The three initial Shuttle/
Spacelab MPS payloads are the:

- Fluids Experiment System
  (FES)

- Vapor Crystal Growth System
  (VCGS)

- Solidification Experiment
  System (SES)

The FES and VCGS are planned
to fly in the manned module of
Spacelab-3 in mid-1984. The SES
will fly on an orbiter pallet in
early 1984.

There are several other efforts
under study for consideration as
first generation Shuttle/Spacelab-
hosted MPS payloads. They are the
Polymer Latex Reactor, Acoustic
Containerless, and Analytical Float
Zone Systems.

The three first generation payloads now under hardware development
are described below.

5. SOLIDIFICATION EXPERIMENT SYSTEM

The SES is a Shuttle bay pallet-
mounted payload which automatically
melts, refines, and resolidifies a
broad range of materials (Figure 8).
It is a co-passenger on a Shuttle mission with communications satellites

which will be deployed early in the mission, allowing the SES payload to use all available electric power from the Shuttle's fuel cells to operate MPS experiments for the remainder of the seven day Shuttle mission.

Figure 8. SES Configuration

The SES operational modes and summary parameter data for each mode are presented in Figure 9.

	SAMPLE SIZE (CM)	SOLIDIFICATION RATE	TEMPERATURE GRADIENT
DIRECTIONAL SOLIDIFICATION	1.27 DIA 25 LENGTH	SAMPLE TRANSLATION .1 to 3600 MM/HR	@ 200 C 20 C/CM 500 70 800 150 1100 200
GRADIENT FREEZE	1.27 to 3.17 D 2.54 to 25.4 L	COOLING RATE 0 - 60 C/MIN	LINEAR 0-100 C/CM OVER 12.7 CM MAXIMUM 180 C/CM OVER 2.54 CM
ISOTHERMAL	1.27 to 3.17 D 2.54 to 25.4 L	COOLING RATE 0 - 1000 C/MIN	TEMPERATURE UNIFORMITY .0 05% OF OPERATING TEMPERATURE OVER 25.4 CM

GENERAL PARAMETERS:
 • TEMPERATURE RANGE 100 C TO 1600 C
 • TIME TO HEAT TO STEADY STATE   1 HR

Figure 9. SES Operating Modes

Diectional solidification processing is designed for the production of high uniform crystalline solids from a melt. A furnace for directional solidification processing must provide a uniformly high temperature environment in one end and a low temperature in the other end. A test specimen of the desired composition is inserted into the hot end, allowed to melt, and is then slowly withdrawn at a controlled rate. Crystal growth occurs at the solid-liquid interface which lies in the gradient zone between the hot and cold ends. The goal of all SES experiments is the production of extremely uniform crystals of a degree of homogeniety which cannot be achieved in a terrestrial environment where the inevitable convection currents distort crystal growth.

Steep axial thermal gradients may be required to prevent constitutional supercooling in the liquid near the interface. Radial gradients, on the other hand, give rise to convective flows at the interface and must be avoided. In space, radial thermal gradients produce a curved interface which distorts crystal growth.

Gradients that can be achieved in a given test specimen will depend on the geometric and physical properties of the specimen, on the properties of its container, and on the thermal characteristics of the furnace. One of the goals of experiment design will be to adjust hot and cold temperatures in the furnace to achieve a liquid-solid interface (at the melt temperature) which lies at a point of maximum gradient within the adiabatic zone. While these principles are common to each of the experiments, each also has its own peculiarities which must be considered in the experiment design.

sitional deviations.

6. FLUIDS EXPERIMENT AND VAPOR CRYSTAL GROWTH SYSTEMS

The FGS and VCGS are being developed to conduct crystal growth studies under weightless conditions with the aid of the payload specialist in the habitable environment of the Spacelab module.

Crystals can be grown from fluids, from vapors, or from melts of solid materials. Payloads are planned to allow each of these phenomena to be investigated in Earth orbit. For growth from fluids, there is an interchangeable experiment cell (Figure 10) that allows crystals to be grown under controlled conditions while being observed by the onboard payload specialist and scientists on the ground. The cell is closely temperature controlled (Figure 11) and has

windows to permit holographic and video recording of the crystal as it is grown. Operation of the experiment is initiated by a payload specialist with seed crystal insertion. The holographic film and grown crystal are returned to Earth for data analysis.

The experiment currently approved for flight on the FES is the Radial Growth Experiment of Dr. R. V. Lal (Alabama A&M University) and Dr. R. Kroes (NASA/MSFC). This is a solution growth of crystals using triglycine sulfate, which has applications as as infrared detector.

To conduct crystal growth from the vapor phase, a sealed ampoule with surrounding heater coils is employed. (Figure 12). The source material (mercuric iodide) is heated to vaporization and recondenses at the cooler seed crystal site. The rate of vaporization and crystal growth is controlled by adjusting the source and seed site temperatures. Crystal growth is observed with the vidicon camera also used by the FES.

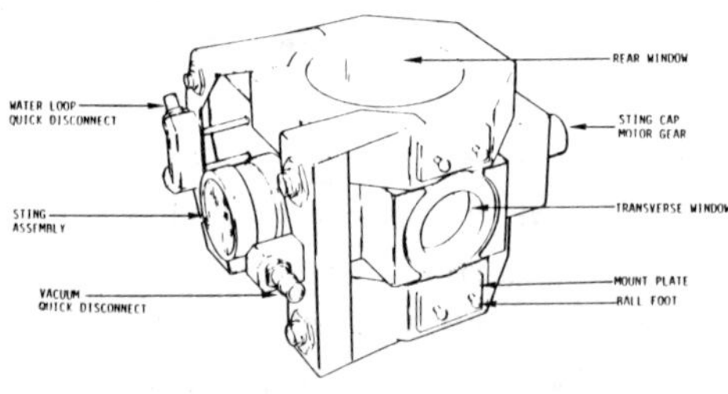

Figure 10. FES Experiment Cell

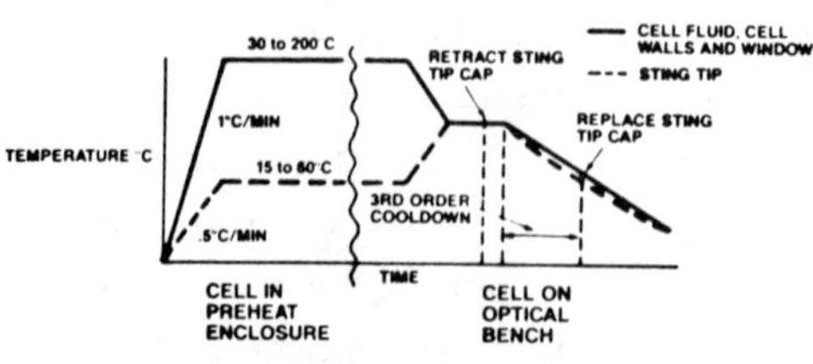

Figure 11. FES Test Cell Temperature Profile

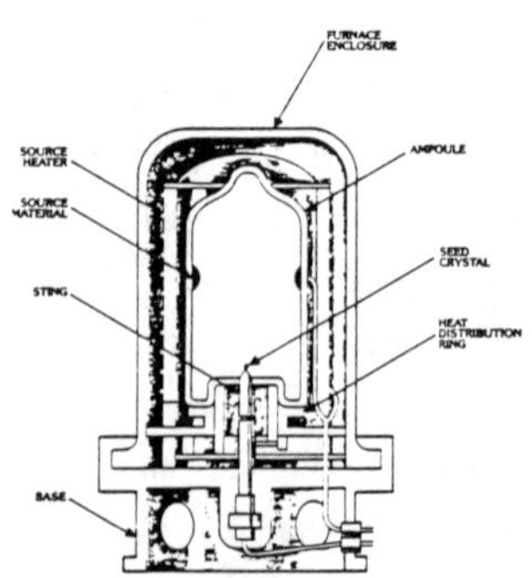

Figure 12. VCGS Module

The SES experiments offer a
valuable opportunity for extended
experimentation in a low-g environment. A first objective of many
investigators will be to study
the effects of crystal structure
of the several processing variables available to them. Specific
experiments are summarized below.

a. Dr. David Larson (Grumman
Aerospace Corporation) - Dr.
Larsons experiments concern the
low-g processing of magnetic materials. Apollo-Soyuz Test Project
experiments with 50 atomic percent bismuth-50 atomic percent
manganese mixtures solidified
in a low-g environment demonstrated substantial improvement
in macroscopic chemical homogeniety. Directionally solidified
Bi/MnBi eutectic samples exhibited
markedly superior magnetic properties. Intrinsic coercive
strengths exceeded previously
published values by 64 percent.
The average value of inductance
was improved by 76 percent and
the energy product by 57 percent.

b. Dr. Robert Naumann (NASA/
MSFC, Huntsville) - Dr Naumann's
experimental objectives are to
determine growth techniques for
the production of high quality
trimetal solid solution crystals
of mercury cadmium telluride for
use in infrared detectors. Military and civilian applications
require larger crystals with
better homogeniety than is currently available. Particular

emphasis will be directed toward
producing materials with extremely high frequency response
for wide bandwidth applications.

c. Dr. Stanley Gelles (S. H.
Gelles Associates) - Dr. Gelles is
interested in three alloy systems:
aluminum-indium, lead-zinc, and
lead-copper. These alloy systems have miscibility gaps in
which the two liquid phases are
immiscible below a particular
temperature called the "consolute
temperature". Such alloys cannot
be formed in Earth gravity because
density differences between the
two liquid phases cause rapid
separation when the melt is cooled
through the miscibility gap during
solidification. Since many other
systems possess miscibility gaps,
perfection of the solidification
technique in the low-g environment of space offers an opportunity to find new alloys with
unusual properties.

d. Dr. Robert Crouch (NASA/
LaRC) - Dr Crouch's interests are
in the production of solid solution
crystals of lead tin telluride.
Detectors using PbSnTe may be superior to those using HgCdTe for
certain NASA applications (such as
pollution minitors to detect HNO_3
and NO_x in the upper atmosphere).
Compared to HgCdTe, PbSnTe is
easier to produce due to lower
vapor pressures and is better
understood in terms of its phase
diagram. PbSnTe is also about
four times less sensitive to compo-

Dr. W. Schnepple (EGG, Santa Barbara, California) is the principal investigator for this mercuric iodide experiment. Mercuric iodide is used as a nuclear detector.

Figure 13 illustrates the Spacelab module racks for conducting both the fluid and vapor phase crystal growth experiments. Five fluid test cells are stored in the double rack on the left and are used for experimentation one at a time. The test cell configuration of Figure 13 has been developed for the specific Spacelab-3 experiment, growth of triglicine sulfate crystals. Other experiments will use test cells individually designed for each investigation. The vapor experiment module configured for the Spacelab-3 mercuric iodide growth experiment is housed in the single rack facility on the right in Figure 13.

Figure 13.   FES/VCGS Configuration

6.1 Holographic Capability

The FES holographic system (Figure 14) will record holograms in flight for analysis on board and later on Earth.

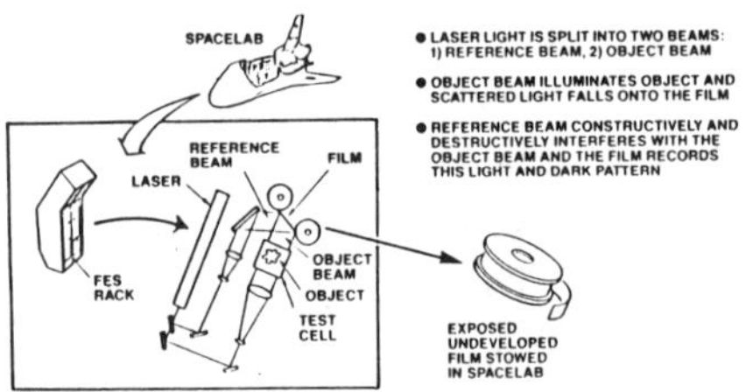

Figure 14.   The FES Holographic Data System is a Major Technical Challenge

7.   SECOND GENERATION PAYLOADS AND THEIR HOST VEHICLES

The next major milestone for MPS payloads is anticipated in the late 1980's. At that time the planned 25 kW Power System is expected to become operational and the projected needs of MPS for samples, processing time, and power required to support sustained, systematic MPS activity will exceed Shuttle capabilities. Thus the 25 kW Power System capability and MPS needs can be matched. In operational terms the 25 kW Power System provides the opportunity to: (1) extend the orbital stay time of the Shuttle/Spacelab pallet while providing a higher power level to the MPS payloads and (2) support experimental and commercial payloads while docked to the 25 kW Power System as a free-flier between Shuttle visits. Both of these capabilities provide significant benefits to the MPS Program.

Previous analysis of the cost of research in space have shown that longer stay time, together with more

power to run experiments, can dramatically reduce the unit cost of experimentation. Ground-based research and
flight experiments to date have shown
that a significant number of samples
must be processed in order to isolate,
characterize and develop MPS processes. In addition, the power requirements of MPS research and processing
apparatus are typically higher than
for other kinds of space activity. It
is already apparent that the electrical power and energy resources of the
Shuttle/Spacelab system will become a
serious limiting factor to the MPS
Program at the levels of activity that
are expected in the late 1980's.

Consequently, the MPS Program will
become a primary user of the 25 kW Power System since it offers increased
orbital stay time, electrical power
and, in the free flying mode, microgravity stability at levels of 10^{-5} g
or better. This system will be used by
MPS in both the advanced Shuttle sortie
and the free-flying modes.

In the advanced Shuttle sortie
mode (Figure 15), second and subsequent
generation MPS payloads that require
manned participation will be accommodated. In this mode, the Shuttle/
Spacelab will dock with the 25 kW Power
System and support manned missions of
up to 30 days.

In the free-flying mode, totally
automated versions of the second generation MPS payloads will be accommodated by the 25 kW Power System on
missions of 60 to 120 days. Planning
for use of the Power System has been
in progress in the MPS Program for over

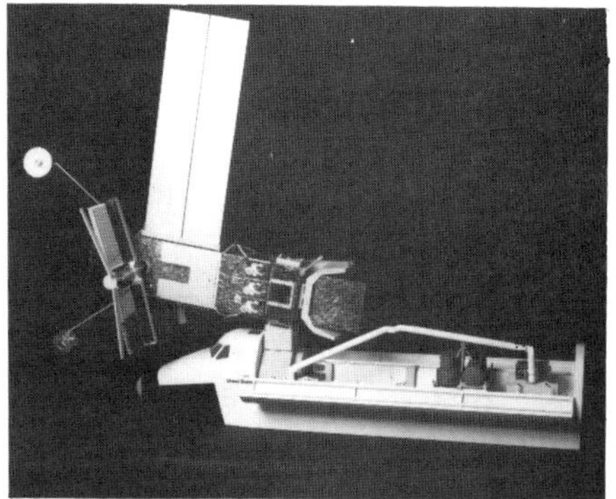

Figure 15. MEC Shuttle Bay Sortie
Mode

a year and it has been determined
that the most useful mode of operation will be to attach experimental equipment to the Power System to be operated while the system
flies freely between visits from
the Shuttle. This mode will require
a structure to support MPS experiments and an associated subsystem
to operate them. The hardware
assembly needed for this purpose
has been named the Materials Experiment Carrier (MEC).

Figure 16 depicts MEC mission
operational elements and requirements. A significant amount of
interface system engineering is
involved since the MEC has major

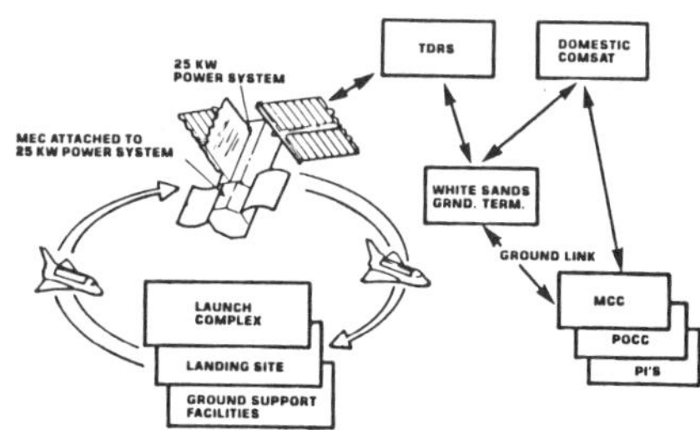

Figure 16. MEC Mission Operational
Elements

interactions with the Shuttle, 25 kW
Power System, Tracking and Data Relay Satellite System, and ground
operational facilities.

8. COMMERCIALIZATION

The ultimate goal of the MPS
Program is to develop commercial
interest in using space to: (1) perform research to improve industrial
technology or to develop new products; (2) prepare research quantities of materials for comparison with
current Earth-based technologies; (3)
manufacture limited quantities of
unique products to test market potentials or to fulfill limited but com-
pelling needs; and (4) produce materials of sufficient quantity and
value to stand on their own economically.

NASA's MPS Program will demonstrate to potential industrial users
that they can learn more about their
processes by conducting experiments
in space and be able to do things in
space that cannot be done on Earth.
NASA is working closely with industries to understand their problems
sufficiently to be able to identify
areas in which space processing can
be used best. It is probably not
realistic to expect major commitments
from industry until a sequence of
spaceflights has been completed to
demonstrate that space offers significant benefits to materials processing.

The processing and manufacturing
of commercial products in space is
attracting considerable interest.
Virtual weightlessness, readily available solar power, and the vacuum of
unlimited capacity offered in space
could yield new, improved or lower-
cost products. At present, the
economics of space processing is a
controversial issue. The space processing experiments of the late
1980's will undoubtedly yield rich
rewards in pure science; it is also
quite probable that this learning
period will point the way toward
commercial opportunities for the
production of selective high-value
space products. These products are
likely to pertain to health, electronics, electro-optics, and optics.
The concept of space manufacturing
is quite new -- the product forms
have yet to emerge, costs are uncer-
tain, very little market analysis
has been conducted, and accurate
cost benefit evaluations are still
required. It seems, nevertheless,
that those companies that are aware
of the rapid advances of technology
can hardly not afford to follow
NASA's experimentation closely;
they should be willing to participate now through their own innovative thinking and planning. The
typical alternative in a free and
competitive society is the risk of
losing out in a profitable commercial venture. "First-to-market"
is always extremely important.

9. SUMMARY

Materials processing in the
space environment will expand our
knowledge of both the science and
the technology of materials.
Greater control over the structure

and properties of materials is the
primary output expected from the
processing investigations to be
first performed in space on Shuttle/
Spacelab 1984 missions. Results
will undoubtedly benefit ground-
based processes. The unique
features of weightlessness are expected to be of ultimate benefit to
a class of new, high value products.
Precursor products, however, must
first be made by the commercial sector to test the risks -- including
the investigation of competitive
ground-based technologies -- as
well as the market demands for such
materials. Cooperative opportunities to explore these possibilities
are currently under consideration by
a growing number of U. S. companies.
The availability of frequent and
economical access to the space environment afforded by the Space Shuttle
and vehicles to follow, like the MEC,
will open new opportunities for
materials research and new product
lines for commercial interests. A
fundamental message is: "As our business looks up, we must keep both feet
on the ground".

With the start of Shuttle/Space-
lab MPS flights, there will be exciting research and experimentation.

But as MPS activities mature from
materials development to materials
applications, it must not be forgot-
ten that there are some very practical things that have to be accomplished to make space productization
a reality.

* * * * * * * * *

Neville J. Barter was educated
in England, receiving his National
Certificate in Mechanical Engineering
at Southhampton University. He has
35 years of aerospace experience and
has been with TRW Defense and Space
Systems Group for the past twenty-
two years.

Mr. Barter is the Project Manager
for the Materials Processing in Space/
Spacelab Project being performed by
TRW for the NASA Marshall Space Flight
Center. Prior to this assignment,
he held key management and design
roles in TRW's commercial and military
communications satellite systems, including management of the design and
development of the Intelsat III and
DSCS II communications satellites.
He led the initial TRW design and
business development activities on
the Tracking and Data Relay Satellite
System (TDRSS) for Western Union.

AVENUES AND INCENTIVES FOR COMMERCIAL

USE OF A LOW-G ENVIRONMENT

Richard L. Brown

and

Lowell K. Zoller

Materials Processing in Space Projects Office

Marshall Space Flight Center, Alabama

Abstract

One of the unique and new technologies which has emerged from the space program is the processing of materials in a low-gravity (low-g) or microgravity (10^{-6}g's to 10^{-2}g's) environment. The reduction or elimination of the pervasive influences of gravity on process mechanisms affords opportunities for understanding and improving ground-based processes and for creating unique materials. The primary goal of NASA's present work in the field is to realize scientific and commercial utilization of the low-g environment for materials research, and for process and product development. For the next several years, any products of commercial interest which necessitate processing in space will likely be low volume, high value items. To encourage the commercialization of materials processing in low-g, NASA , in parallel with establishing and demonstrating the scientific/technological precepts for analyzing and using a low-g environment, is establishing the legal and management mechanisms to share in the cost and risk of early commercial ventures, and is now working with commercial firms on a case-by-case basis to explore applications of this new technology to specific needs of the company.

1. INTRODUCTION

The Space Act directs NASA to conduct its activities to preseve U.S. leadership in aeronautical and space sciences and technology, and their applications. One of the unique and new technologies which have emerged from space effort is the processing of materials in an environment where the effective gravitational acceleration is very low (10^{-6}g's to 10^{-2}g's). This low-gravity (low-g) condition is often called microgravity, zero-g or weightlessness. Due to its origin in and close ties to the space program, this emerging field is commonly referred to as materials processing in space (MPS). For many years, the usefulness of gravity levels above one-g has been recognized. Separators, centrifugal casting operations, pumps and a host of other equipment and industrial processes utilize "variable gravity" above one-g. Only in recent years, however, has much attention been given to "variable gravity" below one-g. Gravity levels below one-g can be achieved by placing an object in free-fall. For short-duration investigations of material properties and process mechanisms (a few seconds to a few minutes), a low-g or microgravity enviroment can be achieved by allowing specimens to free-fall in a drop tube or a drop tower, or by flying aircraft and rockets through carefully prescribed trajectories. A spacecraft in orbit about the Earth may be viewed as being in a state of continual free-fall. Thus, with the advent of the Space Shuttle in 1981, a low-g environment can be sustained for days at a time. By the end of the decade, free-flying spacecraft, which will be serviced by the Shuttle, will be routinely available. These "free flyers" will provide a microgravity environment for long periods of time (months to years). Available and planned NASA owned-facilities and flight systems which can provide a microgravity environment in the range of 10^{-6}g's to 10^{-2}g's, are shown in Figure 1.

2. INFLUENCE OF GRAVITY
ON PROCESS MECHANISMS

From work done to date it is clear that the microgravity environment opens a new frontier for material scientists and processors by providing new insights into the pervasive role of gravity on material properties and process mechanisms and by eliminating some of its disrupting influences that preclude some materials produced on Earth from achieving the theoretical performance characteristics of materials.

2.1. <u>Convection</u>

The elimination of gravity driven convection in molten materials can preclude the sometimes undesirable stirring and mixing encountered during the growth of crystals, the casting or solidification of alloys and composites, chemical reactions, or the separation of biological materials (Figures 2 and 3).

2.2. <u>Sedimentation and Bouyancy</u>

The elimination of gravity induced sedimentation and bouyancy can broaden the spectrum of alloys and composites that may be formed by permitting particles of vastly different density to remain in suspension until solidification occurs. Also, the elimination of sedimentation and bouyancy eliminates the need for mechanical stirring. This is important in instances where the stirring may be detrimental to the materials involved (Figure 4).

2.3. Gravity Induced Deformations

Where hydrostatic pressure controls or limits a process or the force of gravity (weight) causes deformation or fracture of a material, the elimination of gravity can provide opportunties for investigations of unique materials and manufacturing techniques (Figure 5).

2.4. Containerless Processing

The elimination of the necessity to confine liquids and molten materials within a container can open interesting possibilities. Materials, depending upon their electromagnetic characterisitics and the influences of processing in a gaseous environment, may be melted, mixed, manipulated, shaped and solidified in free suspension by use of acoustic, electromagnetic, or electrostatic fields. Surface tension will hold the material together in a mass. Another form of containerless processing which can be enhanced by microgravity is the float zone process; in space, since the molten zone will be confined by surface tension, much more latitude may be available for cyrstal production (Figure 6).

3. LAYING A SOUND FOUNDATION

The scientific and technological benefits which can be derived from the exploitation of materials processing in low-g are fundamental in nature and may result in significant improvements in material utilization and producibility; the potential economic benefits may be both substantive and viable. However, utilization of this technology must be approached with deliberation and realism.

3.1. Technology Base

In early work there was considerable emphasis on investigating material processes in suborbital and orbital experiments that could rapidly lead to the production of commercially viable products in space. While a number of interesting results were obtained, it became clear that much more sophistication was required in process control and diagnostics, particularly with regard to the control and measurement of thermal gradients and quenching rates used in many of the processes. Sample preparation was found to be especially critical when it is necessary to control oxide formation or to completely homogenize a specimen. In containerless processing, much has been learned about the precise positioning and rotational control needed to prevent the sample from contacting the levitating device, as well as disruptive accelerations on and unwanted stirrings within the sample. Better methods for obtaining flow and temperature fields were found to be necessary in order to observe what is happening during a process. Present scientific and technological work sponsored by NASA is concentrating on identification of basic process mechanisms and gravitation influences on these mechanisms. Areas in which NASA-sponsored work is being carried out are shown in Figure 7.

3.2. Commercial Applications

A major difficulty that became apparent in attempting to develop commercial applications from initial experiments in microgravity was the identification of potential products. A number of studies have been carried out on the economic benefits of manufacturing specific items such as silicon ribbon, improved turbine blades, and various pharmaceuticals in space. Although such studies may have had some value in creating interest, stimulating ideas, and developing concepts, they ignored certain realities. It was tacitly assumed that space processing would result in a superior product and that improvements in certain Earth-based processes were limited irrevocably by gravitation effects. As it turns out, many of the identified improvements have been achieved by alternative techniques. In other cases, new technology has supplanted the need for the product. This will always be a problem with trying to identify specific needs for several years hence. Still, materials processing in microgravity can be used to obtain new insights into Earth processes, to make paradigms for comparision with materials made on Earth, and to make new and unique products. With regard to space-made products, present studies show that during the decade of the 1980's, they will likely be limited to low volume, high value items.

3.3. Understanding Commercial Needs

It is clear that careful analyses and close coordination between NASA and prospective users of the technology will be needed to insure that the proper base is laid. To foster proper working relationships with prospective users, the Commercial Applications Office, MPS Projects Office, at the Marshall Space Flight Center, has been created to work exclusively with commercial interests. This team forms a bridge between NASA and the commercial community, serving as a source of information and assistance for the user as well as a focal point for commercial views and a channel by which these views can be articulated to NASA. This team is also working to clarify NASA's policies regarding commercial rights in intellectual property, liabilities, equipment-leasing, and pricing. It is through this effort that NASA believes it can provide a simpler interface to the private sector, develop a better understanding of the incentives needed to elicit private initiatives and stimulate the inventive genius and entrepreneural spirit in this country to fully utilize the benefits to be derived from the MPS program.

4. INCENTIVES FOR COMMERCIALIZATION

Since NASA wishes to encourage the private sector by sharing in the technical and financial risks of early commercialization efforts and

since the regulations for government procurements are not generally
compatible with an entrepreneural
endeavor, NASA does not anticipate
sponsorship of commercialization
work through contracts. Rather
NASA is pioneering a concept of
joint investigations and joint projects with industry.

4.1. Risk Sharing and Incentives

In these joint activites, NASA and
interested, qualified commercial
organizations enter into "construc-
tive partnerships" as equals who
have sufficient motivation toward
common objectives to make independent commitments and to share in the
risk and benefits. Activities are
selected across the spectrum of
materials processes to demonstrate
that the low-g environment is a
valuable tool for isolating and
characterizing gravitational effects
on ground-based material processes,
or for actually producing unique
materials in space for commercial
application. Since market incentives are presently inadequate to
bring about technological innovation
based on low-g technology, under its
joint activities program, NASA can
provide certain tangible incentives
such as: (1) providing flight time
on the Space Transportation System
(Space Shuttle) on appropriate terms
and conditions, (2) providing technical advice, consultation, data,
and use of facilities and equipment,
and (3) entering into joint research
and development programs where each
party funds its own participation.

Joint activities may range from exchanges of technical information
and collaboration on low-g ground-
based and flight experiments to joint
projects (Joint Endeavors) to develop
a marketable product.

4.2. Terms and Conditions

As stated in the "Federal Register"
of August 14, 1979, (pages 47650 and
47651), the terms and conditions of
joint efforts are negotiable; this
includes such factors as the firm's
right to proprietary data and/or
patent ownership, provisions for a
form of exclusivity in special cases,
and recoupment of NASA's contribution
under appropriate cirucmstances. Until such time as normal market incentives can provide a competitive
and self-sustaining condition in the
private sector for materials processing in low-g technology, the
incentives and negotiability of terms
and conditions will be used as a
stimulus to encourage the more technologically advanced, entrepreneur-
ally inclined firms to pursue
applications of low-g technology.

5. NASA/INDUSTRY WORKING
RELATIONSHIPS

NASA has developed three basic levels
of working relationships with private organizations. These provide
the flexibility needed to meet a
wide range of needs from large organizations with strong research departments to small entrepreneural firms
that want to develop a product for
the market. They also provide for
incremental increases in understanding and commitment by the parties.

In all cases, the Government does
not fund any of the work done by
the firm, but rather each party
funds its own activities separately.

5.1. Technical Exchanges

For companies interested in the
application of low-g technology, but
not ready to commit to a specific
space flight experiment or venture,
a Technical Exchange Agreement (TEA)
has been developed. Under a TEA,
NASA and a company agree to exchange
technical information and cooperate
in the conduct and analysis of on-
going ground-based research programs.
In this way, a firm can become familiar with microgravity technology and
its applicability to the company
product line at minimal expense.
Under the TEA, the private company
funds its own participation, and
derives direct access to and results
from NASA facilities and research,
with NASA gaining the support and
expertise of the industrial research
capability. The first two TEA's are
expected to be formalized in the
near future, one with a major metal
producer and another with an equipment manufacturer.

5.2. Industrial Guest Investigators

Another joint activity is the Industrial Guest Investigator (IGI),
where NASA and industry share
sufficient mutual scientific interest that a company arranges for
one of its scientists to collaborate,
again at company expense, with a NASA-
sponsored Principal Investigator
on a space flight experiment. Once

the parties agree to the contribution
to be made to the objectives of the
experiment, the IGI becomes a member
of the investigation team, thus
adding industrial expertise and in-
sight to the experiment. The first
IGI agreement was signed in May of
1980, with TRW Equipment Division
in Cleveland Ohio, for work on solidification experiments.

5.3. Joint Endeavors

Joint Endeavors provide a mechanism
for NASA to share in the cost and
risk of early space projects. The
first Joint Endeavor Agreement between NASA and McDonnell Douglas
Astronautics Company was signed in
January 1980, and illustrates the
key features of this third working
relationship. McDonnell Douglas will
use the Space Shuttle to develop and
demonstrate the technology of contin-
ous flow electrophoresis under low-
gravity conditions and to ascertain
the applicability of that technology
to the production of pharmaceutical
products. The agreement requires a
substantial investment by McDonnell
Douglas and its assocaties in each of
three sequential phases: (1) feasibility studies and planning, (2) flight
experimentation and technology development, and (3) applications demo-
strations. In return for McDonnell
Douglas's promise to make results of the work available to the
U.S. public on reasonable terms and
conditions, NASA agrees to refrain
from entering into similar joint endeavors or international cooperative

agreements directly related to the development of commercial devices and processes which would compete with those resulting from the Mc-Donnell Douglas endeavor. However, NASA is not precluded from selling flight time on the Shuttle to any other organization wanting to conduct the same or similar experiments. Significantly, NASA will not acquire rights in inventions made by Mc-Donnell Douglas or its associates in the course of the joint endeavor unless McDonnell Douglas fails to exploit the inventions or terminates the agreement, or unless the NASA Administrator determines that an emergency exists. Additional joint endeavors are being pursued. For example, detailed discussions have been held with a small business firm regarding electroeptaxial growth of semiconductor cyrstals. Agreements with small business concerns require attention to means for minimizing financial risk and controlling negative cash flow. Joint Endeavor Agreements offer great prospects for commercial exploitation of space technology.

6. CONCLUSION

The unique characteristics of microgravity and existing space technology combine to offer commercial organizations, and through them the public, potential significant benefits in terms of new or improved materials and processes. It is not possible at this time to predict exactly what products will be produced or how they will be produced. This will be largely determined by the ingenuity of the people doing research on materials processing in low-g, the imagination of the entrepreneurs, and the laws of economics. It is NASA's challenge to create the proper "climate" for this to occur by stimulating research on the ground to develop new ideas for flight experiments, by developing ground and flight hardware that is responsive to user needs in controls and diagnostics, by providing easy access and multiple research and product development opportunities in a low-g environment for qualified individuals and firms, and by developing satisfactory patent and data protection for firms to ensure them a reasonable chance of return commensurate with their financial risk. While interest to date by private organizations has been limited, substantial progress is being made under NASA/Industry joint investigations and joint ventures to demonstrate the usefulness of low-g technology from both the technical and business standpoints.

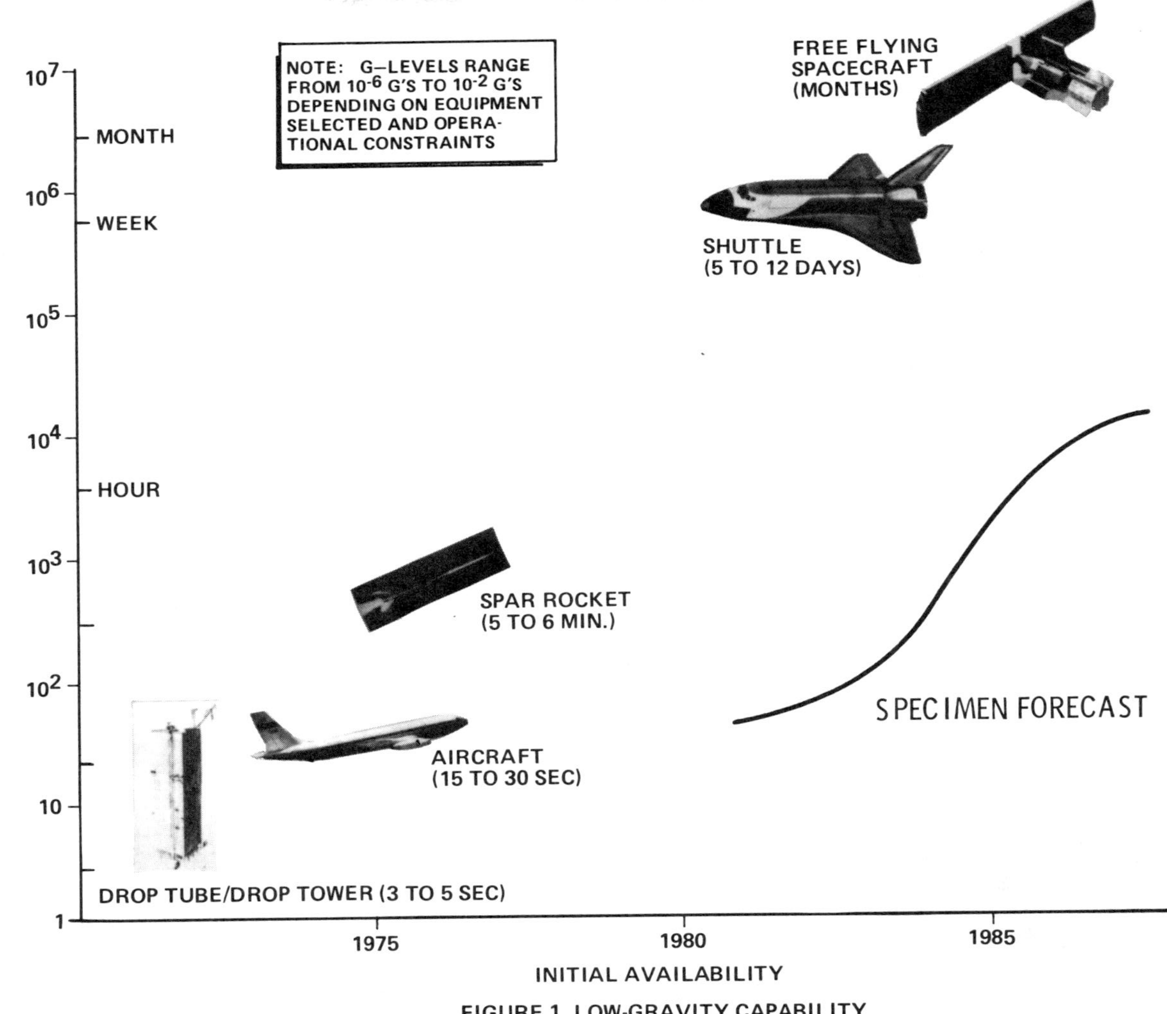

FIGURE 1 LOW-GRAVITY CAPABILITY

This microphotograph (at 350X) of a surface of a germanium crystal shows compositional striations due to convection in the melt when grown in one gravity. It was taken into space, partially melted and resolidified in low gravity. The more homogeneous composition due to the growth in the absence of convection can be seen.

The specimen of lead shown at the right has a small amount of gold incorporated into one end to demonstrate the diffusion of gold into lead when melted. The micrograph taken of the sample melted in one gravity shows that convective flow has dispensed gold throughout the specimen. The micrograph taken of the sample melted in low gravity shows tin diffusion layers progressing through the sample. In microgravity, diffusion controlled processes, undisturbed by convection, are possible.

Crystals grown in one gravity often contain edge defects and dislocations due to imposed stresses. Crystals grown in low gravity appear to have less defects, fewer dislocations, and sharper edges. The material shown is germanium-selenium (GeSe).

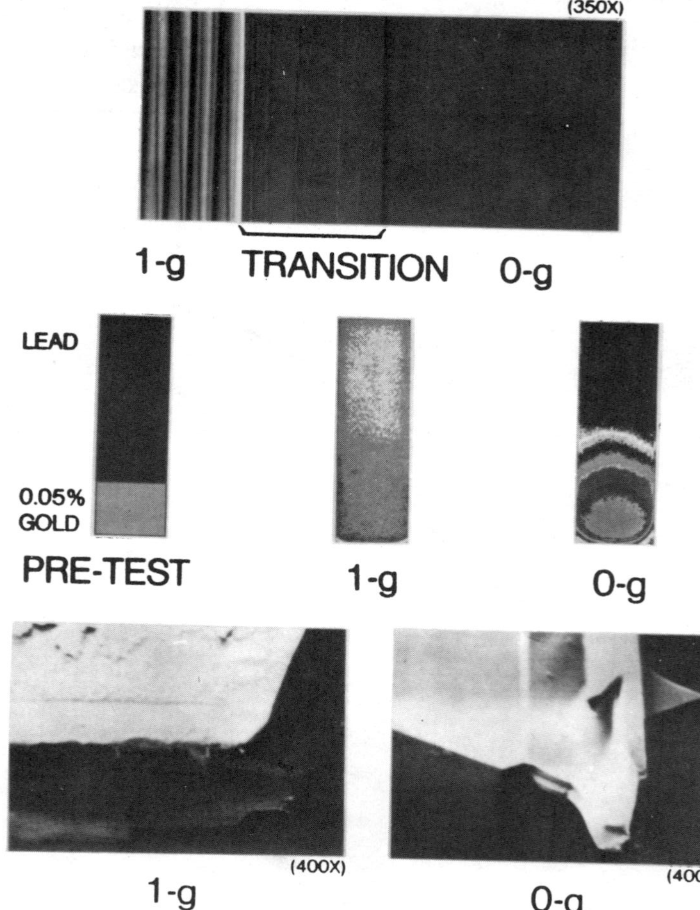

FIG. 2 CONVECTIVE EFFECTS

572

A directionally solidified eutectic composition (natural two-phase solid which grows short fibers) can grow long, continuous fibers in the absence of convection at low gravity. Shown here is a demonstration of this phenomea, using sodium chloride-lithium-fluorine (NaCl-LiF).

In one gravity, convection causes the hot products of combustion to rise and new reactants to be supplied to the flame. In low gravity, a thin layer of flame can be initiated at the surface of the fuel where it contacts the oxidizer and can be carefully controlled, thus providing a unique opportunity to study combustion phenomea.

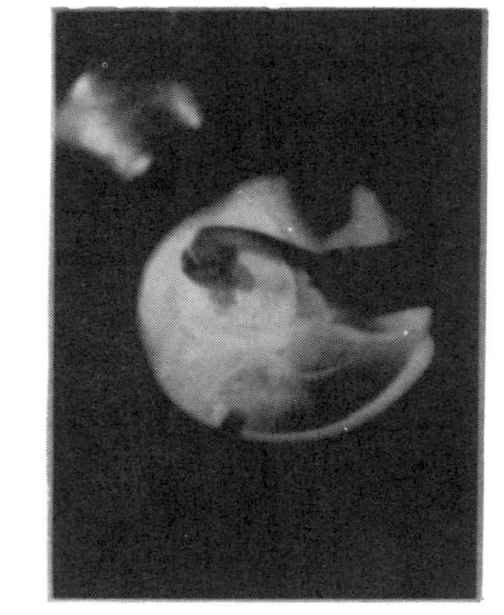

FIG. 3 CONVECTIVE EFFECTS

A number of materials, such as immiscible semiconductors, segregate when solidified in one gravity. In low gravity, phase separation and segregation can be better controlled. This has been demonstrated, using aluminum-antimony (AlSb) as shown here.

Under one-gravity conditions, low density materials will migrate toward the surface of a liquid (or melt). In low g, lighter density materials will remain in suspension for indefinite periods of time, thus removing a key constraint in the processing of composites and alloys where the constitutents have large density differences. Also, materials which may be damaged by mechanical stirring can be processed in low-g.

In this solidification experiment, ammonium chloride (liquid is light and crystals are dark) was used to simulate a casting process. Under one-gravity conditions, the dendrites formed and pieces broke-off and settled to the bottom of the container, thus producing weaknesses in the sample. In low g, the dendrites formed, grew uniformly and did not breakoff. This resulted in significant improvements in the sample.

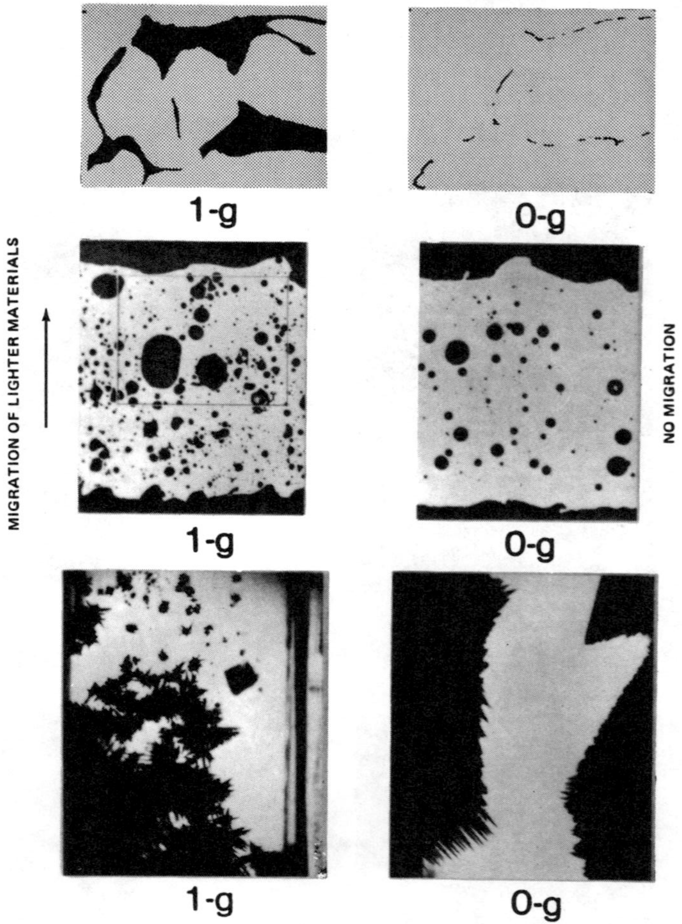

FIG. 4 SEDIMENTATION/BUOYANCY EFFECTS

This highly purified single crystal
deformed along the shear planes
under its own weight during processing in one gravity. Large crystals
which have weak internal bonding
strengths can be made in microgravity
without deformation and with fewer
defects.

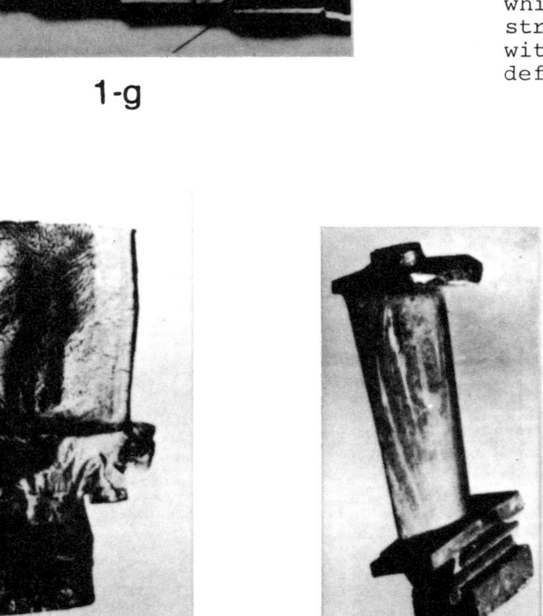

Finished machined parts can be
melted and resolidified in space
without significant deformation
when contained only by a thin
skin of ceramic material as shown
in the photographs at the left.

FIG. 5 GRAVITY INDUCED DEFORMATIONS

In the float zone process a bar of polycrystalline material is moved through the
hot zone of a furnace in such a way that
a thin molten zone moves along the bar
and a single crystal is formed in the
material solidified behind the zone.
In low gravity, surface tension can
hold a much larger molten zone in
place, even if it is rotating rapidly
as shown in the photograph at the far
right.

Gravity necessitates containment of a
liquid (or melt) in a crucible. The
crucible can contaminate the melt or
produce unwanted nucleation sites, resulting in undesirable optical or physical properties in the material as
illustrated in the photograph at the
right.

In microgravity, containerless processing
can be done by using weak acoustic,
electromagentic or electrostatic fields
to position, manipulate and form molten
materials. The picture at the far right
shows a hologram of acoustic standing
waves containing a glass sample being
melted without the use of a container.

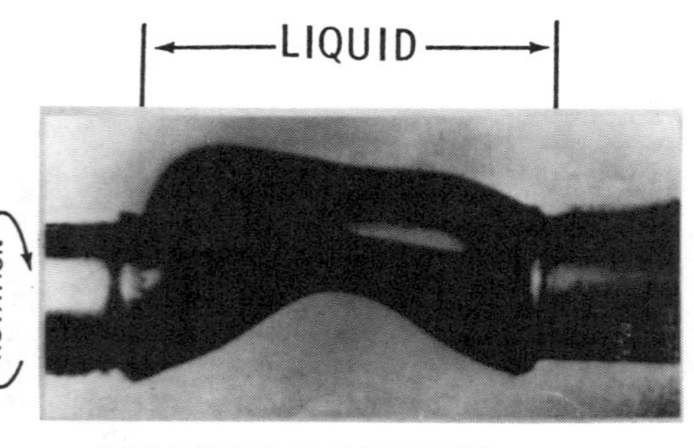

TYPICAL 1—g
FLOAT ZONE PROCESS

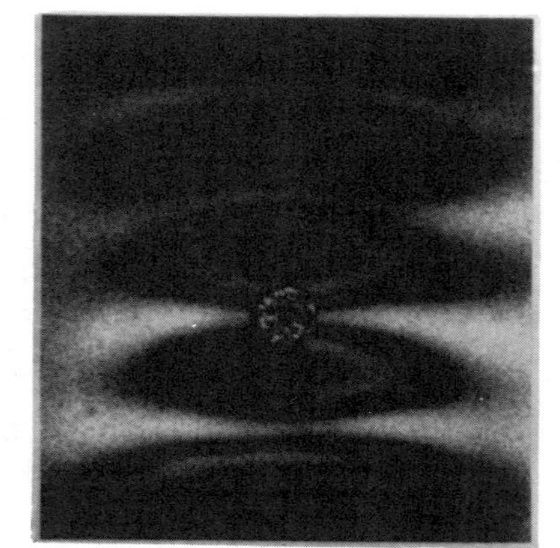

SIMULATED FLOAT ZONE
IN LOW— g

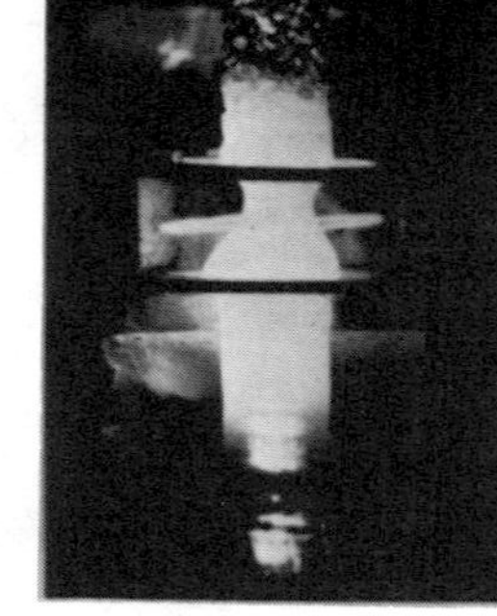

- CRYSTAL GROWTH AND SOLIDIFICATION
  - SOLID SOLUTION IR DETECTORS ($HgCdTe$, $PbSnTe$)
  - VAPOR GROWTH (HgI_2, ALLOY TYPE)
  - SOLUTION GROWTH (TGS, GROWTH ENVIRONMENT VS MORPHOLOGY)
  - FLOAT ZONE (MARANGONI CONVECTION, RADIAL SEGREGATION, INTERFACIAL STABILITY)

- METALLURGICAL MATERIALS AND PROCESSES
  - IMMISCIBLE ALLOYS
  - MAGNETIC COMPOSITES
  - METAL FOAMS
  - HIGH G/R SOLIDIFICATION
  - SOLIDIFICATION AT EXTREME UNDERCOOLING

- COMPOSITES
  - CASTING OF DISPERSION STRENGTHENED ALLOYS
  - SOLID ELECTROLYTES WITH DISPERSED ALUMINA
  - PARTICLE PUSHING BY SOLIDIFICATION INTERFACES

- GLASSES
  - GLASS FINING
  - LASER HOST GLASSES
  - OPTICAL GLASSES WITH UNIQUE PROPERTIES
  - METAL GLASSES

- CHEMICAL PROCESSES
  - MONODISPERSE LATEXES (POLYSTYRENE MICROSPHERES)
  - STABILITY OF FOAMS AND SUSPENSIONS
  - COLLODIAL INTERACTIONS
  - HIGH TEMPERATURE PROPERTIES OF REACTIVE MATERIALS
  - DIFFUSION CONTROLLED SYNTHESIS

- SEPARATION SCIENCES
  - HIGH VOLUME—HIGH RESOLUTION ELECTROPHORESIS CELL SEPARATION
  - PROTEIN PURIFICATION BY CONTINUOUS FLOW ISOELECTRIC FOCUSSING

- FLUID STUDIES
  - NON—BUOYANCY DRIVEN CONVECTIONS
  - WETTING AND SPREADING STUDIES
  - ROLE OF CONVECTION IN PROCESSES (ELECTROKINETIC SEPARATION, ELECTROPLATING, CORROSION, ETC.)

FIG. 7 MATERIALS PROCESSING IN SPACE PROGRAM—CURRENT AREAS OF RESEARCH

26th National SAMPE Symposium
April 28-30, 1981

ICAM "MANUFACTURING COST/DESIGN GUIDE" (MC/DG) BACKGROUND, METALLIC AND NONMETALLIC DEMONSTRATION SECTIONS

Bryan R. Noton
Battelle's Columbus Laboratories
Columbus, Ohio
and
Capt. Richard R. Preston
Materials Laboratory, AFWAL/MLTC
Wright-Patterson Air Force Base, Ohio

Abstract

The "Manufacturing Cost/Design Guide" (MC/DG) is under development for the Computer Integrated Manufacturing Branch, Materials Laboratory, WPAFB, Ohio. The objective is to enable aircraft designers to achieve lowest cost by conducting trade-offs between manufacturing cost and design factors, such as minimum weight. The model of the MC/DG has been developed as well as sections for sheet metal, aerospace sheet metal discrete parts, mechanically-fastened assemblies, advanced composite fabrication, castings, test, inspection and evaluation (TI&E), and also for certain emerging manufacturing technologies. The manufacturing operational sequences have been studied for base parts; i.e., structural elements in their simplest form, and also with designer-influenced cost elements (DICE) such as cut-outs and heat treatment. It is therefore, also possible to identify the cost-driving operational sequences which are candidates for computer-aided manufacturing (CAM). In the case of composites, typical DICE are fiber materials, number of plies, and quality requirements.

The paper presents various designer-oriented formats, for composites and metals, which indicate the influence and magnitude of up to four cost-drivers on a single designer-oriented format. The methodology to utilize these formats will be demonstrated in subsequent papers in these Proceedings. The MC/DG is being developed by a team, consisting of BCL, as prime contractor and the following airframe companies, General Dynamics, Grumman, Lockheed, Northrop and Rockwell International.

INTRODUCTION

The need for the Air Force to arrest
and reduce costs at all levels of
the aircraft system life-cycle is
becoming increasingly important.
Qualitative and quantitative data
and other information on cost drivers useful during the design, manufacturing, operation, and maintenance of aircraft systems are
essential. This is particularly so
because, at the same time, the
number of new aircraft types being
developed for the Air Force inventory is reducing while there is a
need for increased performance
meaning reduce weight, better
quality, lower ownership costs, but
with less energy consumption. The
Air Force must therefore be provided with performance which is affordable (Figure 1).

Strong interaction between design
and manufacturing is essential to
achieve this required advancement
in manufacturing sophistication and
refinements. The "Manufacturing
Cost/Deisgn Guide" (MC/DG) Data
Development Program, a follow-on of
earlier contracts (F33615-75-C-5194
and F33615-77-C-5027) provides an
unprecedented opportunity for the
designer to study a larger number,
e.g., 10, of alternative design
configurations for airframes to
achieve the lowest manufacturing
cost. These contracts are published
in AFML-TR-76-227 and AFWAL-TR-80-
4115.

The MC/DG identifies the cost drivers over which the designer has
control and which he can trade back
for performance once the basic performance requirements of the system
have been exceeded. An example is
composite materials with certain
design configurations; for example,
for complex fuselage sections, it
is possible to show significant
weight savings over traditional
aluminum structures. However,
having met the performance requirements, the designer may desire to
trade back some weight savings to
achieve cost savings. The MC/DG
also provides information to promote interaction between manufacturing and design, for example, regarding alternative facilities due to
shop loading requirements.

OBJECTIVES OF THE MC/DG

The objectives of the MC/DG are to:
- Provide to designers urgently-
  needed, quick, simple, and
  quantitative cost comparisons of
  manufacturing processes
- Emphasize design orientation of
  MC/DG formats and manufacturing
  man-hour data for use at all
  phases of design process, i.e.,
  preliminary and detail design,
  therefore, increasing emphasis
  on cost as a vital design parameter
- Enable more extensive manufacturing cost trade-offs to be conducted on airframe components and
  aerospace electronics fabrication

and assembly

- Emphasize potential cost advantages of emerging materials and manufacturing methods accelerating the transfer to production hardware of these technologies
- Guide the designer to the lowest cost manufacturing process early in the design phase to avoid cost drivers
- Identify cost-driving manufacturing operational sequences which provide targets for future computer-aided manufacturing (CAM) efforts.

Design/Manufacturing interaction is, of course, crucial in the development of the MC/DG and when the MC/DG is complete, it will serve as a unique and valuable tool to achieve and maintain this interaction.

MANUFACTURING COST/DESIGN GUIDE AND ITS UTILIZATION

The contents of the MC/DG volume are shown in Table 1. It will be noted that the costs (man-hours) are presented in formats for the following major categories.

- Procured items
- Material removal
- Detail fabrication
- Material treatment
- Permanent joining methods
- Assembly methods

The contents include both metallic and nonmetallic materials.

A format selection aid for the MC/DG sections is shown in Figure 5.

DESIGNERS' MC/DG WORKSHEET

To determine the total program costs for both discrete parts and assemblies, an MC/DG cost worksheet has been prepared and can be used by industry. This worksheet is shown in Table 2.

MC/DG utilizations being reported by the Air Force are reflecting positive results in the area of cost savings while at the same time allowing the user to accomplish more in the same period of time. Example MC/DG uses being reported are:

Rockwell International's North American Aircraft Division used the MC/DG to conduct a trade study during the replacement of an aluminum isogrid shell structure door that had been designed as a baseline for the MX Missile Stage 4 shell structure. A design engineer, utilizing the cost-driver effect (CDE) and cost estimating data (CED) formats in the MC/DG, performed a cost trade analysis of ten different advanced composite designs. It is estimated that the trade study was conducted in eight hours using the MC/DG. With the normal method of working with manufacturing engineering and cost estimators, the task would have required approximately 40 hours of labor and a calendar span of one week for turn-around.

The MC/DG "Casting Section" was used
by Rockwell International's North
American Aircraft Division to aid in
the purchase of a large sand casting
to be used on a Liquid Rocket Engine
Propulsion System. The use of the
casting format proved to be successful in providing a casting price
which coincided approximately with
the actual price paid.

Lockheed-California Company used the
"Advanced Composites Fabrication
Section" of the MC/DG to make manu-
facturing cost comparisons of Hat vs.
"J" structural members in a compres-
sion panel stiffness study proposed
to the Naval Air Development Center
(NADC).

The Advanced Design Department of
Lockheed-California Company is currently using the MC/DG "Advanced
Composite Fabrication Section" for
determining manufacturing cost trade
off comparisons of flap structured
configurations.

The Air Force states in "The ICAM
Program Report" Vol. 2, No. 4, March/
June 1980 that incoming reports of
successful MC/DG applications are
indicative of savings in both time
and dollars. These reports are
important information sources for
the Air Force as well as industry,
reflecting the performance of the
MC/DG project to date.

DESIGNER-ORIENTED FORMAT
DESIGN CRITERIA

The designer-oriented formats developed were reviewed by interdisciplinary groups in the aerospace
industry.

Furthermore, designer surveys were
conducted and the feedback received
on the MC/DG was as follows:

- Must be simple whenever possible
- Must not be time consuming to
  use in the design process
- Complicated calculations should
  be avoided
- Manufacturing data are urgently
  needed but with designer orientation
- No single airframe company can
  provide all manufacturing cost
  data required due to varying
  expertise
- Designers are more concerned that
  it is the lowest cost rather than
  what it costs, i.e., qualitative
  comparisons are important.

The CDE and CED formats must therefore meet the following criteria:

- Emphasize cost drivers
- Be simple to use
- Use designer language
- Instill confidence
- Be economical
- Be accessible
- Be maintainable.

METHODOLOGIES FOR PRESENTING
MANUFACTURING DATA

The manufacturing man-hour data for

aerospace materials discrete parts
and assemblies, and manufacturing
technologies are presented in two
ways. Firstly, cost-driver effects
(CDE) and, secondly, cost-estimating
data (CED) are shown. The objectives of the CDE and CED methodologies are:

- To develop a simple approach for
  the use of formatted data by designers to achieve lowest manufacturing costs during all design
  phases (CDE and CED)
- To provide qualitative cost guidance to the designer to assure
  lowest manufacturing cost (CDE)
- To provide the designer with the
  capability through quantitative
  guidance to perform simple trade-
  offs on manufacturing costs (CED).

The CDE cost relationships, providing qualitative information, have
the following objectives:

- Identify cost drivers that increase the manufacturing cost of
  the design
- Show relative effects of cost
  elements over which designers
  have control
- Motivate designers to reduce the
  impact of the cost drivers by
  designing around them.

Using the CDE approach, the designer
should realize the lowest cost while
satisfying the performance requirements, e.g., airframe weight and
durability.

The CED cost relationships, providing quantitative information, have
the following objectives:

- Provide designers with manufacturing man-hour data to allow
  trade-offs to be quickly performed to achieve comparative
  cost for candidate structural
  configurations
- Motivate designers to conduct
  trade-offs through the use of
  designer-oriented formats and
  manufacturing man-hour data in
  the MC/DG.

Figure 3 summarizes the above.
Examples of CDE and CED formats are
shown 7 through 12. However, Figure 10 shows the detail of part definition to determine the manufacturing man-hours.

THE MC/DG IN EDUCATION

One of the factors that will
require university graduates to
play an increasingly important part
is shown in Figure 4. This chart,
courtesy of Mr. R. H. Hammer, Boeing
Commercial Airplane Company, shows
the experience distribution of aerospace industry designers as a function of age. The theoretical curve
implies that when an engineer retires
a new person would join the company.
This allows time for the inexper-
ienced designer to develop and gain
knowledge from the seasoned designers
he is associated with. The optimum
curve takes into account early re-
tirement and persons transferring
from the aerospace industry to other

industries. The problem is that
the actual situation is not represented by this optimum curve. This
is caused by basically two factors.
One factor is the large influx of
engineers that occurred durring
World War II and the other is that
during layoffs, such as experienced
during the late 1960's, and to some
extent, in recent years, the last
persons entering the aerospace industry were the first ones released.
As the curve shows, the average age
of designers is approximately 55
years. Furthermore, many experienced
engineers are considering early re-
tirement within the next few years
and unless some method is developed
to transfer the vast amount of knowledge acquired by retiring designers
over the years to less experienced
designers, a valuable resource will
be lost. The MC/DG is one means of
documenting and retaining this
experience thus achieving the
needed transfer of design and
manufacturing knowledge.

Benefits of the MC/DG to university
professors and students are summarized below:

To the Professor

- Provides a realistic, easy-to-
  use source of manufatturing
  cost information for aerospace
  discrete parts and subassemblies
- Provides generally applicable,
  up-to-date source of information,
  as opposed to specific information from the brochures of vendors
- Facilitates the alignment of
  tehoretical courses to industry
  staffing requirements by enabling
  structural performance/manufac-
  turing cost trade studies to be
  conducted in the classroom
- The computerized MC/DG will provide an additional dimension to
  computer activities in engineering schools.

To the Student

- Introduces students to systematic
  methodologies for performing
  trade studies
- Teaches student the impact of
  manufacturing technology selection, comparative costs, and
  manufacturing facility requirements
- Familiarizes students with the
  use of manufacturing cost data
  at all stages of the design
  process
- Aids students in the transition
  from the classroom or laboratory
  environment to industry.

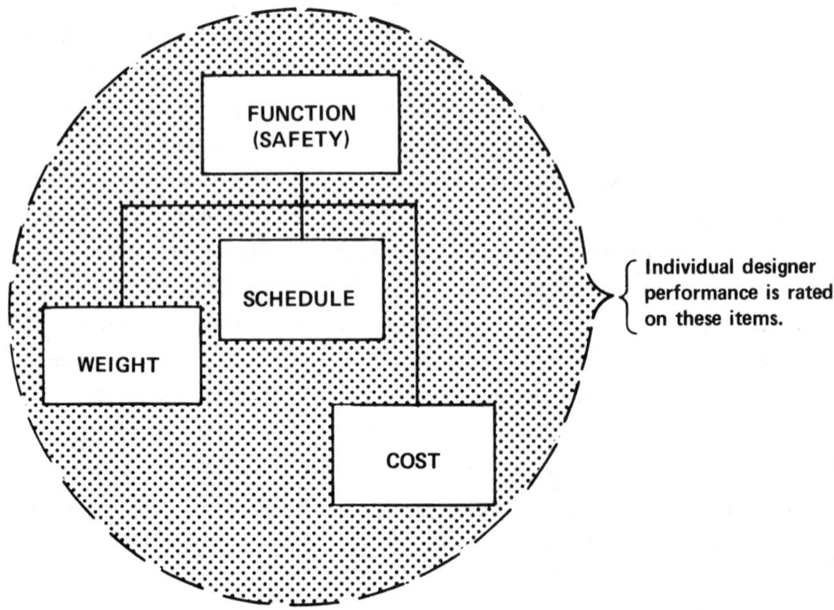

Fig. 1.   Present Aircraft Design Team Priorities

MC/DG VOLUME CONTENTS: "MANUFACTURING TECHNOLOGIES FOR AIRFRAMES"

I	II	III	IV	V	VI
PROCURED ITEM COSTS	MATERIAL REMOVAL COSTS	DETAIL FABRICATION COSTS	MATERIAL TREATMENT COSTS	PERMANENT JOINING* COSTS	ASSEMBLY* COSTS
• Forgings Hand Conventional Blocker Precision • Castings Sand Permanent Mold Investment Die Casting • Extrusions • Materials • Fastener Systems • Emerging Proc. Isothermal Forging Powdered Metal Pultrusion HIP Healing CAM	• Machining Turning Milling Drilling • Chem Milling • EDM • ECM • Emerging Proc. Laser Fluid-Jet CAM EB Cutting	• Metallic Forming Cutting • Non-Metallics Forming Cutting Molding Laminating • Emerging Proc. Superplastic Forming Flow Forming Hydrostatic Forming Thermoplastic Forming CAM	• Heat Treatment • Surface Treatment • Emerging Proc. Laser Treating Nonenvironmental Polluting Treatments	• Welding • Adhesive Bonding • Brazing • Emerging Proc. Diffusion Bonding Weld Bonding Laser Welding Ultrasonic Welding Plasma Arc *Subassembly	• Metallic Assy. Mechanical Fastening • Non-Metallic Assy. Mechanical Fastening • Emerging Proc. Bimetallic Rivets Microwave Curing *Major and Final Assembly

Categories = e.g., Material Removal
Sections = e.g., Machining
Subsections = e.g., Turning and Milling

Table 1. MC/DG Volume Contents "Manufacturing Technologies for Airframes"

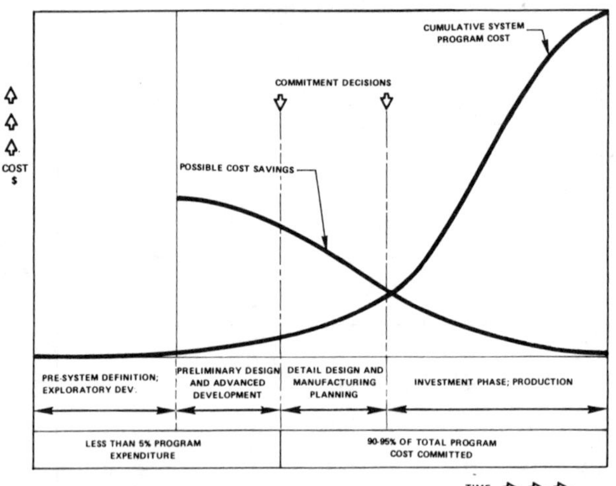

Fig. 2.   Decreasing Cost-Savings Leverage as Program Progresses

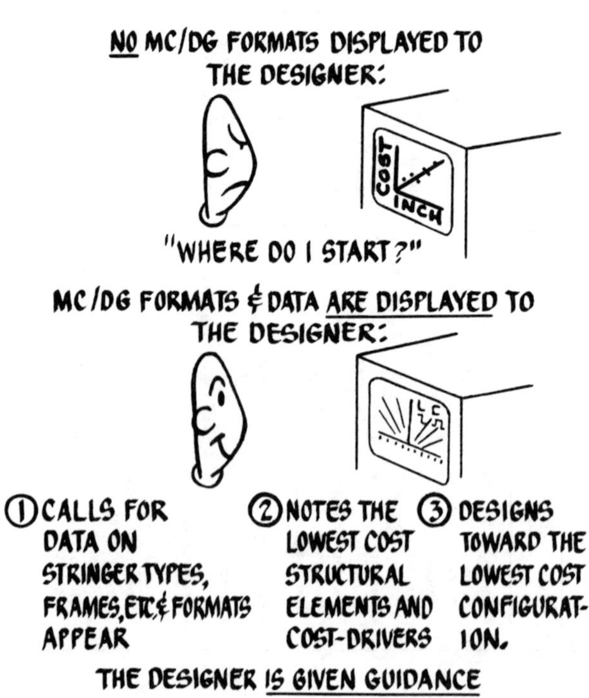

Fig. 3.   Cost-Estimating Vs. MC/DG Designer-Oriented Formats

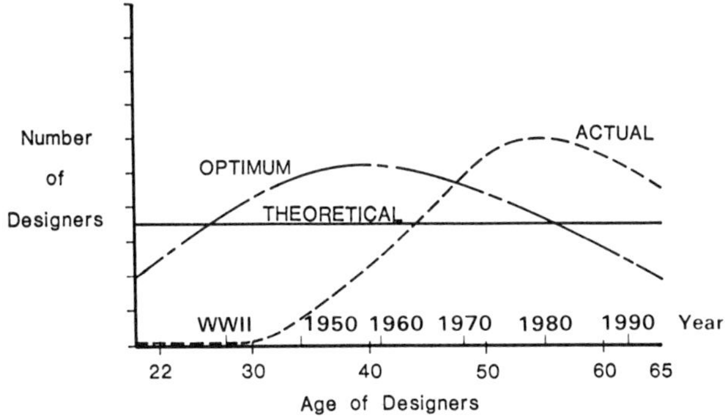

Fig. 4. Experience Distribution of Aerospace Designers

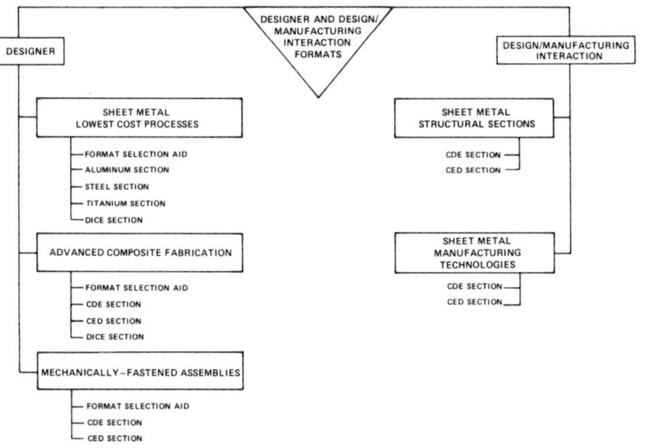

Fig. 5. An MC/DG Selection Aid for Formats

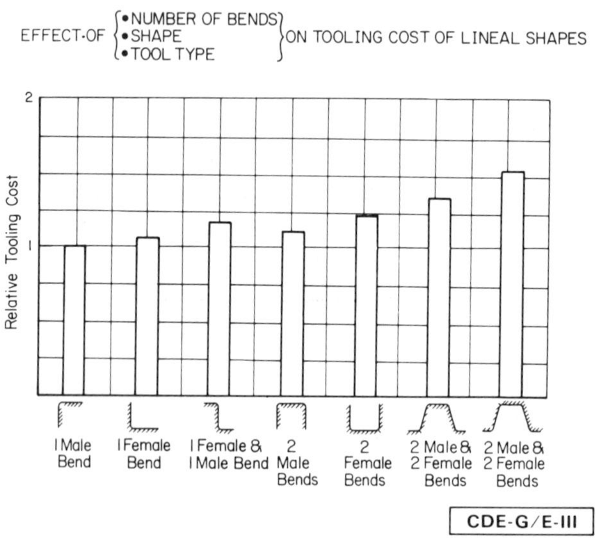

Fig. 11.   Format Showing Cost-Driver Effects
for Composite Tooling

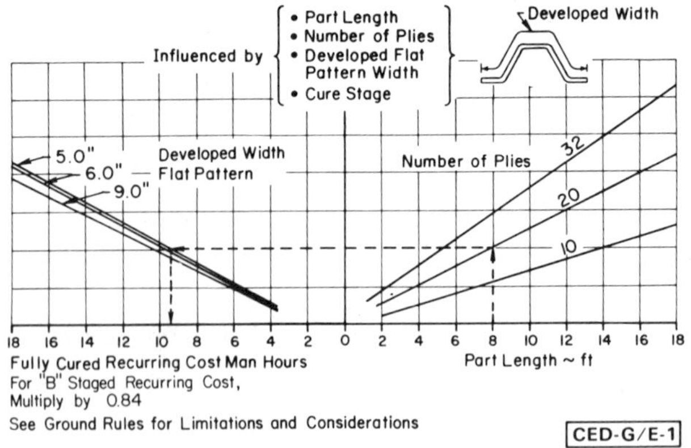

Fig. 12.   Typical Cost-Estimating Data Format
for a Lineal Structural Shape

SAMPLE PARTS USED TO DERIVE COST DATA FOR MC/DG

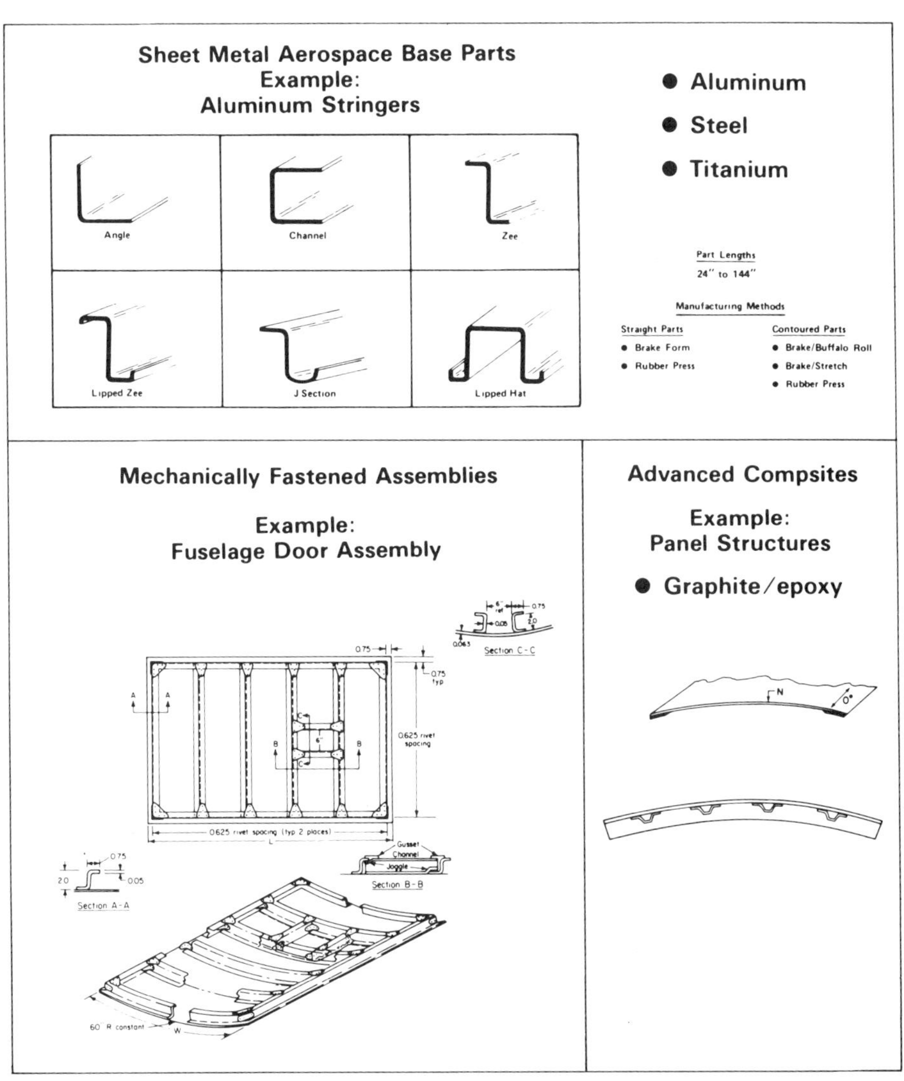

Fig. 6. Sample Parts Used to Derive Man-Hour Data for MC/DG

MATERIAL	BASE PART MANUFACTURING METHOD	STANDARD JOGGLE	FLANGED HOLES	BEADS	HEAT TREATMENT	SPECIAL FINISH	SPECIAL TOLERANCE	LINEAL TRIM	END TRIM	CUTOUTS W/O FLANGES
ALUMINUM	BRAKE FORM	L	L	X	H	L	H	L	L	L
	BRAKE/BUFFALO ROLL	L	L	X	H	L	H	A	L	A
	BRAKE STRETCH	L	L	X	H	L	N	A	A	A
	DIE FORM	N	N	N	N	L	N	L	L	L
	DROP HAMMER	N	N	N	L	L	H	L	X	A
	FARNHAM ROLL	X	L	X	L	L	H	L	X	A
	ROUTED FLAT SHEET	X	L	X	L	L	H	L	X	L
	RUBBER PRESS	N	N	H	N	L	A	L	L	L
	STRETCH FORM	X	L	A	N	L	N	A	X	A
	YODER ROLL	L	L	X	H	L	H	A	A	A
	YODER STRETCH	L	L	H	N	L	N	A	L	A
TITANIUM	BRAKE FORM R.T.	A	L	X	X	L	H	H	H	L
	R.T. BRAKE/HOT STRETCH*	A	L	X	X	L	L	H	H	H
	CREEP FORM*	X	L	X	X	L	L	H	H	H
	FARNHAM ROLL	X	L	X	X	L	H	H	H	H
	HOT PRESS*	N	L	N	X	L	L	N	N	L
	PREFORM/HOT SIZE*	N	L	N	X	L	L	N	N	L
STEEL	BRAKE AND BUFFALO ROLL	A	L	X	N	L	H	H	A	L
	BRAKE FORM R.T.	A	L	X	N	L	H	L	L	L
	BRAKE/R.T. STRETCH	A	L	X	N	L	A	H	L	A
	FARNHAM ROLL	X	L	X	N	L	H	H	L	A
	RUBBER PRESS	N	N	N	N	L	A	L	L	L
	STRETCH FORM	X	L	X	N	L	A	H	A	L

LEGEND

RATING	
X	NOT APPLICABLE
N	NO ADDITIONAL COST INCL. IN BASE PART COST
L	LOW ADDITIONAL COST
A	AVERAGE ADDITIONAL COST
H	HIGH ADDITIONAL COST

Percentage Cost Ranges For Above

L	Up to 10%
A	10–30%
H	Above 30%

*Denotes one or more elevated temperature processing steps.

DICE-0

Fig. 7. Guide to Designer-Influenced Cost-Elements (DICE)

TITANIUM FRAME, LOWEST COST PROCESS
HOT PRESS

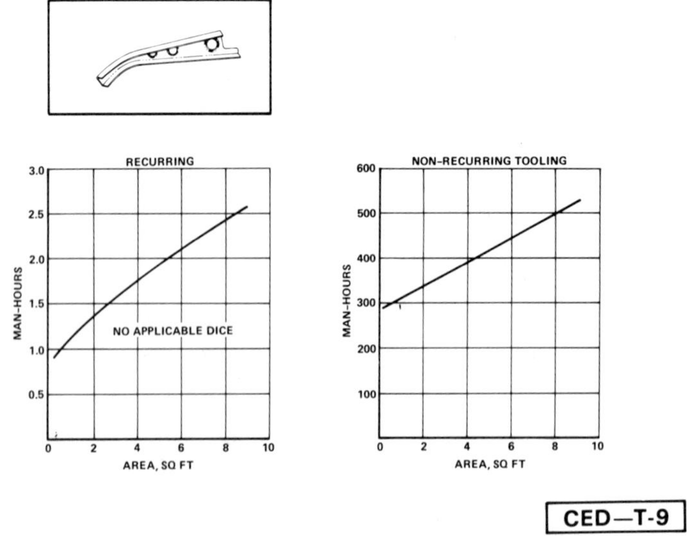

Fig. 8. Typical Format for Sheet Metal Showing Lowest Cost Process

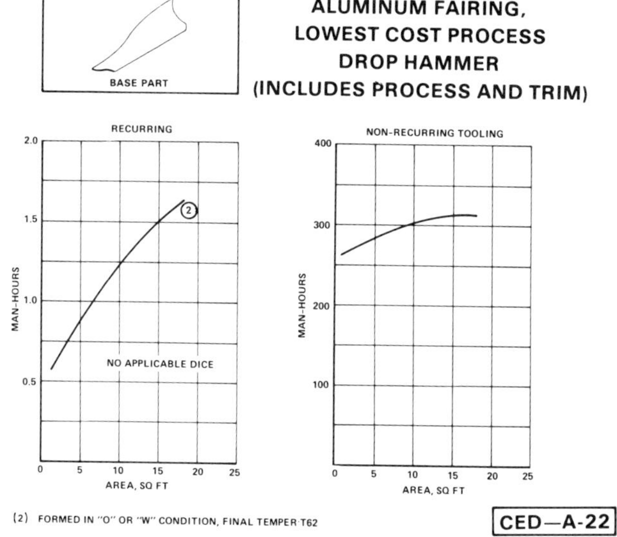

Fig. 9. Typical Format for Sheet Metal Fairing

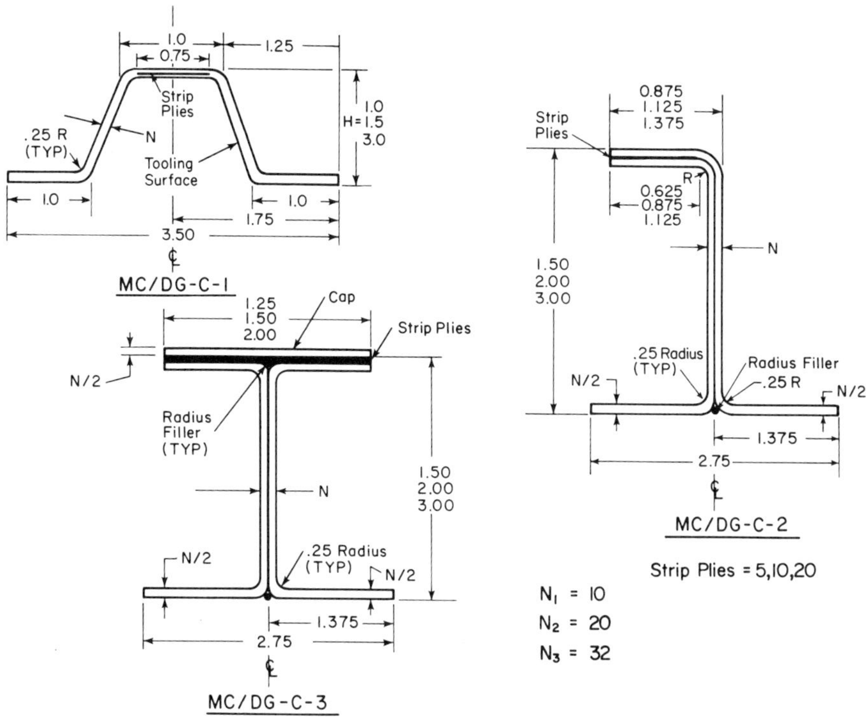

Fig. 10. Composite Parts Studied to Develop Man-Hour Data

591

MC/DG COST WORKSHEET

PAGE ________   ________

| DESIGN CONCEPT | | RECURRING COST (RC) | | | | | | | | | NON-RECURRING COST (NRC) | | | PROGRAM COST | | |
| | | L . LC . LR = L$ + M$ = RC . P/AC . DQ = PRC | | | | | | | | | NRC . LR = PNRC | | | 9 + 12 ÷ DQ = COST/AC | | |
PART NO.	DESCRIPTION	LABOR MC/DG MH/PT (1)	LC FACTOR (2)	LABOR RATE $/MH (3)	LABOR RC $/PT (4)	MAT'L $/PT (5)	REC. COST/ PT. $ (6)	PARTS PER AC (7)	DES. QTY. (8)	PROG. RC $ (9)	NRC MC/DG MH (10)	LABOR RATE $/MH (11)	PROG. NRC $ (12)	PROG. COST $ (13)	DES. QTY. (14)	COST/ AC $ (15)
	TOTALS															

BY: ________________________________

DATE: ______________________________

Table 2.   The MC/DG Designers' Cost Worksheet

UTILIZATION IN INDUSTRY OF
ICAM "MANUFACTURING COST/DESIGN GUIDE" (MC/DG)
FOR COMPOSITE STRUCTURES

Dean S. Klivans
Rockwell International Corp., NAAD, Los Angeles, CA
Bryan R. Noton
Battelle's Columbus Laboratories, Columbus, OH

ABSTRACT

Utilizing the MC/DG, trade-studies have been conducted on advanced composite fuselage shear-panels. The configurations included light weight/high complexity; moderate weight/moderate complexity; and high weight/low complexity configurations. The panels were of single curvature and were assembled using mechanical and cocuring methods. The graphite/epoxy selected was AS/3502-6. The configurations studied included hat and J-sections for the stringers, and J-sections for the frames, and the manufacturing man-hours were determined for each. The results are presented in diagrams indicating the weight and manufacturing man-hours for each configuration. The MC/DG formats were used to develop the design employing a step-by-step approach to support the cost of the concept. The final cost was accurate, and the objectives of the MC/DG program were demonstrated.

1. OBJECTIVES OF AIRFRAME TRADE STUDIES

A series of fuselage shear panels was analyzed with regard to weight and manufacturing cost by three airframe industry team members, utilizing the manufacturing man-hour data presented on designer-oriented formats in the three demonstration sections described earlier in this session, i.e., "Sheet Metal Aerospace Discrete Parts," "Mechanically Fastened Assemblies," and "Advanced Composites Fabrication."

The primary objectives of the fuselage shear panel trade studies were to:

- Demonstrate the use of the MC/DG in an industrial environment designing typical airframe structures

- Determine whether the manufacturing cost (man-hour) formats, providing CDE and CED information, meet the format design criteria established for their development

- Determine whether the CDE and CED
  formats provide the accuracy required by designers in conducting
  comparisons of airframe configurations utilizing both metallic
  and composite materials.

Fuselage panel designs were studied
in the following structural materials by the design departments in
each of the three companies:

- Aluminum alloy--by General Dyna-
  mics Corporation, Fort Worth
  Division
- Titanium alloy--by Lockheed-
  California Company
- Graphite/epoxy--by Rockwell In-
  ternational, Los Angeles Divi-
  sion.

The fuselage panel trade studies
were critically reviewed by:

- Boeing Commerical Airplane Com-
  pany
- Northrop Corporation, Aircraft
  Group.

2. COMPOSITE FUSELAGE SHEAR PANEL TRADE STUDY

This paper presents the details of
the trade-study conducted on the
advanced composite (graphite/epoxy)
fuselage panel. This trade-study
utilized the approach illustrated
in Figure 1 for the cost/weight
evaluation. These six steps are
specifically:

(1) Concept Development
     - Skin panel sizing
     - Frame shape selection
     - Number of frames required
     - Stringer shapes
     - Number of stringers required
     - Candidate manufacturing
       methods to produce each
       discrete part
(2) Determination of manufacturing
     cost for each panel configur-
     ation
(3) Determination of assembly costs
     for each panel configuration
(4) Determination of weight (lbs)
     for each panel configuration
(5) Determination of total manu-
     facturing cost, including mat-
     erials and tooling
(6) Presentation of manufacturing
     man-hours and structural
     weight on design charts and
     tables to facilitate selection
     of the cost-effective designs.

The choice of the fuselage panel for
the demonstration is quite appropriate, as this is a promising
application for advanced composite
materials. The advantages offered
by advanced composites for fuselages
are briefly as follows. Fabrication of fuselage by conventional
methods using metallic materials
has resulted in problems in areas
of cost (acquisition and life-cycle)
weight, maintenance, crashworthi-
ness, and fatigue resistance. Use
of lightweight sandwich panels has
increased stiffness, but complicated corrosion and damage control
and repair. A large quantity of
parts and fasteners typical of met-

allic assemblies also impacts ownership costs (approximately 75 percent) and life-cycle costs (approximately 50 percent). Utilization advanced composite materials and structures has provided both weight and cost savings in primary structures. New approaches for implementation on advanced tactical aircraft have shown large potential reductions in manufacturing cost with significant impact also on the life-cycle cost of structures, such as fuselages.

The material chosen for this study was AS/3501-6, as specified in the ground rules set forth by the BCL/airframe industry team members at the start of the program. The design assumptions for this trade study, specified a panel 36 inches wide by 72 inches long, with single curvature of 60 inch radius. A balanced ply layup with quasi-isotropic skin was selected. The spacing of the structural members was specified as 12 to 24 inches for the frames and 4 inches minimum for the stringers. For assembly, titanium Hi-Lock fasteners and cocuring were used, see Table 1.

The limit loading conditions were:
- $N_{x(comp)}$ = 2000 lb/in
- N_{xy} = 121 lb/in
- Shear buckling was not permitted.

A temperature of 300 F and a dry environment were also specified.

Figure 1 shows that three basic categories of configurations were considered. These categories are:
- light weight/high complexity
- moderate weight/moderate complexity
- high weight/low complexity.

In evaluating the concepts, stringer/frame, stringer/skin, and skin variations were considered. The MC/DG was utilized in analyzing the manufacturing cost of these variations, as indicated by the dashed boxes in Figure 1. Figures 2 through 4 provide examples of the fuselage shear panel configurations evaluated.

Three configurations were analyzed within the category of light weight/high complexity (see Table 2). In the concepts the number of stringers and frames were varied to determine the optimum combination. Once it was determined that 4 stringers with 3 frames were the best combination, the type of stringer and the method of assembly were determined. Table 3 presents a summary of the various configurations showing ply-count, structural weight and cost. Figure 5 is a typical diagram showing the man-hours for the light-weight/high complexity concept. Such plots also show the relationship of each concept to lines representing specific man-hour per pound values. From such figures, Configuration III was chosen as most appropriate to represent the light-weight /high complexity category in the remainder of the trade study.

A similar methodology was followed to select representative configurations from the other two categories. Selecting a configuration for production from those summarized could be accomplished, depending on the relative importance of weight and cost, as well as other design factors for the aircraft under consideration.

In a case where two or more concepts appear to be very close in the cost/ weight trade, a detailed cost estimate would need to be performed by cost estimators. This, combined with other factors, would allow the design team to select the most cost competitive design which still meets all other design parameters.

This trade study provided an opportunity to utilize a number of the designer-oriented formats presented in the advanced composites demonstration section of the MC/DG.

The results of the trade study were independently reviewed by the Boeing Commercial Aircraft Company and by Northrop Corporation's Aircraft Group. These companies studied the results to determine if the accuracy provided by utilizing the MC/DG formats was sufficient to provide a meaningful trade study, and that the trade studies represented the intent of the Air Force for the demonstration of the use of the MC/DG.

The following conclusions resulted from the independent review:

- The practicability of the MC/DG was demonstrated
- MC/DG provides a quick, efficient designer's tool which:
  - Develops costs to identify lower-cost designs
  - Reduces design time for screening candidate design
  - Improves schedule compliance
- Use of MC/DG in obtaining manufacturing costs and performing simple cost estimates was well demonstrated
- Demonstrated selection criteria of dollars/pound weight saved
- Fully demonstrated use of MC/DG in developing cost/weight effective designs
- Wider coverage needed to expand data base for manufacturing technologies, structural configurations, and composite material types.

The following conclusions were also arrived at by the aerospace companies:

- Utilized costing methodology, developed program dollar costs, used material, labor, and tooling costs
- Cost/weight summary chart and recommendations are of particular merit.

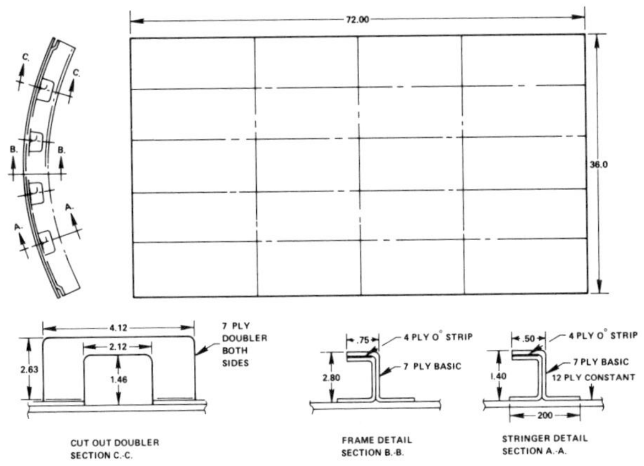

Fig. 2. Moderate Weight/Moderate Complexity
(4 Stringers/3 Frames) Concept

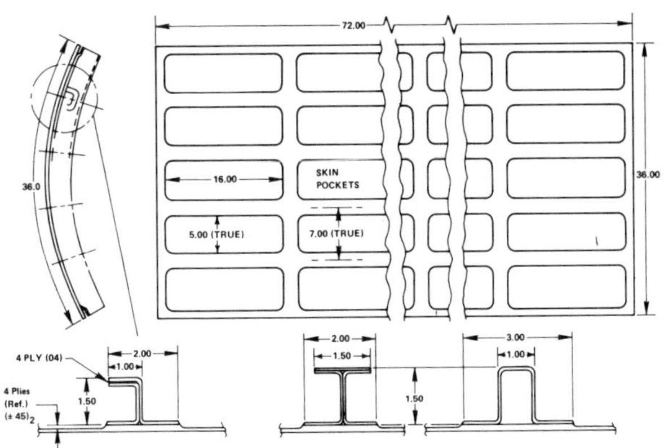

Fig. 3. Baseline Fuselage Shear-Panel

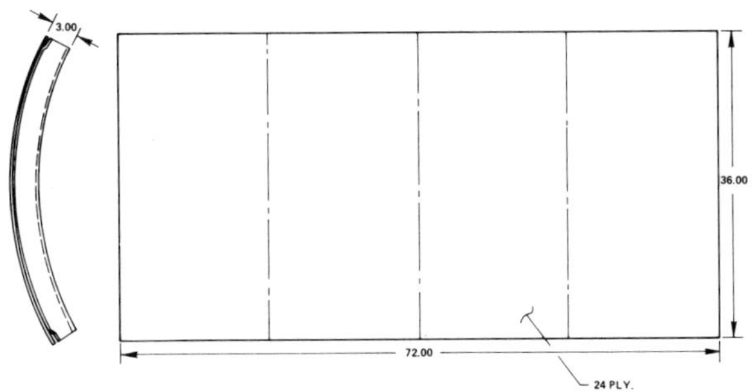

Fig. 4. Minimum Part Count Concept
(Plateskin/3 Frames)

LOADING CONDITIONS	SIZE	CONFIGURATION	ASSEMBLY
** $N_{x(comp)}$ $\underline{2000}$ LB/IN. LIMIT N_{xy} $\underline{121}$ LB/IN. SHEAR BUCKLING NOT PERMITTED ENVIRONMENT • DRY ** • TEMP. 300°F	36 X 72 BALANCED PLY LAYUP • SKIN-QUASI-ISOTROPIC	• SINGLE CURVE R = 60″ • FRAME SPACING 12″ – 24″ • STRINGER SPACING 4″ MIN. • EDGE SPLICE NOT DEFINED 100% CONTINUITY	• MECHANICAL HI-LOCK (Ti) • COCURE

MATERIAL   AS/3501-6[*]

**LOADS AT ROOM TEMPERATURE
300°F LOADS .667 OF R.T. LOADS

TRADE BASED ON COSTS AT 200TH UNIT[*]

(*) GROUND RULE

Table 1.   Design Assumptions for Composite
Fuselage Shear-Panel

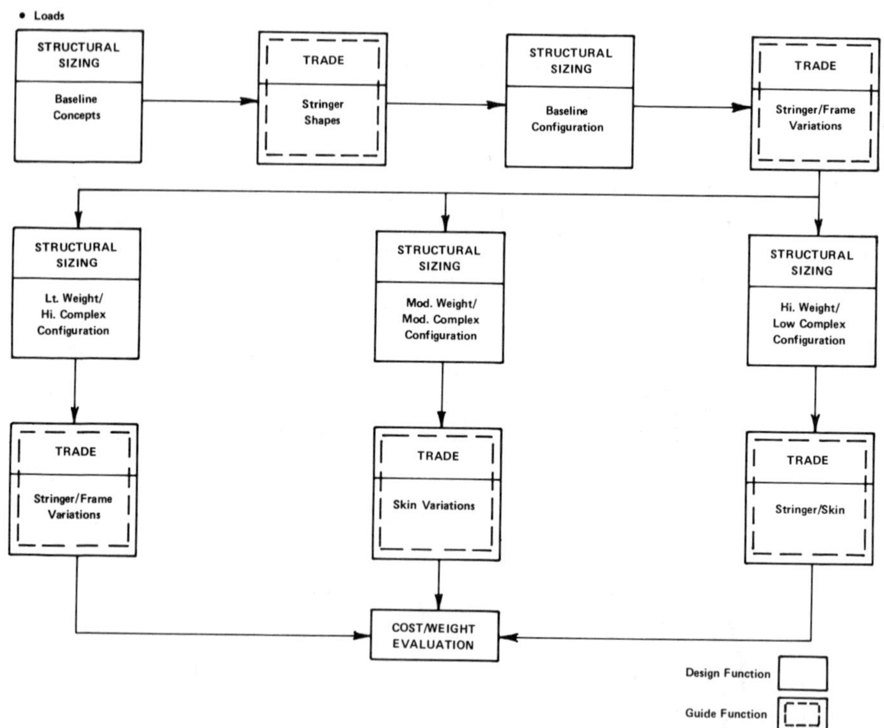

Fig. 1.   Design Function/MC/DG Interaction for Composite
Fuselage Shear-Panel Trade-Study

Table 2. Summary of Light-Weight/ High-Complexity Concept

CONFIG-URATION	STRINGER						FRAME					COST, MAN-HOURS
	TYPE	NO.	a (IN.)	b (IN.)	c (IN.)	ATTACHMENT METHOD	TYPE	NO.	a' (IN.)	b' (IN.)	c' (IN.)	
I	HAT	4	1.0	1.5	3.0	MECHANICALLY FASTENED (100% MANUAL)	J	3	0.75	3.0	2.0	77.9
II	HAT	4	1.0	1.5	3.0	MECHANICALLY FASTENED (80% AUTOMATIC 20% MANUAL)	J	3	≠0.75	3.0	2.0	61.5
III	J	4	1.0	1.5	2.0	COCURED	J	3	0.75	3.0	2.0	51.1

Table 3. Configuration Summary

CONFIG-URATION		STRINGERS				FRAMES				SKIN		*MAN-HOURS	WEIGHT, LB
		NUMBER	a	b	c	NUMBER	a'	b'	c'	t_1	t_2		
A	BASELINE-LIGHT WEIGHT/ HIGH COMPLEXITY COCURED	4	1.00	1.50	2.00	3	.75	3.00	2.00	8 PLY	12 PLY	51.10	17.45
B	MODERATE WEIGHT/ MODERATE COMPLEXITY	4	.50	1.40	2.00	3	.75	2.80	2.00	12 PLY	12 PLY	49.20	17.65
C	MODERATE WEIGHT/ MODERATE COMPLEXITY	3	1.00	1.60	2.00	3	.75	3.20	2.00	12 PLY	12 PLY	48.04	18.80
D	LOW COMPLEXITY	0	0	0	0	3	.75	3.00	2.00	24 PLY	24 PLY	34.79	22.50

$$*\text{RECURRING} + \frac{\text{NON-RECURRING}}{200}$$

599

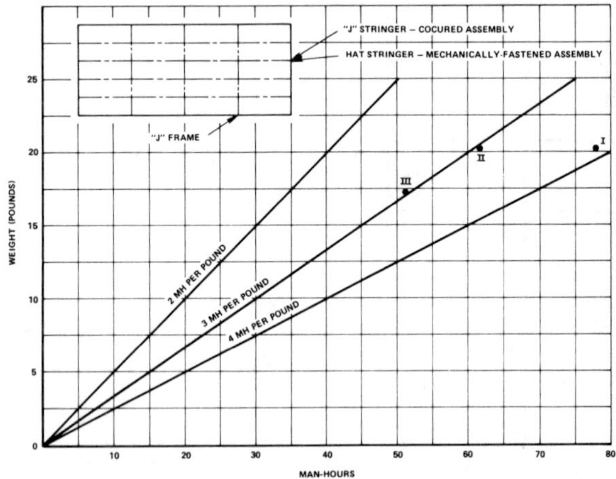

Fig. 5.   Typical Diagram Showing Weight and Man-Hours
(for Light-Weight/High Complexity Concept)

UTILIZATION IN INDUSTRY OF
ICAM "MANUFACTURING COST/DESIGN GUIDE" (MC/DG)
FOR METALLIC STRUCTURES

Anthony J. Pillera
Lockheed-California Company, Burbank, CA
Bryan R. Noton
Battelle's Columbus Laboratories, Columbus, OH

Abstract

The MC/DG has been utilized for structural performance/manufacturing cost studies of seven titanium fuselage panel designs for an SST project. Various structural concepts with varying skin thicknesses, frames, and stringer count, were studied. An engineering team conducting the trade studies, included design, stress, weight, and producibility engineers. Detailed engineering drawings were utilized providing realistic applications of the MC/DG. The study enabled the selection of the most cost-effective concept by determining the cost of weight saved for each panel. The use of the MC/DG was fully demonstrated in developing a cost/weight effective design by utilizing the manufacturing cost methodology developed. This provided program dollar costs (including materials, labor and tooling).

1. OBJECTIVES OF AIRFRAME TRADE STUDIES

A series of fuselage shear panels were analyzed with regard to weight and manufacturing cost by three airframe industry team members, utilizing the manufacturing man-hour data presented on designer-oriented formats in the three demonstration sections described earlier in this session, i.e., "Sheet-Metal Aerospace Discrete Parts", "Mechanically Fastened Assemblies", and "Advanced Composites Fabrication".

The primary objectives of the fuselage shear-panel trade-studies were to:

- Demonstrate the use of the MC/DG in an industrial environment designing typical airframe structures
- Determine whether the manufacturing cost (man-hour) formats, providing CDE and CED information, meet the format design criteria established for their development

- Determine whether the CDE and CED formats provide the accuracy required by designers in conducting realistic comparisons of airframe configurations utilizing both metallic and composite materials.

Fuselage panel designs were studied in the following structural materials by the design departments in each of the three companies:

- Aluminum alloy--by General Dynamics Corporation, Fort Worth Division
- Titanium alloy--by Lockheed-California Company
- Graphite/epoxy--by Rockwell International, North American Aircraft Division.

The fuselage panel trade studies were critically reviewed by:

- Boeing Commercial Airplane Company, and
- Northrop Corporation, Aircraft Group.

2. TITANIUM FUSELAGE SHEAR PANEL TRADE STUDY

This paper presents the results obtained by a design team consisting of design, stress, weight, and producibility engineers, for the trade-study of a titanium fuselage shear panel. The approach used for this trade-study is shown in Figure 1. This commenced with a review of the ground rules and structural sections available.

Figure 2 shows the structural sections selected for this trade-study. In the next step, the structural design premises and the general characteristics of the panel design were specified. Figure 3 shows the panel selected of dimensions 36 x 72 inches with a constant 60-inch radius. The design loads, structural design criteria, and analysis methodology are summarized in Figure 3. These criteria were derived from the first and second generation SST studies; modified to reflect the ground rules set forth by the MC/DG development team.

The third step in the approach to the trade study is to develop candidate design configurations. A generalized drawing of the panel is shown in Figure 4. Table 1 is a summary of the seven design concepts considered. The table provides values of A and B, shown in Figure 4, for each concept. It can be seen from the table that the number of frames, skin thickness, and the number and type of stringers were the variables. Table 2 provides a detailed summary of the concepts, including the number, type and dimensions of each part. Also included in this table are the number of rivets (fastener count) required for assembly of the concept.

The final step, conducted as part
of the trade-study, was to estimate
the cost and weight of each panel
concept. The weight was estimated
by the weight engineer, using
standard weight estimating procedures. To estimate the cost of
each concept, the MC/DG Designer
Worksheet was utilized. The worksheet is included in the first
paper in this session. Utilizing
the man-hours from the MC/DG formats for recurring and non-recurring
tooling cost, the designer can calculate the program cost. The worksheet is a useful aid to the designer when using the MC/DG, as it
provides an orderly outline of what
must be accomplished to determine
the cost of the panel.

Having determined both the manufacturing cost and weight of each
panel, the designer can now organize the data into a convenient
form for selection of the optimum
panel design. Table 3 summarizes
the cost and weight of each concept. For this summary, the least
costly panel, Concept VII, was
chosen as the base design. With a
base design selected, a delta for
weight and cost of each panel can
be calculated relative to the base.
These deltas are then combined to
give a value for the cost of weight
saved, in dollars per pound. These
data are also included in Table 3.
The designer concluded from the
data that Concept II should be the

recommended panel design. In order
to confirm this decision, Table 4
was prepared with Concept II as the
base design. The deltas and cost
of weight saved were again calculated. The results show that
Concept II was the correct choice.

The conclusions, based on a review
of the trade-study, were that the
MC/DG is an effective tool for the
design team, and that the methodology followed in this trade-study
clearly demonstrated the concept of
utilizing the MC/DG in the aerospace industry environment. The
specific conclusions are listed
below:

- Information presented indicative
  of the ultimate function of the
  MC/DG
- Use of MC/DG in obtaining manufacturing costs and performing
  simple cost estimates was well
  demonstrated
- Demonstrated selection criteria
  of dollars/pounds weight saved
- Fully demonstrated use of the
  MC/DG in developing cost/weight
  effective design
- Utilized costing methodology,
  developed program dollar costs,
  material, labor, and tooling
- Cost/weight summary chart and
  recommendations are of particular merit
- Review of each concept given
  with cost-estimating steps
  clearly shown.

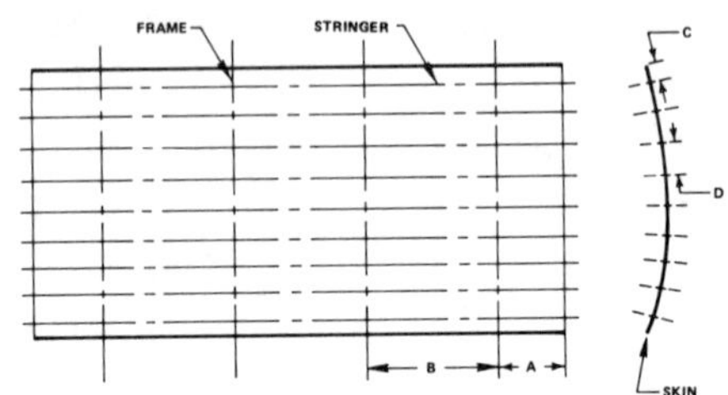

Fig. 4.   General Layout of Titanium Fuselage Shear Panel

SUMMARY OF CONCEPTS								
CONCEPT NO.	SKIN THICKNESS, IN.	NO. OF FRAMES	TYPE OF FRAME	NO. OF STRINGERS	TYPE OF STRINGERS	DIMENSIONS, IN. A	B	DIMENSIONS, IN. A	B
1	0.08	4	ZEE	9	ZEE	9.0	18.0	2.0	4.0
2	0.08	3	ZEE	9	ZEE	12.0	24.0	2.0	4.0
3	0.06	4	ZEE	8	HAT (CLOSED)	9.0	18.0	2.25	4.5
4	0.06	3	ZEE	8	HAT (CLOSED)	12.0	24.0	2.25	4.5
5	0.075	4	ZEE	8	HAT (OPEN)	9.0	18.0	2.25	4.5
6	0.075	3	ZEE	8	HAT (OPEN)	12.0	24.0	2.25	4.5
7	0.190	9	ZEE	0	–	4.0	8.0	–	–

Table 1.   Summary of Structural Concepts

SUMMARY OF CANDIDATE PANEL CONFIGURATIONS											
CONCEPT	SKIN THICKNESS* (IN.)	STRINGERS NO.	THICKNESS (IN.)	TYPE	FRAME ASSEMBLY NO. OF FRAMES	NO. OF RIVETS	FRAME (ZEE) NO.	THICKNESS (IN.)	FRAME ANGLE	CLIP (ANGLE)	NO. OF RIVETS
I	.080	(9)	.050	ZEE	4	29	(4)	.040	(4) .040	(36) .040	761
II	.080	(9)	.050	ZEE	3	29	(3)	.050	(3) .050	(27) .050	690
III	.060	(8)	.040	HAT**	4	13	(4)	.040	(4) .040	32 LH .040 32 RH .040	1380
IV	.060	(8)	.040	HAT**	3	13	(3)	.050	(3) .050	24 LH .050 24 RH .050	1259
V	.075	(8)	.040	HAT***	4	13	(4)	.040	(4) .040	32 LH .040 32 RH .040	1380
VI	.075	(8)	.040	HAT***	3	13	(3)	.050	(3) .050	24 LH .050 24 RH .050	1259
VII	.190	—— NONE ——			— NONE —		(9)	.040	NONE	NONE	261

*SKIN SIZE: 36" x 72".
**FLANGES ATTACH TO SKIN.
***FLANGES AWAY FROM SKIN.

Table 2.   Summary of Candidate Panel Configurations

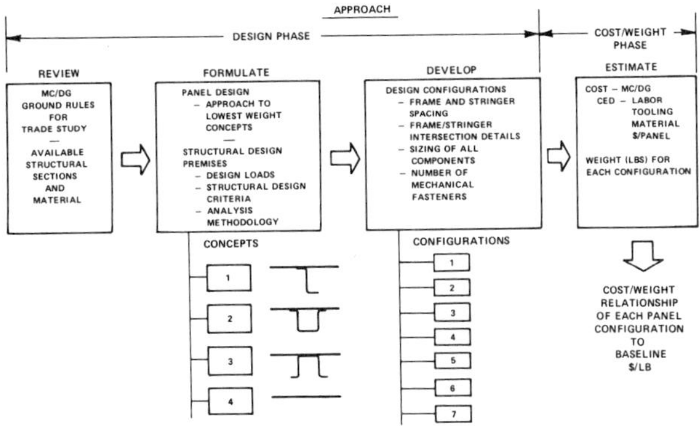

Fig. 1. The Design and Cost/Weight Phases in the Trade-Study

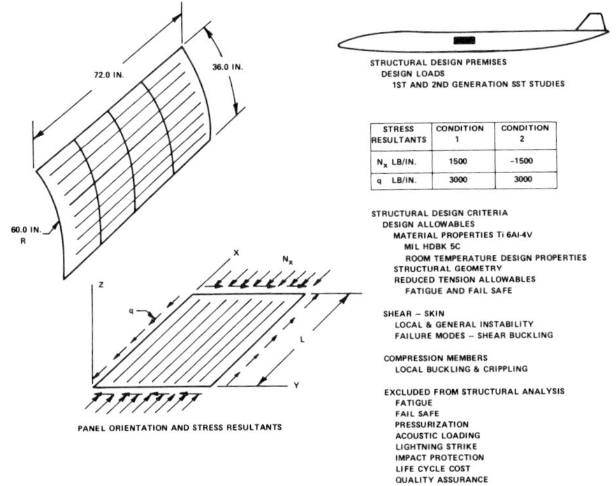

Fig. 2. MC/DG Titanium Structural Sections Utilized

Fig. 3. Loading Conditions and Design Requirements
for Titanium Fuselage Shear Panel

| CONCEPT | COST | | WEIGHT | | COST OF WEIGHT SAVED— |
	$/PANEL	Δ$/PANEL	LBS/PANEL	ΔWT LBS	$/LB
I	2986	994	59.02	-28.77	35
II	2680	688	58.46	-29.33	23
III	4473	2481	58.26	-29.53	84
IV	3915	1923	57.58	-30.21	64
V	4491	2499	64.48	-23.31	107
VI	3933	1941	63.80	-23.99	81
VII	1992	BASE	87.79	BASE	BASE

Table 3. Trade-Off Summary Showing Cost
of Weight Saved ($/Pound)

| CONCEPT | COST | | WEIGHT | | COST OF WEIGHT SAVED— |
	$/PANEL	Δ$/PANEL	LBS/PANEL	ΔWT LBS	$/LB
I	2986	306	59.02	+0.56	①
II	2680	BASE	58.46	BASE	BASE
III	4473	1793	58.26	-0.20	8965
IV	3915	1235	57.58	-0.88	1403
V	4491	1811	64.48	+6.02	①
VI	3933	1253	63.80	+5.34	①

① GREATER WEIGHT AND GREATER COST.

Table 4. Cost-Weight Trade-Off Summary
Showing Recommended Concept

THE ICAM "MANUFACTURING COST/DESIGN
GUIDE" (MC/DG) FOR AVIONICS

Robert Remski
Honeywell, Inc.
St. Louis Park, Minnesota

John G. Vecellio
Rockwell International Corporation
Avionics & Missiles Group
Cedar Rapids, Iowa
and
Bryan R. Noton
Battelle's Columbus Laboratories
Columbus, Ohio

Abstract

A "Manufacturing Cost/Design Guide (MC/DG) for Avionics" is under development for the Computer Integrated Manufacturing Branch (AFWAL/MLTC) Materials Laboratory, WPAFB, Ohio. The MC/DG is being developed by a team; Battelle's Columbus Laboratories (BCL) as prime contractor, Honeywell, Inc., and Rockwell International's Avionics and Missile Group. The data analysis methodologies established for the "MC/DG for Airframes" are being used. General and detailed ground rules are described, which include specifications of electronic discrete parts, materials, assemblies, manufacturing methods, etc. Examples of assemblies are power supply hybrids and chassis. Examples of discrete parts are substrates and printed wiring boards. The paper presents an overview of the aerospace electronic design processes, indicating how the MC/DG will be utilized to conduct trade studies at the conceptual design phase, as well as the detailed design phases for the circuitry and also the chassis. The paper concludes with an overview of the numerous trade-studies and data requirements to achieve affordable performance in avionic systems. This data must be presented to designers so that the cost-trade studies can be conducted with schedule limitations.

The following are examples of cost-drivers in electronic fabrication and assembly:

- Single Source/Proprietary Items
- Insufficient Test Points (Shorten Fault Isolation Testing)
- Unrealistic Tolerances
- Requirement for Use of Special Equipment/Skills/Facilities
- Back to Back Card Modules
- Bussing
- Shielded Wire
- Unique Wire Gages
- Solder Sleeve Terminations
- Flextapes
- Solder Cup Connectors
- Chassis Mounted Components
- Multilayer Printed Wiring Boards
- Components Unable to Be Auto Inserted
- Multirow Interface Connectors
- Masking, Unmasking, and Touch-Up
- Special Tools for Insertion or Extraction
- Multiple Range of Fasteners
- Special Punches
- Excessive Hand Wiring
- Castings
- Transformer Clearances.

THE MC/DG FOR AVIONICS

To enable conceptual and detailed designers to achieve affordable performance in developing avionics systmes, cost-drivers such as listed above, are being addressed by designer-oriented formats presenting manufacturing man-hour data. The list of contents, for the MC/DG is shown in Table 1. It will be noted that this is subdivided into the following categories:

- Procured Items
- Detailed Fabrication
- Assembly
- Test, Inspection and Evaluation.

It will also be noted that both conventional and emerging technologies will eventually be included in this volume of the MC/DG.

The average cost breakdown for different flight control systems desifned in the past and recently, are shown in Figure 1. However, this type of qualitative format, while useful for comparative purposes is not satisfactory for presenting man-hours for manufacturing assembly and inspection. The formats under development to present the man-hour information for discrete parts (base parts with designer-influenced cost elements) must meet the following criteria:

- Emphasize cost drivers
- Be simple to use
- Use designer language
- Instill confidence
- Be economical
- Be accessible
- Be maintainable.

GENERAL GROUND RULES

General and Detailed Ground Rules for Electronics Fabrication and Assembly are under development.

Ground rules are necessary and important as they promote understanding, ensure consistency, uniformity, and accuracy in generating and integrating data into the formats.

The general ground rules are categorized under the following major groupings:
(a) Electronic Assemblies
(b) Discrete Parts
(c) Materials
(d) Manufacturing Methods
(e) Facilities
(f) Data Generation - Recurring Costs
(g) Data Generation - Nonrecurring Costs
(h) Support Function Modifiers
(i) Test, Inspection, and Evaluation (TI&E).

(a) Electronic Assemblies
  (1) The electronic assemblies selected will be those commonly used in the electronic industry. Examples of assemblies are printed wiring assemblies, power supplies, hybrids, and chassis.

(b) Discrete Parts
  (1) The discrete parts selected will be those commonly used in the Aerospace Electronics Industry such as Printed Wiring Board, Wire, Substrate, and Connectors.

(c) Materials
  (1) The materials selected for the electronic assembly will be representative of the range of those more commonly used in the industry.
  (2) Raw material costs for printed wiring board (PWB) or assembly (PWA), power supply, transformer, hybrid, and chassis will not be included in the MC/DG formats, but will be treated by the user at his discretion.
  (3) Material cost of nonrecurring tooling will not be included.

(d) Manufacturing Methods
  (1) Only existing manufacturing methods required to produce the base parts will be considered. No emerging manufacturing methods will be evaluated. However, the potential of new technologies to reduce cost will be highlighted.
  (2) A production, in contrast to a prototype, environment will be assumed for the electronic industry discrete parts.

(3) Manufacturing man-hour data will be developed, where possible, for more than one manufacturing method for each discrete part or assembly. The data will thereby enable the designer, using the MC/DG, to determine the most cost competitive manufacturing method in trade studies.

(4) To generate an effective data base for each selected part, a factory operational sequence for each applicable manufacturing method will be established reflecting the most economical means of fabrication. This standardized sequence will be used by each team member to determine the part cost in man-hours.

(5) Tools required to manufacture the various parts will be identified on the data collection forms.

(e) Facilities

(1) Only present manufacturing facilities, available to the electronics industry, will be considered.

(f) Data Generation - Recurring Costs

(1) All manufacturing labor cost data will be presented in the MC/DG in man-hours and material costs in vendor dollars (1980).

(2) The base part cost (man-hours/vendor 1980 dollars) will be generated for each part type.

(3) Man-hour data will be generated for all manufacturing operational sequences in part fabrication and assembly and will include all hands-on factory direct labor operations prior to entering storage for subsequent assembly.

(4) Setup time (man-hours) will be amortized over the selected lot size and added to the processing time to obtain the base part cost (man-hours).

(5) Recurring tooling costs (tool maintenance, planning, etc.) will not be included.

(6) Lot sizes to be considered for electronic assemblies will be determined.

(7) Lot sizes for procured items will be 50, 200, and 1000.

(g) Data Generation - Nonrecurring
Costs

(1) Tool costs (man-hour) will
be generated for each
fabricated part type. In
addition, tool design and
tool planning costs will
be evaluated with respect
to their impact, to determine whether they should
be included or omitted.

(2) The cost of production
tooling, if included, will
be restricted to contract
or project tools only, for
presentation in the MC/DG.

(3) Nonrecurring vendor costs
(1980 dollars) will be
included and amortized
over the typical lot size.

(4) Nonrecurring tooling material costs will be identified and, when significant,
included in the MC/DG.

(5) Special test equipment
(STE) (i.e., fixtures,
adapters, etc.) nonrecur-
ring costs (man-hours/
vendor 1980 dollars) will
be included and amortized
over the typical lot size.

(6) Test software cost (man-
hours) will be considered
as nonrecurring and will
be amortized over the
typical lot size.

(h) Support Function Modifiers

(1) Additional efforts other
than factory labor, such
as production control,
industrial engineering,
and manufacturing engi-
neering will be excluded
from the part cost data.
These modifiers may be
included later by the
MC/DG users.

(2) Impacts on manufacturing
cost resulting from reli-
ability, maintainability,
and life-cycle cost
requirements included in
product specifications,
will be identified for
the various manufacturing
method alternatives.

(i) Test, Inspection, and
Evaluation (TI&E)

(1) TI&E cost data (man-hours)
will be developed for
"in-process", "func-
tional", and "final
acceptance" inspection/
testing.

DETAILED GROUND RULES

The detailed ground rules are categorized under the following major
groupings:

(a) Material (Purchased Items)

(b) Configuration

(c) Specification Requirements

(d) Manufacturing Methods

(e) Facilities

(f) Test, Inspection, and
     Evaluation (TI&E)

(g) Data Generation - Recurring

(h) Data Generation - Nonrecurring.

(a) Material (Purchased Items)

  (1) All purchased parts will
       be "off-the-shelf" stan-
       dard parts, i.e., no
       screening added or custom-
       designed parts specified.

  (2) Parts will consist of
       integrated circuits (DIP
       packages), resistors,
       capacitors, diodes, con-
       nectors, and extractor(s).

  (3) The printed wiring board
       will be a purchased item.

  (4) No purchased tooling will
       be included.

(b) Configuration

  (1) The PWB will be rectangu-
       lar and with a surface
       area of 20 to 40 in.^{2}.

  (2) The electronic part count
       will be 40-60.

  (3) The PWB board will be two-
       layer, G-10 (MIL-P-18177)
       board material, with
       plated-thru holes.

  (4) A maximum of five (5) cuts
       and jumpers will be
       allowed.

(c) Specification Requirements

  (1) Military standards will
       apply to part selection,
       quality, and workmanship
       practices.

  (2) No assembly level
       screening/debug.

(d) Manufacturing Methods

  (1) Eighty to ninety percent
       (80-90%) of the parts
       will be automatically
       inserted.

  (2) The board will be wave-
       soldered and the appro-
       priate masking will be
       performed.

  (3) The cleaning procedure
       will include vapor
       degreasing and water wash.

  (4) Part locations and assem-
       bly identifiers will be
       silk-screened.

  (5) Conformal coating will be
       applied using standard
       company practices.

(e) Facilities

  (1) Existing production
       facilities will be evalu-
       ated when developing
       manufacturing man-hour
       data.

(f) Test, Inspection, and
     Evaluation (TI&E)

  (1) Normal purchased part
       incoming inspection pro-
       cedures will be employed.

(2) In-circuit testing will
be conducted at the assembly level prior to any
functional testing.

(3) Functional testing will
be conducted using auto-
matic test with the option
of using automatic or
manual testing for repair.

(4) Normal company inspection
procedures will be fol-
lowed in complying with
military standards.

(g) Data Generation - Recurring
(1) The build lot size
selected is 20-30
assemblies

(2) Purchased part costs will
be based on normal lead
times and release quanti-
ties of 200 pieces.

(3) The standard labor hour
will consist of 50
minutes.

(4) Rework hours will be
included in the total
assembly hours.

(h) Data Generation - Nonrecurring
(1) The in-circuit testing
adapter will be amortized
over the lot quantity.

(2) Cost for special test
equipment will be amor-
tized over the lot
quantity.

(3) Software costs for in-
circuit testing and
automatic functional
testing will be amortized
over the lot quantity.

SELECTION OF ELECTRONIC
LOT SIZES ASSEMBLIES

Electronic assemblies for aerospace
products are considered low production (lot sizes less than 25/
month) when compared to production
lot sizes (i.e., 150-300/month).

For small runs (less than 5/month)
or prototypes, the assemblies are
hand assembled and data provided
will be representative of low production quantities.

For medium lot sizes (10-15/month),
the data will be representative of
a combination of both hand assembled and some utilization of
semiautomated assembly.

Average production runs (20-25/
month) are representative of typical lot sizes for electronic assemblies, manufactured on semiautomated assembly lines.

To provide manufacturing cost data
with maximum application, the initial data has been determined for
3 lot sizes, i.e., 50-10-25/month.
This provides the design engineer
with the full range of data for
electronic assemblies and provides

a sufficient base-line of data that can be expanded to large sizes by any company.

TRADE STUDIES AND
DATE REQUIREMENTS

An overview of the design process in aerospace avionics is shown in Figures 2 & 3. It should be noted, however, that this flow diagram will have been modified at the time of the presentation when a new date change will be discussed.

The following are examples of the trade studies which are conducted in the aerospace avionics industry, and also the data requirements to conduct these trade studies are indicated. The trade studies are provided under the following headings:

A. Examples of Conceptual Design Phase Trade Studies

  a. Standard Circuits vs New Technology
- Cost
- Weight
- Size
- Reliability
- Maintainability
- Probability of Being Available
- Factory Capable of Handling
- Multiple Source (for Candidate Circuits)

  b. One Box Vs Two Boxes Vs N Boxes
- Vehicle Configuration
- Redundancy Level Desired
- Vulnerability Level Desired
- Identical Configurations
- Maintenance Philosophy
- Common Functions
- Box Reliability
- Weight

  c. Common Functions Vs Unique Functions
- Multi-Mode Computational Capability
- Reliability
- Space Availability
- Data Transmission
- Fault Isolation Capability
- Flight/Mission Critical

  d. Analog Vs Digital Processing
- Interface
  - I/O
  - Required Conversions
- Part Availability
- Part Costs
- Reliability
- Hardware/Software Integration
- Number of Functions
- Operational Definition
- Test Costs

614

B. Examples of Detailed Circuit
Design Phase Trade Studies
a. Standard Components Vs Custom Components
   - Quantity
   - Number of Gates
   - Non-Recurring Design Cost
   - Tooling Cost
   - Area
   - "I" Level Maintenance Concept
   - Spares Stock Requirements

b. Reliability Vs Part Cost
(MIL Vs Extreme Temperature Vs Commercial)
   - Part Type
   - Part Family
   - Failure Rate of Quality Level
   - Operating Environment
   - Specification Tailoring
   - Part Cost
   - Purchase Quantity
   - Purchase Agreement

c. Part Density Vs Printed Circuit Board Cost Vs Hard-Wired
   - Size
   - Number of Layers
   - Part Types
   - Thermal Density
   - Equipment Mechanical Architecture
   - Field Repairable
   - Test Point Access

d. Discrete Components Vs Hybrid
   - Area
   - Functions/In^2 (Circuit Density)
   - Power Density
   - Cost
   - Non-Recurring Design Cost (Hybrid)
   - Tooling Cost (Hybrid)

e. Automated Test Vs Manual Test
   - Quantity
   - Complexity (Function and Number of Measurements)
   - Fixture Costs (Adapter)
   - Software Costs
   - Number of Parts
   - Type of Parts
   - Function to be Tested
   - Level of Incoming/Receiving Part Testing

C. Examples of Detailed Mechanical
Design Phase Trade Studies
a. Auto-Insertion Vs Hand Insertion
   - Quantity of Assemblies
   - Type of Component
   - Number of Axes - Component Orientation
   - Number of Boards Per Blank
   - Footprint
   - Lot Sizes
   - Assume Production (Firm Design)

b. Adding Cuts/Jumpers Vs Re-
designing Printed Wiring
Board
  ● Quantity of Cuts and Jump-
    ers
  ● Cost per Cut & Jumper
    (Labor)
  ● Cost of Re-Design of PWA
  ● Volume of Assembly Remain-
    ing

c. Soldering: Automatic Wave Vs
Vapor Phase Vs Manual
  ● Density
  ● Type of Component
  ● Type of Lead Form

d. Quantity of Boards Vs Single
Board Vs Multi-Layer Board
  ● Density
  ● Size
  ● Number of PWBs
  ● Number of Interconnects
  ● Package Constraints
  ● Test Considerations

e. Single-Sided Part Mounting
Vs Double-Sided Part
  ● Card Spacing in Chassis
  ● Density
  ● Type of Parts
  ● EMI Shielding
  ● Soldering Process
  ● Recurring Labor Costs
  ● Handling/Storage
  ● Repairability

CONCLUSIONS

Design teams can be motivated into a
design-to-lowest cost attitude by
utilizing the MC/DG in the design
process. Design teams must be
provided with:

● Tools: Identification and documentation of cost-drivers and
  cost reduction methods
● Incentives: Cost targets against
  which performance of design
  personnel can be measured.

The specific objectives of the
"MC/DG for Avionics" are as follows:

● To provide designers with <u>simple</u>,
  relative, and quantitative cost
  comparisons of manufacturing
  processes that can be utilized
  rapidly
● To emphasize <u>design orientation</u>
  of MC/DG formats and manufacturing man-hour data for use at
  <u>all</u> phases of design process,
  e.g., conceptual and detail
  design, therefore, increasing
  emphasis on cost; a vital
  design parameter
● To enable more extensive per-
  formance/manufacturing cost
  <u>trade-offs</u> to be conducted by
  designers on avionic parts and
  assemblies
● To emphasize potential <u>cost advantages of emerging materials
  and manufacturing methods</u>
  accelerating the transfer of
  these technologies to production
  hardware.

I	II	III		IV
PROCURED ITEMS	DETAIL FABRICATION	ASSEMBLY		TEST, INSPECTION, AND EVALUATION (TI&E)

Schematic Parts
 	
• Electronic
• Electromechanical
 	
Interconnect Parts
 	
• PWB
• Connectors
• Signal Interconnect
  – Wire
  – Coaxial
  – Flex Cable
  – Ribbon Cable
  – Wave Guides
 	
Hardware
 	
• Mechanical
• Electrical
 	
Fabricated Parts
 	
• Castings
• Sheet Metal
• Plastic Parts
• Powdered Metal
• Machined Parts
• Extrusions
 	
Emerging Technologies
 	
• VLSI
• Displays (i.e., liquid crystal, electroluminescent, etc.)
• Chip Carriers
• Fiber-optics
• Multilayer Flex Circuit
• Hot Isostatic Processing (HIP)
	Metallic
	
	• Machining
	• Forming
	• Chem Milling
	
	Nonmetallic
	
	• Molding
	• Laminating
	• Machining
	• Forming
	
	Treatment/Coating
	
	• Anodize
	• Paint
	• Silk Screen
	• Plating
	
	Emerging Technologies
	
	• Laser Machining
	• Net Shape
	• Robotics
	• Laser Inspection
	• Video Control Inspection
		Mechanical Assembly
		
		• Connectors
		• Hardware
		
		Part Assembly (Pre-Wave and Post-Wave)
		
		• Automatic
		• Manual
		• Semi-automatic
		
		Cleaning
		
		Solder
		
		• Hand
		• Flow
		• Vapor Phase (batch)
		• Oven
		• Induction
		• Infrared
		
		Sheet Metal/Standoff Assembly (Hard Wiring)
		
		• Standoff Insertion
		• Part Assembly
		• Interconnect
		
		Cable/Wire Harness Assembly
		
		• Coaxial
		• Wire Bundles
		• Ribbon Cable
			Hybrids
			
			Other Assembly Techniques
			
			• Multi-Wire
			• Stitch Wire
			• Wire Wrap
			
			Chassis Assembly
			
			Final Equipment Assembly
			
			• Mechanical
			• Electrical
			
			Post-Assembly Processes
			
			• Conformal Coating
			• Masking
			• Spray/Dip/Brush
			• Sealing
			
			Potting
			
			Adhesives
			
			Inspection
			
			Emerging Technologies
			
			• Chip Carrier
			• Chip/Epoxy
			• Robotics
			• Continuous Vapor Phase Soldering
			• Infrared
				Card/Module Level Test
				
				• Automatic
				• Manual
				• Calibration/Alignment
				
				Testability
				
				Burn-In/Screening Test
				
				• Automatic
				• Manual
				
				Device/Equipment Test
				
				• Automatic
				• Manual
				• Calibration/Alignment
				
				Emerging Technologies
				
				• Automated Assembly Handling
				• Improved CAD/CAM
				• CAT
				• Automatic Optical Comparison
				• Voice Control

Table 1. MC/DG Contents: "Manufacturing Technologies for Aerospace Electronic Fabrication, Assembly, and Test"

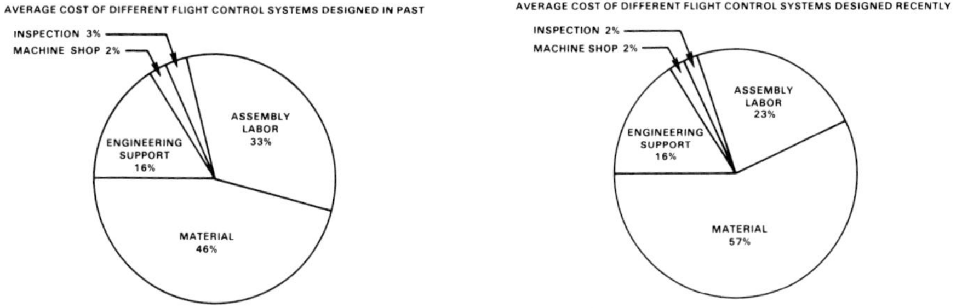

Fig. 1. Typical Cost Breakdown for Aerospace Components

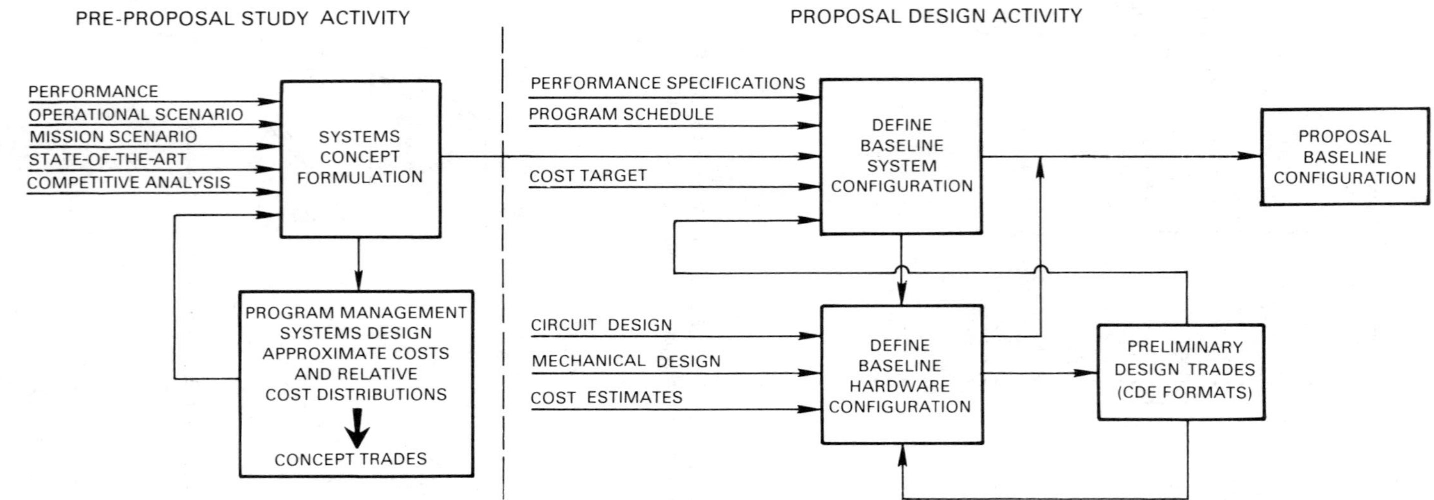

Fig. 2. Typical Design Flow for Aerospace Avionics Industry

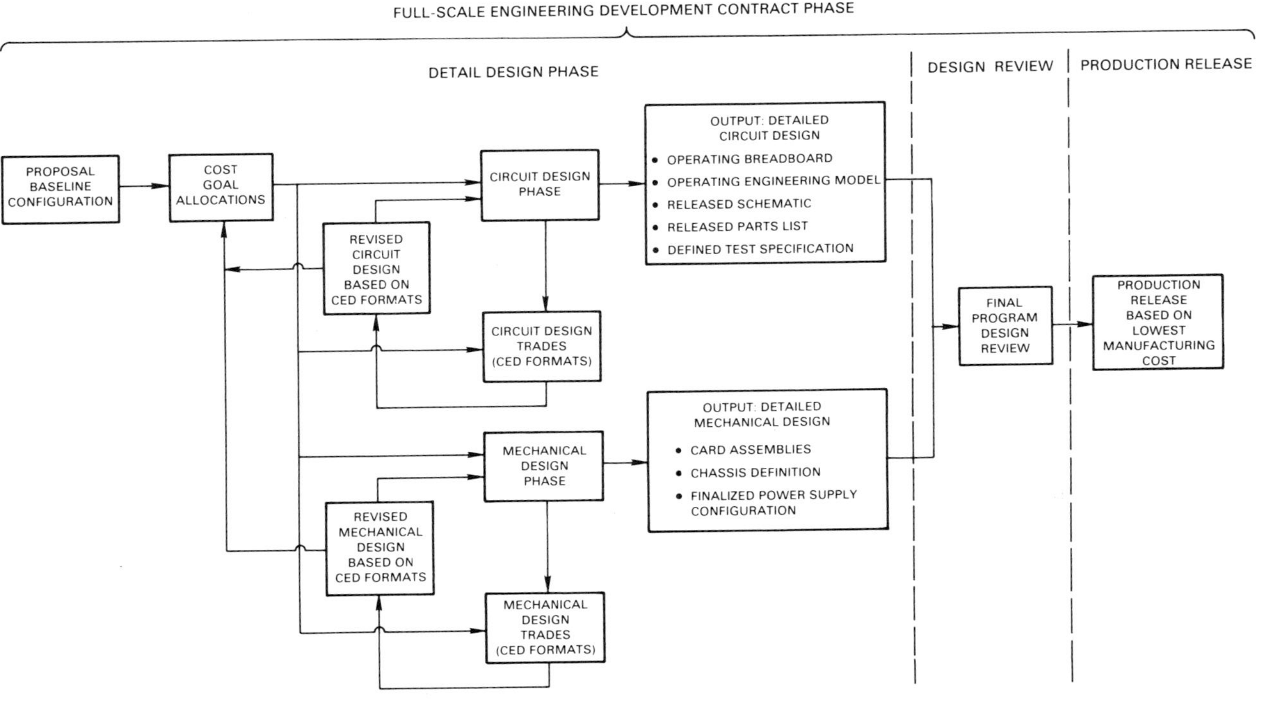

Fig. 3. Typical Design Flow for Aerospace Avionics Industry (Continued)

THE MANUFACTURING COST/DESIGN GUIDE (MC/DG) COMPUTERIZATION PROGRAM
MANUFACTURING COST DESIGN SYSTEM (MCDS)
A PROGRAM UPDATE

Bernard I. Rachowitz and Henry H. Loshigian
Grumman Aerospace Corporation
Bethpage, New York 11714

David Judson
Integrated Computer Aided Manufacturing
(ICAM) Office
Wright-Patterson Air Force Base, Ohio 45433

Abstract

In the early 1970's, the Air Force initiated development of a Manufacturing Cost/Design Guide. The Guide's data development is continuing by the Coalition Team headed by the Battelle Corporation. While the basic guide is useful in a manual sense, the Air Force recognized the need for a computerized system that can be integrated with other ICAM and CAD systems. In order to accomplish this task, the Manufacturing Cost Design System (MCDS) is being developed. This paper provides a brief overview of the development of the MCDS.

The paper will first review the background leading toward the current effort, describing the overall Air Force basic ICAM Program and describing the specific tasks involved in the MCDS development along with a review of concepts and outputs through Phase I.

By the time this paper is given additional data will be available on Phase II implementation and with appropriate approvals will be presented.

DISCUSSION

In recent years the cost of manufacturing has continued to escalate in all industries. The aerospace industry has been no exception. This fact coupled with a shrinking defense budget has made cost reduction and automation to reduce cost a high priority activity. In order to achieve maximum return on investment the Air Force Materials Laboratory has embarked on a coordinated Integrated Computer Aided Manufacturing (ICAM) Program to achieve increased productivity and thereby reduce production cost. The Manufacturing Cost/Design Guide initiated in 1975 by the Air Force provides the basic techniques and formats to allow the design engineer to perform trade studies during the early phases of a program's design activity and thereby establish an optimum design considering the cost to manufacture as a key element in the decision making process.

Figure (1) shows how important it is to conduct these trade-offs early as it is this point in the development of a program when most of the costs are locked in, not only for production but development and Operations and Support as well. It is early in the program's life cycle that the economic leverage to reduce cost is the greatest.

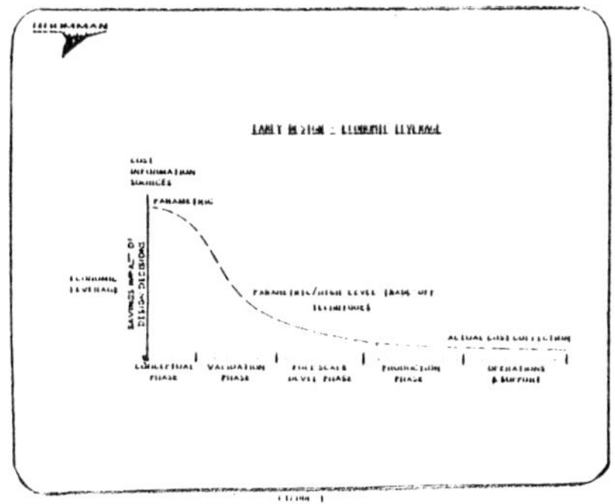

Before proceeding too far into a discussion of the Manufacturing Cost/Design Guide (MC/DG) Computerization Program, let's take a few minutes to review the ICAM program and where the MC/DG and this computerization program fit.

Figure 2 shows the "Basic ICAM Roadmap". Some revisions are currently being made to incorporate composites programs earlier in the schedule but the wedges shown are essentially those currently being pursued by the Air Force. The bar

shown in the first wedge "Design Interaction", is where the MC/DG program is located. ICAM is a major effort in the Manufacturing Technology Division of the Wright Aeronautical Laboratories, covering all aspects of manufacturing including the development of Manufacturing Architectures, automated planning, simulation, language development, fabrication, automated cell/center and factory design, assembly and manufacturing interaction with the design area, etc. Current funding levels are in excess of 100 million dollars, with participation of many industrial firms and universities around the country.

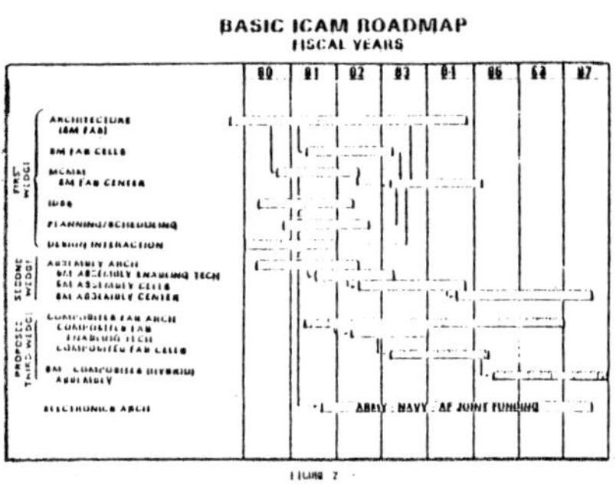

Figure 3 is a matrix showing many of the organizations currently involved in the Air Force ICAM program. With growing competition around the world and the need for increased productivity, Integrated Computer Aided Manufacturing is an important area for the future of the aerospace industry.

With that very brief description of the overall program and its language let us now look at the area of Manufacturing/Design interaction and more specifically at the Manufactur-

ing Cost/Design Computerization
program. To begin with, the development and formatting of the basic
costing and design data is not part
of this particular program, but is
the subject of a separate program
headed by the Battelle Laboratories
under contract to the same ICAM
Program Office in the Air Force
Manufacturing Technology Division.
Therefore this paper will not
specifically address that area but
will outline the computerization of
the outputs of that program and the
interaction of the computerized
MC/DG with other ICAM manufacturing
systems.

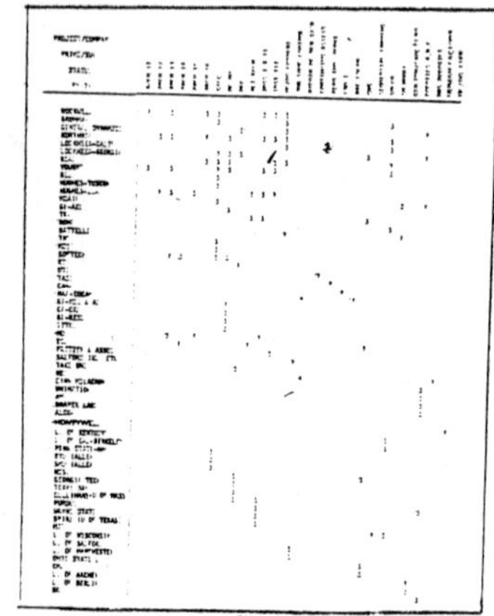

FIGURE 3

With over 160 projects, for the Air
Force to manage in the ICAM area, a
very definite approach has been
established for development of all
ICAM programs. This approach
involves a specific Life Cycle
development that breaks a program
into five major activities (See
Figure 4). They include:
(1) Establish Project Master Plan
and Schedule - this is fairly self
explanatory, but in the case of the

MC/DG Computerization Program
includes development of a number of
detail plans including configuration
management, Software Quality Assurance, etc. (2) Understand the
Problem - this function continues
during the entire Phase I or first
year of the MC/DG Computerization
program and includes:
- Establish MC/DG Environment
- Perform Needs Analysis
- Define System Requirements
- Perform State-of-the-Art Review
(3) Formulate and justify Solution -
during the first year of the program
this function includes formulating a
conceptual design. During Phase II,
the option phase of the program where
actual software is generated, this
function includes performing a preliminary design and detail design.
(4) Construct Integrate and Test -
this function includes:
- Construction and Verification of
  the computer program
- Integration and Validation
- Development of Training Mechanisms

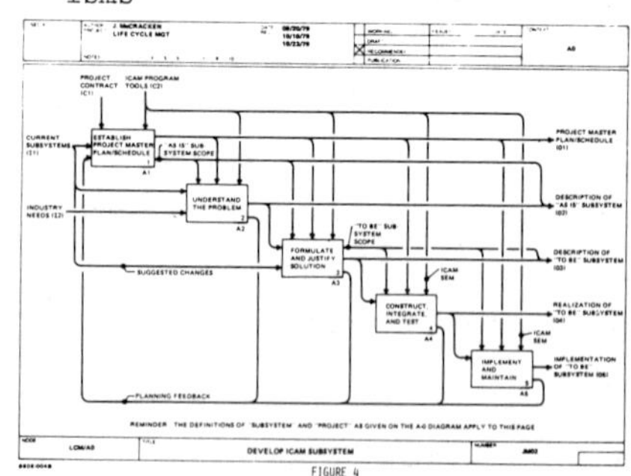

FIGURE 4

(5) Perform Implementation & User
Acceptance - this includes the program demonstration on two host
computers; one at Grumman and one at

Rockwell as well as installation of the program on a National Network.

Figure 5 is a high level overview of the schedule established for the Manufacturing Cost/Design Guide Computerization Program. As you can see the schedule closely parallels the ICAM Life Cycle Plan and as of the first interim report, from which this schedule was taken, the program is proceeding on schedule. Before proceeding through a description of the program concept, and some of the MC/DG coalition's achievements to-date, let us take a minute to review the coalition team.

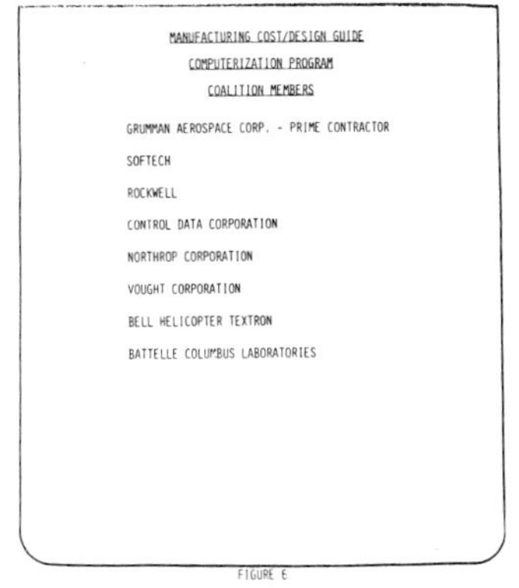

FIGURE 5

Figure 6 shows the coalition member organizations and technical advisors headed up by Grumman Aerospace Corporation as prime contractor.

FIGURE 6

Figure 7 shows the MC/DG Coalition organization. The coalition team members role and cooperation in this type of a program is key to success-ful completion of all program activities.

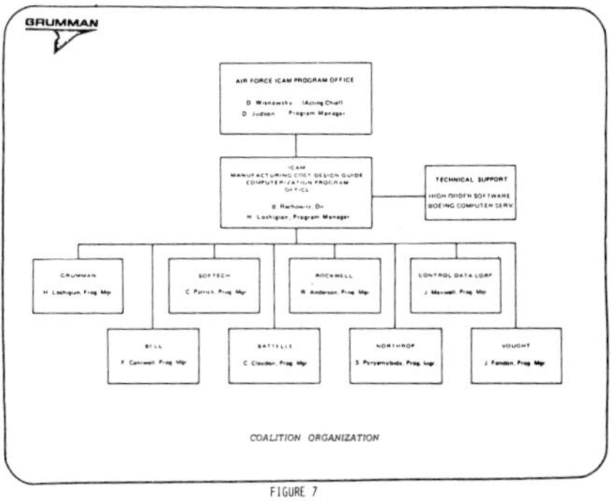

FIGURE 7

Figure 8 is a sample of the initial tasks description matrix that was developed by Grumman and utilized in determining tasks to be performed and the roles individual companies were to play during MC/DG Computer-ization program development.

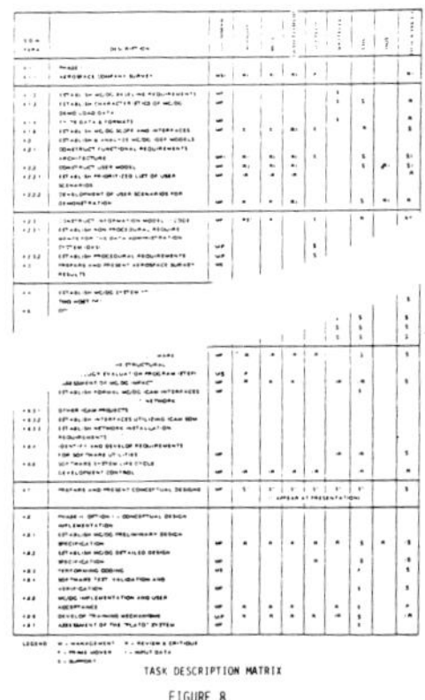

TASK DESCRIPTION MATRIX

FIGURE 8

In order to describe the program and how it fits into the overall scheme of Integrated Computer Aided Manu-facturing in the short time allotted, Figure 9 was prepared as a summary. The lower portion of this figure depicts the MC/DG spanning both the Design and Manufacturing

Architectures. The first triangle
represents the Design Architecture.
This is called the Design Architecture because it includes all the
functions performed in design.
Likewise the second large triangle,
the Manufacturing Architecture
includes all the functions in
manufacturing.

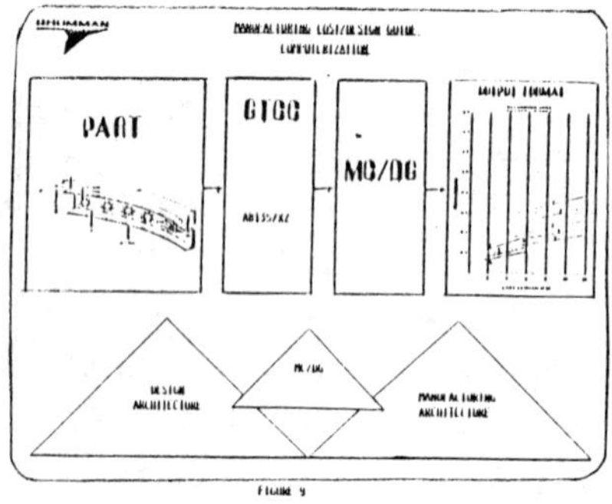

(An example of the Manufacturing
Architecture is shown in the
Figure 10 node diagram "ICAM Composite View of Aerospace Manufacturing"). The triangle that depicts
the Manufacturing Cost/Design Guide
overlaps both architectures demonstrating that the guide plays a
definite role in both Manufacturing
and Design. The computerized guide
will allow the design engineer
access to manufacturing information
relative to production cost for
performing trade studies. Once
integrated with other manufacturing
systems, such data as those determining material and machine availability, etc., will provide the
design engineer with the information
required to make the optimum design
decision the first time, thereby
hopefully eliminating the need for

design changes once the part nears
production. The upper portion of
Figure 9 very simply shows some of
the events that will take place in
developing the part cost and performing the computerized tradeoffs. The
MC/DG system will eventually be provided with the part shape (geometry)
and description from a computer aided
design system. This description will
be provided via the ICAM Group Technology Coding and Characterization
system which will fully describe the
part. Utilizing the data developed
as part of the MC/DG or company data
development activities, the computerized MC/DG will display an output
format for the design engineer.
Depending on the independent variable
(in this case part length) the lowest
cost to produce will be calculated
and noted. As stated previously this
is a very simple example of one of the
possible outputs and how the system
will use other ICAM programs to
develop a required output. The
MC/DG Computerization program is
currently surveying the needs of
industry and developing "as is" current models of various aerospace
companies to determine the many
diverse functions this system will
perform for the Management, the
Designer and Production people alike.

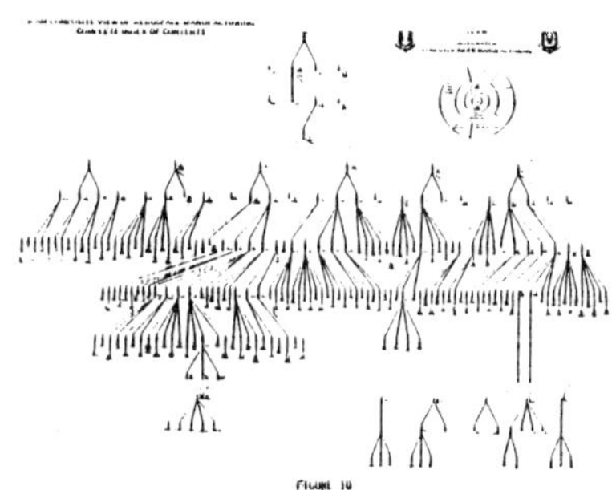

The balance of this paper will review
some of the program activities to
date outlining ICAM project interfaces, the survey being performed
and sample of the factory models
being generated.

Figure 11 is a node diagram showing,
at a very high level, the ICAM project interfaces with the MC/DG program. As you can see the MC/DG
will interface in each of the major
ICAM program areas, having a major
impact in the Application Systems
activity.

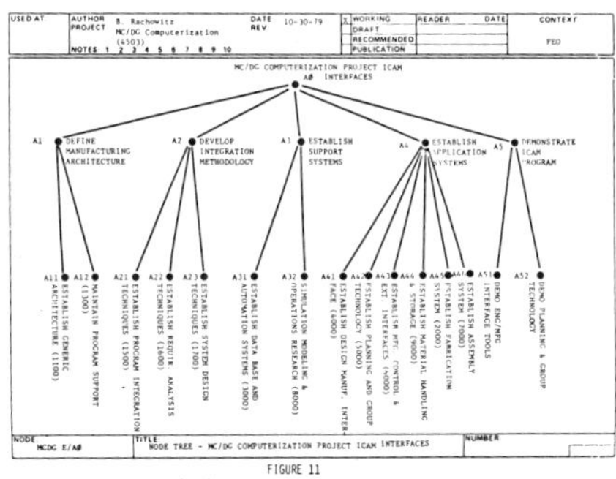

FIGURE 11

As a key element in providing manufacturing information to the design
engineer relative to planning,
manufacturing control, material
handling, etc., the MC/DG will provide the Design/Manufacturing real
time interface that is so important
in reducing costs.

A key task in structuring the overall system is making sure that the
requirements of the user and his or
her supervisors are accounted for in
generating a system that will be both
accepted and useful. Therefore, a
survey has been prepared and distributed to 57 companies, representing small, medium and large manufacturing organizations.

Figure 12 describes the composition
of the survey package. Each survey
administrator is given a complete
package describing; why and what the
survey is all about, stating the
purpose and viewpoint as well as
questions tailored for each of three
types of individuals: (1) Design
Engineer, (2) Design Engineering
Manager and (3) Data Processing
Manager.

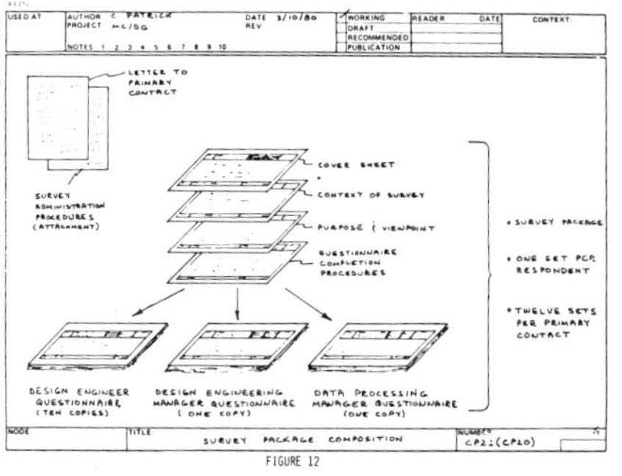

FIGURE 12

Figures 13 through 15 show sample
question pages from each of these
survey areas. At the time of publication of this paper only a few
inputs had been received, but we
have been notified by each of the
companies that a formal response
will be sent by the required due
dates. From this small sample the
desire for such a system has been

clearly indicated.

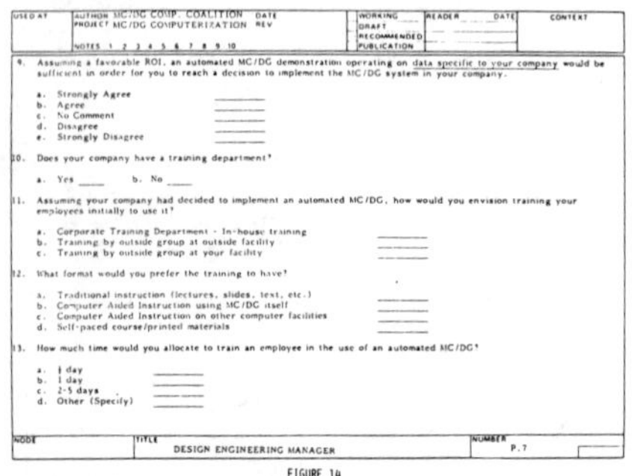

FIGURE 13

Finally as the program develops each
of the major coalition companies
were required to submit an "as is"
factory model of the current methods
for accomplishing the manufacturing
cost/design trade studies using the
Air Force ICAM definition methodology
IDEF.

FIGURE 14

IDEF is a good structured method for
describing functions and their associated inputs, outputs, controls and
mechanisms. During this presentation a further description of this
methodology will be given, but if
you wish to learn more about the
system you can contact the Air
Force ICAM Office (AFWAL/MLTC) at
the Manufacturing Technology
Division, Wright-Patterson Air
Force Base, Ohio 45433.

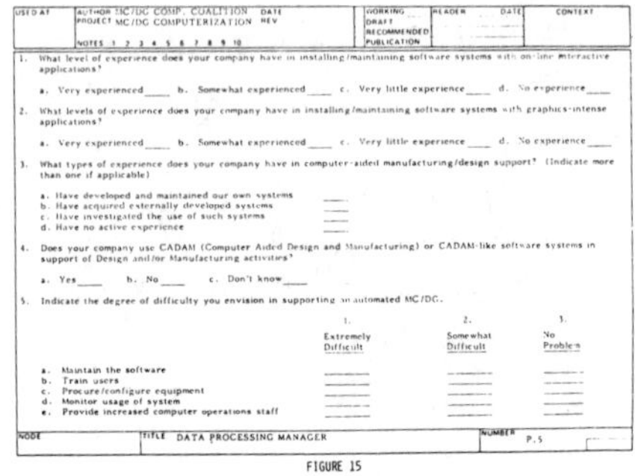

FIGURE 15

Figure 16 is the high level or A-Ø
IDEF chart for "Perform Manufacturing Cost/Design Tradeoffs".

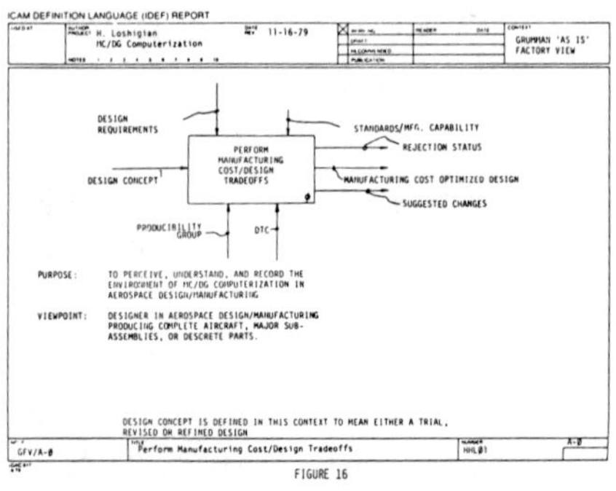

FIGURE 16

Figure 17 is a decomposition of
Figure 16 to the AØ level showing
the functions included in the A-Ø
chart. Further decompositions broke
down each of these functions to the
lowest level possible. Upon completion of each company's "as is"
factory model, a detailed review and
coalition meeting synthesized the
composite model used for development
of the "to be" computerized MC/DG
system. Conceptual designs were
generated to form the basis of final
design selection.

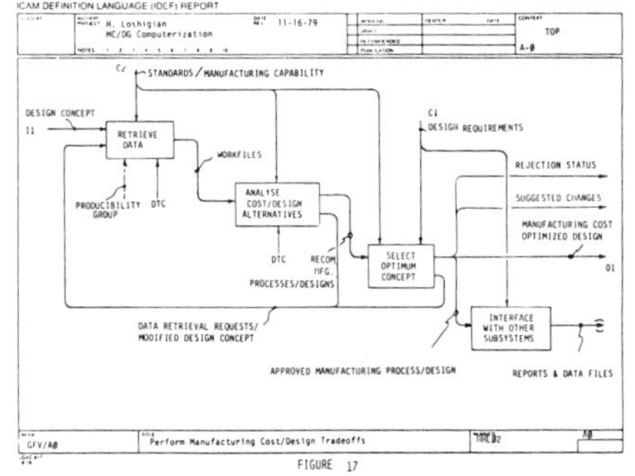

FIGURE 17

The conceptual design activity
resulted in the development of a
Manufacturing Cost Design System
(MCDS) Requirements Document. This
document, Figure 18, provides the
basis of a mutual understanding
between the system developers and
system users in terms of capabilities, functionality, and environment of the system.

BASELINE FUNCTIONAL REQUIREMENTS

0   SUPPORT OF MANUFACTURING COST/DESIGN
    TRADE-OFF ACTIVITIES

0   SUPPORT OF SYSTEM DATA MANAGEMENT ACTIVITIES

0   SUPPORT OF INTERACTIVE USER TRAINING ACTIVITIES

0   SUPPORT OF SYSTEM DEMONSTRATION OF COST DATA
    SYNTHESIS

FIGURE 18

Preliminary Design, (Figure 19),
began before the end of Phase I of
the MC/DG Computerization program,
initiated the functional design,
meeting the requirements with major
emphasis on user interfaces, and
iteratively synthesizes the
alternative architectural design of
the subsystems/modules to effectively perform the functions. At the
time this updated report was being

prepared, the Grumman Coalition had
completed the Phase I activity and
received approval from the Air Force
to begin the Phase II activity.

PRELIMINARY DESIGN - FUNCTIONAL

0   DESIGN OF MCDS USER INTERFACES
    -   HUMAN FACTORS
    -   SYSTEM/USER DIALOGUE
    -   COMPLETENESS
    -   CONSISTENCY
0   ANALYZE MCDS REQUIREMENTS
0   SYNTHESIZE SOFTWARE ARCHITECTURE
0   ALLOCATE SOFTWARE FUNCTIONS TO SUBSYSTEMS/MODULES
0   ANALYZE AND OPTIMIZE HYPOTHESIZED SOFTWARE CONFIGURATION

FIGURE 19

Figure 20 describes the preliminary
information analysis to be performed
during Phase II, to define the
"to be" structure and to transform
$IDEF_1$ definitions into Data Base
(IDBMS) set type definitions.

PRELIMINARY DESIGN - INFORMATION

0   ANALYSIS OF $IDEF_1$ MODEL TO DEFINE "TO BE"
    MCDS INFORMATION STRUCTURE

0   TRANSFORMATION OF $IDEF_1$ ENTITY - RELATIONSHIP -
    ATTRIBUTE DEFINITIONS TO IDBMS SET TYPES,
    OWNER/MEMBER RELATIONSHIPS, ITEM DEFINITIONS.

FIGURE 20

At the completion of this model,
the Design Engineer will have a
tool to interactively perform
timely broad based cost/design trade
studies, resulting in significant
cost effectiveness and productivity
improvements within the Aerospace
Industry.

BIOGRAPHIES

BERNARD I. RACHOWITZ

Education

AAS, Mechanical Technology,
Academy of Aeronautics
BS Cum Laude, C. W. Post College,
Long Island University
MS, Graduate School of Business,
Long Island University

Experience

Mr. Rachowitz has been with Grumman
Aerospace Corporation for 13 years.
He is currently Manager, Product
Technology Development/Software
Systems Department. In this
position he is responsible for all
Product Technology Operations as
well as coordinating all Grumman's
Corporate ICAM activities with the
Air Force as Director of the Grumman
ICAM Interface Office.

Prior to this assignment, he was
Manager of Cost Technology Development, responsible for managing
Grumman's Design-To-Cost activities,
computerized estimating and pricing
development, and implementing and
managing the development of a
number of Air Force and Navy contracts (e.g., Modular Life Cycle
Cost Model for Advanced Aircraft
Systems, Manufacturing Cost Data
Collection and Analysis for Composite Production Hardware, Manufacturing Cost/Design Guide
Subcontract, Composite Material
Factor Development, etc.).

Prior to the above assignments,
Mr. Rachowitz was group leader in
the subcontractor control group of
the Mass Properties Control Engineering section. In this capacity, he
was responsible for preparing specifications, participating in design
trades, establishing and negotiating
subcontractor requirements (both
technical and data), as well as
controlling the efforts of both sub-
and associate contractors. On the
F-14A program, in addition to the
above, he prepared the Grumman
engineering positions and negotiated
many of the changes submitted by the
associate contractors. In addition
to the F-14 program, he was assigned
to a number of proposals such as the
EF-111A, Space Shuttle, Global
Positioning System, etc., where he
developed and exercised a number of
computer programs for both weight
and cost.

Society Members/Papers/Lectures, etc.

Tau Alpha Pi - National Honor
Society for Technical Colleges -
Member.
S.A.W.E. - Senior Member

Mr. Rachowitz has presented a number
of papers for the S.A.W.E., A.I.A.A.,
A.D.P.A., as well as lecturing on
various engineering cost related
topics at the University of Dayton's
Aircraft Design Short Course, Stony
Brook University in New York,
Virginia Polytechnic Institute and
the Air Force Institute.

HENRY H. LOSHIGIAN

Education

BA, Columbia College

BS, Engineering Mechanics,
Columbia Engineering

MS, Engineering Mechanics,
Columbia Engineering

PF, Engineering Mechanics,
Columbia Engineering

Advanced Graduate Study, Polytechnic Institute of Brooklyn

Experience

Mr. Loshigian is currently the Grumman ICAM Program Manager responsible for all ICAM-related activities and the Project Manager for the ICAM Manufacturing Cost/Design Guide Computerization. His most recent position was the Project Manager of the Rapid Aerospace Vehicle Evaluation System (RAVES), which is an integrated system of over 245 analytical computer programs used by over 450 engineering professionals representing 16 technical disciplines on virtually every project in the Grumman Aerospace Corporation. Since 1973, his responsibilities on RAVES have varied from the development of system specifications and integration/acceptance of RAVES into the Engineering Department, to the management of RAVES as a production system during the past four years. This effort included development of all production documentation and user application guides.

Mr. Loshigian is the Grumman representative on the Industry Technical Advisory Board (ITAB) for NASA's Integrated Programs for Aerospace Vehicle Design (IPAD), an ICAM-related program which treats the computerization of engineering processes. He is also a permanent member of the ITAB subcommittee for user scenarios.

Mr. Loshigian's past assignments include NASA Space Shuttle and Lunar Module programs, and a variety of aircraft projects ranging from the Grumman commercial Gulfstream II to the F-14A fighter. Mr. Loshigian was elected to the Grumman Engineering Professional Development Program (EPDP) which is a two-year accelerated career development program structured to provide a broad range of management and design experience.

Society Membership/Papers/Lectures, etc.

American Society of Mechanical Engineers

American Society of Civil Engineers

Licensed Professional Engineer, New York State

Mr. Loshigian has presented a variety of CAD-related papers for the ASME, SAWE and NASA.

MECHANICAL CHARACTERIZATION OF BONDED AND BULK
ADHESIVE SPECIMENS AS AFFECTED BY TEMPERATURE AND MOISTURE

O. Ishai and G. Dolev
Faculty of Mechanical Engineering
Technion - Israel Institute of Technology
Haifa, Israel

Abstract

Extensive tests were conducted on bonded in-situ and bulk FM 73 adhesive specimens under different temperature and moisture conditions.

It was found that the yield strength and the elastic modulus as derived from the torsion test are in correlation for the two adhesive systems. The properties derived from the uniaxial test of the bulk adhesive material in tension and compression were related to those of the adhesive layer in-situ in shear, through a combined stress law which complies with a modified Von Mises yield failure criterion.

The effect of strain-rate and temperature on yield strength in both systems was found to obey a time-temperature superposition law similar to the corresponding behavior of other ductile polymers. Based on this, a master curve was obtained covering the yield behavior of the adhesive for about ten decades of strain-rate.

Moisture in the adhesive specimens reduced both their stiffness and their strength characteristics but had no significant effect on the shift factor, which is an essential parameter in the yield strength-time-temperature relationship.

It was concluded that the mechanical characteristics and the mode of behavior of a bonded in-situ adhesive layer, subjected to a combined state of stress under different environmental conditions, can be approximately predicted on the basis of uniaxial test data for bulk specimens.

1. INTRODUCTION

In an earlier work [1,2] an attempt was made to find the relationship between the mechanical characteristics of an adhesive bonded in-situ and those of the bulk material. It was found that in FM73, the structural adhesive tested, which follows a visco-elastoplastic mode of behavior, shear strength was almost completely unaffected by thickness. This is

supported by other findings reported
in literature for "intermediate-
modulus adhesives"[3-6] which similarly obey a ductile mode in shear.
In the works cited, as well as in
others[7-9], the shear modulus was
found to vary in thin bonded adhesive layers over a ± 20% interval
compared with their thick-layer and
bulk references. This variation is
attributable to a more flexible sub-
layer, which forms close to the
interfaces as a result of physico-
chemical effects in the molecular
structure of the adhesive in the
interfacial zone[10]. The difficulty in accurately determining the
shear strain of a thin bonded layer
may also be responsible for the
high scatter and contradictory results for elastic moduli of the
bonded adhesive[11-13].

Another problem in the mechanical
characterization of a bonded system
is how to relate the basic parameters – such as shear strength and
modulus – to the mechanical behavior of layers bonded in situ,
which are mostly in a complex
multidirectional state of stress –
including the case of an adhesive
layer in a butt-joint under simple
unaxial tensile loading[9,14,15].

The complex experimental character-
isation of an adhesive layer bonded
in-situ on the one hand, and its
interpretation as a design allowable
and the stress analysis of the

bonded joint on the other, justify
treatment of the adhesive layer not
as a distinct phase, but as a regular element of a multi-material
system, which is expected to behave
like other ductile polymers of this
type rather than like FRP laminae
or metal adherents.

The following premises were postulated in the present work:
1. The mechanical and hygrothermal
behavior of an adhesive layer bonded
in-situ is similar to that of the
corresponding bulk polymer.
2. A bonded joint with appropriate
surface treatment is subject to cohesive failure.
3. A simplified elastoplastic model
(characterized by its initial elastic moduli and yield plateau) suffices for the approximate prediction
of the mechanical behavior of the
adhesive[16].

These premises were found to be approximately valid for FM73 at room
temperature[1,2], except for the
hygrothermal aspect. The latter,
dealt with in numerous studies on
environmental effects in structural
bonded joints[17-24], proved a tough
nut to crack, both experimentally
and theoretically. The present
project was accordingly undertaken
with a view to:
1. Establishing a relationship between bulk and in-situ material
characteristics under different
loading conditions;

2. Studying the mechanical behavior
at different states of stress versus
that under simple uniaxial loading -
under different hygrothermal conditions.

3. Studying the effect of strain-
rate, temperature, and moisture, on
the mechanical characteristics of
the bulk adhesive versus a bonded
adhesive layer.

2. EXPERIMENTAL

Structural film adhesive FM73, supplied by Cyanamid Bloomingdale, was
chosen for the representative ductile adhesive system. The adherends were made of aluminum 2023/T3;
the primer was Br-127.

2.1 Preparation of specimens

The specimens were of four types
(Fig.1):

U1 - for uniaxial tension   ) of
U2 - for uniaxial compression) the
) bulk
S1 - for shear   ) ad-
) hesive

S2 - for shear, bonded adhesive (In
     situ ahesive layer). In ad-
     diton, a control specimen
     without an adhesive layer was
     prepared.

The procedure of preparation was as
follows:

- Lay-up of the adhesive films in
  closed mold, in order to prevent
  flow under pressure*.

- Application of 10 at. pressure on
  closed mold.

- Heating to 120°C and holding for
  30 minutes.

- Slow cooling to room temperature.

- Release of pressure.

- Final shaping.

2.2. Environmental variants

The variants were "dry" and "wet".
"Dry" specimens were kept in pro-
longed storage at 50% relative
humidity and 23°C and loaded at four
temperatures, viz.: 23°C, 40°C, 60°C,
and 80°C. "Wet" specimens were
tested after being immersed in distilled water at 60°C until saturation (relative moisture contents
of 3.5% and 2.8% for compressive and
tensile specimens respectively) by
loading at 23°C, 40°C, and 60°C.

2.3 Testing procedure

The specimens of type S2 were tested
by uniaxial loading with the aid of
a special torsion device (Fig. 2);
the shear strain γ was determined
on a conventional Instron extensometer after deduction of the adherend displacement obtained from the
control specimens. The specimens
of type S1 were similarily tested,
while specimens of types U1 and U2
were tested directly on the Instron
tester. A constant strain rate was

---

*
This represents more closely the
real situation of the adhesive
layer region at some distance
from the edges.

maintained throughout the tests.

3. RESULTS AND DISCUSSION

3.1 Environmental effects

Typical tensile stress-strain curves at different temperatures for the dry and wet bulk specimens are given in Fig. 3. The compressive and shear stress-strain behavior of the bulk and the bonded specimens showed similar trends. In all cases the wet specimens showed lower yield and stiffness characteristics compared with their dry counterparts.

The effect of strain-rate and temperature on the yield strength of dry and wet bulk specimens is shown in Figs. 4 and 5, respectively. As can be seen, linearity prevails between yield stress and log strain-rate throughout the temperature and strain-rate ranges involved. This is in agreement with the visco-plastic character of the adhesive, in common with other ductile polymers[25,26]. In these circumstances, the yield stress log-strain rate temperature curves of Figs. 4 and 5 can be merged into a single master curve, using the time temperature superposition shift method. Such master curves are shown in Fig. 6, with strain-rate range being extended to ten decades; the shift factor is plotted as a function of temperature in Fig. 7.

From Figs. 3-7 the following may be concluded:

- Moisture has a significant effect in that it reduces the yield strength but leaves the shift factor almost unaffected.

- At moderate temperatures (40°C and below) the compressive yield strength of specimens is higher than its tensile counterpart, whereas at elevated temperatures there is practically no difference between the two. This is attributable to plasticization of the polymer with the onset of the rubbery mode of behavior, in which the contribution of isotropic stress components becomes negligible.

- Moisture was found to lower the glass transition temperature*, as demonstrated by the predominantly rubbery mode of behavior, which was found to prevail at elevated temperature (60°C) in the case of wet specimens (Fig. 3). Hence no yield data can be obtained in this hygrothermal range, and the failure mechanism is not expected to comply with the ductile strength criteria.

The effects of temperature and moisture on the initial tensile elastic modulus of the bulk adhesive specimens are shown in Fig. 8. The decrease in stiffness with increasing temperature and moisture content is readily seen, although the high scatter somewhat obscures the effect of the strain-rate on this parameter.

3.2 Relationships among mechanical characteristics

Average elastic and yield strength values at the same strain-rate level, derived from the stress-strain curves for the dry bulk and bonded

*This was established for several resins in previous works and during a preliminary stage of the present research for FM73[1,2].

specimens, are given in Table 1 and compared with computed values based on the following theoretical considerations:

Elastic parameters for isotropic materials are related by the classic law

$$G = \frac{E}{2(1+\nu)} \qquad (1)$$

where G, E, and ν are, respectively, the shear modulus, Young's modulus, and Poisson ratio.

Yield strength in shear, tension, and compression, can be related by a modified Von-Mises distortional energy criterion, as follows:

$$K_s \, \tau_{oct} + K_v \, \sigma_m = 1 \qquad (2)$$

where τ_{oct} - octahedral shear stress, given by

$$\tau_{oct} = \frac{1}{3}[(\sigma_1-\sigma_2)^2 + (\sigma_1-\sigma_3)^2 + (\sigma_2-\sigma_3)^2]^{\frac{1}{2}} \qquad (3)$$

σ_m - isotropic normal stress, given by

$$\sigma_m = \frac{1}{3}[\sigma_1 + \sigma_2 + \sigma_3] \qquad (4)$$

$\sigma_1, \sigma_2, \sigma_3$ - the principal normal stresses,

K_s, K_v - material parameters responsible for plastic yield due to the deviatoric and isotropic stress tensors, respectively.

Substitution of the particular cases of uniaxial tension, compression, and shear, in eqs.2-4 yields a direct relationship between the three yield parameters, as follows:

$$\alpha_s = \frac{\sqrt{3} \; (\lambda+1)}{2\lambda} \qquad (5)$$

where

$\lambda = F_c/F_t$ - the ratio between the compressive and tensile yield strength, and

$\alpha_s = F_t/F_s$ - the ratio between the tensile and shear yield strength.

Table 1 shows good agreement between experimental data and elastic and yield values computed with the aid of eqs.1 and 5. The latter equation was also found adequate for predicting the yield behavior of other adhesive systems, in different conditions, as shown in Fig. 11 of Ref. 27.

The yield failure criterion postulated in eqs.2-4 may be described in terms of yield-failure envelopes (Fig. 10). Here again there is good agreement between experimental shear yield values and those derived theoretically from experimental tensile and compressive yield data.

A similar downward trend of the octahedral yield stress with increasing isotropic stress is common to all dry specimens up to 60°C. Above that temperature, however, and in the wet specimens $\lambda \simeq 1.0$, which means that the octahedral stress at yield is constant and unaffected by the isotropic stress, in keeping with the classic Von-Mises criterion.

4. CONCLUSIONS

A program of experimental characterization was conducted on bulk and bonded FM73 adhesive specimens under tensile, compressive, and shear, loading at different temperatures and strain-rates and in different moisture conditions.

The following conclusions could be drawn:

1. The mechanical behavior of the bulk adhesive is similar to that of other ductile polymers.

2. Elastic modulus and yield strength in shear are approximately the same for the bulk and for bonded adhesives.

3. A master yield-strength vs. log-strain-rate curve can be obtained by the time-temperature superposition shift method, as for other ductile polymers.

4. The elastic shear parameter for the bulk and the bonded adhesive may be computed from available experimental tensile elastic parameters with the aid of the classical laws for isotropic materials.

5. The yield strength in shear of the bulk and the bonded adhesive can be computed from tensile and compressive data based on a modified distortional energy criterion. The result is a general law, which predicts the yield behavior of the bonded adhesive under a multiaxial state of stress, as represented by the failure envelope.

6. A similar procedure is proposed for the mechanical characterization of a bonded adhesive as follows:

(i) Uniaxial testing of the bulk adhesive in tension and compression.

(ii) Calculation of shear stiffness and yield strength based on available theoretical relationships.

(iii) Checking of the computed shear data by a napkin-ring, or a thick-adherend, method for in-situ bonded adhesives.

(iv) With (iii) verified, the theoretical formulation embodied in eqs.1–5 may be used to predict the yield and the elastic behavior of the bonded adhesive in a multiaxial state of stress at different temperatures and moisture levels.

(v) For the evaluation of environmental effects on yield it suffices to find the yield vs. strain-rate behavior experimentally at three temperatures, to determine the shift factor as a function of temperature, and to construct the master curve.

(vi) The effect of moisture on the master curve may be similarly evaluated by using the same shift factor as for the dry counterparts.

In order to verify the proposed procedure, further tests must be conducted on bulk and bonded adhesive specimens in direct multiaxial states of stresses at different temperatures and moisture levels.

5. REFERENCES

1. Ishai, O.,and Dolev, G., "Mechanical Characterization of Structural Adhesives", First Year Report, MML-65 Technion - I.I.T., (1979).

2. Dolev, G., "Mechanical Characterization of Structural Adhesives Under Different Loading and Environmental Conditions", M.Sc. Thesis, Technion - I.I.T., (1979)

3. Frazier, T.B., and Seago, R.A., "Effects of Bondline Thickness on Some Shear Parameters of a Film Adhesive", Report No. 599-002-903, Bell Helicopter Company, (1969).

4. Foulkes, H., Shield, J., and Wake, W.C., "The Variation of Bonded Strength with Temperature: A Preliminary Study of Metal-to-Metal Adhesion", J. Adhesion, Vol.2, p.254, (1970).

5. Mccarvil, W.T., and Bell, J.P., "Torsional Test Method for Adhesive Joints", J. Adhesion, Vol.6, pp.185-193, (1974).

6. Tanner, W.C., "Manufacturing Processes with Adhesives Bonding", from: Processing for Adhesive Bonded Structures", Edited by M.J. Bodner, Applied Polymer Symposia, No.19, pp.1-21, (1972).

7. Bryant, R.W., and Dukes, W.A., "The Measurement of the Shear Strength Adhesive Joints in Torsion", Brit.J.Appl.Phys., Vol.16, (1965).

8. Dukes, W.A., and Bryant, R.W., "The Effect of Adhesive Thickness on Joint Strength", J. Adhesion, Vol.1, p.48, (1969).

9. Hughes, E.J., and Rutherford, J.L., "Study of Micromechanical Properties of Adhesive Bonded Joints", Aerospace Research Center, General Precision Systems, Technical Report 3744, (1968).

10. Peretz, D., "Shear Stress-Strain Characteristics of Adhesive Layers", J. Adhesion, Vol.9, pp.115-122, (1978).

11. Renton, W.J., and Jones, W.B., "Instrumental Test Methods for Adhesively Bonded Joints", Proc. of 10th Technical SAMPE Conf., p.939, (1978)

12. Renton, W.J., and Vinson, J.R., "Shear Property Measurement of Adhesives in Composite Material Bonded Joints", Composite Reliability, ASTM STP 580, pp.119-132, (1975).

13. Krieger, R.B., "Fatigue Testing of Structural Adhesives", SAMPE, Vol.24, book 1, p.1102, (1978).

14. Alwar, R.S., and Nagaraja, Y.R., "Elastic Analysis of Adhesive Butt Joint", J.Adhesion, Vol.7, p.279 (1976).

15. Rutherford, J.L., Bossler, F.C., and Hughes, E.J., "On Measuring the Properties of Adhesives in Bonded Joints", SAMPE: "Advanced Techniques for Material Investigation and Fabrications", Vol.14, (1968)

16. Gali, S., and Ishai, O., "Interlaminar Stress Distribution Within an Adhesive Layer in the Nonlinear Range", J.Adhesion, Vol.9, pp.253-266, (1978).

17. Frazier, T.B., and Lajoie, A.D., "Durability of Adhesive Bonded Joints", Technical Report AFML-TR - 74-26, Bell Helicopter Company, (1974).

18. Lohr, J.E., "Factors Affecting the Survivability of Stressed Bonds in Adverse Environments", 18th National SAMPE Symposium: "New Horizons in Materials and Processing", p.418 (1973).

19. Bethune, A.W., "Durability of Bonded Aluminium Structures", SAMPE Journal, Vol.11, No.3, p.4, (1975).

20. DeLollis, N.J., "Durability of Structural Adhesive Bonds: A Review", Adhesive Age, p.41, (1977).

21. Wegman, R.F., Bodnar, M.J., and
    Ross, M.C., "A New Technique for
    Assessing the Durability of
    Structural Adhesives", SAMPE
    Journal, Vol.14, No.1, p.20
    (1978).

22. Althof, W., and Brockman, W.,
    "New Test Methods for the Pre-
    diction of Environmental Resist-
    ance of Adhesive Bonded Joints",
    SAMPE, Vol.21, (1976).

23. Althof, W., Gerhard, K., Neuman,
    G., and Schlothauer, J., "En-
    vironmental Effects on the
    Elastic-Plastic Properties of
    Adhesives in Bonded Metal
    Joints", Royal Aircraft Estab-
    lishment., Library Translation
    1999 (1977).

24. Weitzman, Y., "Stresses in Ad-
    hesive Joints Due to Moisture
    and Temperature", J.Composite
    Materials, Vol.11, p.378, (1977).

25. Ishai, O., "Delayed Yielding of
    Epoxy Resin Under Tension Com-
    pression and Flexure", J.App.
    Poly.Sci., Vol.11, p.963, (1967).

26. Ishai, O., "The Effect of Tem-
    perature on the Delayed Yield
    and Failure of Plasticized Epoxy
    Resin", Pol.Eng. and Sci.,
    Vol.9, No.2, p.177, (1975).

27. Gali, S., Dolev, G., and Ishai,
    O., "An Effective Stress-Strain
    Concept in the Mechanical
    Characterization of Structural
    Adhesive Bonding", Adhesion and
    Adhesives (1981).

Table 1: Average Experimental and computed data on dry specimens.

Temperature	Elastic Characteristics					Yield Strength Characteristic					
	E_t (Kg/m^2)	ν	G (Kg/m^2)			F_t (Kg/m^2)	F_c (Kg/m^2)	$\lambda = \dfrac{F_c}{F_t}$	F_s		
	Bulk	Bulk	Bulk	In-situ	computed eq.(1)	Bulk	Bulk	Bulk	Bulk	In-situ	computed eq.(5)
$23^{\circ}C(R.T.)$	225	0,43	80	75	79	6.6	4.7	1.4	3.0	3.15	3.16
$40^{\circ}C$	185	0.4	–	67	66	5.1	4.0	1.27	–	2.35	2.58
$60^{\circ}C$	145	0.4	60	55	52	3.5	2.8	1.25	1.65	1.70	1.80
$80^{\circ}C$	$\sim$95					1.6	1.6	1.0			0.92

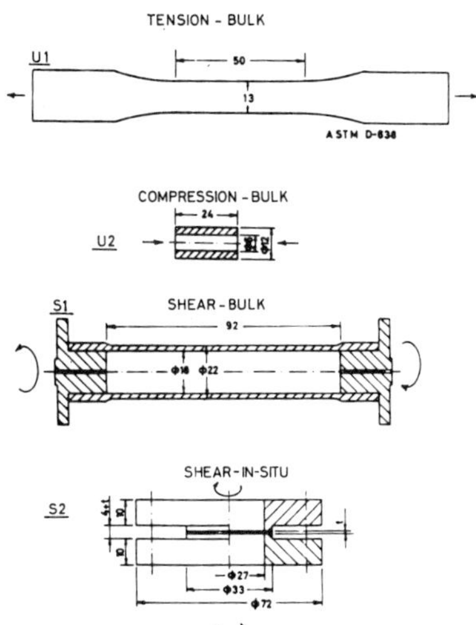

FIG.1. SPECIMEN GEOMETRY

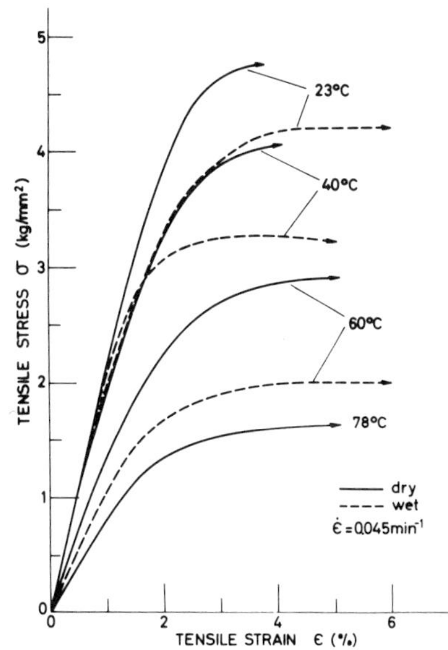

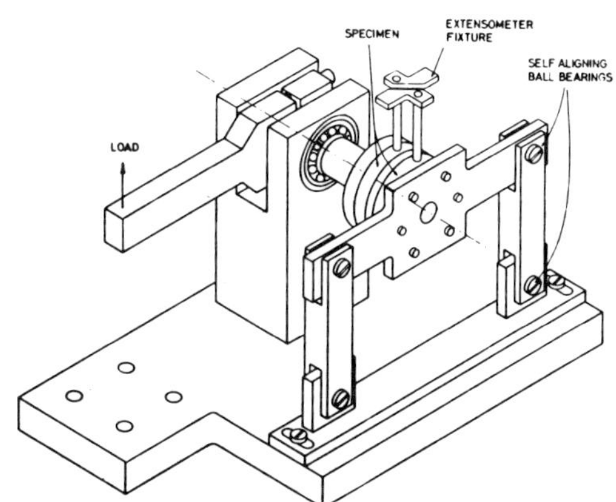

FIG.2. DEVICE FOR THE TORSIONAL
LOADING OF RING SPECIMENS

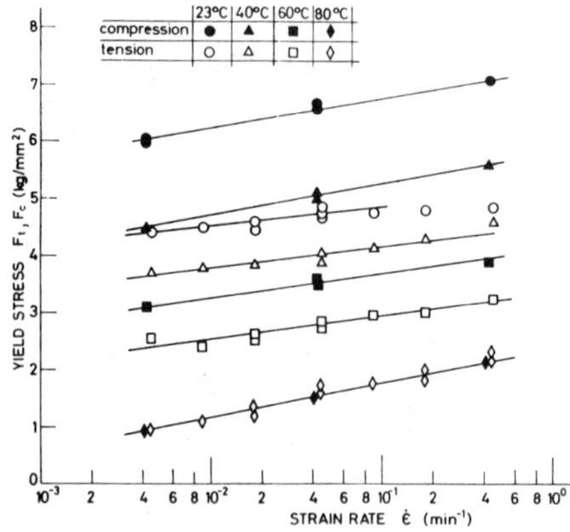

FIG.4. YIELD STRESS – VS. STRAIN
RATE FOR DRY SPECIMENS AT
DIFFERENT TEMPERATURES

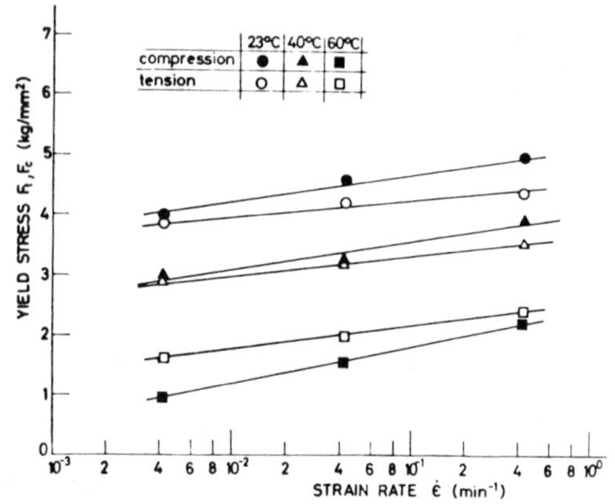

FIG.5. YIELD STRESS – VS. STRAIN
RATE FOR WET SPECIMENS AT
DIFFERENT TEMPERATURES

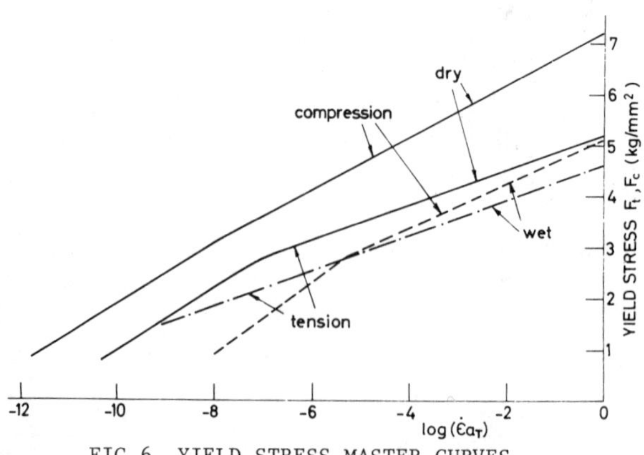

FIG.6. YIELD STRESS MASTER CURVES
FOR DRY AND WET SPECIMENS

640

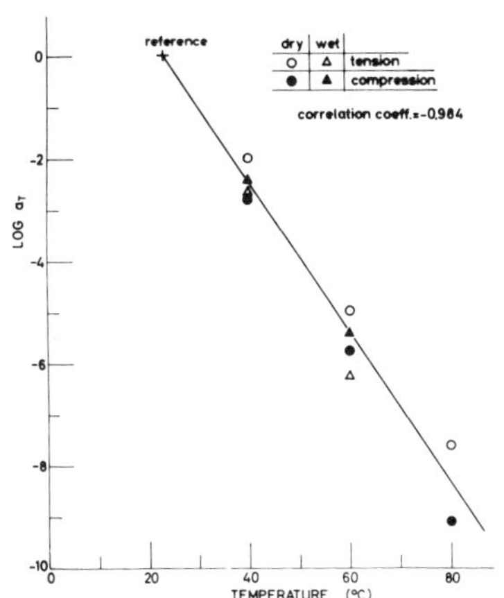

FIG.7. SHIFT FACTOR FOR YIELD STRESS MASTER CURVES AS A FUNCTION OF TEMPERATURES

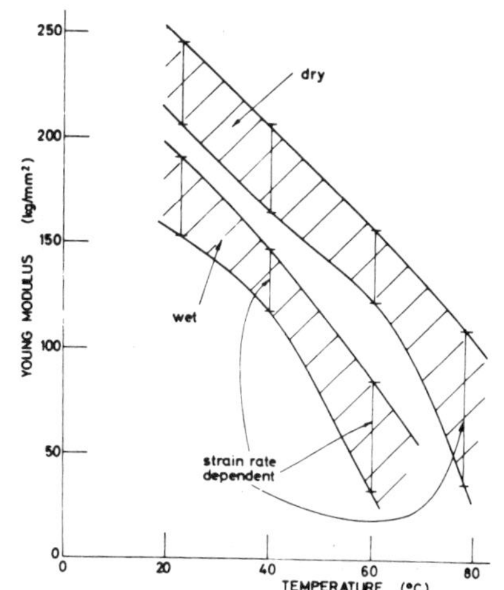

FIG.8. EFFECT OF TEMPERATURE ON THE INITIAL TENSILE ELASTIC MODULUS OF DRY AND WET SPECIMENS

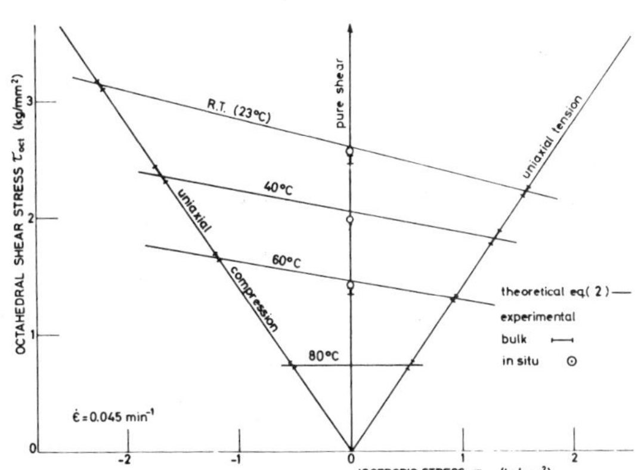

FIG.9. FAILURE ENVELOPES FOR FM-73 ADHESIVE IN DIFFERENT TEMPERATURE CONDITIONS

26th National SAMPE Symposium
April 28-30, 1981

FUTURE APPLICATIONS OF ADHESIVE
BONDING IN THE U.S.AIR FORCE
Theodore J. Reinhart
Materials Laboratory
Air Force Wright Aeronautical Laboratories
Wright-Patterson Air Force Base, Ohio

Abstract

Service experience with structural adhesive bonding on fighter, transport and bomber aircraft as well as recent progress made on Air Force sponsored research and development on structural adhesive bonding is described. Service experience with adhesive bonded aircraft components, both metal to metal and honeycomb construction has since the introduction of corrosion inhibiting primers; the improved adhesive materials and the optimized FPC etch been extremely satisfactory. Bonded construction on the C-5, F-111 and on the F-15 etc., as well as the (so-called) new technology replacement components for the C-130 and C-141 have given outstanding performance. The gradual switch to the Boeing developed phosphoric acid anodize will permit even greater reliability to be built into our adhesive bonded components. Recent R&D work involving the fatigue and fracture characteristics of bonded joints, life prediction methodology the Production Adhesive Bonded Structures Technology, "PABST" all adhesive bonded AMST fuselage, the bonded, laminated structures programs as well as all of the research now being completed on automated chemical composition control, automatic cure monitoring with closed loop control of the cure, electro-depositable corrosion inhibiting primers, have given us an unprecedented technology base from which we can build, to demonstrate that structural adhesive bonding is a cost effective approach to the fabrication of reproducible and reliable minimum weight metallic aircraft structures.

1. INTRODUCTION

Recently very significant progress has been made in advancing the state of the art of structural adhesive bonding. These advances have taken form primarily in the materials and processes area where previously significant technology limitations existed. Briefly, these advances include the following as shown in Table 1; The development of corrosion inhibiting primers, advanced technology adhesive materials, phosphoric acid anodize, chemical composition control procedures, and improved materials and process testing and evaluation procedures. The success and trouble free service provided by this technology prompted the Air Force to fund the very successful "PABST" all bonded fuselage, advanced development program and the successful advanced technology laminated wing structure programs. These efforts vividly demonstrated the outstanding benefits to be gained via bonded metallic structures. Also, during this period other critical areas were being covered by high quality research programs. These areas included (see table 2) work to understand the fatigue and fracture of adhesives, the viscoelastic analysis of adhesives, chemical composition control and automated cure monitoring.

There are still several areas where research is required on critical elements of structural adhesive bonding. These include (See table 3) feed-back control of the cure process, life prediction methodology for adhesive bonded structures, techniques for automated inspection of treated adherend surfaces and the automated application of the adhesive to the surfaces to be bonded, and new concepts in tooling for the bonding process.

It then remains for us to build on the very solid technology base we have been provided. To pull all of these pieces together and demonstrate that structural adhesive bonding is a cost competetive process for producing reliable and reproducible aircraft structures is our ultimate goal.

II. DISCUSSION

We are all more or less familiar with the applications of adhesive bonding on present day aircraft. Aircraft such as the C-141 and C-5 cargo aircraft have extensive quantities of bonded structure as do the T-38, F-111, F-15 and F-16. The extensive and successful use of adhesive bonding on the C-5 is particularly worthy of mention. This aircraft makes extensive use of bonded components in both sandwich and metal to metal assemblies. The entire cargo floor which is a primary fuselage stiffening component is a metal to metal bonded assembly. The under the floor bulkheads are all adhesive bonded honeycomb construction. A total of about 25,000 square feet of honeycomb structure and 10,000 feet of metal to metal bonded structure are performing very successfully on today's C-5 air-

craft.

Many papers have been presented describing the results of the USAF "PABST" ADP. Some of this bears repeating because of the noteworthy accomplishments of this effort conducted by the McDonnell-Douglas Corp. The design (from first principles) fabrication and structural testing of an all bonded AMST fuselage 10 feet in diameter and 40 feet long was accomplished (within funding and on schedule) and was successful far beyond our wildest expections.

The fuselage even though containing many bond line flaws and materials and process irregularities simply refused to fail in fatigue. Huge flaws and damaged areas were inflicted upon this structure with saws and hammers and the propagation of these flaws turned away from the bondlines in every instance thus providing one of the most fail safe structures one can imagine.

It is anticipated that all future fighter, cargo and helicopter fuselage construction will draw and significantly benefit from the technology developed under the PABST - ADP.

Another concept where adhesive bonding will provide improved performing aircraft components is that of the laminated sheet metal composite. Some years ago it was demonstrated that laminated-bonded metallic structures were extremely tolerant to damage and defects and were inherently fail safe materials. Work on all bonded damage tolerant structures has demonstrated that adhesive bonded sheet aluminum specimens have net fracture stresses twice as high as those obtainable on homogeneous aluminum plate specimens of the same thickness. Work performed by the USAF Materials and Flight Dynamics Laboratories has shown the very desirable damage tolerance, and tough failure mode of these laminated sheet metal constructions for use in aircraft. Figure 1 compares the stress strain characteristics of 1/2" thick homogeneous plate and a 1/2" thick laminated plate composed of 5 sheets of 0.1" thick sheet adhesively bonded together.

The alloy utilized was 7075-T6,the adhesive used was FM-1000. Both the bonded sheet and plate specimens were edge notched and fatigued until crack growth was initiated. It should be noted here that at the same net stress level it took 5 times as many stress cycles to produce a crack in the laminated specimen than it did to produce a crack in the plate specimen. The specimens were then tested under static conditions to failure. In the homogeneous plate specimen "pop-in" or catastrophic-unstable crack growth occurred at about 28KS1. The failure was a typical flat fracture expected under the conditions of plane strain, as shown in figure 2 with little evidence of shear lip tearing. As can be seen on figure 1 the laminated specimen did not exhibit "pop-in". The stress strain curve exhibit simple deviation from linearity which is characteristic of a thin section failing by ductile shear, under conditions of plane stress. The laminated specimen failed at about 65 KSI over twice the load that failed the plate specimen. The laminated bonded sheet construction combines thick section load bearing capabilities and thin sheet toughness thus eliminating the brittle failure nature inherent in thick plates. Figure 2 also shows the translation from flat fracture to the mixed mode shear failure as a function of metal sheet thickness. Figure 3 illustrates the predominance of ductile shear failure in the bonded laminated specimens where the sheet thickness is 0.1 inch.

The potential of this type of construction for improving the fatigue life, reducing the cost and eliminating catastrophic failure modes in metal aircraft structure prompted the USAF to fund several advanced technology laminated structure wing box studies.

Figure 4 illustrates an efficient all adhesive bonded laminated box beam. Based on such conceptual structural element work two large demonstration articles were designed, fabricated and structurally tested. Figure 5 shows a demonstration article based on the F-111 root end wing box. Figure 6 shows the demonstration article based on the F-16 root end wing box. Both of the structural components shown have been fatigue tested and statically tested to failure. The results of these programs demonstrated that indeed the fatigue improvement and damage tolerance of a laminated component were significantly increased compared to the conventional plate components. In addition, the fail safe feature of this type of construction was demonstrated by each of the components sustaining over 100% of limit load after complete failure of the spar caps. Also the laminated components were lighter than their mechanically fastened counterparts by about 8% and cost about 10% less to fabricate. The use of the latest state of the art materials and processes in these components as well as design optimization studies to reduce sites for crack initiation in the metal will provide even further increases in component performance.

All of the bits and pieces of the technology of structural adhesive bonding for major aircraft components have been convincingly demonstrated under USAF funded programs. It now remains to put all of these elements together in an optimum manner to demonstrate our ability to produce reliable, reproducible and cost effective adhesive bonded structures in a production atmosphere.

TABLE I

RECENT ADCANCES
IN MATERIALS PROCESSES

Corrosion Inhibiting Primers
Advanced Technology Adhesives
Phosphoric Acid Anodize
Chemical Composition Control

TABLE 2

CRITICAL AREAS OF RESEARCH

Fatigue & Fracture Analysis
Viscoelastic Stress Analysis
Life Prediction Methodology
Cure Monitoring Techniques

TABLE 3

REQUIRED RESEARCH

Cure Control With Feedback
Automated Surface Inspection
Automated Adhesive Application

TABLE 4

PABST ADP CONCLUSIONS

Complete Crack Arrest in Fatigue
Environmental Degradation Negli-
gible
Existing Flaws/Disbonds Did Not
Grow
Bonded Fuselage Extremely Damage
Tolerant

TABLE 5

LAMINATED SHEET METAL CONSTRUCTION

High Toughness
Fatigue Resistant
Damage Tolerant

REFERENCES

(1)   Agard Lecture Series 102
(2)   AFWAL-TR-80-3033
(3)   AFML-TR-79-227
(4)   AFFDL-TR-79-68

BIOGRAPHICAL SUMMARY

Theodore J. Reinhart is presently
Chief of the Composites, Adhesives
and Fibrous Materials Branch in
the Nonmetallic Materials Division
of the AFWAL Materials Laboratory.
The Branch conducts research and
development leading to new and improved materials, processes, and
analytical techniques in the areas
of plastics, composites, structural adhesives and fibrous materials for applications in existing
and future Air Force weapon systems.

3/16"
1/4"
1/8"

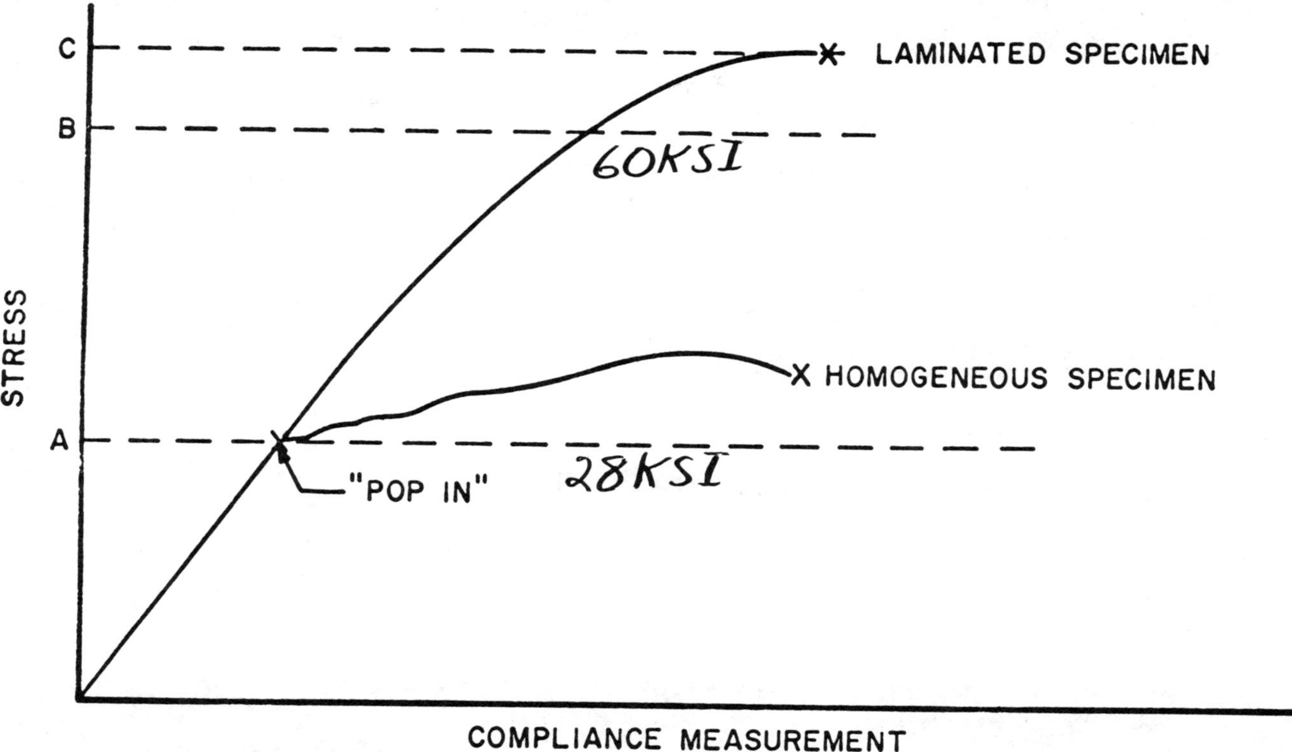

STRESS
COMPLIANCE MEASUREMENT
C
B
A
LAMINATED SPECIMEN
60KSI
HOMOGENEOUS SPECIMEN
"POP IN"
28KSI

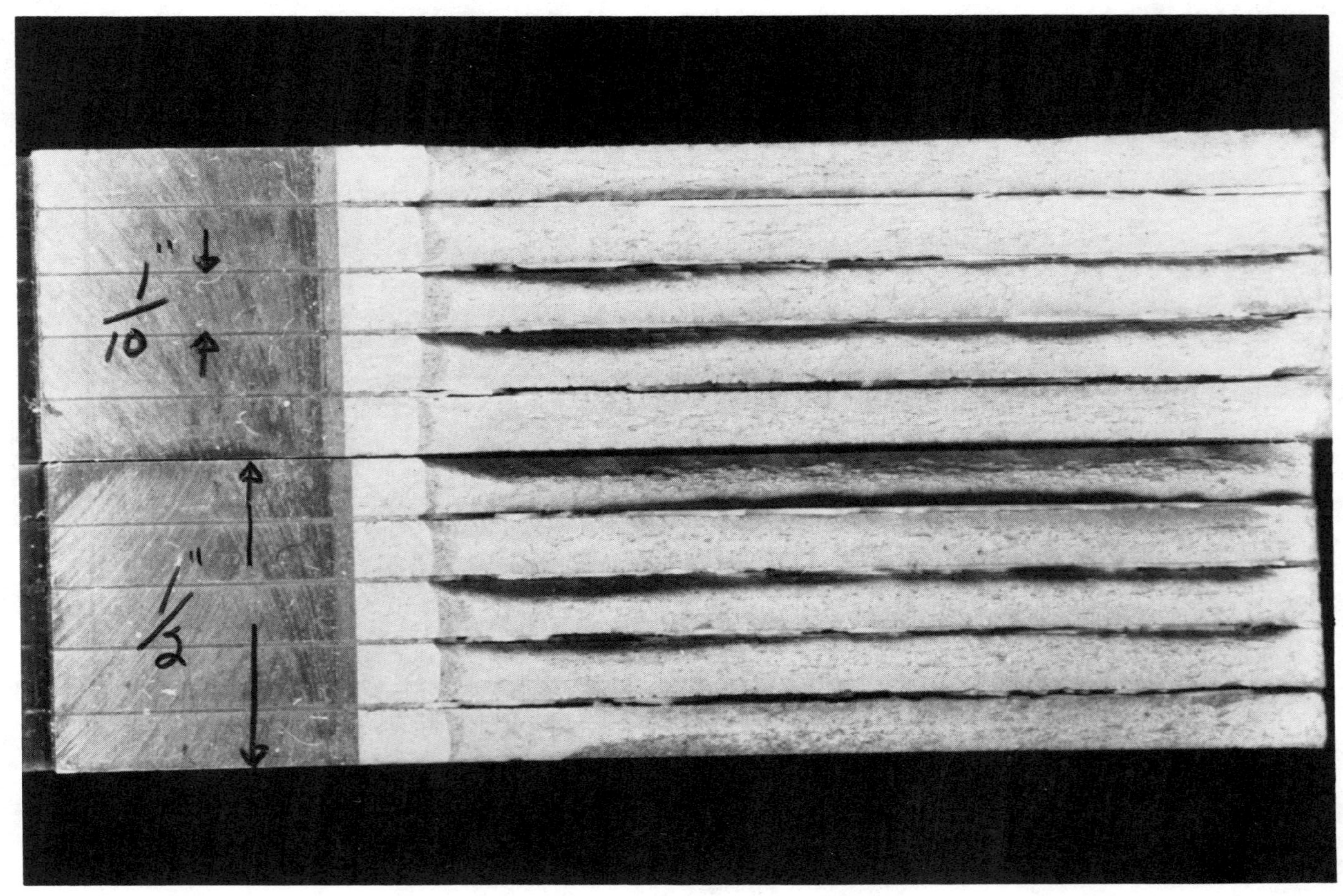

648

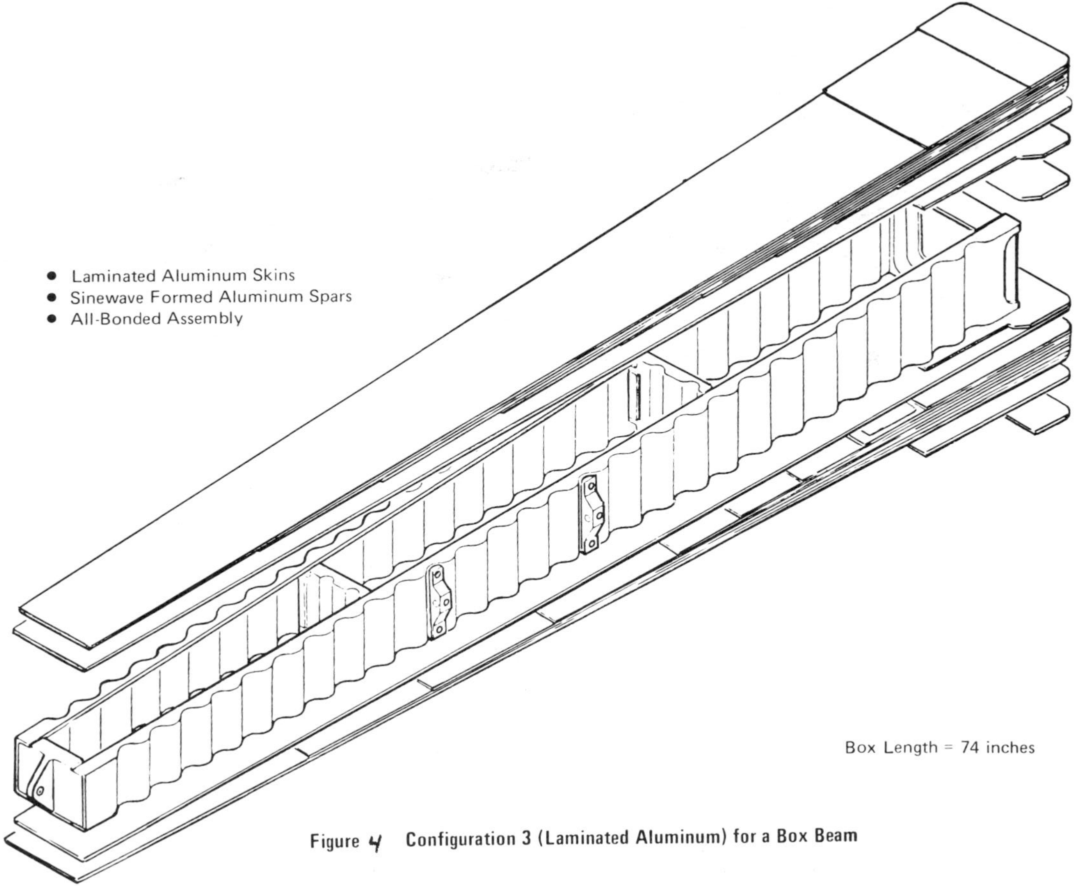

Figure 4   Configuration 3 (Laminated Aluminum) for a Box Beam

649

DEMONSTRATION ARTICLE DEFINITION

F-16 REQUIREMENTS

- o DESIGN DEFINITION
- o CONTOURS
- o CRITERIA/LOADS
- o HARDWARE PROVISIONS

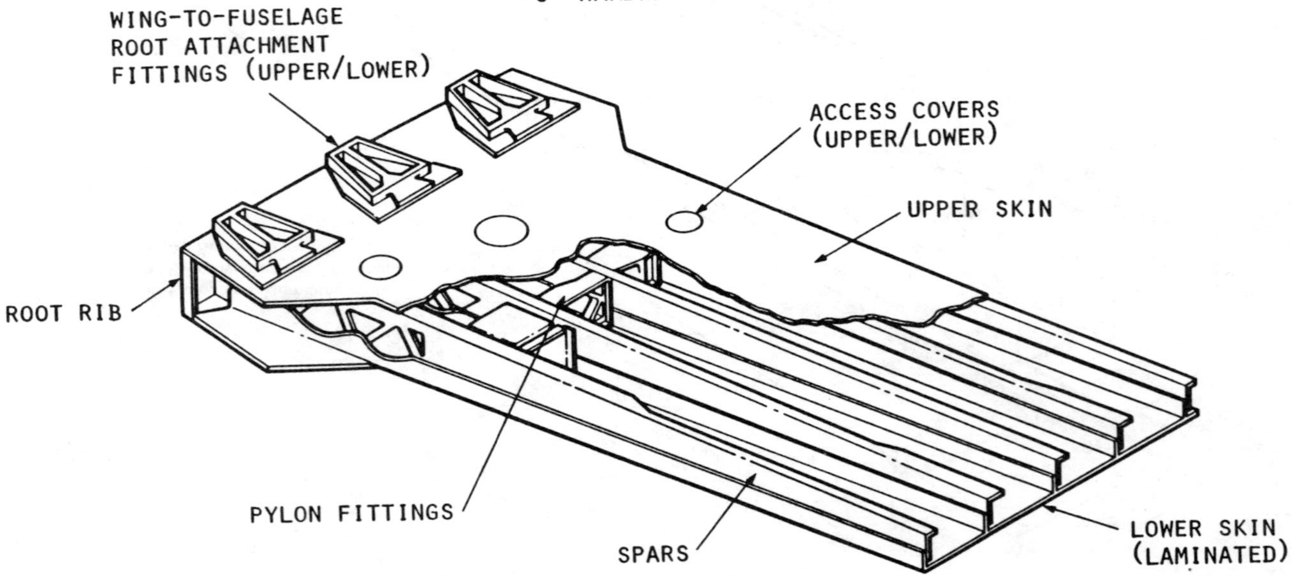

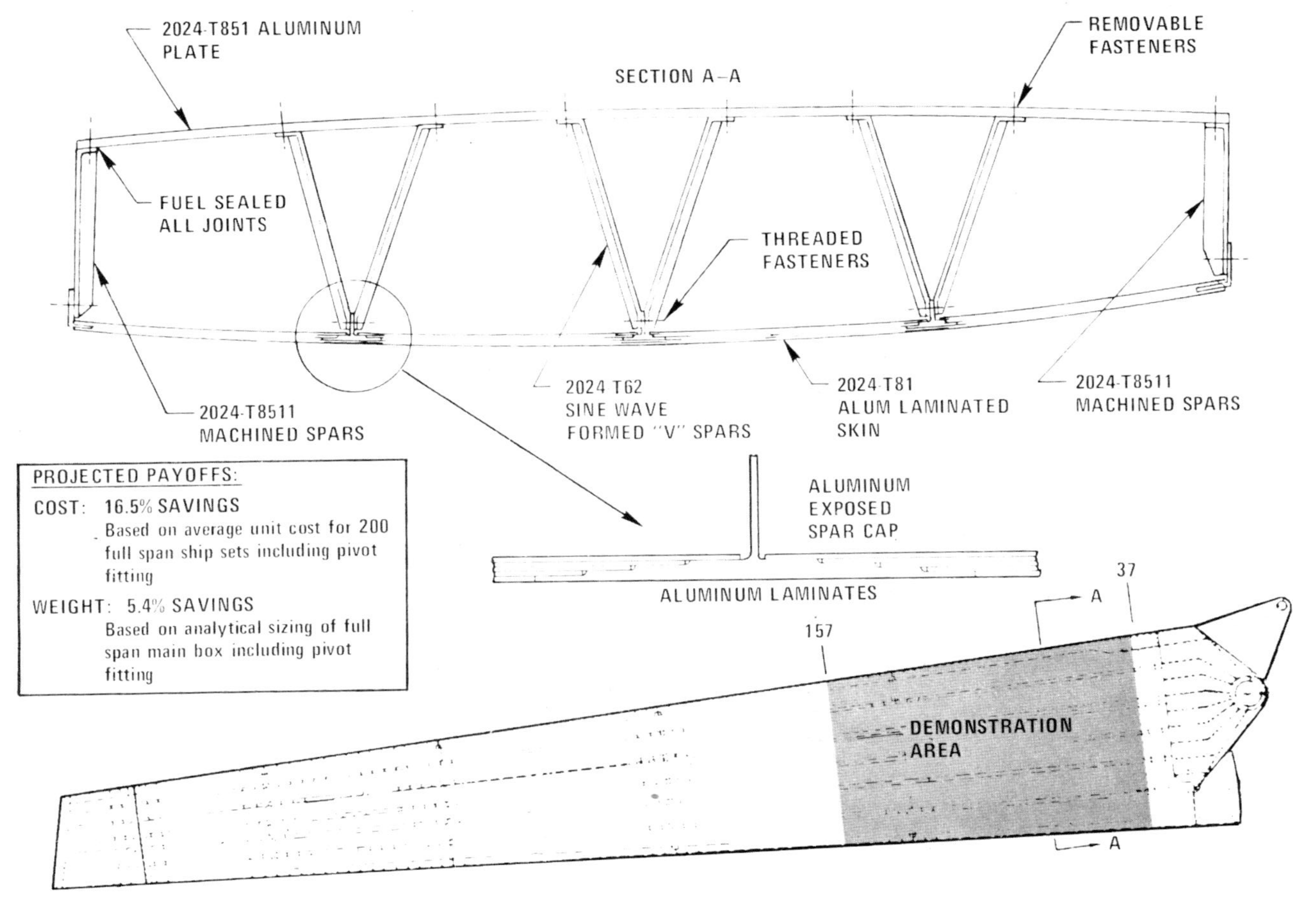

2024-T851 ALUMINUM PLATE
REMOVABLE FASTENERS
SECTION A-A
FUEL SEALED ALL JOINTS
THREADED FASTENERS
2024-T8511 MACHINED SPARS
2024 T62 SINE WAVE FORMED "V" SPARS
2024-T81 ALUM LAMINATED SKIN
2024-T8511 MACHINED SPARS
PROJECTED PAYOFFS:
COST: 16.5% SAVINGS
Based on average unit cost for 200 full span ship sets including pivot fitting
WEIGHT: 5.4% SAVINGS
Based on analytical sizing of full span main box including pivot fitting
ALUMINUM EXPOSED SPAR CAP
ALUMINUM LAMINATES
37
157
A
A
DEMONSTRATION AREA

26th National SAMPE Symposium
April 28-30, 1981

IMPROVED DURABILITY FOR WELDBONDED ALUMINUM STRUCTURES

R.G. Hocker
Northrop Corporation, Aircraft Division,
Hawthorne, California

Abstract

Weldbonding combines the joint efficiency of adhesive bonding with the economy of resistance spot welding. Recent developments in surface preparation have resulted in a weldbond system with adhesive bond durability nearly equal to that of PABST (Primary Adhesively Bonded Structure Technology) and superior to that of the FPL (Forest Products Laboratory) BR-127/FM-123 system used for standard adhesive bonding. Normal thicknesses of anodized layers and primer coatings can provide excellent durability for adhesive bonded joints; however, resistance spot welds cannot be made through these thick insulative layers. A combination of a low voltage phosphoric acid/sodium dichromate anodize (developed at Northrop) and a very thin layer of BR-127 primer provides excellent durability for the weldbonded joint, and this surface can still be resistance spot-welded. Standard wedge test specimens exposed to a five percent salt fog at 95F and to a sea coast environment provided a comparison of durability for the weldbond system with and without the BR-127 primer, and with two other adhesive bonding systems (PABST and FPL/ BR-127/FM-123). Durability comparisons were made using four material combinations: 2024-T3 alclad bonded to 7075-T6 bare, 2024-T3 bare bonded to 7075-T6 bare, 2024-T3 bare bonded to 2024-T3 bare, and 7075-T6 bare bonded to 7075-T6 bare.

Keywords: aluminum, weldbonding, durability, adhesive bonding, anodize, wedge tests.

1. INTRODUCTION

Weldbonding is a joining process which combines adhesive bonding and resistance spot welding. The spot welds are primarily used to hold the component together during the cure cycle in the oven instead of using expensive autoclaves and tooling fixtures. Paste adhesive is placed on the aluminum surfaces to be joined and resistance spot welds are made through this adhesive layer. The component is then placed in an oven and heated to the temperature used to cure the adhesive.

The weldbonding process originated in the early 1960's in the U.S.S.R. where it is still being used to weldbond structural components for military and commercial aircraft. In the mid 1960's, the Lockheed-Georgia Company began extensive investigations to establish methods for weldbonding aluminum structures. During the 1970's under an Air Force Contract, Lockheed weldbonded a large fuselage panel for a C-130 transport. The edges of the panel were well sealed prior to installation. After more than 3,000 hours of flight time, no degradation of the weldbonded joints has been observed. Other production applications for the weldbond joining method include Sikorsky's S67 Blackhawk helicopter and the Centaur Shroud for the Mars-Viking project fabricated at Lockheed, Sunnyvale, California.

Under Air Force Materials Laboratory sponsorship, Northrop Corporation Aircraft Division recently developed a low voltage anodize which provides a weldbond surface which is more durable than all prior weldbond surfaces[1,2]. A follow-on contract with Northrop and Fairchild Republic in 1977-79 established weldbonding as an efficient cost-saving method for fabricating fuselage panels for the A-10 aircraft[3]. A joining process which had been developed in research laboratories has now been transferred into a full scale production line technology at Fairchild Republic for the weldbonding of A-10 panels. This paper discusses the durability of the AFML/Northrop weldbond system as compared to other adhesive bonding joining systems.

2. DURABILITY OF WELDBOND AND ADHESIVE BOND JOINTS

One of the primary concerns in the use of adhesively bonded structures is premature failure due to a time-dependent crack growth in the bond joint under the combined influence of applied stress and a corrosive environment. This type of failure is commonly termed "environmental stress cracking." Resistance to environmental stress cracking is defined as environmental durability or simply durability.

As a result of increased applications for the weldbond process, more and more emphasis is being placed on its durability as compared to the durability of standard adhesive bonding systems. A test program funded by the Air Force Materials Laboratory was conducted for the purpose of comparing the durability of the weldbond system with that of two adhesive bonding systems: PABST (Primary Adhesively Bonded Structure Technology) developed by McDonnell-Douglas, and the FPL (Forest Products Laboratory) system commonly used by aircraft companies to adhesively bond panels. A schematic representation of these systems is shown in Figure 1. The

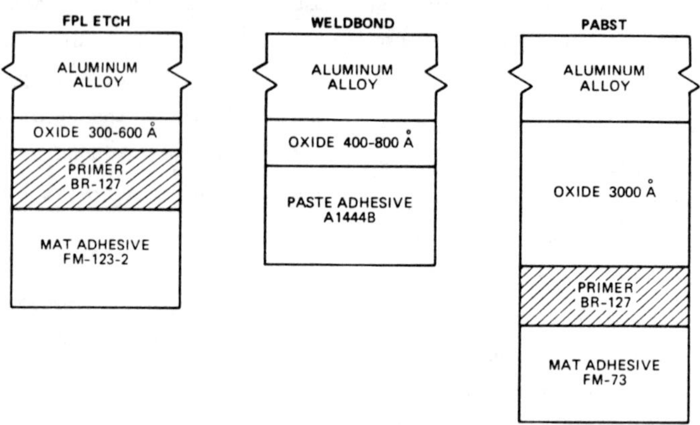

FIGURE 1.   SURFACE/ADHESIVE SYSTEMS FOR FPL, WELDBOND, AND PABST JOINING PROCESSES

PABST system utilizes the 10-volt phosphoric acid anodizing process developed by Boeing which produces a thick boehmite oxide ($\alpha Al_2O_3 \cdot H_2O$) layer (over 3000 angstroms) and includes the BR-127 primer and the FM-73 mat adhesive. The FPL system utilizes a thin (300 to 600 angstroms) boehmite layer, which is formed during the FPL etching process. This bonding system includes the BR-127 primer and the FM 123-2 mat adhesive. The weldbond system utilizes a thin (400 to 800 angstroms) boehmite layer, formed by low-voltage, phosphoric acid/sodium dichromate anodizing, but this system did not originally include a primer. A chromated paste adhesive, Goodrich adhesive A1500B (formerly A1444B), is used for the weldbond system to improve the durability of the weldbond joint and to enable welds to be made through the adhesive. Detailed descriptions of the surface treatments and adhesives used for these three joining systems are presented in Tables 1, 2, and 3.

3.   TEST METHOD, ENVIRONMENT, AND MATERIAL COMBINATIONS

The standard wedge test, Figure 2, was used to evaluate the durability of the three bonding systems. This test method allowed both high stress and the corrosive environment to be applied directly at the bondline. The growth of the crack, or progressive parting of the bond line, is measured as a function of time. This method combined with long exposure periods provided an excellent means for determining small differences in durability in contrast to the "go, no-go" type of wedge test which can be accomplished by the short, one or two day, exposure or the constant load – salt water immersion test for lap shear specimens.

OPERATION	MATERIAL	PROCESS
VAPOR DEGREASE	1-1-1 TRICHLOROETHANE	
ALKALINE CLEAN	ALTREX 1097 5-8 OZ/GAL	10-15 MINUTES 160-190F
SPRAY RINSE	DEIONIZED WATER	5-7 MINUTES 110F
DEOXIDIZE AND ETCH	FPL-ETCH 96% SULFURIC ACID: 38.5-41.5 FLUID OZ/GAL SODIUM DICHROMATE, DIHYDRATE: 8 OZ/GAL 2024 ALUMINUM SCRAP: 0.2 OZ/GAL	12-15 MINUTES 150-160F
PRIME	BR-127, 0.00015-INCH TO 0.00030-INCH THICKNESS SPRAY COAT	AIR DRY: 30 MINUTES AT RT. CURE: 60-120 MINUTES AT 250F
ADHESIVE	FM 123-2 0.045 LB/SQ FT, 2 LAYERS	VACUUM BAG. APPLY AUTOCLAVE PRESSURE OF 50-100 PSI. CURE: 90 MINUTES AT 260F

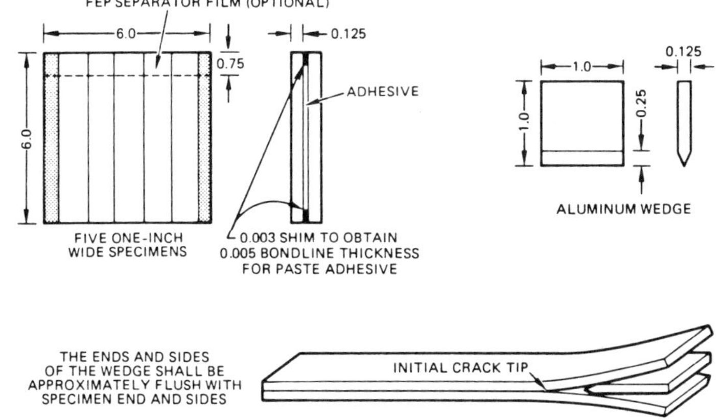

FIGURE 2. WEDGE TEST PANEL, WEDGE, AND COUPON

TABLE 2. NORTHROP WELDBOND PROCESS

OPERATION	MATERIAL	PROCESS
VAPOR DEGREASE	1-1-1 TRICHLOROETHANE	VAPOR 60 SECONDS; CONDENSED FLUID-60 SECONDS; COOL, REPEAT.
ALKALINE CLEAN	TURCO 4215-S, 6-8 OZ/GAL SOLUTION	12-15 MINUTES, 125-165F
SPRAY RINSE	COLD DEIONIZED WATER WITH TAP WATER MIX	5-7 MINUTES
DEOXIDIZE	NITRIC ACID/AMCHEM-7 (MODIFIED) NITRIC ACID — 11-14% BY VOLUME, BE^o 42 (70% HNO_3) AMCHEM-7: 2.9 — 3.3 WT OZ/GALLON SOLUTION ALODINE-45: 11.2 — 11.5 ML/GALLON SOLUTION ALODINE 1200E ACTIVATOR — 13.2 — 13.4 ML/GALLON SOLUTION BALANCE — DI WATER	6-8 MINUTES, ROOM TEMPERATURE IN AGITATED SOLUTION
SPRAY RINSE	COLD DEIONIZED WATER WITH TAP WATER MIX	5-7 MINUTES
ANODIZE	PSD SOLUTION PHOSPHORIC ACID — 1.2 — 1.5 FL OZ/GALLON SOLUTION (85% H_3PO_4) SODIUM DICHROMATE DIHYDRATE: — 1.3 — 1.5 OZ/GALLON SOLUTION	20-25 MINUTES, ROOM TEMPERATURE VOLTAGE — 1.4-1.6 VOLTS, DC FOR BARE ALLOYS 0.9-1.1 VOLTS, DC FOR CLAD ALLOYS
SPRAY RINSE	COLD DEIONIZED WATER WITH TAP WATER MIX	5-7 MINUTES
OVEN DRY	CIRCULATING HOT AIR	30-40 MINUTES, 150±10F
ADHESIVE	A1444B PASTE	APPLY TO BOTH SURFACES
WELD		RESISTANCE SPOT WELD
CURE		250-260F FOR 2 HOURS

Three different environmental exposures were used for the wedge tests: (1) 95 to 100 percent relative humidity at 120F, (2) five percent salt fog at 95F, and (3) the McDonnell-Douglas sea coast beach site at El Segundo, California. Also, four material combinations were evaluated for this program: (1) 2024-T3 alclad bonded to 7075-T6 bare, (2) 2024-T3 bare bonded to 7075-T6 bare, (3) 2024-T3 bare bonded to 2024-T3 bare, and (4) 7075-T6 bare bonded to 7075-T6 bare. The objective of this program was to rank the three bonding systems and the four material combinations in terms of durability behavior.

TABLE 3. PABST PROCESS

OPERATION	MATERIAL	PROCESS
VAPOR DEGREASE OR SOLVENT WIPE	1-1-1 TRICHLOROETHANE	
ALKALINE CLEAN	TURCO 4215-S	4-8 OZ/GAL 15 MINUTES 140F
SPRAY RINSE	DEIONIZED WATER	5-7 MINUTES ROOM TEMPERATURE
DEOXIDIZE	NITRIC ACID/AMCHEM 6-16	15 MINUTES
	AMCHEM 6-16: 6-8% BY VOLUME 8-10% BY WEIGHT	75-95F
	NITRIC ACID: 20 FLUID OZ/GAL, (70% HNO3)	
	DEIONIZED WATER: BALANCE	
SPRAY RINSE	DEIONIZED WATER	5-7 MINUTES ROOM TEMPERATURE
ANODIZE	PHOSPHORIC ACID	20-25 MINUTES ROOM TEMPERATURE AT 10-15 VOLTS
	96% H_3PO_4: 11-16 OZ/GAL DEIONIZED WATER: BALANCE	
SPRAY RINSE	DEIONIZED WATER	5-7 MINUTES ROOM TEMPERATURE
OVEN DRY	CIRCULATING HOT AIR	30-60 MINUTES 150-160F
PRIME:	BR-127, 0.00015-INCH TO 0.00030-INCH THICKNESS SPRAY COAT	AIR DRY: 30 MINUTES AT RT CURE: 1 HOUR AT 225F
ADHESIVE:	FM 73, 0.060 LB/SQ FT	VACUUM BAG APPLY AUTOCLAVE PRESSURE OF 50-100 PSI. CURE: 90 MINUTES AT 250F

4. DURABILITY TEST RESULTS

A total of 144 wedge specimens were exposed to the sea coast environment for one year, i.e., 12 wedge tests for each bonding process and material combination. A total of 72 wedge specimens were exposed to the salt fog, and 72 wedge specimens were exposed to the 95-100 percent relative humidity environment, i.e., 6 wedge tests for each bonding process and material combination. The exposure time for these two environments was 90 days. The results of these tests can be summarized as follows:

1. The PABST system has much better durability than the FPL system, and the FPL bonding system has slightly better durability than the standard weldbond system (without primer).

657

2. Based on wedge test results for the FPL joining system and the weldbond system exposed to the salt fog and sea coast environment, the 2024-T3 bare bonded to 2024-T3 bare had the best durability; 2024-T3 bare/7075-T6 bare and 7075-T6 bare/7075-T6 bare had intermediate durability; and 2024-T3 alclad/7075-T6 bare had the worst durability.

The wedge test crack growth for all three joining systems was less than 0.5 inch after 90 days exposure to the 95-100 percent relative humidity at 120F. Since this environment did not reveal a difference in durability for the three bonding systems, the data curves for this exposure are not presented. However, the sea coast and salt fog environmental exposures did result in considerable differences in durability for the three bonding processes and four material combinations. Average wedge crack growth curves for these exposures are presented in Figures 3 and 4.

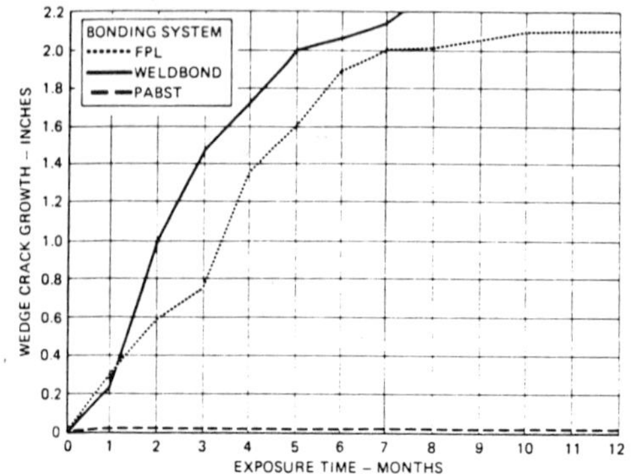

2024-T3 ALCLAD/7075-T6 BARE

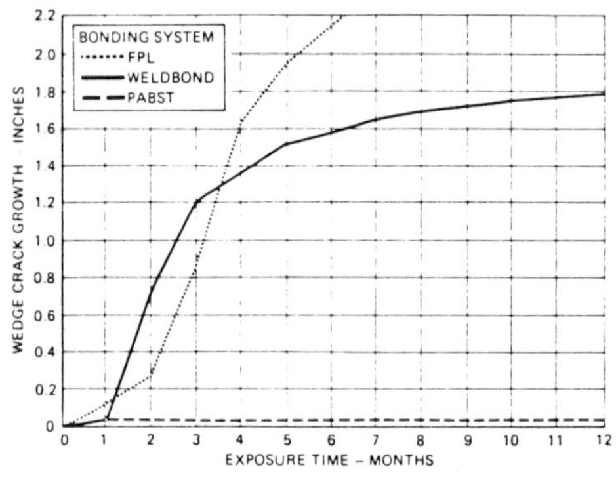

2024-T3 BARE/7075-T6 BARE

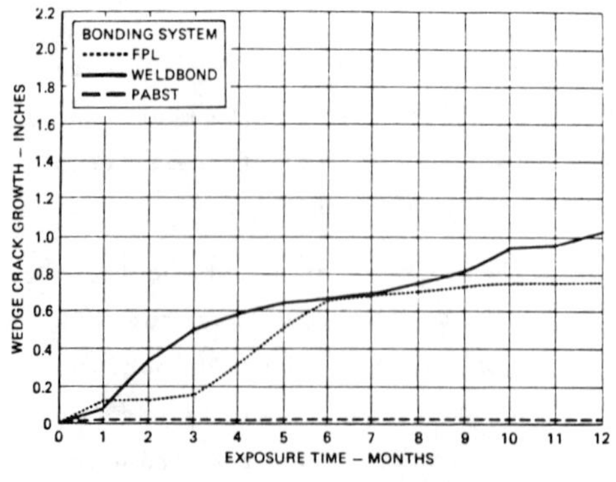

2024-T3 BARE/2024-T3 BARE

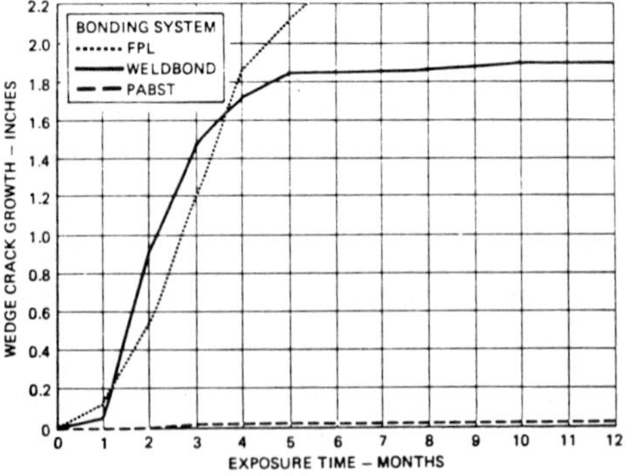

7075-T6 BARE/7075-T6 BARE

FIGURE 3. AVERAGE WEDGE CRACK GROWTH FOR SEA COAST EXPOSURE

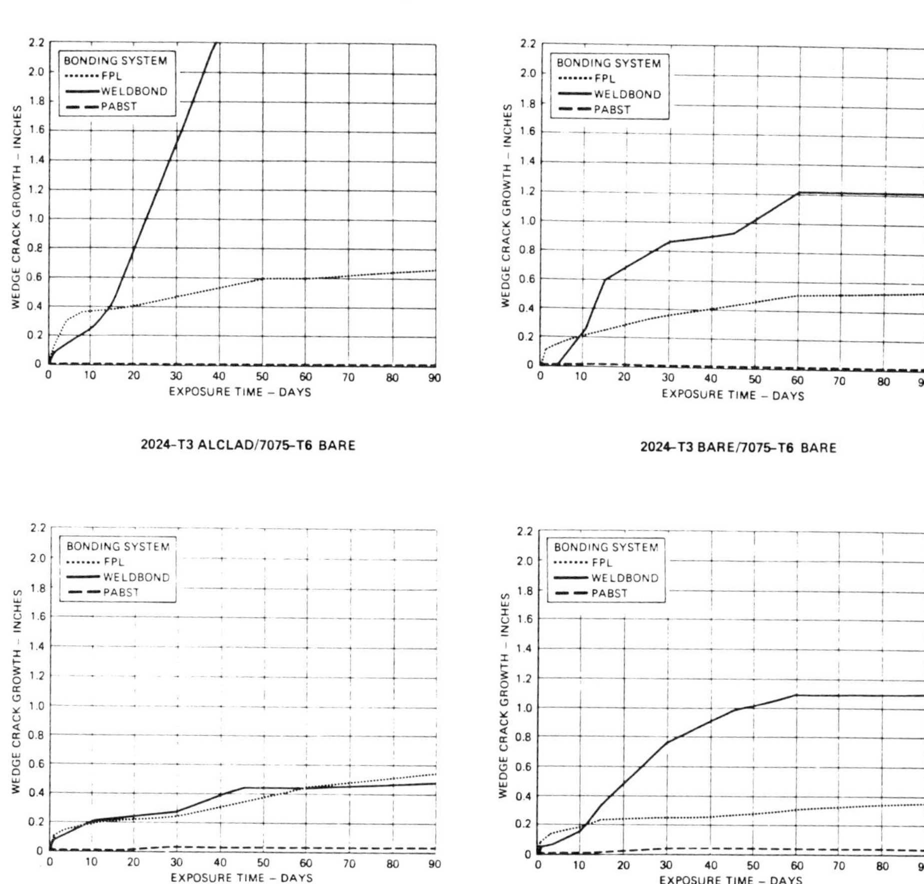

FIGURE 4.   AVERAGE WEDGE CRACK GROWTH FOR SALT FOG EXPOSURE

For the sea coast exposure the weld-bond system showed the most rapid wedge crack growth during the first three to six months compared to the other two joining systems. For three of the material combinations this crack growth exceeded one inch. The durability of the FPL joining system was not much better. Both joining systems had very poor durability in comparison to the "less than 0.05 in." crack growth of the PABST system.

For the salt fog exposure, the weld-bond system again showed the most rapid crack growth with three of the material combinations showing a crack growth exceeding one inch within two months. For this exposure the FPL system showed much better durability than the weldbond system. As for the sea coast exposure, the PABST system revealed the best durability with average crack growths less than 0.05 inch.

659

5. IMPROVED WELDBOND DURABILITY

As a result of this evaluation, additional effort was made to determine methods for improving the durability of the weldbond system. It was obvious that a thick anodized layer with the addition of a primer provided excellent durability for the PABST system. However, for weldbonding, it is not possible to resistance spot weld through thick anodized layers with high resistance. Also, an insulative layer of primer decreases the weldability of this system. Prior development work at Northrop provided a thin anodized layer which could be welded, and provided a surface which had much greater durability than obtained with only the spot weld etch, the surface cleaning process used for weldbonding prior to the 1970's. Therefore, a change in the Northrop anodizing process was not recommended. Using the same logic, it was hypothesized that use of a thinner than standard layer of primer might also provide weldability and greater durability.

A series of weldability tests were conducted using three primer thicknesses; 0.00004, 0.00008, and 0.00015 inch. It was found that the very thin 0.00004 inch (1 micron) layer of BR-127 primer could be welded without much difficulty if the primer were only air dried and not cured at elevated temperature. Weldbond samples for lap shear, T-peel, and fatigue were prepared using the 1 micron thickness of primer which was co-cured with the adhesive after welding. The results for these mechanical tests showed that there was no substantial gain or loss in strength or fatigue life as a result of this thin layer of co-cured BR-127 primer[4]. The average lap shear strength for weldbonded specimens with BR-127 was 4700 psi compared to 4790 psi for unprimed weldbond specimens when tested at room-temperature, and was 3610 psi and 3850 psi, respectively, when tested at -65F. The average T-Peel strength for six specimens was 9.8 and 10.7 pounds per inch width for primed and unprimed weldbond specimens, respectively. The average fatigue life of eight primed specimens was 328,000 cycles compared to the 197,000 cycles average life for five unprimed specimens, for a maximum load of 2,250 lbs and R:0.1. Lap joint dimensions were 1.25 inches wide x 1.25 inches overlap, and a single spot weld was used for these high load-transfer fatigue specimens.

Durability wedge specimens were also prepared with the one-micron layer of BR-127 primer and exposed to the sea coast for one year, and to the salt fog environment for 3 months. The results for these tests are shown in Figure 5. The improvement in durability compared to the standard weld-

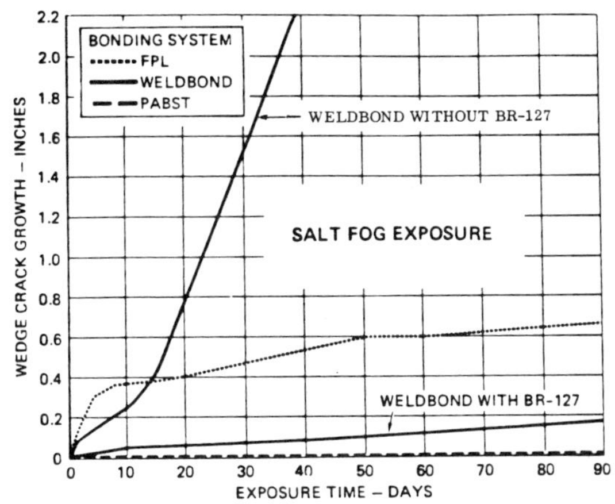

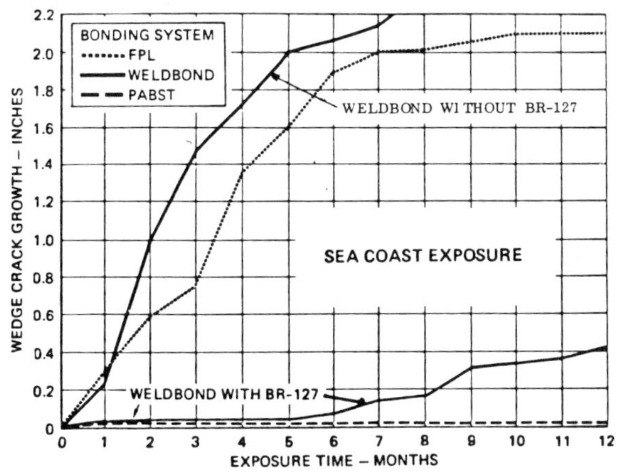

FIGURE 5.   AVERAGE WEDGE CRACK GROWTH FOR 2024-T3
ALCLAD/7075-T6 BARE

bond system was phenomenal. For the most sensitive (least durable) material combination, 2024-T3 alclad bonded to 7075-T6, the standard weldbond system showed an average wedge crack growth greater than 1.5 inches after 1 month's exposure to the salt fog, compared to less than 0.1-inch average crack growth for the weldbond/BR-127 system. For the sea coast exposure, the standard weldbond system showed an average crack growth greater than 1.0 inch after only 2 months exposure, compared to less than 0.05-inch average crack growth for the weldbond/BR-127 system. Even after a 12 month exposure, the average crack growth for the weldbond/BR-127 system was only 0.4 inch compared to 2.9 inches for the standard weldbond system. The durability of the weldbond/BR-127 system is much greater than the durability of the FPL/BR-127/FM-123 adhesive bonding system.

6. SUMMARY

1. The environmental-stress durability of the weldbond system is greatly improved by the addition of a thin (0.00004-inch) layer of BR-127 primer to the anodized surface.

2. The addition of a thin layer of BR-127 primer to the standard weldbond surface preparation does not reduce lap-shear strength, T-peel strength, or high-load transfer fatigue strength for weldbonded panels.

3. High quality welds can be made through a thin layer, 0.00004 inch, of air dried BR-127 primer added to the standard weldbond surface treatment.

7. ACKNOWLEDGEMENT

The work reported in this paper on the durability of the initial three joining systems was performed under AFML Contract F33615-76-C-5412, and is reported in AFML TR-79-4081. Mssrs. S.G. Lee and N.E. Klarquist of the Metals Branch (MLTM) of the Manufacturing Technology Division, Air Force Wright Aeronautical Laboratory (AFWAL), were the project engineers.

The work on the improved weldbond system was performed as part of Northrop's Independent Research and Development programs.

McDonnell-Douglas, Long Beach, California provided the PABST test panels and availability to their sea coast exposure site at El Segundo, California. The cooperation and assistance of the following McDonnell-Douglas personnel is gratefully acknowledged: E.W. Thrall, Jr., PABST, Program Manager; R.W. Shannon, Materials and Process Engineering, Section Chief; R.W. Ochsner, Materials and Process Engineering, Unit Chief; and R.L. Radecky, Materials and Process Engineer.

8. REFERENCES

1. Bowen, B.B., Herfert, R.E., and Wu, K.C., "Development of Corrosion Resistant Surface Treatments for Aluminum Alloys for Spot-Weld Bonding," Report No. AFML-TR-75-69, (March 1975).

2. Wu, K.C., "Advanced Aluminum Weldbond Manufacturing Methods," Report No. AFML-TR-76-131, (September 1976).

3. Croucher, T.R. "Advanced Weldbonding Process Establishment for Aluminum," Report No. AFML-TR-79-4006, (April 1979).

4. Hocker, R.G., "Environmental Durability of the Weldbond, FPL, and PABST Joining Systems," Report No. AFML-TR-79-4081, (July 1979).

9. BIOGRAPHY

Robert Hocker received a B.S. in Metallurgy from Carnegie Mellon, Pittsburgh, Pennsylvania. His experience in the aerospace industry includes 18 years with McDonnell-Douglas in Santa Monica and Huntington Beach, and two years with Fairchild, Stratos Division, Manhattan Beach. He joined the Northrop Corporation in 1976 and has spent the past five years in the field of advanced joining technology including titanium weldbrazing and aluminum weldbonding, and in the field of high strength aluminum powder products for aircraft structures.

26th National SAMPE Symposium
April 28-30, 1981

SURFACE TREATMENT FOR ALUMINUM BONDING
STAB(3)

Tennyson Smith
Rockwell International Science Center
Thousand Oaks, California 91360

Abstract

A new surface treatment for aluminum alloys (particularly Al 2024-T3) has been developed for bonding with 250°F curing modified epoxy adhesives. On a laboratory scale, this treatment has proven to be as effective as any reported thus far, as to bond strength and hydrothermal stress durability. For example, lap shear failure is always cohesive, with the strength reported for the adhesive. The wedge-test crack extension is about 0.2 inch/24 hrs at 60°C and 100% RH. Failure during the wedge test is also cohesive.

Advantages of the treatment are that it is relatively inexpensive, and requires only three steps (3 to 10 min dip, rinse and dry) at room temperature, with no carcinogenic chromates and no degreasing (in most cases).

1. INTRODUCTION

The industry standard surface treatment for aluminum and its alloys in the last thirty years has been the FPL (Forest Product Laboratories) etch. The FPL etch consists of six steps, degrease, alkaline clean, rinse, and sulfuric acid-dichromate etch, rinse and dry. The process removes the as-received contamination and oxide layer, as well as a few microns of metal, and leaves a rough, pitted surface with a clean very thin ($\sim$ 150 Å) oxide (hydroxide) layer (see Ref 1). It is claimed by Venables et al[2] that the oxide is a thin layer about 50 Å thick with thin ($\sim$ 50 Å dia) columns about 400 Å high at the corners of a hexagonal array (this averages to about 150 Å). They suggest that the good adhesion and durability properties of the FPL etch are due to the mechanical interlock between the adhesive and these protruding oxide columns.

More recently Bethune[3] introduced the phosphoric acid anodize (PAA) surface treatment for aluminum alloys. PAA consists of nine steps, the six FPL steps plus anodize (25 min at 10V in 10% phosphoric acid),

rinse and dry steps. The treatment
has proven to be very reliable,
yielding bonds that are strong (fail
primarily in the adhesive, i.e.,
cohesive failure) and durable under
hydrothermal stress. PAA produces
very porous hydroxide films (3000-
5000 Å) into which the adhesive can
penetrate to produce strong mechanical interlocking.

Due to the carcinogenic properties
of the chromates, we have searched
for a non-chromate surface treatment
that is as strong and durable as the
FPL or PAA treatment. In the process of this search various alkali
treatments were tried. A mild
alkaline treatment, referred to as
STAB(1)[4] (STAB stands for surface
treatment of aluminum for bonding)
appeared initially to be successful
but proved later to be difficult to
give consistent results. A second
mild alkaline treatment STAB(2)[5]
appeared promising but proved irre-
producible also. A third alkaline
treatment (STAB(3)), with a highly
concentrated sodium hydroxide solution, has proved very successful and
reproducible on a laboratory scale.
STAB(3) should not be confused with
the usual commercial alkaline cleaner treatments, such as used in the
second step of the FPL etch. Currently used commercial alkaline
cleaners (as for STAB(1) and (2))
produce strong joints, but do not
consistently produce stable films
that are durable under hydrothermal
stress. This is because the

hydroxide concentration is about 20
g per liter of water, which is shown
in this paper to give poor results,
as compared to 200-600 g/l for
STAB(3) which gives excellent results. STAB(3) produces strong,
durable adhesive joints by the formation of a stable porous hydroxide
film. Advantages of STAB(3) are:
no carcinogenic chromates are used;
it is very simple and inexpensive
(only three steps: NaOH dip, rinse
and dry). This paper reports the
experimental results which define
the boundaries within which STAB(3)
can be used.

2. EXPERIMENTAL

2.1. Materials

Primarily Al2024-T3 bare aluminum
alloy was studied. Some results for
Al2219-T37, the 6000 series (Al6061-
T6), 7000 series (Al7075-T6) as well
Ti6Al4V titanium alloy and AM 355
steel are reported. The adhesive
was primarily 121 C°(250°F) curing
modified epoxy (Hysol EA 9628H) with
BR 127 primer, (Hysol EA 9210H primer was also used). Other adhesives
were FM 73 (121°C), AF 163 with XB
3944 primer and 177°C (350°F) curing
PL-729-3 (with PL-728 primer). The
sodium hydroxide was technical grade
flakes and industrial caustic soda,
the water was primarily deionized
although for some experiments tap
water was utilized to check the
difference, if any.

2.2 Surface Preparation

As will be seen, a degrease step is not necessary but for very contaminated panels it is advisable. Degreasing with TMC (trichlorotrifluoroethane, methylene chloride/ethyl alcohol azeotrope) and other organic solvents were used. It was found that aqueous detergent type degreasing is also adequate prior to STAB(3). The detergent used in this study was Alkanox (30 cc/1 of DI water). For 1 in. × 6 in. × 1/8 in. aluminum samples the solution was in glass beakers, for 1 ft × 1 ft × 1/8 in. panels the solution was in a stainless steel tank that held 13 liters.

To be sure that excessive amounts of metal were not being removed, samples were etched for various lengths of time (0-10 min), rinsed, dried and weighed. The weight loss proved linear with time, after a 1 min initiation period. The depth of metal removal (using 2.78 g/cm^3 as the metal density) can be expressed by the equation $d \simeq (5 \times 10^{-5}$ cm/min$)t$ where t is time in minutes after the first minute. STAB(3) removes about 1-3 µm in 3 to 10 min, which is not considered to be excessive but does remove the contamination, oxide and outer roll-worked layer.

For the glass beakers, the small samples were removed after the desired dip time and rinsed with a hard spray of room temperature DI water. The 1 ft × 1 ft panels were hung from a wire frame that could be lowered into and raised from the rinse tank at a controlled speed. The samples were dried by blowing the surface with room temperature dry nitrogen or left hanging in laboratory air to drip dry.

2.3 Bonding

The samples were primed by dipping into a container filled with primer and left in a vertical position for one half hr to dry, prior to placing in an oven to cure at 121°C (250°F) for one hr. Some samples were brushed with primer and some were sprayed to check any differences. The samples were bonded in a press at 50 psi and 121 C°(250°F) for 1 hr.

2.4 Testing

a. Bond Strength

The bond strength was measured with the lap-shear tensile test by bonding two 1 in. × 4 in. × 1/16 in samples with 0.5 in. overlap.

b. Hydrothermal Stress Endurance

Hydrothermal stress endurance refers to the endurance of "wedge" specimens (ASTM D3762179) and "stress durability" specimens (ASTM D2919) in hot humid environments. It was concluded in Ref 6, that although the "wedge test" is very useful for comparing adhesive bond durability, it can be misleading if the rating is made on crack extension alone (i.e., Δa). The correct way is to rate with respect to the initial crack length a_o, plus crack extension Δa, i.e., $(a_o + \Delta a)$.

Although Bethune[3] used $\Delta a < 0.75$ in., at 50°C (122°F) at 100% RH, as the pass criterion, we use a much more stringent criterior in order to compare STAB(3) with literature values of a_o and Δa for FPL etch and PAA at 60 C°(140°F) and 100% RH. For aluminum alloys and 121°C (250°C) curing modified epoxy adhesives, the initial dry crack length $a_o \sim 1.2$–1.6 in. and failure is primarily cohesive, i.e., center of line in the adhesive. The range of crack extension Δa (from various laboratories)[1-10] at 24 hrs is about 0.2 in. to 0.5 in. for FPL or PAA at 60°C and 95% RH. For comparison of STAB(3) with FPL or PAA we choose the wedge test pass criterion to be $(a_o + \Delta a) \leqslant 2$ in.

The stress durability test (ASTM 2919, see Ref. 1, pp. 428–429) uses lap shear joints (1 in. × 4 in. × 1/16 in) that are stressed with a spring to some fraction of the ultimate dry strength and placed in a humidity chamber. The time to failure is recorded. The stress durability results[7,9,11,12] for A12024-T3 with FPL and PAA surface treatment, bonded with 122°C curing epoxies, and tested at 60°C, 95–100% RH at 2000 psi yielded failures from 130 hr to 500 hr and about 38–63 hr at 2470 psi. If STAB(3) yields results in the range 130–500 hr for 2000 psi and 38–63 hr at 2470 psi it

is considered to have passed, although considerably more data are needed for this test, since there are not enough data to make a good comparison of the stress durability test. Testing of STAB(3) was performed by 5 different people over a period of 2 yrs with 3 different batches (purchased at different times) of A12024-T3.

3. RESULTS

Except in the case of deliberate contamination, FPL, PAA and STAB(3) give lap shear values between about 4.5 and 5.5 ksi. The apparent failure mode is cohesive in the adhesive. Less stringent surface preparation, such as, degrease only, or degrease plus alkaline clean also produce strong bonds with cohesive failure but these bonds do not endure under hydrothermal stress.

Thirteen wedge test experiments were performed with different shims to check the effect of bond line thickness. The bond line thickness ranged from 2 to 9 mil. Increasing the bond line thickness increased the initial crack length a_o from about 1.35 to 2.5 in., but had little effect on Δa at 1 hr or 24 hr. The rest of the tests reported have had a bond line thickness of about 4–5 mil.

The surface treatment process parameters are listed in the tables as: type of degrease agent and time of application, sodium hydroxide concentration (as g of NaOH per liter of DI water), time the sample was dipped into the sodium hydroxide solution, type of rinse (DI vs tap water) and time of rinse, type of drying (N_2 blow vs drip dry) and time of drying and temperature of the sodium hydroxide solution. In the tables, the time of N_2 blow dry is given as 1 min; however, after blow drying the samples were left in laboratory air for 0.5 hr prior to priming.

Most of the test results involve Al2024-T3 aluminum alloy, Hysol EA 9628H adhesive and BR 127 primer; however, preliminary results are also given for Al2219-T37, Al6061-T6, Al7075-T6 as well as Ti-6Al4V, titanium alloy and AM 355 steel. Other adhesives and primers tested were: EA 9628H (with EA 9210H), FM 73 (with BR 127), AF 163 (with XB 3944) and PL-729-3 (with PL-728).

3.1 Effect of the Degrease Step

The effect of no degrease and various types of degrease, aqueous detergent solutions (Alkanox and Microclean) and solvents (Gunk and TMC) gave an average initial crack length $a_o \sim 1.3$ in. and the average crack extension in 24 hrs is $\Delta a \sim 0.13$ in. so that $(a_o + \Delta a) \sim 1.4$ in. Therefore, the wedge test for STAB(3) is considered to pass (i.e., comparable with FPL or PAA) with or without a degrease step. It will be shown later that for extremely

contaminated samples a degrease step is desirable.

3.2 Effect of the NaOH Concentration and Temperature

Table 1 shows the effect of NaOH concentration at 23°C (room temperature), 33°C, and 40°C. With a 10 min TMC degrease, and a 10 min NaOH dip, STAB(3) passes the wedge test if the NaOH concentration is above about 140 g/l at RT, but can be considerably below this at high temperatures. Five tests between 23°C and 40°C averaged $(a_o + \Delta a) = 1.5$ in., at 50 C, $(a_o + \Delta a) = 1.70$ in. and at 60 C, $(a_o + \Delta a) = 2.20$ in., without a degrease step, if the NaOH concentration is 562 g/l and the dip time is 10 min.

Table 1

Wedge Test Results: Effect of NaOH Concentration and Temperature

Primer: BR127	Adhesive: EA9628H		Al2024-T3				
Sample	Surface Treatment		Wedge Crack				
	NaOH Concentration g/l	Temp °C	a_o Initial in	Δa 1 hr in	Δa 24 hr in	$a_o + \Delta a$ 24 hr in	Pass-P Fail-F
5-8-80-12	14	23	1.82	1.30	1.35	3.17	F
11	35		1.67	1.08	1.10	2.77	F
10	70		1.58	0.15	0.69	2.27	F
9	140		1.29	0.01	0.09	1.38	P
8	211		1.30	0.11	0.17	1.47	P
7	316		1.51	0.09	0.21	1.72	P
6	351		1.51	0.11	0.20	1.71	P
5	421		1.47	0.15	0.19	1.66	P
3	492		1.32	0.10	0.19	1.51	P
2	527		1.43	0.14	0.11	1.54	P
1	562		1.47	0.15	0.20	1.67	P
5-21-80-1	35	33	1.60	0.10	0.17	1.77	P
2	70		2.00	0.07	0.20	2.20	F
3	140		1.52	0.02	0.22	1.74	P
4	281		1.41	0.03	0.12	1.53	P
5	421		1.30	0.08	0.16	1.46	P
6	562		1.47	0.02	0.12	1.59	P
5-21-80-7	35	40	1.28	0.02	0.16	1.44	P
8	70		1.46	0.00	0.19	1.65	P
9	140		1.37	0.01	0.22	1.59	P
10	281		1.31	0.08	0.17	1.48	P
11	421		1.30	0.01	0.16	0.46	P
12	562		1.50	0.80	0.30	1.80	P

Degrease: 10 min in TMC
NaOH dip time: 10 min
Rinse: 0.5 min in DI water
Dry: 1 min N_2 blow + 30 min ambient

3.3 Effect of NaOH Dip Time, Degrease and Primer

Figure 1 shows the effect of dip time after 10 min TMC degrease, 5 min Alkanox degrease and after no degrease. A 1 min dip is sufficient after the TMC degrease, and a 30 min dip is too long. After the 5 min aqueous detergent degrease (Alkanox) a 5 min dip is required, and after no degrease 10 min dip is required, to pass the wedge test.

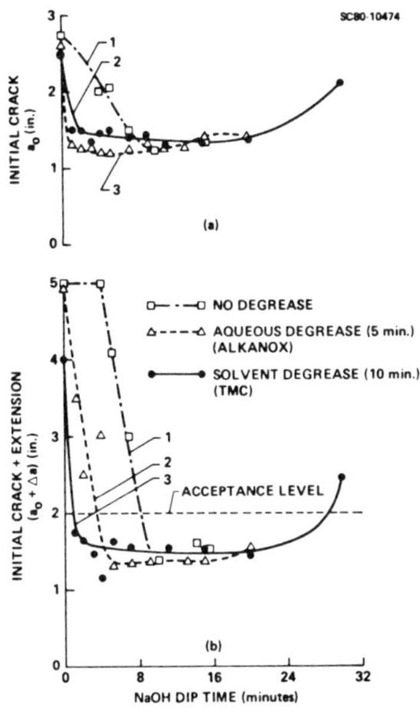

Fig. 1 Effect of NaOH dip time on (a) initial crack a_o, and (b) $a_o + \Delta a$ (24 hr). Al2024-T3, EA 9628H, BR 127, NaOH conc. 562 g/ℓ, DI water rinse, N_2 dry, Room Temp.

With EA 9210H rather than Br 127 the wedge test only passes for dip time between 5 and 11 min, in contrast to 1 and 20 min for BR 127. However, Table 2 gives results for EA 9210H, for a different batch of Al2024-T3 specimens. The results reported in Table 2 are average values for the number of specimens indicated under the sample column (e.g., 12 samples for No. 1-2-79). For this (cleaner) batch of specimens, 98 passed the wedge test without a degrease step and for a 3 min dip. The other 20 samples came within 0.1 in. of passing. The 12 samples that were degreased (MEK) passed.

Table 2

Wedge Test Results: Effect of No Degrease and EA9210H Primer

Primer: EA9210H   Adhesive:EA9628H   Al2024-T3

Sample	a_o Initial in	Δa 1 hr in	Δa 24 hr in	$a_o + \Delta a$ 24 hr in	Pass-P Fail-F
1-2-78 (1-12)	1.40	–	0.30	1.7	P
1-13-79 (1-12)	1.2	0.1	0.9	2.1	F
1-12-19 (13-24)	1.7	–	0.09	1.79	P
1-29-70 (1-12)	1.6	0.07	0.15	1.75	P
2-5-79 (7-12)	1.6	0.10	0.15	1.75	P
1-23-79 (1-8)	1.5	0.08	0.53	2.03	F
2-5-79 (1-12)	1.5	0.0	0.25	1.75	P
2-5-79 (13-24)	1.7	0.13	0.24	1.94	P
2-5-79 (13-24)	1.4	0.08	0.15	1.55	P
1-24-79 (1-6)	1.4	0.04	0.12	1.52	P
1-22-79 (1-6)	1.7	0.14	0.30	2.00	P
2-5-79 (1-12) MEK degrease	1.5	0.08	0.21	1.71	P

Degrease: none, except for 2-5-79 (bottom of table)
NaOH Conc: 562 g/ℓ   NaOH dip time: 3 min
Rinse: DI, 5.0 min   Dry: N_2 blow, 1 min
Temp: 23°C   Delay to Prime: 5 min-24 hr

3.4 Effect of Delay Prior to Rinse

With a degrease step and a 10 min dip, the samples can hang dripping in the air from 0.08 to 10 min and still pass the wedge test, if DI water is used for the rinse. If tap water is used, about half the specimens fail. The average value for 6 tests with DI water was $(a_o + \Delta a)$ = 1.45 in., the average for 6 tests with tap water was $(a_o + \Delta a)$ = 2.1 in.

3.5 Effect Rinse and Rinse Time

As shown above, the use of tap water, rather than deionized (DI) water, can have a dramatic effect. It is crucial that the rinse be hard spray, rather than a water soak. This is because all of the sodium hydroxide must be removed in the rinse step. A number of specimens were rinsed by placing them in a beaker of flowing water. The wedge test $(a_o + \Delta a)$ was found to be directly related to the time of water exposure. It was also found that the pH of the sample surface was directly related to water exposure time. The sample surface was tested for pH by placing a wet litmus paper on the surface and with a micro pH probe on the rinsed sample. A direct correlation exists between the wedge test and the surface pH and if the surface has a pH ~ 6-7 durable bonds result.

In spite of the requirement of complete surface neutralization, a few seconds of water spray is usually sufficient. The small (1 in. × 6 in. × 1/8 in.) samples were spray rinsed about 30 s; the larger 1 ft × 1 ft panels had a few seconds spray rinse at the top of the panel. The rest of the panel was exposed to the water draining from the top, so that the bottom of the panel experienced about 3 min of drain water plus a few seconds of hard spray. The wedge test results appeared independent of the rinse time as long as the surface pH was not basic.

3.6 Effect of Dry and Delay to Prime

There does not appear to be any differences in bond strengths or durability, for Al2024-T3, if the samples are N_2 blow-dried or just left to drip-dry. This is also true for 1 ft × 1 ft panels. As will be seen, there is one set of data for Al7075-T6, for which drip drying dramatically degraded the durability of the joints.

A set of 12 samples was given STAB(3), samples were then left for 5 min, 1 hr, 4 hr, 8 hr, 16 hr, and 24 hr prior to priming. All of the joints passed the wedge test.

3.7 Effects of Scale up to 1 ft × 1 ft Panels

Panels of Al2024-T3, (1 ft × 1 ft × 1/8 in) were surface treated by the STAB(3) process, then cut down the center of the panel to yield two pieces that were bonded together. The 1 in. × 6 in. specimens were labelled A-K for top to bottom in order to investigate the vertical effect of dripping, rinsing, drying and priming on the wedge test. For example, a panel with BR 127 primer was cured at 110°C rather than the specified 122°C. The primer flowed to the bottom of the panel and thus became too thick for proper curing at this temperature. As a consequence the bottom samples failed the wedge test.

Table 3 gives results for 12 panels with varying degrease, dip time, rinse and dry steps. The results from the 11 specimens (A-K) for each panel were averaged and recorded at the right of Table 3. Except for the first panel (degreased, but no NaOH dip) all specimens passed the wedge test. The average $(a_0 + \Delta a)$ = 1.5 in. for 121 tests, standard deviation for a_0 was 0.1 in., for Δa (1 hr) was 0.04 in. and for Δa (24 hrs) was 0.04 in. The large panels give better and more reproducible results than the small specimens.

3.8 Effect of Adhesive

In addition to Hysol EA 9628H, the American Cyanamide 122°C curing modified epoxy (FM 73) (used on the Air Force PABST program), AF 163 (with BX 3944 primer) and 177°C curing PL-729-3 (with PL 728 primer) from B. F. Goodrich, were tested with STAB(3). For FM 73, without a degrease step, STAB(3) passes the wedge test if BR 127 primer is used. STAB(3) also works well with AF 163. The pass-failure criteria used to compare STAB(3) with FPL and PAA for the 122°C curing adhesives, cannot be used for the 177°C curing adhesive because the initial crack

Table 3

Wedge Test Results: Effect of Scale up to 1 ft × 1 ft Panels

Primer: BR127 Adhesive:EA9628H Al2024-T3

Sample	a_0 Initial in	Δa 1 hr in	Δa 24 hr in	$a_0+\Delta a$ 24 hr in	Pass-P Fail-F
4-14-80 (A-K)	2.00	>3	(no dip)	>5	F
4-16-80 (A-K)	1.33	0.09	0.12	1.45	P
5-2-80 (A-K)	1.53	0.07	0.13	1.66	P
5-2-80 (A-K)	1.41	0.09	0.15	1.56	P
4-6-79 (A-K)	1.40	0.09	0.14	1.54	P
4-4-79 (A-K)	1.35	0.05	0.20	1.55	P
3-28-79 (A-K)	1.37	0.04	0.20	1.57	P
3-27-80 (A-K)	1.44	0.05	0.11	1.55	P
3-27-80 (A-K)	1.48	0.06	0.09	1.51	P
4-1-80 (A-K)	1.38	0.05	0.11	1.49	P
4-15-80 (A-K)	1.28	0.08	0.11	1.39	P
4-16-80 (A-K)	1.33	0.08	0.13	1.46	P

Degrease: half of the panels were degreased in Alkanox solution 5 min, the other half were not.
NaOH Conc: 562 g/1
NaOH dip time: top panel (4-14-80) no dip the rest 3 min.

length is consistently larger (i.e., the fracture toughness is less in the dry state) as it is in Ref. 7. For PL-729-3, the average initial crack length was a_0 = 2.0 in. The chosen pass criteria is $(a_0 + \Delta a)$ = 2.4 in., which is comparable with FPL and PAA reported in Ref. 7.

3.9 Caustic Soda

To see if STAB(3) would be satisfactory with inexpensive industrial caustic soda (50% liq) (~ 10¢/lb) rather than technical grade NaOH flakes (~ $2/lb), two panels of Al2024-T3 were prepared and primed with Br 127, bonded with EA 9628H, prior to the wedge test. One panel was dipped in caustic soda (at room temperature) for 10 min, rinsed in hard DI water spray for 3 min, N_2 blow dried plus 0.5 hr ambient dry prior to prime. The other panel had the same treatment but was degreased by soaking in aqueous Alkanox detergent solution for 5 min then sprayed rinse to remove the detergent. The wedge test results for the 22 wedge specimens were essentially identical for both panels (with or without the degrease step). The initial crack length was a_o = 1.38 ± 0.02 in., the crack extension after 1 hr at 60°C, 100% RH was Δa = 0.07 ± 0.02 in. and after 24 hrs Δa = 0.11 ± 0.02 in. Therefore STAB(3) passes the wedge test, $(a_o + \Delta a)$ = 1.49 ± 0.02 in. with even better results for caustic soda.

3.10 Long Time Exposure

An Al2024-T3 panel was given the STAB(3) surface treatment (5 min Alkanox degrease, 3 min NaOH drip, DI water spray rinse, N_2 blow dry) primed with BR 127 and bonded with EA 9628H. The 11 samples A–K were placed in humidity chamber at 60°C, 100% RH for 3834 hr. The average results are recorded in Fig. 2. After 3834 hr the crack extension is only 0.77 in. Crack growth follows a log time relationship which changes slope between 200 and 500 hr. Equations for these two regions are:

$$\Delta a_{1-200} \sim 0.092 \log t + 0.05 \text{ (in.)}$$

$$\Delta a_{500-4000} \sim 0.46 \log t - 0.88 \text{ (in.)}$$

It is of interest to note that after 3834 hr of humidity exposure there was no sign of corrosion on the outer primed surfaces.

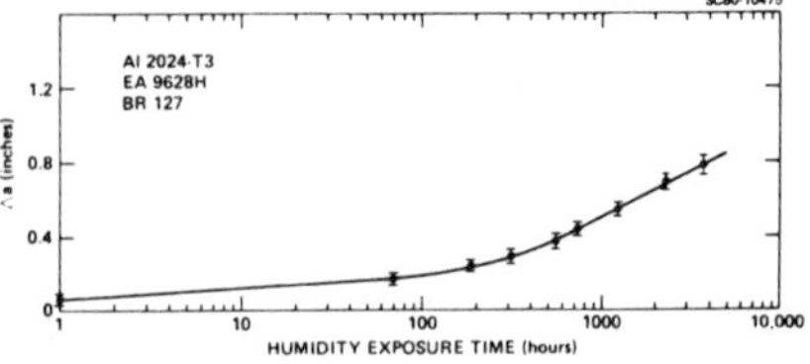

Fig. 2 Long term wedge test, crack extension vs time, RH ~ 100%, T = 60 C, Initial crack length a_o = 1.35 ± 0.02 in.

3.11 Different Alloys

Wedge test results for Al2219-T37, Al6061T6 and Al7075-T6 reveal that dip times of 1 min to 10 min provides passing wedge test results for another 2000 series alloy, Al2219-T37, but not for the 6000 series Al6061-T6 or 7000 series Al7075-T6 alloys at room temperature. Increasing the dip time to 20 min (at RT) provides a passing test for Al6061-T6 but dip times of 60 min is required for Al7075-T6. However, for a 10 min dip, increasing the temperature to 30°C or above, provides passing tests for Al6061-T6 and for a 5 min dip, increasing the temperature to 40°C and

above provides passing tests for
A17075-T6.

3.12 Stress Durability Tests

Some of the lap shear specimens were
used to determine the dry lap shear
strength (~ 4550 psi) and others
were placed in the humidity chamber
at 60°C, 100% RH at stress levels of
2000 psi and 2472 psi. All of the
samples stressed to 2000 psi passed
the test (failure time > 150 hr) if
they were degreased. At 2472 psi
and 3 min dip, half the samples
passed (>50 hr) if they were degreased, none of them passed if they
were not degreased. However, there
are insufficient data here or in the
literature to adequately compare
STAB(3) with FPL or PAA.

3.13 Titanium and Steel

A few preliminary experiments were
performed with Ti6A14V, titanium
alloy and AM 355 steel. For elevated temperature (>40°C) and an
appropriate NaOH concentration there
is a hope that STAB(3) might be useful for titanium and steel. However
the problems of irreproducibility
(as for STAB(1) + (2)) may result.

It is not possible to use the same
pass-fail criterion ($a_o + \Delta a$) for
other metals as for aluminum because
a_o will be different for adherends
of differing dimensions and elastic
constants. Assuming that for the
common adhesive (EA 9628H) the dry
yield strength (G_I) is the same,
then $a_o \sim 1.35$ in. for aluminum, a_o

~ 1.5 in. for titanium and $a_o \sim 0.9$
in. for steel. The pass-fail test
was decided from Δa at 24 hrs. The
dimensions were 1 in. × 6 in. × 1/8
in. for aluminum, 1 in. × 4 in. ×
1/16 in. for titanium and 1 in. × 6
in. × 1/16 in. for steel.

4. ECONOMICS

The economics of any particular surface treatment must be judged on the
basis of needed bond strength and
endurance, initial or additional
capital investments, recurring cost
of maintenance and materials and any
additional costs of meeting OSHA
requirements to meet health and
safety regulations.

The Boeing PAA treatment (BAC 5555)
adds three steps to the standard
Forest Product Laboratories (FPL)
treatment, i.e., the anodize and an
additional rinse and dry. The
Boeing treatment entails 9 steps as
compared to 6 for FPL, and 3 for
Rockwell STAB(3). For badly
contaminated metal a degrease step
may be necessary, increasing the
steps to 4 for STAB(3).

A first approximation as to relative
capital investment was estimated by
giving the value of one unit for
each step, adding a unit to each
step that needed heating and one
unit for the electrical equipment
needed for anodizing. In addition
to relative capital investment,
relative bond endurance, and hazard

were considered. On the basis of lab scale up studies, STAB(3) with only 3 steps yields as durable joints as FPL or PAA with 6 and 9 steps respectively. The crude rating scheme yields cost on the order of 11 units for PAA, 8 units for FPL, 3 units for STAB(3). Of particular importance is the low hazard of STAB(3) as compared to FPL etch or PAA, since one of the main purposes of this research was to eliminate the use of carcinogenic chromates and organic solvents, for factory use.

5. SUMMARY OF RESULTS AND
    CONCLUSIONS

The object of this program was to find a nonacid surface treatment (no carcinogenic chromates) for Al2024-T3 that was inexpensive but would produce strong, durable adhesive joints with Hysol EA 9628H adhesive and Hysol EA 9210H or BR 127 primer. Initial studies[4] indicated that a simple degrease and hot water soak (STAB(1)) would produce strong durable joints. Further investigation revealed great difficulty in reproducing this result, although occasional good results could be obtained. Investigation of surface properties revealed that as-received Al2024-T3 has a surface layer of metal that differs from the bulk. This layer is entirely removed in the standard FPL etch, but is only partially reacted during STAB(1). If the Al2024-T3 is FPL etched then given STAB(1) good bond endurance

with wedge test results. It is concluded that in some instances STAB(1) reacts with the entire outer layer of metal yielding hydroxide that is stable to moisture (hydro-thermal stress). In other instances this layer is only partially re-acted, leaving an unstable layer that can corrode under hydrothermal stress. Also, SEM studies reveal the STAB(1) hydroxide to be layered, allowing fracture between the layers.

In the course of studying STAB(1), a simple degrease and water soak (room temperature) in a water solution containing a commercial cleaning agent (MICRO) gave good bond endur-ance result. This was labeled STAB(2). Further investigation revealed this treatment to suffer from the same reproducibility problems as STAB(1).

In the process of studying STAB(1) and (2) the effect of sodium hydrox-ide concentration was investigated. It was found that 36 w/o solution of NaOH in DI water (STAB(3)) gave excellent bond endurance by the wedge test. Of 322 wedge test specimens, the average crack growth in the first hour was 0.04 in. and after the first 24 hr, 0.2 in.. Failure during the wedge test was primarily cohesive in adhesive or primer. The results for STAB(3) are comparable with the Boeing PAA, and the standard FPL etch treatments. The lap shear strength averages 5.5

ksi, with cohesive failure, and is also comparable with the PAA or FPL results.

Investigation of the surface properties of STAB(3) revealed that the outer unstable layer of metal is removed in 3 min (i.e., ~ 1 μ removal). After rinsing and drying, a very porous hydroxide single layer is formed. These properties are similar to that for PAA in that primer becomes embedded within the hydroxide layer and the hydroxide and underlying metal is stable to corrosion under hydrothermal stress.

Auger electron spectroscopy with Ar^+ sputter etching reveals the chemical constitution of the outer atomic layers for films after the various treatments. As-received and degreased-only Al2024-T3 have aluminum oxide layers rich in Mg whereas FPL etched and PAA do not. It has been hypothesized that the Mg-rich oxide layer causes corrosion instability of the layer, and thus the layer must be removed for improved hydrothermal stress durability. However, STAB(3) leaves films with a high concentration of Mg but also gives good durability. The as-received and degreased-only Al2024-T3 does have less stable oxide, as revealed by the effect of the electron beam. The electron beam reduces Al^{+3} and desorbs oxygen whereas this does not occur for the surface treated alloy.

A number of process variables were investigated to delineate the range for good bonds and the boundaries beyond which poor bonds result. Table 4 lists the parameters, range and remarks for AL2024-T3. The parameters ranges for other alloys are not well established as yet.

Table 4

Process Parameter Boundaries for STAB(3)

Parameter	Range for Good Bonds	Remarks
Glue line thickness	3-7 mil	Normal ~ 4 mil
Primer thickness	0.4-1.4 μ (wt chg) 0.06-0.2 mil (calipers)	
Primer cure	1 hr normal	10 min gave poor peel strength,
	121°C (250°F)	Left overnight before cure gave bad bonds
Degrease	Not needed in most instances	But does not degrade STAB(3)
NaOH concentrate	>140 g/ℓ at RT, can be lowered at higher temp	Best results ~ 600 g/ℓ at RT. Caustic soda works well.
Time of NaOH dip	>3 min	Best results ~ 3-10 min
Delay between NaOH dip and rinse	10 min ok	Probably longer ok
Rinse	Spray rinse with DI water (must remove all NaOH) up to 20 min, ok	Insufficient rinse leaves surface alkaline giving poor durability
Dry	N₂ blow, air blow, drip dry	
Delay before Primer	24 hrs ok	
Delay between Primer and cure	1/2 hr ok, 24 hrs bad	

Although STAB(3) has been optimized for Al2024-T3 and Hysol EA 9628H adhesive and primer, it has been found to be as good for other 121°C (250°F) curing adhesives, e.g., FM 73/BR 127, AF 163/XB 3944. However, the STAB(3) parameters that give good results for Al2024-T3 have not been found optimum for other alloys such as Al7075-T6 or Al6061-T6. Initial results indicate that good results can be obtained by raising the temperature.

STAB(3) produces a pleasing gold color. This plus surface properties measured by ellipsometry, SPD, PEE and water contact angle can be used for quality assurance that the treatment has been properly performed and is the basis for nondestructive inspection techniques.

Care must be taken to ensure adequate spray rinsing. Inspection for inadequate rinsing can be made with a micro pH meter or with litmus paper. The pH should be near 7 by placing wet litmus on the surface. Values of pH > 8 indicate inadequate removal of NaOH, which will result in poor hydrothermal stress durability.

The process times are similar to those for FPL and PAA (i.e., 3-20 min dip, up to 10 min or more delay before rinse, 1-20 min rinse, up to 24 hr or more delay to prime etc.) which are conducive to large scale factory operation.

It is concluded that the laboratory scale tests indicate factory use of STAB(3) will be much less expensive, less hazardous, but as effective as other surface treatments for aluminum alloys. Initial tests of STAB(3) with AM 355 steel and Ti6A14V, titanium alloy indicate possible use for these systems.

ACKNOWLEDGEMENT

This project was partially supported by Department of the Army, U. S. Army Armament R&D Command, Dover, New Jersey, (Contract DAAK10-78-C-0274). Mr. Raymond Wegman was program monitor.

It is a pleasure to acknowledge the help of G. W. Lindberg, L. R. Bivins, R. P. Haak and D. Collins.

REFERENCES

1. T. Smith, "Symposium on Durability of Adhesive Bonded Structures," sponsored by U. S. Army Research and Development Command, Dover, New Jersey, Oct. 1976.

2. J. D. Venables, D. K. McNamara, J. M. Chen and T. S. Sun, SAMPE, 10th National Tech. Conf. 10, pp. 362, Oct. 1978.

3. A. W. Bethune, SAMPE Journal, July/Aug/Sept. 1975.

4. T. Smith, J. Adhesion, 9, 313 (1977).

5. T. Smith, "Surface Treatment for Aluminum Bonding," Final report to U. S. Army Armament R&D Command, Dover, New Jersey, (Contract DAAK10-78-C-0274) 1979.

6. T. Smith, "Adhesive Bond Endurance by the Wedge Test," SAMPE Conference, Hyatt House Hotel, Los Angeles, Feb. 5-6, 1980.

7. H. S. Schwartz, Ref. 1, pp 101.

8. R. W. Malarik, 25th National SAMPE Symposioum, pp. 172, May 6-8, 1980.

9. R. W. Shannon et al, PABST Report, AFFDL-TR-77-107, Sept. 1978.

10. T. Smith, Science Center.

11. W. J. Russel, Ref. 1, pp. 155.

12. R. F. Wegman, Ref. 1, pp. 1.

THE ENVIRONMENTAL DURABILITY OF 2024 T3-KEVLAR[R] 49 FIBER AND 2024 T3-CARBON FIBER LAMINATES

G. H. Koch
BATTELLE
Columbus Laboratories
505 King Avenue
Columbus, Ohio 43201

ABSTRACT

The environmental durability of 2024 T3 bonded with Kevlar[R] 49 and carbon fiber reinforced adhesives has been investigated. The study has indicated that the fibers had no detrimental effect on the durability in humid air and salt spray environments, when non-clad material was used as adherend. When clad material was used severe corrosion of the clad layer at the adhesive-metal interface was observed.

Keywords: durability, corrosion, fiber-reinforced, laminate, Kevlar[R] 49, carbon, aluminum, clad.

1. INTRODUCTION

Laminated metallic structures have exhibited good fracture properties as compared with monolithic structures.[1-4] Further improvement of the fracture behavior of laminates can be obtained by applying high modulus fibers such as Kevlar[R] 49 or carbon fibers in the adhesive.[5,6]

Previous studies[4,5] have shown that in particular carbon fiber reinforced adhesives can significantly improve the fatigue cracking behavior of laminated structures. However, due to a large difference in corrosion potential between carbon and aluminum, it may be possible that the aluminum adherend corrodes rapidly when the laminated structure is in a corrosive environment. Electrochemical studies[7] have reported differences in corrosion potential of approximately 1 Volt.

Although the effect of Kevlar[R] 49 fibers on the fracture behavior of composite-metal laminates is less dramatic, an aluminum-Kevlar[R] 49 combination may be preferred because a better corrosion resistance is expected. Since Kevlar[R] 49 is non-conducting it cannot form a galvanic couple with aluminum. However, Kevlar[R] 49 which is an aromatic polyamide, could readily transport water from a moist environment into the bond line and enhance debonding at the adhesive-metal interface. This process of debonding is generally known as bond line corrosion.

In the present study the corrosion behavior of these laminated materials has been investigated to determine the effect of salt spray and moist air environments on the durability of different carbon and Kevlar$^{(R)}$ 49 reinforced laminates.

2. EXPERIMENTAL

The aluminum-copper-magnesium alloys 2024 T3 bare and clad with pure aluminum were used as adherends, (1.5 mm thick). Both materials were surface treated by one of the following two methods:

- chromic-sulfuric acid pickling
- chromic acid anodizing.

The adhesives which were used to bond the adherends are listed in table 1.

1.	Kevlar$^{(R)}$ 49 prepreg (0.254 mm thick)
2.	Redux$^{(R)}$ 775 and Kevlar$^{(R)}$ 49 prepreg
3.	Carbon fiber prepreg (0.254 mm thick)
4.	Redux$^{(R)}$ 775 and carbon fiber prepreg.

Table 1. The Different Adhesive Bonding Systems

Kevlar$^{(R)}$ 49 aramid is an organic (aromatic polyamide) fiber which combines high modulus with high strength. For the present studies a 181 weave of the fibers was used impregnated with an epoxy resin (Narmco$^{(R)}$ 550). When the Kevlar$^{(R)}$ 49 prepreg weave was used as adhesive, BSL$^{(R)}$ 101 primer, a phenolic formaldehyde resin, was applied to the aluminum adherend and forced dried in an oven at a temperature of 70°C. Subsequently, the laminates consisting of Kevlar$^{(R)}$ 49 prepreg between two aluminum alloy adherends, were assembled to panels from which single lap shear specimens could be machined in accordance with ASTM 1002. The panels were cured in an autoclave for 90 minutes at a temperature of 145°C and a pressure of $3x10^5$ Pa.

When the prepreg Kevlar$^{(R)}$ 49 weave was bonded between the adherends with Redux$^{(R)}$ 775, a two-component phenolic formaldehyde adhesive, the prepreg weave was first cured at a temperature of 125°C at a pressure of $3x10^5$ Pa. The cured sheets were lightly abraded with 400 grid paper and cleaned with methyl ethyl ketone (MEK). Both the cured Kevlar$^{(R)}$ 49 sheets and the pre-treated aluminum alloy sheets were primed with BSL$^{(R)}$ 101 which was dried at room temperature. After the Redux$^{(R)}$ 775 adhesive was applied on both the Kevlar$^{(R)}$ and the aluminum surface, the laminated panels were assembled by curing under vacuum in an autoclave for 30 minutes at a temperature of 145°C. The pressure during curing in the autoclave was held at 3×10^5 Pa.

The high tensile strength carbon fiber prepreg consisted of parallel fibers (unidirectional) impregnated with the Narmco[R] 550 epoxy resin. As in the case of Kevlar[R] 49 the carbon fiber prepreg was both used as adhesive and as precured reinforcement bonded between the aluminum adherends with Redux[R] 775. The laminated panels were prepared under the same conditions.

From the bonded panels lap shear specimens were machined having dimensions in accordance with ASTM 1002. The laminated specimens were assembled such that the carbon fibers and the warp of the Kevlar[R] 49 weave were parallel to the direction of the applied load, see Fig. 1.

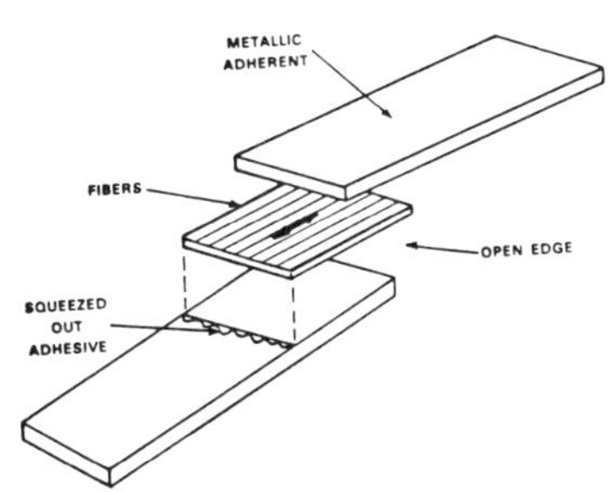

Figure 1.   Schematic of the lap shear specimen assembly, showing the fiber reinforced adhesive between the aluminum alloy adherends. The arrow indicates the direction of the carbon fibers and the warp of the Kevlar[R] 49 weave.

The squeezed out adhesive was not removed so that only corrosive attack could be expected from the sides of the lap shear specimens.

The specimens were exposed in moist air and an aqueous salt spray environment, and removed at different time intervals, viz., 9, 18, 36, 72, and 144 days. The exposure conditions were as follows:

- moist air at a relative humidity of 95% and a temperature of 50°C
- an aqueous salt spray containing 5% NaCl according to ASTM B117.

After removal from the test environment the specimens were loaded to failure in an Instron mechanical testing machine at a cross head speed of 1 mm/min to determine the shear strength.

3.   RESULTS AND DISCUSSION

The effect of exposure time and material variables on the durability was determined for 16 different laminate systems. The material variables were:

- 2 materials, 2024 T3 bare and clad
- 2 surface treatments, pickled and anodized
- 4 adhesive systems.

To determine the effect of these variables an analysis of variance was conducted.[8] The analysis indicated that the effect of surface treatment on the durability in the moist air and salt spray environment was not significant, except for clad material exposed to the salt spray environment. In the latter case the anodized specimens had better resistance to corrosion and consequently higher shear strength after exposure than the pickled specimens. Further, the presence of a clad layer did not influence the durability of the laminates in moist air, while it had a significant effect on the durability in the salt spray environment, Fig. 2, 3, and 4.

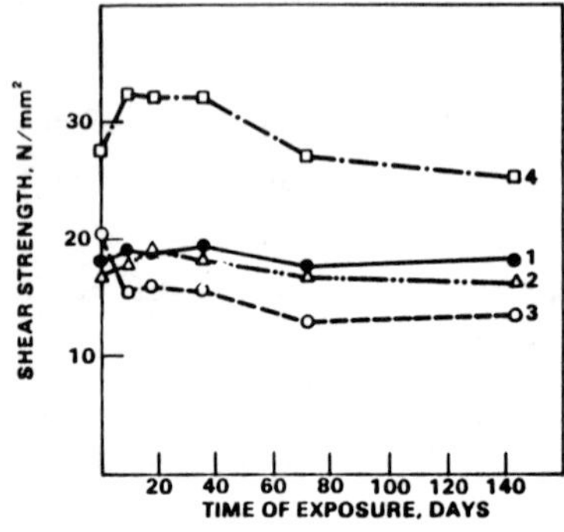

Figure 2. Diagram showing the effect of exposure to the moist environment (RH 95%, T = 50°C) on the shear strength of 2024 T3 bonded with the 4 different adhesives listed in Table 1.

Figure 2 shows the effect of exposure to the moist environment on the shear strength of 2024 T3. Since it was found that the presence of a clad layer and the surface treatment did not influence the durability in moist air, the data points in the figure are the mean values

of the shear strengths determined for bare, clad, pickled and anodized specimens.

The figure shows that the initial shear strengths are approximately the same (near 20 N/mm^2) except for system 4 where precured carbon fiber prepreg was bonded between the aluminum adherends with Redux[R] 775. In some cases an initial increase in shear strength was observed. A possible explanation for this behavior is that the water absorbed from the environment relieves internal stresses in the epoxy, allowing for an overall increase in shear strength. The ability of water to relieve internal stresses has been demonstrated by Brewis et. al.[9]

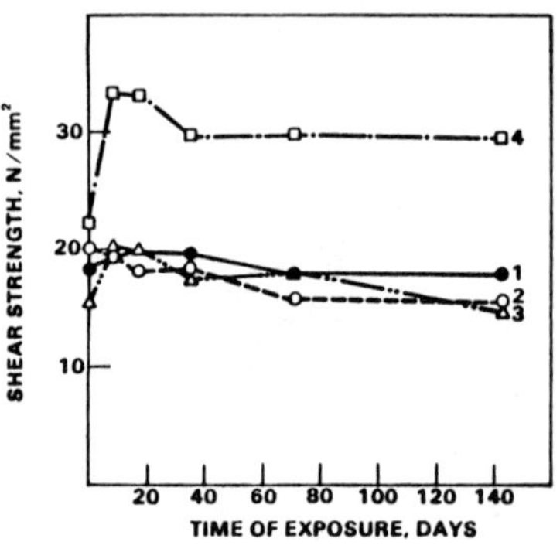

Figure 3. Diagram showing the effect of exposure to a 5% NaCl salt spray on the shear strength of 2024 T3 bare bonded with the 4 different adhesive systems listed in Table 1.

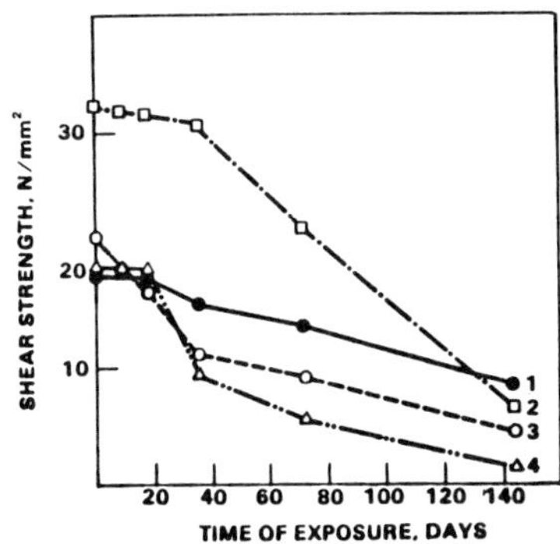

Figure 4. Diagram showing the effect of exposure to a 5% NaCl salt spray on the shear strength of 2024 T3 clad bonded with the 4 different adhesive systems listed in Table 1.

After the initial increase the lap shear specimens suffered a decrease in shear strength. This decrease has generally been attributed to diffusion of water through the adhesive rather than along the metal-adhesive interface,[9-12] although in the case of composite materials there is some evidence of diffusion along the fiber-matrix interface.[13] Experimental evidence supported by thermodynamic arguments[10,11] has suggested that failure occurs along the interface by displacement of the adhesive on the metal oxide by water, which results in debonding. The work of adhesion per unit area is defined by Dupre's equation

$$W_A = \gamma_x + \gamma_y - \gamma_{xy}$$

where γ_x and γ_y are the surface free energies of materials x and y and γ_{xy} is the free energy of the interface between materials x and y. When water is at the interface the work of adhesion often becomes negative indicating that the interface is unstable and debonding will occur. Table 2 shows the detrimental effect of water on the work of adhesion of different interfaces.

Interface	Work of adhesion	
	Inert medium, W_A (mJ/m²)	In water, W_{AL}, (mJ/m²)
Epoxy/ferric-oxide	291	-255
Epoxy/aluminum-oxide	232	-137
Epoxy/carbon-fiber- reinforced plastic	88 90	22 44

Table 2. Values of work of adhesion for some epoxy-metal oxide and epoxy-carbon fiber reinforced plastic interfaces[11,14]

Since the interface between carbon fibers and epoxy is much more stable in the presence of water than that of metal oxides and epoxy, it is expected that failure of the carbon fiber reinforced epoxy bond would occur along the oxide-epoxy interace or through the epoxy.

Since the Kevlar[R] reinforced adhesive bond shows similar behavior in the moist air environment it may be assumed that failure occurs by the same mechanism. However, since Kevlar[R] 49 readily absorbs water, diffusion through the fibers rather than through the epoxy may be rate controlling.

Significantly, no corrosion was observed on the aluminum at the adhesive-metal interface, even in the case of clad material adhesive bonded with carbon fiber reinforced epoxy, Fig. 5

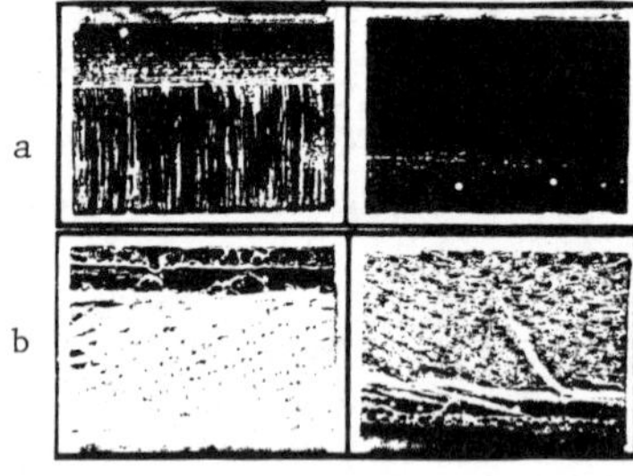

Figure 5. Photographs showing the opposite fracture surfaces of 2024 T3 clad/chromic-sulfuric acid pickled lap shear specimens after 144 days exposure to a moist air environment (RH 95% T = 50°C). The metallic sheets were bonded with (a) carbon and (b) Kevlar[R] 49 fiber reinforced epoxy.

Although carbon forms a significant galvanic couple with aluminum the current flow can be slowed down considerably by embedding the carbon in the non-conducting resin so that contact between the carbon and the aluminum is avoided. When some carbon fibers do make contact with the aluminum, the cathode-anode ratio is so small that the resulting corrosion rate is negligible.

In the more severe corrosive environment of 5% NaCl salt spray a distinct difference in durability of bare and clad adherends was observed, Fig. 3 and 4. The initial increase and subsequent decrease in shear strength found for non-clad materi-

als was similar to that of specimens exposed to moist air suggesting that failure resulted from water diffusion through the bond line and subsequent failure of the epoxy or epoxy-metal oxide interface. Also, no corrosion was observed at the bond lines, but in all cases the 2024 T3 material showed severe pitting.

When 2024 T3 clad was used as adherend severe corrosion was observed on the fracture surfaces both in the case of Kevlar[R] 49 and carbon fiber reinforced adhesives, Fig. 6. Since pure aluminum is more active than alloyed aluminum it will provide protection of the alloy by preferential dissolution. This process is exacerbated by contact with the carbon fibers which resulted in severe corrosion of the interface. Corrosion of the clad layer resulted then in rapid decrease of shear strength with increasing time of exposure, Fig. 4.

4. SUMMARY

The results of this study have indicated that the presence of Kevlar[R] 49 and carbon fibers in the bond line did not have a detrimental effect on the durability of clad and non-clad aluminum alloy laminates when exposed to moist air environments. Also, when a non-clad aluminum alloy was used as adherend the two different types of fibers

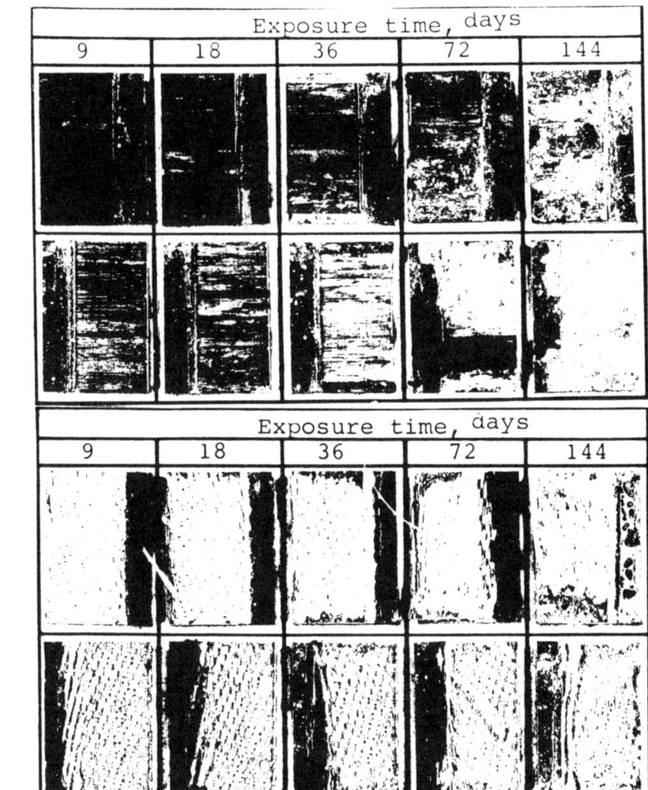

Figure 6. Photographs of fracture surfaces of (a) carbon and (b) Kevlar[R] 49 reinforced adhesive bonded 2024 T3 clad/chronic-sulfuric acid pickled lap shear specimens which were fractured after different exposure times in a 5% NaCl salt spray.

did not appear to have a detrimental effect on the durability in a 5% NaCl salt spray environment. However, when 2024 T3 clad was used severe corrosion of the clad layer at the adhesive-metal interface was observed. The corrosion was particularly severe when carbon fibers were used as reinforcement.

5. ACKNOWLEDGMENT

The study described in this paper was carried out at the Manufacturing Research and Product Development Department of Fokker Netherlands Aircraft Industries. The author is indebted to Ir. R. J. Schliekelmann, head of this department for his comments and permission to publish the results of this work.

6. BIOGRAPHY

G. H. Koch received his BS (1969)
and MS degree (1971) in Aeronautical
Engineering from the Delft Technical
University. After receiving his
PhD degree in metallurgy from the
University of Illinois at Urbana-
Champaign (1976) he joined the Manu-
facturing Research and Product De-
velopment Department of Fokker
Netherlands Aircraft Industries. In
1980 the author joined the Corro-
sion Section of Battelle's Columbus
Laboratories as a research scientist.

7. REFERENCES

1. J. C. Kaufman: Transactions of
    ASME, J. Basic Engineering, Vol.
    87, 1967, p. 303.

2. J. A. Alic and A. Danesh: Eng.
    Fracture Mechanics, Vol. 10,
    1978, p. 177.

3. P.F.A. Bijlmer: Proceedings of
    the 11th ICAS Conference,
    Lissabon, Portugal, September,
    1978.

4. G. H. Koch: SAMPE Quarterly,
    October, 1979, p. 7.

5. G. H. Koch: Proceedings of the
    10th ICAF Symposium, Brussels,
    Belgium, May, 1979, p. 4.4/1.

6. R. Marissen and L. B. Vogelesang:
    Proceedings of SAMPE Symposium,
    Cannes, France, January, 1981.

7. A.R.G. Brown and D. E. Coomber:
    British Corrosion Journal, Vol.
    7, September, 1972, p. 232.

8. O. L. Davies: _Design and
    Analysis of Industrial Experi-
    ments_, Ch. 6 and 7, ed. Oliver.

9. D. H. Brewis, J. Comyn and
    J. L. Tegg: International
    Journal Adhesion and Adhesives,
    Vol. 1, 1980, p. 35.

10. R. A. Gledhill and A. J. Kin-
    loch: Journal of Adhesion,
    Vol. 6, 1974, p. 315.

11. A. J. Kinloch: AGARD Lecture
    Series No. 120, Bonded Joints
    and Preparation for Bonding,
    1979, p. 2.1.

12. A. W. Bethune: SAMPE Journal,
    Vol. 11, 1975, p. 4.

13. P.W.R. Beaumont and B. Harris:
    Journal of Materials Science,
    Vol. 7, 1972, p. 1265.

14. D. H. Kaelbe, P. J. Dynes, L. W.
    Crane and L. Maus: Journal of
    Adhesion, Vol. 7, 1974, p. 25.

26th National SAMPE Symposium
April 28-30, 1981

RESEARCH IN ADHESION AND ADHESIVES: FUTURE DIRECTIONS
IN THE U. S. ARMY

George R. Thomas and Stanley E. Wentworth
U. S. Army Materials and Mechanics Research Center
Watertown, MA 02172

Abstract

Past Army efforts in adhesion and adhesives have been geared to the development of data on the performance of specific adhesives and substrates under a variety of conditions. Although useful to the designer, this approach is inefficient in that it requires that such data be collected for each new system (adhesive and/or substrate). Instrumentation and methodology now exists which permits a much more fundamental approach to the study of adhesive bonding. To this end, a NASA/DOD Committee on Adhesion and Adhesives has been established which will effect interagency coordination in these areas. Army activities will include efforts in shelf life determination for anaerobic adhesives, development of thermally resistant adhesives and improved methods for NDE of adhesive joints. Other studies will address the adhesive properties of interpenetrating polymer networks, the efficacy of coupling agents in improving adhesive bonds and the fundamental aspects of the adhesion of novel block copolymers. These latter investigations are being accomplished thru academic grants. The overall objective of these programs is the development of a sufficient understanding of adhesive bonding to permit bonded joint performance and lifetime prediction and to provide a rational basis for the molecular design of new adhesives.

Over the years, the majority of the Army's efforts in the area of adhesives has been concerned with the evaluation of rather specific combinations of commercial adhesives and substrates under a wide variety of conditions including outdoor environmental exposure. Also extensively studied has been the effect of various processing parameters such as surface treatment on the quality of adhesive bonds. This work has been described in an extensive series of technical reports issued by the Army's Picatinny Arsenal in Dover, New Jersey. A summary of some of this work is given in reference [1].

While this kind of work is of significant value to the design engineer seeking to specify processing conditions or to project useful lifetimes for bonded joints, the approach is rather inefficient. It requires the generation of such data for each new system (adhesive and/or substrate) under consideration. Thus a clear need exists for a more fundamental approach leading ultimately to a sufficient understanding of adhesive bonding to permit the accurate prediction, based on molecular considerations, of performance. In essence, this is the goal of Army research in adhesives and adhesion.

Such an approach obviously involves a multidisciplinary research effort, the cornerstone of which will be materials characterization. Fortunately, the last several years have brought major advances in instrumentation for this purpose. Chief amongst these have been FTIR (Fourier Transform Infrared Spectroscopy) and HPLC (High Performance Liquid Chromatography). These techniques have been applied extensively in our laboratory to a variety of problems including the quality control of glass/epoxy

prepregs [2] and the environmental deterioration of resin matrix composites [3]. Recently, in conjunction with thermal analysis, they have been applied to adhesives problems. One such effort involved determination of the proper cure conditions for an epoxy film adhesive, ultimately leading to the establishment of a simple cure monitoring procedure for the material [4]. Another program utilized these techniques to characterize a series of anaerobic adhesive formulations in preparation for a more ambitious study to be described below [5]. Thus these techniques, together with surface analytical techniques such as ESCA and Auger Spectroscopy, not to mention SEM, STEM and TEM are clearly applicable to a program of fundamental research in adhesion and adhesives. Our initial steps in this direction are the subject of the balance of this paper.

At the present time, work relevant to adhesives and adhesion is being conducted in both Army and academic laboratories. the latter being funded thru the Army Research Office in Durham, North Carolina. In-house, several areas are under investigation. As mentioned above, we have been involved in the characterization of anaerobic adhesive formulations. This work will be the basis for a follow-on program in which we will monitor compositional changes of selected anaerobics as a function of shelf life and correlate them with corresponding changes in mechanical properties. This study will utilize a number of the techniques previously mentioned, especially HPLC and FTIR. In a rather different area, we are engaged in a program of high temperature resin development, one objective of which is the securing of thermally stable adhesives. This program, which has dealt largely with polyphenylquinoxalines, is an example of the sort of advanced materials project that can benefit significantly from a better understanding of the fundamentals of adhesive bonding. In the area of NDE (nondestructive evaluation), a problem exists which will require research for its solution. At present, NDE of adhesive joints can only indicate the presence or absence of a bond. No indication of bond strength is available. Since this information is essential to any realistic NDE procedure to pass or fail a bonded structure, the required research will, in all liklihood, be undertaken. Here a mechanistic approach involving bond fracture mechanics will probably be needed, especially in view of the failure of more empirically based approaches. This area, in fact constitutes one of the better cases for the need for research in adhesive bonding.

The Army funded research in adhesives and adhesion in academic laboratories is, as might be expected, generally of a more basic nature. One of these projects deals with an investigation of the efficacy of coupling agents of the type generally used in composites in promoting adhesion in metal/polymer systems. Another program involves an evaluation at a fundamental level of the adhesive properties of an exciting new class of materials, the interpenetrating polymer networks or so-called IPN's. A third program consists of a major effort in which the adhesion of a series of well characterized novel block copolymers will be examined using a full range of mechanical and surface analytical techniques. A number of other Army grants to academic laboratories deal in varying degrees with various aspects of adhesion and adhesives. Hence the above should not be regarded as comprehesive but representative. Further information about these programs can be obtained from the authors or the Army Research Office.

As this program of fundamental research is implemented, it will be important to co-ordinate the Army's activities with those of other agencies in order to avoid duplication of effort. Such is the objective of the recently established NASA/DOD Committee on Adhesion and Adhesives. This coordination will be accomplished principally thru the conducting of a series of workshops in which both the in-house programs of the various agencies and those of their grantees and contractors are

to be reviewed on a regular basis.
The first of these workshops is
scheduled for Spring, 1981.

While the principle objective of
these research programs is a better
understanding of adhesive bonding,
spin-off benefits can also be
anticipated. Two of these will be a
direct result of the application of
advanced materials characterization
techniques to adhesives. These are
better quality control procedures
for adhesives and better specification documents for adhesives. It
should be possible for instance, to
move in the direction of specifications based on composition
rather than performance. This will
permit the Army to exercise much
tighter control over the uniformity
of the materials it accepts.

In summary, future Army research in
adhesives and adhesion will be of a
much more fundamental nature than in
the past. In concert with other
agencies, a detailed understanding of
the adhesive bonding process will be
sought. This will result not only
in a sound footing for performance
and durability prediction, but will
also provide a rational basis for
the development of new adhesive
systems.

REFERENCES

(1) M. J. Bodnar and R. F. Wegman,
     SAMPE J. $\underline{5}$ (Aug/Sep) 51, 1969.
(2) G. R. Thomas, B. M. Halpin,
     J. F. Sprouse, G. L. Hagnauer
     and R. E. Sacher, Proceedings
     of the 24th National SAMPE
     Symposium, San Francisco,
     May 1979, 458.
(3) J. F. Sprouse, G. A. George
     and R. E. Sacher, Proceedings
     of the International Symposium
     on the Weathering of Plastics
     and Rubber, London, June 1976,
     D 4.1.
(4) R. E. Sacher, Army Materials
     and Mechanics Research Center,
     unpublished results.
(5) S. E. Wentworth, Army Materials
     and Mechanics Research Center,
     unpublished results.

MAINTENANCE AND REPAIR OF
ADVANCED COMPOSITE STRUCTURE
Clark E. Beck, P.E.
Air Force Wright Aeronautical Laboratories
Wright-Patterson Air Force Base, Ohio

Abstract

Advanced composite materials are now proven viable structural materials offering significant performance and economic potential when applied to aerospace vehicle structures. The need for reliable repair methods and techniques is discussed. A review of early composite structure repair technology efforts is given and some advances made since 1975 are presented. It is shown that graphite/epoxy structures, designed in a variety of configurations and for a wide range of load intensities can effectively be repaired to almost all their original strength.

Keywords: Structure, Composites, Repair, Maintenance.

1. INTRODUCTION

The advantages of advanced composites as viable structural materials have been demonstrated by both structural development programs and application to hardware for actual production aircraft. All-composite wing construction has been used on the AV-8B and composite wing skins, as well as several other composite components, are used on the F-18A. Composite spoilers have been flown on B-737 aircraft for several years and composite ailerons and vertical fins have been designed for the L-1011. Ten prototype composite rudders have been fabricated for flight service on the DC-10, and flight testing is planned for a composite vertical stabilizer, floor beams and other components. Composite components including major empennage assemblies and speed brakes are used on F-14, F-15, F-16, and F-18 aircraft. Missile and spacecraft designs and helicopter applications are also taking advantage of the unique properties of advanced composite materials.

As the structural use of advanced composites has increased, a commensurate need has developed for procedures to repair damage which occurs to these structures. The cost of structural repair represents a significant portion of the total operational and maintenance (O&M) cost for an aircraft. For more than two decades, these O&M costs have represented an increasing portion of the total cost of ownership, which currently far exceeds the cost of initial acquisition for an aircraft system.

Table 1 shows the results of a recent study by the Air Force Logistics Command on the Life Cycle Costs (LCC) of a specific transport.

TABLE 1. LIFE CYCLE COST BY PHASE

LIFE CYCLE PHASES	15 YEAR TOTAL COST (IN MILLIONS) 1976 $	PERCENT OF TOTAL 1976 $
(1) R&D	3.221	.04
(2) PROCUREMENT / PRODUCTION	1257.358	17.00
(3) OPERATIONS AND SUPPORT	6134.742	82.95
TOTAL	7395.321	100.00

Approximately 83 percent of the dollars spent went for O&M of that aircraft. Figure 1 shows how that 83 percent of the LCC were spent. Much of this went for maintenance and repair actions. Figure 2 shows a similar story for a helicopter.

It is now economically essential that O&M costs be minimized. One action which will considerably decrease the cost of ownership is

to decrease the cost of structural maintenance and repair. Monetary savings can be realized only if economical procedures for maintenance and repair are developed and implemented.

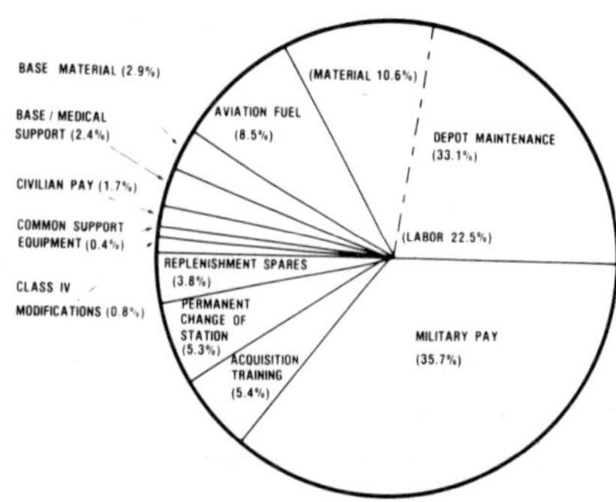

FIGURE 1

OPERATIONS AND SUPPORT 15 YEAR COSTS BY MAJOR CATEGORY

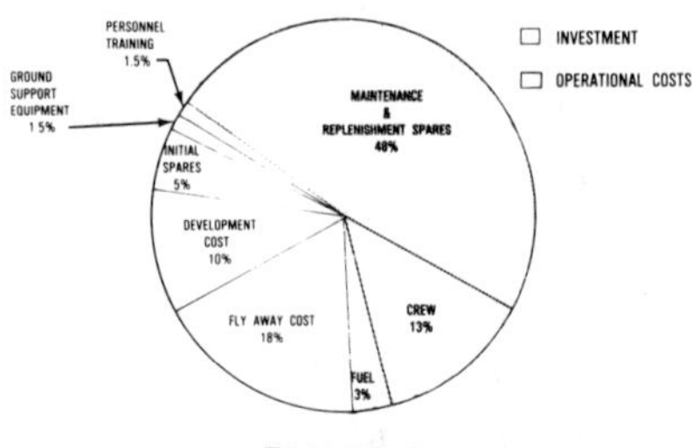

FIGURE 2

TYPICAL LIFE-CYCLE COST BREAKDOWN FOR MILITARY HELICOPTER

Composite structures are susceptible to damage from the same sources which cause damage to metallic structure. Repairs for metallic structure have been developed over many years and have become more or less routine operations in many cases. The rapid increase in the use of advanced composites has created an immediate need for composite repair technology.

2. BACKGROUND

As a result of the various requirements for repair, considerable information has been developed related to advanced composite repair. It is contained in many varied places, some documented and some undocumented. Currently there is a program underway to collect the wealth of existing information pertinent to advanced composite repair, critically evaluate it, identify gaps in the data, develop new information where needed and reduce that information into a form which can be presented in a single concise document.[1]

As early as 1969, Air Force funded research and development work addressed the problems of the repair of small area damage to lightly-loaded boron/epoxy structures.[2] More recently, repair techniques for small damaged areas have been developed for graphite/epoxy and hybrid composite structures, and field repairs for graphite/epoxy laminates have been developed using titanium alloy external patches and mechanical attachments.

Technical Orders for several military aircraft now contain specific repair instructions for small damage areas of composite structure, usually limited to non-strength critical structure. Under a recently completed program, procedures were developed for the repair of large damaged areas, with 80 to 100 percent of the parent laminate strength restored by the repair.[3]

In addition to the documented repair information, there is also a body of knowledge based on actual repair experience which has not been documented. This experience exists as developed by manufacturers of aerospace hardware and by the military and civilian facilities responsible for the maintenance of the hardware. Since much of the most advanced composite repair latest information is not published, personal contact with leaders in the field was very important.

3. REPAIR GUIDE

The current program investigates advance composite structure repair in the categories of fixed wing aircraft, helicopters, and Space and Missile Systems. Some observations in each category follow.

3.1 Helicopters

With the exception of rotor blades, relatively small amounts of advanced composites have been used in helicopters. Almost all advanced composites use on helicopters is on new helicopters which have not been used enough to have incurred damage requiring repair development. Graphite/epoxy is used in limited structural applications on the Sikorsky UTTAS(UH-60), the Sikorsky S-76, the Hughes AAH(AH-64), the Vertol CH-47, the Vertol 234, and the Bell Models 222 and 214ST. As use time is accrued, damage incurred and repairs implemented, the amount of repair data for advanced composites on helicopters will increase.[4]

3.2 <u>Space and Missile Systems</u>
Space and missile systems are experiencing an increase in the use of graphite/epoxy and other advanced composites for critical structural applications. Repair of advanced composites on space systems requires consideration of many factors not associated with aircraft repair. Maintaining aerodynamically smooth surfaces is not a requirement of space system repair. However, the repair materials and processes must be non-contaminating with low out-gassing and survive space environmental conditions including radiation, high vacuum, and wide temperature excursions. Often the maintenance of electrical continuity and the isolation of conductive components are important.

Three categories of space systems employing advanced composites are (a) orbital satellites and planetary probes, (b) the Space Shuttle Orbitor and (c) large space structures. Composite structures on orbital satellites and planetary probes are generally stiff, low weight, intricate, complex, thin walled, high cost, low production quantity items which are easily damaged. Because of these attributes, the repair of this type structure is delicate and highly specialized to meet strict specifications.

The Space Shuttle advanced composite structure repair may be similar to aircraft repair with the added consideration of wide temperature excursions. Repairs to composite beams on large space structures will include considerations of satellite component repair plus possible requirements for repairs to be made in orbit.

For missile applications, retention of high stiffness, high strength, low weight, and high temperature capability are requirements for repaired components. In some cases, selection of repair materials is limited by nuclear vulnerability and hardening requirements and radar observ-ability considerations. Repaired components on missiles must also withstand long-term storage environments.[5]

3.3 <u>Fixed Wing Aircraft</u>
The largest use of advanced composite structure on aircraft has been on fixed wing aircraft. As a result, most work relating to

advanced composite structure
repairs has been done for fixed
wing aircraft. Under Reference 2,
repair procedures were developed
and verified by testing. Only
adhesive bonded repairs were
considered due to bearing allowable
limitations. Two repair configurations, a scarf inclusion and an
external patch were developed. A
portable repair kit was developed
to provide vacuum pressure and
heat in an integral rubber blanket.

Between 1974 and 1976, repair
procedures were developed for
lightly loaded graphite/epoxy
sandwich structure, complex
structural elements and for some
highly loaded structures typical
of those found in a wing or
fuselage.[6] Both flush patches
and tapered external patches were
used, the latter made of either
titanium foil or graphite.

In 1977 and 1978, bolted titanium
external patches were developed
for field repair of laminates up
to 1/2 inch thick.[7] Laminates
used were typical of AV-8B and
F-18A wing skins. Damage up to
4 inches in diameter was considered.

The study of Reference 3 concentrated on the development of repair
procedures for use at military
depot level facilities where
relatively highly skilled personnel
and adequate facilities and equipment are available. Both
monolithic and honeycomb sandwich

construction were considered.
Several variations of an essentially flush repair using a scarf
joint were used to make repairs
with access from one side only,
for partial thickness repairs, and
for use with honeycomb sandwich
construction. As part of this
program, five large panels
(approximately 20 inches by 60
inches test section) were repaired
and subjected to static tests.
Each of these panels represented
a different repair problem. In
addition, an in-service damaged
aircraft speedbrake with graphite/
epoxy skins was repaired and proof
loaded to verify the adequacy of
the repair.[8]

4. ACKNOWLEDGMENT

Much of the work described herein
was sponsored by organizations
within the Department of Defense.
Special mention should be made of
work done by the Northrop Corpora-
tion (J.D. Labor and S.H. Myhre);
Grumman Aerospace Corporation
(G. Lubin); General Dynamics
Corporation (V.J. Studer and
R.M. La Salle); McDonnell Aircraft
Company (J.C. Watson); Bell
Helicopter Textron (S.C. Aker);
Hughes Aircraft Company (R.W.
Seibold) and Lockheed-California
Company (R. Stone) under various
Air Force and Navy contracts.

5. REFERENCES

1. "DoD/NASA Advanced Composite
Structure Repair Guide", Contract

No. F33615-79-C-3217 (AFFDL),
Northrop Corporation.
2. Lubin, G., et. al., "Repair
Technology for Boron/Epoxy
Composites", AFML-TR-71-270,
Grumman Aerospace Corp., February
1972.
3. Labor, J.D. and Myhre, S.H.,
"Large Area Composite Structure
Repair", AFFDL-TR-79-3040,
Northrop Corporation, March 1979.
4. Aker, S.C., Letter 81:SCA
bi-449, Bell Helicopter Textron,
January 1980.
5. Seibold, R.W., Activity
Summary (Contract F33615-79-C-3217),
Hughes Aircraft Company, February
1980.
6. Studer, V.J., and La Salle,
R.M., "Repair Procedures for
Advanced Composite Structures",
AFFDL-TR-76-57, General Dynamics
Corporation, December 1976.
7. Watson, J.C., et. al.,
"Bolted Field Repair of Composite
Structures", NADC-77109-30,
McDonnell Aircraft Company,
March 1979.
8. Myhre, S.H., Activity Summary
(Contract F33615-79-C-3217),
Northrop Corporation, January 1980.

BIOGRAPHY

Clark E. Beck, P.E., has been
involved in aircraft structure
maintenance and repair research
and development since 1973. Since
1975, he has concentrated on the
repair of advanced composite
structures and related research.
His most recent program starts in
this area were the "DoD/NASA
Advanced Composite Structure
Repair Guide" and "Rapid Repair of
Aircraft Structure Battle Damage."

He is presently Technical Manager
of the Experimental Control Group,
Structures Test Branch, Structures
and Dynamics Division of the
Flight Dynamics Laboratory. He is
a registered Professional Engineer
in the State of Ohio and a member
of AIAA, ASME, STC and a past
President of the Ohio Society of
Professional Engineers. He has
degrees in Mathematics, Mechanical
Engineering and Aerospace
Engineering.

26th National SAMPE Symposium
April 28-30, 1981

REVIEW OF NAVY COMPOSITE REPAIR PROGRAM

A. Manno
Naval Air Development Center
Warminster, Pennsylvania

Abstract

Applications of composite materials to current and near-term Navy aircraft have necessitated the establishment of repair methods both for field and depot maintenance levels. Navy requirements are presented for field level repairs as they apply to current repair programs with emphasis on Navy facilities and capabilities. Repair concepts for damages up to 4 inches in diameter have been developed for field level applications for relatively thick wing skins. Results of this work are presented for both bonded and bolted concepts. Additional work is currently being performed as part of a recently initiated program for field level repairs in which a wide range of thicknesses and damage types is being addressed. A survey of existing materials is being made to select the best materials and processes available to meet the Navy requirements. An extensive test and evaluation program is being performed as part of this program.

1. INTRODUCTION

Graphite/epoxy structure have been introduced into current Navy aircraft quite extensively as shown in Figures 1 and 2. To support the use of these materials in the service environment, reliable repair methods are being developed. Initial exploratory efforts within the Navy have addressed the development of field repair methods that are simple enough to be performed by organizational and intermediate level maintenance personnel in shipboard environments. These efforts have been successful in developing concepts which could be implemented immediately. However, the methods are dependent on the availability of suitable materials needed to satisfy the Navy's desire to minimize logistics requirements, time-to-repair, special support equipment and skill levels. Several programs recently initiated are addressing these needs and will help to provide the capability to maintain these structures in both a combat and peace-time environment. Both field and depot maintenance levels are being addressed. The status of the field level repair methods is presented in this paper. The depot level repair program is the subject of another paper being presented in this session. [1]

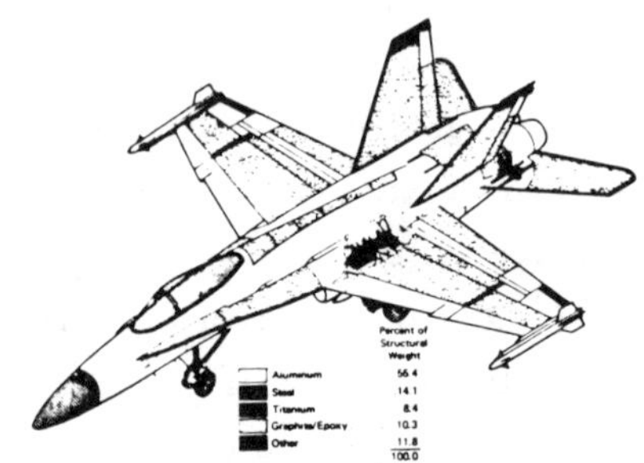

FIGURE 1   F-18 MATERIALS DISTRIBUTION

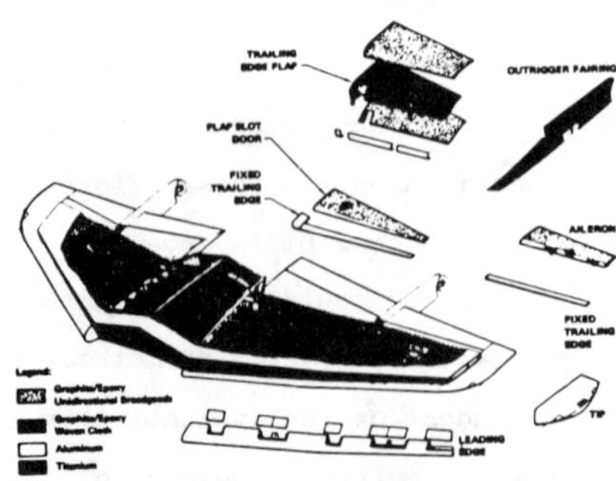

FIGURE 2   AV-8B COMPOSITE WING

2. STRUCTURAL DAMAGE

Navy organizational and intermediate level repair activities are responsible for the repair of damage that can be accomplished on the aircraft with outside access only or on components that are relatively easy to remove. Repair concepts are being developed which will address the types and sizes of damage anticipated to be performed at these levels. Existing damage size limits for field level maintenance of metal structure have been adopted. The construction and damage that are being addressed are the following:

Monolithic Wing Skins - Wing construction having skin thicknesses ranging from 0.1 to 0.4 in. are of interest. Two types of damage are being considered-- (1) penetration with cleaned up hole sizes up to 4 inches and (2) delamination of wing skins resulting from low-velocity impact extending up to 4 inches in diameter.

Full-Depth Honeycomb Sandwich - Skin thicknesses varying from 0.040 to 0.160 in. thick are of interest. Three types of damage are being considered-- (1) facing penetration with hole sizes up to 4 inches, (2) disbonds with core crushing as a result of low velocity impact up

to an area of 12 square inches
and (3) edge damage resulting in
delaminations or chipped edges.

3. DEVELOPMENT OF REPAIR DESIGNS

3.1 Bolted patch

An initial program was completed
demonstrating the application of
mechanically fastened titanium
patches for repair of graphite/
epoxy wing skins. [2], [3] Design
constraints dictated that damage be
accessible from the outside only,
and the repairs be performed with
available field equipment and a
minimum of personnel training.

Skin thicknesses ranged from 0.19
in. to 0.50 in. Damage consisted
of through-the-thickness holes up
to 4.0 in. in diameter in the vicinity of graphite/epoxy sinewave substructure. Five configurations
were fabricated and tested. Static
tension tests to failure were conducted on three specimens with skin
damage near substructure. Two
pressure box specimens were fabricated for internal pressurization
tests. Of these two, one incorporated a rib with cap damage, and the
other a spar/rib inter-section with
accompanying spar/rib cap damage as
shown in Figure 3. The skin was
repaired, but no attempt was made
to repair substructure.

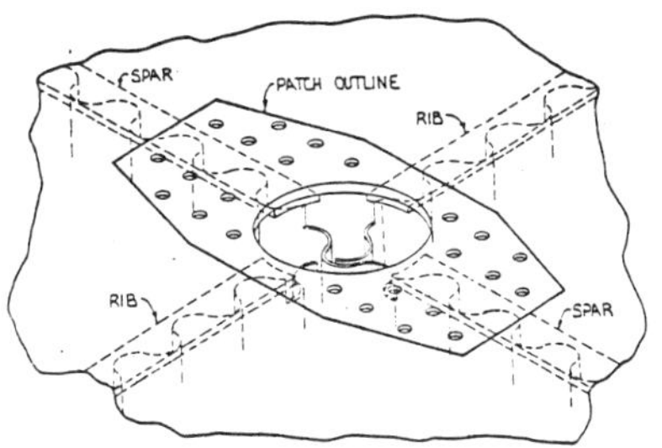

FIGURE 3   DAMAGE AT SPAR AND RIB

3.1.1   Materials. Titanium,
6Al-4V alloy, was selected for the
repair patches because of its high
bearing strength and resistance to
galvanic corrosion. Its stiffness
and thermal expansion properties
are compatible with the parent

composite structure. The 0.16 in.
thick titanium patch is thin
enough to be formed in the field
to mild curvature matching the wing
surface and is the minimum required
for countersinking. Edges were
beveled to minimize aerodynamic
disturbance.

3.1.2   Fasteners. The titanium
patches were designed with 0.25 in.
blind single shear Jo-bolt fasteners. To maintain a minimum patch
size, fasteners were spaced at four
times their diameter (4D). Fasteners were placed as much as possible
in line with the cleaned-up damage
hole and in the direction of load
to provide a direct load path
across the hole. In attaching re-

pair patches over the damage area, care was taken to properly transfer existing fastener hole locations from the skin to the patch. These holes were pilot drilled in the repair plate along with the other fastener holes. A computer program was developed to optimize the number of fasteners and their locations for the final fastener pattern.

3.1.3 <u>Fuel sealing</u>. Repairs in fuel containment areas were sealed with PR 1422 polysulfide per MIL-S-8802. The sealing was accomplished by liberal application of the polysulfide to the faying surfaces of both the patch and laminate. The Jo-bolts were installed wet with sealant. Where damage is confined to a 1.0 in. diameter hole or less, a nonstructural patch (plug) is adequate. This plug is similar to those currently used to seal damaged fuel tanks on existing metal aircraft and has the advantage of simplicity and minimum installation time.

3.1.4 <u>Design criteria</u>. A design allowable strain of 4000μ in/in. in the repaired skins was used since this is the design ultimate strain level to which existing wing skins are designed.

To define the design allowables, unrepaired specimens having hole diameters of 0.25, 1.0, 2.5, and 4.0 in. were instrumented and tested to failure. Gross strain and edge-of-hole strains were measured. The preliminary gross failure strains away from the hole is shown in Figure 4. In general, as the damage hole size increases, the gross strain allowable decreases and levels off at about 2800μ in/in. for a damage hole diameter of approximately 2.5 in. or greater. For diameters larger than 1.0 in., structural repairs are necessary to obtain a gross failure strain in excess of 4000μ in/in.

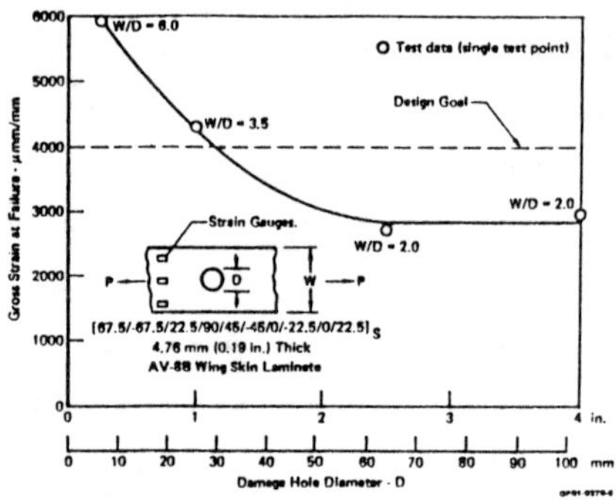

FIGURE 4   GROSS FAILURE STRAIN VS. DAMAGE HOLE DIAMETER

3.1.5 <u>Static tests</u>. Three repair configurations with a 4 inch damage were verified by test to failure. All repairs were performed on the test specimens with access only to the outer surface of the skin. Test specimens included substructural spars and ribs, however, the load plates were attached to the skin only. All three specimens, sustain-

ed design ultimate load based on
4000μ in/in. gross strain. Table
1 summarizes the test parameters
and strain data. In addition to
the tension tests to failure, two
pressure boxes were tested to design limit pressure of 16.7 psig
without damage. Only one leak was
noted on one box and was readily
repaired with sealant.

3.2   Bonded patch

A bonded patch type repair for
primary structure was developed as
part of an in-house Navy program.
The objective was to develop repair
concepts for graphite/epoxy structure and demonstrate that design
structural integrity could be restored with existing Navy equipment
and personnel. Some of the design
constraints included one side access
and minimization of out-of-moldline
protrusion. The repairs were directed to full-depth honeycomb such
as horizontal stabilizers and control surfaces and thin monolithic
skins up to about 0.20 in. thickness such as some areas of the wing
skins and vertical stabilizers.
Damage consisted of through-the-
thickness holes up to 4.0 inches
in diameter.

3.2.1 Materials. Candidate
materials considered for the patch
included titanium and cocured
graphite/epoxy. Both materials

appeared to be acceptable with each
possessing specific advantages and
disadvantages. Advantages of
graphite/epoxy were that special
surface preparation is not necessary before bonding, better conform-
ability, and slightly better stiffness and thermal compatibility with
the parent structure. Disadvantages
are that standard graphite/epoxy
requires special refrigerated handling in transit and refrigerated
storage on site. With refrigeration, present day prepregs have
shelf lives of only six months to a
year. As a result of the initial
screening effort, the graphite/
epoxy repair concept was selected
for further evaluation and
development.

The repair, sketched in Figure 5,
was made by stacking 0.01 in. thick
crossplied discs of AS/3501-6
graphite/epoxy over a 0.015 in.
thick crossplied disc of 7743/2054
fiberglass/epoxy prepreg. The two
ply discs were cut to size from
crossplied sheets of graphite and
fiberglass. A quasi-isotropic
patch was formed by alternately
orienting the discs at $\pm45^{\circ}$ and $0^{\circ}/$
90° to the principal load direction.
The stacked discs were placed on a
sheet of EA 951 adhesive and centered over the damage. EA 951
adhesives was selected because it

was well characterized and best represented an adhesive with medium-modulus and good ductility. An improved current day equivalent would be FM 300 which has improved elevated temperature and environmental performance. The patch was cocured in place under vacuum pressure at 350°F for 120 minutes. The discs were sized such that a taper ratio of 25 to 1 was obtained.

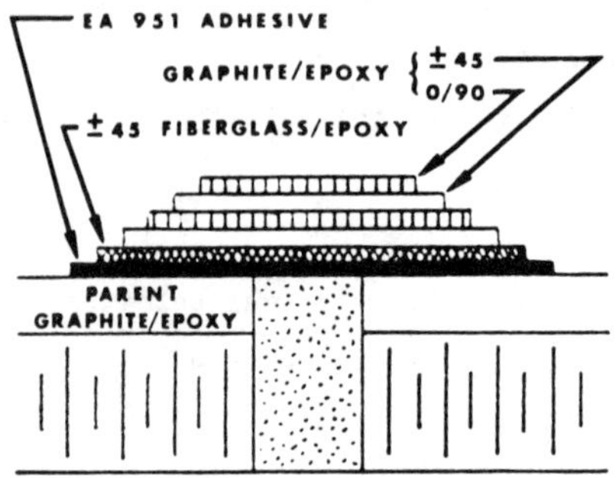

FIGURE 5   EXTERNAL BONDED PATCH

3.2.2   <u>Design criteria</u>. Room temperature static strength restoration of 80 percent "B" basis strength allowable with extensional stiffness restoration was set as the design goal. A design temperature range of -67°F to 220°F was established with representative moisture levels of 1 to 1.2 percent.

3.2.3   <u>Static tests</u>. Both monolithic and full-depth honeycomb beam specimens were statically tested. The repaired skin on the beam specimen (3 in. damage) was subjected to compression loads above 80 percent "B" allowable and then tested in tension to failure. The monolithic specimens were subjected to tension loads only. Table 2 summarizes the major parameters and test results. One specimen achieved 75 percent "B" allowable which does not quite satisfy the design goal but is sufficiently greater than the current ultimate design strain levels currently being used (4000 to 5000μ in/in.) to provide satisfactory margins. For the 4 inch damage specimen, an inadequate bond caused premature failure of the specimen and thus the effectiveness of the patch for this size damage could not be fully evaluated. The poor quality bond may have been caused by outgassing of faulty potting compound. The repair of a 4 inch damage will be evaluated as part of a current contracted effort. [4]

3.2.4   <u>Environmental evaluation</u>. The effects of moisture on the quality and strength of the repair was evaluated. [5] Panels of 24 and 48 ply material were environmentally exposed to absorb 1.2% moisture. Some specimens were dried for 24 hours at 250°F resulting in an average moisture content of 0.7%. The panels were then subjected to a representative cure

cycle with a maximum temperature
of 350°F held for two hours both
with and without a patch applied.
Non destructive inspection and
photomicrographs of the parent
laminates without the patch indicated no blisters, microcracks,
or voids. This was encouraging
since in a previous program [6]
using 3501-5 material, blistering
was observed. Tests performed in
interlaminar shear and flexure
showed reductions in general
agreement with previous investigations. [7] Non destructive inspection and photomicrographs of
the patched laminate indicated the
presence of voids within the patch
and bondline. Patches with relatively high levels of porosity
tended to fail in the vicinity of
the bondline and exhibited significantly reduced properties. Interlaminar shear values at 220°F with
1.2 percent moisture were approximately 55 percent of the room temperature dry values; for 0.7 percent moisture, the values were 65
percent.

Static and fatigue tests of full-
scale repairs at hot/wet conditions were performed. Two repaired
panels, 0.105 in. thick having a
20-ply patch cocured at 350°F over
a 2 in. damage were statically
tested. One panel was tested dry

at room temperature while the
second panel was moisturized to
1.2 percent before and after repair. Failure for both the hot/
wet and the dry panels occurred
through the damage hole. A 17
percent reduction in strength for
the hot/wet condition was obtained.
Although this is a significant
knockdown from room temperature
dry conditions, the reduction in
static strength for the full scale
repairs under hot/wet conditions
is not as drastic as might be
inferred from the short beam
shear tests. For the fatigue test,
a repaired panel 0.19 inch thick
was exposed before and after repair to a moisture content of 1.2
percent. The specimen successfully sustained two lifetimes of
spectrum tension fatigue and 2
lifetimes of compression fatigue
without failure.

Several approaches can be taken to
minimize the effects of moisture.
One involves locally drying the
structure before repair to lower
the moisture content within the
outer plies. Drying may be an
acceptable solution if sufficient
time is available in a field situation. Drying times may take
anywhere from 8 to 24 hours depending on the drying temperature
and thickness. A second approach

is to design the repair for the reduced properties expected as a result of the service environment. This will minimize the time to repair but result in less efficient patches.

TABLE 1. BOLTED REPAIR TEST SUMMARY

Laminate Thickness (in.)	Damage Configuration	Test Ultimate Load (lb.)	Failure Load (lb.)	Gross Strain at Failure (μin./in.)
3/16	Near Spar and Rib	108,320	124,000	4,380
1/2	Near Spar	305,900	360,000	4,280
1/2	Near Spar and Rib	305,900	381,000	4,480

TABLE 2. BONDED REPAIR TEST SUMMARY

Laminate Thickness (in.)	Damage Size (in.)	Failure Strain (μin./in.)	"B" Allow. (%)	Failure Mode
0.10	2	7050	85	Net Tension
0.15	2	6125	75	Net Tension
0.15	3	>7550	>90	Skin-core bond
0.15	4	5500	66	Patch unbonding

4. CURRENT DEVELOPMENT

4.1 Development/validation

Although much work has been accomplished in laying the ground work for the development of repair designs for field use, the number of full-scale validation tests has remained relatively few. A recently initiated program [4] includes significant numbers of specimens both in the screening and validation test phases to provide the confidence in the reliability of the methods developed. In addition to providing a larger data base, the program is addressing

areas of concern within the Navy. Some of these concerns described below were touched upon in previous work but will receive additional attention.

4.1.1 <u>Materials</u>. Logistics problems inherent with using materials which have relatively short shelf life (less than one year) present hardship on board ship and in uncontrolled environments. These materials require refrigeration and specialized handling storage and resupply. It is desirable to eliminate the need for refrigeration. Modifications to existing material systems and processing procedures are being investigated. Promising methods currently under investigation are those to achieve "latency" and thereby make ambient storage of materials possible. These include a new class of stable hardening agents to resins; introducing blocking mechanisms into resins to bind up their hardening agents until driven off by vigorous heating; and advancing standard prepregs beyond their "B" stage but not to the extent that they become boardy.

4.1.2 <u>Repair time</u>. Problems posed by relatively long times to repair damaged structure in combat conditions are of concern to the

Navy. Bonded repair techniques currently available require about eight hours to accomplish under ideal conditions. It is desirable to reduce this time to about four to six hours. Therefore, resins and adhesives having cure times of one to two hours are being sought to meet these goals.

4.1.3 <u>Moisture</u>. Potential problems associated with the bonded repair of structure that may have significant amounts of moisture in the laminate is being evaluated. In the case of honeycomb construction, attention is being given to insure that the procedures developed will not cause damage to the surrounding structure.

4.1.4 <u>Simplicity/Ease of Repair</u>. The use of mechanically attached titanium plates to repair thick wing skins may have some drawbacks. These include the problems associated with the machining of titanium plates, application to curved surfaces and complexity of fastener spacing. Simpler methods are being sought.

4.2 <u>Ground Support Equipment</u> A number of problems were identified with the existing Navy repair kit. As a result, the heating/vacuum blanket was modified. The efficacy of these modifications is being evaluated. Lack of portability of the existing repair kit is also an area of concern as is the need for special ground support equipment. Specifications for an improved kit are being prepared.

5. SUMMARY

The development and validation of new and improved methods for repairing graphite/epoxy structures being used extensively on next-in-line aircraft are being addressed. Initial development efforts have been instrumental in demonstrating that composite wing skins can be repaired by both bonded and bolted methods. Design development efforts have recently been initiated for field level and depot level repairs. Areas of concern for each maintenance level are being addressed. Improved methods for repair of primary and secondary structure will be developed in addition to addressing other construction and damage types not previously considered.

6. ACKNOWLEDGEMENT

The author wishes to acknowledge Mr. E. Rosenzweig of the Naval Air Development Center who is the technical project engineer on the repair program and whose work was used in the preparation of this paper.

7. REFERENCES

1. Labor, J.D., "Composite Repair Concepts For Depot Level Use," 26th National SAMPE Symposium, 28-30 April 1981.

2. Watson, J.C., et. al., "Bolted Repair of Composite Structures," NADC-77109-30, McDonnell Aircraft Company, 1 March 1979.

3. Bohlmann, R.E., Glaeser, D.A., and Watson, J.D., "Field Repair of Composite Structural Components," NADC-78073-60, McDonnell Aircraft Company, February, 1980.

4. NADC Contract N62269-80-C-0283, "Field Level Repair for Composite Structure," Lockheed-California Company.

5. Rosenzweig, E. L., "Environmental and Fatigue Evaluation of Field Repair Concepts," NASC Workshop on Shipboard Repair of Composite Structural Materials," 9-11 September 1980.

6. Labor, J.D. and Myhre, S.H., "Large Area Composite Structure Repair," AFFDL-TR-79-3040, Northrop Corporation, March 1979.

7. Hedrick, G. and Whiteside, J., "Effects of Environment on Advanced Composite Structures," AIAA Reprint 77-463, March 1977.

8. BIOGRAPHY

The author, A. Manno, is a Senior Engineer with the Aero Structures Division. Mr. Manno is the program manager for the Exploratory Development Program and has been active in the field of composite structures for many years. Within the past several years, he has been deeply involved with the development of repair methods for composite structure.

COMPOSITE REPAIR CONCEPTS FOR DEPOT LEVEL USE

James D. Labor
Northrop Corporation
Hawthorne, California

Abstract

Concepts are described which are being developed for use at Naval Air Rework Facilities to repair various graphite epoxy structural components which are typical of emerging Navy aircraft. Repairs have been considered for three structural configurations: (1) monolithic skin panels up to 0.4 inch (10.2 mm) thick, (2) full-depth honeycomb sandwich construction and (3) sine-wave spar or rib substructure.

The repair concepts have been developed to be generic in nature, i.e., applicable to a variety of structure, rather than for specific hardware configurations. However, to ensure that the concepts can be used on current and future Navy aircraft, criteria were established to govern the development of the repair concepts. The criteria cover service requirements such as operating temperature range, strength, fatigue loading and environmental exposure. In addition, criteria were established for the materials and processes which could be used, including maximum cure temperature, cure pressures and accessibility to the structure being repaired.

A review of available materials was made, and the repair concepts were then developed to meet the criteria, drawing upon available materials information. Repair concepts which had been used previously were reviewed and either adapted to the current criteria or modified as necessary. Small scale testing was conducted on representative structural elements to evaluate concepts and materials. Large scale testing will be conducted to verify the use of the most attractive materials and concepts based on the small scale testing program.

1. INTRODUCTION

Graphite/Epoxy materials are being used extensively for structural applications on emerging Navy aircraft. As these aircraft go into service, procedures will be needed to repair various kinds of damage which are expected to occur.

1.1 Background

Repair procedures that are applicable to advanced composite structures have been developed and tested in several R&D programs over the past several years.[1-4] One of the earliest repair configurations was developed by Grumman Aerospace Corporation[1] for use on boron/epoxy face sheets of honeycomb sandwich structure. It used 8 mil titanium foil interleaved with layers of film adhesive to form a bonded patch, external to the outer surface, as shown in Figure 1. This repair, limited to a 2-inch diameter damage, has been used for many repairs on the empennage structure of F-14 aircraft.

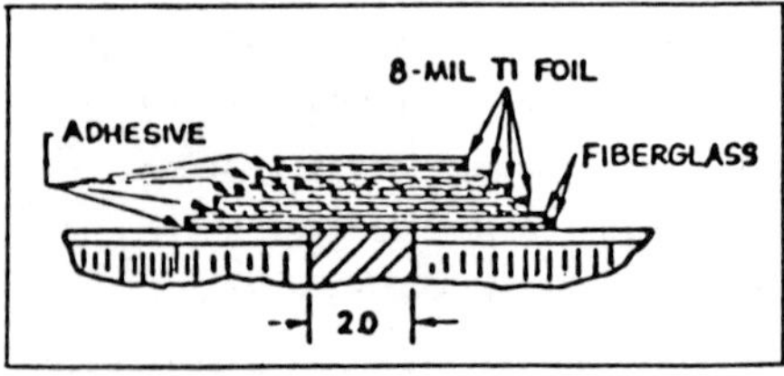

Figure 1. External Patch

Other repair configurations that have been developed include bonded external patches made of aluminum, titanium and precured composite materials. Most external bonded patches have been used on relatively thin skins, on structure that is lightly loaded, and for small damage areas.

Repairs for which metal patches are attached over damaged skins by mechanical fasteners have been developed for field repairs to graphite/epoxy skins up to 0.50 inch (12.7 mm) thick.[2] Strength restoration that will develop 4000 μ m/m was obtained for holes up to 4 inches (102 mm) in diameter. Both blind fasteners and backup plates with nutplates have been used. For higher strength repairs, especially for thick skins, recessed or semi-flush patch configurations have been developed,[3,4] using both precured and cocured patches with scarf joints and step-lap joints. A cocured configuration which restored 100 percent of the parent laminate strength for most test conditions is shown in Figure 2.[4]

1.2 Navy Depot Level Repair Program

A program is currently being sponsored by the Naval Air Development Center to develop and validate depot level repair procedures for graphite/epoxy structure for use at Naval Air Rework Facilities (NARF's). The procedures are intended to be generic in nature, i.e., useful for a variety of hardware and not based solely on the requirements for a specific aircraft. However, procedures will be applicable to structural configurations typical of emerging Navy aircraft, including thick monolithic wing skins, full depth honeycomb sandwich and sine wave shear web substructure.

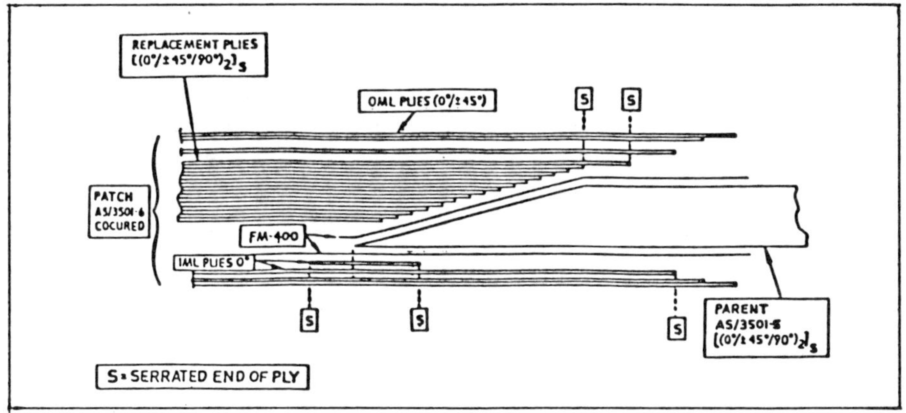

Figure 2. Typical Scarf Joint Repair Configuration

The first task in the program has defined the criteria that repairs must meet. The second task is devoted to evaluation and selection of materials and processes to be used for repairs. The third task is concerned with the development and evaluation of repair concepts. A fourth task will verify the selected materials and concepts on realistic large scale test hardware. The initial portions of this program are discussed in the following paragraphs.

2. REPAIR CRITERIA

Of particular interest in the current program is the realism of the criteria which the repair procedures are required to meet. An objective of the program is to develop repair procedures that are as simple as possible, e.g., they are to minimize the facilities, equipment, skill levels and the time required to make the repair.

In general, a trade-off must be made between making a repair simple and making it meet high performance standards. The approach has been to define realistic criteria, and then develop concepts as simple as possible which will meet those requirements. Criteria have been defined so that the generic repair procedures developed will be applicable to the types of aircraft expected to go into Navy service use in the near future.

2.1 Service Usage Requirements

The ultimate strength to be developed by repairs must allow for 4000 μ m/m in tension and 5000 μ m/m in compression for monolithic skins, and 5000 μ m/m in both tension and compression for honeycomb sandwich structure. For sinewave substructure, full web shear allowables are to be restored, and cap pulloff loads due to fuel tank pressurization must be reacted.

While no decrease in the original
stiffness is permitted, moderate
local increases in stiffness due
to the repairs are acceptable.
Weight increase is to be minimized,
especially on honeycomb sandwich
structure, which is frequently used
on moveable control surfaces for
which flutter properties may be
sensitive to weight and balance
changes. Repairs are not required
to be aerodynamically smooth,
although forward facing steps beyond
the outer moldline are to be ramped
or chamfered, and are not to exceed
0.10 inches (2.5 mm) for most
repairs.

Repairs are to be adequate for
service temperatures from -65°F
(-54°C) to 220°F (104°C), except
that the maximum service temperature
for substructure is 180°F (82°C),
with 150°F (66°C) for the web. All
repairs will be assumed to be made
on composite materials which have
absorbed one percent moisture prior
to repair. Repairs to monolithic
skins and substructure will be
exposed to fuel after the repair is
completed. All repairs will be
considered permanent, and must
be designed for fatigue loading
containing both tension and compression loadings, representing 2
lifetimes of use after the repair is
completed.

2.2 Material and Processing
     Requirements

Procedures for depot level repair
are expected to involve curing of
adhesives and prepreg materials,
so it has been important to define
any limitations to be imposed on
materials or processes. So that
repairs can be made in place on the
aircraft, without removing structural components, cure cycles will
not use autoclave pressure. Either
vacuum pressure or pressure from
mechanical means will be used.

The maximum acceptable temperature
for other materials adjacent to
the repair was reviewed. Cure
temperatures up to 350°F (177°C)
can be used on the composite
material without exceeding the
permissible temperature for the
adjacent materials. To permit
repairs without removal of skins,
all procedures will assume access
from the outside only. To the
extent that it is practical, a
repair to substructure also will be
made without removing skins.

Any materials selected for use in
the repair procedure must be readily
available to the Navy, and should
be in general use and have been
previously qualified to a major
aerospace manufacturer's specification.

3.0 MATERIALS AND PROCESSES
Repair procedures for depot level
are generally assumed to use bonded
or cocured repairs for most composite material applications.
Selection of the best available
materials and curing procedures is
essential to meet the criteria for

repairs, while maintaining the desired simplicity. A review of the available information on prepreg and adhesive materials was made early in the program. While many materials were identified that could be used, on all but a few, information was limited to the manufacturer's data and recommended processing procedures. Information on non-standard processing, such as vacuum pressure or reduced temperature cures that may be needed for repairs, was not available, except for a few materials that have been used previously under repair conditions.

No materials were identified that offered any proven advantages over those used previously for repairs. The material and process selection was therefore limited to coupon testing of previously-used materials under varying conditions of temperature and pressure to determine their properties under the cure cycle parameters to be used during repair.

As a part of the material and process portion of the program, an investigation was made of the tendency for a parent laminate that had absorbed moisture, to delaminate or "blister" during the repair cure cycle. Groups of 4-inch square, 50 ply specimens approximately 0.26 inches (6.6 mm) thick were moisture-conditioned by both water immersion and humidity exposure to obtain some specimens having a steep moisture gradient through the thickness, and others with a nearly uniform moisture distribution.

Specimens were bagged and shop vacuum was applied as specimens were heated to simulate repair cure cycles. In addition to the different moisture conditioning procedures, parameters investigated included the heat-up rate, the maximum temperature, the maximum pressure and the duration of the simulated cure.

Specimens were inspected with through-transmission ultrasonic C-scans, and some specimens were cross sectioned and inspected under magnification and with opaque penetrant X-ray. This testing and specimen evaluation will be completed early in 1981, and the final results will influence the materials and processing variables to be selected for repairs.

4. REPAIR CONCEPTS

A variety of repair concepts have been considered in several previous programs. These have used many materials, many processes and many configurations. The concepts being investigated in the current program, therefore, tend to be adaptations or modifications of earlier concepts, with an effort made to simplify them whenever possible while still meeting the criteria established for the repairs.

The approach has been to start with
existing concepts, simplify where
analysis or judgement indicates it
is possible, and then evaluate
the critical parts of the simplified
concepts with small scale testing.
The concepts which appear best will
then be verified on large scale
test hardware later in the program.

The following paragraphs discuss
some of the concepts which are
being considered for monolithic
panels, full-depth honeycomb sandwich and sine wave substructure.

4.1 Monolithic Panel Concepts

The requirement that repairs be made
with access from one side only
has a major influence on monolithic-
panel repair concepts. The thickness of the parent laminate, up to
0.4 inches (10.2mm), together
with the significant strength
requirements, tends to require a
recessed patch and a joint that
will develop adequate strength.
Either a scarfed (sloping surface)
joint or a stepped lap joint could
be considered. Two configurations based on a scarf joint are
shown in Figures 3 and 4, using
cocured and precured patches.

Figure 3 shows a concept for repair
of a monolithic laminate with
access from the top side only. A
thin precured laminate or metal
sheet, with a chamfer to match a
scarf joint in the skin, is bonded
in place to provide a surface on
which a typical cocured scarf joint

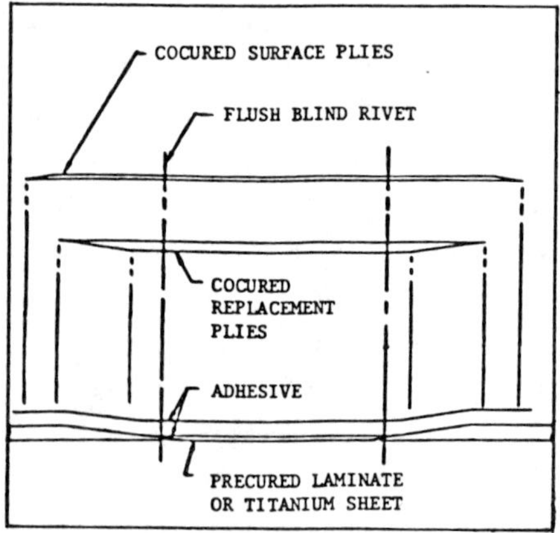

Figure 3. Monolithic Laminate
Cocured Repair

repair is laid up. A non-structural
bond could be used to hold the
precured laminate in place, and no
structural strength would be assumed
for it. The balance of the patch
would be designed to carry the
required loads, with allowance
made for the eccentricity of the
load-carrying patch plies. Small
countersunk blind rivets could be
used if needed to resist peel
forces near the tip of the scarf
joint. This concept is an alternative to bonding a precured
laminate or metal doubler on the
inside surface to provide a solid
surface, which is difficult to do
with access from one side only. The
concept shown is considered a
simplification of that approach,
although some reduction in strength
is expected.

Figure 4 shows two variations on the
use of precured patches used with a
scarf joint configuration. To

710

obtain adequate strength in the scarf joint bond line, it is essential that the mating surfaces of the precured patch and the parent laminate be carefully machined to match each other, so that the bond line is of uniform thickness. A shop-aid type of guide would be used during machining to obtain well-machined surfaces. The advantage of the precured patch is that it eliminates the need for providing a separate surface on which to layup prepreg for a cocured patch. As with other repair configurations involving scarf joints, small blind rivets and/or external cocured plies can be used to increase the overall strength and reduce the tendency for peeling at the ends of the bondline.

4.2 Full Depth Honeycomb Sandwich Concepts

For full-depth honeycomb sandwich skin repairs, bonded external patches have often been used, and are probably the simplest method available. For relatively thin skins, patches can also be relatively thin, and the eccentricity of the patch can be accepted. However, in this program, the sandwich skins to be repaired have been specified to be 0.16 inches (4.1 mm) thick, with repaired strengths of 5000 μ m/m. For this strength criterion, the thickness of external patch would exceed the out-of-moldline smoothness considered reasonable for depot level repairs.

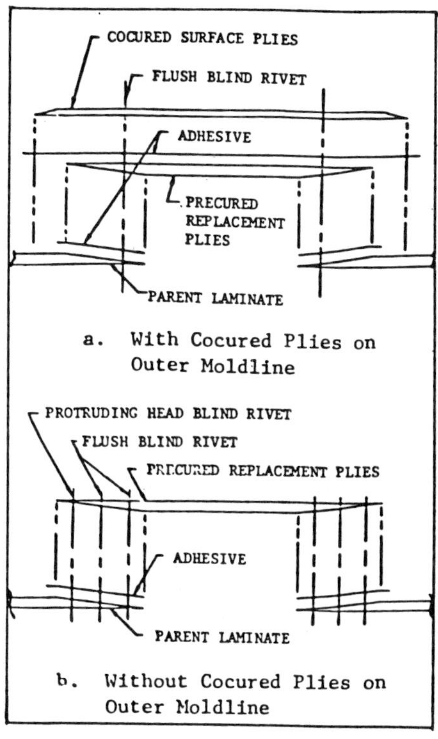

Figure 4. Monolithic Laminate Precured Repairs

Because of these strength and smoothness considerations, scarf-joint (or other semi-flush) configurations are recommended. As a result, concepts for sandwich skins tend to resemble those for monothlic laminates. The primary difference is that once the damaged core is replaced, it provides a surface on which to lay up prepreg for a cocured patch. A typical configuration is shown in Figure 5. Materials other than metallic honeycomb core will be considered to simplify the core cleaning and machining required. Various precured or

711

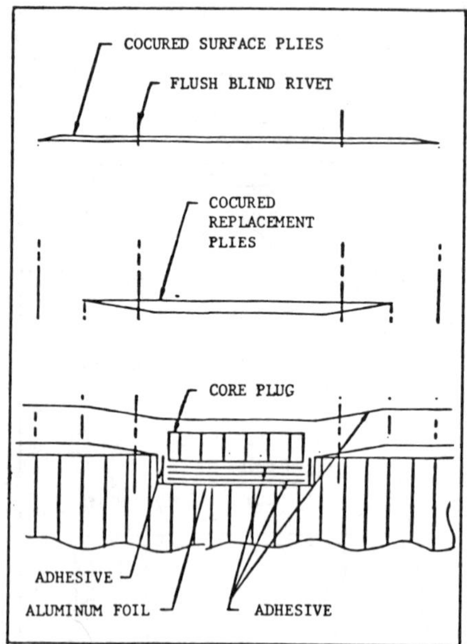

Figure 5. Honeycomb Sandwich Repair

foam-in-place materials may be available which can be used, and these are being considered.

4.3 Sine Wave Substructure Concepts

A logical repair concept for sine wave structure is to adhesively bond a precured or cocured doubler using vacuum bag pressure, with the bag placed over the repair area inside the wing. Bridging of the bag material across the bottom of each corrugation is felt to be a potential source of inadequate bonding pressure, especially when both web and cap material are being repaired so that the cap-to-web corner also provides a potential location for bag bridging. The concepts shown in Figures 6 and 7 were developed as alternatives to the use of a vacuum bag inside the structure. The concept shown in

Figure 6 replaces the damaged web with an aluminum honeycomb sandwich element fabricated separately and bonded to the undamaged web. The top of the graphite/epoxy channel replaces the damaged cap material, while the core-stablized vertical legs replace the damaged sine wave web.

The repair segment subassembly would be fabricated, then slipped over the cutout in the web being repaired, and joined to it by a bonded scarf joint at the cap interface and to the web corrugations with fasteners. The top unsupported edge of the sine wave web is stabilized by the bonded graphite/epoxy laminate, which is in turn bonded to the bottom of the core of the spar repair segment. The bonding pressure for the cap splice joints would be provided by the use of clamps. Doublers may be required on the undamaged sine wave in the area where they are mechanically attached to the vertical legs of the repair channel. These doublers would be cocured onto the undamaged sine wave web, with the bonding pressure obtained with preformed rubber blocks and clamps.

A possible simplification of this configuration omits the core and replaces the precured channel with a similar bent sheet metal channel, thick enough to be non-buckling without stabilization.

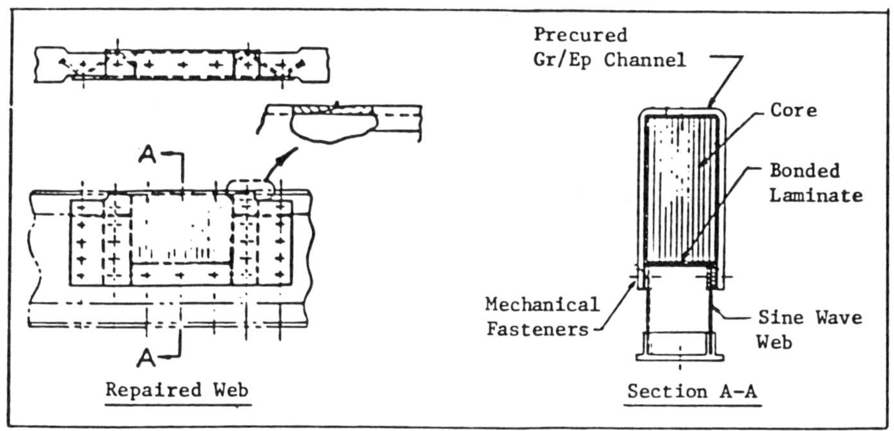

Figure 6. Sine Wave Substructure Repair

The concept shown in Figure 7 is intended to be less complicated than the concept in Figure 6. In Figure 7, a precured sine wave web segment is used, either obtained in stock sheets from a vendor or fabricated from tooling available at the Navy depot. The bonding pressure required for the web splices can be provided by mechanical methods to avoid the difficulty of working with a vacuum bag through limited access in a small hole in the skin. Three mechanical methods for applying pressure are being considered; (1) rivet bonding, (2) spring washers with screws or bolts, and (3) mechanical screw clamps.

An important consideration for the sine wave substructure repair is the size of the opening required in the skin to provide access, since the repair is to be made without removing the skin when possible. Access size has been determined for

the tools required for drilling fastener holes, installing clamps, etc.

A special tool which could be used to obtain pressure is shown in Figure 8. The rubber block is

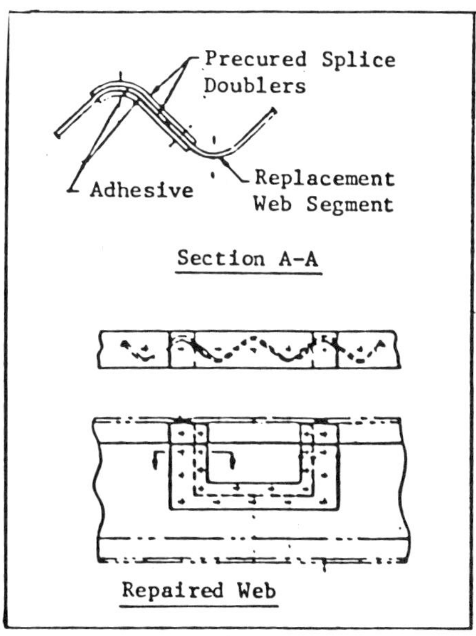

Figure 7. Sine Wave Web Repair

molded to match the sine wave shape.
A heating grid is molded into the
rubber to provide heat for curing
the bond. The rubber block has a
metal backing plate to which clamp-
up forces are applied with clamps.

5. SUMMARY

A program is being conducted to
develop and validate repair procedures for use at Navy depot level
facilities. While repairs will be
generic, i.e., useful for a variety
of hardware, they will be applicable
to structural configurations typical
of emerging Navy aircraft, including
thick monolithic wing skins, full
depth honeycomb sandwich and sine
wave substructure.

Realistic criteria that the repairs
must meet have been defined,
including service useage requirements and material and processing
requirements. Among the more
significant criteria are that no
autoclave pressure will be used, so
that repairs can be made on the
aircraft, and that access will be
limited to one side only.

Materials and processes to be used
for the repairs were investigated,
and data are being obtained for
non-standard cure cycle parameters,
such as vacuum pressure, that may be
more applicable to the repair
environment. A separate investigation was made of the tendency
for a parent laminate, which had
absorbed moisture, to delaminate or
"blister" during the repair cure
cycle.

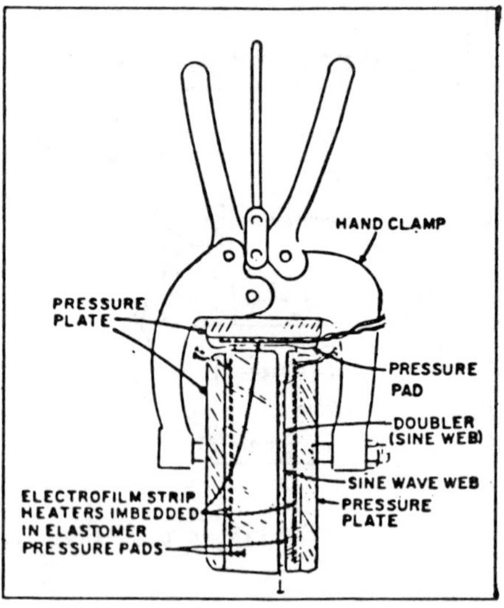

Figure 8.   Tooling For Sine Wave
Web Repair

Concepts have been described
for repairs of monolithic skin
panels up to 0.4 inches (10.2 mm)
thick, full depth honeycomb sandwich
structure, and sine wave substructure. The emphasis has been to
simplify the concepts as much as
possible while still meeting the
established criteria. Concepts
will be evaluated with small test
specimens and later verified on
large scale specimens.

6. REFERENCES

1. Lubin, G., et. al., "Repair
    Technology for Boron/Epoxy
    Composites," AFML-TR-71-270,
    Grumman Aerospace Corp.,
    February 1972.

2. Watson, James C., et.al.,
    "Bolted Field Repair of Compos-
    ite Structures," NADC-77109-30,
    McDonnell Aircraft Co., 1 March
    1979.

3. Studer, V. J. and La Salle, R.
    M., "Repair Procedures for
    Advanced Composite Structures",
    AFFDL-TR-76-57, General Dynamics
    Corporation, December 1976.

4. Labor, J. D. and Myhre, S. H.,
    "Large Area Composite Structure
    Repair," AFFDL-TR-79-3040,
    Northrop Corporation, March
    1979.

7. ACKNOWLEDGEMENT

The work reported was conducted as a part of U.S. Naval Air Development Center Contract No. N62269-80-C-0232, "Development and Validation of Depot Level Repair Methods for Composite Structure." Mr. Ed Rosenzweig, the Navy project engineer, has contributed valuable information and helpful suggestions to the program.

8. BIOGRAPHY

The author, James D. Labor, is a Senior Technical specialist with the Northrop Structural Mechanics Research Organization. Mr. Labor is the program manager for the Navy depot level repair program described, and he has been active in the field of composite structure repair for several years.

ADVANCED COMPOSITE REPAIR -- RECENT DEVELOPMENTS AND SOME PROBLEMS

S. H. Myhre*

NORTHROP CORPORATION
Aircraft Division
Hawthorne, California

ABSTRACT

A team of investigators at Northrop -- supported by others at Lockheed-California, Hughes Aircraft and Bell Helicopter -- has completed a one year review of more than 400 documents for information relevant to advanced composite repair and allied subjects. Before proceeding with the DOD/NASA Advanced Composite Structure Repair Guide, a pause has been taken to evaluate the current status of the survey to identify specific areas of observed deficiencies. This paper will present the major developments of the past decade, including the seven programs now in progress. The problem areas will be frankly discussed to point out where useful work may still be done. The paper is concerned with generic problems rather than those related to specific hardware.

---

*Senior Technical Specialist, Structural Mechanics Research Department

Keywords: Composite Materials, Structural Design, Repair Technology, Maintainability and Logistics Support

1. INTRODUCTION

The use of advanced composite materials has grown in recent years to be included in nearly every production program of flight structure. Pioneered in the laboratory in the 60's and extended by experimental hardware programs through the 70's, advanced composite materials have found a place for themselves in the production lines of the 80's. The avante garde of these programs, the USAF F-16 and the U.S. Navy F/A-18A already have aircraft in service and flight test use. The U.S. Army UH-60A and the commercial S-76 helicopters are also ready with composite structure at the beginning of the decade. In business aircraft, the Lear Fan 2100 has made apparently the largest percentage commitment to composite materials, using them for primary

structure in both wing and fuselage.
The largest single components
constructed to date are the payload
bay doors of the NASA Space Shuttle.
Impressive as these programs are,
they are dwarfed by the decision of
the Boeing Company to commit to
production in composite materials
essentially all of the secondary
structural components of the 767 and
the 757 commercial transports.[1]
The maintenance and repair of these
structures is now becoming the
concern of the users, as well as the
manufacturers of the aircraft.

Components needing repair will
be largely honeycomb sandwich wing
tips, control surfaces, fairings,
panels and doors. These light
structures are subject to being
easily damaged and may be the only
components on the aircraft made with
composite materials. As the use of
composites in primary structure
expands, the need to repair heavier
structure will grow.

A background of research has
already been completed which is
adequate to affirmatively answer the
basic question of repairability of
composite structure. Each of five
contractors, working specifically on
composite structure repair, has
explored in depth at least one
unique approach to the repair
problem. The Grumman program
conducted by Lubin[2] and others,
developed a fairly successful
external patch repair using 0.008-
inch titanium alloy (6AL-4V)
foils adhesively bonded to the skin

and to each other, as shown in
Figure 1. Two plies of fiberglass
prepreg, oriented at ±45 to the
major loading direction, are included in the bondline next to the
skin. Such a repair is applicable
to flat or simply curved skin and
has been used in service. An
important contribution of the
Grumman work was to develop and
provide a field repair kit which
is self-contained, requiring only
external electrical power. The kit
contains a vacuum pump and a combined heating-pad/vacuum-cover
suitable for use with many kinds of
bonded repair.

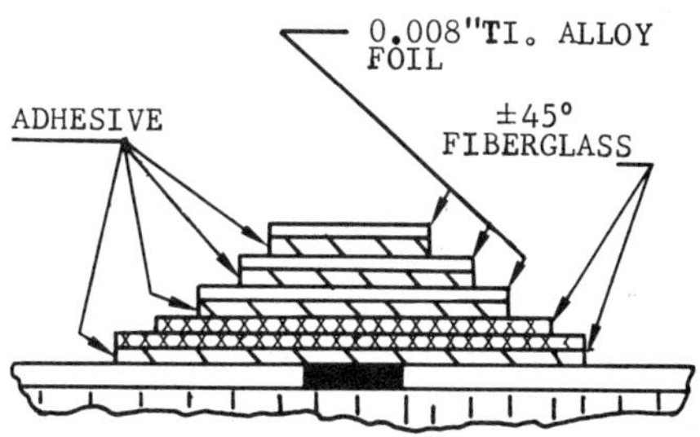

FIGURE 1.   Bonded Titanium Foil
Patch

The work of Christian, Miller
and Doyal[3] on the repair of
boron/ aluminum composites was
particularly interesting because of
the amount of information these
investigators extracted from
relatively few test components.
Each component was tested first in
the undamaged condition by the
application of design-type loads to
failure, then repaired and loaded to
failure again in the same manner. A

summary of these results is given in Figure 2. Since all tests were conducted at room temperature, it is not known how successful these repairs would be at elevated temperatures which would be typical of boron/aluminum components.

Component	Weight Change %	Failure Load Change %
Sandwich Beam	6.5	+16.5
Elastic Buckling Tube 25" long, L/D=25.4 .	6.0	-13.
Elastic Buckling Tube 30" long, L/D=30.3 .	11.0	-10.
Skin-Stringer (Bending)	12.0	+20.
Tension/Bending Tube 2.25" Dia x 40" long	15.0	+76.
Web Spice	4.0	+58.
Tension Field Panel ..	10.0	+41.

FIGURE 2. Summary Boron/Aluminum Repairability Results

The General Dynamics program conducted by Studer and LaSalle[4], investigated repair techniques for several kinds of damage in a variety of structures. They had good success with the exception of resin injections into voids and disbonds. This was ineffective in restoring compression strength and in some cases caused extension of the original damage. Similar observations have been made by English[5] in a more recent study. For reliable repair, the laminate must be removed down through the discrepancy and replaced with a patch of compatible material.

Exploratory work on mechanically fastened repairs has been performed by Watson[6] and others at McDonnell Aircraft. These repairs, utilizing titanium alloy patches and backing plates, were developed for application on fuel cell composite wing surfaces of graphite/epoxy. The repairs were generally capable of strains at failure of 0.0040, which is representative of ultimate strain allowable for laminates with loaded holes.

The most recently completed program in advanced composite repair was reported by Labor and Myhre[7] at Northrop. The major design approach developed a scarf joint using a patch of cocured graphite/ epoxy with thin internal and external doublers, as shown in Figure 3. Serrating ply ends with pinking shears effectively prevented patch peeling, which was recognized to be a problem with the relatively brittle FM 400 adhesive. No difficulty was found in making a long scarf (L/T = 18/1 or 36/1) with hand held tools and no fit problems were experienced using the cocured patch. The basic Northrop modified scarf joint was adapted to many different repair problems in graphite laminates from 16 to 50 plies, a 64-ply boron/graphite hybrid, sandwich panels with 8-ply faces and a damaged F-15 speedbrake. A total of 252 tests were conducted on coupons and small beams under adverse conditions of temperature (-65F, R.T. and 265F), moisture (dry and wet to 1.0%) and loading (static and 2-life spectrum fatigue). The

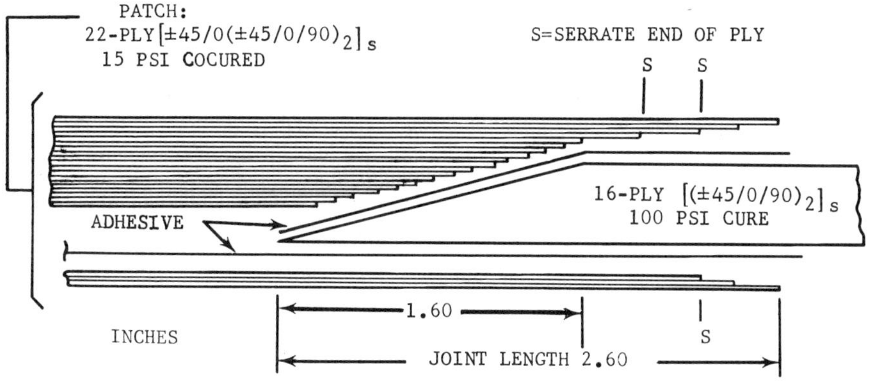

Figure 3. Single Scarf Flush Repair

conclusion is reached that reliable repairs can be made which will restore at least 80% of the parent laminate static strength and will retain this strength for at least one design lifetime. In terms of strain, this represents a restored strain capability of 0.0088 using the AS-3501-5 material. Since current aircraft design allowables utilize only 50% or less of the laminate strength (because of fastener hole strength reduction), it follows that structures can be repaired to a "good-as-new" condition. These conclusions were verified on four intermediate size panels (12 by 48 inches) each with a repaired 4-inch diameter damage hole and on five large panels (19 by 60 inches) each with a different repair problem, ranging from a 4-inch diameter hole to a hole 6.6 inches across by 12 inches long.

2.   RECENT DEVELOPMENTS

Seven funded programs involving repair of advanced composite structures are currently in progress. Each of these contractors is exploring some remaining concerns about advanced composite repair.

The NASA program at Lockheed "...for Commercial Transport Applications"[8] in Phase I has conducted two surveys: (a) airline damage experience and maintenance/ repair capabilities and (b) defect sensitivity of composites. These surveys indicated a need for a wide range of repair procedures suitable for both maintenance base and line station operations. Phase II work has experimentally evaluated several approaches to repair. The test results have indicated that aerodynamically flush repair, similar to the Northrop modified scarf joint, using structural grade adhesive (M329) and pre-preg (Narmco T300/5208) provided the greatest strength recovery for both lightly loaded and highly loaded structures. External graphite patch repairs using structural grade adhesive provided adequate strength restoration and are less complex and expensive. Cold-bonded, wet lay-up repairs and bolted repairs with blind fasteners provided only limited strength recovery, which may, however, be adequate for many lightly loaded components. The Phase III activities, now in progress, will include repairs to more complex skin-stiffener structures.

The Navy sponsored program "Depot Level Repair for Composite Structures"[9] at Northrop has thus far been involved with evaluating adhesives, patch materials and repair concepts. The requirements include repairing sine-wave spars typical of the AV-8B. Repairs will be made through the skin damage hole. The problem of blistering a moisture-laden laminate by rapid heating is being investigated using T300/5208 and AS/3501-6 materials.

Another Navy sponsored program "Field Level Repair for Composite Structures"[10] has begun recently at Lockheed-California. The goal is to repair the typical damaged area within 4 hours, so that the aircraft can return to operational status for a reasonably long time, i.e., several months. The Lockheed approach is to screen a large number of repair techniques including bonded, bolted and bolted/bonded repairs and several delamination injection repairs suitable for field (and shipboard) implementation. Thick skins, as well as honeycomb sandwich panels, will be repaired.

Rockwell International, North American Aircraft, has recently begun work on a program of "Rapid Repair of Battle Damaged Aircraft Structure".[11] The objective of this 36-month program is to develop and demonstrate rapid field repairs for composite or metallic structure so that one more mission can be flown. The limitation on down-time is of utmost importance, while durability, smoothness and appearance are of no importance.

Another program recently awarded to Rockwell International, Space Systems Group, concerns "...Repair Techniques and Processes for ... Graphite/Polyimid Composite Structures"[12] The polyimid matrix can raise the maximum temperature limitation to the 400-500°F range. Hence, the repair would presumably be required to perform in the same temperature range. Without autoclave curing, the challenge appears formidable indeed.

In addition to these funded programs in industry, the University of Delaware has a NASA grant on "Damage Repair Technology for Composite Materials,"[13] now in the fourth year. Stress analysis of scarf joints has been an important focus of this work, illuminating the high stress area at the tip of each adherend. This work would now be expected to expand to explore practical methods of stress alleviation.

The publication of "Introduction to Composite Materials" by Tsai and Hahn[14] and the packaging of the computational procedures involved into a Composite Materials Module I[15] by the AF Materials Laboratory will prove a boon to the repair depot structures engineer, as well as many others. The text is a clear presentation of laminate theory and the module (installed in the TI-58 or TI-59 calculator) facilitates convenient, quick, accurate and inexpensive computation in the areas of laminate in-plane or bending analysis involving mechanical and environmental strains or stresses. The nine programs are designed to be used interactively with the user or a user-written program. Using the calculator on the printer and suppressing the printing of those items not required, the user has a tape showing all input data and the precise output desired. An example, using module programs 01 and 06 to solve for the D-matrix and then use these values to determine the longitudinal compression and shear stability allowables for a simply supported long panel is shown in Figure 4.

Structural grade adhesives have shown steady improvement over the last decade. The availability of FM 300, FM 400, EA 9647, M 398 and others, together with peel resisting configuration techniques,[7] gives the repair team the ability to make excellent repairs. No substitute has yet been found for training, craftsmanship and eyeball-to-eyeball communication for every member of the team.

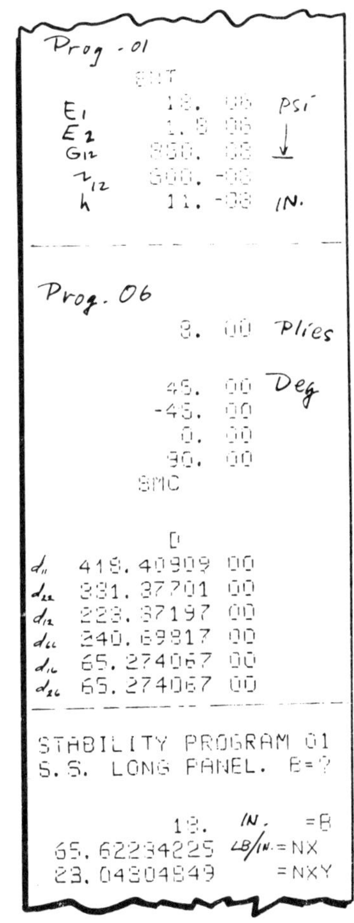

Figure 4. Example Calculation Using TI-59 Composite Materials Module and User Programs

These have been the principal developments observed in a survey of the literature and current practice conducted for the "DOD/NASA Advanced Composite Structure Repair Guide,"[16] conducted by Northrop with the help of

Lockheed-California, Hughes Aircraft and Bell Textron Helicopters. The study began in August 1979 and will be completed with the publication of the first edition of the Repair Guide in March 1972. Periodically through the study we have paused for an overview of what we had found and especially to identify what seemed to be missing.

3. SOME PROBLEMS REMAINING

3.1 Bonding on Wet Laminates

No practical technique has been found to measure, in a nondestructive manner, the moisture content of a laminate on the airplane. Some experience of wet laminate blistering during 350°F cure has been reported· on thick AS/3501-5 laminates.[7] Adhesive bond strength can only be degraded by moisture from the parent laminate. The process of drying is simple enough, but slow. Hence, the questions: Does the part need to be dried? If so, how can drying be done quickly, and when is it dry enough? The approach of monitoring the effluent gasses of the drying process is being developed in the Repair Guide program. Drying would begin immediately at 200°F over an area somewhat larger than the patch cure heating area. Drying temperatures would be raised as drying progresses, until reaching the adhesive cure temperature.

3.2 Nondestructive Inspection (NDI)

The subject of NDI, while not part of the repair process as such, is an important related subject, particularly NDI for repair verification. A really practical technique has yet to be discovered, though several methods are available which can help. A vacuum pressure cocured repair presents the most difficult verification problem, especially in the thick laminates. The difficulty arises because of laminate porosity resulting from the relatively low pressure cure. In thinner laminates, tapping can detect a flaw but it requires a quiet place and is subjective. This deficiency is a real concern, since the vacuum pressure cocured repair method is a versatile technique which has been shown to produce excellent repairs. [4, 7, 8]

3.3 Environmental Effects

Every repair technique will be affected by moisture, temperature and fatigue environment and should be evaluated for strength and durability under a broad range of these conditions. This results in a great deal of work which could be recommended in this area.

Bonded repairs with 250°F curing adhesives are of special interest, since these structural adhesives (e.g., FM73, AF 127) are well established for bonded repairs on metal structure with service temperatures to 180°F.[17] Limitations of the hot/wet strength of these adhesives in advanced composite repair joints is unknown. Adhesive elastic and plastic strains

to failure both increase with
increasing temperature as strength
is declining. For this reason, the
strength of a properly made repair
cannot be extrapolated to higher
temperatures on the basis of simple
lap shear tests. It is expected
therefore, that the service temperature limitation of 250°F curing
adhesives can be significantly
increased. Because of the possible
requirement of a 250°F maximum
repair cure temperature together
with good strength retention at
temperatures from 200°F to 250°F,
the real potential of a selected
250°F curing adhesive and prepreg
should be determined. A small test
program has been initiated in the
Repair Guide program to explore this
problem.

3.4 Simplified Field Repair

Simplified field repairs are
not adequately explored and documented. Since a quick repair is
always desirable and at times
essential, many such methods should
be evaluated for strength, durability and repair time.

Repair research to date has
generally begun with a clean damage
hole, i.e., all damaged material is
removed. In the process, much good
material is also removed and must be
replaced. The material removal
takes time and may be unnecessary.
In the case of many low-energy
impact damage situations, the
simpler approach of leaving the
damaged material in place, vacuum
injecting a low viscosity resin and

cocuring an external patch may yield
adequate results and would certainly
be quicker. In many cases, the need
for a repair may be considered
marginal, and for such cases, the
simplified repair has the advantage
of being an acceptable compromise
choice between no repair and a more
costly repair. The interim quick-
fix, which will accomplish its
limited objective and yet does not
complicate the later permanent
repair, would serve a valuable
purpose. If research and experience
can demonstrate that the quick-fix
is adequate for permanent repairs,
the costs of repairs could be
reduced significantly.

The use of ambient temperature
curing resins and adhesives to date
has not proven to be generally
satisfactory, especially if even
moderate temperature (180°F) service
is required. For repairs for which
durability and elevated temperatures
are not important criteria, these
materials may still find a use.
The use of elevated temperature and
vacuum pressure for bonded repairs
are becoming more widely accepted.
Hence, the use of quick repair
techniques developed with ambient
temperature curing materials could
now be adapted to elevated temperature materials where necessary.
The conventional structural adhesives requiring two hours to cure
have been used with good success on
both simplified and high performance
repairs. For the reduced cure
time required for combat repair

situations, the modified epoxy
family should be explored. Indications are that curing time for some
of these would be as low as three
minutes at 250°F. Therefore,
simplified field repairs which can
be made in less than one hour and
can be considered permanent now
appear possible. It is expected
that the two active field repair
programs[10, 11] will adequately
investigate this area.

3.5 Joint Repair

Composite material bonded
to metal substructure appears
to be a type of configuration for
which repair research and experience is deficient, although some
work has been done. Studer and
LaSalle[4] addressed this problem
for the repair of a step-lap bonded
graphite-to-titanium joint, but the
performance of this graphite patch
repair was obscured by several
factors affecting the test configuration. In addition to those factors
mentioned by the authors, some
processing variables may have also
affected the results. Specimens
were prepared for repair by removing
all composite material down to the
stepped titanium surface on the
fitting end. Unsolved problems with
this type of repair include how
to remove the composite without
damaging or contaminating the clean
surface of the metal, and how to
reclean exposed metal, when required, without harm to the adjacent
composite material.

While cleaning and anodizing
are generally accomplished in the
factory using tanks for immersion of
the parts, a non-tank anodizing
process which can be used on-
aircraft has been demonstrated
successfully for bonded aluminum
repairs.[17] A similar process
is expected to be possible for
titanium but has not yet been
demonstrated. Since titanium
fittings and substructure are
expected to be used frequently with
graphite epoxy structure, the lack
of adequate procedures for repair of
such elements is considered to be a
significant deficiency at this time.
A small effort in this area has
been planned for the Repair Guide
program.

A specific deficiency in the
area of joint repair is that of worn
fastener holes. Simply filling with
an epoxy potting plug and redrilling
will generally not be adequate for a
permanent repair. Since the wear
indicates that the original design
was inadequate, some improvement
should be made, with due regard
for weight and cost.

Fastener hole wear is especially likely to occur at removable
fasteners, such as at door edges,
and results in a difficult, unsolved
repair problem. Any plug installed
and redrilled, or any type of
grommet used must be positively
retained so that it will stay in
place when fasteners are loosened or
removed. Concepts for accomplishing

positive retention tend to either
interfere with adjacent structure or
reduce the fastener edge distance.
Development and validation of
acceptable repairs for fastener hole
damage are needed.

3.6 Kevlar/Epoxy Repair

Repairs for Kevlar/epoxy are
noticeable by their absence in the
technical literature. It is expected that Kevlar composites would
be repairable by the same techniques
developed for graphite but demonstration of this has not been found.
Kevlar's greater affinity for
moisture could make blistering
problems more significant for Kevlar
composite repairs. The moisture
problem also suggests that bonded
field repairs on Kevlar should not
utilize Kevlar patch material,
but rather fiberglass or a hybrid of
fiberglass and graphite.

3.7 Substructure Repairs

Repairs involving substructure
in advanced composites are inade-
quately developed, primarily because
of the great variety of possibilities which exist. Current
composite wing designs use skins
either bonded or mechanically
attached to substructure or integrally cured skins and substructure. Similar integrally stiffened
skins are used on helicopters,
missiles, and a variety of vehicles.
Each individual design creates its
own repair problems: sharp radii,
lack of access, irregular shapes,
etc. Some of these detailed repair
problems may be sufficiently generic

to be studied outside of a specific
hardware project.

A significant amount of
repair development work has been
done in this area, however, which
can be applied to many of the new
designs. The work of Studer and
LaSalle[4] contributed significantly to this area with repairs
to hat stringers and skin. The
work of McCarty and Horton[17]
on bonded metal structure can also
be applied to bonded composite
structure.

One of the principal deficiencies of concern is the repair of
a bonded or integrally cured skin/
spar joint. While generous margins
are required for these joints, and
in some designs failsafe pins are
installed, the joint is a critical
structural detail with complex load
carrying requirements which must be
considered in making a repair. Fuel
pressure results in pressure on the
skin/spar joint and local chordwise
bending in the skin. In addition,
the joint transmits the spar shear
and the skin carries its basic
biaxial combined loading. Repair
studies have been performed on such
joints but no test work has been
found to verify repair concepts.[18]

Another deficiency in available
substructure repair procedures deals
with accessibility. The difficulties associated with any type
of substructure repair are substantially increased if access to
both sides of the damaged parts
is not available. With skin and

substructure damaged in a wing or
empenage surface of skin-stringer or
multiple spar construction, the
problem becomes so acute that
current design requires that at
least one skin be removable. Yet
the removal of a wing skin to
accomplish a repair is in itself
such a large task as to be undesirable. Continued development of
techniques for repairs which
could be made with limited access
could remove the requirement for a
removable skin and reduce the cost
of a major repair. Such work should
be done on curved surfaces to
further demonstrate the adaptability
of advanced composite repair
techniques to real conditions.

4. CONCLUSIONS

A wealth of information has
been found on repair of advanced
composite structures. The development of repair technology throughout
the decade of the 70's has demonstrated the repairability of composite structure through at least five
independent investigations. Current
efforts are addressing special
problems of particular users,
circumstances and material. The
remaining problems, as we see them,
have been identified here.

5. ACKNOWLEDGEMENT

The support of the Air Force
Wright Aeronautical Laboratories
through Contract F33615-79-C-3217,
DOD/NASA Advanced Composite Structure Repair Guide, is gratefully
acknowledged.

7. REFERENCES

1. Hammer, Robert, "The Advanced-
Composites Hurdle for 767
Production," Boeing Commercial
Airplane Company, ASTRONAUTICS
AND AERONAUTICS, pg. 40,
Oct. 1980.

2. Lubin, G., Dastin, S., Mahon,
J., Woodrum, T., Paa, C.,
"Repair Technology for Boron/
Epoxy Composites," AFML
TR-71-270, Final Report,
Grumman Aerospace Corp.,
Bethpage, NY, 276 pp, February
1972.

3. Christian, J. L., Miller,
M. F., Doyel, F. H., "Repair
of Boron/Aluminum Composites,"
General Dynamics, Convair
Division, SME Paper EM75-709,
given in Los Angeles, CA,
12 pp, March 1975.

4. Studer, V. J., and LaSalle,
R. M., "Repair Procedures
for Advanced Composite Structures" AFFDL TR-76-57-Vols.
I & II, General Dynamics,
Ft. Worth. Div., December
1976.

5. English, F. C., "Advanced
Composite Aileron for L-1011
Transport Aircraft," LR 29352,
Quarterly Tech. Rpt. No. 9,
NASA Contract NAS 1-15069,
Lockheed-California Company,
22 Jan. 1980.

6. Watson, James B., Glaser,
D.A., Harvey, F. L., Fukimoto,
W.T., Padilla, V. E., "Bolted
Field Repair of Composite
Structures," NADC 77109-30,

McDonnell Aircraft Co., St. Louis, Mo, Report No. MDC-A5583, 217 pp, March 1979.

. Labor, J. D., and Myhre, S. H., "Large Area Composite Structure Repair," Final Report AFFDL-TR-79-3040, Northrop Corporation, Aircraft Division, March 1979.

. Stone, R. H., "Development, Demonstration and Verification of Repair Techniques and Processes for Graphite/Epoxy Structures for Commercial Transport Applications," NASA Contract NAS 1-15269, Lockheed-California Company.

. Labor, J. D., "Depot Level Repair for Composite Structures," Contract N62269-80-C-0232, Naval Air Development Center, Northrop Corporation, Aircraft Division, Jan. 1980 start, 36 month duration, Bimonthly Progress Reports.

0. Stone, R. H., "Field Level Repair for Composite Structures," Contract N62269-80-C-0283, NADC Lockheed-California Company, July 8, 1980.

1. Sanders, A. L., "Rapid Repair of Battle Damaged Aircraft Structure," Contract F33615-80-C-3235, AFFDL, Rockwell International, Los Angeles Division, Sept. 9, 1980.

2. Jones, Jack, "Development Demonstration and Verification of Repair Techniques and

Processes for Celion 6000/Lark 160 Graphite/Polyimid Composite Structures," Contract NAS 1-16448, NASA, Rockwell International, Space Systems Group, Sept. 24, 1980.

13. Pipes, R. B., et. al., "Damage Repair Technology for Composite Materials," NASA Grant 1304, University of Delaware, Newark, Delaware.

14. Tsai, S. W., and Hahn, H. T., INTRODUCTION TO COMPOSITE MATERIALS, Technomic Publishing Co., Westport Ct, 06880, July 1980.

15. Knoll, K. T., "Instructions for Use of the Composite Materials Module I," AFWAL-TR-80-XX, February 1980.

16. Labor, J. D., and Myhre, S. H., "DOD/NASA Advanced Composite Structure Repair Guide," Contract F33615-79-C3217, AFWAL, Northrop Corporation, Aircraft Division, Aug. 1, 1979.

17. McCarty, J, E., and Horton, R. E., "Repair of Bonded Primary Structure," AFFDL-TR-78-79, Boeing Commercial Airplane Co., Seattle, WA, Final Report, 110 pp, June 1978.

18. Butler, B. M., et.al., "Wing/Fuselage Critical Component Preliminary Design (Northrop)," AFFDL-TR-78-174, Northrop Corporation, Aircraft Division, March 1979.

26th National SAMPE Symposium
April 28-30, 1981

LIFE SCIENCES FLIGHT EXPERIMENTS PROGRAM:
OVERVIEW

W. E. Berry
Chief, Life Sciences Flight Experiments Project Office
NASA-Ames Research Center
Moffett Field, California

C. C. Dant
Experiment Support Scientist
Management and Technical Services Company
General Electric
NASA-Ames Research Center
Moffett Field, California

Abstract

The Life Sciences Flight Experiments (LSFE) program at NASA-Ames Research Center, responsible for Life Sciences flight investigations using nonhuman specimens, is establishing the capability through the E.S.A. Spacelab and the Space Transportation System (STS) for conducting life sciences research and development in space, and is exploiting the uniqueness of the space environment to study the effects of weightlessness on a variety of biological systems. For the first time in NASA's history, the capability for scientists to fly in space without having to be astronauts will be provided on a frequent basis. Opportunities for the life sciences community to participate in the STS flight program are extensive. The Spacelab provides the capability to fly a number of life sciences experiments, to retrieve and reuse experimental equipment, and to undertake sequentially related studies. This paper presents an overview of the Ames LSFE program, which is working to take advantage of the opportunities available on Spacelab.

Key Words

Life Sciences

Spacelab

Space Biology

Comparative Physiology

Shuttle

Space Transportation System

Animal Research

1. INTRODUCTION

Once operational, the European Space Agency's (ESA) Spacelab, carried by the NASA Space Transportation System's (STS) Shuttle, will provide new and unique opportunities for the Life Sciences community to conduct research into the effects of spaceflight and hypogravity on living organisms under conditions approximating those of ground-based laboratories. The experiences of the Skylab, Salyut, and Cosmos missions have demonstrated the feasibility of performing a wide variety of general experimental manipulations in weightlessness. This experience has provided a foundation for the concept of a manned space laboratory in which more sophisticated biological and medical studies can be conducted. The unique opportunity to manipulate and control experimental procedures and animals directly rather than remotely is essential to Life Sciences research. The capability for immediate follow-up of new experimental findings on reflights will be an important facet of biomedical study in Spacelab.

The NASA Life Sciences Flight Experiments Program (LSFEP) focuses on Spacelab life sciences missions planned for the 1984-5 time frame. Life Sciences Spacelab payloads, launched at approximately 18-month intervals, will enable scientists to test hypotheses from such disciplines as vestibular physiology, developmental biology, biochemistry, cell biology, plant physiology, and a variety of other life sciences. Building on experience gained from ground-based mission simulations, orbital flight test experiments, and the first three Spacelab missions, NASA will be able to progressively develop the scientific, engineering, and management capabilities necessary for the first Spacelab dedicated to the life sciences. Development of experiments for these missions will require implementation of human/ animal compatible life-support systems and research equipment not previously flown in space.

One of many steps leading to a flight mission is the issuance of an Announcement of Opportunity (AO). In 1978, NASA's Office of Space Sciences issued an AO for Life Sciences flight investigations on the Shuttle/Spacelab system for the first mission dedicated to the life sciences. Over 360 proposals received were evaluated by a team of scientists and engineers. Of these, 88 studies were selected as candidates for a rigorous definition study; 44 of these candidate proposals deal with nonhuman subjects (plants and animals) and are managed from the LSFE Project Office at Ames. Because only a few experiments may be flown on any one mission, definition of each experiment requires accurate detailed planning before final selection, which will be made in Spring of 1981. Once selected, each investigation will undergo final development and fabrication of hardware in preparation for flight. As experiment requirements are defined, the necessary ground support systems and flight laboratory equipment items will be prepared by teams of NASA and contractor scientists and engineers.

Responsibility for management of these experiments is divided between NASA's Ames Research Center and Johnson Space Center, according to experimentation of nonhuman or human test subjects, respectively. Ames, in its role as lead research center for studies involving comparative physiology and basic biology will be responsible for the science management of nonhuman proposals as well as the development, fabrication, test, and flight operations support for all hardware and software systems required to conduct these experiments. Hardware systems currently under development and planned for use over multiple missions, include the Research Animal Holding Facility (RAHF), Life Sciences General Purpose Work Station (GPWS), Plant Growth Unit (PGU), Dynamic Environment Measuring System (DEMS), Biotelemetry System (BTS), and a complement of small standard laboratory instruments and supplies.

In this paper, we present an overview of the LSFE program at Ames that will take advantage of the unique opportunities for biological experimentation possible on Spacelab.

2. PROGRAM STRUCTURE, SCHEDULES, AND STATUS

2.1 Structure

The LSFE program is organized through three levels at NASA. Level I, which operates out of NASA Headquarters in Washington D.C., is directed from the Office of the Associate Administrator for Space Sciences by the Director for Life Sciences. The NASA Lyndon B. Johnson Space Center (JSC) in Houston is the Level-II LSFEP management office and lead NASA center for coordinating the functional efforts at three Level-III offices.

The JSC level-III LSFE project office is responsible for human experimentation in a parallel and similar effort with the Ames level-III LSFEP office. The Kennedy Space Center (KSC) Spacelab Payload Mission office directs integration, launch, and postflight operations of Spacelab payloads.

2.2 Schedules and Missions

To date, there have been 88 studies selected for Definition Phase from an original 360 Life Sciences proposals submitted to NASA's AO in 1978. Half of these comprise a wide variety of animal and plant studies managed from Ames. These studies are currently undergoing detailed definition that will identify each experiment's scientific, engineering, and cost requirements as well as precisely define how each study will be managed so that final selection can be made. This phase, begun in December 1979, will end by Spring 1981, when experiment selection will be made. Detailed development of each experiment will occupy the majority of Ames efforts from 1981-1984.

Prior to the first Life Sciences Dedicated Lab, three Spacelab (SL) missions are planned: SL-1 is an international mission composed of space and atmospheric physics, materials technology, astronomy, and a select group of life sciences studies; SL-2 is a pallet mission for which Ames is developing a plant lignification study; SL-3, now scheduled for launch in April 1984, is planned to verify the design and function of the RAHF for rodents and squirrel monkeys, as well as to test biotelemetry systems planned for later flights.

2.3 Program Status

In planning for the scheduled launch dates for SL-3 and dedicated missions, Ames has issued contracts

to 44 principal investigators
through April 1981 for experiment
definition. As planning for
experiment selection and

development continues, the Ames
LSFE project office is engaged in
several major efforts. Planning
for the design and fabrication of
major Ground Support Equipment
(GSE), Data Systems Integration
Support Equipment, Life Sciences
Laboratory Equipment (LSLE), and
facilities at KSC and Ames have
been a major thrust of recent
activities. To help manage the
program, a data base management
system is being implemented for the
normally voluminous data required
for experiment implementation.

These efforts are supported at Ames
and JSC through a prime contract,
the Development Engineering and
Operations (DEO), with Management
and Technical Services Company
(MATSCO), a major General Electric
(Space Division) subsidiary. The
DEO program at Ames has grown in
one year to over 50 personnel that
include separate Life Sciences,
Engineering, Operations, and Data
Systems groups. This DEO effort
provides the manpower to plan,
produce and coordinate essential
documentation for the Definition
Phase. MATSCO plays an integral
part with NASA in supporting
mission management decisions.

3.0 SCIENCE ROLE

3.1 Structure

The life sciences research effort
at Ames has historically held a
vital place in NASA's programs. On
this basis, the Ames project has
established the need to maintain a
strong role in the science
management of the LSFE program
under its responsibility.

A project team of scientists and
engineers, under the direction of a
NASA Project Scientist, interacts
with the principal investigators
throughout the experiment's history
from proposal selection through
development and flight operations.
They help define scientific
requirements and protocols,
identify hardware needs, and
support development throughout
NASA's implementation of the flight
study.

3.2 Selection

The selection of 44 studies for
definition encompasses a lengthy
and ardent process involving
several peer reviews from
scientists and engineers outside
and inside NASA. Of these studies,
11 science disciplines involving
studies of several animal and plant

species, were proposed from 29 major research institutions and universities, as well as 11 originating from the Biomedical Research Division at NASA-Ames. Three proposals were selected from foreign institutions in England, Wales, and Australia. The major science disciplines that involved joint animal/human experimentation are: vestibular physiology, cardiovascular and cardiopulmonary physiology, renal and endocrine balance, hematology and immunology and bone/muscle physiology.

3.3 Science Investigator Working Groups

During experiment definition and development, advisory interactions between NASA and MATSCO scientists, engineers, and managers, and the principal scientists are vital to the LSFE program. A Science Working Group (SWG) patterned after the Investigator's Working Group (IWG), which will grow from the SWG following final experiment selection, is a forum for interaction of investigators with Ames project personnel. Some of the many tasks accomplished by the SWG include: providing recommendations on combining and sharing of specimens and experiment resources; making recommendations for the development of the animal

holding facilities, work station, and common research equipment planned for Spacelab; playing an advisory role in NASA animal use planning; and helping to define and develop training requirements for the Spacelab crew (payload and mission specialists).

4. LIFE SCIENCES LABORATORY EQUIPMENT (LSLE) PROGRAM

Standard research equipment common to many life sciences experiments is provided by NASA from several major contracts managed and implemented from the Ames project office and MATSCO. Ames contracts major hardware items to be designed and built to be compatible with Spacelab instrumentation, stowage, electrical and data systems, and meet specifications for safe and proper functioning in zero-g and one-g.

There is a potentially overwhelming list of standard biological laboratory equipment that could be modified for Spacelab use. Many of these items are obviously mandatory for a space-borne biological laboratory and are now being developed; for other equipment, we are waiting for the final selection of specific experiments before committing NASA's resources for hardware development. Typical items under development or being considered for development include: hematology kits, solution handling kits, refrigerated centrifuges, microscope systems, a rodent sacrificing system, rodent metabolic cages, refrigerators, and cryogenic freezers.

4.1 General Purpose Work Station (GPWS)

The GPWS is a Biohazard Cabinet designed to support life science laboratory experiments aboard the Space Shuttle. Its primary functions are both to protect both the astronauts and animals and to control liquids and chemical vapors released within the work station in a zero gravity environment. The station consists of a laminar-flow cabinet, a trace contaminant-control system, a thermal-control system, control electronics and lighting, an avionics-air cooling system and several utility functions such as magnets, camera mount, water and vacuum.

The unit fits into a Spacelab double rack. The cabinet is mounted on four rails so that the assembly can be slid in and out of the rack. As planned now, the assembly can also be fully extended for two operator access.

In October, 1980, a functional test of the GPWS was completed. The objective was to demonstrate the capabilities of the GPWS Design Verification Unit and verify its suitability for life science experimentation. The test pointed out areas of concern in human factors for the design of the GPWS, which are currently undergoing review by NASA.

4.2 Research Animal Holding Facility (RAHF)

A major objective of the NASA Life Science Flight Experiments Program is provision of functioning laboratory conditions approximating those in ground-based facilities. The Research Animal Holding Facility (RAHF) supports this objective by providing life-support housing of small research animals on Spacelab missions for various life science investigations.

The facilities to maintain and house small experimental animals is of primary importance to successful attainment of experiment goals. To assist the Principal Investigator to this end, NASA has developed the modular RAHF for use on Spacelab missions. In addition to providing general housing for various animal species, RAHF was designed to

minimize disturbance of the
specimens by vehicle and missions
operations. Life-sustaining
capabilities, such as provisions
for food, water, and waste removal
as well as environmental control,
are provided. By necessity, RAHF
is reusable and is of modular
construction to accommodate a
variety of small animals.

The RAHF system will accommodate a
combination of twenty-four 500-gram
rats, 144 mice, or a mixed number
of rats and mice. An alternative
design accommodates four squirrel
monkeys. The RAHF system is
independently housed in a single
ESA rack, each measuring
approximately 262 x 50 x 76 cm
(103 x 19.5 x 30 in.). Because
Spacelab is launched vertically,
the cages are installed on their
sides to provide proper orientation
for operation of the waste
management system in the prelaunch
g environment.

Lockheed Missile and Space Company
provided the design and will build
the first Spacelab RAHF system that
together with NASA-built animals
cages will be tested on the
Spacelab-3 mission with rats and
squirrel monkeys. Currently, the
RAHF is to be built and delivered
to Ames Research Center by early
1982.

4.3 Plant Growth Unit (PGU)

The Space Shuttle Plant Growth Unit
(PGU) has been designed and built
by Lockheed and NASA to investigate
the effect of zero gravity on the
formation of plant lignin. Lignin
is a product of metabolism in
plants and forms the woody
structure that stiffens the plant
walls, aiding the plant in growing
upright against gravity.
Weightlessness affords a unique
tool to further define the
biochemical pathways vital to
lignin synthesis.

The PGU is a self-contained
carry-on experiment designed to be
loaded into the Shuttle late in the
launch countdown and mounted in
place of one of the storage lockers
located in the Spacelab Orbiter
middeck. The only other interface
required of the Orbiter is
electrical power. The experiment
is completely automatic requiring
no flight crew interaction, and all
data are stored within the
experiment package. The hardware
holds six plant growth chambers,
each containing approximately 16
plant seedlings, a total of 96
plants. The experiment hardware
also includes a thermal-control
system, lighting, instrumentation,
and a data-management system.

NASA plans to first fly the
experiment hardware on Operational
Flight Test 4 (OFT-4). This flight
will involve species from different
plant classes; the outcome will
provide basic data on lignin
formation and will also help plant
scientists finally select the plant
species to be used in the second
flight experiment, scheduled to fly
on Spacelab 2. In this flight, all
of the plants are scheduled to be
of the same species, now
tentatively pine.

5. CONCLUSION

The life science community has
shown a strong interest in the LSFE
program. Several thousand letters
indicating interest were received
to an "Invitation to Participate in
Planning the NASA Life Sciences
Program in Space." Responses were
received from all segments of the
life sciences community,
representing more than 500
institutions and private
organizations.   In response to the
most recent Announcement of
Opportunity for the first dedicated
Life Sciences Spacelab missions,
over 360 experiment proposals were
received from the life sciences
community.

Ames personnel are committed to
planning the detailed definition
and implementation of the science,
engineering, and resource
requirements that are necessary to
develop optimum payloads for
Spacelab and to identify equipment
and facilities necessary in
conducting nonhuman experiments.
As the Definition Phase efforts
conclude and experiment selection
becomes eminent, NASA is beginning
tasks for detailed payload
definition. Such concerns as
sharing of animals between
experiments, prelaunch and
postlanding access needs,
constraints on Spacelab power,
crewtime, and storage space are
being analyzed to accurately
synthesize and implement Spacelab
payloads.

With a successful Shuttle Program,
the LSFE program at Ames will
continue through experiment
selection and development for the
1984-5 Spacelab missions.

BIOGRAPHY – William E. Berry

William E. Berry holds a BS degree
in Mechanical Engineering from
Drexel University, Philadelphia, PA
and has done graduate work at
Columbia University, New York City,
and Stanford University.

His professional experience during
14 years at NASA-Ames Research

Center, Moffett Field has included:
Project Engineer for the
Biosatellite Thermal Control and
Gas Management Systems, Project
Engineer for Development of the
Viking Mars Mission Life Detection
Experiment, and Principal Engineer
and Project Manager for the Pioneer
Venus Gas Chromatograph Atmospheric
Analyzer Experiment. At present
Mr. Berry is Chief, Life Sciences
Flight Experiments Project Office,
Biosystems Division. Life Sciences
Flight Experiments Program Ames
Level-III Manager. Responsible for
all non-human investigations on the
Shuttle.

BIOGRAPHY - Christopher Dant

Christopher Dant received his BS
(Zoology, 1965) and MA (Genetics,
Endocrinology, 1968) from Indiana
University. In 1968, he took a
position as research scientist and
fellow in biochemical genetics in
the Human Genetics section of the
University of Michigan's Medical
School. In 1970, he was awarded a
Ph.D. teaching fellowship in Cell
Biology at the University of
Vermont, where he went on to a
research position in cellular
biochemistry after completing his
coursework (without dissertation).
In 1973, he left the academic
community to work as a biomedical
writer, first at Syntex and
Ceba-Geigy, and later at the

University of California as a
liaison between biomedical research
and technical publications.
Christopher joined G.E. as an
experiment support scientist to
NASA in 1979, after spending a
year's work on contracts to NASA
Biosystems Division.

LDEF EXPERIMENT: CRYSTAL GROWTH BY SOLUTE DIFFUSION PROCESSES

M. D. Lind and N. G. Taylor
Rockwell International
P. O. Box 1085
Thousand Oaks, California 91360

Abstract

Experiments in crystal growth by solute diffusion processes will be performed on the first flight of the NASA Long Duration Exposure Facility (LDEF) in 1983. In near-zero gravity, reactant solutions will be allowed to diffuse slowly toward each other through a region of pure solvent in which they will undergo a chemical reaction to form crystals. The processes are analogous to the growth of crystals in gels in normal gravity. Crystals tentatively selected for investigation are PbS and $CaCO_3$. The LDEF experiments are an extension of experiments performed during the Apollo-Soyuz Test Project (ASTP) and others now being developed for the first Spacelab flight. The LDEF flight will be the first space flight to completely satisfy the processing time requirements of these processes. Another important feature of the LDEF experiments is that they will incorporate precise temperature control, which was not possible in the ASTP experiments. The LDEF experiments are being developed in cooperation with the Technical University of Denmark and two Danish companies, Danmarks Rumfartsselskab AS and Terma Elektronik AS.

1. INTRODUCTION

1.1 Crystal growth in space

Experimentation in solidification and single crystal growth processes in earth-orbiting space laboratories is now receiving much attention. In nearly all solidification and single crystal growth processes, convection, sedimentation, and contamination by containers cause structural defects and compositional inhomogeneities in the solids produced. It is likely that these imperfections impair the performance of devices (e.g., electronic, magnetic, optical, acoustic, and structural) made from the materials. In an orbiting space laboratory, the gravitational force is nearly zero, so that convection, sedimentation, and the need for containers are eliminated, and better control of the quality and properties of solids, and improvements in performance, yields, and costs may be achievable.

1.2 Solute diffusion processes

Certain kinds of single crystals, which are slightly soluble in a particular solvent, can be grown from solutions by solute diffusion processes. In such processes, two or more reactant solutions diffuse together and react chemically at controlled rates according to the following chemical equation to form the desired single crystals:

$$A(soluble) + B(soluble + \ldots \rightleftharpoons C\downarrow$$
$$(slightly\ soluble) + D(soluble) + \ldots$$

Two kinds of experimental arrangement for solute diffusion crystal growth processes in normal gravity are shown in Figure 1. In the u-tube arrangement (Figure 1a), the crystals normally grow within the gel or at its interfaces with the reactant solutions. In the second arrangement (Figure 1b-d), the crystals can grow in the pure solvent at the center.

In these processes, the gel separates the two reactant solutions and suppresses the rapid convective mixing that would be caused by their difference in density. Diffusion is the predominant mechanism for the mixing of the reactant solutions. This allows control of the material transport by variation of solution concentrations and/or reactor shape.

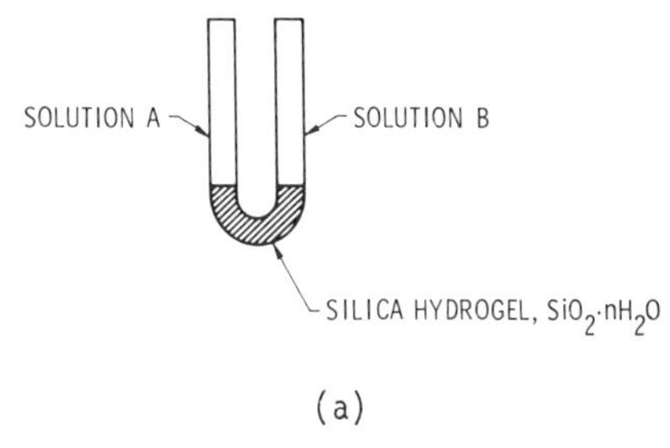

(a)

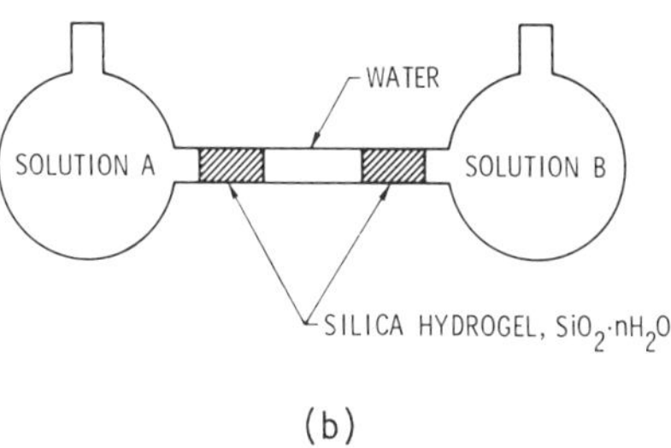

(b)

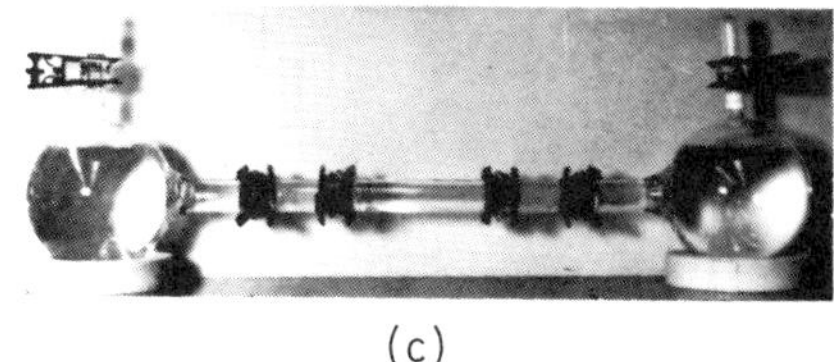

(c)

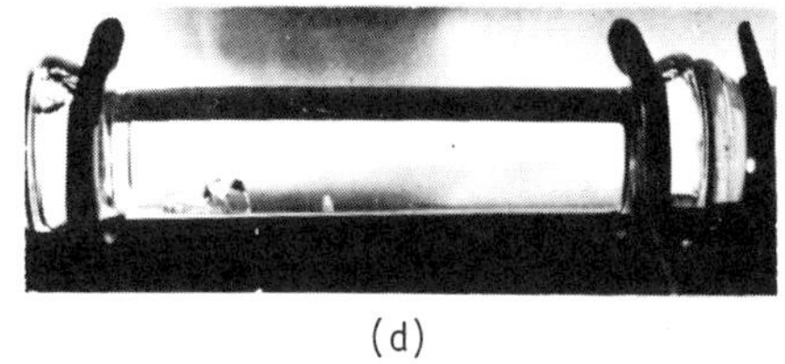

(d)

Figure 1. Typical experimental arrangements for crystal growth in gels: (a) u-tube; (b) double; (c) calcium tartrate crystal growth reactor; (d) close-up view of crystals in center section of (c).

The gel also suppresses sedimentation, and in the u-tube arrangement

keeps the growing crystals suspended
in the solution. Typical gels are
approximately 95% solvent. As the
crystals grow, they are exposed to
solution on all sides and have no
contact with container surfaces
except those of the flexible gel.

1.3 Solute diffusion in space

When the gravitational force is small
enough to make convection and sedimentation negligible, as in an orbiting space laboratory, the gel is not
needed and can be replaced by a
region of pure solvent. Diffusion
will be the predominant mechanism of
mixing even with no gel present.

Solute diffusion processes at ambient
temperatures ($\sim$25°C) are advantageous
because they minimize problems of
thermally induced strain, phase
transformations, volatility of components and container contamination
usually inherent in high temperature
crystal growth methods. Solute diffusion processes using gels have some
disadvantages, including excessive
nucleation resulting in small crystals, contamination of the crystals
by gel constitutents or impurities in
the gel, and instability of the gel.
The space flight experiments can
overcome all these disadvantages
while retaining the advantages.
Therefore, solute diffusion processes in near-zero gravity may be
expected to yield large, high-quali-
ty crystals of a variety of materials.

1.4 ASTP experiments

The first experiments in solute diffusion crystal growth under near-
zero gravity were performed on the
ASTP flight in 1975.[1-3] The simple reactor used in these experiments is shown schematically in
Figure 2. Six such reactors
(Figure 3) were carried on the ASTP
flight. The crystals investigated

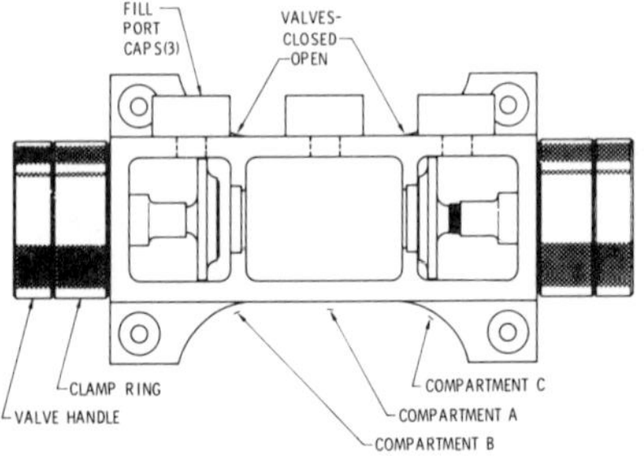

Figure 2. ASTP crystal growth
reactor schematic.

Figure 3. ASTP flight reactors.

were calcium tartrate, calcium carbonate, and lead sulfide. After the
flight, the reactors were returned
to our laboratory for analysis of
the results. The experiments were
successful, and the results provide
ample justification for undertaking

further, more sophisticated experiments. An example of the results is shown in Figure 4. Further details of the ASTP experiments and their results can be found in the cited publications.[1-3]

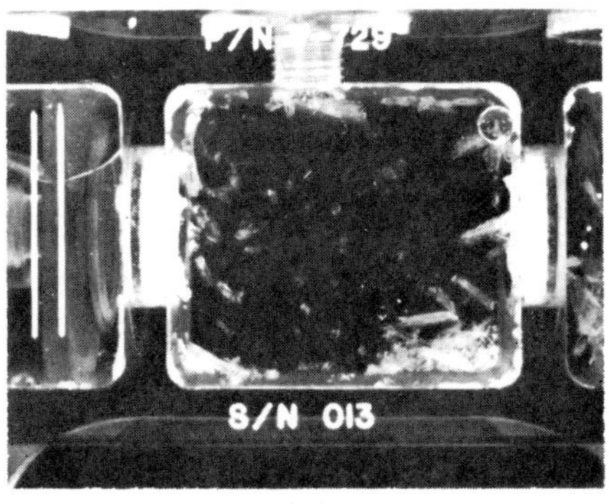

(a)

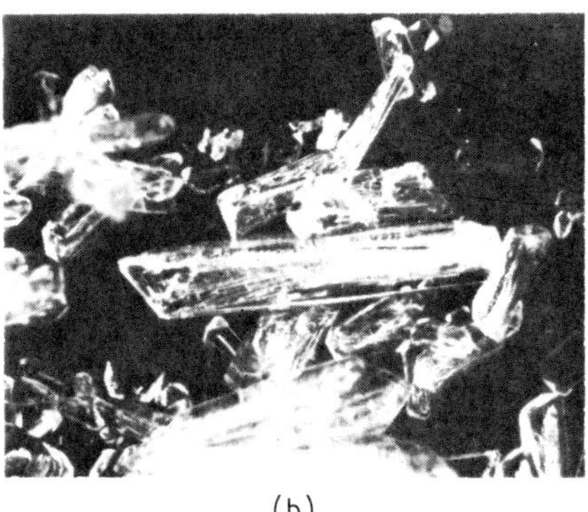

(b)

Figure 4. ASTP results: (a) calcium tartrate reactor; (b) calcium tartrate crystals under magnification.

1.5 Spacelab experiments

Some similar experiments to be performed on the first flight of the Spacelab are now being prepared by two groups in Europe. K. F. Nielsen and co-workers at the Technical University of Denmark are investigating tetrathiofulvalene-tetracyanoquino-dimethane (TTF-TCNQ) with acetonitrile as the solvent.[4] A. Authier and co-workers at the University of Paris are studying carbonates, including calcium carbonate and others, grown from aqueous solutions.[5] The first Spacelab flight will be in 1983.

2. LDEF EXPERIMENTS

2.1 Rationale

Ideally, solute diffusion crystal growth processes require times of the order of weeks or months, rather than the 4 to 5 days provided by the Apollo-Soyuz and Spacelab flights. The first LDEF flight will be the first space flight to satisfy this requirement completely, and will provide the first opportunity for fully exploring the potential of the process. Accordingly, in cooperation with the Danish group, we have designed and built a solute diffusion crystal growth experiment package to be carried in the first LDEF flight.

2.2 LDEF description

The LDEF is a passive, unmanned, free-flying, gravity gradient stabilized spacecraft, developed by the NASA Langley Research Center to be deployed by the Space Transportation System Orbiter in a near-Earth

orbit, and retrieved after a period
of 6 months or more. The first
LDEF, also planned for 1983, is
shown in Figure 5.

Figure 5. First LDEF.

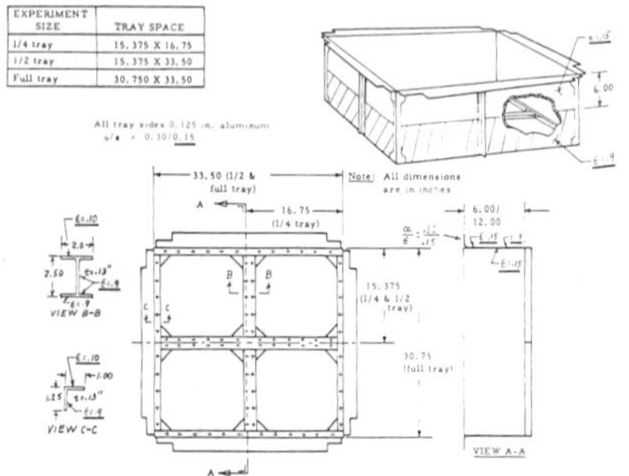

Figure 6. LDEF experiment tray
schematic.

The LDEF will accommodate self-contained experiments packaged in standardized trays such as that shown
schematically in Figure 6. The
trays are furnished by NASA.

2.3 Crystals to be investigated

The reactions tentatively chosen for
investigation in the first LDEF
flight are:

(1) Lead sulfide:

$$PbCl_2 + CH_3CSNH_2 + H_2O \rightleftharpoons$$
$$PbS\downarrow + CH_3CONH_2 + 2HCl$$

(2) Calcium carbonate:

$$CaCl_2 + (NH_4)_2CO_3 \rightleftharpoons$$
$$CaCO_3\downarrow + 2NH_4Cl$$

These crystals were chosen because
they can be grown by solute diffusion methods in normal gravity, and
because of their importance in
research and technology. Lead sulfide is a semiconductor useful as an
infrared detector and laser diode.
Calcium carbonate has useful optical
properties, including birefringence.

The LDEF experiments are expected to
yield crystals of each of these
materials superior in size, structural perfection, and compositional
homogeneity to those obtainable in
normal gravity. The availability of
such crystals will make possible
better determination of their properties and, perhaps, investigations
of possible device applications.

Although the experimental apparatus
is designed for these materials, it
incorporates sufficient flexibility
for substituting alternatives if
warranted by circumstances before
the flight.

The Danish group is planning similar
experiments in the growth of TTF-
TCNQ crystals for the first LDEF
flight.

2.4 LDEF reactors

The solute diffusion crystal growth
process to be investigated is the

diffusion of two reactant solutions into a region of pure solvent in which they react chemically to form single crystals of the desired material. The process will utilize specially designed reactors (Figure 7) having three compartments separated by valves for keeping the reactant solutions and solvent separated until the LDEF has been deployed in orbit. The two outer compartments are 7.6 cm in diameter and will contain the reactant solutions. The central compartment is 5.1 cm in diameter and will contain pure solvent. The lengths of the compartments will be different for the various crystals; the maximum length of each reactor is 71 cm. Each compartment has three ports to facilitate loading and to accommodate a pressure-equalizing manifold. Each compartment contains a teflon bag to provide for volume changes caused by temperature or phase changes.

The valves are two circular plates each having an identical pattern of three holes with "O" ring seals (Figure 7c); they are opened and closed by rotation in the same way as a salt shaker, except that the shafts operating them have threads which apply pressure to the "O" rings in the closed position. The valve openings contain glass frits of a porosity and thickness calculated to provide for damping of fluid flow induced by opening the valves. The valves will be opened

and closed by electric motors. By means of a very low gear ratio, the valves will be opened very slowly over an eight-hour period to minimize the induced fluid motion.

There will be four reactors in the LDEF package. Two of these, for our lead sulfide and calcium carbonate experiments, are made of stainless steel with Viton "O" rings; the other two, furnished by the Danish group for the TTF-TCNQ experiments, are made of teflon, with fused quartz valve plates and Kalrez "O" rings.

The electric motors for all four reactors are being furnished by the Danish participants.

2.5 Reactor enclosure

The four crystal growth reactors are mounted in an insulated, hermetically sealed cylindrical enclosure, 30 cm in diameter and 82 cm long, which, in turn, is mounted in the $85 \times 78 \times 30.5$ cm^3 LDEF tray (Figure 8). The reactor enclosure is a double-walled aluminum pressure vessel sealed by Viton "O" rings in the end caps. The region between the two walls contains multilayer aluminized mylar super-insulation and is evacuated to 10^{-6} torr for thermal isolation of the reactors. The mounting brackets of the reactor enclosure are designed to be either insulating or conducting, depending on the outcome of further thermal analysis and testing.

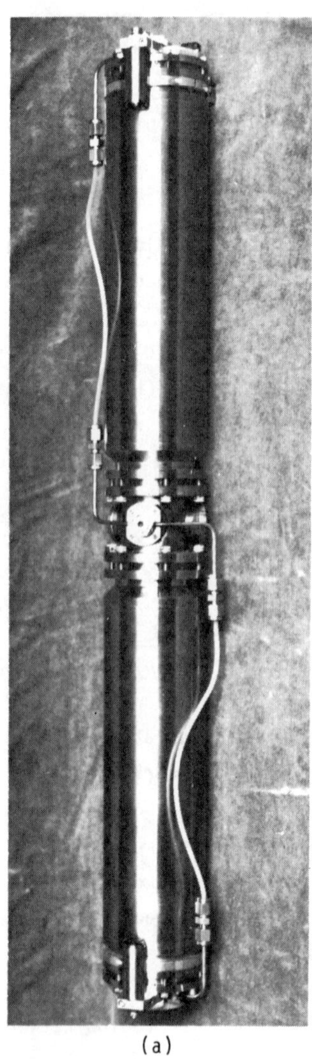

(a)

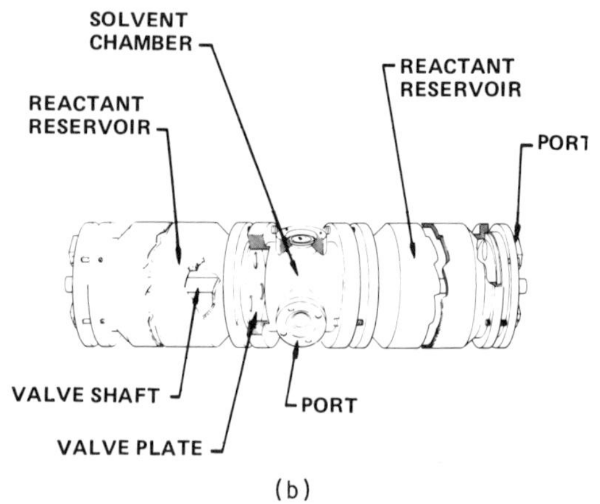

(c)

Figure 7. LDEF crystal growth reactor: (a) photograph of complete reactor; (b) schematic; (c) valve schematic. Reactor dimensions are given in the text.

Within the inner tube of the enclosure, the four reactors are surrounded by a cylindrical resistance heating element and argon at a pressure of 1 atm. Temperature sensors are mounted on each reactor. The heater will maintain the temperature inside the reactor enclosure at a preset value in the range of 25 to 45°C. The heating element and temperature sensors are being furnished by the Danish participants.

2.6 Electrical power

Electrical power will be provided by batteries in an adjacent tray. The processing time will be limited by both the weight and the cost of the batteries and is currently estimated at about two months.

2.7   Process controller

The process will be controlled and monitored by an electronic controller designed and built by Terma Elektronik AS in Denmark. It is a modification of apparatus they are furnishing for the European Spacelab experimental package. The controller is housed in a single-walled hermetically sealed enclosure mounted in the LDEF tray as shown in Figure 8.

Throughout the experiment, the controller will regulate the temperature of the reactors and record any deviations from the desired value. Precise control of the temperature is an important improvement over the ASTP experiments, because they had to be performed at the Apollo cabin temperature, which varied more than we expected.

After deployment of the LDEF, the controller will open the valves to start the diffusion and growth processes. It will monitor the battery power and terminate the experiments by closing the valves just before the batteries are depleted.

The controller is being furnished by the Danish participants.

2.8   Thermal design

A thermal model of the LDEF experiment package has been derived by

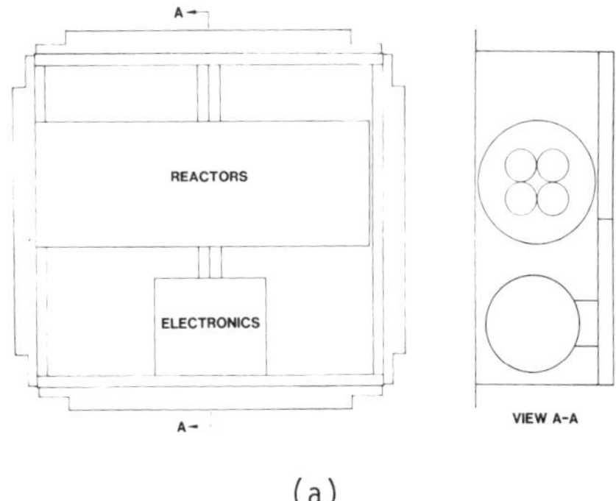

(a)

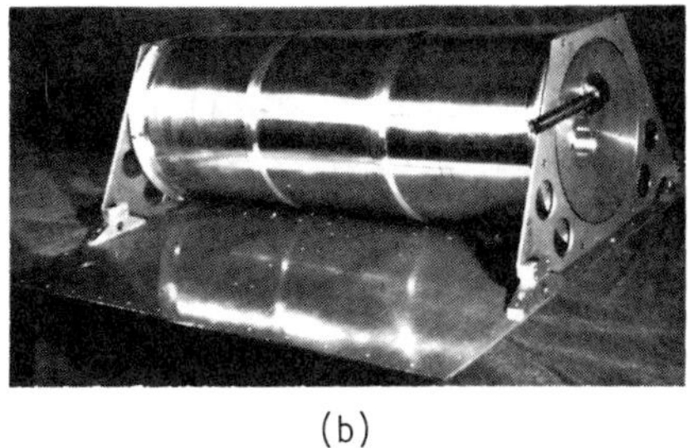

(b)

Figure 8.   Tray layout: (a) schematic; (b) photograph of reactor enclosure mounted on base plate with space for electronics enclosure in foreground. Dimensions are given in the text.

modifying an empty-tray thermal model, provided by NASA, to include the experimental apparatus added to the tray. Using the electrical circuit analogy, the hardware components and boundary elements were considered as nodes in a thermal network; the thermal capacitance of each node and the conduction resistances and radiation connections between them were calculated. These data, combined with the thermal boundary conditions, were used to calculate heat fluxes and temperatures by conventional thermal network analysis procedures. The results indicate that approximately 1 watt of power will be required to maintain the reactors at a temperature of 35°C.

The thermal model will be refined as the results of functional tests of the apparatus become available. An accurate thermal model is required both to assure the compatibility of our package with the overall LDEF thermal design and to determine the power requirements for regulating the temperature of the crystal growth reactors.

2.9 Critical aspects of the design

The most critical aspects of the apparatus design are the valves and the thermal design. The valves must remain closed with no leakage until the start of the process and then must operate without stirring the fluids. The thermal design must provide for temperature regulation throughout the experiment, despite the temperature extremes to be encountered during the flight. Both aspects are being subjected to thorough functional testing.

2.10 Processing variables

The processing variables include the reactant solution compositions, reservoir dimensions, temperature, and duration. For the LDEF flight, these will be based primarily on the results of analogous solute diffusion crystal growth experiments under normal laboratory conditions and on the power requirements as determined by functional tests and thermal analysis.

REFERENCES

1. M. D. Lind, "Crystal Growth, Experiment MA-028" in Apollo-Soyuz Test Project Summary Science Report, NASA SP-412, 1977, p. 555.

2. M. D. Lind, "Crystal Growth from Solutions in Low Gravity," Paper 77-197, Proceedings of the AIAA 15th Aerospace Sciences Meeting, Los Angeles, CA, January 24-26, 1977.

3. M. D. Lind, "Crystal Growth from Solutions in Low Gravity," AIAA Journal 16, 458 (1978).

4. K. F. Nielsen, "Diffusion Growth of Organic Charge-Transfer Crystals," in European Space Agency Special Publication No. 114, 1976, p. 255.

5. A. Authier, private communication.

ACKNOWLEDGEMENTS

The contributions of the Danish participants and Langley Research Center personnel are gratefully acknowledged. We thank K. F. Nielsen, G. Galster, O. Lundorff, S. Thomsen, A. Kloster, J. Jones, C. Kiser, W. Kinnard, R. Greene, C. Saunders, G. Haynes, C. Moore, D. Raleigh, and M. Manoff for helpful discussions.

M. D. Lind is a Member of Technical Staff, Microelectronics Research and Development Center, Rockwell International. He received a B.S. in Chemistry from Otterbein College in 1957 and a Ph.D. in Physical Chemistry from Cornell University in 1962. After 1-1/2 years of post-doctoral work at Cornell University, he was employed at the Union Oil Company Research Center from 1963 to 1966 and has been with Rockwell International since 1966. His principal research interests are in crystal growth and X-ray crystallography. He has worked in the field of crystal growth in near-zero gravity recently and has been Principal Investigator for crystal growth experiments in the Apollo-Soyuz Test Project, the Space Processing Applications Rocket Project, and the Long Duration Exposure Facility.

N. G. Taylor is Supervisor of the Research Shop at Rockwell International Science Center. He received a B.S. in Mechanical Engineering from the University of Southern California in 1970. He has been with the Science Center since 1962. Prior to assuming supervision of the Research Shop in 1978, he worked as a design engineer. He has provided support for many scientific research projects including design of crystal growth hardware for the Apollo-Soyuz Test Project, the Space Processing Applications Rocket Project, and the Long Duration Exposure Facility.

DEVELOPMENT OF A CURE-IN-PLACE
ABLATOR FOR SPACE SHUTTLE INSULATION
Blevins, C. E. and Gonzalez, R.
McDonnell Douglas Astronautics Company
5301 Bolsa Avenue
Huntington Beach, California 92647

Abstract

The probability of damage to the Shuttle Orbiter Thermal Protection System (TPS) during ascent has made it necessary to provide the Shuttle crew with a TPS flight repair kit. This paper covers the development and evaluation of an ablator and its dispensing gun for repairing the Shuttle Orbiter TPS while in orbit. A two-component RTV silicone ablator was developed that could be mixed, dispensed, and cured in a space environment. It was demonstrated during this investigation that certain catalysts will cure silicones in a hard vacuum at low temperatures. The ablator materials developed were tested and found to meet the in-space processing criteria and to have the mechanical and thermal properties required to provide protection to the orbiter during reentry. The ablator dispensing gun was specially designed to be used in a space environment by an astronaut working in a manned maneuvering unit. It was shown that with proper viscosity adjustment and a combination of selected catalysts a two-component RTV silicone ablator can be efficiently mixed in a static mixing head.

1. INTRODUCTION

This paper summarizes the work accomplished in the development of a flight repair kit for on-orbit repairs of the Shuttle Orbiter tile insulation. The project consisted of two main tasks: (1) development of ablator materials for applying to damaged areas and (2) design and fabrication of a gun for dispensing the uncured ablator.

1.1 Ablators

Two ablator materials were developed: a cure-in-place silicone compound for use in repairing small areas and precured ablator tiles for repairing large areas. The cure-in-place silicone ablator was also used for bonding the precured ablator tiles.

The development program was undertaken because, at the time, there were no ablation compounds available which could be applied in-flight, would cure in vacuum, and had the required properties.

Various commerical silicone elastomers were examined. Because one
of the objectives was to produce a
low density material, the addition
of hollow microspheres was investigated. Of seven cure-in-place
ablators considered, RTV-560-55 and
RTV-560-32 were developed and fully
tested. Except for char retention
properties, these materials satisfy
all the requirements listed under
Section 2. Char retention properties of these two materials can be
improved by adding refractory fibers
to the formulations. Of the precured ablators considered, three
materials were found to meet the
criteria: Purple Blend Mod. 5 (an
old NASA formulation), precured
RTV-560-32, and the ceramic tiles
already used as insulation on the
Shuttle Orbiter.

1.2 Material Applicator

Several applicator concepts were
studied and evaluated to the system
requirements. From this effort two
concepts were selected.

a) A self-contained applicator
(Figure 7).

b) A three-part applicator design
(Figures 8 and 9).

Both material applicators operate
on the same concept, using gas
pressure to expel the material and
static mixing heads to mix the
catalyst and resin. Twin concentric pistons or bellows are used to
extrude the catalyst and resin. A
functional mockup (Figure 5) was
constructed and used to verify the
basic functions of the proposed gun
concepts.

2. PROGRAM REQUIREMENTS

The cure-in-place ablator had to
meet the following requirements:

Thermal performance: maintain
orbiter temperature below 350°F during reentry.

Cure properties: cure in vacuum at
10^{-5} torr and temperatures from
40° to 125°F.

Work life: 1 hour after application at 40°F.

Cure time: cure within 18 hours
at 40°F.

Bond strength: greater than 40 psi
when bonded to RTV 560.

Tensile strength: greater than
40 psi.

Shelf life (unmixed): minimum of
6 months of 80°F.

Viscosity: sufficient to remain
in place after application.

Low density: to minimize thermal
stress on adjacent tile and meet
repair kit requirements of 300 lb
maximum weight.

3. SCREENING AND SELECTION
OF CANDIDATE MATERIALS

An effort was made to select cure-
in-place in-flight ablation materials for which a good data base
was available. However, no reports
on applying and curing ablation
materials in space existed. Therefore, a material had to be developed.

The main factor limiting the candidates to the RTV silicones was the
requirement to bond to an RTV-560
substrate. Addition-cure silicones
were ruled out because they will not
adhere to RTV-560, which being

metallic-soap cured poisons the addition-cure catalyst.

RTV-560, 566, 88, 577, and PR 1977 were candidates for cure-in-place ablators. The RTV-88, 577, and PR 1977 had densities and viscosities that were too high for mixer/applicator suitability. An obvious candidate for cure-in-place ablator was the substrate to which the ablator would be bonded, RTV-560. RTV-560 is a methyl-phenyl-type polymer that provides the best low-temperature flexibility and is formulated with an iron-oxide filler for improved high-temperature stability.

RTV-560 was chosen as the base elastomer for the cure-in-place ablative compound because it would adhere to the RTV-560 substrate used to attach the tiles and also had the best overall properties of the commercial RTV silicones. This material is basically a sealant, so the incorporation of additional fillers is required to provide improved ablative properties and to increase viscosity.

To decrease the density and thermal conductivity, hollow microspheres were added (silica Eccospheres, phenolic Microballoons). The Eccospheres increased oxidation resistance and viscosity. At high temperatures, the Eccospheres yield a protective high viscosity melt. The phenolic Microballoons increased char yield and viscosity. Refractory fibers were not included in the cure-in-place ablators formulated

for this study, because of their potential for clogging the static mixers in the applicators.

Two of the many formulations investigated were more promising than the others. In the first, hollow microspheres were added to RTV-560 until the viscosity was estimated to be the maximum compatible with the mixer/applicator. This compound was named RTV-560-55, the "55" being roughly equal to its final density in lb/ft^3.

The second RTV-560 ablator, designated RTV-560-32, was made by adding a very large amount of hollow silica microspheres. Incorporation of the microspheres and lowering of the viscosity to a workable level was accomplished by adding a low viscosity silicone fluid diluent. The compositions of the two cure-in-place ablators are given in Table 1.

Table 1. Composition of Cure-in-Place Ablators[1]

Material	RTV-560-55	RTV-560-32
RTV-560	82.6 parts by weight	61.0 parts by weight
RTV-9811 Catalyst	4.6	3.4
RTV Catalyst F	4.6	3.4
Eccospheres Si	4.1	20.0
Microballoons, phenolic	4.1	---
Silicone fluid SF-99	---	12.2

4. DEVELOPMENT OF CATALYST SYSTEMS

The curing agent had two basic requirements:

a) A means to adjust the curing speed of the ablator formulation.

b) Proper volume and viscosity of the curing agent so that it could be metered and mixed uniformly into the base compound.

Metallic soaps are normally used as curing agents for the condensa-

tion-cure silicone elastomers.
While some early work was done with
mixtures of commerically available
catalysts (as indicated in Table 1),
the cure studies and the final product employed dilutions of concentrated catalysts. Dibutyl tin
dilaurate is a frequently used curing agent that provides a moderate
cure rate (18 to 36 hours at 293°K).
Stannous octoate provides a very
rapid cure (4 to 12 hours at
293°K) and cures at low temperatures.
Because our materials had to cure
within 18 hours at 40°F, it was
reasoned that a mixture of dibutyl
tin dilaurate and stannous octoate
would be satisfactory. This mixture, suitably diluted with hollow
silica microspheres (Eccospheres)
and silicone oil, was used.

The cure rate of RTV-560-55 was
investigated as a function of temperature at two selected concentrations of stannous octoate (diluted
1/1 with dibutyl tin dilaurate).
The extent of cure was measured by
testing the hardness with a Shore A
durometer. The results are shown
in Figure 1. At the lower concentration of 1.25 parts by weight

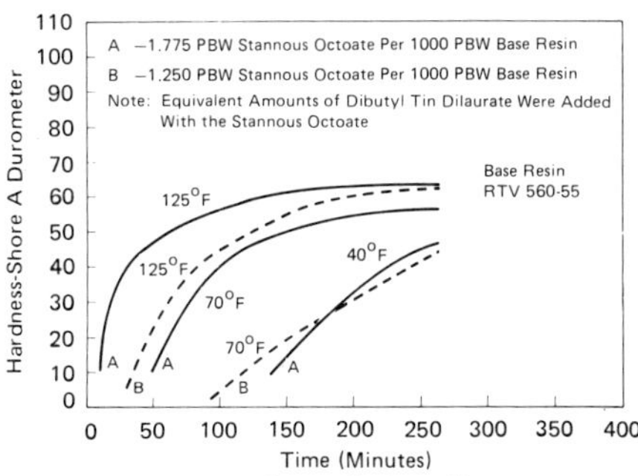

Figure 1. Cure Rates Versus
Temperature

(PWB) of stannous octoate to 1000
parts of RTV-560-55, cure at 278°K
(40°F) is very close to meeting the
mission requirements of a one hour
work life and is fully cured within
18 hours. The 1.775 PBW concentration of curing agent provides a cure
within 18 hours at 40°F. However,
the work life at 125°F is too short.
A cure rate study of the type described above was not done for RTV-
560-32.

Tests indicated that RTV-560-32
cured slower than RTV-560-55 and
would require a higher catalyst
concentration.

5. TESTING THE MATERIALS FOR PHYSICAL AND THERMAL PROPERTIES

The specimens used to evaluate the
cure-in-place compounds were made
as follows: the hollow microspheres
(Eccospheres SI and phenolic microballoons) were dried 16 hours minimum in a vacuum oven at 325°K
(140°F). The RTV-560 was degassed
under vacuum, the fillers were
added, and the mixture was stirred
under vacuum. Then, while the
mixture was exposed to atmospheric
conditions, the curing agent was
mixed in. The trace amounts of
water needed for cure were supplied by not degassing or drying
the catalyst mixture.

5.1 Viscosity

The viscosities of the RTV-560-55
and the RTV-560-32 are listed as
a function of temperature in
Table 2.

5.2 Tensile Strength

Vacuum-cast tensile specimens were

Table 2. Viscosity of Cure-in-Place Ablators

Material	Temperature °F (°K)	Spindle no.	Speed RPM	Viscosity, cp (N sec/m^2)
RTV 560-55	40 (278)	7	20	344,000 (344)
	70 (294)	7	20	277,000 (277)
	125 (326)	7	20	108,000 (108)
RTV 560-32	70 (294)	6	20	136,000 (136)
	70 (294)	6	10	140,000 (140)

cut from the cured sheets using a die conforming to ASTM D412.[2] Testing was done at 0.5 in/min, and the results are presented in Table 3. The tensile test results indicate that the modified RTV-560 formulations had sufficient strength to meet program requirements.

Table 3. Tensile Strength of Cure-in-Place Ablators

Material	Cure	Tensile strength psi (10^6 N/m^2)
RTV-560-55	Room temperature in vacuum	198, 151 (1.36, 1.04)
RTV-560-32	125°F (325°K) in vacuum	158, 131, 164, 126, 174 (1.09, 0.903, 1.13, 0.869, 1.20) Average = 151 (1.04)

5.3 Tensile Adhesion of Vacuum-Cast Specimens

Tensile adhesion tests were conducted using round aluminum tensile blocks with a 2 in^2 cross-sectional area conforming at ASTM D429.[3] A brush coat of catalyzed RTV 560 silicone rubber was then applied to each DC 1200-primed block and cured.

Mixtures of the cure-in-place ablator materials were prepared as previously described and applied to both surfaces of the RTV-560-coated tensile blocks. All specimens were cured in a vacuum and were tested in tension at a load rate of 0.0021 m/sec (0.5 in/min). The results of the tensile adhesion tests are shown in Tables 4 and 5. Unless otherwise indicated, all failures were cohesive. The results indicated that both cure-in-place ablators meet the tensile bond strength requirements. Tensile adhesion samples to the ceramic tile failed by cohesive failure in the ceramic tile.

5.4 Densities

The densities of the cure-in-place ablators are listed in Table 6.

5.5 Thermal Expansion

The thermal expansion characteristics of the ablative materials were

Table 4. Adhesion of RTV-560-55 Cure-In-Place Ablator

Material	Substrate	Cure temperature °K (°F)*	Tensile bond strength 10^5 N/m^2 (psi)**	
RTV-560-55	RTV-560	278 (40)	3.70, 4.20, 3.30, 3.42 (53.7, 60.9, 48.0, 49.6	Avg = 3.65 Avg = 53.0)
RTV-560-55	RTV-560	Room temperature	4.20, 4.92, 3.95 (61.0, 71.3, 57.3, 43.9	Avg = 4.36 Avg = 58.4)
RTV-560-55	RTV-560	325 (125)	3.03*, 3.67, 4.03*, 4.3* (43.9*, 53.2, 58.4*, 63.0*	Avg = 3.77 Avg = 54.6)
RTV-560-55	Purple blend and RTV-560	Room temperature	5.91, 6.48, 6.48, 5.23 (85.7, 94.0, 94.0, 75.9	Avg = 6.03 Avg = 87.4)
RTV-560-55	HRSI Tile	Room temperature	1.08, 2.28 (15.7, 28.6	Avg = 1.18 Avg = 17.2)
RTV-560-55	Emittance coating on HRSI	Room temperature	2.42, 2.56 (35.1, 37.2	Avg = 2.49 Avg = 36.2)

*All specimens cured under vacuum.
**All values identified thus * consist of adhesive failure between the RTV-560 substrate and the aluminum tensile block.

752

Table 5. Adhesion of RTV-560-32 Cure-In-Place Ablator

Material	Substrate	Cure temperature ^{o}K (^{o}F)*	Tensile bond strength 10^5 N/m^2 (psi)**	
RTV-560-32	RTV-560	278 (40)	3.83, 4.91, 4.67, 4.81 (55.5, 71.3, 67.7, 69.7	Avg = 4.55 Avg = 66.0)
RTV-560-32	RTV-560	Room temperature	8.82*, 4.52*, 5.16* (128*, 65.6*, 74.9*	Avg = 6.17 Avg = 89.5)
RTV-560-32	RTV-560	325 (125)	5.56, 5.70, 6.52, 4.59 (80.6, 82.6, 94.5, 66.6	Avg = 5.59 Avg = 81.1)
RTV-560-32	HRSI Tile	Room temperature	0.731, 0.586 (10.6, 8.5	Avg = 0.659 Avg = 9.6)
RTV-560-32	Emittance coating on HRSI***	Room temperature	8.20, 2.56 (119***, 37.2	5.38 Avg = 78)

*All specimens cured in vacuum
**All values identified thus * consist of adhesive failure between RTV-560 substrate and the aluminum tensile block
***Refer to text for discussion of anomalous high value

Table 6. Cure-in-Place Densities

Material	Method of Cure	Density, lb/ft^3
RTV-560-55	Cured in air	58.9
	Cured in vacuum	50.4
RTV-560-32	Cured in vacuum	32.8

determined with a Perkin Elmer Thermal Mechanical Analyzer using a quartz expansion probe. The measurement was conducted from ambient temperature to 200°C. The results are shown in Table 7.

Table 7. Properties of RTV-560-55 and RTV-560-32

Property	RTV-560-55	RTV-560-32
Density, lb/ft^3	50	32
Tensile strength, psi	135	145
Hardness "A" shore	70	66
Adhesion to RTV-560, psi	87	89.5
Viscosity, cps		
40°F	344,000	– –
70°F	263,000	136,000
125°F	109,000	– –
Thermal conductivity, Btu/hr-ft-^{o}F	0.031	0.025
Coefficient of thermal expansion, in/in/^{o}F	1×10^{-5}	1×10^{-5}
Specific heat, Btu/lb-^{o}F	0.34	0.34
Thermal diffusivity, ft^2/hr	0.00175	0.00230

5.6 Summary of Cure-in-Place Ablator Properties

The properties of the cure-in-place ablators are summarized in Table 7.

5.7 Thermal Analyses

Thermal analyses were made to evaluate the performance of candidate thermal protection repair materials. The critiera were maximum aluminum skin temperatures of 450°K (350°F) for honeycomb structure areas and 478°K (400°F) for other locations. Body points 213 and 1702 (Figures 2 and 3) were selected to represent areas requiring thick and thin thermal protection material. The heating environments for these two body points were determined from the data furnished by NASA Johnson Space Center (JSC).

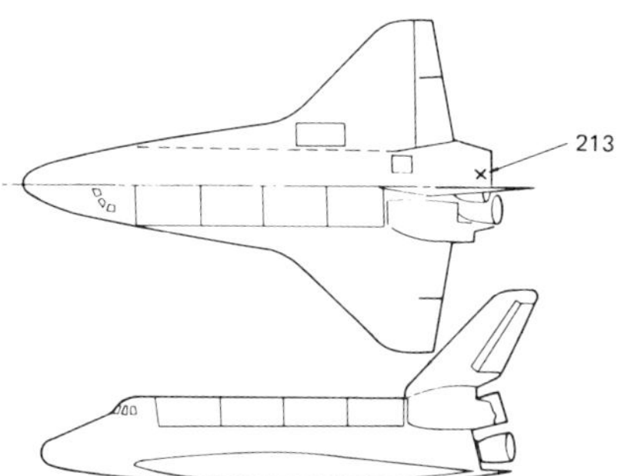

Figure 2. Body Point 213

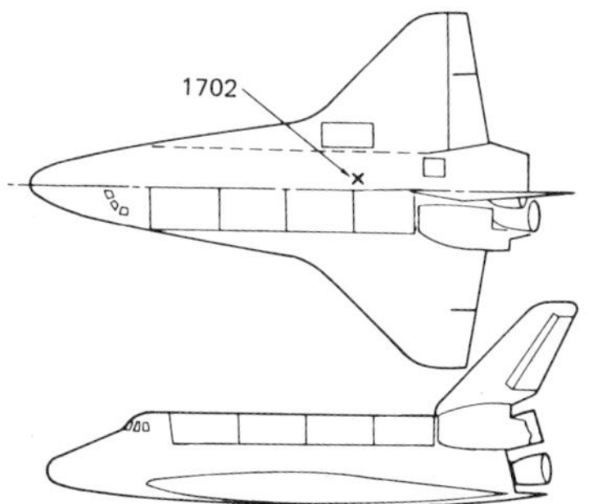

Figure 3. Body Point 1702

Cure-in-place material performance
was evaluated at body point 1702
(Figure 3). The design thermal
protection is provided by 0.0206 m
(0.81 in) of silica tile, 0.00406 m
(0.16 in) of Nomex felt, and
0.0038 m (0.015 in) of RTV adhesive,
for a total thickness of 0.0250 m
(0.985 in) over the 0.00457 m
(0.18 in) aluminum structure.

The Arrhenius decomposition rate
equation used in the CHAMP (charring
ablator) code used rate parameters
determined from TGA measurements of
specimen weight versus temperature
at measured heating rates. Due to
problems encountered with the CHAMP
code (slow convergence and an indicated significant weight loss at
only 400°F), the analysis was continued with the MDAC aerodynamic
heating code AERO HEAT. Because of
the lack of ablation data, the heat
transfer through the thermal protection material and structure was
modeled as transient conduction
using properties based on results
from the CHAMP attempts. The results
for the cure-in-place analysis are
shown in Figure 4. A linear relationship between the minimum thickness (0.25-in. recess) and full
thickness results is assumed. Note
that the aluminum temperature change
is given. Consequently, for an
initial temperature of 322°K (120°F),
which is the maximum application
temperature, the minimum thickness
of 0.01867 m (0.735 in) is sufficient for RTV-560-55 to meet the
450°K (350°F) limit for the alumi-

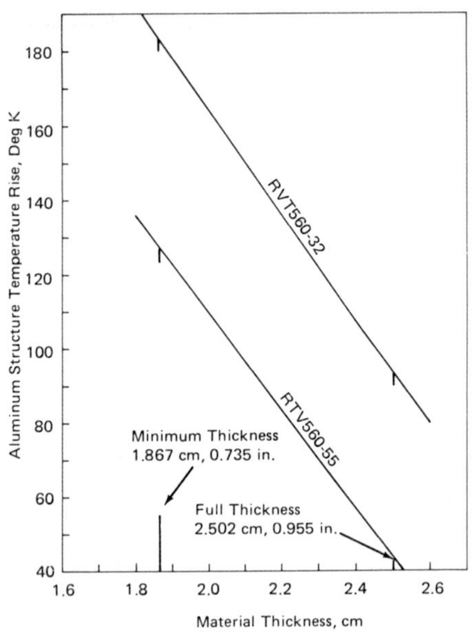

Figure 4. Body Point 1702 Cure-
in-Place Repair

num; whereas 0.02261 m (0.89 in) of
RTV-560-32 is needed to satisfy this
requirement.

6. ARC JET TESTING AND SIMULATION
Two sets of cure-in-place specimens
were submitted for NASA JSC Arc
Jet Testing. In the initial set,
specimens consisted of 0.051 m
(2 in) thick by 0.051 m (2 in) diameter samples of the ablative material contained in tubular holders
of the HRSI tile and bonded to a
substrate representative of an on-
orbit repair substrate. Thermocouples were inbedded at 0.25,
0.50, 1.0 and 2.0-in depths
(0.00635, 0.0127, 0.0254, 0.0508
m). The ablative materials in
this initial set of specimens were
cast into a cylindrical mold under
atmospheric pressure. The temperature versus depth results for the
cure-in-place ablators are given in

Table 8. A second set of arc jet specimens that were vacuum cast was prepared using RTV-560-55 and RTV-560-32. A final set of arc jet specimens was prepared to evaluate char adherence. These specimens consisted of RTV-560-55 cast at atmospheric pressure into rectangular blocks 0.15 by 0.15 m by 0.051 m (6 by 6 by 2 in).

6.2 Arc Jet Results

All ablative materials included in the arc jet testing exhibited a thermal performance in the tests sufficient to meet the mission requirements to maintain the Shuttle structure temperature below 350°F during reentry conditions, assuming that initial structure temperatures are 325°K (125°F) or less.

Char separation occurred in the 0.15 by 0.15 by 0.05 m (6 by 6 by 2 in) RTV-560-55 samples. Modification of the RTV-560 formulations by refractory fiber addition may therefore by required to maintain ablator integrity.

6.3 Arc Jet Computer Simulations

Analytical simulations of the arc jet tests of RTV-560-55 and RTV-560-32 were attempted, using the properties given in Table 7 and the AERO HEAT computer code. The arc test and simulation analysis results at 600 sec are given in Table 8. The poor agreement between test and analysis temperatures at the the 0.00635 and 0.0127 m nominal depths probably arises from inaccurate overall representation of blocking effects and thermal diffusivity changes over the decomposition temperature range by the analytical model. Comparisons for the 1-in depth indicate that the solid conductivity component was undervalued for the RTV-560 materials model.

6.4 Arc Jet Simulation Conclusions

For thick thermal protection sections, the density, specific heat, and thermal conductivity of the repair material are of secondary importance to the ability to maintain structural integrity in the entry environment. For thinner sections, the thermal diffusivity becomes increasingly important. Because the specific heat values

Table 8. Arc Test Data and Simulation Analysis Temperature Rise at 600 sec

Material	Model	Depth m($\times 10^{-2}$)	in.	Test K	F	Analysis K	F	Model	Depth m($\times 10^{-2}$)	in.	Test K	F	Analysis K	F
Purple Blend	1A	0.00	0.00	1340	2410	1449	2608	1B	0.00	0.00	1389	2500	1449	2608
		0.74	0.29	1066	1919	1262	2271		0.61	0.24	1095	1971	1297	2335
		1.45	0.57	725	1305	1032	1858		1.40	0.55	822	1480	1046	1883
		2.67	1.05	68	123	366	658		2.59	1.02	90	162	414	746
MDAC S-10	2A	0.00	0.00	1311	2360			2B	0.00	0.00	1378	2480		
		0.71	0.28	1059	1906				0.74	0.29	1078	1940		
		1.22	0.48	891	1604				1.24	0.49	876	1578		
		2.59	1.02	101	182				2.62	1.03	101	181		
RTV-560-55	3A	0.00	0.00	1367	2460	1454	2617	3B	0.00	0.00	1450	2610	1454	2617
		0.69	0.27	951	1711	1150	2070		0.66	0.26	999	1798	1164	2095
		1.37	0.54	428	770	679	1222		1.35	0.53	477	858	696	1253
		2.67	1.05	60	108	6	11		2.62	1.03	74	132	7	13
RTV-560-32	4A	0.00	0.00	1355	2440	1456	2620	4B	0.00	0.00	1444	2600	1456	2620
		0.64	0.25	980	1765	1247	2245		0.66	0.26	944	1699	1241	2234
		1.32	0.52	617	1111	942	1696		1.50	0.59	453	815	842	1516
		2.54	1.00	142	255	124	224		2.74	1.08	94	170	61	109

do not differ significantly among
the materials under consideration
and their conductivities vary little
over the range of densities, the
density primarily determines the
diffusivity. Thermal diffusivity
is inversely proportional to the
density. Because low values of
diffusivity are desired, the higher density materials give better
thermal protection, other factors
being equal.

The class of silicone-based composites included in this study having
a density of 513 kg/m^3 (32 lb/ft^3)
or more appear satisfactory with
respect to their thermal protection
properties.

7. MATERIAL APPLICATOR AND
EXTRUSION TESTS

Early design and fabrication phases
were needed to ensure completion of
the functional mockup in time to
perform tests before the delivery
date. Designs for the self-
contained and three-part applicators followed and drew upon the
mockup development experience.

7.1 Functional Mockup

The working unit needed to contain,
mix, and deliver the resin in a
reliable and efficient manner.
Because of the delivery schedule no
attempt was made to optimize weight
or develop detailed mechanisms
(such as a trigger valve control).
The functional mockup is shown in
Figure 5.

The design selected has an aluminum
containment body with a CO_2 cylinder
pressure source. The design used a

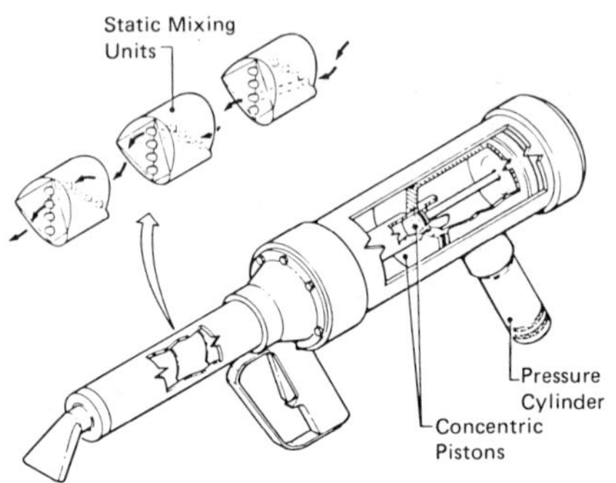

Figure 5. Functional Mockup

thin aluminum burst disk to prevent
contact of the catalyst and resin
prior to activation. The mix ratio
of resin/catalyst is controlled by
the use of concentric pistons, which
are sized to a 9.59/1 area ratio.
The functional mockup would deliver
72 to 80 in^3 ablator material.
Three standard 2-in-dia static mix-
heads (Charles Ross & Son Co.),
were used in the nozzle extension
tube and provided adequate mixing.

7.1.1 Review Requirements. The
functional mockup and the designs
proposed for flight applications
must meet the following requirements:

- Reliable mixing and dispensing.
- One-hand operation (for appli-
  cation).
- Minimum operator fatigue.
- Ease of assembly.
- Good visibility of repair area.
- Safe to handle.

The repair concepts for cure-in-
place and precured materials are
shown in Figure 6. These concepts
involve using cure-in-place material for light damage and using

756

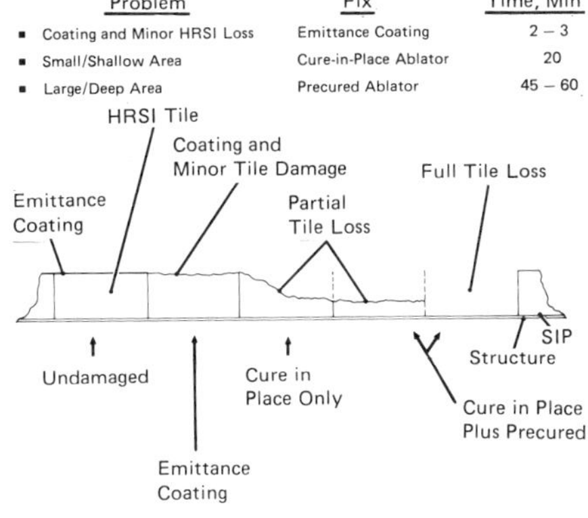

Figure 6. Repair Concepts precured tiles to fill heavily damaged areas.

The bridged, concentric pistons approach was selected because it was judged simple and reliable.

7.2 Hardware/Test

The S. L. Morse Company fabricated the mockup, using aluminum alloy. After proof testing, the gun was filled with RTV-560-55, and the extrusion rate was measured as a function of pressure. The extrusion rate varied in a nearly linear manner with pressure.

7.2.1 Demonstration.

The functional mockup was demonstrated at the NASA facility in Houston, Texas, on November 30, 1979. Three samples were expelled, a sample tile cavity was filled, and an adhesive bed was placed for precured tile. Prior to adjournment, the material had hardened to a satisfactory degree, with very little marbling.

7.3 Self-Contained Applicators and Three-Part Applicators

Several applicator concepts were studied and evaluated to the NASA

requirements and two were selected:

- A self-contained applicator shown in Figure 7.
- A three-part applicator design shown in Figures 8 and 9.

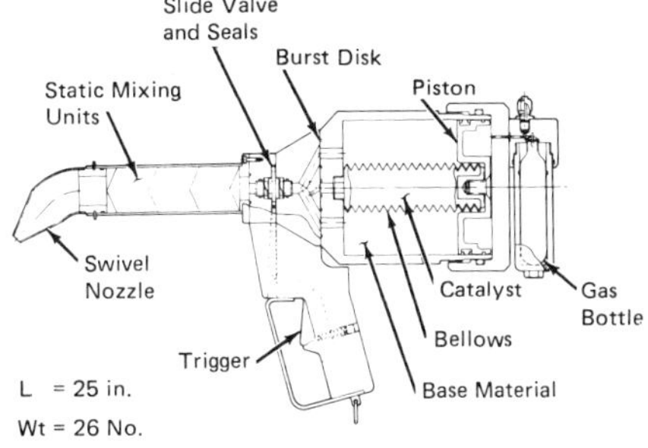

Figure 7. Self-Contained Applicator (Bellows)

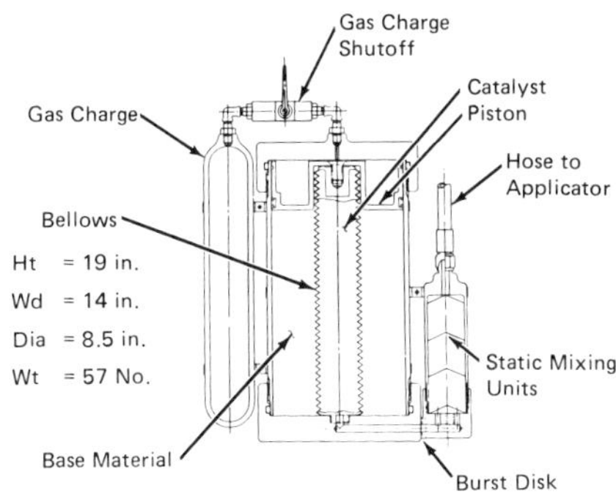

Figure 8. 3 Part Applicator Bellows Reservoir

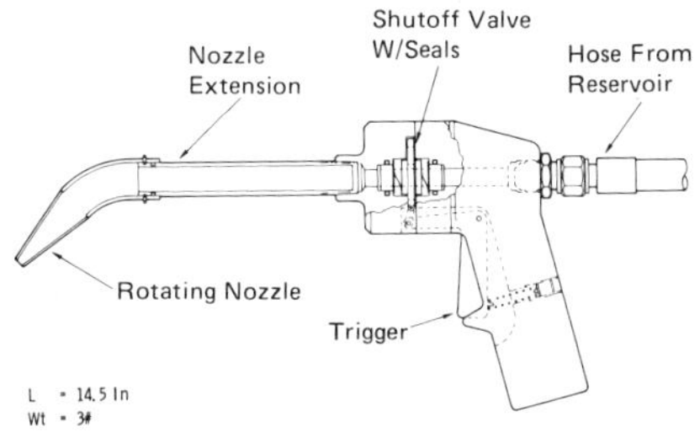

Figure 9. 3 Part Applicator

Both material applicators operate on the same concept, using gas pressure to expel the material and static mixing heads to mix the

catalyst and resin. The basic difference between applicator concepts
is in the containment of the material.

8. CONCLUSIONS

1. Cure-in-place ablators were
developed, tested, and analyzed and
were found to possess the required
thermal properties.
2. The cure-in-place ablators were
formulated with a combination of
curing agents to meet NASA cure
rate requirements.
3. The adhesion of the cure-in-
place ablators to the Orbiter TPS
materials easily met the 40 psi
requirement.
4. The cure-in-place ablators were
found to dispense from the functional mockup at low pressure (<200
psi) with efficient mixing of the
catalyst and resin.
5. The functional mockup was successfully tested in the laboratory
and demonstrated during the final
presentation at NASA JSC. The dispensed material cured to the
required hardness within 5 hours.

9. REFERENCES

1. MDAC G8261 Shuttle Orbiter
TPS Flight Repair Kit Development -
Final Report. Prepared under
Contract No. NAS9-15971 and modi-
fication to Contract No. CCA No. 1
by McDonnell Douglas Astronautics
Co., for NASA JSC, December 1979.
2. ASTMD412-75 Standard Test
Methods for Rubber Properties in
Tension.
3. ASTMD429-73 Standard Test
Methods for Rubber Property -
Adhesion to Rigid Substrates.

10. ACKNOWLEDGEMENT

The authors acknowledge the assistance of the MDAC personnel in
Mechanical Design, Structural Analysis, and Thermal Analysis who significantly contributed to this program. We also acknowledge the
direction of Dr. L. Leger, NASA
JSC.

11. BIOGRAPHIES

C. E. Blevins is an Associate Engineer in the Plastics and Elastomers
Groups at McDonnell Douglas
Astronautics Company-Huntington
Beach. He received his B.S. degree
from the University of Kentucky in
1975 and his M.S. Ch.E. degree
from the University of Kansas in
1979. Before joining MDAC-HB, he
did research in tertiary oil recovery at the University of Kansas. He
also wrote catalytic combustor
control programs and viscoelastic
data reduction programs for
Wright-Patterson AFB, Ohio.
Rodolfo Gonzalez is a Materials
Research and Development Specialist
at the McDonnell Douglas
Astronautics Company - Huntington
Beach, where he has been employed
since 1952. He is Unit Chief of
the Plastics and Elastomers Groups.
He is responsible to design and
manufacturing departments for
nonmetallic materials technology in
support of major company programs
and for internal research activites.
Mr. Gonzalez has been instrumental

in the development and application
of structural plastics, acoustical
and thermal insulation, and elastomeric components in the Company's
space vehicle, missile, and aircraft
projects. He was graduated from
the University of Texas at El Paso
in 1952.

EFFECT OF FILLER AND FABRIC ON THE THERMAL DURABILITY OF A 350°F SERVICE EPOXY ADHESIVE

Andrew C. deLeon
Daniel L. Paradis
Alex D. Alderman
Hysol Division/The Dexter Corporation
P. O. Box 312, Pittsburg, CA 94565

Abstract

Support fabric and filler have considerable effect on the performance of 350°F service structural epoxy adhesives. Inclusion of fabric support may lead to five-fold improvement in peel performance. Large differences in thermal durability due to fabric have been found.

Filler incorporation brings about dramatic changes on these adhesive systems. In contrast to state-of-the-art 180°F service systems, these filled systems often out-perform non-filled systems. In particular, peel strength can be increased two-fold in comparison to unfilled systems. Many factors influence performance of fillers; among these are generic type, particle size, and particular geometry.

1. INTRODUCTION

First generation 350°F service epoxy adhesives were known for their brittleness. Fabric support were used mainly to give integrity to the tape and to facilitate handling more than for improvement of physical properties.

Second generation 350°F service epoxy adhesives, i.e, toughened, rely upon fabric and fillers for improvements in physical properties. This is in contrast to conventional 180°F service epoxy adhesives whose physical properties are usually degraded by inclusion of filler or fabric.

The purpose of this study was to determine the effect of various fillers and fabrics on mechanical and thermal properties of a second generation epoxy adhesive.

2. MATERIAL SELECTION

2.1 Fabric Support

Two widely used support fabrics, glass and nylon tricot(6-6 nylon) and unsupported film were chosen for evaluation.

2.2 Fillers

As herein used, a filler is defined as a solid material used at a reasonably large level. Fillers have mainly been considered as extenders and not primarily used to improve intrinsic adhesive properties.

Most fillers are ground minerals. As such they contain various size particles. Within a particular generic type, fillers can be characterized by particle shape, particle size distribution, and packing density.

Average particle size and a measure of polydispersity are necessary for a description of particle size distribution. Different particle sizes of a particular filler can be compared only if the degree of polydispersity is approximately the same. The packing density is primarily a function of particle size distribution (mixture of particle sizes pack tighter than monodisperse particle sizes).

Filler chemical composition is a primary consideration for all epoxy systems. Reactivity due to chemical composition, contaminant, or surface effects are possible. The mechanisms are not always straight-forward and care must be taken to exclude fillers which lead to undesirable reactions with resins or other modifiers commonly used in epoxy formulations.

Fillers increase the viscosity of epoxy formulations: regularly shaped particles the least, fibrous the most. Smaller particle sizes of the same generic type generally increase the viscosity more than larger particle sizes due to greater surface area.

The thermal effects of a filler upon a typical epoxy formulation are usually considered as follows: thermal conductivity, specific heat, coefficient of thermal expansion, and thermochemical.

2.3 Materials

The following fillers were selected for evaluation:

    (1) Fused Silica
    (2) Precipitated Silica Gel
    (3) Amorphous Silica
    (4) Microcrystalline Silica
    (5) Clay
    (6) Calcium Carbonate
    (7) Mica
    (8) Barium Titanate
    (9) Zirconium Silicate
   (10) Powdered Aluminum
   (11) Tabular Alumina
   (12) Microtalc

Several of the silicates tested were silane treated.

3. EXPERIMENTAL

3.1 Basic Formulation

A state-of-the-art epoxy formulation was standardly used. Filler was added to resin until viscosity reached the upper working limit of this formulation. Asbestos was used as a thixotrope to insure proper flow control in honeycomb structures.

3.2 Preparation of Test Specimen

Film adhesives were prepared with conventional hot melt equipment.

Unsupported tapes .06 psf were reticulated onto 3/8" honeycomb core by use of an air knife. This method, widely used in aerospace honeycomb bonding technology, results in the adhesive positioned on the edge of the core. In this geometry the core may be bonded to perforated aluminum skins with two advantages: first, the adhesive is present only at the bonding site (fillet) and second, most of the perforated holes are not plugged with adhesive. This allows environmental interchange of the bonded structure and increased strength.

Supported tapes .10 psf were handled in standard fashion. A state-of-the-art formulation incorporating powdered aluminum filler was used to prepare nylon supported and glass supported tapes.

Lap shear and honeycomb test specimen were prepared to MMM-A-132 or MIL-A25463 specifications. Adhesives were autoclave cured by heating at 3°F/minute up to 350°F and holding for 60 minutes at 350°F. A positive pressure of 30 psi was applied during entire cure cycle. A vacuum of 25-30 inches Hg was applied to over-pressure. Bags were then vented.

4. RESULTS AND CONCLUSIONS

4.1 Effect of Fabric Support

Initial adhesive tapes were prepared to study the following variables:

a) Unsupported and supported tapes - nylon and glass fabric

b) Inclusion of aluminum filler

Both these variations are typical of commercially available products.

Baseline data (shown in Table 1) developed included flatwise tensile, honeycomb climbing drum peel (HCCD), and tensile lap shear at ambient temperature and 300°F. Aluminum filler improved both the room temperature lap shear and flatwise results. The flatwise averaged 850 psi for the filled tapes versus 670 psi for the unfilled tapes. Lap Shear increased from 2800 to 3700 psi. The 300°F results were relatively constant.

The honeycomb peel strength was improved by both fabrics and also improved by aluminum filler. Values ranged from 4.6 in-lb/3 inch for the unfilled, unsupported tape to 30 in-lb/3 in for the aluminum filled/nylon supported tape. The nylon fabric yielded approximately twice the peel strength as the glass fabric for both the filled and unfilled variations. By visual inspection, both the filled and unfilled systems failed adhesively from the glass fabric.

This type of testing is typical of that required of 350°F service adhesives. An important consideration of these systems is their thermal durability. Since this requires many thousands of hours of aging, its use as a standard quality control test is ruled out. Model adhesive systems were prepared to test the influence of the fabric on the thermal durability of this adhesive system.

Large differences in the thermal stability of the model adhesive system due to the fabric support have been found. As shown in Figure 1, the nylon supported tape retained only 25-30% of its original strength after 10,000 hours at 350°F. The unsupported and glass supported tapes were virtually unchanged during this time period.

Another consideration in the determination of thermal stability is the test temperature. The generally accepted theory for reduced strength after thermal aging is that polymer degradation suppresses the glass transition (Tg) temperature of the adhesive system. If the adhesive is tested at a temperature well below its Tg, very little loss in strength might be found. High temperature is more sensitive for measuring this property.

4.2 Effect of Fillers

The silica fillers examined can be classified into several subsets: precipitated, fused, amorphous, and microcrystalline. Because of the size of this group, they were considered separately. Although silicas are all chemically silicon dioxide, different generic classifications affected adhesive properties differently.

The addition of various precipitated silicas increased the viscosity of the resin system at a very high rate. System viscosities became unworkable at 20 parts per hundred parts resin level. This is a much lower level than other silica fillers and may be due to the large surface area or to an undemonstrated chemical interaction with the resin (these fillers have an acidic surface). The other silica fillers were formulated to approximately 120 parts filler to 100 parts resin.

A word of caution about the "amorphous" silicas. True amorphous silicas consist of the fumed or pyrogenic variety commonly used as thixotropes, precipitated silica and silica gel. All these silicas are produced by chemical reaction. These particular "amorphous" silicas (this is a manufacturer's term) are more closely associated with microcrystalline silicas. Their particle sizes are finer than typical microcrystalline silicas so that crystalline structures are not easily seen.

The baseline data of the silica fillers is presented in Table 2.

This testing was performed at ambient temperature to screen the various filler systems for later high temperature aging. The particle size distributions of these fillers were fairly symmetrical (equal amount of dispersity). The combination of tensile shear and peel results yielded some interesting results. To generalize, both 250°F and 350°F epoxy formulations are toughened at the expense of lowered shear values. Second generation technology allows us to toughen a system without degrading shear values. However, to increase both is difficult. The microcrystalline silica fillers impart a great deal more toughness and a moderate increase in shear strength over the unfilled formulation.

There are two unique features to microcrystalline silica. First, both peel and shear properties optimize at approximately 6.5 micron particle size; second, the use of silane coupling agent improved baseline properties by 10% across the entire range of particle sizes.

The improvement with silane coupling agents has been noted previously. The generally accepted mechanism is that the coupling agent improves wetting and therefore promotes adhesion. However, it has been postulated that coupling agents interact with metal oxide films to promote adhesion.

Baseline data for other fillers is presented in Table 3. Aluminum is deserving of special attention and will be treated separately. The tensile shear strength of all these filled systems was lower than the unfilled formulation. The HCCD peel was not significantly affected except by calcium carbonate which increased from 4.6 for the unfilled to 10 in-lb/3 in.

Note the different amounts of each filler which was used. The clay, talc, and mica fillers imparted higher viscosities than typical silicas, titanates, aluminum and calcium carbonate and could not be loaded as highly.

Contrary to these fillers, aluminum improves both tensile shear and honeycomb peel properties. The peel was increased threefold. Powdered aluminum has long been used in epoxy formulations to impart toughness. It has been postulated that this has been due to the better thermal expansion match and the resultant lower internal strain.

The best baseline fillers were chosen for further investigation. Flatwise tensile specimen were prepared using performated aluminum skins. The perforated skins allowed for environmental interchange so that oxidation would not be limited by design. Table 4 presents the rate of strength loss at an aging temperature of 420°F.

In general all the filled systems had a greater strength retention than the unfilled system. The lower values with the 3.6 μ micro-crystalline silica may be due to poor reticulation. Note that the aluminum filled system retained the largest amount of its original strength after 2500 hours at 420°F.

This resin system was designed for intermittant exposure up to 420°F. We went beyond this and aged flatwise specimen at 500°F. By conventional kinetic calculations, this should increase the rate of oxidation by approximately one order of magnitude. As with the 420°F aging, the flatwise tensile strength was determined at 300°F. This results again confirm that aluminum filler retains a larger amount of its original strength (34%) than the other filled and unfilled systems. The silica fillers behaved similarly to the unfilled system retaining 5-10% of their original strength.

This action might well be physical in nature rather than chemical. It is believed that the higher temperature induces stress cracking or crazing in all these systems but aluminum filler, because of its closer thermal expansion match, relieves these forces more than the other systems.

It is recognized that this testing is harsh and might not be encountered in the intended use of the adhesive system. It was not attempted to correlate this data with other thermal durability tests. In particular, tensile shear (or any other overlapped specimen) results would not be expected to correlate with this data. The exact testing geometry should be carefully studied by material engineers so that the results truly measure the adhesive's expected environment.

5. SUMMARY

Second generation 350°F epoxy adhesives rely on fillers and fabric for improvement of their baseline adhesive properties. The choice of such fabrics and fillers must be tempered by their effects upon the thermal durability of these adhesive systems.

Nylon fabric increases the peel strength of the model adhesive system by a two or three fold magnitude. However, thermal durability of these systems are not as good as glass supported adhesives.

Fillers can often increase baseline material properties and thermal durability properties. In particular aluminum powder has been found to both toughen this brittle 350°F system and improve its thermal durability.

6. REFERENCES

1) "Epoxy Resins, Chemistry and Technology", C. A. May and Y. Tanaka, 1973

2) "Handbook of Fillers and
   Reinforcements for Plastics",
   H. S. Katz and J. V. Milewski,
   1978

3) "Treatise on Adhesion and
   Adhesives", R. L. Patrick,
   1967

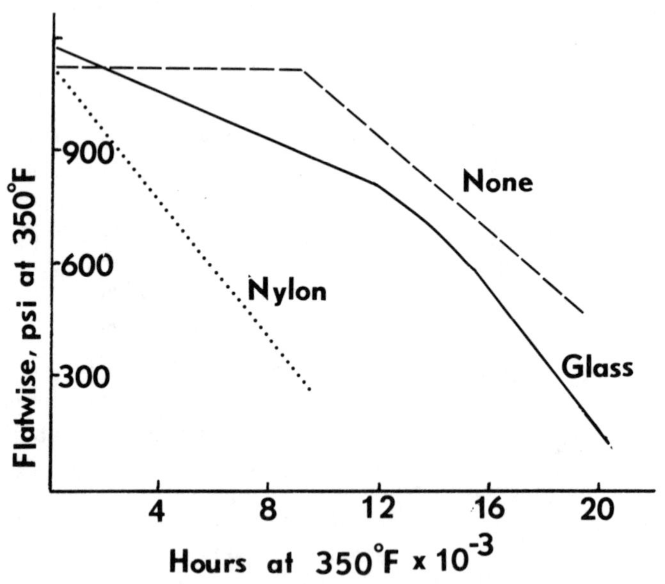

FIGURE 1

EFFECT OF FABRIC ON THERMAL DURABILITY

TABLE 1 - BASELINE PROPERTIES OF 350°F SERVICE TAPES [a]

Carrier	Nylon	Glass	None	Nylon	Glass	None
Aluminum Filler	Yes	Yes	Yes	No	No	No
Flatwise Tensile psi @ RT	870	804	885	658	619	724
Flatwise Tensile psi @ 300°F	685	604	723	632	632	691
Tensile Lap Shear psi @ RT	3510	4010	3400	2600	2640	3180
Tensile Lap Shear psi @ 300°F	3430	3920	3350	3450	3800	4110
HCCD -3/8" Cell in-lb/3 in @ RT	30.2	9.1	13.0	17.5	8.3	4.6

TABLE 2 - BASELINE DATA OF SILICA FILLERS [a]

Type	Average Particle Size, μ	Tensile Shear @ RT	HCCD In-lb/3 Inch Width @ RT
Fused	2.0	3290	4.5
Fused High Purity	6.5	2730	9.5
Fused Trace of Carbon	6.5	2600	12.0
Silica Gel	4.0	4170	9.2
"Amorphous"	2.0	2680	9.0
"Amorphous"	2.8	2380	7.3
"Amorphous" + Silane	2.8	2420	8.7
Microcrystalline	3.6	2730	8.0
Microcrystalline + Silane	3.6	3390	9.3
Microcrystalline	6.5	3370	10.7
Microcrystalline + Silane	6.5	3720	11.2
Microcrystalline	11.5	2420	4.7
Microcrystalline + Silane	11.5	2610	6.0
Microcrystalline	>12.0	2490	4.8
Unfilled		3180	4.6

a) Per MMM-A-132 or Mil-A-25463

TABLE 3 - BASELINE DATA OF FILLERS

Type	Average Particle Size, μ	Tensile Shear @ RT	HCCD In-lb/3 Inch Width @ RT
Calcined Clay	2.0	2300	4.7
Attapulgus Clay	2.9	2050	2.7
Attapulgus Clay	5.3	1980	3.0
Attapulgus Clay	18.0	2340	3.8
Powdered Aluminum	17.0	3400	13.0
Mica	< 2.0	1660	4.5
Zirconium Silicate	3.9	2980	5.0
Tabular Alumina		3030	6.8
Talc	Platy,> 10.0	2420	3.0
Barium Titanate	2.0	2850	5.3
Barium Titanate	1.4	2640	5.3
Calcium Carbonate	3.2	2220	8.5
Calcium Carbonate	17.2	2550	10.0
Precip. CA CO_3	5.7	2780	5.7
Unfilled		3180	4.6

TABLE 3A - FILLER CONCENTRATION

Filler	Parts Filler/100 Parts Resin
Fused Silica	108
Precipitated Silica Gel	14
"Amorphous" Silica	108
Microcrystalline	130
Aluminum Powder	120
Microtalc	50
Mica	66
Clay	65
Barium Titanate	108
Calcium Carbonate	108

TABLE 4 - 420°F AGING
Flatwise Tested at 300°F, psi

Filler	0 Hrs	680 Hrs	1000 Hrs	2500 Hrs	% Retained
Fused Silica 6.5µ	712	753	690	522	73
Silica Gel 4.0µ	631	678	632	-	-
"Amorphous" 2.8µ	756	686	632	557	74
Microcrystalline + silane 6.5µ	635	656	662	422	66
Microcrystalline 6.5µ	621	547	515	344	55
Microcrystalline 3.6µ	422	411	375	174	41
Unfilled	651	548	297	312	48
Aluminum Powder	770	813	734	642	83

(Column header: Time at 420°F)

TABLE 5 - 500°F AGING
Flatwise Tested at 300°F, psi

Filler	0 Hrs	680 Hrs	1000 Hrs	2500 Hrs	% Retained
Fused Silica 6.5µ	749	401	283	51	7
Silica Gel 4.0µ	760	362	309	64	8
"Amorphous" 2.8µ	831	398	366	47	6
Microcrystalline + silane 6.5µ	784	387	326	50	6
Microcrystalline 6.5µ	735	357	329	29	4
Microcrystalline 3.6µ	530	371	224	22	4
Unfilled	723	349	273	65	9
Aluminum Powder	770	410	414	258	34

(Column header: Time at 420°F)

TECHNIQUES OF MEASURING
THE INTERFACIAL AGING MECHANISMS
IN METAL BONDS

Dr. Hansgeorg Kollek and Dr. Walter Brockmann
Fraunhofer-Institut für angewandte
Materialforschung (IFAM), Bremen, Germany

Abstract

Measurements of adhesion and its aging mechanisms by mechanical, physical and chemical tests were compared. Especially the wedge test seems to give no good results on testing quality of adhesion.

Another problem in mechanical testing of adhesion is, that a real failure of adhesion doesn't occur, but a failure in the adhesive near the metal surface. This is pointed out on hand of photometry and scanning electron microscopy.

I. INTRODUCTION

In highly loaded adhesive bonds under humid environmental conditions delamination may be observed. Responsible for this is water imigrating into the glue line. It will not only change the mechanical properties of the adhesive itself, but also decrease adhesion. So lifetime of bonded structures often depends only on the velocity of destruction of adhesion.

The mechanisms of aging are closely connected to the mechanisms of the reactions of the adhesive with the metal surface.

For the description of adhesion a combined theory of chemical reactions and micro-mechanical linking is discussed .[1-3] To proof this, techniques for measuring adhesion have to be checked for their evidence.

2. FAILURE OF ADHESION IN MECHANICAL TESTS

Several mechanical tests for measuring adhesion are known. Combining them with an aging procedure, e.g. storing in a hot and humid environment, should give information to the decrease of adhesion forces.

In mechanical testing of adhesion a crack propagates at the interface between adhesive and oxide layer of the metal surface. This can be realized macroscopically in most peel-tests using a not optimal surface preparation of the metal. But it has been shown previously, using a C^{14}-labelled adhesive, that in the case of adhesional failure mostly there is still some adhesive remaining, fig. 1.[4]

So there exists not really a failure of adhesion, because the crack propagates in the adhesive or the primer near the metal surface.

For other specimens also failure in the oxide layer occurred.[5]

Even in lap shear specimens with a macroscopically absolute failure of adhesion, small dots of adhesive are still on the metal surface.

In this case the determination of the adhesive on the metal was made by microscope photometry. Adhesives have a high absorption coefficient for light in the ultraviolet range at a wavelength of 270 nm. At this wavelength the reflection of the metal surface was determined within an area for measuring of 10 x 10 μm. Using a scanning stage measurements were made over a whole area of 200 x 200 μm² or 20 x 20 of the above mentioned area of measuring. The values of reflection in extinction units were plotted by a computer in form of a map containing lines of equal extinction. The lines below the average of all values are dotted and the lines above the average are solid.

In fig. 2 a computed map of a titanium surface from a lap shear specimen is given. Although macroscopically there was an absolute failure of adhesion, some dots of adhesive could be detected with the microscope photometer. These dots were marked by black colour in fig. 2.

In trying to obtain a real adhesional failure in lap shear specimens the CSA-etching (tab. 1) process with an etching time of 3 min was used for 2024 clad aluminium alloy. This surface preparation with its short etching time should give a not well adhering surface. The surface was scanned partially with the microscope photometer before bonding. After bonding, aging and destroying the lap shear specimen the same area was scanned again. The area of measurement was in the area of failure of adhesion, as it is shown in fig. 3. The microscope-photometry showed a film of adhesive of primer covering the whole area of "adhesional failure". The amount of adhesive is at about 10 - 20 μg/cm². But the lap shear specimen is especially in combination with high strength adhesives an unsufficient tool for evaluating the adhesional behaviour in short time aging tests. So other test methods like peel- and fracture mechanic specimens have been developed, from which the so called wedge test [6] is a simple version easy to handle.

To learn more about the capability of this test method wedge specimens from 2024 clad and bare alloy were prepared with the adhesive/primer system FM 73/ BR 127. Surface preparation was made by CSA-etching over various times. This process is, except etching time, quite similar to the optimized FPL-process, as shown in tab. 1.

Times chosen for the CSA-process were 1, 2, 3, 4, 5, 6, 7, 10, 15, 20, 25, and 30 min. The CSA-process over 10 minutes leads in micromorphology to similar structures like FPL-etching. [7] Fig. 4 shows some of these wedge-specimens of 2024 clad alloy. After an etching time of more than four minutes no failure of adhesion will occur. In the case of 2024 bare alloy there is no failure of adhesion even after shorter etching times, fig. 5. For these specimens aging time was 30 days at 40 °C and 95 % rel. humidity although normal-ly an aging time of only two days is recommended. [8] On the other hand it is known, that surfaces etched over less than 25 min may lead to metal bonds of less durability. From this point of view the wedge test seems to be only a test of adhesional properties even when small durability of adhesion is required. And at least photometric measurements showed, that also in this test there is no real failure of adhesion, too.

For measuring adhesion in the unaged and aged state, when exactly a fracture between adhesive and metal surface or oxide layer is required, non of the commonly used mechanical tests is satisfactory. They all test strength and durability of the adhesive, sometimes near to the zone of adhesion or strength of the oxide layer. This is no reason for neglecting these techniques as practical tests for comparing adhesional properties. But they give only indirect information to adhesion and its durability itself.

3. SOME CHEMICAL ASPECTS OF
   ADHESION

The observations described above do not affect the theory of adhesion. While most of the mechanical tests are orientated at the failure of adhesion only on the metal and its oxide layers, they neglect the fracture surface of adhesive. When an adhesive is cured at elevated temperatures, curing will start at the metal surface. The energy passes through the metal and reaches the centre of glue line at last. So temperature of activation will be reached first in the adhesion near the metal surface. This adhesive has been chemisorbed by the oxide layer. Chemisorption was detected by remission photometry [9] and by microscope photometry using a phenolic resin. The major compounds in a phenolic resin are given in fig. 6. Especially the substituted phenols are able to form with aluminium chelate-complexes of relative high stability against hydrolysis. This may be a reason for the high stability of adhesion, when phenols are used as adhesives or in primers.

On anodized aluminium surfaces, on which today the best results in aging stability or adhesion is obtained, the reaction of the resin with the oxide is detectable by UV-photometry. The amount of chemisorbed resin was roughly calculated to 10 to 20 µg/cm².

On etched surfaces the amount is about ten times lower than on anodized and a chemical reaction was not clearly detectable by photometry.

Another fact, detectable by photometry is the inhomogeneity of the technical metal surface. Its influence on adhesion and strength of a metal bond is still unknown.

Fig. 7 shows the UV-images of a CSA-etched surface before and after adsorbing the resin, measured as mentioned above. The difference of both is given in the lower part. This map represents the distribution of the adsorbed resin on the surface. It is easy to be seen, that this adsorption is not homogeneous. There are dots of high amounts of adsorbate, which may become after curing and destroying adhesion the dots of remaining adhesive.

This inhomogeneity is connected to the surface pretreatment of the metal, giving a certain rate of polymerisation to the adhesive. This leads at last to the effect, that quality of adhesion can be measured mechanically, even when fracture occurs in the adhesive near the boundary layer. On the other hand micromorphology of the metal surface will influence the orientation of the

adsorbed molecules and therefore may be some steric effects
on the polymerisation.

While in the foreground of
these investigations stood
only the metal surface and its
reactions with the adhesives,
further investigations were
done to get more information
to the structure of adhesive
in the glue line.

Fig. 8 shows two cross-sections through a glue line
made with cold curing epoxi-
resin. On the left the distribution of the amine hardener
is indicated by the silver
distribution. Silver ions react chemically with diamine-
groups and silver enriched
areas represent amine-rich
parts of the adhesive. This
image gives no information to
the boundary layer because of
its low magnification.

At a higher magnification,
fig. 9, there seems to be no
good contact between the metal
in the lower part of the fig.
and the adhesive in the upper
part. This may be caused by
the technique of preparation,
in which water and alcohol
was used. These fluids may
have washed out parts of the
adhesive.

Because a well cured epoxi-
resin should not be soluble in

water or alcohol, this effect
could be explained by an incomplete curing of the adhesive near the metal surface.
Using a hot curing epoxi-resin,
and using aluminium in the
ground state to be bonded such
effect may occur again. Fig.10
shows the image of the adhesive surface after etching
away the aluminium parts in a
dilute alcaline aqueous solution.

This porous structure was observed also by using a cyano-
acrylate. There seems to be a
dependence of the porosity to
the cleanliness of the surface.
Degreasing gives a high porosity, grinding a lower one,
and a well done etching or
anodizing process gives no
porosity.

The pores again may occur by
the preparation process, but
the partial solubility of adhesive gives hints to an incomplete cured system near the
boundary layer. The porous
structures were not observed
in bulky adhesives.

3. CONCLUSIONS

In trying to interpret the
experimental results mentioned
above there is an important
fact, which has to be pointed
out: Adhesion in metal bonding
by adhesives and its aging

depends on the environmental
conditions, the properties of
the metal surface and on the
adhesive or primer.

The metal surface will alter
the grad of polymerisation of
the adhesive near to it. So
in most cases of so called
"adhesional failures" not the
adhesion itself fails, but
the adhesive near the adhesional zone. This gives some restrictions to mechanical testing of adhesion, especially
in using the wedge test.

5. REFERENCES

(1) Brockmann, W., O.-D. Hennemann and H. Kollek: Reakti-
vität und Morphologie von Me-
talloberflächen als Basis für
ein Modell der Adhäsion.
farbe + lack 86, p. 42o/25
(198o)

(2) Brockmann, W. and H.Kollek:
Possibilities of Increasing
The Stability of Adhesion In
Metal Bonds, Proc. of 23rd
Nat. SAMPE Symp., p. 1119/
1125 (1978)

(3) Venables, J.D. et al.:
Oxide Morphologies On Aluminium Prepared For Adhesive
Bonding. Appl. Surf. Sci. 3
(1979), p. 88/98

(4) Brockmann, W.: Über Haft-
vorgänge beim Metallkleben.
Adhäsion (1969) p. 335/42,
p. 448/6o (197o) p. 52/66,
p. 25o/52

(5) Bethune, A.W.: Die Bestän-
digkeit geklebter Aluminium-
Konstruktionen. Adhäsion (1976)
p. 347/51 and (1977) p. 25/28

(6) Bethune, A.W.: Durability
of Bonded Aluminium Structure,
SAMPE-Journal, July-September
1975

(7) Hennemann, O.-D.: A Compa-
rison of Anodize Processes On
The Durability And Strength Of
Adhesively Bonded Aluminium,
Proc. of 12. ICAS Conf.,
Munich 198o, p. 723/728

(8) Thrall, E.W.: Primary Bonded Structure Technology
(PABST) Douglas Paper 6539

(9) Brockmann, W. and H.Kollek:
Haftprobleme beim Metallkleben.
defazet (1978), p. 166/7o

	CSA	FPL
$Na_2Cr_2O_7$	8o g/l	88 g/l
H_2SO_4	28o g/l	3o5 g/l
Al	3,o g/l	9 g/l
Cu	o,3 g/l	*
Temperature	6o °C	65 °C
Etching time	3o min	1o min

Tab. 1: Comparison of CSA and
FPL processes
(* not specified, but
comparible to CSA)

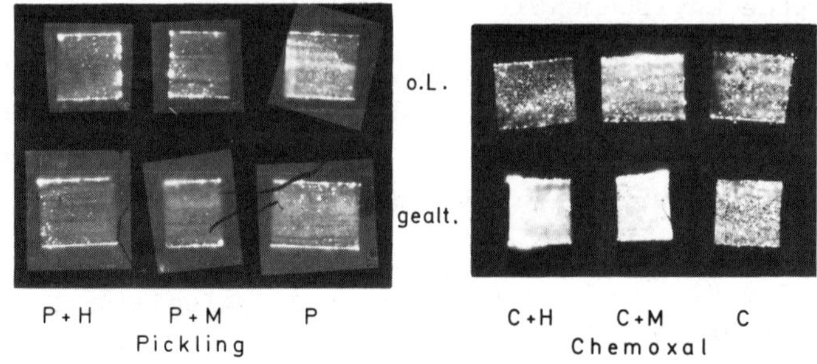

Fig. 1: Autoradiographs of aluminium surfaces after peel test. The light dots are the still remaining adhesive.

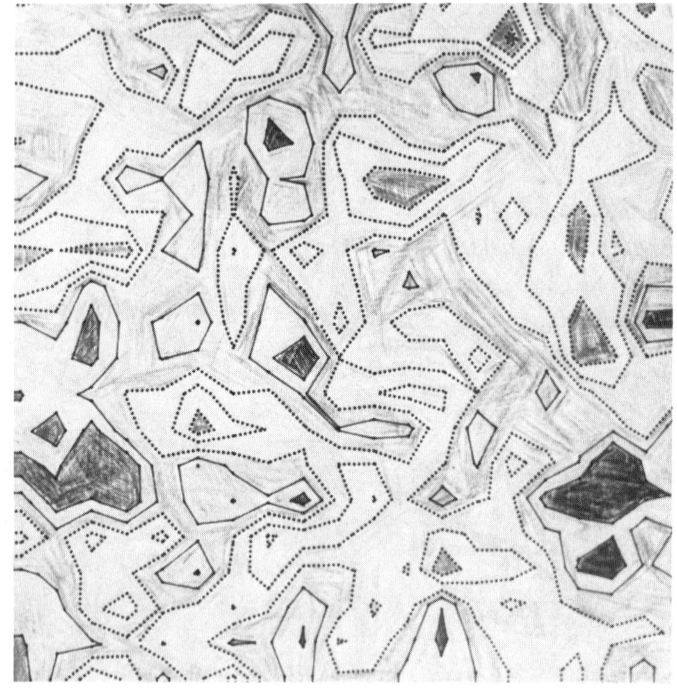

Fig. 2: Computed image of the UV-reflection of titanium surface

(Area: 2oo x 2oo μm^2)

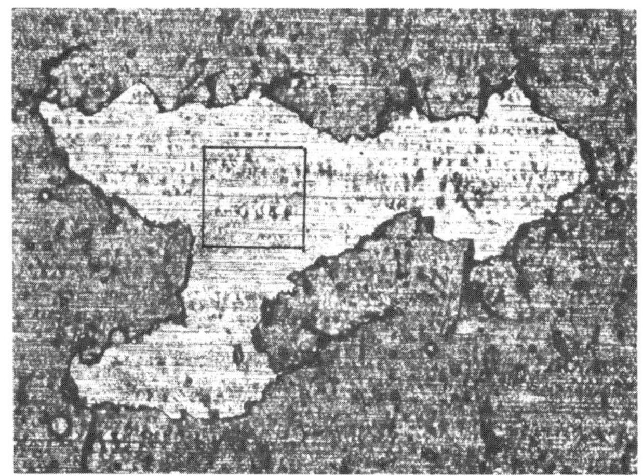

Fig. 3: Area of a "pseudo" failure of adhesion. (Marked area: 2oo x 2oo μm²)

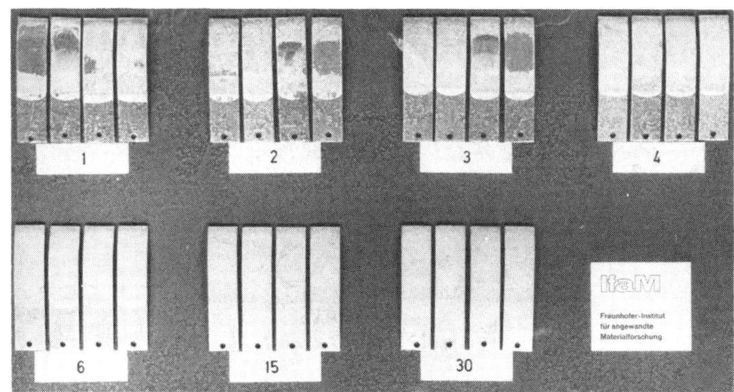

Fig. 4: Fracture surfaces of wedge specimens from 2o24 clad alloy. Figures mark the various etching times in minutes.

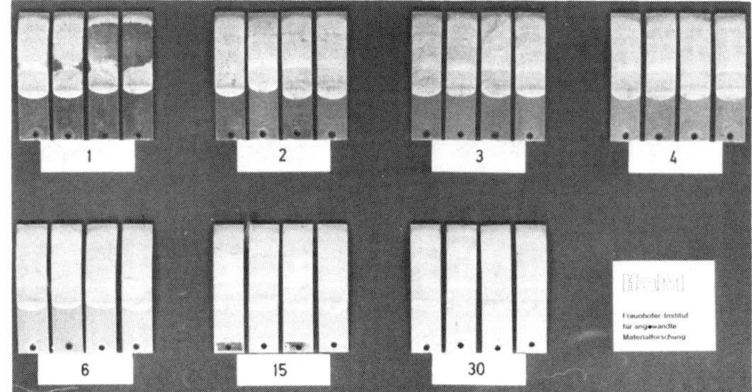

Fig. 5: Fracture surfaces of wedge specimens from 2o24 bare alloy. Figures mark the various etching times in minutes.

Phenol | 2-Hydroxibenzylalcohol | Bis-(2-hydroxphenyl)-methane

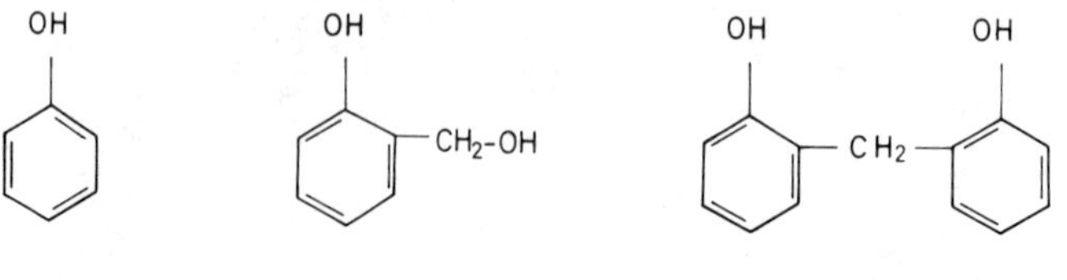

Fig. 6: Major compounds in a phenolic resin

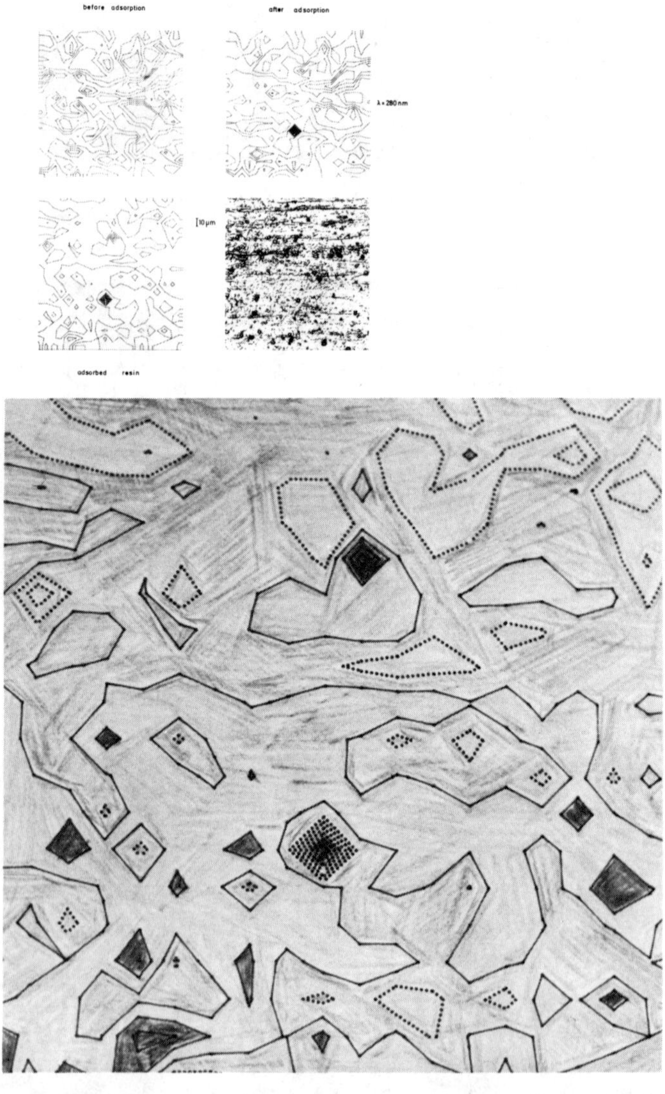

Fig. 7: Computed UV-images of CSA-etched 2o24 alloy surface before and after adsorption of a phenolic resin with inhomogeneity of adsorption in the lower part

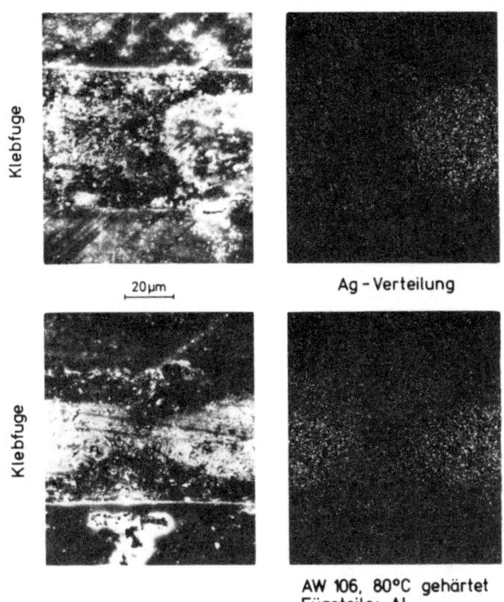

Fig. 8: Cross sections through a glue
line (left: SEM-image, right:
Ag-distribution)

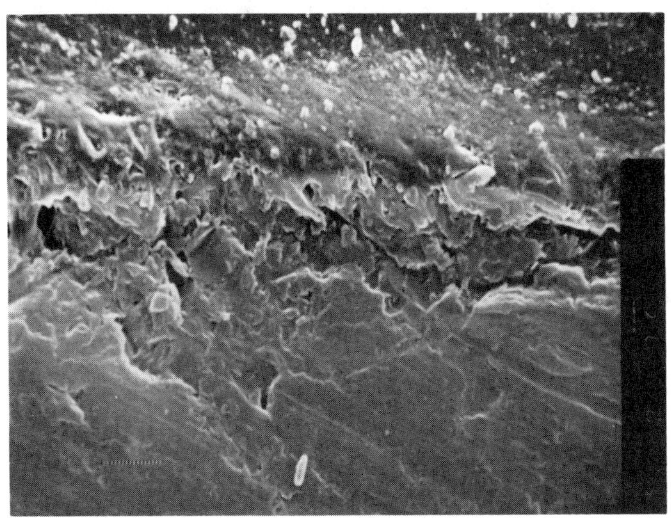

Fig. 9: SEM-image of a boundary layer in a glue line
(cross section, lower part: aluminium, upper
part: adhesive)

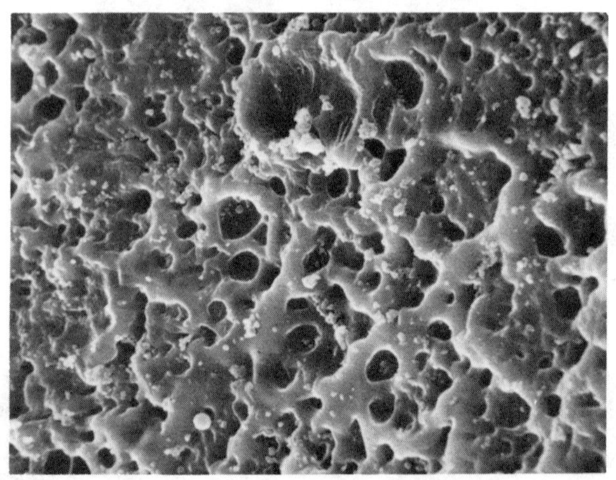

Fig. 1o: SEM-image of adhesive after etching
away the metal surface

MOISTURE EFFECTS IN EPOXY ADHESIVES

Joseph Boutilier, Edward J. Hughes, and John L. Rutherford
The Singer Company, Kearfott Division
Little Falls, N.J.   07424

Abstract

This paper describes the effects of moisture on the dimensional stability of gyro grade epoxy adhesives using a simplified method developed to evaluate the behavior. Measurements were made on epoxy adhesives conditioned under a range of temperatures and humidities. The diffusion coefficient for moisture in the adhesives was determined together with the stress and strain induced by the moisture. This relatively low-cost method can be used over a wide range of environmental conditions and can be easily adapted to accommodate a variety of resins.

1.   INTRODUCTION

It is well-known that epoxy adhesives absorb and desorb moisture when the temperature and/or the surrounding relative humidity change.[1,2,3] Absorption of moisture usually causes the epoxies to expand, while desorption leads to a volume decrease. Although the strains associated with these volume changes are small, they are significant when the adhesive is used to bond components in precision applications, such as inertial sensing instruments or optical systems. The moisture weight-gain studies in the literature provide a data base and primary equations with which to determine the amount of moisture absorption and the diffusion constants which govern the absorption. However, the designer of precision instruments must know the strains associated with moisture in epoxy adhesives so that com-

pensation can be made for the dimensional instability that results from adhesive swelling or shrinking.

This paper describes a newly-developed test method that determines the strains developed in adhesives as a function of moisture content. The test results can be used to determine residual stresses developed within the adhesive. Of greater importance to the design engineer, however, is the time-dependence of the strain changes that can be recognized by the test method.

2.   EXPERIMENTAL PROCEDURES

This newly-developed test apparatus consists of a cantilever metal beam that has a slot that contains the adhesive being studied (see Figures 1 and 2). A resistance-type strain gauge is bonded to the beam opposite to the adhesive, and a compensating strain gauge is mounted on an inactive part of the beam. When the temperature is changed, the beam bends because of the difference in the values of the coefficients of thermal expansion of the metal beam and the adhesive. More important, however, the beam will also bend at constant temperature, as the adhesive swells or shrinks when it absorbs or desorbs moisture.

The beam is positioned in a small volume,high thermal mass oven. The specimen temperature is established using flexible tape heaters with a current-proportioning controller and a fine-wire chromel-alumel thermocouple is attached to a mechanically inactive part of the beam to monitor the specimen temperature. The test specimen and oven is used inside a chamber which can be evacuated and in which the relative humidity (RH) can be controlled by the use of constant humidity solutions (see Figure 3). With this arrangement, the beam can be heated to the test temperature and thermal equilibrium established within a few minutes.

The experiments described in this paper were made with copper-beryllium beams and two structural adhesives: Bacon Laboratories LCA-4LV with BA9 hardener and Epoxy Technology H-74.(6,7)

After the epoxy adhesive was cured in the beam slot, it was conditioned by exposing it to a selected constant humidity at room temperature until the strain gauge output stabilized; i.e. until the adhesive was saturated with moisture and stopped swelling or shrinking (usually taking about seventy-two hours). Then the constant-humidity solution

was removed and the chamber was evacuated using a mechanical fore-pump. At the same time, the specimen was heated to the desired test temperature. Due to the differential expansion of the epoxy/metal beam, a large strain gauge output was observed while the specimen was being heated. After thermal equilibrium is established, any changes in the strain gauge reading are due to a shrinking of the epoxy as it gives up moisture. The equilibrium moisture content of the epoxy will be lowered because of the increased temperature and because of the absence of moisture in the test chamber environment.

This apparatus can also be used to monitor the swelling of the adhesive as it abosrbs moisture at constant temperature; and it can be used to monitor shrinkage strains as the adhesive is cured. There are several desirable features of this test apparatus:

a) it uses adhesives in thin film form, and not in bulk

b) the adhesive is bonded to a metal substrate and is not freestanding

c) it provides strain values as a function of moisture content and not just weight gain

d) the thermal characteristics closely approximate those of the precision inertial sensing instruments that use adhesives in critically important areas

e) the temperature and environmental conditions can be altered very quickly

f) the test results are recorded easily

3. RESULTS AND DISCUSSION

The strains developed as a result of desorption of moisture in two adhesives are shown in Figure 4. Each specimen had been conditioned (after curing in air) for seventy-two hours at room temperature in a 58 percent RH. The constant temperature solution was then removed and the specimen was heated to 160°F (taking about two minutes) while the chamber was evacuated using the fore-pump. As each adhesive gave up its moisture, it contracted causing the beam to bend. The initial rate of desorption was high. With time, however, the rate decreased and approached zero after about 15 hours for the LCA-4 and about 24 hours for the H-74. This latter adhesive showed a much greater desorption strain than did the LCA-4, indicating that it had absorbed more moisture.

The initial desorption strains for LCA-4 are shown in Figure 5 for four moisture contents. Test specimens were individually conditioned at room temperature at four RH levels: 58, 31, 20, and 0 percent RH. The test chamber was then evacuated and the temperature raised to 160°F. After thermal equilibrium was established, the 58 percent RH conditioned specimen showed the greatest desorption strain; with the 31 and 20 percent RH specimens having proportionately less. The slight shrinkage strain in the 0 percent RH specimen is attributed to additional curing that took place at 160°F.

The strains measured in these tests can be used to determine the diffusion coefficient for moisture in adhesives. From Fick's laws, the moisture content (M) in an adhesive is given by

$$G = \frac{M-M_i}{M_m-M_i} \qquad (1)$$

where

G = time dependent factor

M = instantaneous moisture content

M_m = maximum moisture content

M_i = initial moisture content

Assuming that the strains developed in these tests are proportional to the moisture content of the epoxy, the measured values of strain can be substituted for M_m, M and M_i in Equation (1). The time-dependent function G has been shown by Shen and Springer[4] to be approximated by

$$G = 1 - \exp\left[-7.3\left(\frac{DT}{4t_p^2}\right)^{0.75}\right] \quad (2)$$

where

D = diffusion coefficient

T = time

t_p = epoxy thickness

Which can be rewritten

$$DT = 4t_p^2\left[-\frac{\ln\left(1-\frac{\varepsilon-\varepsilon_i}{\varepsilon_{max}-\varepsilon_i}\right)}{7.3}\right] \quad (3)$$

where

ε = instantaneous strain

ε_{max} = maximum strain

ε_i = initial strain

In Equation (3), the only unknown is D. From a plot of DT vs T, the value of D can be obtained from the slope of the resulting approximate straight line (Figure 6). The values of D obtained in this manner were 0.28 and 0.52 x 10-6 in2/min for Epotek H-**74** and LCA4/BA9, respectively. These values are comparable with those found for similar epoxy systems by other workers. (5)

An important feature of the experiments described here is that they enable the stress developed in a bonded epoxy due to moisture absorption to be determined directly. Moisture absorption in these adhesives results in a swelling of the epoxy which leads to deformation of the beam. The total strain developed and measured in the beam (ε_m) is made up of two components: one stretching and one bending and is given by

$$\varepsilon_m = \frac{\Delta}{S}\left[\frac{1}{K_m\ell} + \frac{(-t_m+d)3t}{2WQ}\right] \quad (4)$$

where

Δ = net change in length of epoxy

t_m = thickness of metal

ℓ = length of slot

d = distance of neutral axis from metal/epoxy interface

t = total thickness of metal and epoxy

$$S = \frac{\ell}{W}\left(\frac{1}{t_mE_m} + \frac{1}{t_pE_p}\right) \quad (5)$$

$$K_m = \frac{Wt_mE_m}{\ell} \quad (6)$$

$$Q = E_m\left[d^3-(-t_m+d)^3\right] + E_p\left[(t_p+d)^3-d^3\right] \quad (7)$$

where

W = width of slot

t_p = thickness of epoxy

E_m and E_p = Young's modulus of metal and epoxy, respectively

Combining Equations (4), (5), (6) and (7) and rearranging gives

$$\Delta = \frac{X}{Y+Z} \quad (8)$$

where

$$X = \varepsilon_m\ell\left(\frac{1}{A_mE_m} + \frac{1}{A_pE_p}\right)$$

$$Y = \frac{3t(d-t_m)}{2W\{E_m\left[d^3-(d-t_m)^3\right] + E_p\left[(d+t_p)^3-d^3\right]\}}$$

$$Z = 1/(A_mE_m)$$

A_m = area of metal (W x t_m)

A_p = area of epoxy (W x t_p)

In Equation (8), the only unknown is d, the distance of the neutral axis from the metal/epoxy interface, which is given by

$$d = \frac{(E_m/E_p)t_m^2 - t_p^2}{(2E_m/E_p)t_m - t_p} \quad (9)$$

Using Equations (8) and (9), the net change in length of the epoxy and the strain can be calculated if the Young's modulus of the metal and epoxy are known. These values were measured for the two epoxies using a capacitance-type extensometer(8,9) ,handbook values were used for the modulus of the Cu-Be beam. The results are presented in Table 1, which also includes the stresses corresponding to the epoxy strains. Depending on the temperature and humidity, the stress developed varies from about 1000 psi to 250 psi in the LCA4 system and from 2100 psi to 1100 psi in the Epotek H-74 system. In

the calculation, no allowance
was made for any creep which
might occur in the epoxies.
However, previous work(8,9)
has shown that at these
stresses little or no creep
should occur. The magnitude
of the stresses developed
illustrates that moisture absorption can considerably reduce the load-bearing capacity
of a bonded joint, since the
stresses developed are additive to the applied stress.

In order to make the data more
useful to the design engineer,
the strain developed by moisture absorption was reduced to
a basic material property expressed in terms of strain
per percent change in RH.
Values of this Coefficient of
Moisture Expansion are included in Table 1 and can be
used in much the same way as
the Coefficient of Thermal
Expansion. The temperature
dependence of the Coefficient
of Moisture Expansion is shown
for the LCA4 system in Figure
7.

4. SUMMARY

a) A test method has been developed which provides a
   means for determining
   stress and strain levels in
   adhesives bonded to metal
   substrates.

b) The time dependence of
   strains in adhesives caused
   by moisture can be characterized.

c) The method can be used to
   evaluate the moisture sensitivity of adhesives to
   aid in the selection of
   materials.

d) A Coefficient of Moisture
   Expansion can be formulated,
   to be used in a manner
   similar to that of the
   Coefficient of Thermal
   Expansion.

ACKNOWLEDGMENTS

The original concept for the
need and for this approach to
studying strains associated
with moisture absorption in
bonded adhesives was due to
Mr. W. Krupick, Senior Member,
Technical Staff. We are indebted to Professor N. Brown,
of the University of Pennsylvania, for many valuable suggestions during the course of
the study and for developing
the equations used to calculate the epoxy strains.
Finally, we would like to
thank Mr. L. Dunham and
Mr. J. Chien for their help
in performing some of the
experiments.

REFERENCES

1. "Prediction and Verification of Moisture Effects
   on Carbon Fiber-Epoxy
   Composites," J.M. Augl,
   Tech.Rept. NSWC TR-79-43
   (1979).

2. "The Mechanisms of Elevated
   Temperature Property Losses
   in High Performance Struc-
   tural Epoxy Resin Matrix
   Materials after Exposure to
   High Humidity Environments,"
   C.E. Browning, Tech.Rept.
   AFML TR-76-153 (1977).

3. "Moisture Effects in Epoxy
   Matrix Composites," C.E.
   Browning, Tech.Rept. AFML
   TR-77-17, (1977).

4. "Moisture Absorption and
   Desorption of Composite
   Materials, C.Shen and G.S.
   Springer, J. of Comp.Mat.
   10, (1976).

5. "The Diffusion of Water
   Vapour in Humid Air in the
   Bond Lines of Adhesive
   Bonded Metal Joints,"
   W. Altof, 11th Nat.
   S.A.M.P.E. Tech.Conf.,
   Boston (1979).

6. Bacon Industries, Inc.
   192 Pleasant Street
   Watertown, MA 02171

7. Epoxy Technology, Inc.
   14 Fortune Drive
   Billerica, MA 01821

8. "Capacitance Methods for
   Measuring Properties of
   Adhesives in Bonded
   Joints," E.J. Hughes,
   J.L. Rutherford, and
   F.C. Bossler, Rev. Sci.
   Instr., 39 (1969).

9. Unpublished work,E.J.Hughes
   and J.L. Rutherford.

BIOGRAPHIES

Dr. Edward J. Hughes,Principal
Scientist, has been with
Singer-Kearfott for over fifteen years where he has conducted extensive research into the microstrain behavior
of metals and adhesives. Prior
to that he was a Senior Scientist at the Materials
Research Corporation, Orange-
burg, N.Y. Dr. Hughes received
his B.Sc. (Honors) and Ph.D.
degrees in Physical Metallurgy
at Birmingham University in
England. He is a member of
S.A.M.P.E. and ASM.

Joseph Boutilier, Research
Associate, has been a materials and processing technician
with Singer-Kearfott for
eleven years. He has worked
with the development of experimental instrumentation for
the testing of materials. He
also has certification from
the American Chemical Society
in liquid chromatography,
Vishay Intertechnology in
strain gauge technology and
Singer-Kearfott in microelec-
tronic analysis. Mr. Boutilier
has a certificate in Electronic Technology from Newark
College of Engineering, NJ,
and an Associates Degree in
Applied Science, Electronic
Technology, from the County
College of Morris, Randolph,
NJ.

Dr. John L. Rutherford is
Director of the Materials and
Processes Laboratory which
supports Engineering and
Operations in Kearfott. Before
joining Singer-Kearfott over
seventeen years ago, he worked
at the Franklin Institute
Laboratories in Philadelphia
where he helped develop
floating zone-refining techniques for iron and titanium.
Dr. Rutherford hold a B.A.
degree in Physics and M.S.
and Ph.D. degrees in Metallurgy, all awarded by the
University of Pennsylvania,in
Philadelphia. He is listed in
American Men of Science.

TA3LE 1 - Effect of Temperature and Relative Humidity on Values
of Stress and Strain Generated by Moisture Absorption

Adhesive	Relative Humidity (percent)	Test Temp. (^{o}F)	Measured Beam Strain (x10-6)	(x10^{-6} inch)	Calculated Epoxy Strain (x10-6)	Calculated Epoxy Stress (psi)	in/in/% RH (x10^{-6})
LCA4/BA9	58	160	148	152.44	1020	1020	17.59
	31	160	85	87.55	584	584	18.84
	58	140	124	127.72	852	852	14.69
	31	140	61	62.83	419	419	13.52
	58	120	77	79.31	529	529	9.12
	31	120	36	37.08	247	247	7.97
Epotek H-74	58	160	407	450.96	3010	2100	51.9
	58	140	375	415.5	2770	1940	47.76
	31	140	210	232.7	1550	1090	50.0

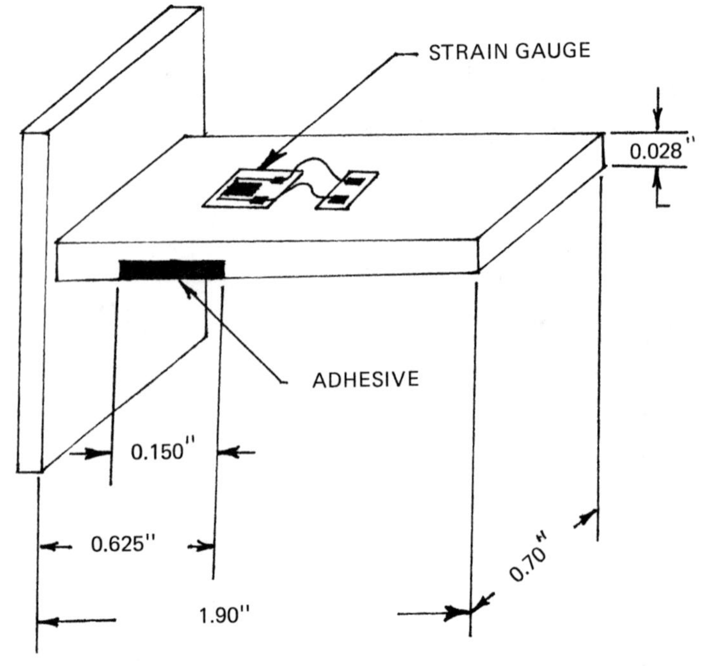

DIMENSION OF CANTILEVER BEAM SPECIMEN
FIGURE 1

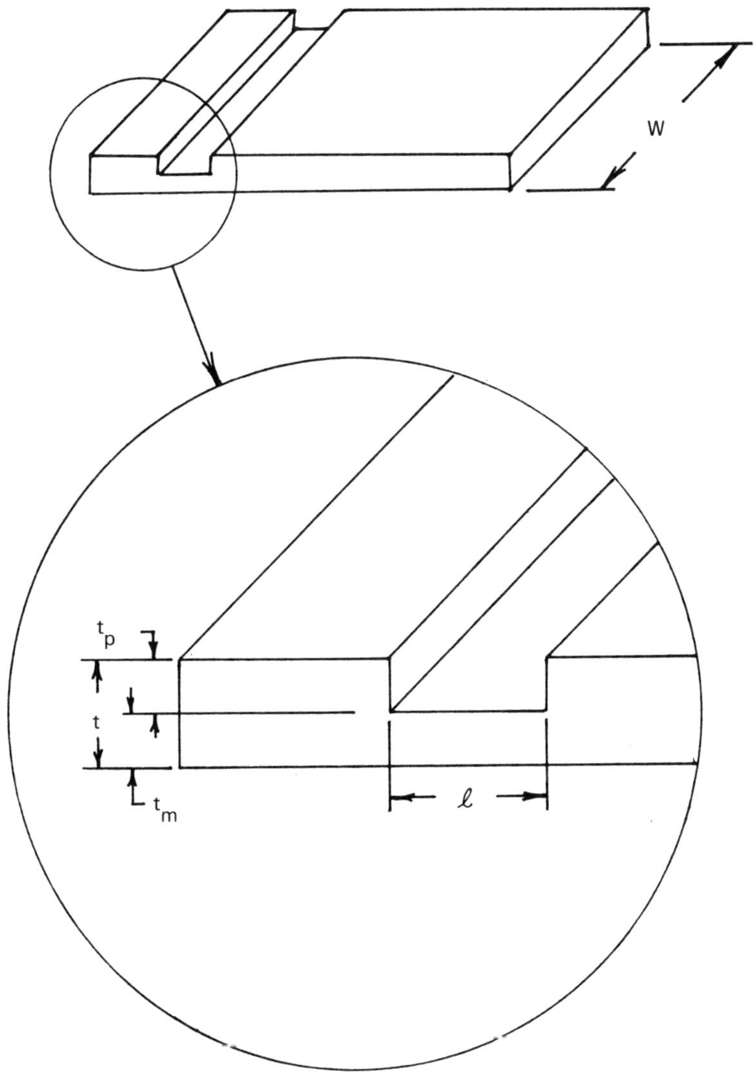

SCHEMATIC OF COPPER-BERYLLIUM BEAM

FIGURE 2

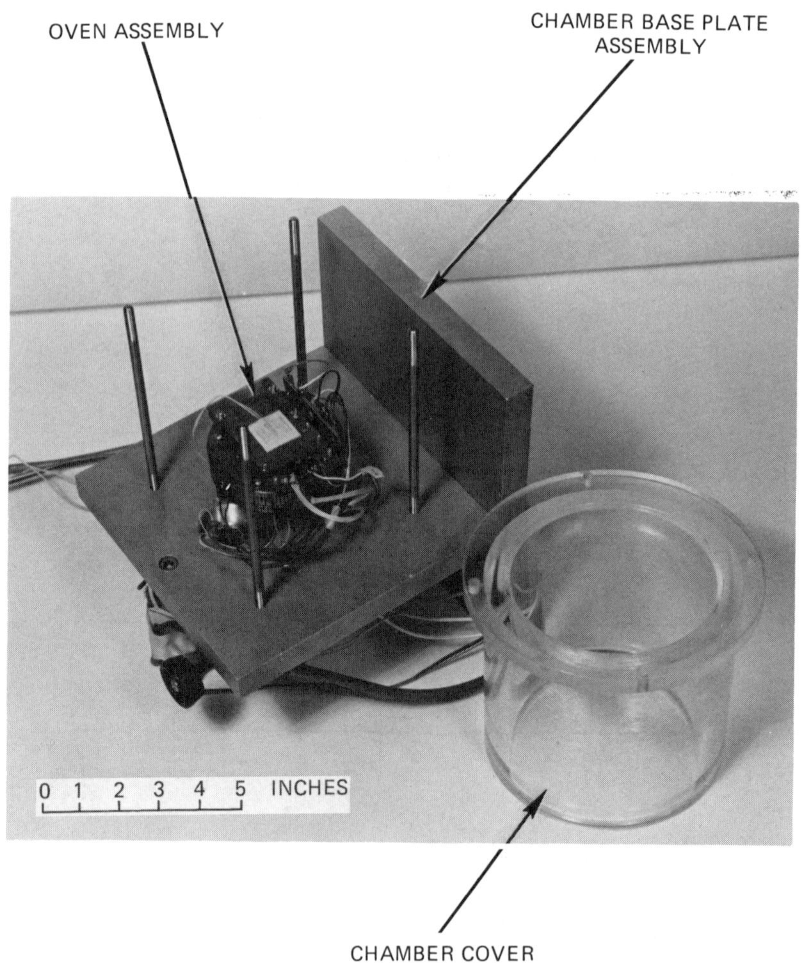

ENVIRONMENTAL CHAMBER WITH LOW VOLUME
HIGH THERMAL MASS OVEN ASSEMBLY

FIGURE 3

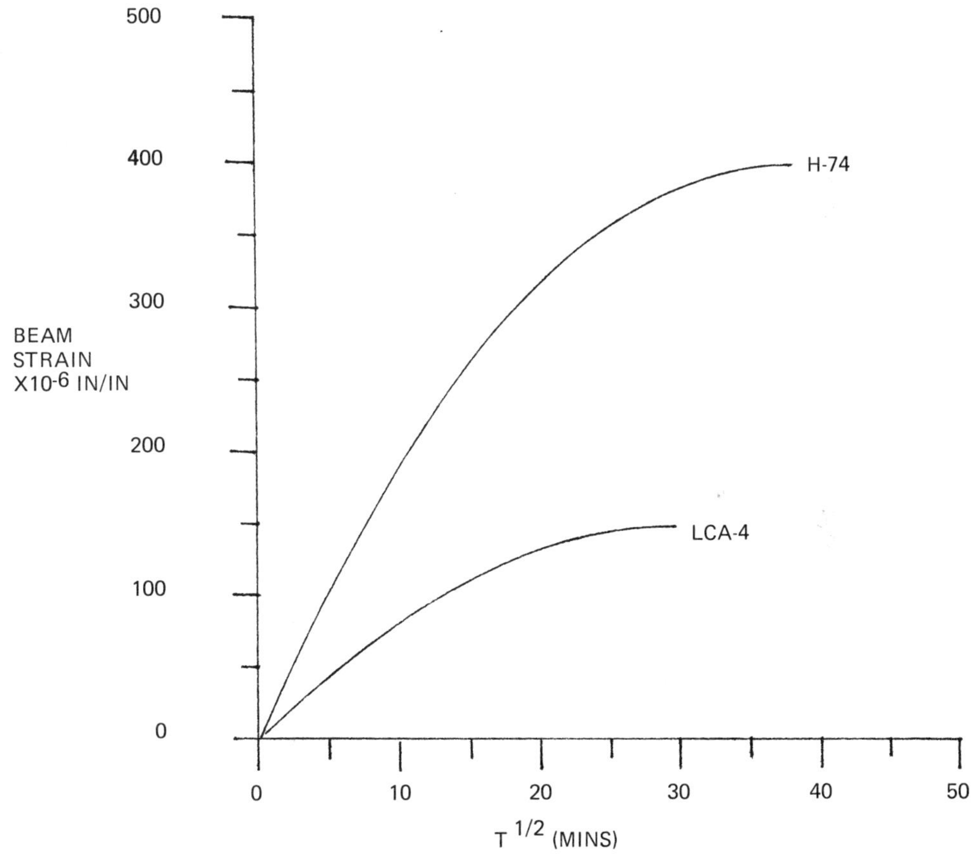

COMPARISON OF MOISTURE DESORPTION STRAINS
IN EPOTEK H-74 AND LCA4/BA 9 AT 160°F

FIGURE 4

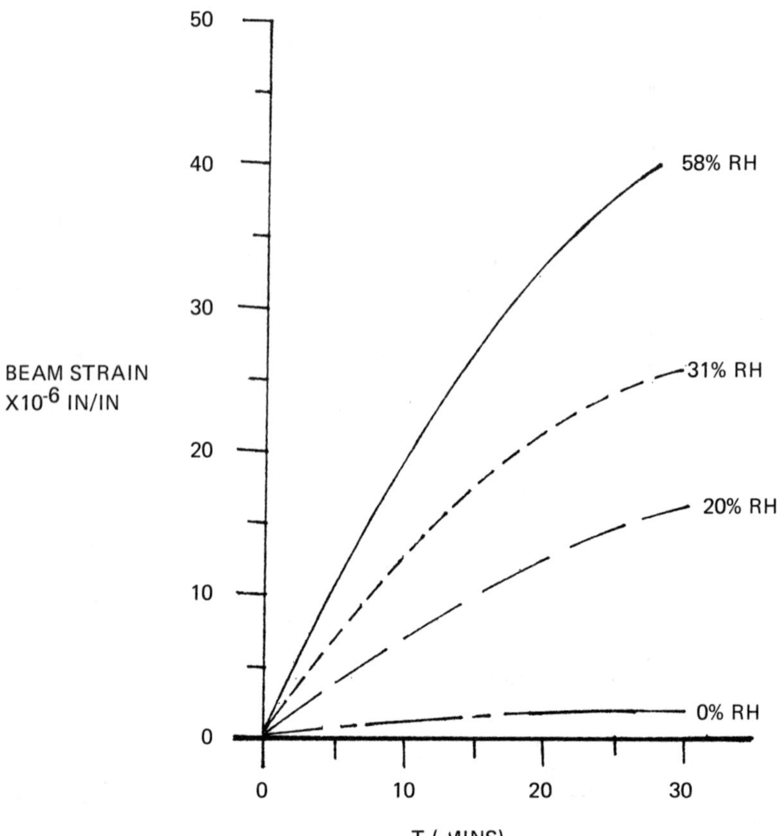

EFFECT OF CONDITIONING HUMIDITY ON
DESORPTION STRAIN IN LCA4/BA9 at 160°F
FIGURE 5

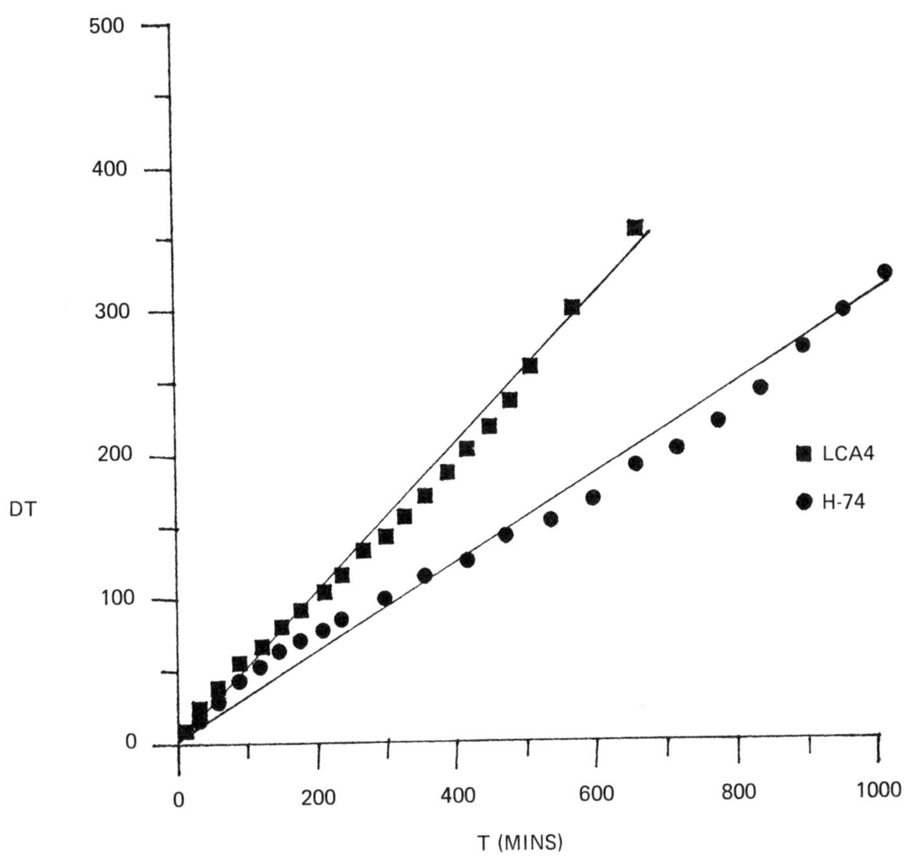

DETERMINATION OF DIFFUSION COEFFICIENT FROM BEAM STRAINS

FIGURE 6

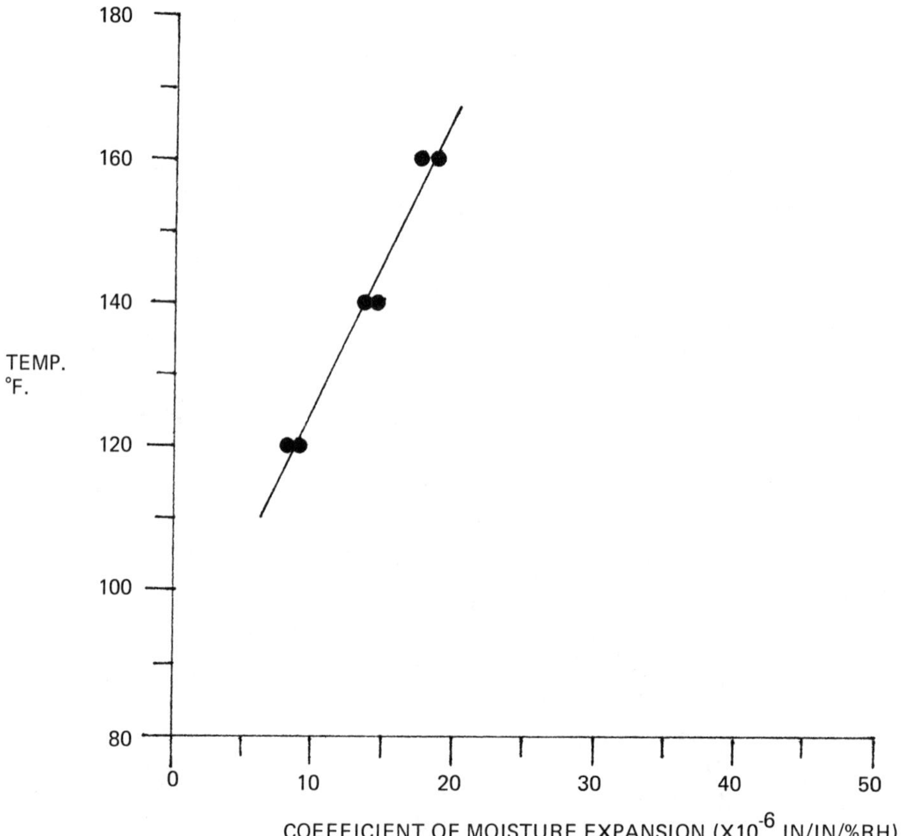

TEMPERATURE DEPENDANCE OF COEFFICIENT OF MOISTURE ABSORPTI
OF LCA4/BA9

FIGURE 7

SYNTACTIC CORE - A NEW COMPOSITE MATERIAL
David K. Klapprott
Dr. David Landman
Hysol Division/The Dexter Corporation
P. O. Box 312, Pittsburg, CA 94565

Abstract

A new, convenient form of syntactic core tape material for use in sandwich construction is described. Two thicknesses of core were evaluated per MIL-STD-401 using E-glass, graphite and Kevlar [R] skins. A 350°F resin system was used as the composite and core matrix material. Initial results show that syntactic core composite structures can provide substantial strengths. Also, sandwich properties vary considerably with skin reinforcement type and that the level of adhesion by the resin to the fiber is a controlling variable. Sandwich shear properties, in most cases, are inversely proportional to core thickness.

1. INTRODUCTION

A syntactic core is a low density foam supplied in film form. It combines an epoxy resin system with high strength glass microballoons. It can be used to lighten structure, reduce the cost of assembly, increase the stiffness to weight ratio (when compared to other foams) and improve the machinability of the finished part. Compared to other core materials, it has the disadvantage of being more notch sensitive and having a greater density. However, syntactic cores have higher strength to weight ratios (Table 1). Presently it is used in designs where core sections of 0.1 inches or less are desired, and honeycomb core is difficult to form reliably. Syntactic foam pastes are available from various sources, but present the problem of poorly controlled cured part thickness. The syntactic core, produced as a controlled thickness film, avoids many of these problems. The structural properties of ADX-819, the core material described in this presentation, are shown in Table 2. Use of syntactic core systems also eliminates the need for a separate adhesive system for bonding the facing material to the core.

Syntactic cores are currently being investigated or used in several aerospace applications. Examples of these are the Lockheed L-1011 aileron where a graphite skin/syntactic core structure replaced an all metal part with a 20% weight saving. This work was done by Lockheed under a NASA contract [1]. Lockheed/Georgia is using a syntactic core/Kevlar skin structure to build the "Aquilia" remotely-piloted vehicle [2]. The entire vehicle uses this construction including all interior rib structures.

This report covers the preliminary results of a program, undertaken by Hysol, to acquire a data base on sandwich structures using syntactic core and various high performance composite skins.

2. TEST PROGRAM

The objective of the test program was to acquire strength data on ADX 819 as a core material. Consequently, data was developed per MIL-STD-401A using the following skin reinforcements:

- Style 1581 Woven E-Glass
- Style 285 Woven Kevlar[R] *
- Unidirectional Thornel[R] **
  T300/3000 Graphite Tape
- Style A193P (Hercules)
  Woven Graphite

These fabrics were impregnated with the same resin system as used in ADX-819. The skins were constructed in either of two ways:

- Woven fabrics - six layers, alternating 0°,90°
- Unidirectional tape- six layers, 45°/0°/135°/45°/0°/135°

The syntactic foam was tested in both 20 and 50 mil thicknesses.

Figure 1 shows a typical lay-up configuration for making the laminates. The size of the laminate was 11" x 13" from which coupons were cut for testing. Curing was done in a vented autoclave using a one hour cure at 350°F and 55 psi schedule. The Kevlar structure was cut by using a band saw with the teeth in reverse direction, followed by sanding with fine sandpaper to remove frayed edges. All other panels were cut using a diamond slab saw. Test specimens were then assembled using a secondary adhesive to bond the shaped core specimens to the metal loading blocks required by the test.

The core shear, flatwise tension, flatwise compression, edgewise tension and compression strengths of structure using 20 mil core have been determined to date. Several difficulties were experienced in acquiring the data. Initially a room temperature curing adhesive was used for

* Kevlar[R] is a registered trademark of E. I. Du Pont de Nemours & Co.
** Thornel[R] is a registered trademark of Union Carbide Corporation

bonding the metal loading pieces.
This was done to avoid any additional heat exposure, especially
on those specimens exposed to a
moist environment. The strength
of the sandwich assemblies ex -
ceeded in many cases that of the
secondary adhesive. In these
cases, the specimens were remade
with a 250°F curing film adhesive.

In most cases, the loading fixtures normally used with MIL-STD-
401 testing were severely strain-
ed or broken in testing the syntactic core structures. The
specimen size for core shear
strength had to be downsized by
one-half to give breaking loads
within the capabilities of the
test fixtures.

Even with these modification, the
fractures observed, especially in
connection with the core shear
and flatwise tension tests, were
a mixture of cohesive, secondary
bond and primary bond (i.e.
between core and facing skins)
failures. In some instances, the
failure fracture was a mixture of
the latter two and it was not
possible to determine what type
of failure was represented by the
failure stress. Therefore, the
data presented is the average of
those specimens which failed in
either a cohesive, primary bond
failure or a mixture of both. It
is probably safe to say that the
actual strengths are in excess of

the reported values.

3. RESULTS

Table 3 shows the core shear
strengths of the various sandwich
constructions containing 20 mil
core, at several temperatures.
Strengths at 75°F vary from a
minimum of 1550 psi using Kevlar
skins to a maximum of 2000 psi for
the glass structure. A sandwich
using 2024-T3 aluminum skins was
included to investigate the poten-
ial strength of the core. These
specimens failed at 2400 psi. The
results from all types of composite skins showed an equivalent
amount of primary bond and cohesive failure while the aluminum
skinned specimen failed in a cohesive mode. A similar relationship
of failure stress and skin type is
seen at both -65°F and 180°F. The
greatest strength occurs with
glass skinned structures followed
by woven graphite, unidirectional
graphite and, finally, Kevlar
constructions.

Similar core shear testing was
conducted using 50 mil core in
place of the 20 mil material.
The 75°F strengths (Table 4) for
these specimens were, in most
cases, less than the 20 mil core
sandwichs. The differences ranged
from a 9% decrease for the
unidirectional graphite to a 46%
decrease for the woven graphite.
Interestingly, the specimens using
Kevlar skins exhibited a 13%

797

increase in strength when 50 mil core was used. The strength of the aluminum skin sandwich decreased by 21% (to 1900 psi) with this change. Apparently both specimen geometry and skin type play a significant role in determining core shear strength.

The flatwise tensile strengths found are shown in Table 5. A ranking of strength with facing type similar to the core shear tests was found. The sandwich structure using woven glass face sheets exhibited the maximum strength at all temperatures followed by the unidirectional graphite, the woven graphite and finally the Kevlar skinned specimens. The Kevlar specimens also failed in an interlaminar manner, a phenomenon not shown by the other composites. Interestingly enough, the strengths shown by the glass skinned specimens were greater than the aluminum construction. An attempt was made to determine the effect of exposure to 180°F and 100% R. H. on core shear and flatwise tensile strength. Unfortunately, effective secondary bonds have yet to be made to the exposed specimens. This testing is currently being repeated and will be reported in subsequent reports.

The flatwise compressive strengths are shown in Table 6. As in the core shear tests, the size of these specimens had to be reduced so that the failure loads were within the range of the Instron testing machine. The failure point was difficult to discern when reading the Instron chart. However, all failures were verified by microscopic analysis. The trend of strength versus skin types is similar to the other testing. The glass sandwichs were the strongest, failing at 9500 psi. The woven graphite was next, failing at 8600 psi. Interestingly, the Kevlar specimen was stronger (7300 psi) than the unidirectional graphite specimen.

The edgewise tension and compressive strength results are shown in Tables 7 and 8. The trends of strength with skin types are those which would be expected from observation of the fiber properties (the low edgewise tensile strength of the unidirectional graphite structure are due to the lay-up construction).

4. CONCLUSIONS

From the results presented here, it can be seen that syntactic cores can be combined in sandwich structures with composite skins and provide substantial strength properties. The level of strength depends not only on the core but also on the type of fiber and the construction of the skin reinforcement. Adhesion to the fiber appears to play some controlling

part in this phenomenon. Thus,
the highest levels of strength
are found using glass reinforcements. Historically, epoxies
have had the highest level of
adhesion. Generally, the lowest
level of strengths were exhibited
by the Kevlar sandwiches. This
is the fiber to which cured epoxy
has poor adhesion. Sandwich
structures with syntactic core
containing both unidirectional
or woven graphite skins, generally, had strength properties
intermediate to the glass and
Kevlar. Preliminary results
indicate that the core shear
strength is inversely proportional
to the thickness of the core
(Table 4).

Work is continuing at Hysol both
in refining the test data presented here and in defining the
strength to weight ratios of the
sandwich materials in comparison
to the pure composite materials
themselves. Results from both
this and earlier work suggest
that the syntactic core/composite
skin structure may have some
unique energy absorption properties. This implies that
syntactic core construction may
provide materials that do not
fail catastrophically and are
therefore "fail-safe".

5. ACKNOWLEDGEMENTS

It is a pleasure to acknowledge
the work of my associates, Mr.
Chuck Dean and Mr. Paul R.
Schreiner who have done much of the
work described in this paper.
This work was done under the aus-
pices of Hysol Division/The
Dexter Corporation at Pittsburg,
California.

6. REFERENCES

1. Lockheed-California Co./
   NASA contract NASI-15069.
2. "Mini-RPV Being Developed for
   Army", Aviation Week and Space
   Technology, 54, Jan. 7, 1980.

TABLE 1 - PROPERTIES OF VARIOUS CORE MATERIALS

Material	Density(lb/ft^3)	Shear Strength to Weight Ratio	Compressive Strength to Weight Ratio
Honeycomb	1.1 - 8.0	480 - 1080	360 - 2280
Balsa	4.8 - 10.1	360 - 420	170 - 240
Urethane Foam	1.8 - 28	180 - 360	120 - 780
Syntactic Core	29 - 50	840 - 1440	1560 - 2520

TABLE 2 - PROPERTIES OF ADX-819 SYNTACTIC FOAM

Cure Schedule - 1 Hour at 350°F

Flow at 350°F & 85 psi - 35-55% (w/w)

Cured Density - 0.023 lb/in^3

Tensile Strength (lb/in^2)
@ 75°F - 2200
@180°F - 2200

Compressive Strength (lb/in^2)
@ 75°F - 7430
@180°F - 6460

Shear Strength (lb/in^2)
@ 75°F - 2110
@-65°F - 2060
@180°F - 2250

TABLE 3 - CORE SHEAR STRENGTH OF SEVERAL SANDWICH STRUCTURES
Syntactic Core - ADX-819, 0.020 in. thick

| Facing Material | Minimum Strength (lb/in^2) | | |
	@ 75°F	@ -65°F	@ 180°F
Style 1581 Glass	2000	1750	1900
Style 285 Kevlar	1550	1450	950
Unidirectional T-300/Graphite	1750	1550	1250
Woven Hercules A193P Graphite	1950	1700	1700
2024-T3-Aluminum	2400	-	-

TABLE 4 - EFFECT OF SYNTACTIC CORE THICKNESS ON CORE SHEAR STRENGTH

Facing Material	75°F Core Shear Strength (lb/in)using	
	20 mil	50 mil
Style 1581 Glass	2000	1800
Style 285 Kevlar	1550	1750
Unidirectional T-300 Graphite	1750	1600
Woven A193P Graphite	1950	1050
2024-T3-Aluminum	2400	1900

TABLE 5 - FLATWISE TENSILE STRENGTH OF SEVERAL SANDWICH STRUCTURES

Syntactic Core Thickness - 0.020 in.

Facing Material	Minimum Strength (lb/in^2)		
	@ 75°F	@ -65°F	@ 180°F
Style 1581 Glass	4200	3600	3250
Style 285 Kevlar	2000	1600	1560
Unidirectional T-300 Graphite	3050	2000	2200
Style A193P Graphite	2900	1750	2900
2024-T3-Aluminum	3400	-	3150

TABLE 6 - FLATWISE COMPRESSIVE STRENGTH OF SEVERAL SANDWICH STRUCTURES

Syntactic Core Thickness - 0.020 in.

Facing Material	Strength (lb/in^2) @ 75°F
Style 1581 Glass	9500
Style 285 Kevlar	7300
Unidirectional T-300 Graphite	7100
Style A193P Graphite	8600

TABLE 7 - EDGEWISE TENSILE PROPERTIES

Test Temperature - 75°F

Facing Material	Strength (lb/in^2)	Modulus ($lb/in^2 \times 10^{-6}$)	Elongation @ Failure(%)
Style 1581 Glass	41,600	2.93	1.8
Style 285 Kevlar	50,400	3.80	1.4
Unidirectional T-300 Graphite	20,000	2.36	1.0
Style A193P Graphite	86,700	7.73	1.1

TABLE 8 - EDGEWISE COMPRESSIVE PROPERTIES

Test Temperature - 75°F

Facing Material	Strength (lb/in^2)	Modulus ($lb/in^2 \times 10^{-6}$)
Style 1581 Glass	51,600	3.04
Style 285 Kevlar	16,100	3.04
Unidirectional T-300 Graphite	64,600	5.61
Style A193P Graphite	48,600	9.53

FIGURE 1

TYPICAL LAY-UP

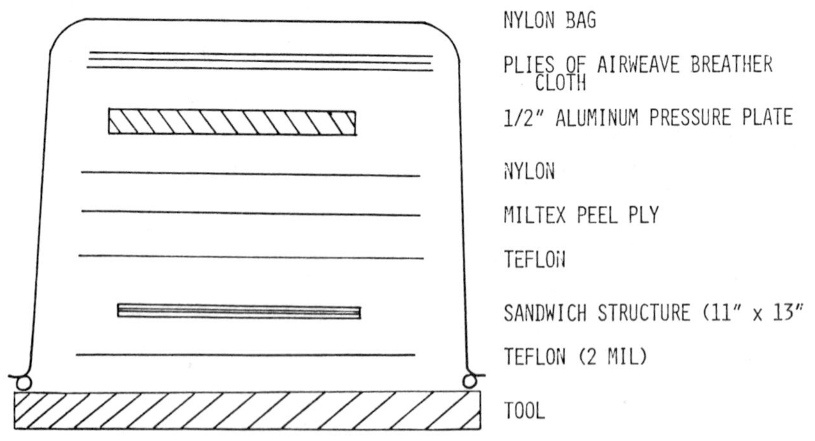

BIOGRAPHIES

David K. Klapprott is the R & D
Group Manager for Film Adhesives
for the Hysol Division of the
Dexter Corporation. His main
areas of effort have been the
development of high strength
adhesives of widely varying temperature requirements and
structural adhesive prepregs. He
received his B. S. degree in
chemistry from Arizona State
University in 1966.

David Landman is an Associate
Scientist at Hysol Division, The
Dexter Corporation located at
Pittsburg, California. He is
responsible for development of
syntactic foam materials. Prior
to joining Hysol, he was employed
by Milliken Chemical where he was
responsible for technical
development of epoxy curing agents.
Dr. Landman received his B. S. and
Ph.D. in chemistry at the
University of Sydney, Australia.

ADVANCED STRUCTURAL REPAIR FOR LOWER COST
A. L. Sanders, Jr
Rockwell International Corporation
North American Aircraft Division, California
H. C. Croop
Air Force Wright Aeronautical Laboratories/FIBCB
Wright-Patterson Air Force Base, Ohio

ABSTRACT

This paper summarizes the results of the initial effort of an Air Force Flight Dynamics Laboratory (AFFDL) sponsored 40-month study to apply current and advanced structures technology to the reduction of costs of aircraft structural maintenance and repair. Discussion is provided of maintenance data obtained from sources such as AFM66-1 and visits to the Air Force Logistic Centers. Results are provided of maintenance data analyses that were conducted to establish a list of major maintenance cost contributors. A description is included of the process used to reduce the major maintenance cost contributor list to the five candidates used in detailed design studies. The Air Force and contractor then selected three of the major maintenance cost contributors to be used in subsequent program effort, which will consist of developing, designing, and demonstrating the feasibiligy of maintenance concepts devised to significantly reduce maintenance costs. The major cost contributors are identified, the basis for their selection is summarized, and the phase III activities are identified.

1. INTRODUCTION

The objectives of defense department reliability and maintainability (R&M) activities (established in the 8 July 1980 DOD R&M directive 5000.40) may be summarized as follows: (1) increase operational readiness and mission success, (2) reduce demand for maintenance and logistic support, (3) use available skills and training, (4) provide specific types of essential R&M data, and (5) ensure that cost and schedule investment in R&M contributes significantly to the preceding objectives. One measure of the importance of these objectives is the attention being given to operational and support problems in military and civilian publications. This attention and the serious nature of maintenance-related problems are illustrated in table I, which appeared in the 6 October 1980 issue of Aviation Week and Space Technology.

Table I

FIGHTER AIRCRAFT READINESS INDICATORS

Aircraft	Complexity	Not Mission Capable (%)	Mean Flight Hr. Between Failure	Maint. Events Per Sortle	Maintenance Man-Hours Per Sortle
Air Force					
A-10	Low	32.6	1.2	1.6	18.4
A-7D	Med	38.6	0.9	1.9	23.8
F-4E	Med	34.1	0.4	3.6	38.0
F-15	High	44.3	0.5	2.8	33.6
F-111F	High	36.9	0.3	9.2	74.7
F-111D	High	65.6	0.2	10.2	98.4
Navy					
A-4M	Low	27.7	0.7	2.4	28.5
AV-8A	Low	39.7	0.4	3.0	43.5
A-7E	Med	36.7	0.4	3.7	53.0
F-4J	Med	34.2	0.3	5.9	82.7
A-6E	High	39.3	0.3	4.8	71.3
F-14A	High	47.1	0.3	6.0	97.8

As one contribution to the overall
solution to achievement of DOD's R&M
objectives, Rockwell International's
North American Aircraft Division
(NAAD), under Air Force contract
F33615-78-C-3208, is conducting a
study of possible applications of
newer technologies to reduce structural maintenance costs of inservice
Air Force aircraft. This is the
second of two similar programs
wherein it is acknolwedged that
while research and development are
frequently directed towards future
systems applications, numerous
opportunities exist to impact today's
problems with current technologies.
Several computerized data sources
were first consulted to identify
high-cost structural maintenance
items for a number of Air Force aircraft. These listings then, provided
a basis for discussions with maintenance personnel at Air Force and
other DOD maintenance facilities.
The onsite discussions frequently
resulted in a considerably revised
high-cost-maintenance item list.
From these lists, a systematic evaluation was then initiated to select
a limited number of items for further
detailed studies.

The primary objectives of the
program are to:

1. Identify chronic, high-cost
    maintenance structural and
    corrosion problems of 16 Air
    Force operational aircraft

2. Develop innovative, cost-
    effective repair solutions or
    new designs using advanced
    materials and processes

3. Conduct feasibility demon-
    strations of repair-solution
    specimens by manufacturing and
    testing to provide proof-of-
    concept validation.

The program is structured to be
completed in four separate phases:

- Phase I - Data Research and
  Acquisition

- Phase II - Repair Concept
  Development

- Phase III - Feasibility
  Demonstration

- Phase IV - Technology Transfer
  and Documentation

2. PHASE I - DATA RESEARCH AND
    ACQUISITION

The phase I objective was to
identify chronic high-cost structural maintenance problems of 16
operational Air Force aircraft
currently in the Air Force inventory.

Also, through all program phases,
Rockwell will maintain a responsive
posture regarding customer requests
for developing improved solutions for
for specific repairs. An example
of this was reflected in initiation
of repair procedures for the F-4
stabilator inboard trailing edge
panel, which occurred during phase I
of the program.

An initial step in the program consisted of selection of the study
aircraft in accordance with the following criteria:

- Have an estimated remaining
  inventory life in excess of 10
  years.

- Major aircraft types will be
  represented in the initial steps
  of the selection process; i.e.,
  attack, bomber, cargo, fighters,
  observation, and trainer.

- Each selected model will be in
  inventory quantities sufficient
  to provide significant logis-
  tical support cost (LSC)
  savings.

Initially, a maximum of 15 aircraft
was considered; however, the number

was increased to 16 so as to include the FB-111. This aircraft was added because of the large quantity involved and its structural similarity to the F-111 (which was one of the original 15 selected). The aircraft selected for study are listed in table II.

Table II

SELECTED CANDIDATE AIRCRAFT

Type	Designator
Attack	A-7, A-10
Bombers	B-52, FB-111
Cargo/Transport	C-5, C-130, C-141, KC-135
Fighters	F-4, F-15, F-106, F-111
Observation/Utility	OV-10
Trainers	T-37, T-38, T-39

Electronic and utility aircraft or helicopters were not considered to meet the preceding selection requirements. However, forthcoming repair or redesign proposals may be generic in scope and could be applicable to aircraft not included in the program under discussion in this paper.

Maintenance data related to these aircraft, such as AFM66-1, AFM65-110, and KO51 maintenance, operations, and LSC computerized data tapes were obtained from the responsible Air Force agencies and transferred to Rockwell's data bank. Rockwell-developed computerized analysis programs were then used to extract the desired high-cost structural data for purposes of analysis and field audit study. As with any large computerized data base these tapes are subject to limitations, such as not being current and complete. However they were suitable for developing initial lists of "top maintenance contributors" for the study aircraft. A brief sample from the F-4 list is shown in table III. These lists, in conjunction with a

Table III

F-4 SUMMARY - TOP MAINTENANCE CONTRIBUTORS (ASR/LC)

WUC	Nomenclature	LSC X $100	Total MMH	Struct MMH[*]	% Struct	Major Struct HMC's (%)	Action Taken
111HC	Door 96 L-R (eng acc)	6,060	56,391	38,786	68.7	Worn 30.7, lost hdw 16.4, crack 11.8, missing hdw 3.4	A-G-G-D
111GU	Door 92 L-R (eng acc)	3,401	30,788	18,950	61.4	Lost hdwr 21.3, crack 20.9, worn 8.1, missing hdw 4.4	G-G-A-G
111HA	Door 81 L-R	2,630	24,783	18,150	73.1	Lost hdw 20.9, crack 19.0, worn 10.3, torn 4.9	G-G-G-G
111G4	Door 74 L-R ctr fus	2,007	19,022	15,974	83.8	Lost hdw 47.9, crack 18.3, missing hdw 10.9, worn 3.0	G-G-G-G
111FO	Center fus sec	544	18,569	14,452	77.7	Crack 33.2, lost hdw 21.5, worn 5.3, core 5.2	G-G-G-2
111GS	Door 83 L-R (eng acc)	6,280	58,806	13,236	22.4	Lost hdw 25.3, crack 9.9 missing hdw 6.2, worn 2.5	G-G-G-G
111GR	Door 82 L-R (eng acc)	2,621	24,298	13,162	54.1	Lost hdw 28.2, crack 11.3, missing hdw 6.9, worn 2.8	G-A-G-G
1122F	Door 101 L-R (ctr wing)	2,687	24,684	11,210	45.4	Lost hdw 26.5, missing hdw 7.2, crack 4.9, deteriorate 3.4	G-F-G-G
111CH	Door 6 L-R	1,765	28,150	10,868	38.6	Lost hdw 27.1, missing hdw 3.6, broken 3.1, crack 2.1	G-G-F-G
111C3	Door 22 L-R	2,972	27,589	9,757	35.3	Lost hdw 16.9, crack 10.7, missing hdw 3.1, broken 0.8	G-G-G-G
111CP	Door Data Link acc 19	1,297	16,880	9,670	57.2	Lost hdw 39.6, missing hdw 8.9, broken 3.2, worn 1.8	G-G-G-G
1122B	Door 75 L-R (ctr wing)	173	10,462	9,321	89.0	Lost hdw 47.6, missing hdw 15.0, crack 14.0, worn 8.4	G-G-G-A
111KD	Tail cone	1,053	13,773	8,967	65.0	Crack 34.9, lost hdw 10.7, worn 8.9, deteriorate 3.7	G-G-F-F
1122A	Door 102 L-R (ctr wing)	2,167	20,392	8,343	40.8	Lost hdw 23.8, missing hdw 8.5, crack 3.8, deteriorate 2.5	G-G-G-G
111F2	Web assy keel FS 303.62	933	12,424	7,174	78.9	Missing hdw 33.2, lost hdw 31.5, crack 10.9, worn 1.6	G-G-G-G
1132M	Bellmouth cables	328	12,424	7,163	57.6	Broken 31.0, worn 23.4, crack 1.3, binding 0.9	R-R-R-L
111FG	Fairing Aft eng	158	7,760	6,404	82.5	Crack 40.0, worn 19.5, lost hdw 14.8, missing hdw 4.2	A-A-G-G
1122J	Door 88 L-R (ctr wing)	1,150	14,251	6,295	44.1	Lost hdw 31.5, missing hour 6.4, crack 3.8, broken 0.6	G-G-F-G
1132C	Variable bellmouth ring	2,081	15,345	5,678	36.9	Worn 21.1, crack 8.0, lost hdw 2.6, torn 1.5	A-G-G-A
111CM	Door Oxy bay acc	388	8,607	5,646	65.5	Lost hdw 35.2, missing hdw 11.7, crack 8.4, worn 4.4	G-G-G-G
111GB	Door fuel all 48	2,109	20,207	4,884	24.1	Lost hdw 11.8, crack 5.9, missing hdw 3 .7, deteriotrate 1.0	G-G-G-G
111DL	Door 9 L-R (fwd fus)	265	8,326	3,394	40.7	Lost hdw 31.7, missing hdw 4.4, crack 1.2, worn 0.8	G-G-G-G
11320	Bypass bellmouth sys	888	11,549	3,182	27.5	Worn 13.7, broken 6.5, crack 2.2, lost hdw 1.6	F-F-F-G
112BL	Door 141 L-R (ctr wing)	496	4,595	2,928	63.7	Lost hdw 24.9, crack 23.7, missing hdw 8.1, rent 1.7	G-A-G-A
11230	Outer wing sect	858	7,989	2,764	34.6	Crack 16.6, lost hdw 6.5, rent 2.3, broken 1.8	R-G-R-R
1211G	Main instr panel (C/P)	1,188	13,507	1,923	14.2	Lost hdw 6.3, missing hdw 2.6, corr 1.4, crack 1.0	G-G-2-G
12240	Pilot's eject. seat MK117	2,170	19,147	1,841	9.6	Corr 5.8, stress corr 1.3, lost hdw 0.9, broken 0.6	Z-Z-G-G
1226F	Bucket, seat	24,456	247,371	1,670	0.5	Corr 0.2, lost hdw 0.1, crack 0.1, broken 0.1	Z-G-A-G
112BJ	Door 88 L-R 32-11035-( )	416	3,710	1,542	41.5	Lost hdw 29.9, missing hdw 7.4, crack 1.5, worn 1.2	G-G-G-G
12250	Radar pilot's eject.seat	3,421	29,399	1,348	4.5	Corr 2.2, stress corr 0.7, lost hdw 0.5, broken 0.7	Z-Z-G-G
1121J	Fuel tank RH int wing	6,893	72,381	711	0.8	Lost hdw 0.6, missing hdw 0.1, worn 0.1, deteriorate	G-G-G-G
12265	Container drogue	701	26,369	280	1.0	Worn 0.2, lost hdw 0.2, crack 0.1, broken 0.1	G-G-A-A
1225A	Bucket, seat (radar pilot)	3,873	27,625	181	0.5	Lost hdw 0.2, broken 0.1, crack 0.1, worn 0.1	G-G-G-G
1224A	Bucket assy (pilot's seat)	176	2,740	62	2.1	Stress corr 0.9, lost hdw 0.3, crack 0.2, scored 0.2	G-G-G
11327	Bellmouth cab	388	53	27	50.0	Broken 44.4, missing 3.7, worn 1.9	R-G-G
1224E	(Door 111 L-R)	1,551	36	13	34.8	Missing hdw 21.2, lost hdw 13.6	G-G

[*]Rank in descending order/struct MMH

Action taken codes:
A - bench chk & repair
F - repair
G - repair/replace minor parts
L - adjust
R - remove and replace
S - remove and reinstall
X - test, inspect, service
Z - Corrosion repair

field audit questionnaire (table IV), were used to structure interviews with maintenance personnel at the field audit sites listed in table V.

Table IV

STRUCTURAL FAILURE AND REPAIR FIELD AUDIT QUESTIONNAIRE

```
                                                    DATE ______________
   ORGANIZATION ______________________________  LOCATION ______________
   AIRCRAFT TYPE, MODEL, BLK NO _______________  SERIAL NO. ______  A/C HR ______

 1.  PART NO. _____________________  NOMENCLATURE ______________________
 2.  WORK UNIT CODE ________________  PART HOURS _________________________
 3.  PARTS CATALOG NO. _____________  PARTS CATALOG FIGURE & INDEX ________
 4.  REPAIR MANUAL NO. _____________  REPAIR MANUAL REFERENCE _____________
 5.  PART LOCATION (FS, WL, BL, ETC) _____________________________________
 6.  STRUCTURE          PRIMARY          SECONDARY            OTHER
 7.  WHEN DISCOVERED _______________  HOW DISCOVERED _________________
 8.  DESCRIPTION OF FAILURE _______________________________________________
     _______________________________________________________________________
     _______________________________________________________________________

 9.  TYPE/FORM OF PART MATERIAL/FASTENERS ________________________________
     _______________________________________________________________________

10.  CAUSE OF FAILURE ____________________________________________________
11.  REPAIR OR REPLACEMENT ______________  PERMANENT OR TEMPORARY REPAIR _____
12.  DESCRIPTION OF REPAIR _______________________________________________
     _______________________________________________________________________
     _______________________________________________________________________

13.  TYPE OF REPAIR MATERIAL/FASTENERS ___________________________________
     _______________________________________________________________________
14.  FABRICATION METHOD __________________________________________________
15.  MATERIAL PROCESSING _________________________________________________
16.  REPAIR INFORMATION SOURCE ___________________________________________
17.  REPAIR OR REPLACEMENT MAN-HOURS ____________  LABOR COST _____________
18.  PREREPAIR MAN-HOUR EXPENDITURE (DEFUEL, JACK, DEPANEL, ETC) __________
19.  POSTREPAIR MAN-HOUR EXPENDITURE (RETURN A/C TO FLT STATUS) __________
20.  MATERIAL, PARTS, KIT COSTS __________________________________________
21.  SPEC TOOL RQMT ___________________________  COST _____________________
22.  DEPOT OR FIELD-LEVEL REPAIR ___________  MEAN TIME BETWEEN FAILURES ___
23.  HAS TCTO, I-TIME INSP, ETC. BEEN ISSUED ____________  NUMBER _________
24.  ANY FLT RESTRICTIONS IMPOSED BY THIS FAILURE ________________________
25.  HAS A/C BEEN SUBJECTED TO SPECIAL FLT OPS RESULTING IN HI-G, ETC, DUE TO NONSTD LOADS, WEATHER,
     TWG, ETC ____________________________________________________________
     _______________________________________________________________________
26.  COMMENTS, SUGGESTIONS _______________________________________________
     _______________________________________________________________________
     _______________________________________________________________________
     _______________________________________________________________________
     _______________________________________________________________________
     _______________________________________________________________________
```

Table V

FIELD AUDIT SITES AND AIRCRAFT

Site	Aircraft
● **Air Logistics Centers**	
-Sacramento ALC, McClellan AFB, CA	A-10, FB-111, F-111, T-39, F-4
-Ogden ALC, Hill AFB, Utah	
-Oklahoma City ALC, Tinker AFB, Oklahoma	A-7, B-52, KC-135
-San Antonio ALC, Kelly AFB, Texas	C-5, F-106, OV-10, T-37, T-38
-Warner-Robins ALC, Warner-Robins AFB, Georgia	C-130, C-141, F-15
● Air Force Operational Facility	
-35th Tactical Fighter Wing, George AFB, CA	F-4
● Air National Guard Facility	
-146th Tactical Airlift Wing, Van Nuys, CA	C-130
● Naval Air Rework Facilities	
-NARF, Alameda NAS, CA	A-6, A-7, F-4, P-3, S-3A
-NARF, North Island NAS, CA	F-4
-NARF, Jacksonville NAS, Florida	A-7, P-3, RA-5C, S-2
-NARF, Cherry Point MCAS, North Carolina	A-8, C-130, F-4, OV-10
● Marine Corps Operational Unit	
-Tactical Reconnaissance Squadron 3, El Toro MCAS, CA	RF-4

Following field audit, a total of 241 potential candidate items had been identified; of these, 76 had been acquired through the field audit process, while the remaining 165 items were holdovers from the Rockwell-prepared lists. Many of the latter were retained because field audit information was not adequate either to substantiate or remove them from consideration. Thus, they were retained solely on the basis of field maintenance data (AFM66-1).

Of the 241 candidates, some were deleted from consideration because they had no significant time/cost-saving potential or their repair opportunities would provide no permanent corrective solution. Other candidate problems were dropped because the failure frequency and/or aircraft numbers did not justify the expenditures involved. As a result of this initial screening, the candidate list was reduced to a quantity of 72. The 72 items were reviewed in detail by specialists in the fields of design, structural analysis, manufacturing, manufacturing engineering, and materials and processing.

An evaluation of the specialists review comments, cost data information, field audit information, Air Logistics Center priorities resulted in the selection of 25 phase I candidate study items. From this list of 25 candidates, the five items shown on table VI were submitted by Rockwell for Air Force consideration as those which most closely fill the contractural requirements and therefore should be carried into phase II for repair design development.

Table VI

PHASE II CANDIDATE STUDY ITEMS

Item No.	Aircraft	WUC	Description
1	KC-135	14AHO	Outboard aileron tab
		14ADO	Inboard aileron tab
		14BHO	Rudder tab
2	F-4	1431B	Stabilator inboard trailing edge panel (repair)
3	F-4	2375E	Constant-speed drive gearbox access doors
4	F-4	1431B	Stabilator inboard trailing edge panel (redesign)
5	Generic		Pylon-to-wing attach points

2.1 <u>Summarized Conclusions and Comments From Phase I</u>

● A list of five recommended

phases I candidate study items
was established (table VI).

* Facilities visited were genu-
  inely interested in this program and its purpose of
  reducing life cycle costs by
  the application of innovative
  materials/processes.

* The corrosion problem is as
  much an Air Force problem as
  with the Navy. Preventive
  maintenance is the answer to
  today's inventory; future
  design efforts must be made
  to minmize or hopefully
  eliminate the problem in the
  next generation of aircraft.

* The maintenance data collection system does not fully
  satisfy the purposes
  intended: (1) the work unit
  codes are often too broad in
  scope for proper problem
  identification of the part in
  question, (2) little depot-
  level maintenance information
  finds its way into the AFM66-1
  system, and (3) the existing
  cost accounting systems do
  not provide detail data sufficient to identify and analyze
  specific problem areas. Field
  visits provide the only means
  of obtaining specific information associated with a given
  aircraft system.

3.   PHASE II - REPAIR CONCEPT
          DEVELOPMENT

The activities of phase II were
structured by two objectives: (1)
selected a primary repair/replacement
candidate, and (2) develop innovative
cost-effective repair solutions or
new designs, using advanced materials
and/or processes.

The five candidates selected in
phase I (table VI) were used as the
point of departure for the activities
of phase II.

Evaluation of each candidate started
with definition of the operational
environment and analysis of the
required specifications, together
with their impact on the system. High
maintenance cost items and their
causes were identified. Required
original design criteria, applicable
specifications, and field reports
were reviewed and drawings acquired.

System analysis was performed,
emphasizing the use of advanced technology for defect detection and
repair or replacement. Alternate
design concepts were evaluated and
trade studies conducted to determine
the most cost-effective design philosophy. The results of this effort
are briefly summarized in the following list and will be discussed in subsequent pages in the order listed.

* A C-130 aileron trim tab repair
  is presented, as practiced by
  commercial carriers.

* Wing to pylon joint corrosion
  prevention or detection systems analyzed were found not
  cost effective. A revision
  and maintenance plan is
  recommended.

* The F-4 constant-speed drive
  gearbox door repair using
  mechanical attachments was
  developed for bench test and
  early operational introduction.

* The design of F-4 stabilator
  replacement part using the
  titanium SPF/DB four-sheet
  technology was completed.

3.1 KC-135 Aileron and Rudder Trim
     Tabs

These tabs are shown in figure 1.
Damage of these parts appear to fall
into two categories: (1) water
intrusion and its effects, and (2)
mechanical damage such as dehts,
scratches, punctures, and delamina-

tions. Fixes for the mechanical
damage are presently made by generally accepted honeycomb repair
techniques. These are not the cost
driver for this item. The chronic
problem is water intrusion and the
corrosion and delamination problems
it causes. To try to seal such a
structure interrupted by hinges,
actuator balance weights, end fitting, and lightening strike provisions is nearly impossible. The
present method of dewatering the
parts does not cure the original
structural problems, which is the
perforated core which allows moisture to invade the interior.

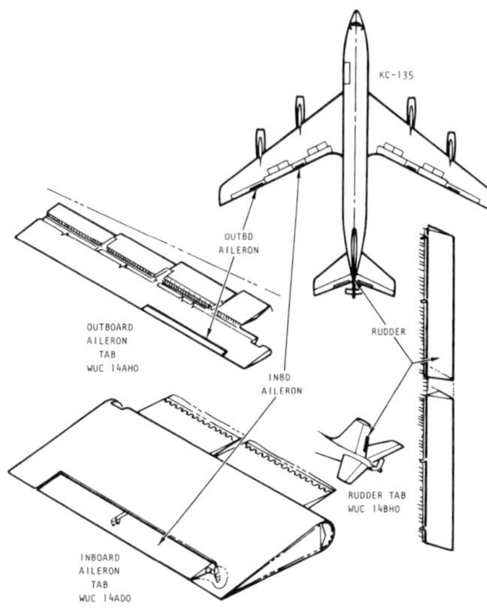

Figure 1. KC-135 Aileron
and Rudder Tabs

Because the commercial Boeing 707
airplane is similar in design to the
military KC-135, the project team
made an effort to benefit from
experiences of the airlines. A visit
was made to the TWA maintenance
facility at Los Angeles International
Airport to discuss this part with
their maintenance engineers. They
were generous in sharing their
experiences with flight control tab
repairs. Currently, they are not
having corrosion problems with 707

tabs. The reasons are: (1) this
airplane is moving out of inventory;
therefore, there are only a small
number in service, and (2) they have
a rebuilding technique developed by
their Midcontinent International
repair facility at Kansas City. A
subsequent call to the director of
this facility, Mr. Floyd Mielke,
provided more definition. Corroded
tabs are rebuilt from the fittings
up, using unvented core, nonoutgas-
sing adhesives, and new skins. This
repair service has been provided for
a number of airlines and appears to
work. The rework effort has gone
from over 100 tabs per year to two
to three at present.

Use of this or a similar repair service to resolve this problem on the
KC-135 is Rockwell's recommendation.

3.2 <u>Wing-to-Pylon Corrosion</u>

The structural interface on the wing
to pylon of the C-5, B-52, C-130,
and C-141' is a special structural
problem (figure 2). In all cases,
the close tolerances required for
strength and the installation techniques used eliminated the inclusion
of protective materials to prevent
corrosion. One possible factor promoting rapid corrosion of the pylon
attach fittings is fretting. These
unlubricated joints are subjected to
both high loads and significant
vibration which promote fretting.

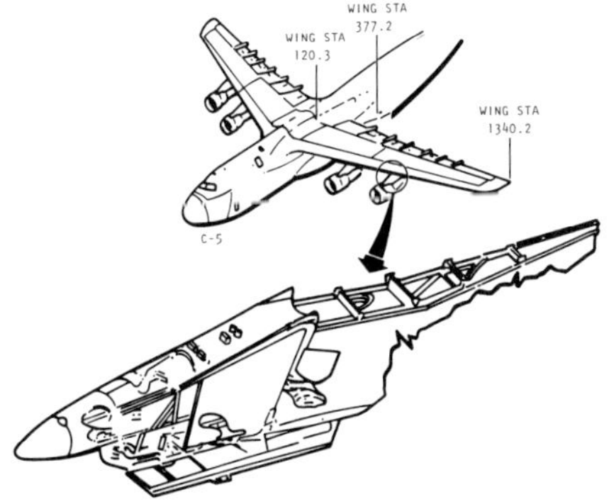

Figure 2. Generic Illustration of
Corrosion Area Wing-to-Pylon

When fretting occurs, the small
amount of debris generated rapidly
corrodes whenever water is present,
such as condensation from weather
changes. Once this corrosion
starts, the spongy residues formed
retain condensed water and further
accelerate the corrosion process.

Because perfect sealing is seldom,
if ever, achieved, attempts to
exclude water by application of
sealant almost always fails.
Schedule lubrication to prevent
fretting and reduce wetting of the
surface is far more practical and
usually the most effective method
for protecting working joints
against corrosion. It is recommended that future designs incorporate other methods of attachment
which prevent fretting and eliminate
trapped fluids which promote coro-
sion.

A "fix" which deserves further evaluation is to replace the present
bolt-bushing combinations with one
having periodic lubrication capability, together with a maintenance
schedule to insure that water is
regularly flushed from the joint
by fresh lubricant.

3.3 F-4 Constant-Speed Drive (CSD) Gearbox Door

The failure mode of this high-
maintenance cost contributor is
failure of the spot welds joining
the inner and outer sheets of titanium. (See figure 3.)

The anti-icing air comes from the
17th stage bleed on the engine at
1,200° R and 190 psi pressure. If
this pressure is reduced to one-
fourth (45 psi) in the door, a load
tending to fail the spot weld in
tension is that pressure times the
area around each spot. Because
spot welds are approximately 2
inches apart, the load is $45 \times 2^2 =$

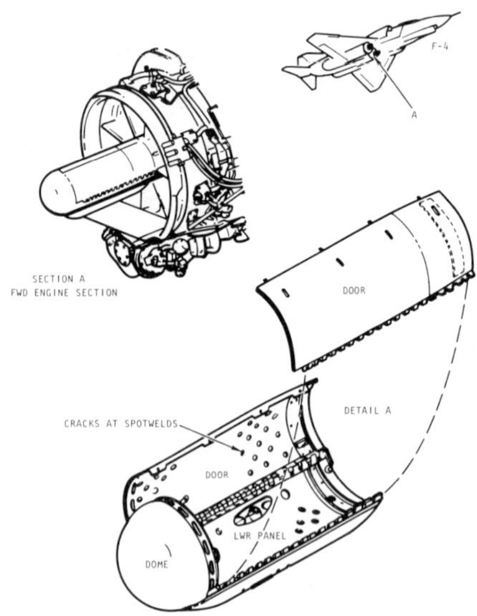

Figure 3. Constant-Speed Drive Door

180 pounds per spot. The allowable
shear strength of spot weld in titanium is 312 pounds. (Reference MIL-
HABK-5C table 8.2.2.3.4 (b) for
0.020 thickness.) The design allowable in tension is 25 percent or 78
pounds. The tension load imposed
by pressure will quickly fail the
spot weld loading. Thus, this
method of joining the sheets should
be replaced by one having higher
tension capabilities, particularly
in fatigue. This analysis is somewhat conservative inasmuch as there
are no shape factors involved.
Design studies resulting in possible
repair methods are identified in
table VII. These methods are all
configurations which would improve
tension fatigue capability.

3.4 F-4 Stabilizer Inboard Trailing Edge Repair

Concurrent with the studies described
in the preceding paragraphs, studies
were in progress on the F-4 stabilator trailing edge. As these stabilator studies progressed, they began
to indicate that a cost-effective

810

Table VII

F-4 CONSTANT-SPEED DRIVE DOORS REPAIR

	Crack Crack Size	Hole Dia	Washer Spacer	Fastener	O-Ring or Spring	Comments
SH 1	All	3/16	53-014044-3 1 CRES	M520615-6M	-	Seal
SH 2	.25	1/4	1 CRES	CSR925-8	-	Seal
SH 3	.31	5/16	3 CRES	M520615-4M	-	Seal
SH 4	.31	5/16	2 Titanium	M520615-6M weld	-	Seal
SH 5	.43	7/16	3 Titanium	M520615-6M weld	-	Seal
SH 6	.31	5/16	2 Ti or CRES	CSR925-10	O-Ring silicone	Self-seal
SH 7	.37	3/8	1 Ti or CRES	CSR925-10	O-Ring silicone	Self-seal
SH 8	.31	5/16	1 Ti	Weld		Self-seal
SH 9	.31	5/16	-	Explosive Rivet	O-Ring silicone	Self-seal
SH 10	.20	.21	None	Slug rivet	None	Hole filling

repair method (or a durable replacement) promised greater savings in
logistics costs than could be expected from any of the other three
candidates. Therefore, emphasis for
the remainder of phase II was concentrated on the F-4 stabilator
which is shown in figure 4.

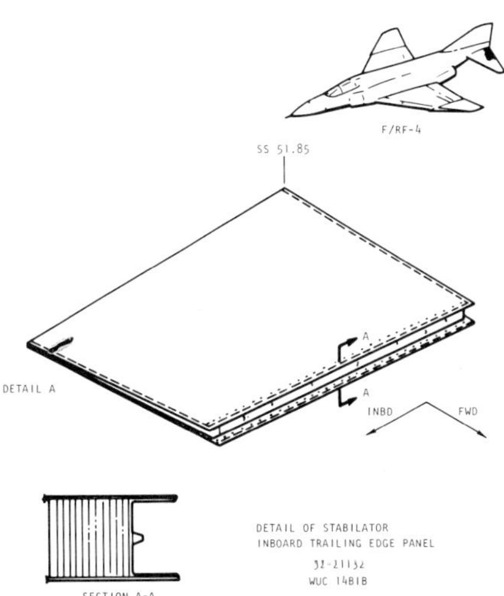

Figure 4. F-4 Stabilator

3.4.1 _Description_. The F-4 stabilator trailing edge is one of the
high-cost maintenance items identified under phase I of the contract.

Present configuration is a full-
depth honeycomb assembly. (See
figure 4.) Face sheet and core
material are 15-7 stainless-steel
silver brazed. The failure mode is
face sheet separation from the core,
followed by cracking in the cover.
Repairs consisting of a doubler on
both the top and bottom surface
attached by pins passing through
the core and welded to the
doublers were not successful. The
disbonded area continues to grow,
while weld failures at the edge of
the doubler open the interior to the
environment and can become a source
for cracks. This has required
replacement of trailing edges at
such a rate that shortages have
developed in replacement parts. Tactical readiness of the F-4 aircraft
is threatened by this situation.

Evaluation of a condemned part was
made by Rockwell and it was found
possible to peel the entire upper
face sheet from the core. The fracture of the joint had occurred in
the core to face sheet bond without
tearing the core. No contamination
was evident on the face sheet or
within the core. This type of
failure is attributed to lack of
sufficient fillet brazing alloy to
cause the failure to propagate into
the core. Further examination
revealed that the core cell nodes
were completely brazed. The faying
surfaces within the core represent
a large capillary reservoir that is
capable of taking brazing alloy away
from fillet areas. As a result, the
fillets are undersized and incapable
of transferring sufficient load to
the core to cause core failure,
hence the delaminations.

Past experience has shown that the
face sheet, which is on the upper
side during the brazing process,
will usually have smaller fillets
than the bottom side. This is due
to the additional affect of gravity
on the molten brazing metal. The

lack of brazing alloy of the upper
side fillets is the most likely
reason for failure in service of the
F-4 stabilizer inboard trailing
edge parts.

3.4.2 Design Criteria for the Trailing Edge.

* Form - Any repair or replacement will not significantly
  modify form, fit, or function.

* Weight - Weight distribution
  will be the same as the original part.

* Acoustics - Part will withstand a sound pressure level
  of 165 db.

* Temperature - Range will be
  from -67° to 400° F.

* Loads - The stabilator trailing edge will be 7-1/2 psi
  ultimate (down) and evenly
  distributed over the surface.

* Application - Application of
  the repair will be within the
  capability of the existing
  Air Force maintenance facility.

Work done during this contract to
define the stabilator environment
consisted of several actions to
verify temperatures and loads.

Design temperatures were at one time
thought to be 1,100° F. Because the
stabilator is above and aft of the
engine nozzle, it could be expected
that the upper surface would be
considerably cooler than the bottom.
A thermodynamic investigation showed
that upper surface temperatures with
aerodynamic cooling would be one-
half to one-fourth those of the
heated surface underneath. Figure 5
is a graph of upper skin temperature
as a function of two specific lower
skin temperatures and core depth.
This encouraged us to look at repairs

that could be used on the cooler
portions of the surface. During
this time, Hill AFB provided facilities for flight test of a stabilator
striped with temperature-sensitive
paint.

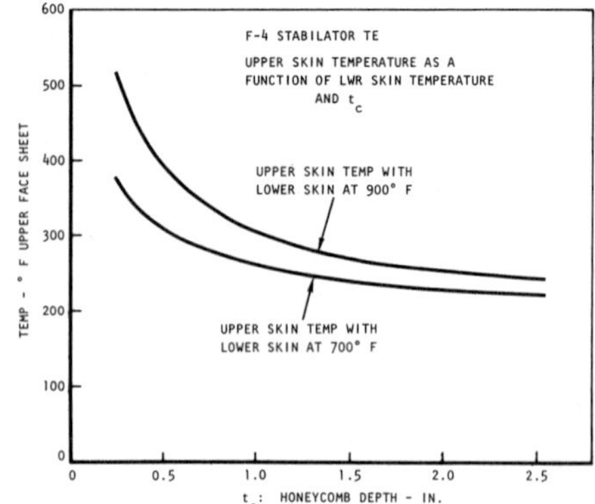

Figure 5. Stabilator Skin
Temperature Plots

The aircraft was flown on a mission
profile deemed by the airframe manufacturer to produce maximum heating
on the stabilator. Maximum temperature on the stabilator trailing
edge exceeded 400° F over approximately one-third of the lower surface, but was below 475° F maximum.
On the upper surface, temperatures
did not exceed 300° F. Immediately
adjacent to the part, on the outboard side, the 200° F paint was not
affected. On the basis of these data,
an organic repair of the stabilator
was proposed, using epoxy novilac
adhesive. Subsequent to selection
of the F-4 stabilator inboard trailing edge as the prime repair/replace-
ment candidate, Rockwell had assisted
the Advanced Structure Technology
Group (FIBCB), Structural Concepts
Branch of the Wright Aeronautical
Laboratories to develop a comprehensive instrumentation and flight-test
plan to define strains, acoustic
noise, and surface temperatures and
establish gradients over the entire
surface.

3.4.3 Solutions. On 24 October
1978, a request was received from

the Air Force Project Office for an
emergency repair procedure to prevent the trailing edge from becoming a non-mission-capable supply
item. A series of alternate repair
concepts was developed and defined
on engineering sketches.

Four concepts were of the pin
doubler type, one of which is shown
in figure 6.

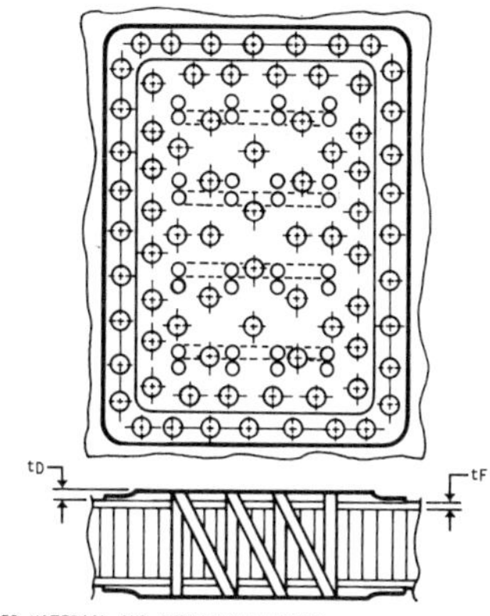

Figure 6. External Doubler
and Pin Repair

Another repair method investigated
was to reattach the face sheets by
applying an adhesive between face
sheet and core. Tests were run to
determine the strength of repairs
on a stainless-steel honeycomb
assembly with core-to-face voids.
Repairs were made by injecting an
adhesive through holes drilled into
the voided area. Specimens were
2-1/2 by 12 inch portions of a
separated F-4 stabilator trailing
edge, which were tested in three-
point beam bending. This determined
whether they could withstand the
bending fatigue requirement of the
part. Two adhesive candidates were
evaluated. One, a savreisen ceramic cement, would be compatible
with steel in temperature capability
and could be injected. The second,
HT 424 epoxy adhesive, would sustain
temperatures as high as 400° F and
thus would be applicable for repairs
over most of the trailing edge.
Both systems failed. The ceramic
cement did not develop enough adhesion and separated during loading,
and the epoxy did not inject well
because it was too viscous to pass
through the small gap between face
sheet and core. Because void adhesion was incomplete, failure
occurred prematurely.

Void injection of brazed steel
honeycomb does not appear to be a
feasible method of repair. Stiffness of the face sheets makes spreading difficult, and adhesion of the
high-temperature adhesives investigated was below expectations for
this application.

Establishment of the trailing edge
temperature environment showed that
organic repair was generally applicable. Therefore, a repair involving
the replacement of a disbonded area
by bonding became a promising solution. Such repair would involve
removing the cover in the disbonded
area, cleaning and stabilizing the
core with foaming adhesive, and
applying an external patch. A development test and demonstration plan
was initiated to provide this repair.

Data from this test consist of lap
shears of several lengths of skin
joints and flatwise tensile, edgewise compression, and core shears of
sandwich specimens made from material procured from a scrapped part.
Results of lap shear tests are
plotted in figure 7. As shown in
this figure, a minimum overlap of
1 inch would marginally meet the
ultimate load requirements. However, because it is likely that

biaxial forces exist and to give
more statistical confidence, an overlap length of 2 inches was chosen.
This length will develop 90 percent
of the face sheet yield strength and
should make the structural joint not
critical. Specimen fabrication and
testing will continue in order to
qualify this repair for application
in service. This effort is planned
to provide supporting data for a
subsequent F-4 stabilator repair
demonstration.

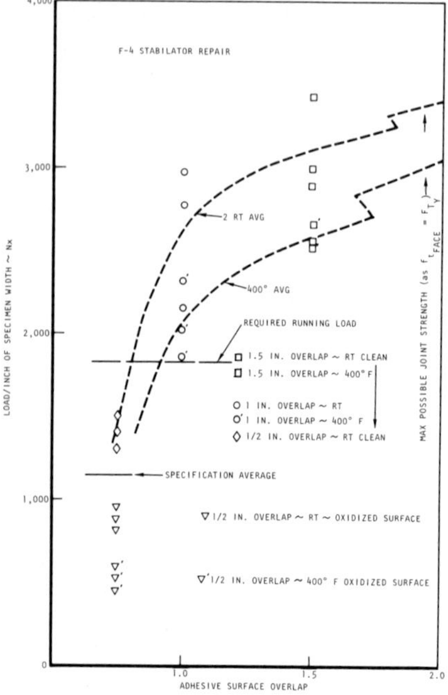

Figure 7. Lap Shear Coupon Results

3.5 Stabilator Trailing Edge Replacement

Description of the problem and
design criteria for a replacement
part are similar to that used in the
repair study (paragraph 3.4).

The existing trailing edge is made
of stainless steel. However, the
reductions in design temperature
allow a flexibility of material
choices. Good strength in the 400 –
600° F range is believed to be adequate. Titanium performs well in
this range, has a good corrosion
resistance, and can be diffusion
bonded (DB) and superplastically
formed (SPF). SPF/DB is a process
patented by NAAD and has been demonstrated to make complex parts with
integral internal support members.
Though several components are used
during the assembly process, the
resulting part is a single piece
where interfaces have metallurgically
been eliminated. Because interfaces
are a problem with the F-4 stabilator trailing edge, a study of SPF/DB
designs for the part was made.

During phase II, conceptual design
studies were conducted as a basis
for selection of a configuration for
an SPF/DB replacement stabilator.
Six concepts having various core
configurations were defined on preliminary design drawings. These
concepts are to be evaluated in
phase III. This evaluation process
will culminate in an Air Force/
Rockwell joint selection of the
most cost-effective concept. The
selected concept will then be detail
designed.

Preliminary evaluation indicates
that the double truss core concept
may prove to be an attractive concept for this use. Double truss
core was used successfully by NAAD
in fabrication of a cooler duct for
the British Tornado. (See figure
8.) Two variations of this concept
are shown in figures 9 and 10.

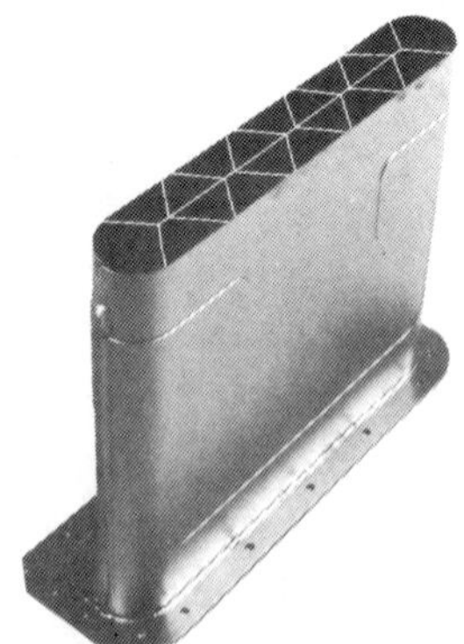

Figure 8. Double Truss Core
Cooling Duct for British
Tornado

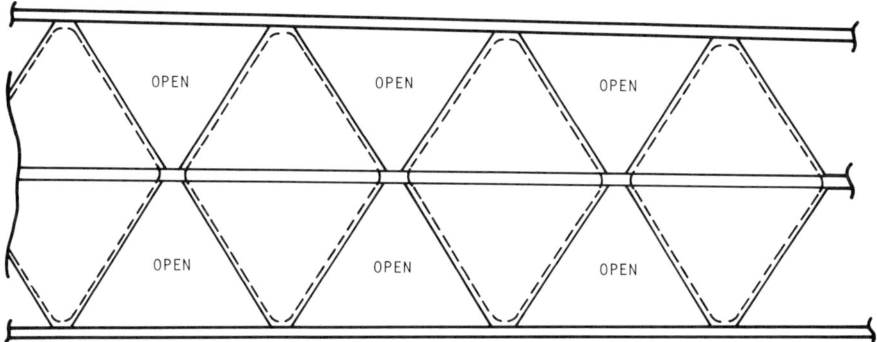

Figure 9.   Option 1 View Looking Aft at Core Termination, Showing
Webs Formed by Bringing Core Out of Part at Centerline

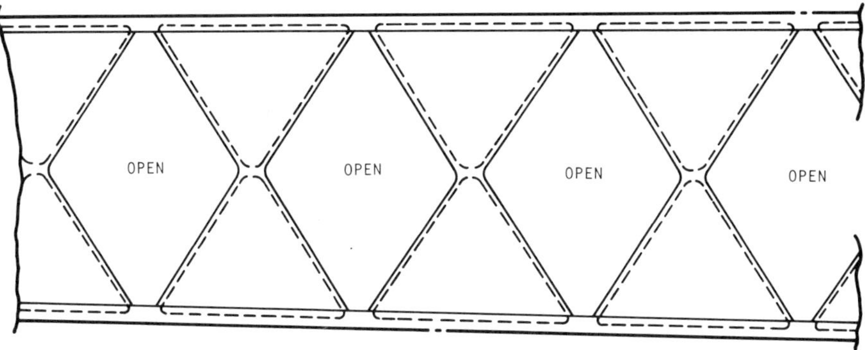

Figure 10. Options 2 and 3 View Looking Aft at Core Termination, Showing Webs
Formed by Bringing Core Out of Part with Skins

4.   PHASE III - FEASIBILITY
     DEMONSTRATION

In phase III, the following
activities are planned:

- Detail test plans

- Fabrication and instrumentation
  of test specimens

- Test and evaluation

- Test report

Successful completion of this program is expected to make a significant contribution to the DOD
objectives summarized in paragraph 1
of this paper.

REPAIR OF ADHESIVE BONDED STRUCTURE
ON COMMERCIAL AIRCRAFT
Tony Seidl
United Airlines Maintenance Operations
San Francisco, California

Abstract

This presentation provides a general review of United Airlines' experience in the maintenance and repair of adhesive bonded aircraft parts. Both the metallics and the nonmetallics will be discussed. United's efforts in this field span two decades, beginning with modest wet lay-up resin/fiberglass patches and culminating with complete rebuilds to O.E.M. standards utilizing state-of-the-art technology. United's decision to engage in repair processes of such magnitude was based on economic considerations: high unit replacement costs; long lead times; high inventory costs.

Once philosophically committed, the airline had to overcome a number of obstacles: training of mechanic personnel in "white glove" techniques and attitudes; designing and constructing a bonding facility capable of reproducing O.E.M. standards; designing and building bonding tooling; developing both engineering and technical specifications; testing and qualifying materials, processes, and tooling. United's approach has been and is governed by practicality: our goal has always been to develop the most cost effective repairs consistent with safety and legal requirements. Much of the progress achieved by United Airlines would have been impossible without the generous assistance of the original equipment manufacturers and material suppliers who, in the interest of continued product improvement, cooperated with us at many critical junctures.

1. INTRODUCTION

Without structural adhesives, modern aircraft would be inconceivable. The advantages of bonded over conventional structure require no elaboration. When bonded parts were first introduced, however, maintenance was ill-prepared for them. And even though there existed a sizeable body of theoretical and scientific literature on the subject, little of it was of much practical use. Maintenance inspectors and mechanics soon discovered the disadvantages of bonded, particularly honeycomb structures: they required special handling, since they were so easily dented or punctured; they were subject to delamination, water contamination, corrosion, and weight gain. The latter phenomenon was of special concern when it affected weight sensitive flight control surfaces.

Operational and economic necessity soon generated a new skill category: The Bonding Repair Specialist.

2. INSPECTION

One of the most common problems associated with bonded parts is their propensity to absorb water, the effects of which are by now well known. The presence of water may be inferred if the unit weight has increased over a previously established weight, or it may be verified by an X-ray examination or, in the case of nonmetallic structure, by means of a moisture detector. Delaminations, often directly resulting from ingested water, may be identified by coin-tapping. External and/or filiform corrosion are easily spotted visually.

2.1 ESTABLISHING WEIGHT CRITERIA

It is prudent policy to weigh and tag all flight control tabs, wing spoilers, and similar types of sandwich structure at the earliest possible opportunity. At future removals, quick "wet-or-dry" decisions can be made on the basis of weight alone, thus avoiding costly and time consuming X-ray procedures. Figure 1 illustrates this principle.

2.2 RADIOGRAPHY

X-rays provide a useful means of discovering liquids in honeycomb cells. Using the developed film as a "map", it is relatively easy to spot and evacuate the contaminants. Occasionally, anomalies occur that are mistaken for water: heavy resin concentrations, fillers injected at previous repair visits, or metal particles left over from the original core machining operation.

2.3 MOISTURE DETECTORS

These electronic devices are extremely accurate in locating even small traces of moisture inside a honeycomb panel. Unfortunately, their use is restricted to the nonmetallics. Metallized coatings (anti-static paints or flame-spray coatings) must first be removed from the fiberglass surface before any moisture readings may be taken. After initially calibrating the instrument, the operator simply scans the panel and "maps" the wet/ dry boundary, thus isolating the water-contaminated areas.

2.4 COIN-TAPPING

This relatively simple procedure is widely used despite the availability of much more sophisticated instrumentation. When coupled with an experienced ear, the coin-tap test is remarkably accurate. Its limitations are that it detects only actual disbonds, and that it gives no qualitative analysis. The tap test is used most reliably on metallics, but can be used on nonmetallics as well.

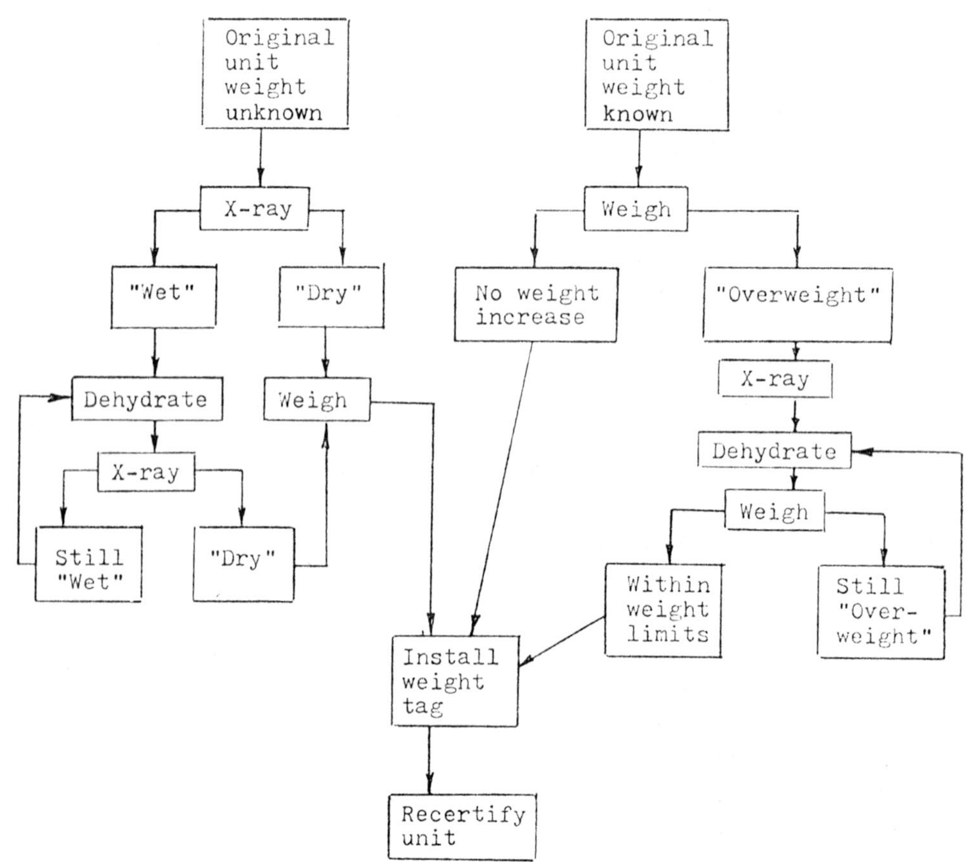

DECISION DIAGRAM

Establishing Weight Criteria for Honeycomb Sandwich Structures

Figure 1

3. REPAIR

Repair action is called for whenever established serviceability criteria are no longer met or certain limitations are exceeded. These criteria and limitations are a function of the service environment of a given part and therefore subject to considerable variabilities. Generally, parts are removed from aircraft because of physical damage, delamination, corrosion, or a combination of the preceding.

3.1 PHYSICAL DAMAGE

A relatively new and otherwise sound bondment damaged by external causes is often the easiest to repair, since it has not suffered the gradual degradation characteristic of older units. Provided no major structural parts are damaged, core plugs and skin patches are usually sufficient to restore the part to service. Large size repair bonds are usually accomplished in an autoclave. Small patches may be applied using heating blankets and vacuum bags. In all cases, the repair must be inspected for potential leak paths and then sealed to prevent moisture entry. Polyurethane paint and/or polysulfide sealant are used for this purpose.

3.2 DELAMINATIONS

In aluminum honeycomb bondments, delaminations are generally associated with corrosion at the metal/adhesive interface, although delaminations caused by excessive temperatures are not uncommon. Unlike physical damage, delaminations are a sign of degradation and delamination repairs are by definition life-limited.

In nonmetallics, delaminations are mostly the result of water ingestion followed by freeze-and-thaw cycles. Overheating can also precipitate disbonds.

Hydraulic fluids are less likely to cause delaminations since they cannot freeze; they have nevertheless a most deleterious effect on nonmetallics, for they seem to become absorbed by the fiberglass and subsequently plasticize the resin matrix. "Skydrol" soaked panels have a characteristic spongy appearance.

3.2.1 CORROSION REMOVAL

Interfacial corrosion occurs when water enters the bondment through the adhesive layer from where it is free to attack the adherend surfaces. Corroded adherends are routinely removed and discarded.

Core corrosion occurs when water enters either through the adhesive layer, or through potted inserts, once the protective finishes have been allowed to deteriorate. Once entrapped inside the core, water can in time destroy the entire detail. Core splice adhesives, especially low density types, absorb water and transfer it over large areas. Contaminated and corroded core is routinely removed and discarded.

Corrosion on structural members (spars, hinge fittings, etc.) is amenable to conventional treatment. Such parts are reusable after they have been etched in a hot sulfuric acid/sodium dichromate solution, anodized in phosphoric acid, and coated with a corrosion-inhibiting adhesive primer baked after application.

Surface corrosion and minor pitting of the face sheets, provided no disbonds or internal/interfacial corrosion exist, can be cleaned up using conventional methods. Such parts should, however, be protected against further degradation by an appropriate surface finish.

Filiform corrosion under existing paint finishes is generally observed along fastener rows or panel edges. Left untreated, filiform corrosion can result in severe degradation, especially on panels with thin gage outer face sheets. The fasteners are usually removed from the affected area, all finishes stripped, the corrosion removed, and an appropriate corrosion inhibiting paint system applied. The fasteners are then reinstalled "wet" with polysulfide sealant to provide dissimilar metals protection.

3.2.2 LIQUID REMOVAL

Liquid removal through evacuation should only be attempted if there are no disbonds already in evidence, and the contaminated areas are small and localized. The actual evacuation path is created by drilling small holes into the contaminated areas, then perforating the adjacent cell walls with a needle tool. At low vacuum pressure and moderate heating, the contaminants are then drawn off, the evacuation

holes sealed up and the part refinished.

The preceding is generally used only on metallic sandwiches; on nonmetallics, water removal most often requires the removal of the face sheets. In some cases, however, water has been successfully removed by low heat and vacuum pressure alone.

Removing hydraulic fluid ("Skydrol") from a nonmetallic panel is practically impossible. The nature of this fluid is such that it prevents adhesion and thus renders contaminated panels non-bondable. Attempts to effect resin bonds on such panels have had disappointing results. At United Airlines, "Skydrol"-contaminated nonmetallics are normally scrapped.

3.2.3 REPLACEMENT OF FACE SHEETS, DOUBLERS, AND CORE MATERIAL

Delaminated materials are normally removed and discarded. As a standard practice, the replacement core is of the corrosion-resistant type. Replacement face sheets and doublers are routinely cleaned in an alkaline solution, etched in sulfuric acid/ sodium dichromate, anodized in phosphoric acid, and spray-primed with corrsion-inhibiting adhesive primer followed by oven bake.

On partial face sheet/doubler repairs, the lap splice area of the remaining structure (which cannot be submerged in the solution tanks) is etched with an acid paste, rinsed and spray-coated with CIP. A bench type phosphoric acid anodizing process, which utilizes wire mesh cathodes and acid jelly has been rejected by United Airlines by United Airlines as too cumbersome, at least for the present time.

All repair bonding is accomplished in an autoclave or (depending on patch size, availability of tooling, or operational constraints) by means of heating blankets under vacuum bags.

At the completion of the bonding repair all bondlines are inspected for possible leak paths before being sealed with polyurethane enamel and/or polysulfide sealant.

3.2.4 REPLACEMENT OF STRUCTURAL MEMBERS

Corroded spars, ribs, closeouts, hinge fittings, pans, etc., that yield unacceptable cross-sectional dimensions after corrosion clean-up are generally scrapped. Replacement parts receive state-of-art pre-bond treatments (Alkaline – FPL – PAA – CIP).

Replacement of structural members cannot normally be accomplished without tooling and fixtures.

3.2.5 REPLACEMENT OF FACE SHEETS AND CORE IN NONMETALLICS

In the repair of nonmetallic structure, the operator has a number of options. 1 – Wet Lay-Up; 2 – Prepreg Lay-Up; 3 – Application of Precured Patches and/or Core Plugs. Wet lay-up repairs are the most common and are highly effective provided that damage is small and some weight gains can be tolerated. As a standard practice, wet lay-ups should be cured under vacuum pressure.

Prepreg lay-up repairs are more desirable from the standpoint of durability, moisture resistance, and weight savings. These repairs, however, require heat input and therefore tooling and fixtures.

Precured patches and plugs are sometimes used in the repair of large structures that do not lend themselves to being fixtured and autoclaved.

The precured parts may either be hot-bonded to the existing structure utilizing film adhesives and heating blankets, or cold-bonded under vacuum pressure.

All repairs to nonmetallics require protective coatings and paint finishes. Some require the restoration of anti-static or flame-spray coatings also. In such cases, it is important to make sure a ground path exists to allow for static bleed-off via bonding fasteners.

4. FINISHES AND SEALANTS

Of equal or perhaps even greater importance than the bonding repair itself is the selection, application, and maintenance of finishes and sealants. With the advent of nonmetallics and, more recently, the advanced composites, protective coatings must be maintained on an ever growing percentage of an aircraft's exterior surface. The maintenance of these coatings will be costly but unavoidable.

It cannot be overemphasized that
the initiation of most bond failures starts when a protective finish becomes damaged or is allowed
to deteriorate. A high proportion
of disbonds are initiated when fasteners are installed dry rather
than wet, or when edges of panels
are allowed to remain unprotected
by sealants, or potted edges develop cracks through which moisture
enters the core.

4.1 POLYURETHANE ENAMEL

An easily applied bondline sealant
and protective coating is polyurethane enamel over a compatible
fluid-resistant primer. Used on
both metallic and nonmetallic
structure, this system has become
the preferred finish for most bonded parts. It is also used as a replacement for the bondable Tedlar
film used by some manufacturers as
a moisture barrier.

4.2 POLYSULFIDE SEALANTS

Various types and consistencies of
these sealants exist. The following are among the more common uses:
1 - Filling gaps between panels, or
between panels and other structure,
thus providing the necessary edge
protection against moisture attack;
2 - Wet installation of fasteners,
thus providing both a moisture seal
and dissimilar metals protection;
3 - Sealing of cavities, inside corners, hinge boxes, etc., thus providing an important moisture barrier; 4 - Bonding of rubstrips, bus
strips and static dischargers, thus
preventing the inevitable corrosion

found on such installations when
bonded with epoxy resins instead.

4.3 SILICONE SEALANTS

Silicone sealants are excellent heat
barriers. They have been used with
good results on bonded structure
located in close proximity to heat
sources, both actual and potential,
(thermal anti-ice ducts, valves,
duct joints, engine exhaust air,
etc.). Silicone adhesives provide
thermal protection when used to
bond titanium fire shields to bonded
structure, in contrast to the epoxy
resins sometimes used during manufacture.

5. TOOLING AND FIXTURES

Major repairs to bonded structure
are impossible without proper fixtures, contour molds, honeycomb
cutters, etc. Since much of this
equipment must be tailored to specific parts, it is therefore not
commercially available and operators
are constrained by necessity to design and manufacture their own. The
following paragraphs are intended to
give a general review of United Airlines' experiences and practices in
this phase of the operation.

5.1 USING THE PART AS
     ITS OWN TOOL

Small repairs can be applied successfully by envelope-bagging a
part and heating the entire assembly in a pressurized autoclave.
While saving tooling costs, this
method can be very risky: weaknesses in the existing adhesive can
lead to additional disbonds while
the repair is being cured; weaknes-

ses in the existing core may mani-
fest themselves in the form of node
bond separations, although some of
this problem may be overcome by the
use of pressure cauls; worst of all
the risks, however, is the chance
of warping the part out-of-contour.
Bagging a part "to itself" should
only be contemplated if more conventional alternatives are not
available.

5.2 BONDING PLATENS

Flat panels of any composition are
most often bonded directly on an
aluminum table of sufficient thickness to yield warpage-free parts.
These autoclave tables are generally provided with a wheeled under-
carriage that allows them to be
rolled directly into an autoclave.
Individualized platen fixtures designed to hold specific bonded
units, composed of a platen base,
spar and rib support blocks, and
various alignment provisions, are
in common use. At the completion
of the part lay-up, they are normally bagged to the autoclave table.
Since many of these fixtures are
quite large, and many of their constituent parts are quite massive,
heat-up rates must be carefully
monitored to achieve proper curing
of the adhesives located next to
spar and rib supports, which tend
to represent "cold spots".

5.3 CAUL PLATES

Curved panels are often cured
against caul plates -- metallic
sandwich type structure against
aluminum cauls, nonmetallics

against nonmetallic cauls.
Aluminum platens can be locally
manufactured to match the contour
of the original part either by step-
braking a thick sheet of aluminum,
or by laminating several thin sheets
that have been rolled individually
and are then bonded together with
structural film or paste adhesives.
Nonmetallic platens are most often
made from several plies of glass
fabric and backed by overexpanded
honeycomb core. Both fiberglass
cloth/resin combinations as well as
epoxy prepregs have been used with
success. Caution must be exercised
when designing such caul plates, to
select thermally compatible materials, and to make sure that the tool
is of sufficient thickness to prevent part warpage during the cure
cycle.

5.4 BONDING MOLDS

Panels exhibiting compound curvatures, regardless of composition
(metallic or nonmetallic) have been
successfully restored, using tooling
molds made from nonmetallic laminates backed by either overexpanded
or flex core. Large assemblies such
as engine cowling or thrust rever-
sers may require either an "egg
crate" back-up or spreader bars in
order to maintain the required dimensional stability, particularly if
the part is being cured at elevated
temperatures.

5.5 CORE CUTTING AND
MACHINING

Slabs of expanded honeycomb core are
normally cut using high speed band-

saws with specially designed blades.
The final machining takes place after the completion of the first
cure cycle.

Wedges of regular geometry are generally "shaved" to final core dimension on a vacuum chuck, using an
adjustable milling machine. Parts
having irregular geometries or variable contours require fixtures especially designed to yield the proper shapes.

6. FACILITY REQUIREMENTS

A typical bonding repair facility
should contain the following equipment: 1 - A cleaning facility for
the pre-bond treatment of aluminum,
consisting of hot alkaline and FPL
solution tanks with appropriate
rinse provisions; 2 - A phosporic
acid anodizing tank with appropri-
ately sized DC rectifier; 3 - A
spray booth and associated equipment for the application of corrosion-inhibiting adhesive primer;
4 - An oven, or infrared heat
source, for the pre-curing of the
CIP; 5 - A contamination-free, temperature and humidity-controlled,
positively pressurized adhesive
lay-up room; 6 - An autoclave of
appropriate pressure and temperature capability, and of the necessary size; 7 - Frozen storage capability for adhesives and prepregs;
8 - Lab facilities for the testing
and evaluation of adhesives and
other materials used in the bonding
processes.

7. DISCLAIMER

United Airlines offers the preceding information as helpful suggestions only. United makes no guaran-
tee of results and assumes no obli-
gation or liability in connection
with this information. Anyone using
information contained in this paper
concerning equipment, techniques, or
materials, should satisfy him/her-
self that the recommendations are
suitable for his/her specific needs.
This paper is not intended to suggest an infringement of any existing
patent or proprietary information.

8. BIOGRAPHY

Mr. Seidl is currently employed by
United Airlines at the San Francisco
California, Maintenance Operations
Center. His current assignment is
as a Staff Maintenance Specialist in
the Sheet Metal Component and Adhesive and Bonding Shops. For the
past 15 years, his primary responsibility has been in the development
of repair procedures and specifications used in the repair and maintenance of adhesive bonded assemblies. Previously, he served in the
United States Air Force as an aircraft maintenance mechanic, following which he joined United Airlines
working in Cabin Interior, Paint,
and Sheet Metal skills. Mr. Seidl
attended Georgetown University,
Washington D.C. and San Jose State
University, San Jose, California.
He holds B.A. and M.A. degrees, respectively. He is a member of the
Northern California Chapter of
S.A.M.P.E.

26th National SAMPE Symposium
April 28-30, 1981

DEVELOPMENT OF LARGE-AREA DAMAGE
REPAIR PROCEDURES FOR THE F-14A
HORIZONTAL STABILIZER
PHASE I - REPAIR DEVELOPMENT

John Mahon
Grumman Aerospace Corporation
Bethpage, New York
and
James Candella
NARF, Norfolk, Virginia

Abstract

This paper describes a large-area repair approach that was developed and successfully demonstrated to repair structural damage up to 6-in. in diameter to the boron/epoxy covers of the F-14A horizontal stabilizer. This repair technique, which is applicable to those areas of the covers up to and including 24 plies thick, involves application of an autoclave cocured and adhesive bonded boron/epoxy flush patch to a damage site that has been stepped to accommodate the patch. The stepped repair concept is based on the STEPS 7 computer program that analyzes multiple-step bonded joints in which one or both adherends consists of advanced composite materials. This program, which performs a modified elastic analysis of the advanced composite joint, compares combined uniaxial tensile or compressive loads with in-plane shear loads using interactive equations. The large-area repair was verified by successful static testing of three unconditioned, damaged, and repaired test elements at both ambient and elevated temperatures. A single repaired element also successfully withstood spectrum fatigue loading for the number of cycles corresponding to 18,000 equivalent flight hours (EFH). This element, which was subsequently subjected to static loading at 300°F, failed at 139% of design ultimate load (DUL).

Keywords: Large-Area Damage
           Repair, Advanced
           Composites, Boron/
           Epoxy Flush Patch,
           Stepped Site

1. INTRODUCTION

To date, over 430 F-14A aircraft have been manufactured by Grumman; a large percentage of these aircraft have been in active service for several years. As a result of this fleet deployment, many of these aircraft have sustained damage to the boron/epoxy stabilizer. Much of this damage has been less than 2-in. in diameter and was effectively repaired by cognizant NARF agencies. The damage to the horizontal stabilizers of some aircraft, however, has been far more extensive, approaching 6-in. in diameter. At present, there are several stabilizers in storage at NARF-Norfolk with damage that could be repaired by application of the large-area stepped patch.

Typical stabilizer damage is most likely the result of impact between the stabilizer and sharp corners, the tines of a forklift truck, or protruding parts on other moving vehicles. This type of impact can result in:

- Rupture and delamination of the stabilizer covers
- Rupture of the adhesive bond between the composite cover and the aluminum honeycomb substructure
- Crushed honeycomb core.

Experience with the above damage (primarily due to forklift impact) is that the resultant damage is almost always restricted to one cover and the underlying honeycomb core. The cover opposite the impact site is usually unaffected by the collision. Most of the presently available scrapped stabilizers have damage to one side only. The damage typically consists of an elongated puncture approximately 6 x 2-in. surrounded by an area of localized disbonds and delaminations. The total area of the stabilizer cover degraded by the forklift impact is seldom in excess of 10 x 6 in.

The Large-Area Repair Program (Development of Large-Area Damage Repair Procedures for the F-14A Boron/Epoxy Horizontal Stabilizer) was initiated to develop structural repairs for damage up to 6 in. in diameter to the F-14 horizontal stabilizer boron/epoxy covers (Fig. 1). The repairs are to be applicable to those areas of the stabilizer covers which sustain relatively moderate loads during flight, that is, areas having thicknesses up to and including 24 plies boron/epoxy prepreg. The repair approach involves application of an autocalve cocured, adhesive bonded, boron/epoxy flush patch to the damage site. The repair site is stepped to accommodate the flush protion of the patch. The external doubler portion of the repair patch is covered with a ply of fiberglass/epoxy. A stepped repair site is shown in Fig. 2. A tested repair patch is presented in Fig. 3.

The Large-Area Repair Program is divided into four phases:

- Phase I - Repair Development
- Phase II - Environmental and Fatigue Resistance
- Phase III - Design/ Manufacturing Iteration - Production Repair Procedures
- Phase IV - Large-Area Repair Verification.

This paper presents the technical results of the Phase I effort.

During Phase I, the large-area repair was developed and verified by the successful static testing of three unconditioned, damaged and repaired

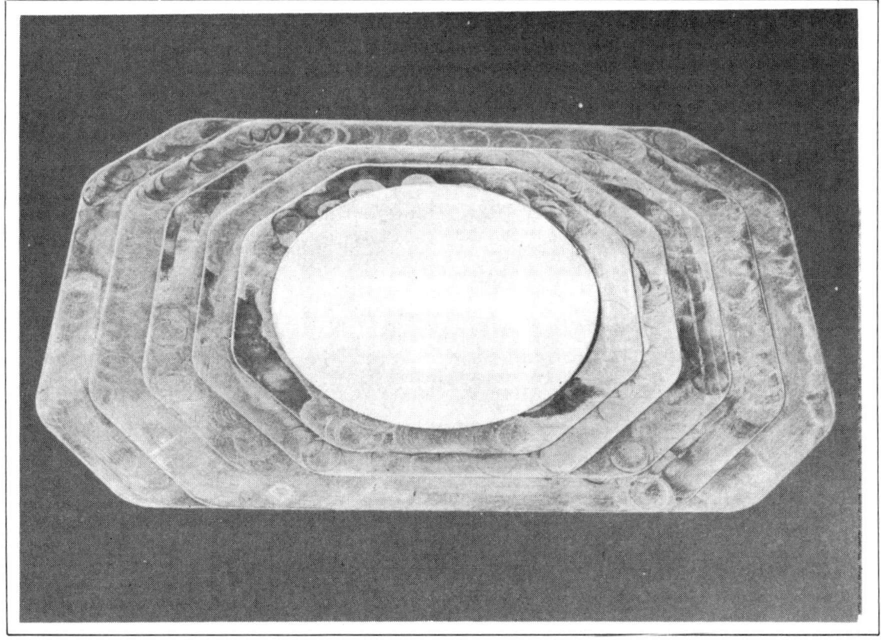

Fig. 1 F-14A Boron/Epoxy Horizontal Stabilizer

Fig. 2 Repair Site Terraces

Fig. 3 Repaired Test Element

test elements at ambient and elevated temperatures. In addition, a single repaired element was tested in spectrum fatigue loading (maximum load-design limit load) for a number of cycles corresponding to 18,000 equivalent flight hours (EFH). Following fatigue testing, the repaired element was successfully tested under static loading at 300°F. The repaired test element failed at 139% of design ultimate load. Prior to the above testing of the repaired specimens, a control element was evaluated under static loading at ambient temperature. The Phase II effort will validate the environmental durability and fatigue resistance of the stepped/flush repair. This phase, which will be performed concurrently with the Phase III effort, will include the residual static testing of repaired elements in tension and compression after spectrum fatigue loading at 170°F and 95% RH. The duration of fatigue exposure will be equivalent to three times the design life (18,000 EFH).

Phase III will be performed concurrently with Phase II. This effort will include a design/manufacturing iteration to establish a repair methodology that is cost-effective and compatible with the material and configuration requirements of full-size stabilizers.

The Phase IV technical effort will validate the large area repair methodology. Three fleet stabilizers damaged in the repairable areas defined in Phase I will be repaired and statically tested at 300°F in a final demonstration of the performance of the large-area repair.

2. F-14A HORIZONTAL STABILIZER

2.1 Design Criteria

A fully reversible, limit air load of 40,000 lb, acting at any point within a specified center-pressure envelope, was established for the stabilizer. An appropriate amount of fully reversible inertia that produces a net-limit design load equal to 42,040 lb. was also considered. Horizontal stabilizer temperatures corresponding to critical design conditions are 335°F for the covers and 300°F for the substructure.

2.2 General Description

The stabilizer is comprised of four major assemblies: the torque box,

leading edge, trailing edge, and tip
cap. The main load-carrying component is the torque box; the other
three components serve to support
and distribute the aerodynamic
pressures to the center box.
The leading edge, trailing edge, and
tip cap assemblies are fabricated from
conventional aluminum skin and full-
depth 2024-T81 aluminum honeycomb
core. The main box structure consists of the root rib, forward and aft
bearing support beams, outboard
bearings, outboard rib, front and
rear beams, tip rib, honeycomb core,
and two covers. The main torque
box is filled with full depth, 2024-
T81 aluminum honeycomb core which
acts as the principal shear-carrying
structure in addition to stabilizing
the skins. The torque box upper
and lower covers consist of boron/
epoxy skins, a titanium peripheral
boundary forming the edge splice at
the root rib, front and rear beams,
and tip rib, and an integral titanium
splice plate covering the pivot region
from the vicinity of the outboard rib
to the root rib. The stabilizer has
been designed so that there are no
mechanical fasteners through the
boron skins.
The torque box boron/epoxy composite covers are composed of a basic
0°, 90° and ± 45° laminate configuration using Avco 5505 prepreg
material. The 0° layers are parallel
to the 50% chord line except in the
"apex" region, i.e., that portion of
the surface bonded by the front beam,

outboard rib, forward bearing support
beam and root. Spanwise cover
loading is primarily carried by the
0° layers, chordwise loads by the 90°
layers, and in-plane shear loads by
the ± 45° layers.

3. STEPS 7

COMPUTER ANALYSIS PROGRAM
STEPS 7 is a computer program that
estimates the strength of multiple-
step, lap-bonded joints built with
advanced composite materials. This
program is based on a previous
investigation and includes considerations for thermal effects.
The STEPS 7 program can be applied
to joints having one or both of the
adherends made of advanced composite materials. In the case where
only one of the adherends is composed
of advanced composites, the other
adherend is assumed to be homogenous and isotropic. In addition, each
composite laminate can be made up
of three different materials oriented
at 0, 90, $\pm \theta$ degrees, respectively.
The elastic properties of the composites are determined by using classical lamination theory while the
strength is determined by examining
the filament failure, the Hill-Von
Mises, and the interlaminar shear
cutoff criteria. The program performs a modified elastic analysis of
the composite joint by including a
correction of the elastic solution
which accounts for the effects of the
non-linear behavior of the adhesive,
which is of major importance in determining the strength of the bonded

joints. The program is capable of analyzing uniaxial tensile or compressive loads and/or in-plane shear loads. STEPS 7 is available in both FORTRAN and BASIC computer languages for more flexibility of usage on various computers. The program utilizes interaction features allowing the user to make design changes in joint geometry and laminate configuration while the program is running.

4. REPAIR PATCH DESIGN

An analysis effort using STEPS 7 was performed to develop a repair patch design, which if properly installed would provide at least 100% of the B-basis axial and shear strength of the unnotched parent laminate in the respective repair zone. The ease of manufacture as well as the relationship of repair strength to fabrication tolerances were major considerations in this analysis. A one-ply, depth-of-cut tolerance was established for the repair concepts under evaluation. It was assumed that this tolerance could occur as a short or long cut at any depth with any of the ply orientations made possible by the parent laminate stacking sequence.

Prospective repair patch design concepts must be compatible with the standard composite fabrication techniques currently in use for production hardware. For example, it is required that, as nearly as possible, the internal portion of the patch be a balanced symmetrical layup. Additional considerations include the requirements that there be no 0° - 90° fiber overlaps on the step faces and the patch design be such that there are no critical stress intensities in the adhesive joint. Working within these guidelines, it was found that a five-step internal repair patch, in combination with a five-step/12-ply (2-2-8) external repair patch, would provide the smoothest load transfer from the parent laminate to the repair patch.

The above repair patch configuration was modified to tradeoff test specimen geometry requirements for minimum step lengths (i.e., the proximity of load points to the edge of the patch) versus increased step length requirements to satisfy reduced adhesive bond strengths at 300°F.

Minor changes were made in the configuration of the external repair. The original first step was an 0-0-4 layup. Based on subsequent analytical refinements, the first step was reconfigured to two steps (0-0-2 and 0-0-4). This modification accommodated minimum ply dropoff requirements. Analysis of the modified configuration indicated a repair laminate allowable which is less than the B-basis shear allowable of the parent laminate. This situation violated original design goals (i.e., 100% retention of B-basis shear strength). The modified design, however, is considered conservative because the adhesive and fiberglass layers will cause load introduction to be relatively gradual. Finally, the original external patch layup configuration was modified to provide a balanced symmetrical laminate unto itself. Schematic cross-sections of

the finalized patch are shown in
Fig. 4 and 5.

In summary, the advantages of the proposed large-area repair patch are as follows:

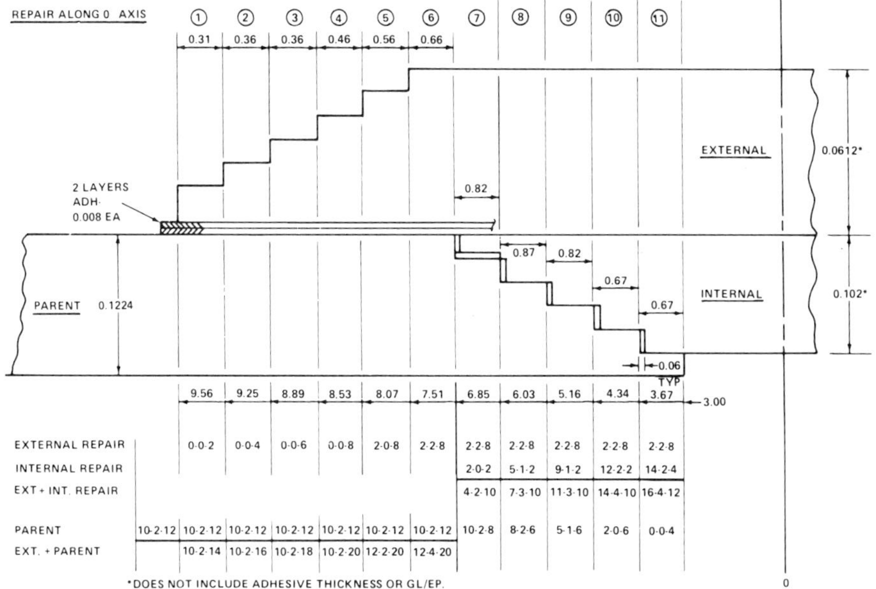

Fig. 4 Cross-Section of Repair Along Zero-Degree Axis

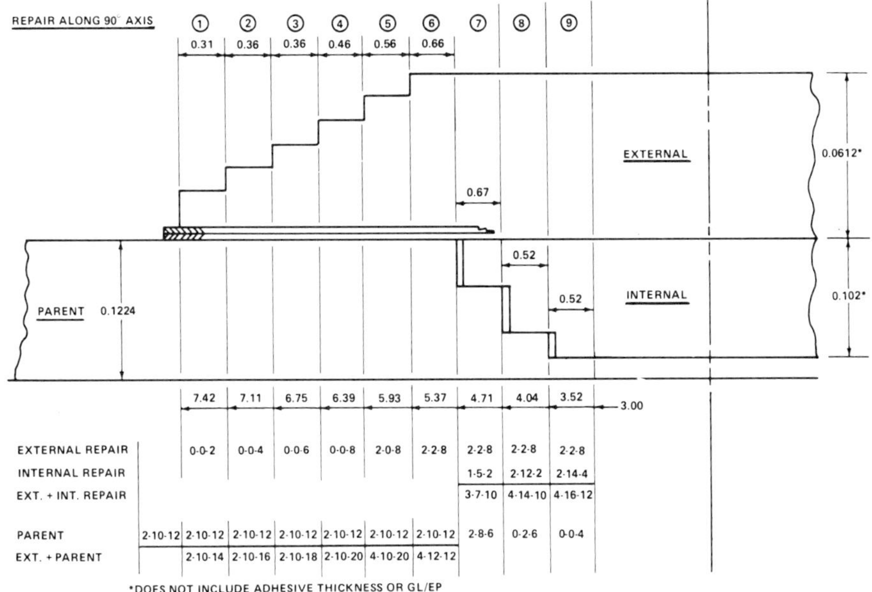

Fig. 5 Cross-Section of Repair Along 90-Degree Axis

831

- The proposed repair patch can develop full parent axial strength in both the longitudinal and transverse directions at ambient temperature and at 300°F

- Transverse axial and shear strengths provided by the repair patch are at least three times greater than the highest loads developed in areas of the stabilizer up to 24 plies in thickness

- Relatively less parent material is removed from a composite detail part repaired using an octagonal patch, compared to the amount of material removed when a circular patch is applied

- The patch utilizes minimum internal and external step lengths and provides the smallest possible overall patch which is capable of maintaining parent laminate axial strength.

5. TEST ELEMENTS

5.1 Control

The test elements designed to evaluate the Phase I large-area repair patch were 3-in.-thick, 18 by 60-in. sandwich structures. The tension facings of these sandwich structures were 24-ply, 10-2-12, boron/epoxy laminates machined from larger laminates. The compression facings of the elements were 0.125-in.-thick, 4130 heat-treated steel (180 ksi) sheets. The metal and composite facings were bonded to 1/8-5052 H39-22.1 honeycomb core sections with Metlbond 329-1A adhesive.

5.2 Static Test

The static test elements were instrumented with seven axial strain gages and, in the case of the elements tested at 300°F, with four thermocouples. An instrumented static test specimen is shown in Fig. 6.

5.3 Fatigue Test

The fatigue test element, which was identical to the static test elements, was instrumented as shown in Fig. 6.

6. STATIC TESTS

6.1 Procedure

Calibrated load cells at each load application point were used to control and monitor the load to the specimen. Deflections were monitored in three places: one in the center of the specimen and one at each loading point. Environmental conditioning for static testing at 300°F was provided by two Thermatron portable temperature conditioning units. The static test specimens were mounted in a loading fixture. Loads were applied by means of a hydraulic actuator. A calibrated load cell was installed between the actuator and the test article. The specimens were failed statically by a four-point loading distribution to develop pure bending in the test specimens. Design ultimate load (DUL) for the sandwich elements was 74,279 lb and design limit load (DLL) was 49,519 lb. Test loads were applied in 10-

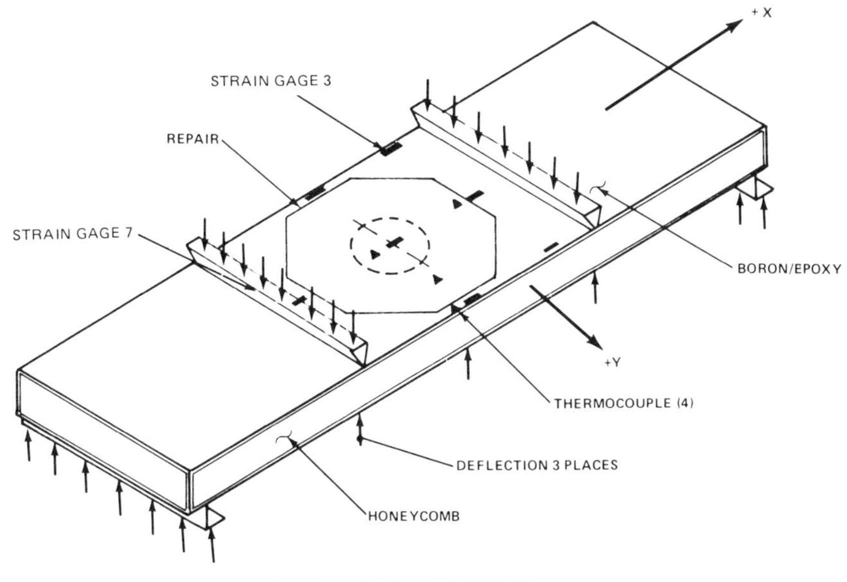

Fig. 6 Location of Strain Gages and Thermocouples

percent increments until DLL was reached. Loading was then continued in 5-percent increments until failure.

6.2 Results

6.2.1 Test Element 1A. Test Element 1A was tested initially undamaged. The beam was loaded to 89,440 lb (122% of DUL) without failure. At peak loading, the average measured strain in the boron/epoxy facing was 7148 μin./in. This value compared favorably with the predicted strain at failure (6562 μin./in.). Test results are presented in Fig. 7.

A 6-in.-diameter, centrally located hole was machined into the boron/epoxy facing of Test Element 1A. This specimen was then loaded to failure at ambient temperature. The unrepaired element failed at 46,096 lb (62% of the DUL of the undamaged element or 72,279 lb).

6.2.2 Test Element 2. Test Element 2 failed at an applied load of 105,411 lb (or 139% of specimen DUL). The measured strain at failure was 7965 μin./in. which was 136% of the strain at DUL (5832 μin./in.) The axial load at failure was 15,014 lb/in. The axial load corresponding to DUL was 10,993 lb/in. The failure mode for Test Element 2 was net tension along the forward edge of the external doubler of the patch.

6.2.3 Test Element 3. During the repair of the facing for Test Element 3 (terracing operation) the innermost step of the terraced repair site was gouged. The gouged area was filled with four plies of graphite/epoxy

TEST SPECIMEN		TEST TEMP. °F	EXTERNAL APPLIED LOADS ~ LBS		STRAIN AT FAILURE μ IN/IN			BORON/EPOXY FACING STRAINS AND LOADS						FAILURE MODE
								STRAIN ~ μ IN/IN				AXIAL LOAD ~ LBS/IN		
NO.	DESCRIPTION		DESIGN ULTIMATE LOAD (DUL)	FAILURE	LOCATION NO. 5	AVERAGE-LOCATIONS 4 AND 6	AVERAGE-LOCATIONS 1, 2 AND3	DUL	PREDICTED FAILURE	FAILURE	% DUL	DUL	FAILURE	
1A	UNDAMAGED CONTROL	RT	74,279 [1]	89,440	–	–	7148	5832	6552	7148	122	10993	–	NO FAILURE
1A	CONTROL WITH 6 IN. HOLE UNREPAIRED	RT	35,323	46,046	–	4857	3574	2684	3050 [2]	3574	133	5059	6737	NET TENSION B/EP FACING
2	6 IN. HOLE REPAIRED	RT	75,550 [3]	105,411	5602	6548	7965	5832	6552	7965	136	10993	15014	NET TENSION B/EP FACING
3	6 IN. HOLE [4] GOUGE-TERRACE REPAIRED	RT	74,279	100,200	5649	6451	7827	5832	6552	7827	134	10993	14754	NET TENSION B/EP FACING
4	6 IN. HOLE REPAIRED	300	64,817	97,226	6080	7152	7835	5626	6322	7835	140	9572	13330	NET TENSION B/EP FACING
5	6 IN. HOLE [5] REPAIRED (FATIGUE)	300	64817	90,036	5294	6320	7240	5626	6322	7240	129	9572	12318	NET TENSION B/EP FACING

NOTES:

1. Design ultimate load at ambient temperature - 74,279 lbs. (Technical Report AC SM ST-8085-B-Basis Value).

2. Failure prediction based upon net tension across minimum width of boron/epoxy laminate

3. Predicted failing load is based on correction for additional 0.125-inch thick steel facing (increased moment of inertia consideration).

4. Lowest repair site terrace contained a gouged area

5. Specimen subjected to Spectrum Fatigue Loading for 18,000 Equivalent Flight Hours (EFH). Maximum load was Design Limit Load (DLL), 48,912 lbs. Maximum strain at DLL was 3,888 μ in/in. Prior to above fatigue exposure, Specimen 5 was subjected to 15,000 EFH at loads and strains that were approximately 19% lower than that achieved at DLL.

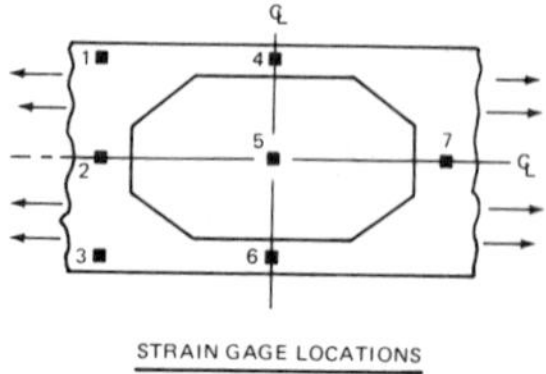

STRAIN GAGE LOCATIONS

Fig. 7 Static Test Results

prepreg and the stepped damage site repaired using the manufacturing procedures previously utilized for Test Element 2. This specimen failed in net tension of the boron/epoxy facing (same pattern as that observed for Test Element 2) at an applied load of 100,200 lb (135% of specimen DUL). The measured strain at failure was 7827μ in./in. which was 134% of the strain at DUL (5832 μin./in.). The axial load at failure was 14,754 lb/in. which compared favorably with the axial load at DUL (10,993 lb/in.).

6.2.4 Test Element 4. Test Element 4 was tested to failure at 300°F. The "predicted failing load" and "predicted strain at failure" (based on average values) for the test elements are 72,828 lb and 6322 μin./in., respectively. Design utlimate load and strain (B-basis values 89% of average values) for these elements are 64,817 lb and 5626 μin./in., respectively. The measured load at failure was 97,226 lb; 150% DUL and the calculated strain was 7835 μin./in. Failure mode was net tension in the boron/epoxy facing at the extreme edge of the external patch. The axial load at failure was 13,330 lb/in. which compared favorably with corresponding axial load at DUL (9,572 lb/in.).

6.2.5 Test Element 5. Test Element 5 was successfully tested in spectrum fatigue loading for 18,000 EFH. Maximum load was DLL (48,912 lb). Maximum strain at DLL was 3,888 μin./in. Prior to the above failure exposure, Test Element 5 was subjected to 15,000 EFH at loads and strains that were approximately 19% lower than

those achieved at DLL. Test Element
5 failed at 90,036 lb (139% of DUL).
The measured strain at failure was
7240 μin./in. The axial tensile load
of failure was 12,318 lb/in. This
value was approximately 8% less than
the axial tensile strength at 300°F
developed by Test Element 4 (13,330
lb/in.) which was not subjected to
spectrum fatigue loading.

7. CONCLUSIONS

The cocured, internally and externally stepped, boron/epoxy repair patch
is a structurally efficient configuration for the repair of 6-in.-diameter
damage to areas of F-14A horizontal
stabilizer covers with thicknesses up
to and including 24 boron/epoxy
plies.

Repaired elements, tested in four-
point beam bending, with the restored
boron/epoxy facing in tension, failed
at strain levels that were significantly
greater than corresponding values
at DUL (135% of strain at 73°F and
300°F).

A repaired element was subjected
to 18,000 EFH spectrum fatigue loading at ambient temperature (maximum
load was design limit load) without
failure. The static residual strain of
the above fatigue element at 300°F was
129% greater than corresponding strain
at design ultimate load.

The large-area repair patch departs
from the air-passage surface by only
0.080 in.

8. BIOGRAPHY

Mr. John Mahon is Group Head of the
Chemical Processes and Composites
Group of Grumman's Advanced
Materials and Manufacturing Department. He is a graduate of Brooklyn
College with a BS in Chemistry (1952)
and MA in Inorganic Chemistry (1961)
Mr. Mahon served as Lieutenant
(Junior Grade) in the U.S. Navy in
the Atlantic Fleet-Mine Force from
1952 to 1956. After completing his
military service, he was employed by
Sperry Gyroscope, Lake Success,
N.Y., as a Process Engineer from
1956 to 1964, and by Kollsman
Instrument as a reliability Engineer
from 1964 to 1965. Mr. Mahon has
been with Grumman since 1965,
serving as the Project Engineer on
such major programs as the Development of Large-Area Repair Procedures for the F-14A Boron/Epoxy
Horizontal Stabilizer and Serviceability Evaluation of F-14A Composite
Main-Landing-Gear-Strut Doors and
Overwing Fairings. Mr. Mahon also
served as Secretary of the Metro-
politan-New York Chapter of SAMPE
in 1968.

NONHONEYCOMB COMPOSITE SKIN STABILIZATION
A SOLUTION TO HIGH COSTS OF
MAINTENANCE

Ralph L. Miller and Walter C. Cooper
Vought Corporation
Dallas, Texas

Abstract

Many aluminum honeycomb secondary structures have experienced high maintenance costs in the field. These high costs are associated with frequent inspection, repair and replacement due to corrosion. Substitution of nonhoneycomb composite structures for these components is a viable solution to this problem. Manufacturing technology was demonstrated for such a composite skin stabilized secondary trailing edge structure using a low-cost nonautoclave fabrication method. A return on investment (ROI) of 2.1:1 was projected for application of this technology to 800 in-service aircraft over a 10-year period. The program included demonstration of elastomeric bladder molding (EBM) processes for a typical aircraft aileron composite trailing edge structure. Graphite/epoxy and fiberglass/epoxy fabrics were hybridized to effect lower material costs. The generic technology demonstrated an acceptable alternative to the metallic construction at a competitive replacement cost and with a reduced life cycle cost. Employing all features of the developed technology, several complete composite trailing edge articles for the Air Force T-38 aircraft were fabricated to verify the selected tooling approaches and to finalize processing criteria limits. A total of 32 T-38 aileron composite trailing edges were manufactured under Air Force (AFWAL and ALC) sponsorship. The full-scale static test article withstood a load equivalent to 404% of design limit. The production readiness of the approach and earlier cost predictions were verified. A total of 30 articles were delivered to the San Antonio Air Logistics Center (SAALC) for flight test evaluations. A repair manual was also provided.

Key Words: Maintenance; Repair; Composite Skin Stabilization; Elastomeric Bladder Molding; and Life Cycle Cost.

1. INTRODUCTION

The Air Force's operational experience with adhesively bonded metallic honeycomb sandwich structures has spanned a quarter of a century. The recurring problems are well characterized and documented at the Air Logistic Centers for many types of aircraft. The principal problem is corrosion and/or debonding of metallic sandwich structures leading to excessive inspection, maintenance and replacement costs. Dynamic instability of control surfaces due to entrapped water is potentially dangerous, and repeated inspection is costly. Additionally, nondestructive inspection techniques of the inaccessible interiors are unreliable. The operating costs attributed to deficiencies in metallic honeycomb structures have been, and continue to be, of serious concern.

Replacement of these metallic articles with composite components employing low-cost manufacturing methods results in cost-competitive structures of equivalent weight that exhibit significantly improved in-service durability.

Vought's solution to the high maintenance cost consisted of designing and developing a hybrid fiberglass-graphite/epoxy component to replace the baseline metallic structure. The approach employed the aileron trailing edge of the Air Force T-38 aircraft as a baseline. Detailed engineering analyses and structural optimization resulted in a composite structure equivalent in weight to the metallic article it replaced.

The manufacturing methodology demonstrated during this program was predicated upon a nonautoclave, elastomeric bladder molding procedure.

2. COMPONENT SELECTION AND DESIGN

A variety of aircraft secondary structure applications were considered in selecting the component to be demonstrated. The T-38 aileron trailing edge, Figure 1, was chosen because:

- It is typical of a variety of trailing edge structures.
- It has a history of frequent failure due to corrosion and debonding of honeycomb and skins.
- There are a large number of T-38 aircraft in service with severe maintenance problems.

Design of the composite T-38 aileron trailing edge was based on the criteria of Northrop Aircraft, Inc., Report NAI-57-56, T-38 Wing Control Surface Stress Analysis. The bending stiffness of the metallic counterpart was matched or exceeded. The materials chosen were nonmetallic for corrosion avoidance, and interfaces were treated with corrosion barriers at the point of attachment to the aileron main body. Also, 350°F curing composite materials were used for maximum resistance to moisture degradation. Several candidate configurations were evaluated. In simplified form, the most promising configurations which were considered are illustrated in Figure 2.

Detailed trade studies were performed on each of these concepts, assuming the use of 100% graphite/epoxy materials. As a result of these studies, configuration type 2 (Figure 2) was selected for further development efforts.

Further engineering analyses were conducted to effect a lower article cost through substitution of fiberglass for a portion of the graphite material originally considered. Since weight saving was not a design criterion, the design concept was revised to include 38% by weight of fiberglass/epoxy fabric resulting in a lower material cost. The revised design retained a projected weight savings of 7% over the metallic baseline. Based on these studies and results of design verification tests, a feasibility component was designed and fabricated using prototype tooling. Figure 3 illustrates the component construction.

A static test of this component was performed verifying the design. The component was uniformly loaded, as a cantilever beam, to a pressure of 14.4 psi before failure. Design ultimate was 8.47 psi. Failure resulted from combined shear/compression buckling of the hat section substructure.

Success of this effort led to the design of a full-scale functional replacement for the T-38 aileron metallic trailing edge.

3. TECHNOLOGY DEVELOPMENT

The manufacturing approach developed during the feasibility program was based on previous Vought programs and related government-sponsored programs. The specific method sought was one which would produce the desired component using a minimum number of cure cycles and one that would not require autoclave processing. The methodology would also be applicable to similar structures on other aircraft.

Prototype EBM tools were used to produce articles with each of three candidate materials. The materials used were rated in order of article appearance and ultrasonic inspection results. The most promising candidate materials, based on these results, were Hexcel's F263 epoxy prepregs (graphite and fiberglass). Specimens were processed to determine the optimum cure cycle suitable for the nonautoclave curing approach. Specimens B-10, B-13 and B-15 (Table 1) were found to meet all specification requirements but the longer, lower temperature staging cycle of B-10 was found to permit greater latitude in processing time and was the cure cycle selected.

The basic tooling approach, illustrated in Figures 4 and 5, was to produce an outside mold line (OML) metallic tool controlling exterior part dimensions and an internal elastomeric bladder applying densification forces through air pressurization. The elastomeric bladder was produced by casting Dow Corning's Silastic E silicone rubber into the aluminum OML tool using a dummy part and a pattern to control bladder cavity configuration.

The tooling philosophy for the front spar was essentially the same as for the wedge; a rigid external tool and an expandable internal bladder. See Figures 6 and 7.

4. COMPONENT FABRICATION

A total of 32 articles were produced under Air Force sponsorship (12 for AFWAL and 20 for SAALC), the last 24 of which were fabricated at a rate of 2 per week. The basic approach to fabrication of the articles is illustrated in Figure 8 and is summarized as follows:

- Lay up multiple-ply fabric broad-goods on mechanized ply cutting machine.
- Cut prepreg patterns to net size.
- Assemble uncured patterns in kits, each kit containing all the details for one trailing edge assembly.
- Preform the core pattern into a trapezoidal corrugation using vacuum pressure and a forming fixture.
- Position all details in the elastomeric bladder mold, inserting the bladder into the component cavity.
- Oven cure the components, wedge and front spar, using bladder pressurization for densification forces.
- Trim peripheries of the cured components.
- Apply structural paste adhesive and assemble the spar with the wedge.
- Oven cure the adhesive completing the trailing edge assembly.

During fabrication of the first article, intentional defects were positioned in the layup to provide an ultrasonic inspection reference standard for the remaining articles. The construction of this standard is illustrated in Figure 9. Before its acceptance, each cured detail and assembly was compared ultrasonically to this standard. The only significant defects were minor delaminations, which occurred during bladder removal. These damaged areas were successfully repaired by injecting EC 3445 paste adhesive into the delaminated area, clamping with "C" clamps, and oven curing. Field repairs of similarly damaged areas may be repaired in like manner using EC 1614 adhesive, a room temperature curing epoxy paste.

5. STRUCTURAL TESTING

A full-scale static test was conducted. The composite component was assembled with a test fixture which simulated the T-38 aileron main structure, and loads were applied to the upper surface through 10 flexible-foam compression pads (Figure 10). The panel was loaded continuously to failure with temporary pauses for observations at 150% and 300% of design limit load. Failure to the stiffness critical structure occurred at 404% (6,814 pounds) of design limit load. Failure mode was compression instability of the lower skin. See Figure 11.

6. COST ANALYSIS

Comprehensive cost tracking, recording and reporting were conducted. Material utilization was closely monitored and controlled. Labor-hour and cost element distribution experienced during the program are graphically illustrated in Figures 12 and 13. All manufacturing costs were analyzed, learning curves predicted and comparisons made to actual labor-hours. This analysis showed that a net savings of $400,000 could be achieved by replacing metallic with composite trailing edge articles. The noncorrodible nature of the composite skin stabilized article, as well as the reduced inspection requirements, achieves reduced life cycle costs through lower replacement rates and less frequent inspection. These reduced costs, when projected over a 10-year service life of 800 aircraft and considering a nonrecurring tooling implementation cost of $189,000, result in an ROI of 2.1:1.

For comparison purposes, it is estimated that the 1980 spares cost of the current metallic article is $612. The average unit recurring cost for a production quantity of replacement composite articles is estimated to be $673 (in 1980 dollars), or within 10% of the metallic article.

7. REPAIRABILITY MANUAL

A manual describing repair requirements and procedures was prepared for the composite skin stabilized articles. Damage types were defined and damage limits were established as shown in Table 2. Repair procedures include descriptions for:

- Nonstructural surface repair
- Structural external patch repair
- Edge wrap for edge damage
- Adhesive injection for delamination

Figure 16 illustrates the vacuum bag method of fabricating a structural patch for subsequent bonding to the composite article using room temperature curing paste adhesive (EC 1614).

8. CONCLUSIONS

The conclusions reached from the successful completion of this program were:

- Nonhoneycomb skin stabilized composite structures are a viable alternative to the continued high maintenance cost of many metallic honeycomb structures.
- These composite articles are direct replacements for their metallic counterparts.
- The composite structures can be produced cost-competitively.
- Nonautoclave processing methods are applicable to this type of structure.
- The manufacturing technology and article design concept is generic to many aircraft secondary structures.
- Life cycle projections indicate a significant reduction in cost through lower replacement rates and less frequent and less complex inspection requirements.

9. CREDITS

The efforts described herein were sponsored in part by the Air Force Wright Aeronautical Laboratories, Contract Number F33615-78-C-5108 (H. S. Reinert, project engineer) and by the Air Force San Antonio Air Logistics Center, Contract Number F34601-77-A-0176-SA47 (J. C. Smith, program manager).

10. BIOGRAPHIES

Walter C. Cooper is a Manufacturing Technology engineer specialist at Vought Corporation in Dallas, Texas, where he has been engaged since 1968 in the development and demonstration of new manufacturing techniques for composite structures. He has been directly involved in all major manufacturing technology programs including the A-7 Composite Outer Wing Panel, Air Force Tape Laying Machine Demonstration and Navy S-3A Composite Spoiler and was principal investigator for the Composite Skin Stabilization program. Prior to joining Vought, he developed tooling and procedures for fabrication of ablative coating heat shields and was instrumental in the development of reinforced plastic materials and processes. Mr. Cooper received an A.S. degree from Southern Technical Institute in Marietta, Georgia, and currently has 22 years of experience in nonmetallic materials manufacturing technology.

Ralph L. Miller received a B.A. (cum laude) degree in Chemistry and Mathematics from Texas Christian University with post-graduate work performed at TCU, California State/Fullerton, and the University of Texas at Arlington. At the Vought Corporation, Mr. Miller directs development programs for improving producibility and reducing the cost of advanced composite structures. While at Vought, his technical involvements have included all major composite programs such as the Air Force Tape Laying Machine Demonstration, A-7 Composite Outer Wing Panel, Composite Fasteners, Navy S-3A Spoiler, Navy V/STOL Aft Fuselage and the Air Force Composite Skin Stabilized Structures. Prior to joining Vought in 1968, Mr. Miller was employed at Philco-Ford Corporation, Rohr Corporation and General Dynamics/Fort Worth where he obtained an extensive background in all facets of nonmetallic materials and processes.

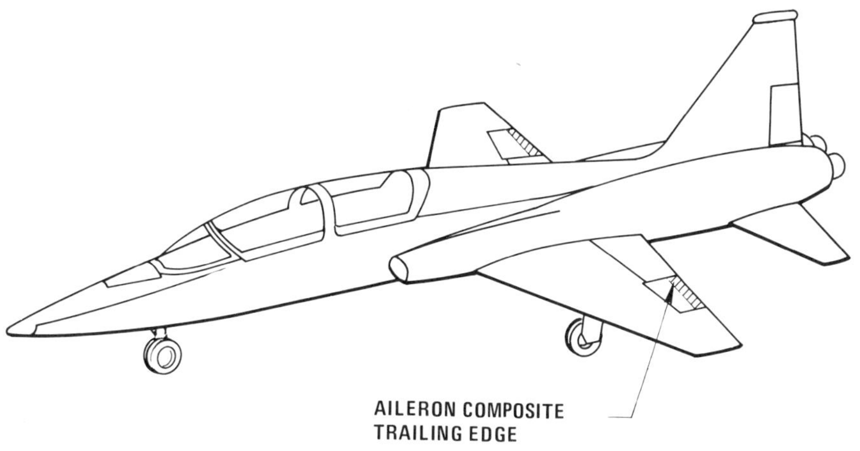

Figure 1 Air Force T-38 Aircraft

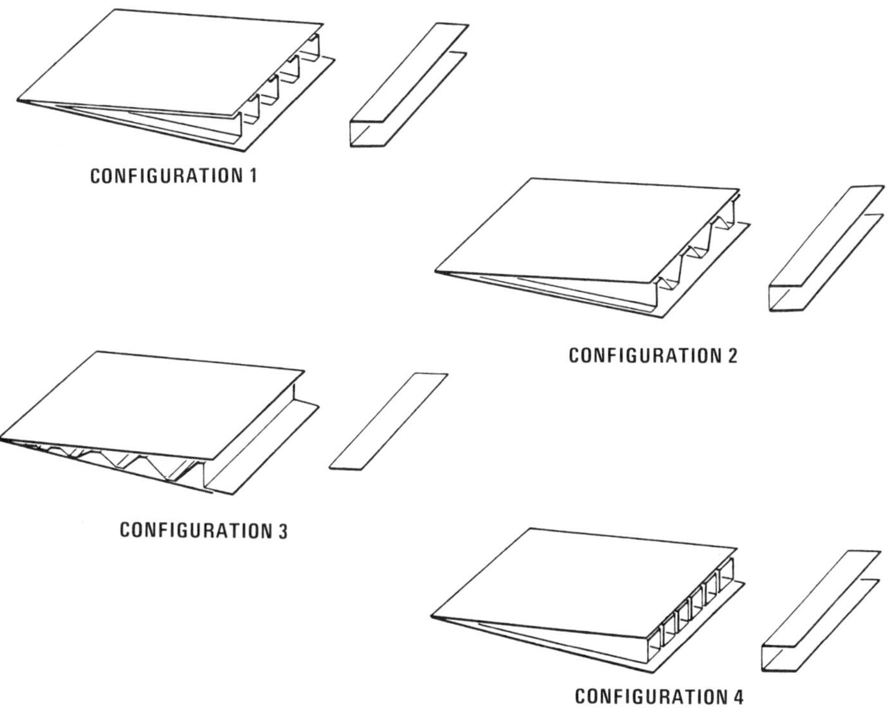

Figure 2 Candidate Composite Configurations

TABLE 1 PROCESS STUDIES

SPECIMEN NUMBER	STAGING TIME (1) (min)	TEMPERATURE °F	CURE PRESSURE (2) (psi)	FIBER VOLUME (%)	SBS (ksi)	FLEX STRENGTH (ksi)	FLEX MODULUS (msi)	ULTRASONIC INSPECTION	COMMENTS
SPECIFICATION REQUIREMENTS			75-85	59-65	6.8	90	8.5-11.5		VOUGHT PROCESS SPEC NO. 208-7-25
CONTROL			85	65.8	9.9	127	9.9	GOOD	AUTOCLAVE CURE
B-1	30	245	90						EXCESS FLOW
B-2	45	225	90						EXCESS FLOW
B-3	15	300	90						APPEARANCE GOOD
B-4	30	275	90						BLADDER RUPTURE
B-5	30	275	90						EXCESS FLOW
B-6	60	275	80	—	—	—	—	POOR	APPEARANCE POOR
B-7	60	275	80	56.7	8.7	138	8.2	POOR	APPEARANCE GOOD
B-8	60	275	50	57.5	10.7	124	9.3	POOR	APPEARANCE POOR
B-9	50	275	80	—	7.8	135	9.0	POOR	APPEARANCE GOOD
B-10	45	275	85	62.3	11.3	127	9.8	GOOD	APPEARANCE GOOD
B-11	60	275	95	—	10.9	117	9.9	GOOD	
B-12	45	275	50	—	9.5	129	8.8	POOR	OPERATOR ERROR
B-13	30	290	80	61.2	11.3	137	9.2	GOOD	APPEARANCE GOOD
B-14	30	290	60	—	—	—	—	POOR	
B-15	15	300	60	61.9	8.8	140	9.2	GOOD	APPEARANCE GOOD

NOTES:

(1) ALL STAGING PERFORMED AT 5 PSI BLADDER PRESSURE
(2) ALL FINAL CURE CYCLES 60 MINUTES AT 350°F

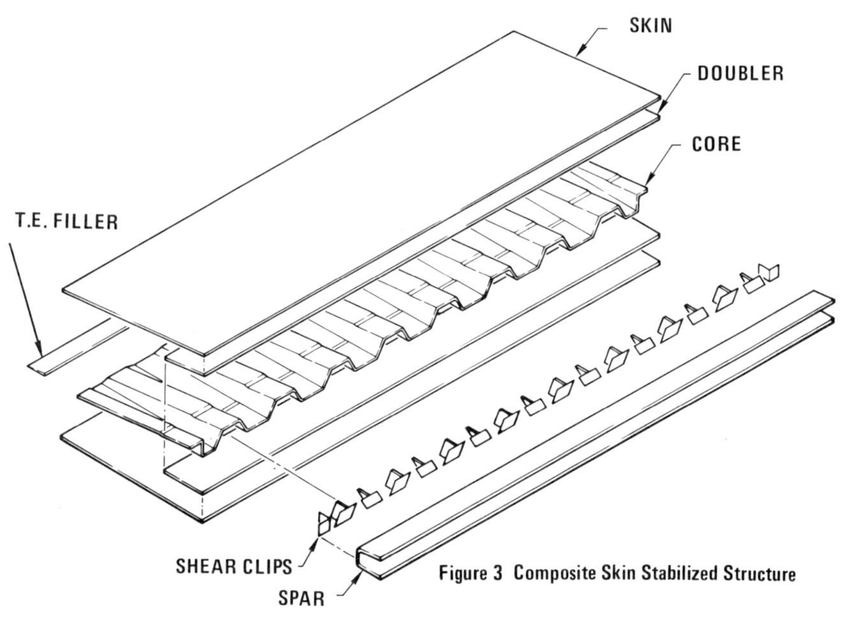

Figure 3 Composite Skin Stabilized Structure

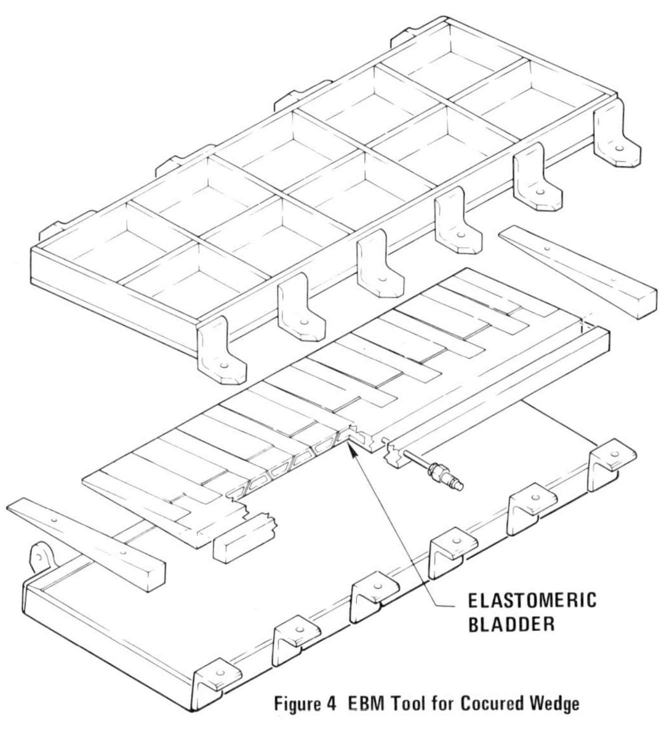

Figure 4 EBM Tool for Cocured Wedge

Figure 5 EBM Tool Shown with Cured Wedge on Bladder

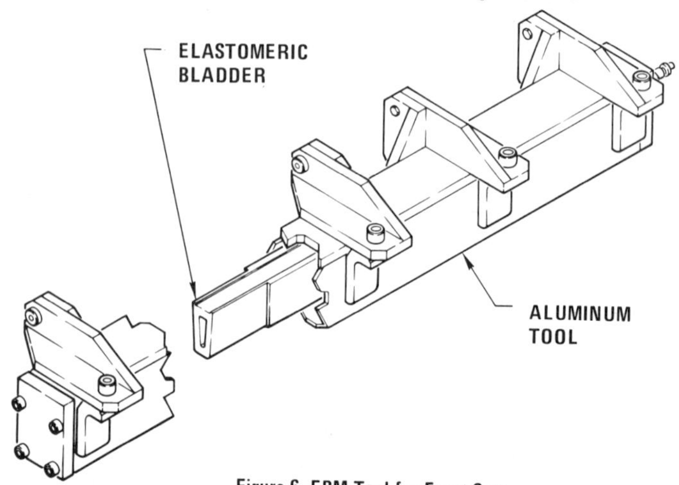

Figure 6 EBM Tool for Front Spar

Figure 7 EBM Tool for Front Spar

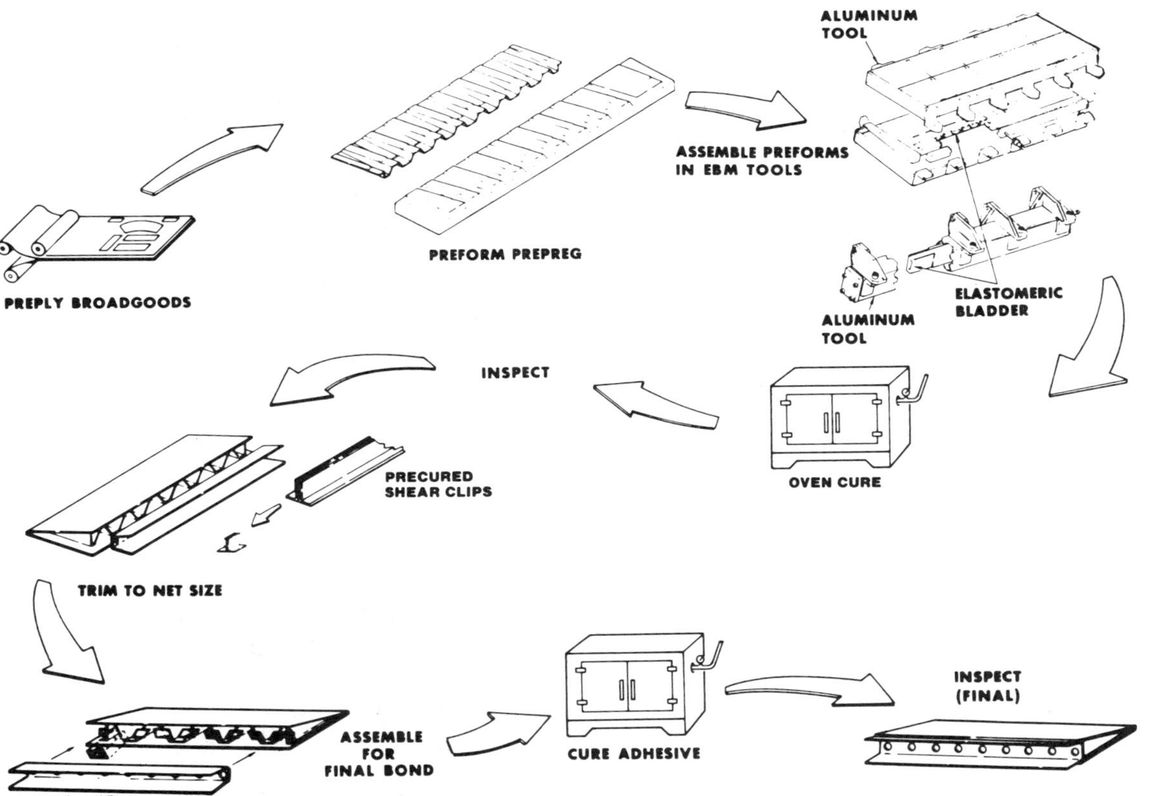

Figure 8 Manufacturing Flow Plan

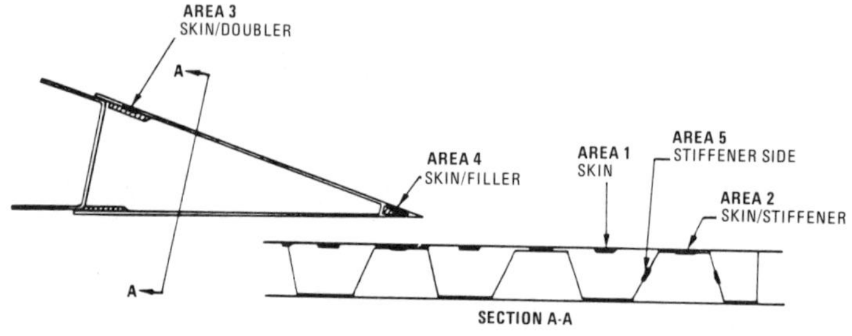

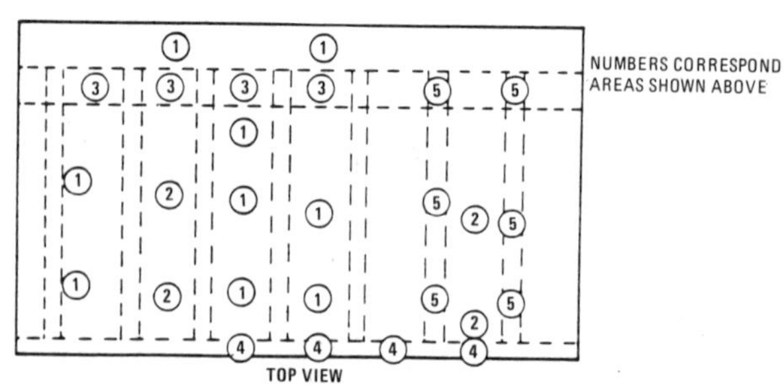

Figure 9 Location of Defects in Standard

Figure 10 Static Test Setup

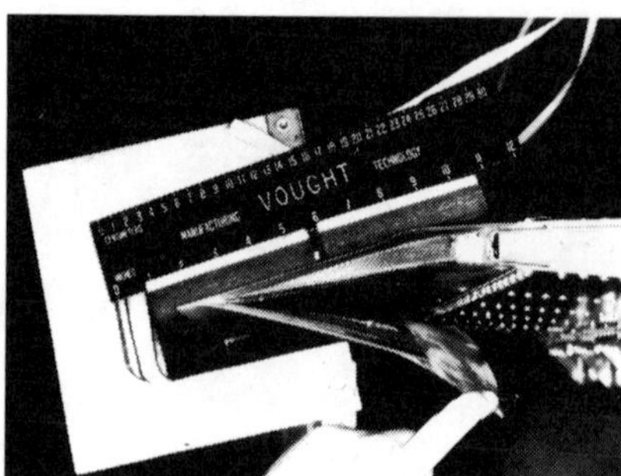

Figure 11 Static Test Article Failure

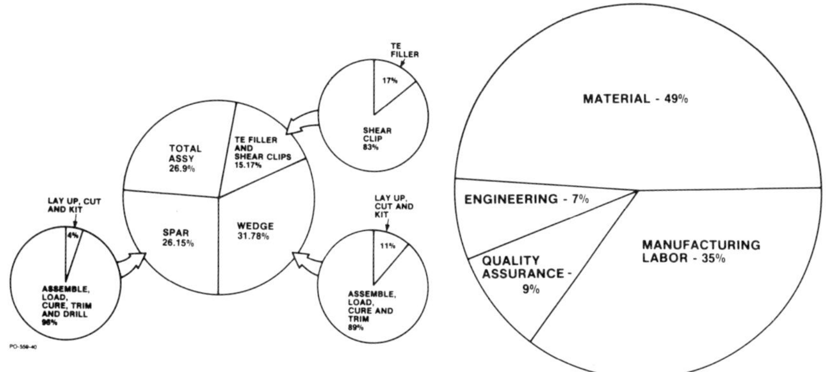

<table>
<tr><td>Figure 12 Labor-Hour Distribution</td><td>Figure 13 Distribution of Cost Elements</td></tr>
</table>

TABLE 2 DAMAGE CLASSIFICATION

DAMAGE TYPE	MAXIMUM ALLOWABLE DAMAGE SIZE FOR NONSTRUCTURAL REPAIR MINOR DAMAGE	MAXIMUM ALLOWABLE DAMAGE SIZE FOR STRUCTURAL REPAIR MAJOR DAMAGE
SCRATCH (SURFACE ONLY)	ANY LENGTH	NO STRUCTURAL REPAIR REQUIRED
SCRATCH (THROUGH SKIN)	3.0 in. (76.2 mm) LENGTH	12.0 in. (304.8 mm) LENGTH
DENT	2.0 in. (50.8 mm) DIAMETER	12.0 in. (304.8 mm) DIAMETER
PUNCTURE	1/4 in. (3.18 mm) DIAMETER	4.0 in. (101.6 mm) DIAMETER 12.5 in.^{2} (8,065 mm) AREA
EDGE CRUSHING	3/4 in. (19.05 mm) LENGTH 0.5 in.^{2} (322.6 mm^2) AREA	6.0 in. (152.4 mm) LENGTH 12 in.^{2} (7,742 mm^2) AREA
DEBOND	1.0 in.^{2} (645 mm^2) AREA	12.5 in.^{2} (8,065 mm^2) AREA
FIRE EXPOSURE	(REMOVE AND REPLACE)	(REMOVE AND REPLACE)

MAXIMUM ALLOWABLE DAMAGE SIZES ARE INDEPENDENT OF REPAIR ZONE.

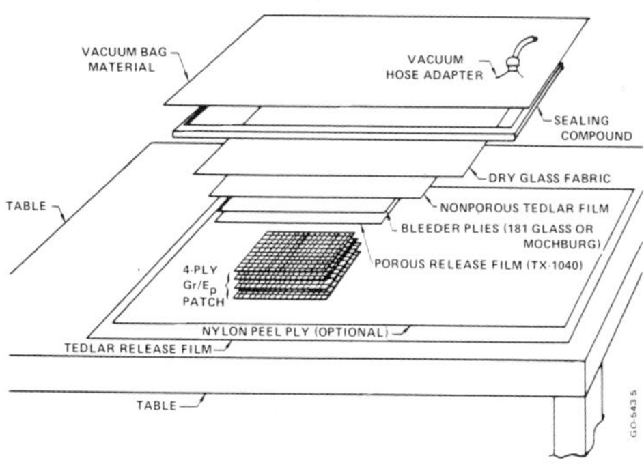

Figure 14 Patch Preparation Description

HIGH PRECISION GRAPHITE / EPOXY ANTENNAS
FOR COMMUNICATIONS SATELLITES

Bernd Abt and Hermann Rieger
Dornier System GmbH
D 7990 Friedrichshafen, Germany

Abstract

This paper deals with the development, manufacture and test of a precision graphite/epoxy antenna structure (reflector and tower), a program that was performed in view of the German direct-to-home TV broadcasting satellite. The TV-SAT has two deployable antennas of the same type as described, mounted to a common tower. The reflectors are sandwich shell construction supported by tubular frameworks. Ultra high modulus carbon fibres are used for the skins yielding a high thermal stability and structural stiffness, and in particular this design does not require an additional metallic coating, even for an application beyond 18 GHz. High precision tooling equipment assures best contour quality. The thermal control design is passive using thermal tape for the framework, and depending on the spacecraft concept, thermal coating of the dish may be required or not. Data of the reflectors mechanical, thermal, and electrical performances are given. In parallel to the reflector a thermally stable CFRP antenna tower truss consisting of filament wound tubes has been developed.

1. INTRODUCTION

Since 1974 the German industry has carried out programs, funded by the German Minister for Research and Technology, to develop the technology for critical components of direct TV broad-casting satellites. High-performance satellite antennas are required to provide this intended service to individual homes and/or communities.

The most important requirements
these antenna structures have to
meet are:

- high contour accuracy of the large
  parabolic reflector
- low reflection loss and high cross
  polarization attenuation of the
  reflecting surface
- high thermal stability
- low mass
- high stiffness and strength in
  the launch configuration

The development should be oriented
towards a 11/14 GHz antenna to be
launched by ARIANE. Futhermore, for
the configuration it should be assumed that there were two antennas
mounted to a common tower and folded during launch due to shroud limitations.

2. REFLECTOR

.2.1 Mechanical Concept

The large reflector to be built
should meet the following requirements

- offset reflector
- suitable for 11/14 GHz, i.e. reflection loss < 0.1 dB
- focal distance 1.2 m
- aperture 2 m in diameter
- offset angle 60°
- contour accuracy including thermal deformations rms < 0.5 mm
- mechanical loads for an ARIANE
  launch

Different solutions for the reflector design, all employing the carbon
fibre technology, were considered.
Several reflectors (0.9 m and 1.3 m)
were built and tested including solar simulation.

In parallel reflectivity measurements
on CFRP samples were taken.

The evaluation of the test results,
and backed-up by analyses found the
sandwich shell with supporting truss
to be the most suitable solution
(Fig. 1). The shell has two 4 layer
$/0^{\circ}, 90^{\circ}/_{s}$ 0.06 mm GY 70/Code 95[+] prepreg face sheets with a 20 mm alumi-
nium "Flexcore"[++] between them. The
supporting truss consists of filament wound (M 40/CA 209)[+++] CFRP tubes
with titanium end fittings. The fibre
lay-up $(/0^{\circ}_{2} \pm 30^{\circ}/)_{s}$ was chosen to
provide the tubes a negative CTE in
order to compensate for titanium.

2.2 Contour Accuracy of the 2 m Reflector

The contour accuracy of reflector
shells very much depends on the accuracy of the tooling used. Therefore
the surface of the cast iron laminate
tooling was produced on a numerically
controlled vertical turret boring machine.

Contour measurements showed that the
difference in CTE between cast iron
and the manufactured sandwich is neglectible. For manufacturing the CFRP
parts the autoclave technique was employed.

[+]Fothergill & Harvey   [++]Hexcel   [+++]Torayca-Ciba

A special measurement device was used
to determine the contour of the reflector at different times of the
development program.

1. after manufacture
2. after vibration and acoustic tests
3. after thermal tests
4. 7 months past the first one to
   monitor any time dependent
   (creeping) effects.

Measure-ment	Δf^+ [mm]	rms [mm]
1	0.11	0.12
2	0.28	0.15
3	0.25	0.14
4	0.25	0.13

The measurements show excellent results, i.e. the reflector could be
used at frequencies even above
14 GHz. Past all qualification tests
and the storage after, the reflector contour remained unchanged -
the small differences of the individual measurements are neglectible -
and it can be assumed that it would
perform perfectly.

2.3 Analysis and Test Results

2.3.1 Structural

Eigenfrequencies and eigenmodes were
calculated and verified in a sine
vibration test.

The low level run showed the fundamental frequencies:

	analysis [Hz]	test [Hz]
symmetric	32.9	34
antimetric	41.8	42

The respective eigenmodes are shown
in Fig. 3 and Fig. 4. There was almost no difference in the modes between analysis and test.

The subsequent qualification level input was in all axes 7 g maximum.

To avoid an additional load test a
quasi-static loading was performed
on the vibrator at frequencies far
off the fundamental ones.

The outputs at the centre of gravity
are pretty close to the values taken
for the static and stress analyses.

axis	analysis [g]	test [g]
x	25	24
y	25	27
z	35	36

Subsequently an acoustic noise test
was performed.
This test simply was to prove survival. Neither an analysis was performed nor measurements except the sound
pressure levels were taken. The acoustic load spectrum corresponded at
that time, to the ARIANE qualification requirements. This meant an
overall sound pressure level of 147 dB
for 120 seconds.

No deterioration on the reflector
could be noted.

$^+\Delta f$ = change in focal distance

2.3.2 Thermal

Investigation on white thermal paint concerning the adherence to CFRP and the degradation during a 7 years orbital life time did not provide confidence in the use of it. Moreover an additional loss in gain would have to be taken into account. Therefore the thermal design was to leave the reflector black as manufactured and to thermally tape the supporting truss.

A solar simulation test was performed, in order

1. to stabilize the antenna by low temperature cycles
2. to measure the temperature distribution at different thermal load cases
3. to measure the thermal deformations at extreme load cases
4. to adapt the analytical models and calculate the in orbit deformations.

The thermal cycling was performed between the temperatures $+20^{o}$ and -150°C. A total of 15 cycles was applied and after every 5 cycles a contour check by photogrammetry was taken at 20°C to detect permanent deformations (none could be observed).

The solar simulation was made for two cases, the one that gives the highest temperatures (sun incidence $\varphi = 0^{o}$) and the second that gives maximum temperature gradients across the shell surface ($\varphi = 75^{o}$). The measured temperatures compared with the calculated ones are shown in Fig. 5 and Fig. 6. Photogrammetric contour measurements were taken for the two thermal load cases to be compared with the calculated deformations. But the accuracy of the photogrammetric method to measure elastic or permanent deformations was less than expected. This accuracy now is given to $\pm$ 0.3 mm and lies in the range of the deformations to measure. Taken this into account no differences between analysis and test could be noted.

For the two sun load cases the in orbit thermal deformations were computed with an adapted thermal model and are with regard to the ideal contour.

load case	max.de- formation [mm]	rms [mm]
Fig.7: $\varphi = 0^{o}$	0.56	0.26
Fig.8: $\varphi = 75^{o}$	0.21	0.04

If the rms values are computed with regard to a best-fit paraboloid the values are

load case	pointing error	rms [mm]
$\varphi = 0^{o}$	0.03^{o}	0.05
$\varphi = 75^{o}$	0.006^{o}	0.02

2.3.3 Electrical

The reflection loss measurements on the selected CFRP lay-up showed excellent values (Fig. 9 and Fig. 10). So no further reflective material is needed.

The reflection loss is

  at 12 GHz   0.05 $\pm$ 0.03 dB

  at 18 GHz   0.07 $\pm$ 0.05 dB.

After the performance of all mechanical tests the electrical performance of the reflector was evaluated at the frequency 12.2 GHz by MBB on their far field antenna range (1000 m).

The measurements confirmed the reflection measurements on samples, and in addition that the reflector shows despite the UD-fibre orientation an excellent symmetry and polarization purity at low side-lobe levels (Fig. 11)[1], and checked very well with the predicted values.

Characteristic data of the antenna are [1]:

	ideal surface	CFRP
linear polarized		
gain	45.0 dBi	44.94 dBi
1. sidelobe	-24.2 dB	-24.0 dB
cross pol.	-19.0 dB	-18.3 dB
circular polarized		
gain	45.0 dBi	44.94 dBi
1. sidelobe	-23.6 dB	-23.5 dB
cross pol.	-50.0 dB	-35.2 dB
beam squint	0.09 deg	0.09 deg

3. TOWER

3.1   Mechanical Concept

Besides the reflector, the technology for the before mentioned common tower was developed. The main requirements on this tower are to take loads of the antennas and the electrical antenna components during launch and be thermally stable in orbit.

- height   2.8 m

- feed horn alignment
  axis half-cone angle   < 0.05°
  phase center   < 0.2 mm

- feed horn alignment stability
  axis half-cone angle   < 0.2°
  phase center   < 0.2 mm

The use of a tower with thin CFRP-sandwich sidewalls is mostly prohibited by integration requirements and by mass limitations. Therefore the development concentrated on a CFRP-tower truss. The problem that had to be solved was to provide good load transfer at the intersection points in combination with low thermal expansion.

The truss members consist of filament wound CFRP tubes. The diagonals and circumferencials are adhesively bonded to the undivided longerons via CFRP plates and clips.

The tubes all have a negative CTE and the plates and clips a pseudo isotropic lay-up with a low CTE. A test article, a triangular truss 30 cm wide and 70 cm high, were designed and built using the high modulus M40A fibres and CY 209 resin system.

3.2   Analysis and Test Results

3.2.1 Structural

To the technology test article a mass dummy was eccentrically attached to provide a nonsymmetric loading. Ana-

lysis showed that during the sine
vibration test the input shall be limited such, that in the center of
gravity of the mass dummy the output
will not exceed 26 g in the x-direc-
tion in order to prevent rupture.

The measured eigenfrequencies compared to the analysis are

	analysis [Hz]	test [Hz]
x	66	65
y	49	49
z	226	132

We assumed the mismatch in the z-
direction to stem from the attachment of the mass dummy. A retesting
was not possible because at 14 g in
the y-direction a failure - interlaminar shear - occured at the attachment of one diagonal to the plate
penetrating the longeron. The test
article then was refurbished and a
static load test performed. At a
load of 10100 N, applied in the y-
direction at the top, the same failure as during vibration testing
occured again.

The corresponding interlaminar shear
strengths were

vibration $\quad \tau = 15.5$ N/mm^2

static $\quad \tau = 15.8$ N/mm^2

The two values are identical but
slightly below the predicted one
$\tau = 18.3$ N/mm^2. A program to improve
the load transfer at the joints has
been undertaken in the meantime.

3.2.2 Thermal

To stabilize the thermal expansion
behaviour the truss was subjected to
15 temperature cycles from $+20^{\circ}$C
down to -170°C.
Subsequently the thermal stability
was measured using the holographic
method.

Converted into CTEs the analytically
predicted values compared to the test
were

	analysis	test
	[10^{-6} K^{-1}]	
α_x	0.39	0.19 ± 0.21
α_z	-0.11	-0.09 ± 0.05

The measurement inaccuracy in the
shorter transverse direction is necessarily higher, but the values are
acceptable.

5. GERMAN TV-SAT ANTENNA

The foregoing described program has
been the basis for the German TV-SAT
antenna being presently developed.
Many of the assumed requirements are
still true.
During Phase C for the German TV-SAT
Dornier System has developed the mechanical concept for the antenna subsystem (Fig. 12).
The TV-SAT antenna module is mounted
to the earth oriented panel of the
spacecraft. There are two offset fed
paraboloid reflector antennas, the
Receive Antenna (2.7 x 1.4 m) and the
Transmit Antenna (ϕ 2.1 m). Both,
reflectors and their individual feeds
are mounted to one common tower

(2.8 m high), thus providing the required modularity. Other components like waveguides, RF-cables, IF-sensors, RF-sensor and S-Band Antenna to provide telemetry, telecommand, tracking and ranging links, are attached to the tower as well.

During launch and transfer orbit both reflectors are swivelled and linked to the tower at their upper rims. To achieve the operational configuration pyrotechnical devices release the reflectors and then identical Antenna Pointing Mechanisms (APM) by taking upon the deployment task rotate them into their final positions. Subsequently the APMs perform the fine pointing about two axes each with an accuracy of 0.01°. The TV-SAT overall pointing error must be less than 0.1°. - The APM is installed between tower and reflector.

To take advantage of the modularity the antenna subsystem can be thermally insulated from the satellite. Except for some areas on the rear of the reflectors they remain black as manufactured and the tower is completely wrapped in Multi Layer Insulations. Depending on their individual temperature requirements the integrated electrical components have SSM-foils, OSRs and/or heaters attached.

6. REFERENCES

(1) L. Heichele, MBB
Influence on Antenna Gain and Polarization Purity of Reflectors Manufactured from Carbon Fibre Composite Materials AGARD, Lisbon, 1980.

7. BIOGRAPHIES

Within the Structures Department of Dornier System Bernd Abt is program manager concerning satellite antenna structures and thermally stable structures.

Hermann Rieger is the head of the Structures and Mechanisms Department of Dornier System, where the development work - design and analysis - for space structures, mechanisms, and antennas (earthstations and satellite) is done.

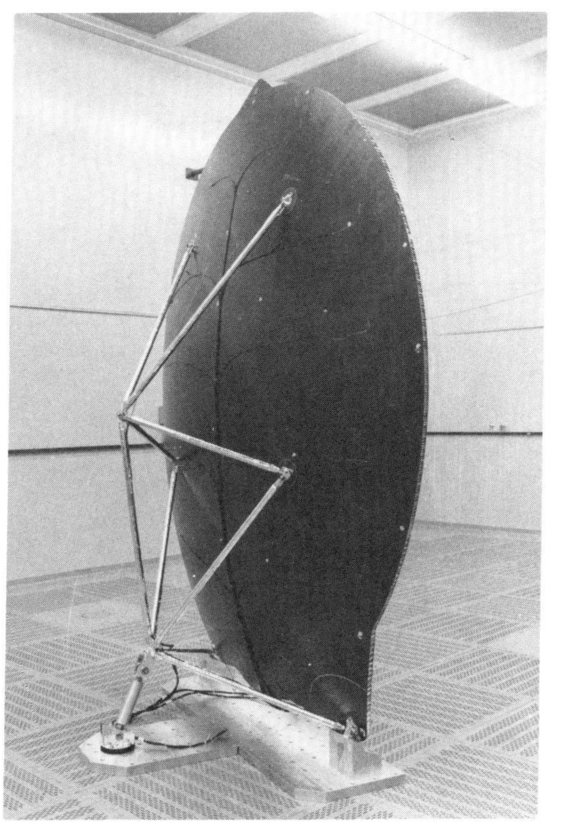

Fig. 1 Precision CFRP Offset Reflector
        Aperture 2 m

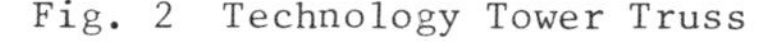

Fig. 2 Technology Tower Truss

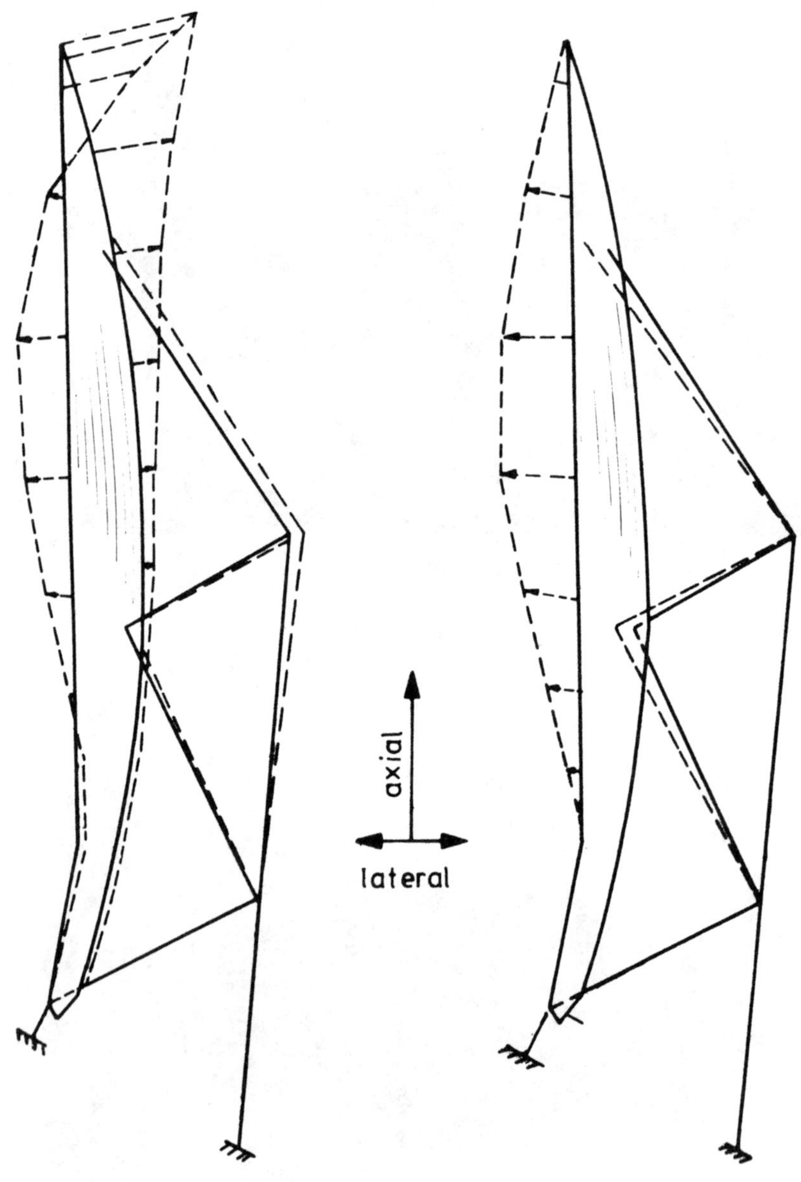

Fig. 3 1. Symmetric Eigenmode
        at 34 Hz

Fig. 4 1. Antimetric Eigenmode
        at 42 Hz

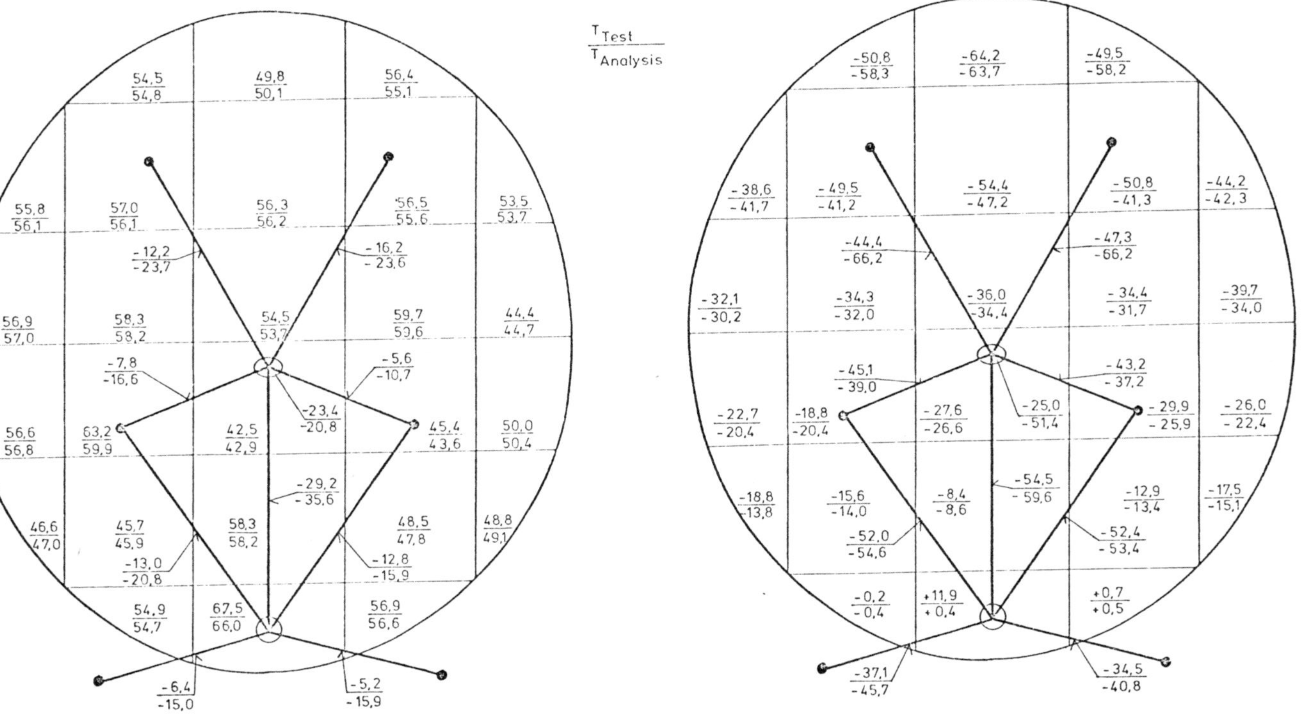

Fig. 5 Analysis and Test Temperatures
for a Sun Incidence $\varphi = 0^\circ$

Fig. 6 Analysis and Test Temperatures
for a Sun Incidence $\varphi = 75^\circ$

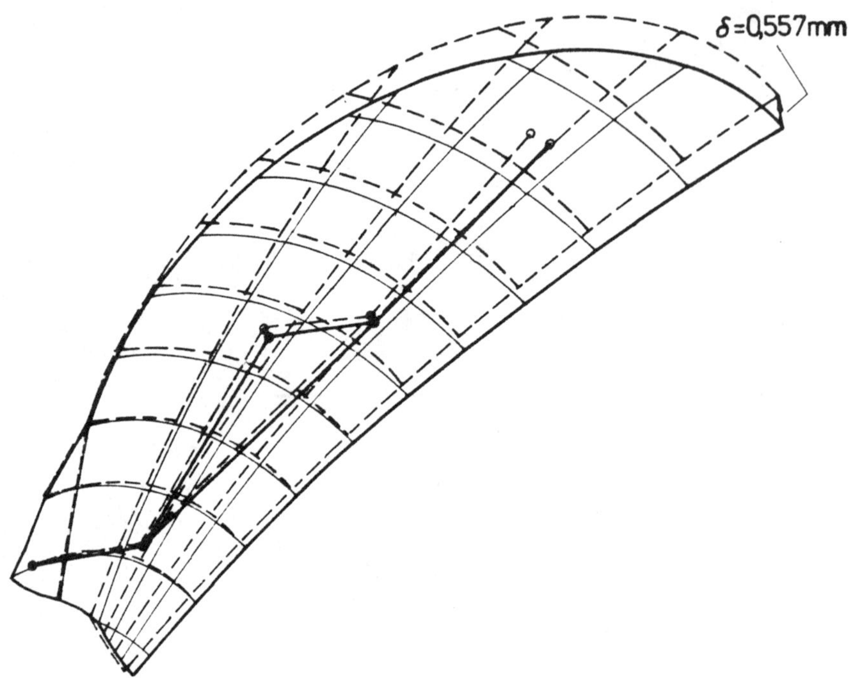

Fig 7   Thermal Deformations for $\varphi = 0^o$

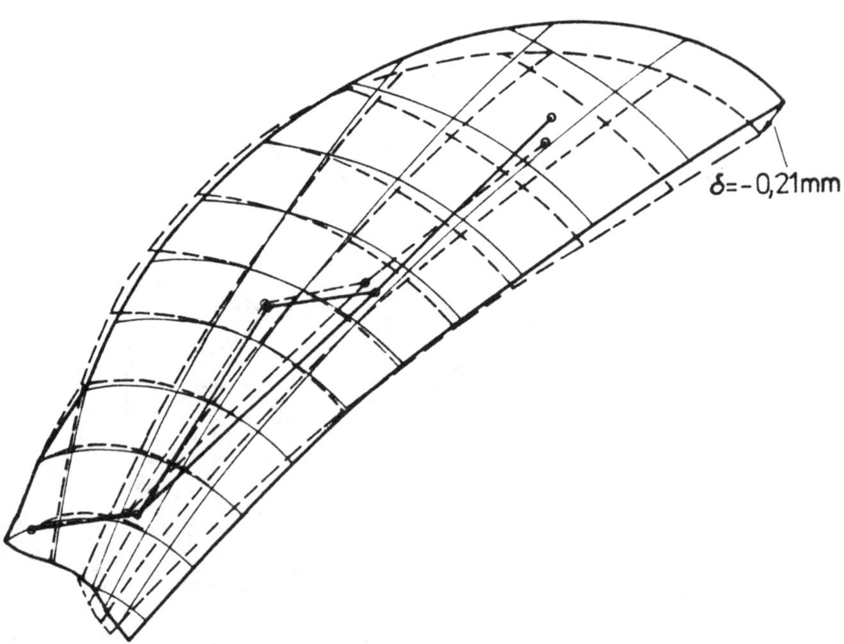

Fig. 8   Thermal Deformations for $\varphi = 75^o$

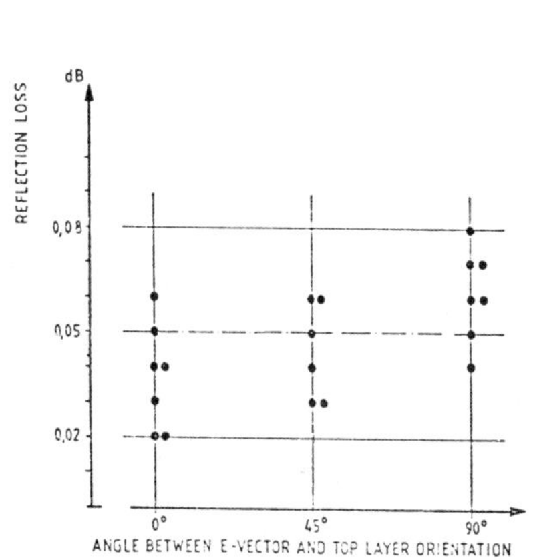

Fig. 9 Reflection loss at 12 GHz
UHM Carbon Fibre Laminate
$(0^{o}, 90^{o})_s$; 0.25 mm thick

Fig. 10 Reflection loss at 18 GHz
UHM Carbon Fibre Laminate
$(0^{o}, 90^{o})_s$; 0.25 mm thick

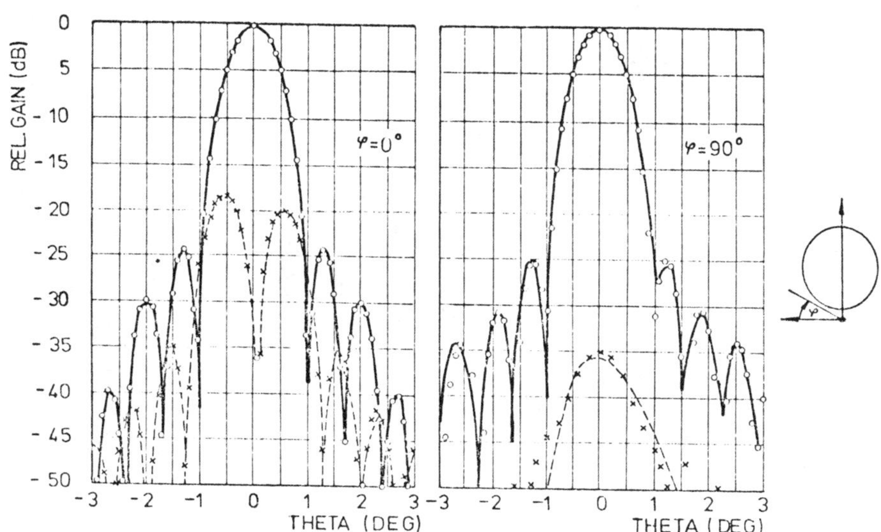

Fig. 11 Co- and Crosspolar Far Field Patterns

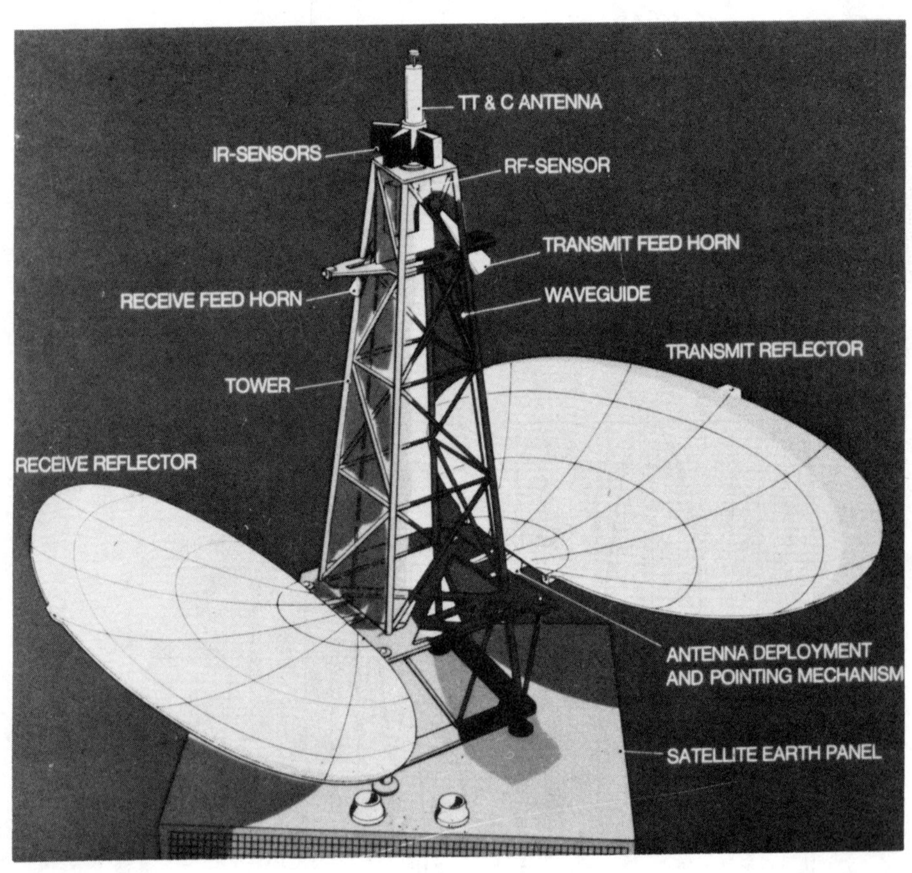

Fig. 12   German TV-SAT Antenna Module

SKIN-STABILIZED HORIZONTAL
STABILIZER FOR THE T-38 AIRCRAFT

D. L. Stansbarger
Northrop Corporation
Aircraft Division
Hawthorne, California

Abstract

Under an on-going program sponsored by the Air Force
Materials Laboratory, Manufacturing Technology Division,
Northrop Corporation is developing an innovative manufac-
turing/design approach to the production of advanced
composite empennage structures which utilize skin stiffened
stabilization techniques in lieu of full depth honeycomb
core. This technology will allow a direct replacement
to be made for existing metallic honeycomb assemblies,
resulting in a reduction in the current high maintenance
cost (life cycle cost) associated with existing metallic
honeycomb sandwich structures due to corrosion and water
entrapment, as well as improving the durability and
maintainability of the resulting structures.

1. INTRODUCTION

In recent years, the Air Force has
experienced high maintenance,
inspection, and replacement costs
associated with aluminum honeycomb
structures. These costs, in
many cases, can be traced to
water entrapment and corrosion
as well as an overall lack of
durability of aluminum honeycomb
components. In the case of full
depth honeycomb empennage assemblies, the spare replacement costs
and frequency of replacements
become extremely high as a direct
result of the water entrapment and
corrosion problems. As such, it
becomes imperative to find

alternatives to today's existing
metallic honeycomb sandwich structures in order to reduce the
extremely high life cycle cost to
the Air Force while providing
structurally sound manufacturable
replacements.

One viable solution to this problem
is the utilization of advanced
composites with innovative manufac-
turing/design approaches which
will allow a direct replacement to
be made for the metallic honeycomb
assembly. Recent development
efforts sponsored by the Air Force

in concert with industry sponsored
independent research efforts have
established the feasibility of
composite skin stiffened stabilization techniques with improved
durability and maintainability as
alternative to full depth honeycomb
metallic structures. These new
skin stiffened stabilized design
concepts appear to result in structures which can achieve weight
parity with today's metallic honeycomb assemblies, while potentially
reducing the life cycle costs.
The utilization of the composite
skin stiffened components will not
only eliminate the honeycomb
corrosion problems and associated
inspection costs but, as replacement parts for the spares inventory,
they will provide an excellent
opportunity to gain service
experience with innovative new
composite design/manufacturing
approaches.

This paper presents the design,
manufacturing and associated
technology developed to date on the
composite skin stiffened horizontal
stabilizer which will lead to
a structurally sound assembly which
offers life cycle cost reductions
and weight savings over the
metallic honeycomb sandwich baseline.

2. DESIGN AND MANUFACTURING
APPROACHES

Design Approach. Through a continuing commitment to design concept
development, evaluation, and validation, Northrop brought to this

program a validated design concept
for an advanced composite skin
stabilized horizontal stabilizer
assembly for an operational T-38
aircraft. Realizing the need for
design simplification, Northrop
approached the design of the composite horizontal stabilizer with the
primary idea of reducing acquisition
and life cycle cost, even at the
expense of maximum weight savings
if need be. Utilizing this design
philosophy, the skin stiffened
horizontal stabilizer component
design was driven by low cost innovative manufacturing/assembly
approaches in order to provide a
structurally sound component that
could be produced at the lowest
possible costs while having the
potential of going into service on
an operational T-38 Air Force aircraft.

The graphite multi-rib horizontal
stabilizer assembly selected for this
program is of sufficient size (five
feet in the spanwise direction and
tapering from five feet to three
feet in the chordwise direction)
and complexity to demonstrate the
proposed concept in all the required
technologies (design, manufacturing,
assembly, and cost) one must assess
for the efficient, low cost fabrication of a skin stiffened composite
horizontal stabilizer assembly.
The stabilizer is a two-cell,
single spar bonded and mechanically
fastened assembly with graphite/
epoxy substructure corrugations,

graphite/epoxy skins, ribs and outboard spar. The existing production
steel integral torque tube/in-
board spar is utilized with minor
modifications. Bending and shear
loads are transferred from the
aerodynamic surface to the torque
tube through the spar with the
torsional loads being transferred
from the skins to the torque tube
fitting through the graphite inboard closing ribs.

The composite skin assemblies are
contoured aerodynamic surfaces,
tapered in both the spanwise and
chordwise directions with skin
thickness ranging from 4 plies at
the trailing edge, building up to
36 plies at the center of the skin
and again tapering to 8 plies at
the leading edge of the assembly.
The skin contains composite
plies of various orientations
(0°, 90°, and $\pm45°$), sized and
stacking sequences, which must be
accurately located to provide the
required tapering, dimensional
tolerances and structural capability associated with the design
and manufacture of the assembly.
The zero-degree bending material
is concentrated along the main spar,
while plies of 90° and $\pm45°$
orientations are provided in the
section forward of the main spar
as this section acts as the primary torque box of the stabilizer.
The aft section of the stabilizer
is relatively inefficient as a
torque box with the primary func-

tion of this region being to
transmit surface pressure loading
to the spar. As such, a four-ply
orientation is used in this area.
The skins were designed for complete automated fabrication
procedures using the techniques
and results developed under the
Air Force funded Automated Composite Material Transfer Program
(Contract F33615-77-C-5121) and
Northrop's Gerber reciprocating
knife evaluations. As such, not
only is the lowest possible cost
obtained, but repeatability
between ply tolerances is assured,
which is important in the load
transfer from the composite skins
into the hinge fitting through a
bonded/bolted joint.

The graphite substructure corrugations are representative of low
cost concepts to skin stiffened
design approaches and, as such,
provide viable information from a
design/manufacturing standpoint
through the fabrication and
assembly of these components during
the program. The corrugations
consist of two separate details
(upper and lower half) which are
cured and sucsequently bonded
together to form the forward
and aft corrugations. Each half of
the corrugations are comprised of
two plies each of bidirectional
woven graphite oriented at $\pm45°$.
In designing the corrugations, some
of the weight savings was sacri-
ficed (by shortening the webs) to

provide ease and cost reductions
in the manufacture as the graphite
material does not have to be
formed as deep into the tool.
Also in designing the corrugations,
"cell" spacing was extremely
important as it not only controlled
skin thicknesses for buckling
considerations, but it also
played a big part in the manufacturing considerations as the
closer the spacing the more difficult the layup and forming operations. After many considerations,
a two inch cell spacing was
selected as the optimum balance
for skin buckling and ease of
manufacturing considerations.
After the upper and lower corrugation halves are bonded together,
they are foam bonded to the main
spar and hinge fitting transferring the corrugations shears
into these assemblies.

The aft and forward close-out
ribs which are cocured to the
hinge fitting are comprised of
tapering bidirectional woven
materials oriented at $\pm 45°$.
These ribs take the torsional
shears and close out the stabilizer assembly. The outboard spar
extension consists of two back-
to-back woven graphite channels
which are cocured to the hinge
fitting using a structural film
adhesive with the joint in double
stress. The loads from the
graphite outboard spar to the steel
hinge fitting are transferred
through a bonded/bolted joint.
The bonded joint is designed to
take ultimate load while the
fasteners are designed to take
limit load. The leading edge and
tip rib caps are metal which are
bonded to the horizontal stabilizer
during the final assembly operation.

Manufacturing Approach. The
proposed manufacturing and assembly
approach combines the "best of all
worlds" through the utilization of
large area broadgood materials
(preplied and bidirectional woven),
innovative cost reducing manufacturing approaches (cocuring, zero
resin bleeding, reusable rubber
vacuum bags, vacuum pressure
bonding, etc.), and automation
(Gerber reciprocating knife cutter,
robotic material transfer mechanism
and automated material dispensing
and transfer). Through the
utilization of these manufacturing
approaches, in concert with the
low cost producible design,
Northrop can offer a generic
approach to the lowest possible
cost skin stiffened horizontal
stabilizer assembly. The baseline
production approach for fabricating
and assembling the horizontal
stabilizer is discussed in the
following paragraphs.

Cocured Graphite "Cross". The
graphite plies for the outboard
spar extension channels and the
aft and forward rib assemblies are
comprised of bidirectional woven

materials which will be cut
automatically using the Gerber
reciprocating knife cutter. The
cut plies for the rib assemblies
are draped over male forming tools
to initially "shape" the assemblies
and subsequently removed from these
tools and placed on the substructure curing tool at the
interface of the steel "shotgun"
fitting. The cut plies for the
outboard spar extensions are also
draped over male forming tools to
"pre-shape" the assemblies.
After the forming operation, the
channels will be placed back-to-
back forming an "I" beam and
draped on either side of the adhesive coated steel "shotgun"
fitting assembly. The top half of
the cocuring tool will then be
positioned on top of the graphite/
steel shotgun "cross" assembly
and the top flanges of the
graphite forward and aft ribs and
outboard spar extension channels
will be formed to the upper
cocuring tool to establish and
dimensionally control the bondline
of the substructure for subsequent
adhesive bonding operations.
After the components have been
draped to shape, the entire
assembly is vacuum bagged using a
reusable rubber bag and cocured (to
form the required joint for load
transfer between the graphite and
metallic assemblies) in the autoclave utilizing ion graphing
techniques to control the time-
temperature and pressure profile,

with no resin bleeding being
accomplished during the cure cycle.
Upon completion of the cocuring
cycle, the mechanical fasteners
will be installed and the subassembly will be subjected to nondestructive evaluations to insure composite
and bondline integrity, configuration stability, and dimensional
tolerances. The assembly will then
be trimmed to final dimensions and
set aside for subsequent joining
operations.

Graphite Corrugations. The
graphite plies for the upper and
lower halves of the aft and forward
corrugations will be cut automatically by the Gerber reciprocating
knife cutter and layed up in the
flat by the automated robotic
transfer mechanism. The flat layup
will then be transferred to the
"female" corrugation curing tool
where it is formed into the tool
through segmented pressure devices
insuring conformation with the tool
in the web and radius areas. The
corrugation will then be prepared
for autoclave curing by vacuum
bagging with a reusable rubber
bag and rubber densifier which is
molded to the shape of the
corrugations. The assemblies are
then autoclave cured using ion
graphing techniques to control the
time-temperature and pressure profiles. The composite corrugations
will not be resin bled during the
cure cycle. After curing the
assemblies will be subjected to

nondestructive testing evaluations
(ultrasonic and in-motion radiography) to insure composite
integrity. Dimensional and configuration stability checks will
also be performed on the corrugation halves to evaluate these
critical parameters. After
inspection, the corrugations
will be prepared for adhesive
bonding to mate the upper and
lower corrugation halves. A
layer of film adhesive (FM-73)
will be draped on the faying surfaces of the upper and lower
corrugations, with the lower
corrugation being placed back into
the curing fixture. The upper half
of the corrugation is then located
on top of the lower half and
adhesively tacked in place. The
assembly is then vacuum bagged
using a reusable rubber vacuum
bag and placed in an oven to bond
the "mating" halves of the
corrugations at vacuum bag pressure,
completing the substructural
corrugations (aft and forward
components). The bonded assemblies
are again inspected to insure bondline integrity, final trimmed
and set aside for subsequent
joining operations to the steel
shotgun/graphite outboard spar
subassembly (cocured cross).

Graphite Upper and Lower Skins.
The skin assemblies for the
horizontal stabilizer component
were designed with automated
fabrication techniques in mind.

The internal drop offs, tapering
and stacking sequence associated
with the skins allows them to be
laminated with Northrop's existing
robotic transfer mechanism. The
graphite plies will be cut with the
Gerber reciprocating knife and
picked up, transferred, and layed
over the "male" curing tool with the
robotic transfer mechanism. After
the skins have been laminated,
they will be vacuum bagged using
a reusable rubber vacuum bag and
cured in the autoclave, using ion
graphing techniques to control the
time-temperature and pressure
profiles, with no resin bleeding
taking place during the cure cycle.

After curing, the skin assemblies
will be inspected with ultrasonic
and low K-V radiography to insure
composite integrity. After inspection, they will be not trimmed and
set aside for subsequent assembly
operations.

First Stage Assembly. The first
stage assembly operation joins the
graphite corrugations (aft and forward) to the graphite outboard
spar extension/steel shotgun
subassembly through foaming adhesive.
The corrugations containing foaming
adhesive in the joggle areas (for
joining to the outboard spars
and steel shotgun fitting) and
faying surfaces (where they "mate"
to the aft and forward ribs) are
located and positioned into the
subassembly tool. The assembly
will be "dead weight" loaded and

placed in an oven allowing
the foam to expand under heat
creating the required structural
joint between the corrugations,
and spar extension/steel shotgun
fitting subassembly. After cure,
the completed subassembly will
be inspected through ultrasonics
and radiography to insure foam
bondline integrity, and set
aside for the final bonding
operation.

Second Stage Assembly. The precured graphite upper and lower
skin assemblies containing a
layer of film adhesive (FM-73),
in the areas which will join the
substructure corrugations and
cocured cross, are located and
adhesively tacked to the substructure assemblies. The
metallic leading edge cap and
outboard tip rib, which have
been preformed and adhesively
coated, are then positioned on
the horizontal stabilizer assembly.
The entire assembly is then
vacuum bagged using a reusable
rubber vacuum bag and placed
in an autocalve where the final
adhesive bonding cycle takes
place at 30 psi pressure. After
the bonding cycle, the stabilizer
assembly is subjected to ultrasonic and radiography inspection
techniques to insure bondline
integrity. After inspection,
the upper and lower skin assemblies
are mechanically fastened in the
area of the steel "shotgun"

fitting, completing the horizontal
stabilizer component.

3. CRITICAL MANUFACTURING/ ASSEMBLY PARAMETER STUDIES

To evaluate, demonstrate, and
verify critical manufacturing and
assembly techniques as associated
with the proposed skin stiffened
horizontal stabilizer prior to the
fabrication of the tool proofing
article, a representative subcomponent assembly (spar splice) was
fabricated and structurally tested.
The spar splice assembly shown in
Figure 1 is representative of the
full size horizontal stabilizer
assembly in that it contains all
the detail parts (graphite corrugations, outboard spar extensions,
upper and lower skins and a simulated steel "shotgun" fitting)
which make up the stabilizer
assembly and, as such, provided the
associated problem areas that would
be encountered when fabricating and
assembling the proposed full size
stabilizer component utilizing the
proposed manufacturing and assembly
methods. The design of the selected
subcomponent was such that the
many complexities (dimensional
tolerances, configuration stability,
assembly interactions) experienced
in manufacturing and assembling
such structures can be evaluated
and proven prior to the fabrication of the tool proofing horizontal stabilizer. The spar splice
represents the most critical
section of the horizontal stabilizer component both from a load

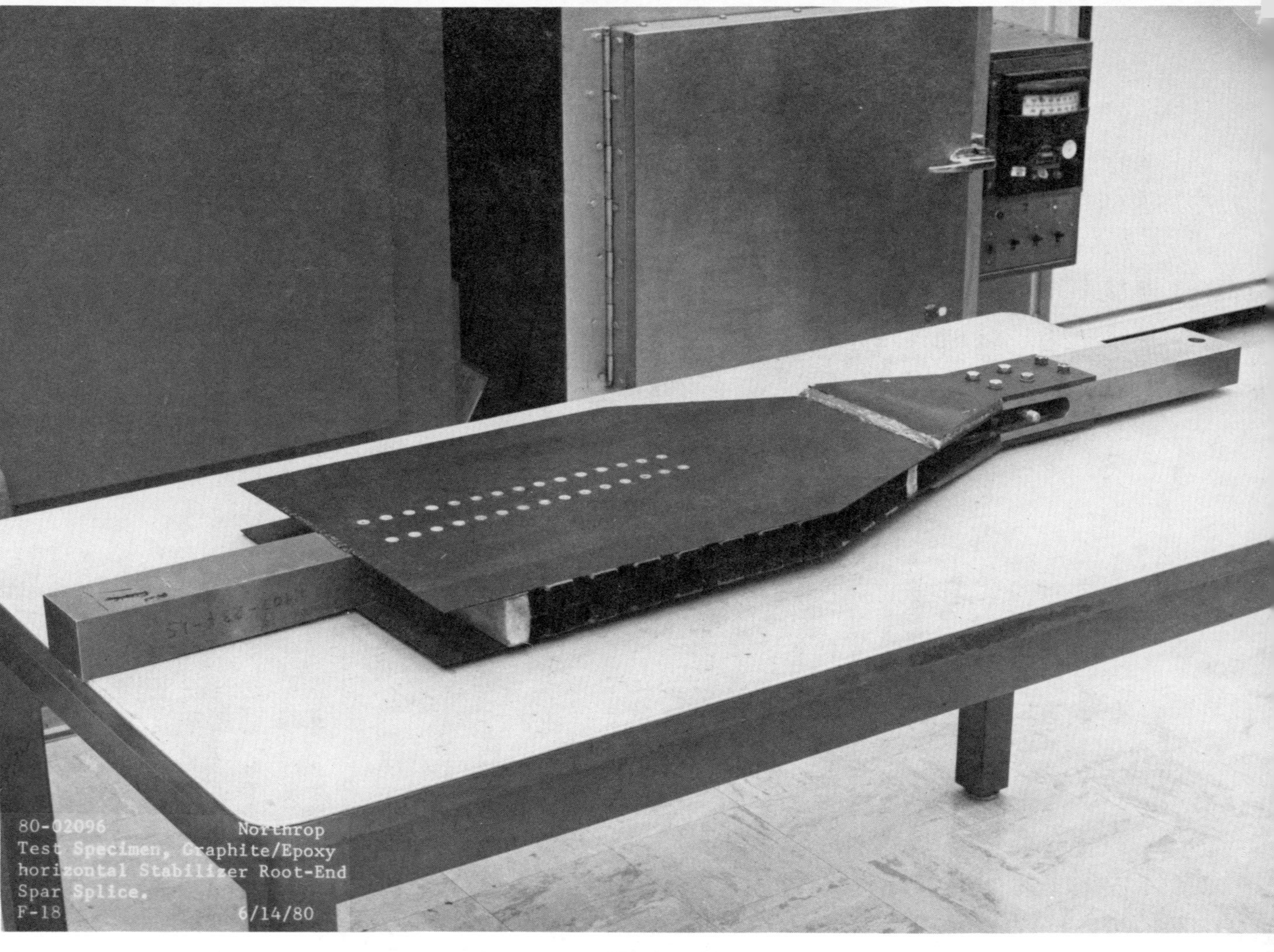

Figure 1. Spar Splice Assembly

carrying and manufacturing and
assembly standpoint. As shown
previously in Figure 1, the spar
splice assembly is basically a
section through the center of the
full size stabilizer representing
the main load carrying path
(graphite outboard spar/steel
shotgun fitting assembly and
graphite skin assemblies)
associated with the horizontal
stabilizer. As such, the
structural integrity of the
manufacturing approach was
evaluated through the static
testing of this assembly.

The assembly was mounted in a
rigid test fixture as a cantilever beam as shown in Figure 2
with the simulated air loads
distributed to the component by
means of a whipple tree arrangement. The applied shear and
moment loadings applied to the
spar splice assembly closely
paralleled those applied to the
full size T-38 horizontal stabilizer. Three test loadings
were applied to the spar splice
including:

- Loading to limit loads
  (67% ultimate) at increments
  of 10%
- Loading to 100% test ultimate
  load at increments of 20%, and
- Loading to 100% test ultimate
  load at increments of 20% and
  150% test ultimate load at
  increments of 10%.

The spar splice assembly was

loaded in the first test condition
(67% of design ultimate) with test
loads being applied in 10% increments to limit load, after which
time the load was reduced to 20%
test ultimate load and then to
zero. After the load had been
removed from the specimen, the
assembly was visually inspected for
damage and/or permanent set. The
test load was then re-applied and
increased to 20% test ultimate load,
limit load and 100% test ultimate
load in 20% increments. The load
was again reduced and the specimen
was visually inspected for damage
and/or failure. After visual
inspection, the spar splice was
again loaded to 100% ultimate load
without incident, with the load
steadily increased in 10% increments
to 150% ultimate load. As the spar
splice reached 130% ultimate load
a definite cracking noise was heard
but no problems were encountered in
reaching the 150% ultimate load.
Subsequent inspection of the
assembly revealed that a fastener
head had sheared off in the upper
(compression) graphite skin at the
extreme inboard end near point "G",
and an unbond had occurred between
the composite skin and the steel
shotgun fitting at the extreme
inboard end. Other fastener
holes adjacent to the failed
fastener also showed evidence of
bearing failure in the graphite
skin assembly. Visual inspection
of the total assembly revealed no

Figure 2. Spar Splice Assembly in Test Jig

additional apparent or significant damage. The component is currently being subjected to low K-V x-ray evaluations to ascertain any additional damage, with the x-ray analysis to be used as a baseline for subsequent development of repair techniques for the full size horizontal stabilizer. With the successful static testing of the spar splice assembly, work was initiated on the full size tool proofing horizontal stabilizer.

4. FABRICATION OF HORIZONTAL STABILIZER

At the present time, three of the required four corrugations for the tool proofing horizontal stabilizer have been fabricated. Bidirectional woven graphite was utilized in the construction of the corrugations. The woven pre-preg was layed in the flat on top of the corrugation tools and subsequently formed into the shape of the tool using preformed pressure bars. In order to assure that the woven graphite material was "intimate" with the tool in the many radiuses required, the forming operation initiated in the center corrugation with subsequent corrugations on either side being formed progressively while maintaining pressure on the previously formed corrugations. After fabrication, the corrugations were subjected to non-destructive evaluations to insure structural integrity. Subsequent to this operation, they were trimmed and set aside for assembly operations.

Fabrication of the substructural cross which consists of the forward and aft grpahite ribs and the graphite outboard spar extension cocured to the steel shotgun fitting has also been initiated. The tooling consists of a "female" brake formed steel inner moldline surface with steel rib and outboard spar tools for controlling the faying surfaces for subsequent bonding operations. At the present time, the graphite ribs and outboard spar assemblies are being layed up and inserted into the steel shotgun fitting for cocuring operations. After the layup operations, the "cross" will be vacuum bagged and cocured in the autoclave.

The completed graphite skins for the tool proofing horizontal stabilizer are shown in Figure 3.

5. SUMMARY

The development and validation of the skin stiffened composite empennage design and manufacturing techniques will provide, to the industry, an innovative approach to fabricating near term and next generation aircraft structures which will offer; (1) reduced maintenance and inspection costs, (2) increased durability, and (3) cost and weight parity with today's composite honeycomb sandwich designs.

Figure 3. Upper and Lower Skins for Tool Proofing Component

6. BIOGRAPHY

Don Stansbarger received a B.S.
from California State Polytechnic
University at Pomona and an LL.B
in Law from LaSalle University.
His experience includes 20 years
in the aerospace industry working
with reinforced plastics of
which the past 15 years have been
concentrated on Advanced Composite
materials. He joined Northrop
Corporation in 1968 and has been
involved with all facets of
advanced composites technology
including; design, materials
and processing, fabrication and
assembly, testing and for the
past five years with the development of automated techniques of
fabrication. During the past
five years, he has been the
Manager of Advanced Composites at
Northrop Corporation. Don is also
an active member of the State Bar
of California practicing both
criminal and civil law.